THIAZOLE
AND ITS DERIVATIVES

IN THREE PARTS

PART ONE

This is the thirty-fourth volume in the series
THE CHEMISTRY OF HETEROCYCLIC COMPOUNDS

THE CHEMISTRY OF HETEROCYCLIC COMPOUNDS
A SERIES OF MONOGRAPHS
ARNOLD WEISSBERGER and EDWARD C. TAYLOR
Editors

THIAZOLE
AND ITS DERIVATIVES

PART ONE

Edited by

Jacques V. Metzger

UNIVERSITY OF AIX-MARSEILLES
FRANCE

AN INTERSCIENCE ® PUBLICATION

JOHN WILEY & SONS

NEW YORK · CHICHESTER · BRISBANE · TORONTO

An Interscience® Publication
Copyright © 1979 by John Wiley & Sons, Inc.

All right reserved. Published simultaneously in Canada.

Reproduction or translation of any part of this work
beyond that permitted by Sections 107 or 108 of the
1976 United States Copyright Act without the permission
of the copyright owner is unlawful. Requests for
permission or further information should be addressed to
the Permissions Department, John Wiley & Sons, Inc.

Library of Congress Cataloging in Publication Data:

Main entry under title:

Thiazole and its derivatives.

   (The Chemistry of heterocyclic compounds; v. 34,
  pt. 1)
   "An Interscience publication."
   Includes indexes.
   1. Thiazoles. I. Metzger, Jacques V.
QD403.T53   547'.594   78–17740
ISBN 0–471–03993–4
Printed in the United States of America

10 9 8 7 6 5 4 3 2 1

Contributors

J. P. Aune, *University of Aix-Marseilles III, France*

J. Chouteau, *University of Aix-Marseilles III, France*

J. Crousier, *University of Aix-Marseilles I, France*

H. J. M. Dou, *University of Aix-Marseilles III, France*

L. Forlani, *University of Bologna, Italy*

J. Metzger, *University of Aix-Marseilles III, France*

R. Meyer, *University of Aix-Marseilles III, France*

G. Mille, *University of Aix-Marseilles III, France*

P. Todesco, *University of Bologna, Italy*

G. Vernin, *University of Aix-Marseilles III, France*

E. J. Vincent, *University of Aix-Marseilles III, France*

To the memory of
ARTHUR RUDOLF HANTZSCH
*great chemist and teacher
and pioneer of thiazole chemistry*

The Chemistry of Heterocyclic Compounds

The chemistry of heterocyclic compounds is one of the most complex branches of organic chemistry. It is equally interesting for its theoretical implications, for the diversity of its synthetic procedures, and for the physiological and industrial significance of heterocyclic compounds.

A field of such importance and intrinsic difficulty should be made as readily accessible as possible, and the lack of a modern detailed and comprehensive presentation of heterocyclic chemistry is therefore keenly felt. It is the intention of the present series to fill this gap by expert presentations of the various branches of heterocyclic chemistry. The subdivisions have been designed to cover the field in its entirety by monographs which reflect the importance and the interrelations of the various compounds, and accommodate the specific interests of the authors.

In order to continue to make heterocyclic chemistry as readily accessible as possible, new editions are planned for those areas where the respective volumes in the first edition have become obsolete by overwhelming progress. If, however, the changes are not too great so that the first editions can be brought up-to-date by supplementary volumes, supplements to the respective volumes will be published in the first edition.

Research Laboratories ARNOLD WEISSBERGER
Eastman Kodak Company
Rochester, New York

Princeton University EDWARD C. TAYLOR
Princeton, New Jersey

Preface

Given their theoretical as well as practical interest, five-membered aromatic rings occupy a position of particular significance in the enormous field of heterocyclic chemistry. Thiazole is one of the important members of this family and thus merits a comprehensive study. The purpose of this book is to condense into a volume of reasonable size the chemistry of thiazole, covering the literature of approximately one century up to December, 1976. For technical reasons this present work has been limited to the study of monocyclic thiazoles, excluding thiamine and partially reduced thiazoles but including selenazoles. Though most of the important material has been published in the last twenty years, all the literature concerning thiazoles has been surveyed, and it is of special interest to see with what energy Arthur Hantzsch was obliged to defend his historical discovery of thiazole.

In the first chapter, devoted to thiazole itself, specific emphasis has been given to the structure and mechanistic aspects of the reactivity of the molecule: most of the theoretical methods and physical techniques available to date have been applied in the study of thiazole and its derivatives, and the results are discussed in detail. The chapter devoted to methods of synthesis is especially developed and traces the way for the preparation of any monocyclic thiazole derivative. Three chapters concern the nontautomeric functional derivatives, and two are devoted to amino-, hydroxy-, and mercaptothiazoles: these chapters constitute the core of the book. All discussion of chemical properties is completed by tables in which all the known derivatives are inventoried and characterized by their usual physical properties. This information should be of particular value to organic chemists in identifying natural or synthetic thiazoles. Two brief chapters concern mesoionic thiazoles and selenazoles. An important chapter is devoted to cyanine dyes derived from thiazolium salts, completing some classical reviews on the subject and discussing recent developments in the studies of the reaction mechanisms involved in their synthesis.

The importance of this work, which was begun by Dr. J. M. Swan of Monash University, Melbourne, Australia, was very quickly recognized, and in 1964 I joined him in his endeavor. Three years later, Dr. Swan was obliged to abandon, and, for the last decade, 17 distinguished scientists have labored to realize this book. I acknowledge with sincere thanks the cooperation and perseverance of all of them, but I am especially indebted to Michel Chanon and his collaborator, René Barone, for the untiring efficiency in management that they have exhibited throughout the realization of this book. I acknowledge also the help of the numerous heterocyclic chemists of the world who sent so many of their valuable reprints to Marseilles. My thanks are also due to Mrs. J. de Caseneuve and Mrs. G. Formanek, who carried out the tedious task of typing the manuscript, and to Thomas Murphy for his help in adjusting the poor original English of most of the manuscript to a hopefully acceptable one. Finally, grateful thanks are due to the University of Aix-Marseilles for financial support and library facilities.

Marseilles, France JACQUES V. METZGER
December 1978

Contents

PART ONE

Introduction   1

I. Properties and Reactions of Thiazole   5

J. V. METZGER and E. J. VINCENT, with the collaboration of J. CHOUTEAU and G. MILLE

II. General Synthetic Methods for Thiazole and Thiazolium Salts   165

G. VERNIN

III. Alkyl, Aryl, Aralkyl, and Related Thiazole Derivatives   337

J. P. AUNE, H. J. M. DOU, and J. CROUSIER

IV. Thiazolecarboxylic Acids, Thiazolecarboxaldehydes, and Thiazolyl Ketones   519

R. MEYER

V. Halo- and Nitrothiazoles   565

L. FORLANI and P. E. TODESCO

Subject Index   587

PART TWO

General Introduction to Protomeric Thiazoles

M. CHANON

VI. **Aminothiazoles and Their Derivatives**

R. BARONE, M. CHANON, and R. GALLO

VII. **Mercaptothiazoles, Hydroxythiazoles, and Their Derivatives**

C. ROUSSEL, M. CHANON, and R. BARONE

Subject Index

PART THREE

VIII. **Mesoionic Thiazoles**

M. BEGTRUP and C. ROUSSEL

IX. **Cyanine Dyes Derived from Thiazolium Salts**

H. LARIVE and R. DENNILAULER

X. **Selenazole and Derivatives**

R. GUGLIELMETTI

Subject Index
Cumulative Author Index

THIAZOLE
AND ITS DERIVATIVES

IN THREE PARTS

PART ONE

This is the thirty-fourth volume in the series
THE CHEMISTRY OF HETEROCYCLIC COMPOUNDS

Introduction

This volume is intended to present a comprehensive description of the chemistry of thiazole and its monocyclic derivatives, based on the chemical literature up to December, 1976. It is not concerned with polycyclic thiazoles, such as benzo- or naphthothiazole, nor with hydrogenated derivatives, such as thiazolines or thiazolidines: later volumes in this series are devoted to these derivatives. The chemistry of thiamine has also been excluded from the present volume because of the enormous amount of literature corresponding to the subject and is developed in another volume. On the other hand, a discussion of selenazole and its monocyclic derivatives has been included, and particular emphasis has been given to the cyanine dyes derived from thiazolium salts.

The first chapter discusses thiazole itself, its structure, and its physical and chemical properties. The second chapter describes the general synthetic methods for thiazoles and thiazolium salts.

Chapters III to VII discuss the general properties of thiazoles having hydrocarbon and functional substituents, respectively. A special chapter (Chapter VIII) is devoted to *meso*-ionic thiazoles, and Chapter IX describes the thiazolium salts and their numerous cyanine dyes derivatives. The last chapter concerns the monocyclic selenazoles.

Although unsubstituted thiazole is not found in nature, naturally occurring thiazoles are numerous and are described in the chapters corresponding to their structure. Their origin has often been related to the cyclization of a peptide chain at cysteine residue with formation of a thiazoline ring and thence, by dehydrogenation, of a thiazole ring: Several antibiotics, such as althiomycine (1) and micrococcin (2, 3), contain a thiazole ring, as do many metabolic products of living organisms (4, 5); numerous natural aromas contain thiazole derivatives: tomato (6), roasted coffee (7, 8), roasted peanuts (9), the basic fraction of Scotch whisky and Jamaica rum (10) and so on. Thiazoles have also been separated from nitrogen bases of some petroleums (11). Many thiazole derivatives possess biological activity, and numerous medicaments contain a thiazole ring in their structure.

$$\text{R'—NH—}\underset{\underset{O}{\|}}{C}\text{—}\underset{\underset{CH_2\text{—SH}}{|}}{CH}\text{—NH—}\underset{\underset{O}{\|}}{C}\text{—R} \xrightarrow{-H_2O} \text{R'—NH—}\underset{\underset{O}{\|}}{C}\text{—}\underset{\underset{H_2C}{|}}{CH}\underset{S}{\diagdown}\underset{}{\overset{N}{\underset{\|}{C}}}\text{—R}$$

$$\xrightarrow{-2H} \text{R'—NH—}\underset{\underset{O}{\|}}{C}\text{—}\underset{HC}{\overset{\|}{C}}\underset{S}{\diagdown}\underset{}{\overset{N}{\underset{\|}{C}}}\text{—R}$$

Among the pentaatomic heterocyclic rings, thiazole is one of the most intensively investigated, and its chemistry maintains steadily its intensive development. The number of annual publications dealing with monocyclic thiazoles is continuously growing, and the excellent but condensed review published in 1957 by Sprague and Land (12) is now quite obsolete. The Card Index of Thiazole Compounds edited by Prijs in 1952 (13) described the 2100 individual thiazolic compounds known at that time; unfortunately more recent editions have not been published, and the present number of known compounds is greater than 12,500. Since 1970 a condensed but comprehensive biannual review of the published work concerning thiazole appears in the collection, "Organic compounds of S, Se and Te," edited by The Chemical Society under the signature of F. Kurzer (14–16): this excellent review calls with insistence for "the volumes concerning the chemistry of thiazole, in the series of monographs edited by A. Weissberger." This book is an attempt to fill this gap.

Most chapters contain tables giving an inventory of all known derivatives of the family concerned. The classification of substituents in these tables is made according to the following hierarchy:

1. Nature of the heteroatom bonded to the ring: C⟩N⟩O⟩S⟩P.
2. Simply bonded⟩doubly bonded substituents.
3. Size of the substituent: –H⟩–CH$_3$⟩–CH$_2$R⟩–CHR$_2$⟩–CR$_3$⟩–CH=⟩ CR=⟩aryl⟩heteroaryl⟩–C≡⟩halogeno⟩–NH$_2$⟩–NHR⟩–NR$_2$⟩–N=⟩–OH⟩ OR⟩–SH⟩–SR⟩–PR$_2$⟩. . . .

In these tables most of the reference numbers are accompanied with small letters, the significance being that the paper concerned deals with one or several of the following topics:

a. Practical application
c. Cotton effect, circular dichroïsm
d. Dipole moment
e. Medicament, pesticid
f. Fluorescence, phosphorescence

g. Vapor-phase chromatography
h. HMO and other theoretical evaluations
i. Infrared spectrometry
k. Kinetic data
l. Living material
m. Mass spectrometry
n. Nuclear quadrupole resonance
o. Microwave spectrometry
p. Polarography
q. Raman spectrometry
r. Nuclear-magnetic-resonance spectrometry
s. Electron-spin resonance spectrometry
u. Ultraviolet spectrometry
v. Other physical determinations
w. Thin-layer chromatography
x. X-ray diffraction
z. Optical properties

REFERENCES

1. H. A. Kirst, E. F. Szymanski, D. E. Dorman, J. C. Occolowitz, N. D. Jones, M. O. Chaney, R. L. Hamill, and M. M. Hoehn, *J. Antibiot.*, **28,** 286 (1975).
2. P. Brookes, A. T. Fuller, and J. Walker, *J. Chem. Soc.*, 689 (1957).
3. M. N. G. James, and K. J. Watson, *J. Chem. Soc. C*, 1361 (1966); *Chem. Abstr.* **65,** 12189H.
4. J. Jadot, J. Casimir, and R. Warin, *Bull. Soc. Chim. Belges*, **78,** 299 (1969).
5. Y. Yamada, N. Seki, T. Kitahari, M. Takahashi, and M. Matsui, *Agric. Biol. Chem. Japan*, **35,** 780 (1970); *Chem. Abstr.* **73,** 35265.
6. R. Viani, J. Bricout, J. P. Marion, F. Mueggler-Chavan, D. Reymond, and R. H. Egli, *Helv. Chim. Acta*, **52,** 887 (1969), *Chem. Abstr.* **71,** 11875
7. O. G. Vitzthum, and P. Werkhoff, *J. Food Sci.*, **39,** 1210 (1974).
8. O. G. Vitzthum, and P. Werkhoff, *Z. Lebensm. Forsch.*, **156,** 300 (1974).
9. J. P. Walradt, A. O. Pittet, T. E. Kinlin, R. Muralidhara, and A. Sanderson, *J. Agric. Food Chem*, **19,** 972 (1971); *Chem. Abstr.* **75,** 108666.
10. H. J. Wobben, R. Timmer, R. Ter Heide, and P. J. De Valois, *J. Food Sci.*, **36,** 464 (1971); *Chem. Abstr.* **75,** 3997.
11. S. L. Gusinskaya, V. Y. Telly, and A. Aidogdyev, *Uzbek. Khim. Zhur.*, **11,** 21 (1967), *Chem. Abstr.* **67,** 4551.
12. J. M. Sprague, and A. H. Land, "Heterocyclic Compounds," R. C. Elderfield, ed., Wiley, New York, 1957, Vol. 5, p. 485.

13. B. Prijs, "Kartothek der Thiazol Verbindungen," Karger, Basel, 1952, Vols. 1–4,
14. F. Kurzer, "Organic Compounds of Sulphur, Selenium and Tellurium," D. H. Reid, Ed., Chemical Society, London, 1970, Vol. 1, p. 378.
15. F. Kurzer, "Organic Compounds of Sulphur, Selenium and Tellurium," D. H. Reid, Ed., Chemical Society, London, 1973, Vol. 2, p. 587.
16. F. Kurzer, "Organic Compounds of Sulphur, Selenium and Tellurium," D. H. Reid, Ed. Chemical Society, London, 1975, Vol. 3, p. 566.

I

Properties and Reactions of Thiazole

JACQUES V. METZGER

Laboratoire de Chimie Organique, Faculté des Sciences et Techniques de Saint-Jérôme, Marseille, France

EMILE-JEAN VINCENT

Laboratoire de Chimie Organique Physique, Faculté des Sciences et Techniques de aint-Jérôme, Marseille, France

with the collaboration of

JACQUES CHOUTEAU AND GILBERT MILLE

Centre de Spectroscopie Infra-rouge, Faculté des Sciences et Techniques de Saint-Jérôme, Marseille, France

```
I. General Aspects . . . . . . . . . . . . . . . . . . . . . . .    7
   1. Introduction . . . . . . . . . . . . . . . . . . . . . . .    7
      A. Numbering the Thiazole Ring . . . . . . . . . . . . . .    7
   2. A Brief History of the Discovery of the Thiazole Ring . . .   8
   3. The Long-Known Derivatives of Thiazole . . . . . . . . . .   14
      A. Thiocyanoacetic Acid . . . . . . . . . . . . . . . . .    15
      B. Thiohydantoin . . . . . . . . . . . . . . . . . . . . .   16
```

	C. Rhodanine	19
	D. 2-Mercaptothiazole	21
	E. 2-Aminothiazole	22
	4. Preparation	24
	5. Theoretical Model, Electronic Structure	26
	A. Influence of Substituents	42
II.	Physical Properties	45
	1. Geometrical Structure	45
	2. Ultraviolet Absorption Spectroscopy	46
	A. Quaternary Salts	50
	B. Phosphorescence Spectra	50
	3. Photoelectron Spectroscopy	51
	4. Infrared Absorption and Raman Diffusion	53
	A. Numbering, Activity, and Nomenclature of the Vibrations	53
	B. Assignments	56
	C. Thiazole Derivatives	63
	5. Nuclear Magnetic Resonance	66
	6. Mass Spectrometry	81
	7. Electronic Paramagnetic Resonance Spectrometry	84
	8. Thermodynamic Data	85
	9. Magnetic Susceptibility	89
	10. Dipole Moment	89
	11. Refractivity	90
III.	Chemical Reactivity	90
	1. Theoretical Aspects	90
	2. Basicity	91
	3. Additions to the Thiazole System	94
	4. Electrophilic Substitution	99
	5. Free Radical Substitution	108
	6. Nucleophilic Substitution	113
	A. Hydrogen-Deuterium Exchange Reactions	113
	B. Hydrogen-Metal Exchange Reactions	119
	C. Nucleophilic Amination	124
	7. Thiazole as an Electron Donor	125
	A. Quaternary Thiazolium Salts Formation	125
	B. Thiazole Complexes with Metal Atoms	126
	8. Miscellaneous Reactions	130
	A. Oxidation	130
	B. Reduction	132
	C. Ring Cleavage	134
	D. Photochemically Induced Transformations	136
	E. Ring Expansion to 1,4-Thiazines	139
IV.	Effects of the Thiazole Ring on Substituents	143
	1. Alkylthiazoles	143
	2. α-Chloroethylthiazoles	146
	3. Halothiazoles	147
	4. Hydroxy-, Mercapto-, and Aminothiazoles	147
V.	Linear Free-Energy Relationships	147
VI.	References	149

I. GENERAL ASPECTS

1. Introduction

This chapter follows the general pattern of the series, starting in Section I with a historical survey of the most important points in the discovery of thiazole and the long-known derivatives of this molecule, followed by its preparations and the description of its electronic structure. The second section presents the most interesting physical properties of thiazole and some of its simple derivatives. The third section is devoted to the chemical reactivity of the thiazole molecule; it is followed by a short section relating the effect of the thiazole ring on substituents, and finally a brief discussion of free-energy relationships in this system is given. Most of the general aspects of thiazole chemistry are only briefly discussed here: they are developed in the following chapters.

A. *Numbering the Thiazole Ring*

First described by A. Hantzsch and J. Weber (1) as "the pyridine of the thiophene series," thiazole (**1**) received a nomenclature conforming to this definition (2): the letters α and β were justified by the great analogy of both positions with the corresponding ones of pyridine, while the *meso* position, to which μ was assigned, had no equivalent in the pyridine ring. In Richter's *Lexikon*, as well as in the second edition of Meyer-Jacobson's *Lehrbuch der Organischen Chemie*, the sulfur atom was numbered 1 and the nitrogen 3 (**2**). Other nomenclatures have also been used: T. B. Johnson (3) proposed the numbering **3**, whereas J. Metzger et al. (4–14) used another system (**4**) then recommended by Grignard's *Traîté de Chimie Organique* (15). The correct numbering is that given by The Ring Index (16) and Chemical Abstracts, and corresponds to **2**.

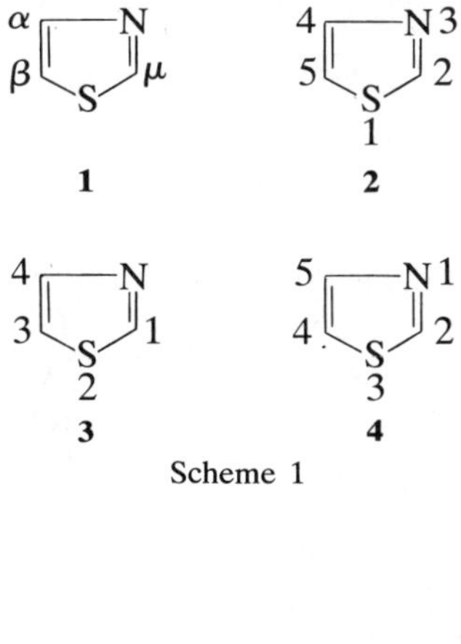

Scheme 1

2. A Brief History of the Discovery of the Thiazole Ring

On November 18, 1887, Arthur Rudolf Hantzsch with his collaborator, J. H. Weber, signed the birth certificate of thiazole. Under the title, "Ueber Verbindungen des Thiazoles (Pyridins der Thiophenreihe)," Hantzsch and Weber (1) gave the following definition: "Unter Thiazolen verstehen wir diejenigen Stickstoff und Schwefel in ringförmiger Bindung enthaltenden Substanzen von Formel $(CH)_3NS$, welche sich zum Pyridin verhalten, wie das Thiophen zum Benzol." They proved the existence of both thiazole (**5**) and isothiazole (**6**), although neither of these was then known in the free state, compared them to glyoxaline and pyrazole, respectively, and proposed to name benzothiazole the "Methenylamidophenylmercaptan" just discovered by A. W. Hofmann (17). Extending this definition they proposed naming the three azole derivatives of thiophen, pyrrole, and furan: thiazole, imidazole and oxazole, respectively, and their benzo-derivatives, benzothiazole, benzimidazole, and benzoxazole. Making allowance for the high reactivity of the groups attached to the carbon atom placed between sulfur and nitrogen they proposed the prefix *meso* for this position in the normal azole series. They noted that numerous derivatives of thiazole could already be known without their constitution being correctly described: they ascribed all the so-called α-thiocyano derivatives of ketones and aldehydes (**7**) ("Rhodanketone") to being thiazole derivatives, especially *meso*-oxythiazoles (**8**).

Scheme 2

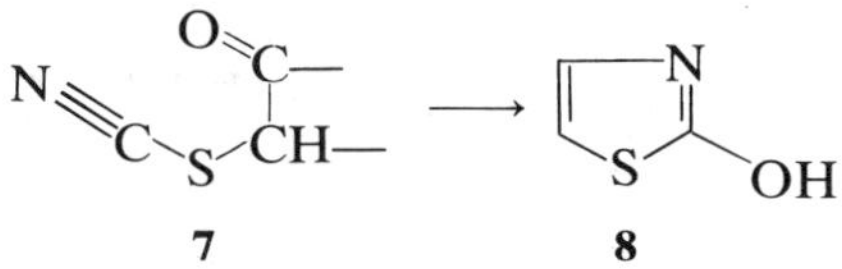

Scheme 3

Hantzsch and Weber began their description with the compound which led them indirectly to the discovery of the thiazoles: the α-thiocyanoacetone imine ("Rhodanpropimin") of J. Tcherniac and C. H. Norton, $C_4H_6N_2S$, obtained by reaction of ammonium thiocyanate with chloroacetone. After Tcherniac and Norton (18), the α thiocyanoacetone

(**9**) first formed would react with a second molecule of ammonium thiocyanate to give the thiocyanate of α-thiocyanoacetone imine (**10**). The physical properties of the new base prompted Hantzsch and Weber to study its structure carefully, and they discovered it to be cyclic and derived from the thiazole nucleus (**11**):

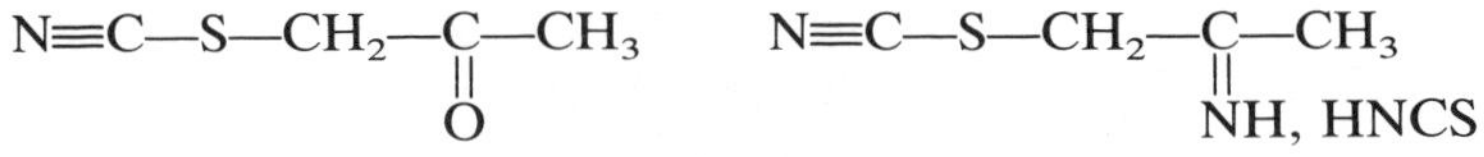

$$N\equiv C-S-CH_2-\underset{\underset{O}{\|}}{C}-CH_3 \qquad N\equiv C-S-CH_2-\underset{\underset{NH,\ HNCS}{\|}}{C}-CH_3$$

9   10

Scheme 4

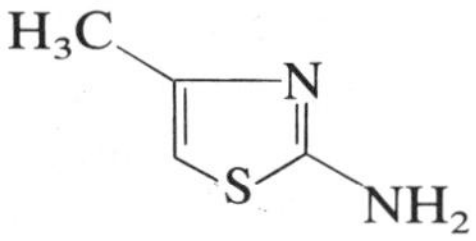

[structure of 2-amino-4-methylthiazole]
11

Scheme 5

m-aminomethylthiazole. Although most of the classical reactions of primary amines were unsuccessful, reaction with methyl iodide showed the presence of two (and only two) hydrogens fixed at nitrogen, and the acetylation proved them to be attached to the same nitrogen atom. The dimethylamino derivative (**12**) thus obtained reacted with bromine to give a monobromo compound (**13**) the constitution of which was only compatible with the formula of a trisubstituted thiazole.

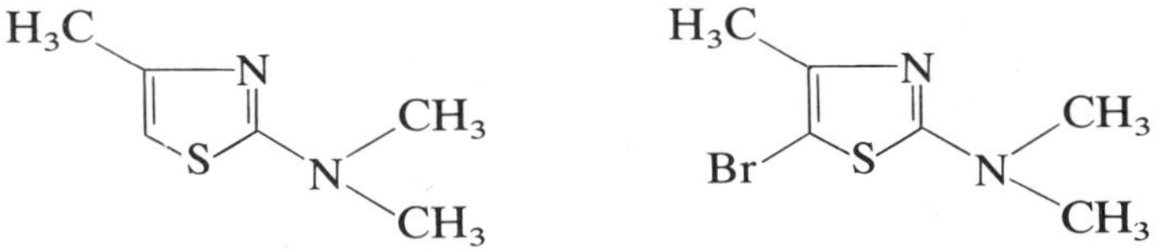

[structures 12 and 13]
12   13

Scheme 6

Looking back to the precursor of the aminomethylthiazole (**11**), Hantzsch and Weber studied the so-called thiocyanoacetone (**9**). This compound had been already prepared by Tcherniac and Hellon (19) by the action of barium thiocyanate with chloroacetone and was described by these authors as an oily product ("Rhodanacetone"). Hantzsch and Weber, introducing a *slight* modification to the preparation given by Tcherniac and Hellon (salting out the product from its aqueous solution by adding Na_2CO_3), obtained a white solid (m.p. 98°) that did not show the classical properties of ketones but contained an hydroxyl group exchangeable with chlorine under the action of PCl_5, and to which they attributed the structure of m-oxymethylthiazole (**14**). The cyclic nature of

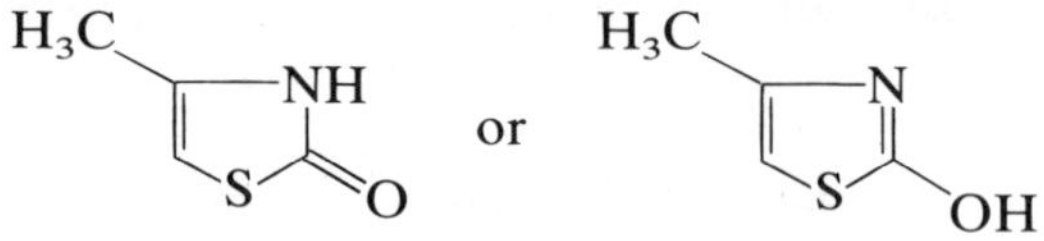

14a   or   **14b**

Scheme 7

the compound was compatible with the analogous formula attributed eight years earlier by C. Liebermann and A. Lange (20) to the product called "Senfölessigsaüre" (senevolacetic acid) (**17**), obtained, according to Claesson (21) and Nencki (22), by hydration of sulfocyanoacetic acid (**15**) to give 'carbaminthioglycolic acid" (**16**) and its subsequent dehydration: Hantzsch and Weber proposed the name dioxythiazole for **17**. In the

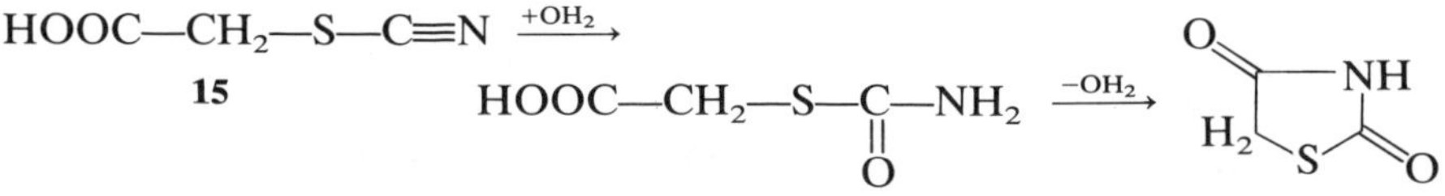

HOOC—CH$_2$—S—C≡N $\xrightarrow{+OH_2}$

15

HOOC—CH$_2$—S—C(=O)—NH$_2$ $\xrightarrow{-OH_2}$

16

17

Scheme 8

same way one could consider the similar structure of the "rhodaninacetic acid" of M. Nencki (22), the constitution of which was demonstrated by Liebermann and Lange (20) to be that of a sulfur derivative of dioxythiazole (**18**). Finally, in a subsequent paper (23), Hantzsch and Weber recalled the thiazolic structure of the long-known thiohydantoïne of Maly (24) to which Liebermann and Lange (20) attributed the cyclic formula **19**. Shortly after, Arapides (25), a collaborator of Hantzsch,

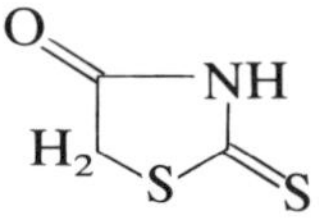

18

Scheme 9

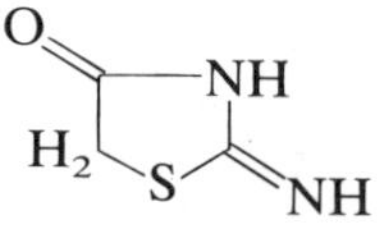

19

Scheme 10

I. General Aspects 11

prepared thiocyanoacetophenone (**20**) (m.p. 74°C) (cf. R. Dyckerhoff (26) and proved it reactive in hot concentrated hydrochloric acid, forming the soluble "carbaminthioacetophenone" (**21**) (m.p. of the chlorhydrate, 175 to 180°C), which by further heating of the solution eliminated 1 mole of water to give 2-oxy-4-phenylthiazole (**22**) (m.p. 204°C). He isolated thiocyanoacetone (**9**) working in accordance with the data of J. Tcherniac and

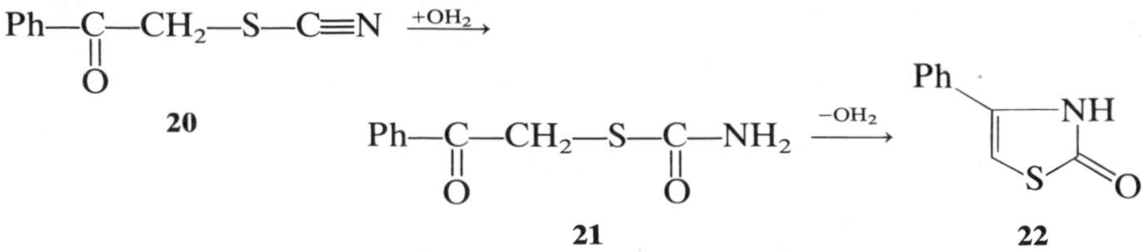

Ph—C(=O)—CH$_2$—S—C≡N $\xrightarrow{+OH_2}$ Ph—C(=O)—CH$_2$—S—C(=O)—NH$_2$ $\xrightarrow{-OH_2}$ [2-oxy-4-phenylthiazole]

20 **21** **22**

Scheme 11

R. Hellon (19) but using anhydrous reagents. He obtained an oil that reacted exothermically with hydroxylamine (oxime of m.p. 135°) and that isomerized to 2-oxy-4-methylthiazole (**14**) upon heating with diluted hydrochloric acid. The thiazolic nature of oxymethylthiazole was clearly demonstrated by its reduction by zinc powder distillation into 4-methylthiazole (**23**), the first free thiazole ever described.

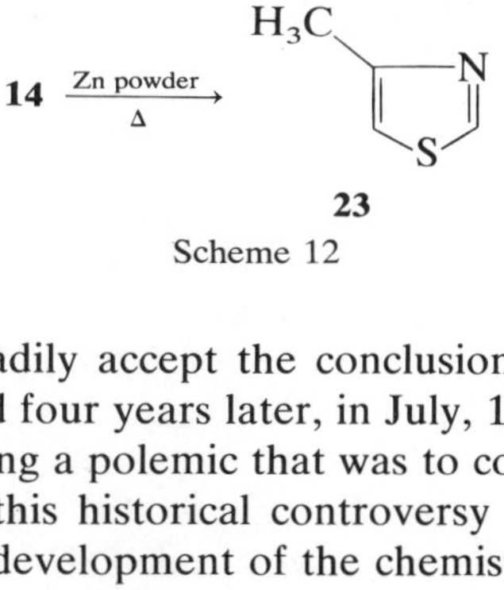

14 $\xrightarrow[\Delta]{\text{Zn powder}}$ [4-methylthiazole]

23

Scheme 12

Tcherniac did not readily accept the conclusions of Hantzsch and his collaborators (1, 25) and four years later, in July, 1892, published a series of papers (27–30) opening a polemic that was to continue for 36 years! A few details concerning this historical controversy merit being related to give an insight into the development of the chemistry of thiazole. In July, 1892, Tcherniac condemned the unsparing criticism of Hantzsch, who asserted (1) that the product obtained by Tcherniac and Hellon (19) was an impure sample of 4-methyl-2-oxythiazole; the impudence of Arapides (25) who claimed the discovery of the true thiocyanatoacetone; and explained the failure of the method of Tcherniac and Hellon by the ability of thiocyanoacetone to isomerize into methyloxythiazole by a hydration-dehydration process, catalyzed by strong acids. Tcherniac, again taking up his preparation of thiocyanatoacetone, used highly purified material and careful conditions of extraction of the product. He asserted that he had obtained an analytically pure sample of thiocyanoacetone that resisted isomerization either by contact with water or even by heating in a waterbath with dilute HCl. On the contrary, isomerization took place upon contact with a basic reagent, like the Na_2CO_3 used by Hantzsch and his collaborators to extract the product of reaction, or $CaCl_2$ (always containing traces of $Ca(OH)_2$) used to dehydrate their etheral solutions. To determine the purity of these samples of thiocyanoacetone Tcherniac measured the elevation of temperature caused by their isomerization to methyloxythiazole under the influence of Na_2CO_3.

Three months later, in October, Hantzsch published an impetuous response (31) to the series of papers by Tcherniac, in which his work with J. Weber (1) and Arapides (25) had been severely criticized. He emphasized the correctness of the structure proposed for "Rhodanpropimin" and, as for the thiocyanoacetone of Tcherniac, he recalled the scarcity of the data given by Tcherniac and Hellon concerning the purity and the properties of the so-called "Rhodanacetone." He defended both his collaborators, showing how Arapides was able to discover the isomerization of thiocyanoacetone to thiazole. Finally he protested vehemently against "the attitude of that chemist who had not been able to note the inconsistency between the wrong formula he attributed to Rhodanpronimin and its chemical behavior, who had not been able to pick up oxymethylthiazole, though well crystallized, and who accused of crude errors conclusions that corrected, in several essential points, his own statements and thus allowed opening the field, then new, of thiazole." Two months later, December 5, 1892, Tcherniac, in an overbearing paper (32), described with self-satisfaction an accurate method using $NaHCO_3$ as a basic reagent to induce the isomerization of thiocyanoacetone into methyloxythiazole. Operating on large quantities he was able to obtain, with a yield of 40%, a sample of 2-oxy-4-methylthiazole melting at 105 to 106°C, whereas Hantzsch reported a melting point of 98°.

The controversy seemed then to be closed. In 1890 Hantzsch had already started his work on the structure of oximes, and his synthetic work on heterocycles was practically ended. However, 27 years later, in July 1919, Tcherniac published a new paper entitled 'Thiocyanoacetone and its derivatives as isomerides" (33), where, after the description of improved and generalized methods for the preparation of thiocyanoacetone he came to the explosive conclusion that "the substance which has been known since 1887 as hydroxymethylthiazole is not a thiazole at all. It might be called 2-imino-4-methylthioxole, but for the sake of simplicity, and in view of the now proved existence of two other isomerides of thiocyanoacetone, it seems preferable to adopt the generic name of *rhodim* for this class of compounds." The strongest arguments of Tcherniac against formula **14a, 14b** for hydroxymethylthiazole were (1) that it behaved neither as an alcohol nor as a ketone (!), (2) that under the action of PCl_5 the oxygen atom was left intact while chlorine was introduced into the molecule, and (3) that neither of the two methyl derivatives (**14** and **25**) could give methylamine on hydrolysis.

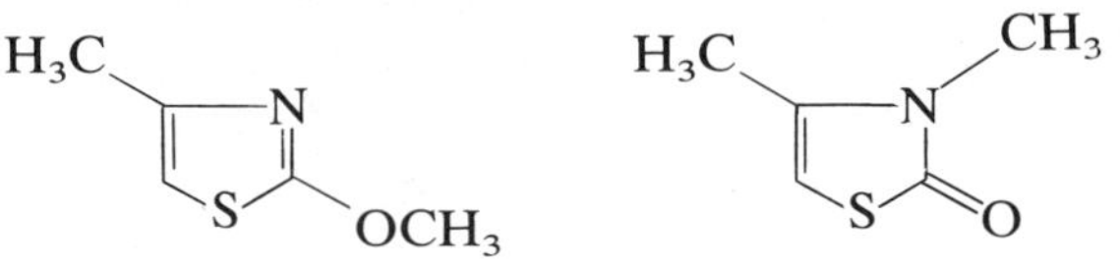

Scheme 13

He then concluded that the NCH_3 group occupied the 2-position, bringing him to the formula (**27**) for "hydroxymethylthiazole," the formation of which could easily be derived by the following scheme from thiocyanoacetone.

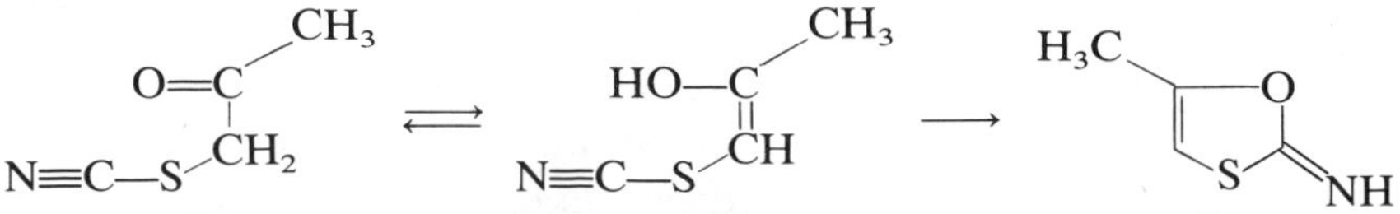

Scheme 14

He named compound **27**, α-methylrhodim. He further described two other isomers named β-methylrhodim and isomethylrhodim, the first being a dimer of the α-isomer (**27**), the second probably a tetramer of the oxymethylthiazole of Hantzsch (**14a**). This paper was apparently ignored by Hantzsch until 1927 when he discovered it and refuted without difficulty the arguments alleged in favor of the rhodim formula (**27**) (34). He demonstrated particularly the tautomerism of oxymethylthiazole (**14a, 14b**) through its spectrochemical behavior and its reaction towards diazomethane as a function of solvent, emphasized the amidic nature of the oxygen atom, and gave the correct formula (**28**) of the chloroderivative. Two months later, Tcherniac published a self-sufficient answer (35)

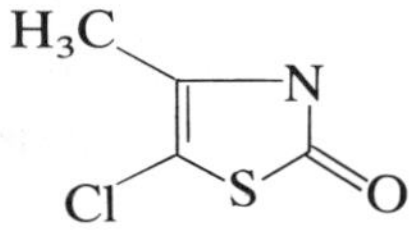

Scheme 15

in which he disputed once more the different points concerning the structure of α-methylrhodim and concluded by proudly confirming his authorship for the formula and the name of the isomeric rhodims.

The finishing stroke to the wrong theories of Tcherniac was given in July 1928 by the last paper of Hantzsch on thiazoles (36): he methodically refuted the erroneous denials of Tcherniac regarding the acidic cyclization of thiocyanoacetone, its poorer basic cyclization, and its reaction with anhydrous NH_3 or methylamine to give aminothiazoles. He conclusively demonstrated the nonexistence of the supposed rhodim, by the tautomerisation of 4-methylthiazolone (**14**), the easy substitution of its 2-OH group by $POCl_3$, the well-known reduction of thiazolone into thiazole. He proved that both cyclic-*N*-methyl (**29**) and *exo-N*-methyl (**30**) isomers of methylated aminothiazole give ammonia and methylamine

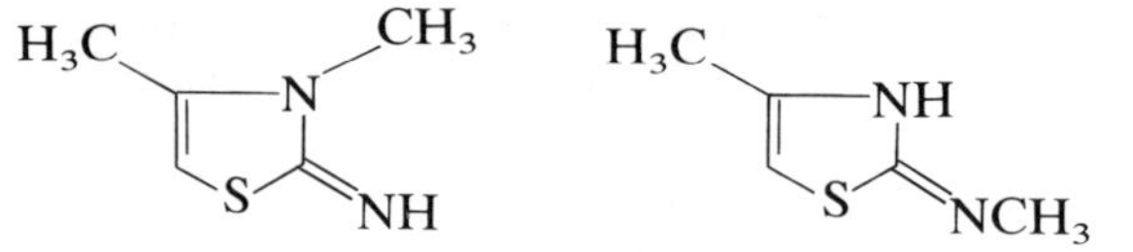

Scheme 16

on hydrolysis with HCl. He clearly demonstrated the two supposedly isomers, β- and *iso*- of the methylrhodim, to be dimeric 4-methylthiazolone and cyclic trimeric thiocyanoacetone (triacetonylthiocyanurate) (**31**), respectively, the supposedly dimethylrhodim to be 3,4-dimethylthiazolone (**25**) and the supposedly chloromethylrhodim to be 5-chloro-4-methylthiazolone (**28**).

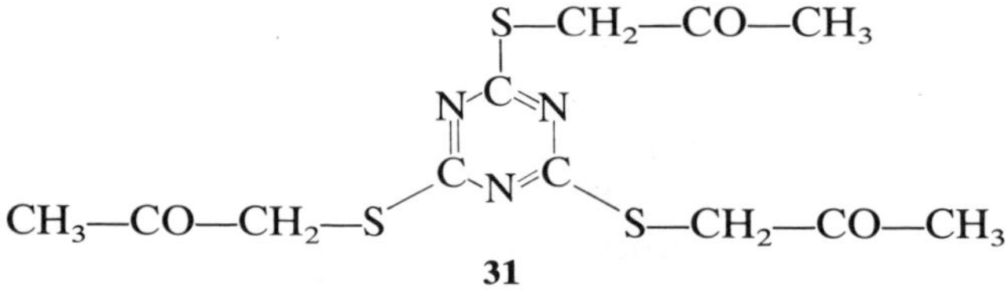

Scheme 17

3. The Long-Known Derivatives of Thiazole

Among the thiazole derivatives known for a long time, but whose structure was not at once acknowledged as such, leading to some controversies, are so-called thiocyanoacetic acid, thiohydantoine, rhodanine, mercaptothiazole, and sulfuvinuric acid.

A. Thiocyanoacetic Acid

In 1865, Heintz (37) prepared ethylthiocyanoacetate (**32**) from potassium sulfocyanate and ethyl monochloroacetate. By saponification he obtained white crystals (m.p. 128°C), which he claimed to be thiocyanoacetic acid (**33**). In 1874, Volhard (38, 39), condensing thiourea with chloracetic acid, attributed to the product a formula (**34**) similar to that of hydantoin (prepared from urea and chloroacetic acid) and named it glycolylthiourea or thiohydantoin. Thiohydantoin was hydrolyzed by aqueous HCl to give a white solid (m.p. 100°C), which he supposed was isothiocyanoacetic acid (**35**) (Senfölessigsäure). In 1877, Claesson (21) improved the reaction described by Volhard, and Nencki (22) developed the reaction of free thiocyanic acid with chloroacetic acid to give the supposed carbamylthioglycolic acid (**36**). By action of ammonium

$$KSCN + Cl-CH_2-CO_2Et \longrightarrow$$

$$N{\equiv}C{-}S{-}CH_2{-}CO_2Et \xrightarrow[2)ClH]{1)KOH} N{\equiv}C{-}S{-}CH_2{-}CO_2H$$

32 **33**

Scheme 18

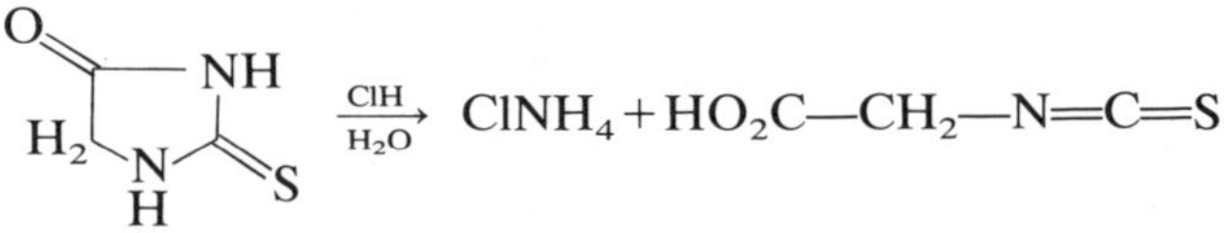

[structure **34**: thiohydantoin ring] $\xrightarrow[H_2O]{ClH}$ $ClNH_4 + HO_2C-CH_2-N{=}C{=}S$

34 **35**

Scheme 19

$$H{-}SCN + Cl{-}CH_2CO_2H \xrightarrow{H_2O} H_2N{-}\underset{\underset{O}{\|}}{C}{-}S{-}CH_2{-}CO_2H$$

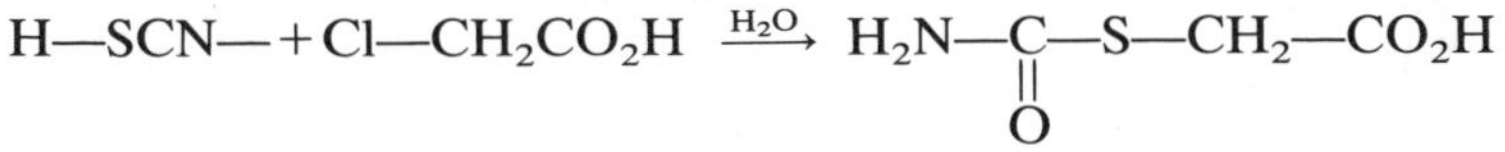

36

Scheme 20

thiocyanate he obtained a product he called rhodaninic acid. In 1879, Liebermann and Lange (20) by alkaline hydrolysis of the so-called thiohydantoine (**34**) obtained thioglycolic acid and demonstrated that the compound was derived from a *S,N*-pentaatomic ring instead of the *N,N*-ring of hydantoin). The new formulation of thiohydantoin (**37**) had to be extended:

1. To the case of the isothiocyanoacetic acid (**35**) of Volhard (39), to which formula **38** had to be attributed.

2. To the rhodaninic acid of Nencki (22), for which he proposed formula **39**.

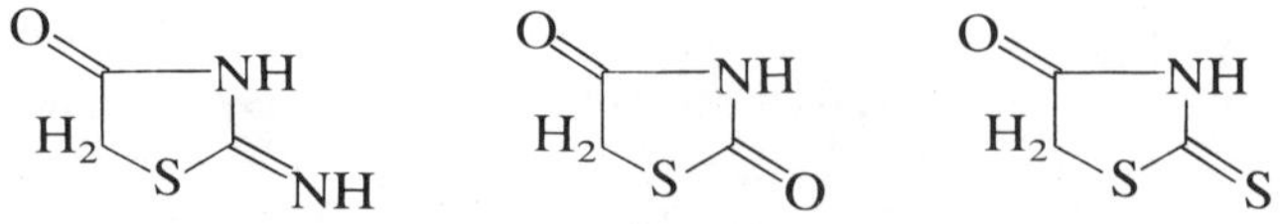

37 **38** **39**

Scheme 21

In 1880, Liebermann and Völtzkow (40), and then Völtzkow (41), condensing chloroacetic acid with both ethyl N-phenylthiocarbamate and p-tolylisothiocyanate obtained homologous compounds to which they attributed formulas **40** and **41**, whereas their structure probably derives from that of **38** by substituting an aryl group on the cyclic nitrogen.

In 1887, Hantzsch and Weber (1) confirmed the thiazolic structure of the so-called isothiocyanoacetic acid (**38**).

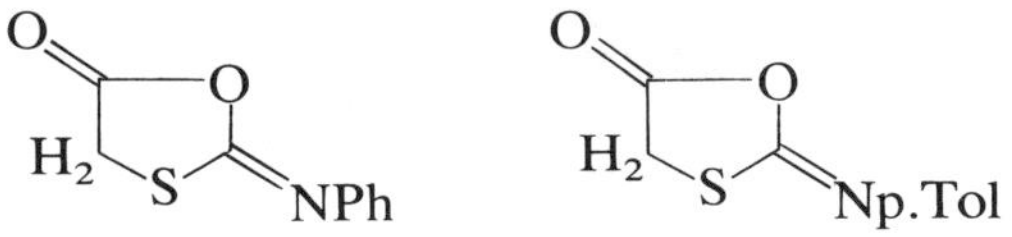

40 **41**

Scheme 22

B. Thiohydantoin

In 1873, almost simultaneously, Maly (24), Volhard (38), and Nencki (42) studied the action of thiourea on chloroacetic acid. As mentioned previously, they believed the product to be the thioanalog of hydantoin and called it thiohydantoin with formula **34**.

In 1874, Volhard (39) showed that thiohydantoin resisted desulfuration, proving its sulfur atom to be more firmly bonded than formula **34** indicated.

In 1875, Mulder (43) extended the synthesis reaction of thiohydantoine to the ethyl ester and amide of chloroacetic acid. Claus (44) demonstrated the acidic properties of thiohydantoin and its ability to form metallic salts.

In 1877, Maly (45) discussing formula **34** applied to thiohydantoine found it unable to explain the basic properties of the compound. He preferred a structure in which the $-CH_2-CO-$ group would be bonded to only one nitrogen atom. Meyer (46) prepared a monophenyl thiohydantoin (m.p. 178°C) by condensing chloroacetanilide with thiourea and proposed **42** for its structure.

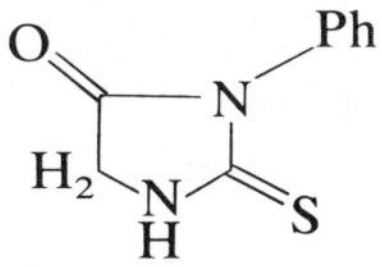

I. General Aspects

42

Scheme 23

In 1879, Lange (47) prepared N,N'-diphenylthiohydantoin (m.p. 178°C) by condensing chloroacetic acid with N,N'-diphenylthiourea and attributed to it structure **43**, noting however that it was difficult to derive from I-**43** the structure of the monophenyl derivative (m.p. 148°C)

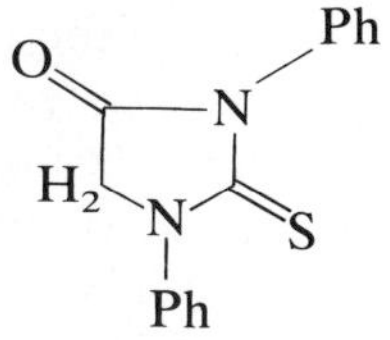

43

Scheme 24

resulting from its acidic hydrolysis. The same year Andreasch (48) hydrolyzing thiohydantoin in alkaline medium obtained thioglycolic acid, a result difficult to reconcile with structure **34**. The credit for the rationalization of all the experimental data concerning the acid hydrolysis of thiohydantoine and the basic hydrolysis of its derivatives, went to Liebermann and Lange (20), who proposed a new structure for this compound (**37**), and for its diphenyl derivative (**44**).

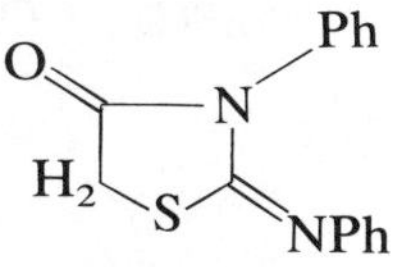

44

Scheme 25

The formation of a sulfur-containing ring was justified by the attack of the halogenated carbon of the chloroacetyl derivative by the sulfur atom of thiourea, a fact in accordance with the results just discussed by Wallach (49, 50) and Claus (51). The new formula (**37**) of thiohydantoine explained why, contrary to thiourea, its desulfuration was difficult.

In 1880, R. Andreasch (52) confirmed the new formula (**37**) by preparing thiohydantoine through condensation of thioglycolic acid with cyanamide (the reverse reaction of the basic hydrolysis of thiohydantoin).

In 1881, Liebermann and Lange (53) confirmed their previous conclusions (20) and proposed formula **45** for the monophenyl derivative obtained by Meyer (**46**) (m.p. 178°C).

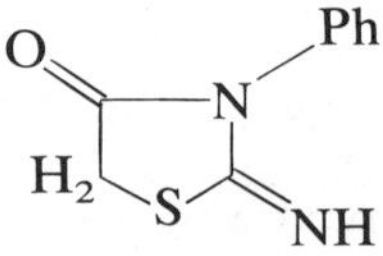
O
Ph
N
H2
S
NH

45

Scheme 26

Then followed a twenty-year period during which several laboratories contributed to the development of syntheses in the thiohydantoin series (54–63) but could not unambiguously solve the structural problem of the *N*-monosubstituted derivatives (64–67) until 1902, when Wheeler and Johnson (68) examining the structure of the two isomeric monophenylthiohydantoin proved the inability of acid hydrolysis to determine the location of the phenyl group either on the ring nitrogen (**46**) or on the exocyclic nitrogen (**47**). They demonstrated that the reaction proceeded via ring opening, to phenylthiohydantoïc acid (**48**), followed by ring closure leading to a mixture of two 2,4-diketotetrahydrothiazoles (**49**) and (**50**). They demonstrated that when chloroacetanilide was gently warmed

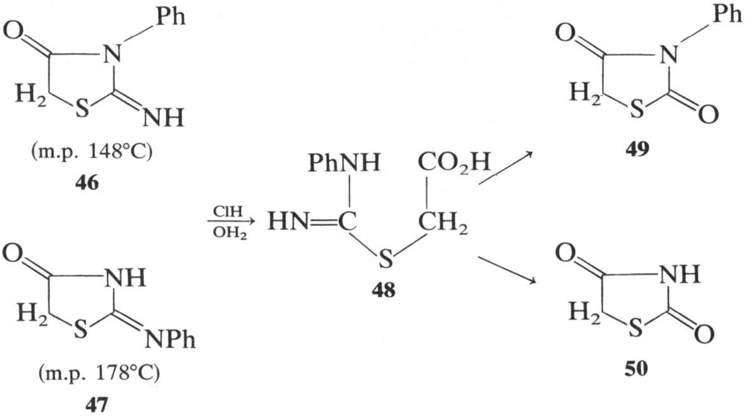
(m.p. 148°C)
46
(m.p. 178°C)
47
ClH
OH2
PhNH
CO2H
HN=C
CH2
S
48
49
50

Scheme 27

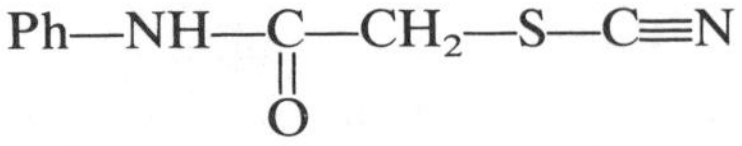
Ph—NH—C—CH2—S—C≡N
O

(m.p. 91°C)

51

Scheme 28

with KSCN in EtOH for ½ hr the first product to be formed was a true sulfocyanoacetanilide (**51**) (m.p. 91°C). When **51** was melted in a water bath for 15 min, it isomerized to a compound of m.p. 148°C identified as the labile 3-phenylthiohydantoin (**46**). By heating **46** or **51** at a higher temperature or for a longer time, they changed to the stable phenylthiohydantoin (**47**) of m.p. 178°C, first obtained Meyer (46). The confirmation of the attack on the sulfur atom of thiourea on the $ClCH_2$–group of chloroacetyl derivatives was brought by Dixon and Taylor (69).

C. Rhodanine

In 1877, Nencki (22) condensing ammonium thiocyanate with chloroacetic acid, attributed the name rhodaninic acid (Rhodaninsäure) to the compound he obtained. He noted the ability of rhodaninic acid to give colored derivatives with ferric salts.

In 1879, Liebermann and Lange (20) demonstrated the formula of rhodaninic acid to be **39**.

In 1884, Nencki (70), studying the properties of rhodaninic acid, confirmed the formula proposed by Libermann and Lange (**39**) and noted its ability to condense with aldehydes.

In 1886, Ginsburg and Bondzynski (71) and Berlinerblau (72) developed the reaction of rhodanine with aldehydes, but proposed a chain formula for the condensation product.

In 1887 (57) and 1889 (73) Andreasch confirmed the cyclic formula proposed by Liebermann and Lange (20). Freydl (74) obtained rhodaninic acid by condensing thioglycolic acid with thiocyanic acid.

In 1891, Miolati (75) confirmed the cyclic formula of Liebermann and Lange by preparing the compound by three new pathways: (1) reaction of CS_2 on thiohydantoin, (2) condensation of ammonium dithiocarbamate with chloroacetic ester, (3) reaction of H_2S on thiocyanoacetic acid.

In 1902, von Braun (76) showed that ammonium N-phenyldithiocarbamate condensed at low temperature to give ethyl N-phenyldithiocarbamylacetate (**52**) which, on heating, led to N-phenylrhodanine (**53**).

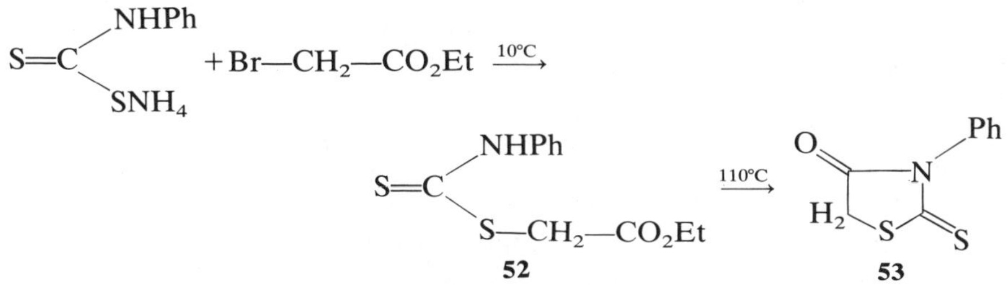

$$S=C\begin{smallmatrix}NHPh\\SNH_4\end{smallmatrix} + Br-CH_2-CO_2Et \xrightarrow{10°C}$$

$$S=C\begin{smallmatrix}NHPh\\S-CH_2-CO_2Et\end{smallmatrix} \xrightarrow{110°C}$$

52   **53**

Scheme 29

In 1902, Wheeler and Johnson (63) obtained 5-substituted rhodanines (**54**) by condensing substituted bromomalonic esters with postassium thiocyanate, then thiolacetic acid, the cyclization resulting from an alkaline treatment.

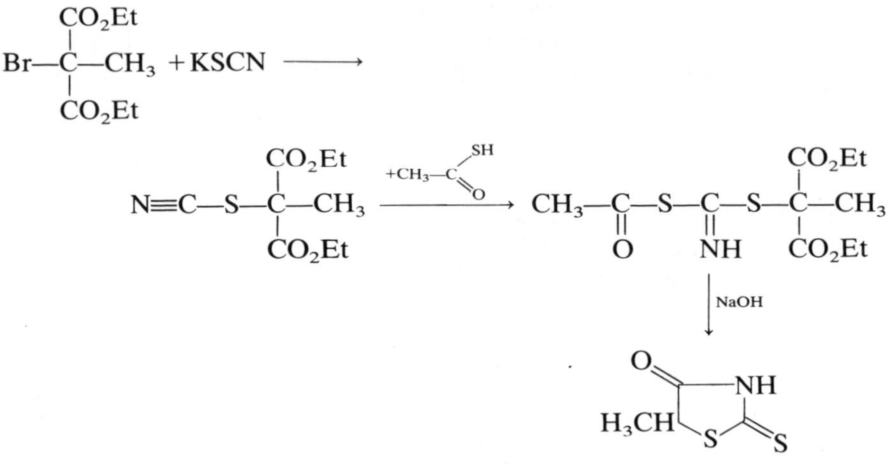

Scheme 30

In 1908, Körner (77) developed the synthesis of *N*-substituted rhodanines (**55**) from dithiocarbaminacetic acid:

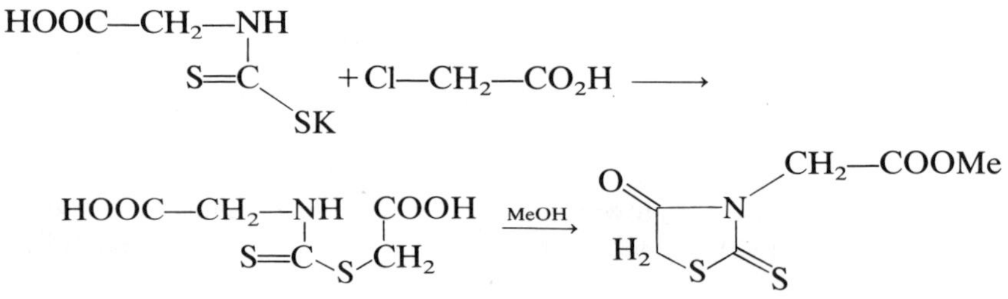

Scheme 31

During a relatively long period, the condensation of rhodanines with aldehydes was developed, especially by Andreasch's group (78–87). Finally, Holmberg (88, 89) described the best method to obtain rhodanines: the condensation of ammonium dithiocarbamate with a sodium or potassium salt of an α-chloro acid.

D. 2-Mercaptothiazole

In 1882, Will (90), by reacting CS$_2$ with the product (**56**) resulting from the condensation of dibromoethane with N,N'-diphenylthiourea, obtained the first derivative of thiazolidine-2-thione (**57**). He observed the reaction of **57** with methyl iodide to afford an addition compound (**58**).

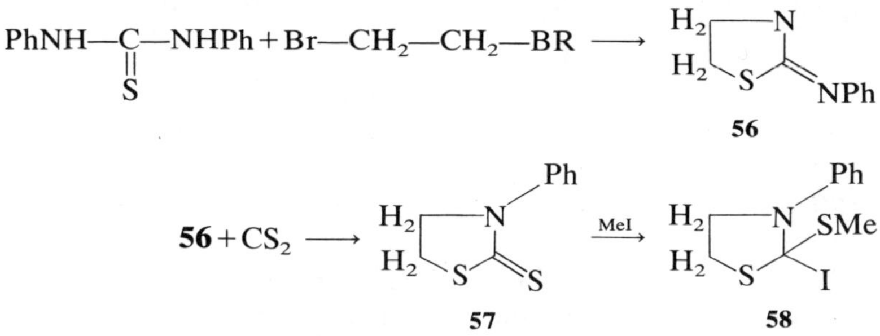

Scheme 32

In 1888, Foerster (91), reproducing the same reaction with dianisylthiourea, demonstrated that the compound he obtained (**59**) could lose a sulfur atom by reduction with tin and hydrochloric acid to form a product analogous to N-phenylpiperidine (**60**).

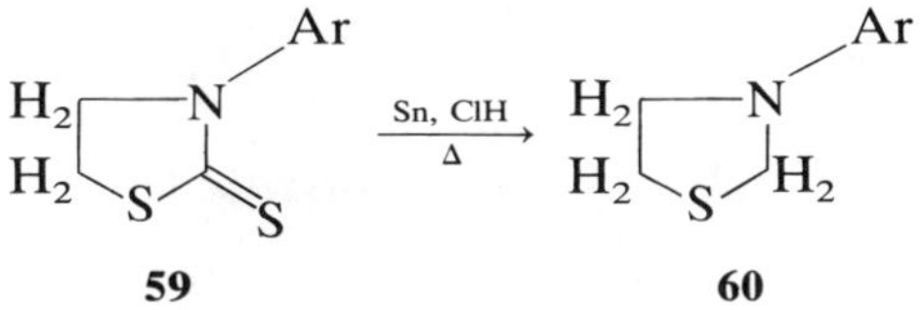

Scheme 33

In 1890, Gabriel and Lauer (92) established that α-bromoamines react with CS$_2$ to give 2-mercaptothiazolines (**61**). In the same laboratory, Hirsch (93) reacted μ-mercapto β-methylthiazoline (**61**) with various alkyl iodides and obtained the corresponding S-alkyl derivatives.

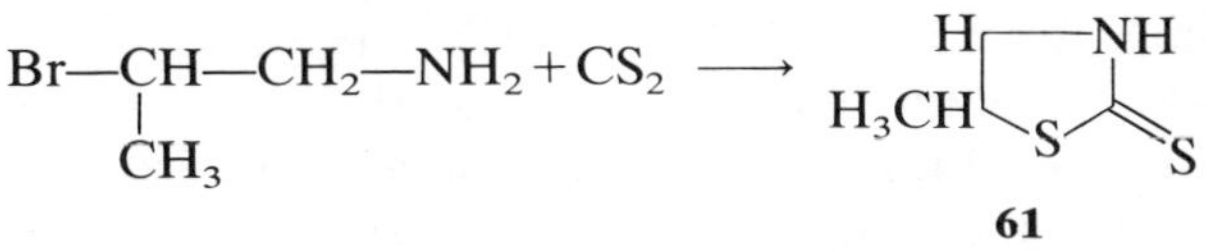

Scheme 34

In 1892, Marchesini (94) described the synthesis of 4-phenyl-2-mercaptothiazole (**62**) by condensing bromoacetophenone with ammonium dithiocarbamate.

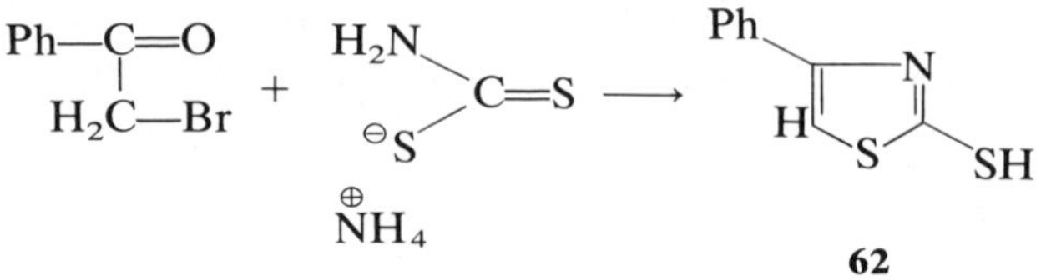

$$Ph-\underset{H_2C-Br}{\overset{\displaystyle C=O}{|}} + \underset{\overset{\oplus}{NH_4}}{\underset{\ominus S}{H_2N}}C=S \longrightarrow \underset{\mathbf{62}}{\overset{Ph}{\underset{H}{\bigsqcup_{S}^{N}}}}SH$$

Scheme 35

In 1893, Miolati (95) generalized the reaction of Marchesini to α-halocarbonyl compounds.

In 1896, Gabriel and Freiherr von Hirsch (96) condensed CS_2 with what they believed to be 1-aminopropylene (2-methylaziridine) and obtained isomeric 4- and 5-methyl-2-mercaptothiazoline (**63** and **64**).

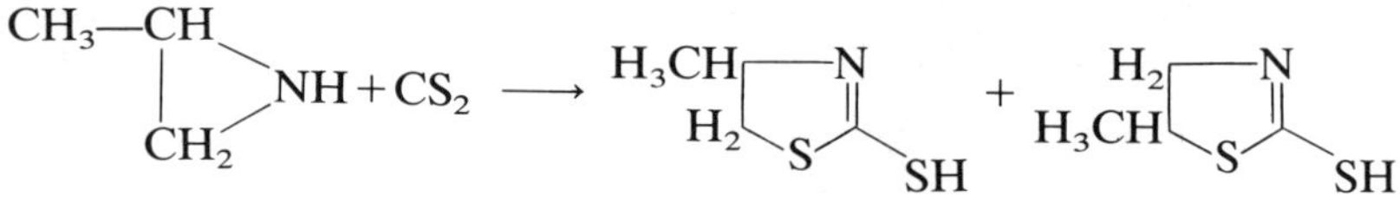

$$\underset{CH_2}{\overset{CH_3-CH}{|}}NH + CS_2 \longrightarrow \underset{\mathbf{63}}{\overset{H_3CH}{\underset{H_2}{\bigsqcup_{S}^{N}}}}SH + \underset{\mathbf{64}}{\overset{H_2}{\underset{H_3CH}{\bigsqcup_{S}^{N}}}}SH$$

Scheme 36

E. 2-Aminothiazole

In 1881, Will (97) obtained 2-phenylamino-*N*-phenylthiazolidine (**56**) by condensing 1,2-dibromoethane with *N,N'*-diphenylthiourea.

In 1882, Nencki and Sieber (98), condensing dibromopyruvic acid with thiourea, obtained a compound they named sulfuvinuric acid (Sulfuvinursäure), which was later demonstrated (99) to be a derivative of 2-aminothiazole. The same year Will (100) observed that the sulfur atom of **56** is masked to $Pb(NO_3)_2$ and to alkalis.

In 1883, Andreasch (101) oxidized Will's compound **56** and identified the oxidation product as $Ph-NH-CO-N(Ph)CH_2-CH_2SO_3H$.

In 1887, Hantzsch and Weber (1) proposed the structure of μ-amino-α-methylthiazole (**66**) for the "Rhodanpropimin" prepared by Tcherniac and Norton (18), and, until then, believed to be thiocyanoacetoneimine (**65**).

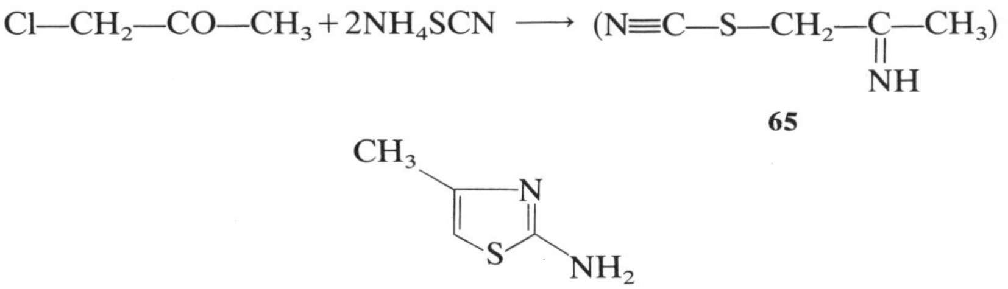

$$Cl-CH_2-CO-CH_3 + 2NH_4SCN \longrightarrow (N\equiv C-S-CH_2-\underset{NH}{\overset{\|}{C}}-CH_3)$$
65

[Structure **66**: 4-methyl-2-aminothiazole]

Scheme 37

In 1888, Hantzsch and Traumann (102) described a general method of synthesis for 2-aminothiazoles (**67**) by condensation of thiourea with

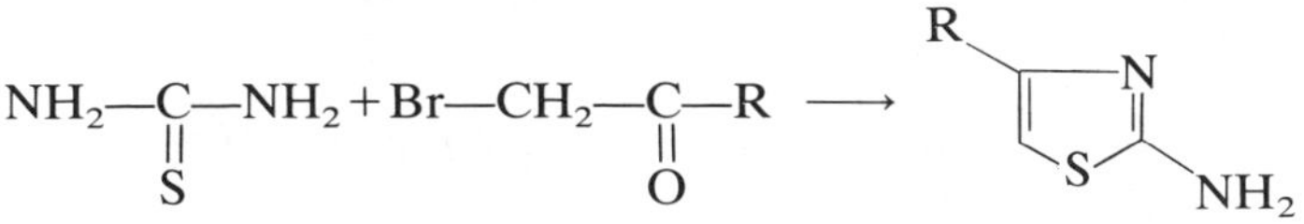

$$NH_2-\underset{S}{\overset{\|}{C}}-NH_2 + Br-CH_2-\underset{O}{\overset{\|}{C}}-R \longrightarrow \text{[thiazole **67**]}$$

Scheme 38

α-halocarbonyl derivatives. Traumann (103) developed this method and showed that monoalkylthiourea (**68**) led to 2-alkylaminothiazole (**69**),

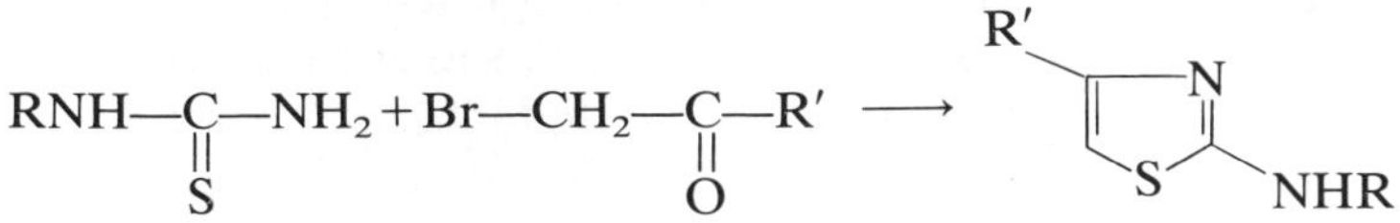

$$RNH-\underset{S}{\overset{\|}{C}}-NH_2 + Br-CH_2-\underset{O}{\overset{\|}{C}}-R' \longrightarrow \text{[thiazole **69**]}$$
68   **69**

Scheme 39

whereas 3-alkyl-2-iminothiazolines (**70**) were obtained by alkylation of 2-aminothiazoles with alkyl iodides. Addition of *sym*-dialkylthiourea led to 3-alkyl-2-alkyliminothiazolines (**71**).

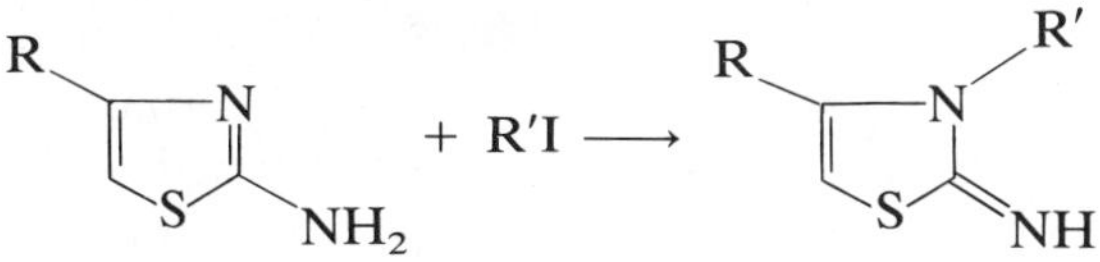

[Structure **67** + R'I → Structure **70**]

Scheme 40

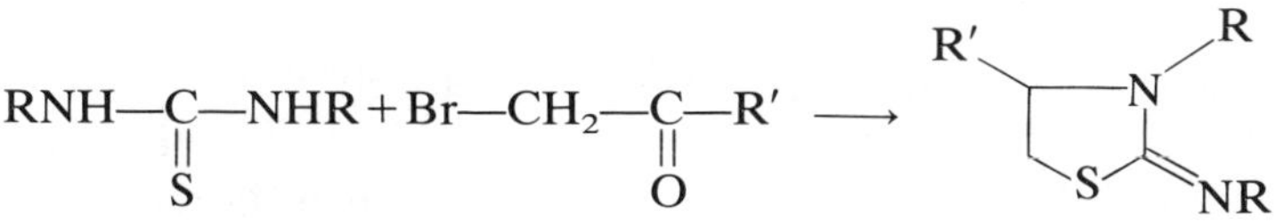

$$RNH-\underset{S}{\overset{\|}{C}}-NHR + Br-CH_2-\underset{O}{\overset{\|}{C}}-R' \longrightarrow \text{[thiazoline **71**]}$$

Scheme 41

In 1889, Popp (104) confirmed the structure of 2-aminothiazole (**72**): by diazotization and reduction he obtained thiazole itself (**73**).

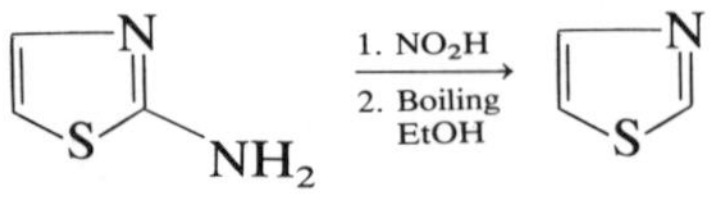

Scheme 42

4. Preparation

Probably first obtained by Hantzsch and Arapides (105) by condensation of α,β-dichlorether with barium thiocyanate, and identified by its pyridine-like odor, thiazole was first prepared in 1889 by G. Popp (104) with a yield of 10% by the reduction in boiling ethanol of thiazol-2-yldiazonium sulfate resulting from the diazotization of 2-aminothiazole, prepared the year before by Traumann (103). The unique cyclization reaction affording directly the thiazole molecule was described in 1914 by Gabriel and Bachstez (106). They applied the method of cyclization, developed by Gabriel (107, 108), to the diethylacetal of 2-formylaminoethanal and obtained thiazole with a yield of 62%. Thiazole was also formed in the course of a study on the ease of decarboxylation of the three possible monocarboxylic acids derived from it (109). On the other

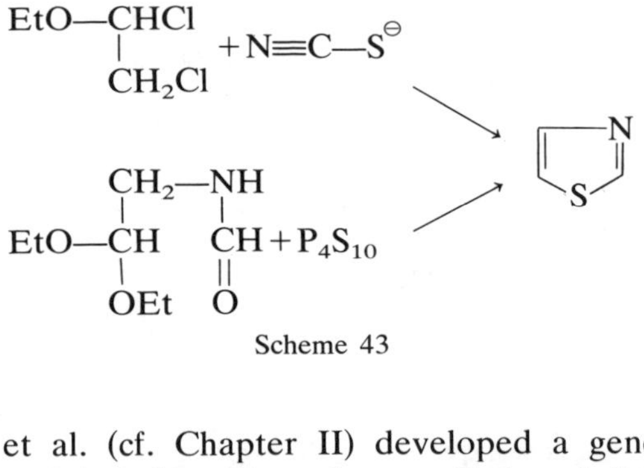

Scheme 43

hand, Asinger et al. (cf. Chapter II) developed a general method of dehydrogenation of Δ-3-thiazolines that, applied to Δ-3-thiazoline itself, led to thiazole. More recently thiazole was obtained by photorearrangement of isothiazole (110). Various isotopically labeled thiazoles have been synthezised for physicochemical purposes: 2- and 5-deutero and

2,5-dideuterothiazoles were obtained by reduction of the corresponding mono- and dibromo derivatives by zinc in deuteroacetic acid (111);

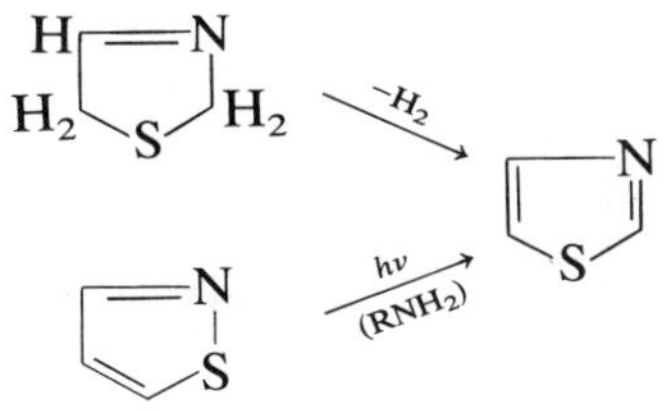

Scheme 44

2-deuterothiazole resulted from the deuterolysis of thiazol-2-yl lithium (111); 2,4,5-trideuterothiazole was afforded by two exchange reactions between thiazole and D_2O: the first was performed by recycling a mixture of the vapors of both constituents through a column of 10% platinum on asbestos at 340°C for 15 days, the second by heating a solution of thiazole in $5.6N$ D_2SO_4 at 118° for 8 days (111). The 4- and 5-deutero derivatives were obtained by the reduction with hypophosphorous acid of 4- and 5-deutero-2-aminothiazoles resulting from the Hantzsch cyclization of 1-deutero and 2-deutero-2-bromoethanal with thiourea (112); 4-deuterothiazole, mixed with 0.5 mole of 2-deutero- and 3.5 moles of nondeuterated thiazole, resulted from the thermal decarboxylation of thiazol-4-yldeuterocarboxylic acid (113). ^{15}N-Thiazole was prepared from labeled thiourea through the Hantzsch cyclization with 1,2-dichloroethoxyethane, Sandmeyer bromination of the 2-aminothiazole, and reduction by zinc and acetic acid of the ^{15}N-2-bromothiazole obtained (112).

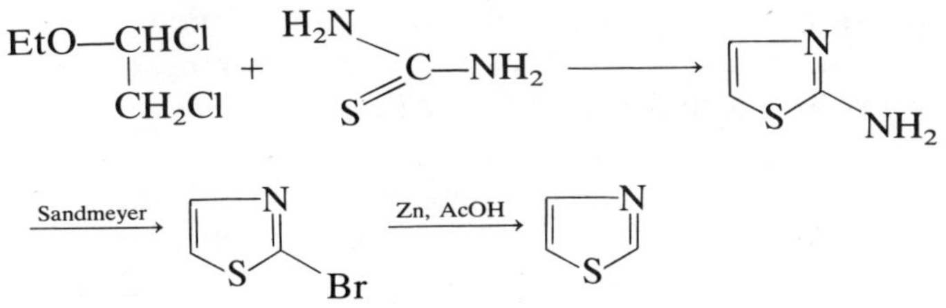

Scheme 45

2-^{13}C- (112, 113), 4-^{13}C- and 5-^{13}C-thiazoles (112) were prepared by the same general pathway from the corresponding ^{13}C-labeled 2-aminothiazoles synthesized by the Hantzsch cyclization of ^{13}C-thiourea, 1-^{13}C-2-bromoethanal, and 2-^{13}C-2-bromoethanal, respectively (112). A mixture of 4- and 5-^{13}C-thiazoles was prepared by the same general pathway from the corresponding mixture of 2-aminothiazoles resulting from the condensation of thiourea with a mixture of 1- and 2-^{13}C-1,2-dibromoethoxyethane (113). 2-^{14}C-thiazole has been prepared from ^{14}C-labeled thiourea (114).

5. Theoretical Model, Electronic Structure

The first empirical and qualitative approach to the electronic structure of thiazole appeared in 1931 in a paper entitled "Aspects of the chemistry of the thiazole group" (115). In this historical review, Hunter showed the technical importance of the group, especially of the benzothiazole derivatives, and correlated the observed reactivity with the mobility of the electronic system. In 1943, Jensen et al. (116) explained the low value observed for the dipole moment of thiazole (1.64 D in benzene) by the small contribution of the polar-limiting structures and thus by an essentially dienic character of the π system of thiazole. The first theoretical calculation of the electronic structure of thiazole, benzothiazole, and their methyl derivatives was performed by Pullman and Metzger using the Hückel method (5, 6, 8).

As part of the same HMO method but with various approximations, appeared the calculations of Zahradnik and Koutecki (117), Vincent and Metzger (118), Vitry-Raymond and Metzger (119), Bonnier and Gelus (120), and Bonnier et al. (121). In 1966, Vincent et al. applied to thiazole the iterative methods restricted to the π system in the following approximations: ω'' (122), ω^* (123), and P.P.P. (124, 123). This last method was later employed with different approximations and for various purposes by Chowdhury and Basu (125), J. Devanneaux and Labarre (126), Yoshida and Kobayashi (127), Witanowskiet et al. (128), and E. Corradi et al. (129).

More elaborated treatments have also been applied: *ab initio* methods by Bouscasse (130) and Bernardi et al. (131); then the all-valence-electrons methods, derived from PPP, by Gelus et al. (132) and by Phan-Tan-Luu et al. (133) and CNDO methods by Bojesen et al. (113) and by Salmona et al. (134).

Table I-1 lists the various theoretical treatments published on the thiazole molecule; for each the type of approximation, the mode of parametrization, and, eventually, the geometry employed are given net charges and bond orders for various theoretical calculations are listed in Tables I-2 and I-3.

TABLE I-1. SYNOPSIS OF VARIOUS THEORETICAL METHODS UTILIZED IN STUDY OF THE ELECTRONIC STRUCTURE OF THIAZOLE

Method	Parameterization		Ref.
HMO	$\delta_S = 1$; $\delta_N = 2$; $\lambda = 0, 1$; $\rho_{CC} = \rho_{CS} = \rho_{CN} = 1^a$		8
HMO	$\delta_S = 0$; $\delta_N = 0.5$; $\rho_{CC} = \rho_{CS} = \rho_{CN} = 0.5$		117
HMO	$\delta_S = 0$; $\delta_N = 1$; $\lambda = 0.1$; $\lambda' = 0.01$; $\rho_{CC} = 1$; $\rho_{CS} = 0.77$; $\rho_{CN} = 0.9$	Overlap $S_{ij} = 1$ for $i = j$ and 1calculated for $i \neq j$ (Mulliken)	118
HMO	$\delta_S = 0$; $\delta_N = 0.5$; $\lambda = 0.1$; $\lambda' = 0.01$; $\rho_{CC} = 1.1$; $\rho_{CS} = 0.6$; $\rho_{CN} = 1$		119
HMO	$\alpha_r = (IP)^{(1)}$ for atom r with simply charged core $\alpha_r = (EN)$ for atom r with doubly charged core β_{rs}: Wolfsberg-Helmholtz (135)		120
ω''	$\delta_S^{(0)} = 1.5$; $\delta_N^{(0)} = 0.5$; $\lambda = 0.1$; $\rho_{CC} = 1$; $\rho_{CS} = 0.5$; $\rho_{CN} = 0.8$; for the nth iteration: $\delta_r^{(n)} = \delta_r^{(0)} + \omega q_r^{*(n-1)} + \sum_s \omega' q_s^{*(n-1)} + \sum_t \omega'' q_t^{*(n-1)}$	$\omega = 1.4$; $\omega' = 0.93$; $\omega'' = 0.64$ σ-system calculated by Del Re method (136)	122
ω^*	−id−	$\omega^* = 0.6$ (mean of various terms in $\frac{1}{2}(p^2, q^2)$ of the P.P.P. method)	123
PPP	$\rho_{rs} = \rho_{rs}^{(0)} + \omega^* 1_{rs}$	Ideal geometry; configurations' interaction	124
PPP	γ_{rr}: Pariser (137) γ_{rs}: Pariser and Parr (138) (139) β_{rs}: Kon (140)		123
PPP	−id−	Approximative geometry calculated from $J_{13_{C-H}}$ repartition corrected versus σ-system organization (141)	126
PPP	Parametrisation of (124)	Ideal geometry	125
PPP	γ_{rr}: Pariser (137)		
PPP	γ_{rs}, β_{rs}: Pariser and Parr (138, 139) $\gamma_{rr} = Z_\alpha$ (α = constant) $\gamma_{rs} = 0.5(\gamma_{rr} + \gamma_{ss}) + AR^2 + BR$ for $R < 4$ Å and ponctual charges approximation for $R > 4$ Å	Optimisation by integrals' variation	127

TABLE I-1 (Continued)

Method	Parameterization		Ref.
PPP	$\beta_{rs} = KS_{rs}(W'_r + W'_s)$ with $W'_r = W_r + (q_{rr} - 1)\gamma_{rr}$ γ_{rr}; γ_{rs}: Nishimoto and Mataga (142–144)		128
PPP	γ_{rr}: Paloni (145) γ_{rs}: Pariser and Parr (138) $\beta_{rs} = KS_{rs}$ (147)	Geometry deduced from $J_{^{13}C-H}$	146
PPP	γ_{rr}: Pariser (137) γ_{rs}: Nishimoto and Mataga (144) β_{rs}: empirical determination (148, 149)	Ideal geometry Refinement of heteroatomic parameters from self-consistent perturbation McWeeny (150); σ-charges calculated by Del Re method (136)	129
PPP	Dewar's variant (151–153)		154
$\sigma - \pi$	γ_{rr} and γ_{rs} estimated from spectroscopic values β_{rs}: Wolfsberg-Helmholtz (135)	Approximated geometry (155); no invariance by rotation	132
$\sigma - \pi$	γ_{rr}: Katagiri and Sandorfy (156) γ_{rs}: Pariser and Parr (138, 139) β_{rs}: Wolfsberg and Helmholtz (135)	No invariance by rotation; geometry estimated from $J_{^{13}C-H}$; configurations' interaction	133
CNDO/2		Approximated geometry	134
ab initio		Slater's orbitals; approximative geometry	130
ab initio		STO-3G level	131

[a] λ = auxiliary inductive parameter (A.I.P.); λ' = A.I.P. for atoms in β position of heteroatom. δ_X = coulombic parameter of atom N; ρ_{XY} = exchange parameter between atoms X and Y.

TABLE I-2. NET CHARGES ON THE THIAZOLE RING[a]

Method	System	$S_{(1)}$	$C_{(2)}$	$N_{(3)}$	$C_{(4)}$	$C_{(5)}$	$H_{(2)}$	$H_{(4)}$	$H_{(5)}$	Ref.
HMO	π	0.665	0.020	−0.630	0.001	−0.056	—	—	—	8
										6
HMO	π	0.034	0.131	−0.194	0.040	−0.011	—	—	—	117
HMO	π	0.185	0.122	−0.409	0.043	0.059	—	—	—	118
HMO	π	0.047	0.087	−0.168	0.020	0.014	—	—	—	119
HMO	π	0.244	0.232	−0.438	−0.055	0.017	—	—	—	120
ω''	π	0.140	0.146	−0.255	0.012	−0.042	—	—	—	157
	σ^b	−0.180	0.076	−0.116	0.050	0.029	0.054	0.029	0.054	
ω	π	0.153	0.179	−0.275	0.046	−0.103	—	—	—	123
PPP	π	0.246	0.076	−0.238	0.025	−0.110	—	—	—	124
PPP	π	0.152	0.180	−0.291	0.044	−0.086	—	—	—	123
PPP	π	0.246	0.105	−0.296	0.054	−0.110	—	—	—	126
PPP	π	0.230	−0.059	+0.003	−0.266	0.092	—	—	—	125
PPP	π	0.1713	0.2276	−0.3599	0.1006	−0.1395	—	—	—	154
PPP	π	—	−0.031	—	−0.050	−0.064	—	—	—	129
	σ^b									
$\sigma-\pi$	π	0.152	0.100	−0.196	0.029	−0.086	0.053	0.039	0.054	133
	σ	0.208	−0.034	−0.260	0.117	−0.194	0.060	0.050	0.060	
$\sigma-\pi$	π	0.400	−0.380	−0.050	0.010	0.020	—	—	—	132
	σ	−0.350	−0.370	−0.290	−0.030	0.070	—	—	—	
CNDO/2	π	0.180	0.009	−0.089	−0.024	−0.076	—	—	—	134
	σ	−0.204	0.107	−0.055	0.085	0.047	0.005	−0.004	0.019	
ab initio	π	0.053	0.330	−0.479	−0.001	0.105	—	—	—	130
	σ	0.535	−0.349	1.179	−0.186	−0.176	−0.330	−0.343	−0.336	

[a] In some cases, the calculated values reported in this table are not explicitly given in the referenced paper; they were calculated by this author on the basis of the published data.
[b] Calculated by the Del Re method; this σ-charges' repartition was also used in Refs. 123 and 124.

TABLE I-3. BOND ORDERS OF THE THIAZOLE RING[a]

Method	System	$S_{(1)}-C_{(2)}$	$C_{(2)}-N_{(3)}$	$N_{(3)}-C_{(4)}$	$C_{(4)}-C_{(5)}$	$C_{(5)}-S_{(1)}$				Ref.
HMO	π	0.716	0.528	0.448	0.775	0.553	—	—	—	8
										6
HMO	π	0.475	0.773	0.544	0.791	0.470	—	—	—	117
HMO	π	0.642	0.647	0.569	0.729	0.602	—	—	—	118
HMO	π	0.469	0.787	0.527	0.806	0.454	—	—	—	119
ω''	π	0.352	0.831	0.454	0.873	0.282	—	—	—	122
ω	π	0.369	0.856	0.388	0.900	0.270	—	—	—	123
PPP	π	0.446	0.836	0.412	0.881	0.346	—	—	—	124
PPP	π	0.371	0.854	0.371	0.908	0.268	—	—	—	123
PPP	π	0.376	0.824	0.517	0.782	0.433	—	—	—	125
PPP	π	0.351	0.861	0.421	0.886	0.291	—	—	—	133
							C_2-H_2	C_4-H_4	C_5-H_5	
	σ	0.938	0.972	0.952	0.993	0.917	0.994	0.980	0.986	
CNDO/2	π	0.361	0.883	0.391	0.895	0.313	—	—	—	134
ab initio	π	0.027	0.737	0.385	0.783	0.015	—	—	—	130

[a] In some cases the calculated values reported in this table are not explicitly given in the referenced paper; they were calculated by this author on the basis of the published data.

I. General Aspects

Table I-4 gives some calculated reactivity indices: free valence or Wheland atomic localization energies for radical, electrophilic, or nucleophilic substitution. For each set of data the order of decreasing reactivity is indicated. In practice this order is more reliable than the absolute values of the reactivity indices themselves.

TABLE I-4. REACTIVITY INDICES

Nature of index	$C_{(2)}$	$C_{(4)}$	$C_{(5)}$	Reactivity order	Ref.
Free valence	0.436	0.417	0.352	2>4>5	6
Free valence	0.484	0.397	0.471	2>5>4	117
Free valence	0.443	0.434	0.401	2>4>5	118
Free valence	0.476	0.399	0.472	2>5>4	119
Free valence	0.547	0.409	0.517	2>5>4	120
Polarization energy[a]					119
Radical	2.007	2.675	2.164	2>5>4	
Electrophilic	2.134	2.679	2.670	2>5>4	
Nucleophilic	1.879	2.670	2.104	2>5>4	
Polarization energy[a]					117
Radical	1.986	2.509	2.007	2>5>4	
Electrophilic	2.189	2.552	2.007	5>2>4	
Nucleophilic	1.789	2.467	2.007	2>5>4	

[a] In β_0 units.

Whatever the method employed and its degree of sophistication some common trends can be noted in the electronic properties of the thiazole molecule.

1. In all cases the π net charge of sulfur is positive, whereas its σ net charge is sometimes positive (133, 130) and sometimes negative (122, 132). The all-electrons methods, like *ab initio*, give a positive total net charge with the exception of the CNDO/2 method for which it is negative (134).
2. In all cases the total net charge of nitrogen is negative. In only one PPP calculation is it slightly positive (125).
3. Of the three carbon atoms of the ring only C-2 shoes a practically clear situation: its total net charge is generally positive or vanishing, its π net charge being usually also positive. In the all-valence-electron methods the absolute value of this charge is considerably lowered (132–134), being even negative in the all-electron approximations (130). The charges at C-4 and C-5 are not so uniformly described; the π net charge at C-4 is

generally small, whereas that at C-5 varies considerably, being mostly slightly negative.

4. From their σ net charge the three hydrogen atoms have decreasing acidity in the order, H-2≥H-5>H-4.

The evolution of the mean π net charge of the five atoms of the ring as a function of the calculation method is reported in Fig. I-1. The sophistication of the method corresponds roughly to a leveling of the charges except for sulfur. The *ab initio* model, being unique, is tentatively reported in Fig. I-1.

5. The discussion of the π-bond orders is interesting because it gives a picture of the distribution of π-electrons along the σ-frame of the ring

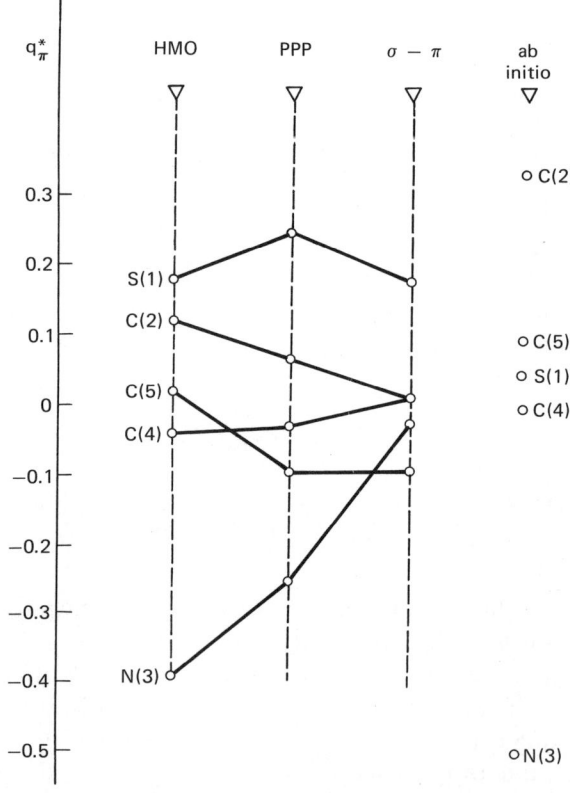

Fig. I-1. Variation of the mean π net charge of the five atoms of thiazole ring as a function of the calculation method employed.

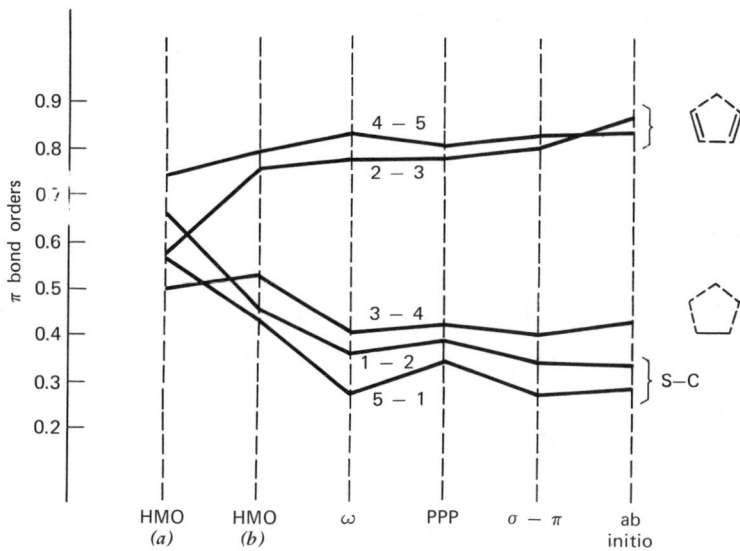

Fig. I-2. Variation of the π-bond orders of thiazole as a function of the calculation method employed.

and, therefore, of its aromaticity, which can be accounted for by NMR and ultraviolet spectroscopy. The HMO results can be divided into two groups, the ones giving thiazole a slightly aromatic character (8, 6, 118) whereas the others are indicative of a more dienic structure (117, 119). The iterative methods, being less sensitive to the choice of parameters, are also more reliable with respect to the bond order results: they largely confirm the dienic character of thiazole. This aspect of the electronic distribution in the ring is yet more exaggerated in the *ab initio calculation*. Figure I-2 gives the evolution of π bond orders as a function of the calculation method.

6. The free valencies calculated by the first group of HMO methods are in an incorrect order (2>4>5), whereas those of the second group, as well as the radical polarization energies, conform to the experimental reactivity (2>5>4). Only one of the second group methods predicts the right order of reactivity of thiazole with electrophilic (5>2>4) and with nucleophilic (2>5>4) species, from calculation of the corresponding polarization energies (117). A useful comparison of the electronic structure of thiazole with those of thiophene and pyridine can be made from the results of three methods where the same approximations have been made: in the π system, the ω and PPP methods (122, 123), and in the

TABLE I-5. COMPARISON BETWEEN ELECTRONIC DIAGRAMS OF THIOPHENE AND THIAZOLE

Method (Ref.)	Position	π Net charges		Bond	π-Bond orders		Position	σ Net charges		Bond	σ-Bond orders	
	r	Thiophene q_r^*	Thiazole q_r^*	r–s	Thiophene $1rs$	Thiazole $1rs$	r	Thiophene q_r^*	Thiazole q_r^*	r–s	Thiophene $1rs$	Thiazole $1rs$
ω" (122)	1	0.126	0.140	1–2	0.298	0.352	—	—	—	—	—	—
	2	−0.029	0.146	2–3	0.849	0.831	—	—	—	—	—	—
	3	−0.034	−0.255	3–4	0.499	0.454	—	—	—	—	—	—
	4	−0.034	0.012	4–5	0.849	0.873	—	—	—	—	—	—
	5	−0.029	−0.042	5–1	0.298	0.282	—	—	—	—	—	—
PPP-π (123)	1	0.140	0.152	1–2	0.300	0.371	—	—	—	—	—	—
	2	−0.038	0.180	2–3	0.896	0.854	—	—	—	—	—	—
	3	−0.032	−0.291	3–4	0.398	0.371	—	—	—	—	—	—
	4	−0.032	0.044	4–5	0.896	0.908	—	—	—	—	—	—
	5	−0.038	−0.086	5–1	0.300	0.268	—	—	—	—	—	—
PPP-σ-π (133)	1	0.145	0.152	1–2	0.309	0.351	1	0.284	0.208	1–2	0.939	0.938
	2	−0.049	0.100	2–3	0.877	0.861	2	−0.203	−0.034	2–3	0.992	0.972
	3	−0.024	−0.196	3–4	0.435	0.421	3	−0.044	−0.260	3–4	0.964	0.952
	4	−0.024	0.029	4–5	0.877	0.886	4	−0.044	0.117	4–5	0.992	0.993
	5	−0.049	−0.086	5–1	0.309	0.291	5	−0.203	−0.194	5–1	0.939	0.917

$\sigma - \pi$ system, the PPP $\sigma - \pi$ method (133). The π and σ net charges and bond orders of thiophene and thiazole are compared in Table I-5. Whatever the method considered the variation of the indices occurs in the same sense when passing from thiophene to thiazole: the replacement in the 3-position of a carbon atom by a nitrogen induces

1. A large increase in σ- and π-electronic density that is not unexpected, allowance being made for the increase in electronegativity of the considered atom.
2. A decrease of σ- and π-electronic density in both adjacent positions. For the σ system this decrease is approximately the same at the 2- and 4-positions, which expresses an equivalent electron withdrawing from nitrogen in both positions. On the other hand, the decrease in π-electronic density is twice as large at C-2 as at C-4.
3. A decreasse in σ net charge at C-5 with, as a corollary, a slight increase in π-electronic density at that position.
4. A clear decrease of σ net charge at the sulfur atom, whereas the π-electronic density remains stationary.

Summarizing, the introduction of nitrogen at the place of C-3 in thiophene does not deeply disturb the electronic environment of the sulfur atom, but it induces in the rest of the molecule some alternating modification of the electronic density (Figs. I-3 and I-4). The perturbations induced by the nitrogen in the π bond order of thiophene are

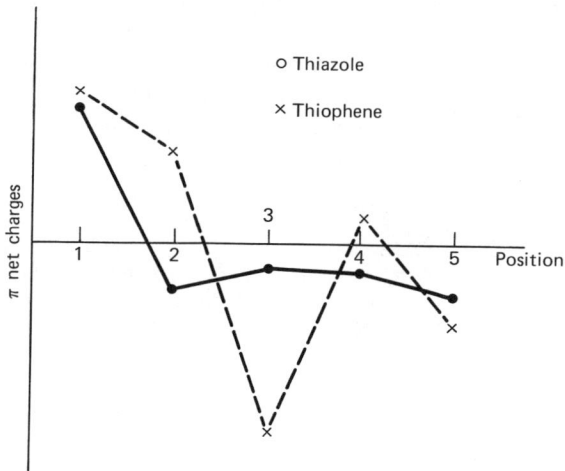

Fig. I-3. Variation of the π net charges by introduction of nitrogen in place of C-3 in thiophene.

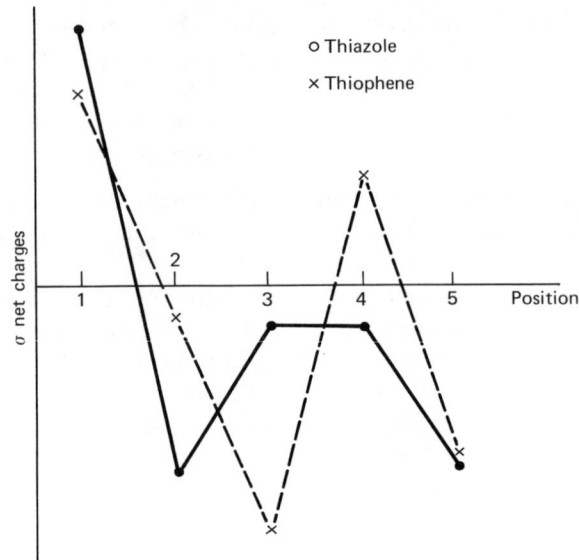

Fig. I-4. Variation of the σ net charges by introduction of nitrogen in place of C-3 in thiophene.

relatively unimportant, both heterocycle keeping a pronounced dienic character.

The comparison of thiazole and pyridine is possible through the PPP-π calculations, but owing to a slight difference in the parameterization, the results are hardly comparable. Nevertheless, one can observe that the contraction of the six- to the five-membered ring gives an increase in the negative charge on the nitrogen atom, while in both systems the π net charges at the α- and β-positions are alternating (Table I-6). The dissymmetry of the ring introduced by the replacement of C-3 and C-4 of pyridine by sulfur corresponds to a π net charge at C-2 slightly higher than at C_4.

TABLE I-6. COMPARISON BETWEEN π NET CHARGES OF NITROGEN AND CARBON ATOMS IN THE α- AND β-POSITIONS IN PYRIDINE AND THIAZOLE (PPP-π METHOD)

	Net charges			Ref.
	N	$C_{(\alpha)}$	$C_{(\beta)}$	
Pyridine	−0.234	0.109	−0.012	160
Thiazole	−0.291	(2) 0.180	−0.086	123
		(4) 0.044		

TABLE I-7. CALCULATED DIPOLE MOMENT OF THIAZOLE[a]

Methods	Ref.	$\mu_{tot\,calc}$(D)
ω'' Del Re	122	1.4
ω^* Del Re	123	1.3
PPP Del Re	124	1.78
PPP Del Re	123	1.41
$(\sigma-\pi)$ PPP	133	1.86

[a] $\mu_{exp} = 1.61$ (116, 158).

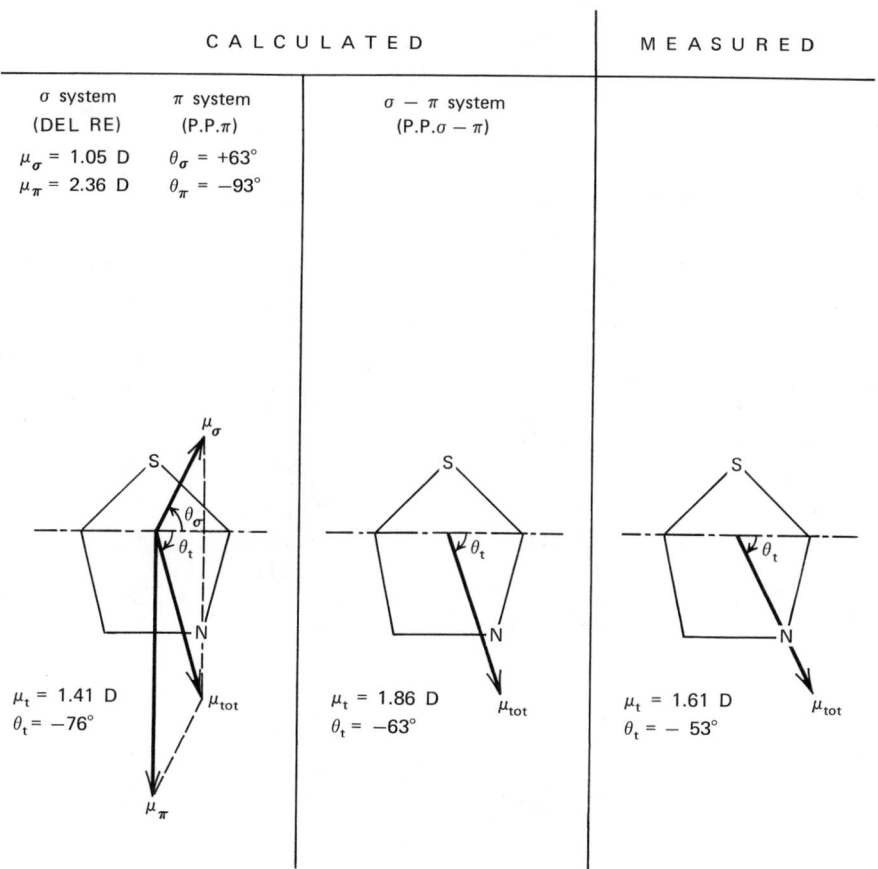

Fig. I-5. Dipole moment of thiazole.

TABLE I-8. DIPOLE MOMENTS CALCULATED FOR VARIOUS THIAZOLE DERIVATIVES

Thiazole derivatives	Calcd	(Debye)	Ref.	Measured	(Debye)	Ref.
Thiazole	1.5	HMO	118	1.64	C_6H_6	116
	1.4	ω''	122	1.61	gazeous	158
	3.38	$\sigma - \pi$	132	1.04		132
	1.78	PPP	124	1.72	CCl_4	146
	1.41	PPP	123	1.62	C_6H_{12}	146
	1.30	ω^*	123			
	1.86	$\sigma - \pi$	133			
	7.49	ab initio	130			
	1.86	HMO	146			
2-Methyl	1.5	ω''	122	1.5	C_6H_6	116
	1.66	PPP	123			
	1.51	ω^*	123			
4-Methyl	1.2	ω''	122	1.51	C_6H_6	116
	1.07	PPP	123			
	1.02	ω^*	123			
5-Methyl	1.5	ω''	122	1.82	C_6H_6	116
	1.70	PPP	123			
	1.47	ω^*	123			
2-Phenyl	1.42	HMO	146	1.21	CCl_4	146
				1.18	C_6H_{12}	146
4-Phenyl	1.45	HMO	146	1.33	CCl_4	146
				1.27	C_6H_{12}	146
5-Phenyl	1.62	HMO	146	1.89	CCl_4	146
2,5-Diphenyl	1.18	HMO	146	0.96	CCl_4	146
				1.02	C_6H_{12}	146
2,4-Diphenyl	0.98	HMO	146	0.70	CCl_4	146
				0.69	C_6H_{12}	146
4,5-Diphenyl	1.67	HMO	146	1.69	CCl_4	146
				1.68	C_6H_{12}	146
2,4,5-Triphenyl	1.29	HMO	146	0.71	CCl_4	146
				0.71	C_6H_{12}	146
2-Amino	1.51	PPP	123	1.75	C_6H_6	116
	1.33	ω^*	123			
4-Amino	1.48	PPP	123	—	—	—
	1.56	ω^*	123	—	—	—
5-Amino	1.43	PPP	123	—	—	—
	1.27	ω^*	123	—	—	—
2-Chloro	1.88	PPP	123	—	—	—
	1.76	ω^*	123	—	—	—
4-Chloro	2.74	PPP	123	—	—	—
	2.78	ω^*	123	—	—	—
5-Chloro	1.65	PPP	123	—	—	—
	1.78	PPP	123			

I. General Aspects

The calculated electronic distribution leads to an evaluation of the dipole moment of thiazole. Some values are collected in Table I-7 that can be compared to the experimental value of 1.61 D (158).

The orientation of the dipole moment, experimentally established by the Stark effect (158), can also be compared to the calculated ones (Fig. I-5).

Finally, the CN and CC π-bond orders can be determined by use of a relation between bond orders and bond length (122) and the experimentally measured bond length (159).

Dipole moments were calculated for a large number of thiazole derivatives; the corresponding results are reported in Table I-8.

TABLE I-9. π-BOND ORDERS AND BOND LENGTHS IN ALL-VALENCE-ELECTRONS CALCULATIONS

Methods (Ref.)	Bond				
	d_{12}	d_{23}	d_{34}	d_{45}	d_{51}
	Bond orders				
PPP, $\sigma - \pi$ (133)	0.351	0.861	0.421	0.886	0.291
CNDO/2 (134)	0.361	0.883	0.391	0.895	0.313
ab initio (130)	0.218	0.779	0.370	0.891	0.165
	Bond lengths (Å)				
Experimental (159)	1.724	1.304	1.372	1.367	1.713
	Bonds orders[a]				
	0.26	0.66	0.42	0.83	0.30

[a] Bond orders calculated from experimental bond lengths (122).

In Table I-9 we have collected only the π-bond orders calculated by all-valence-electrons methods and compared their values with those deduced from experimental bond lengths. Both data are indicative of an aromatic molecule with a large dienic character. The 2–3 and 4–5 bonds especially present a large double-bond character, whereas both C–S bonds are relatively simple.

TABLE I-10. ELECTRONIC STRUCTURE OF METHYL

Methods	Position	System	Net Charges				
			q_1^*	q_2^*	q_3^*	q_4^*	q_5^*
HMO	2	π	0.633	0.068	−0.643	−0.014	0.062
HMO	2	π	0.029	0.122	−0.167	0.012	0.004
ω''	2	π	0.129	0.195	−0.286	0.014	−0.052
ω	2	π	0.143	0.193	−0.308	0.057	−0.115
PPP (π)	2	π	0.144	0.186	−0.324	0.050	−0.098
PPP ($\sigma-\pi$)	2	π	0.153	0.095	−0.186	0.026	−0.079
Del Re	2	σ	−0.181	0.120	−0.120	0.049	0.027
PPP (π)		σ	0.205	0.003	−0.268	0.121	−0.196
PPP ($\sigma-\pi$)	2	$\sigma+\pi$	0.358	0.098	−0.454	0.147	−0.275
HMO	4	π	0.040	0.066	−0.160	0.062	−0.007
ω''	4	π	0.135	0.149	−0.266	0.045	−0.067
ω	4	π	0.154	0.197	−0.295	0.064	−0.147
PPP (π)	4	π	0.152	0.184	−0.299	0.053	−0.119
PPP ($\sigma-\pi$)	4	π	0.154	0.096	−0.191	0.026	−0.073
Del Re	4	σ	−0.180	0.076	−0.117	0.092	0.028
PPP ($\sigma-\pi$)	4	σ	0.211	−0.029	−0.263	0.148	−0.195
PPP ($\sigma-\pi$)	4	$\sigma+\pi$	0.365	0.067	−0.454	0.174	−0.268
HMO	5	π	0.048	0.070	−0.165	−0.006	0.064
ω''	5	π	0.132	0.135	−0.253	−0.009	−0.004
ω	5	π	0.146	0.170	−0.267	0.017	−0.084
PPP (π)	5	π	0.147	0.168	−0.284	0.013	−0.071
PPP ($\sigma-\pi$)	5	π	0.153	0.103	−0.196	0.039	−0.081
Del Re	5	σ	−0.181	0.075	−0.117	0.047	0.072
PPP ($\sigma-\pi$)	5	σ	0.140	0.078	−0.260	0.119	−0.156
PPP ($\sigma+\pi$)	5	$\sigma+\pi$	0.293	0.181	−0.456	0.158	−0.237

THIAZOLES ACCORDING TO VARIOUS METHODS

Bond Orders					Miscellaneous	Ref.
l_{12}	l_{23}	l_{34}	l_{45}	l_{51}		
0.700	0.523	0.472	0.784	0.542	Free valencies: $F_2 = 0.263$; $F_4 = 0.424$; $F_5 = 0.345$	6
0.420	0.824	0.480	0.832	0.430	$F_2 = 0.488$; $F_4 = 0.420$; $F_5 = 0.470$	119
0.343	0.824	0.454	0.873	0.271	$F_4 = 0.405$; $F_5 = 0.588$	122
0.361	0.839	0.376	0.907	0.256		123
0.361	0.832	0.367	0.940	0.257		123
0.347	0.848	0.425	0.883	0.295		133
—	—	—	—	—	CH_3: $q^*_{(c)} = -0.105$; $q^*_{(H)} = 0.041$; $H_{(4)}$: $q^* = 0.039$; $H_{(5)}$: $q^* = 0.049$	122
0.927	0.960	0.951	0.993	0.917	CH_3: $q^*_{(c)} = -0.044$; $q^*_{(H)} = 0.020$; $H_{(4)}$: $q^* = 0.040$; $H_{(5)}$: $q^* = 0.055$	133
—	—	—	—	—		132
0.502	0.739	0.599	0.754	0.491	Free valencies: $F_2 = 0.491$; $F_4 = 0.379$; $F_5 = 0.487$	119
0.351	0.833	0.445	0.876	0.277	$F_2 = 0.548$; $F_5 = 0.579$	122
0.378	0.850	0.375	0.891	0.260		123
0.374	0.852	0.366	0.894	0.260		123
0.343	0.863	0.415	0.874	0.294		133
—	—	—	—	—	CH_3: $q^*_{(c)} = -0.118$; $q^*_{(H)} = 0.039$; $H_{(2)}$: $q^* = 0.054$; $H_{(5)}$: $q^* = 0.049$	122
0.938	0.972	0.942	0.985	0.917	CH_3: $q^*_{(c)} = -0.058$; $q^*_{(H)} = 0.020$; $H_{(2)}$: $q^* = 0.055$; $H_{(5)}$: $q^* = 0.050$	133
—	—	—	—	—		132
0.465	0.876	0.530	0.804	0.448	Free valencies: $F_2 = 0.481$; $F_4 = 0.398$; $F_5 = 0.480$	119
0.342	0.830	0.458	0.871	0.277	$F_2 = 0.560$; $F_4 = 0.403$	122
0.357	0.861	0.378	0.899	0.269		123
0.362	0.857	0.372	0.898	0.266		123
0.353	0.859	0.424	0.869	0.287		133
—	—	—	—	—	CH_3: $q^*_{(c)} = -0.109$; $q^*_{(H)} = 0.040$; $H_{(2)}$: $q^* = 0.054$; $H_{(4)}$: $q^* = 0.038$	122
0.938	0.972	0.953	0.986	0.900	CH_3: $q^*_{(c)} = -0.043$; $q^*_{(H)} = 0.980$; $H_{(2)}$: $q^* = 0.950$; $H_{(4)}$: $q^* = 0.960$	133
—	—	—	—	—		132

A. Influence of Substituents

a. METHYLTHIAZOLES

The introduction of a methyl substituent into the empirical calculations may be performed according to two main different models: the pseudoheteroatomic model and the hyperconjugated model (161–166). Both approximations have been used in π-electron methods (HMO, ω, PPP). On the other hand, in the all-valence-electrons

TABLE I-11. SUBSTITUTION EFFECTS: QUALITATIVE VARIATIONS OF NET CHARGES INDUCED BY A METHYL SUBSTITUTION

	Method	Positions					Ref.
		1	2	3	4	5	
Thiazole → 2-methylthiazole	HMO	↑[a]	↑	↓	↓	↑	6
	HMO	↑	↑	—	(↓)	(↓)	119
	ω''	↓	↑	↓	—	↓	122
	ω	↓	↑	↓	↑	↓	123
	PPP (π)	↓	(↑)	↓	(↑)	↓	123
	PPP $\sigma-\pi(\pi)$	—	(↓)	(↑)	↑	(↑)	133
	Del Re (σ)	—	↑	(↓)	—	—	122
	PPP $\sigma-\pi(\sigma)$	—	↑	(↓)	(↑)	—	133
	$\sigma-\pi$	—	↑	—	—	—	132
Thiazole → 4-methylthiazole	HMO	(↓)	↓	(↑)	↑	↓	6
	HMO						119
	ω''	(↓)	—	(↓)	↑	↓	122
	ω	—	↑	↓	↑	↓	123
	PPP (π)	—	—	(↓)	(↑)	↓	123
	PPP $\sigma-\pi(\pi)$	—	—	—	—	(↑)	133
	Del Re (σ)	—	—	—	↑	—	122
	PPP $\sigma-\pi(\sigma)$	—	—	—	↑	—	133
	$\sigma-\pi$	—	—	—	—	—	132
Thiazole → 5-methylthiazole	HMO	—	↓	—	↓	↑	6
	HMO						119
	ω''	(↓)	(↓)	—	↓	↑	122
	ω	(↓)	(↓)	(↑)	↓	↑	123
	PPP (π)	—	↓	(↑)	↓	↑	123
	PPP $\sigma-\pi(\pi)$	—	—	—	↑	(↑)	133
	Del Re (σ)	—	—	—	—	↑	122
	PPP $\sigma-\pi(\sigma)$	↓	↓	—	—	↑	133
	$\sigma-\pi$	↓	↓	—	↑	↑	132

[a] Symbols show the variations of net charges relative to unsubstituted thiazole: —: $|\Delta q_r| < 0.005$. (↑) or (↓) are, increase and decrease respectively, of q_r so that $0.005 < |\Delta q_r| < 0.010$, ↑ or ↓ are, increase and decrease respectively, of q_r more important than 0.10.

I. General Aspects

TABLE I-12. SUBSTITUTION EFFECTS: QUALITATIVE VARIATIONS OF π-BOND ORDERS INDUCED BY A METHYL SUBSTITUTION[a]

	Method	\multicolumn{5}{c}{Positions}	Ref.
		1	2	3	4	5	
Thiazole → 2-methylthiazole	HMO	(↓)	—	↑	((↑))	((↓))	6
	HMO	↓	↑	↓	↑	↓	119
	ω''	((↓))	((↓))	—	—	(↓)	122
		((↓))	(↓)	(↓)	((↑))	(↓)	123
	PPP (π)	((↓))	↓	((↓))	↑	(↓)	123
	PPP-σ-$\pi(\pi)$	—	(↓)	—	—	—	133
	PPP-σ-$\pi(\sigma)$	(↓)	(↓)	—	—	—	133
Thiazole → 4-methylthiazole	HMO	↑	↓	↑	↓	↑	119
	ω''	—	—	((↓))	—	((↓))	122
	ω	((↑))	((↓))	(↓)	((↓))	(↓)	122
	PPP (π)	—	—	((↓))	(↓)	((↓))	122
	PPP-σ-$\pi(\pi)$	((↓))	—	((↓))	(↓)	—	133
	PPP-σ-$\pi(\sigma)$	—	—	((↓))	((↓))	—	133
Thiazole → 5-methylthiazole	HMO	—	↑	—	—	((↓))	119
	ω''	((↓))	—	—	—	((↓))	122
	ω	(↓)	((↑))	((↓))	—	—	123
	PPP (π)	((↓))	—	—	((↓))	—	123
	PPP-σ-$\pi(\pi)$	—	—	—	(↓)	—	133
	PPP-σ-$\pi(\sigma)$	—	—	—	((↓))	(↓)	133

[a] Symbols show the variations of bond order relative to unsubstituted thiazole: —: $|\Delta l_{rs}| <$ 0.005. ((↑)) or ((↓)) are increase and decrease, respectively, of l_{rs} so that $0.005 < |\Delta l_{rs}| <$ 0.010. (↑) or (↓) are increase and decrease, respectively, of l_{rs} so that $0.010 < |\Delta l_{rs}| <$ 0.020. ↑ or ↓ are increase and decrease, respectively, of l_{rs} more important than 0.020.

methods, the methyl group is introduced with the aid of its classical C- and H-atomic orbitals. The results concerning the σ and π net charges, the σ- and π-bond orders, and some reactivity indices are collected in Table I-10 for the three monomethylthiazoles. The qualitative variations of both charges and π-bond orders are reported in Tables I-11 and I-12, respectively. The replacement of a H atom by a methyl group induces, a decrease in π- and σ-electronic density on the carbon atom of the ring; on the nitrogen atom the perturbation is relatively small, whereas on the sulfur atom a slight increase in electronic density is observed. The bond orders are weakly perturbed by methylation.

b. CHLORO- AND AMINOTHIAZOLES

The π net charges and π-bond calculated by the PPP π and ω methods (123) are collected in Table I-13, and the qualitative variations of both quantities are reported in Table I-14.

TABLE I-13. ELECTRONIC STRUCTURES OF CHLOROTHIAZOLES AND AMINOTHIAZOLE (PPP-π AND ω METHOD (123))

Compounds	Method	Net charges						Mobile bond orders				
		q_1^*	q_2^*	q_3^*	q_4^*	q_5^*	$q_{(Cl)}^*$	l_{12}	l_{23}	l_{34}	l_{45}	l_{51}	l_{CCl}
Chlorothiazole (chlorine position)													
2	PPP-π	0.145	0.186	−0.319	0.049	−0.096	0.035	0.363	0.836	0.368	0.910	0.258	0.204
	ω	0.144	0.192	−0.304	0.056	−0.114	0.026	0.362	0.841	0.377	0.907	0.218	0.175
4	PPP-π	0.152	0.183	−0.298	0.052	−0.114	0.025	0.374	0.852	0.367	0.896	0.261	0.160
	ω	0.154	0.195	−0.292	0.062	−0.142	0.023	0.377	0.851	0.376	0.893	0.261	0.155
5	PPP-π	0.148	0.170	−0.285	0.018	−0.072	0.022	0.363	0.857	0.372	0.899	0.267	0.142
	ω	0.147	0.171	−0.266	0.020	−0.086	0.015	0.359	0.860	0.378	0.900	0.269	0.115
Aminothiazole (amino position)													
2	PPP-π	0.142	0.176	−0.330	0.051	−0.100	0.061	0.357	0.824	0.367	0.911	0.255	0.267
	ω	0.142	0.191	−0.311	0.058	−0.117	0.036	0.360	0.836	0.376	0.907	0.255	0.206
4	PPP-π	0.152	0.185	−0.301	0.048	−0.125	0.042	0.375	0.851	0.364	0.889	0.258	0.208
	ω	0.155	0.200	−0.298	0.064	−0.153	0.032	0.380	0.849	0.374	0.888	0.259	0.183
5	PPP-π	0.146	0.166	−0.283	0.009	−0.074	0.036	0.360	0.857	0.372	0.893	0.265	0.183
	ω	0.145	0.168	−0.264	0.015	−0.084	0.020	0.356	0.862	0.378	0.898	0.268	0.133

TABLE I-14. SUBSTITUTION EFFECTS: QUALITATIVE VARIATIONS OF π NET CHARGE INDUCED BY THE SUBSTITUTION OF A CHLORINE OR AN AMINO GROUP (123)[a]

Substituent	Substitution position	Method	Position				
			1	2	3	4	5
Chlorine	2	PPP	(↓)	(↑)	↓	—	(↓)
		ω	(↓)	↑	↓	(↑)	↓
	4	PPP	—	—	(↓)	(↑)	↓
		ω	—	↑	↓	↑	↓
	5	PPP	—	(↓)	(↑)	↓	↑
		ω	(↓)	(↓)	(↑)	↓	↑
Amino group	2	PPP	(↓)	—	↓	(↑)	↓
		ω	↓	↑	↓	↑	↓
	4	PPP	—	—	↓	—	↓
		ω	—	↑	↓	↑	↓
	5	PPP	(↓)	↓	(↑)	↓	↑
		ω	(↓)	↓	↑	↓	↑

[a] Symbols: see Table I-11.

As with methylation, substitution by –Cl or –NH$_2$ induces a decrease in π electronic density on the substituted carbon atom and a slight increase in both adjacent positions. The perturbation of an –NH$_2$ group is slightly larger than that of a –Cl group.

II. PHYSICAL PROPERTIES

1. Geometrical Structure

The geometrical structure of thiazole was first approached (167) by the combination of bond angles deduced from a correlation between ^{13}C–H proton NMR coupling constants and interorbital and internuclear bond angles, C–H bond length deduced from a correlation between the same coupling constants and C–H bond length (168), and C–C, C–N, and C–S bond length obtained from bond orders calculated with the HMO iterative ω method (112). More recently, a complete determination of the geometrical parameters of the molecule was performed by microwave spectrometric study of thiazole and eight isotopically labeled isomers

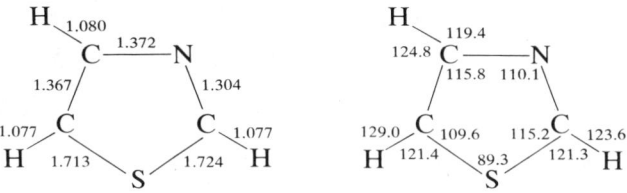

Fig. I-6. Molecular structure of thiazole; bond lengths in Å (left), bond angles in degrees (right).

(159) (Fig. I-6). The structure obtained for thiazole is surprisingly close to an average of the structures of thiophene (169) and 1,3,4-thiadiazole (170) (Fig. I-7). From a comparison of the molecular structures of thiazole, thiophene, thiadiazole, and pyridine (171), it appears that around C(4) the bond angles of thiazole C(4)–H with both adjacent C(4)–N and C(4)–C(5) bonds show a difference of 5.4° that, compared to a difference in C(2)–H of pyridine of 4.2°, is interpreted by L. Nygaard (159) as resulting from an attraction of H(4) by the electron lone pair of nitrogen.

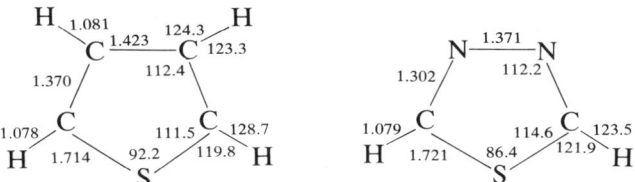

Fig. I-7. Molecular structures of thiophene and 1,3,4-thiadiazole; bond lengths in Å (left), bond angles in degrees (right).

From the direction of the quadrupole axis of nitrogen it is concluded that its lone pair is symmetrically placed outside the ring, along the bisector of angle C(2)–N–C(4) (159).

2. Ultraviolet Absorption Spectroscopy

The ultraviolet absorption spectra of most new thiazoles currently synthesized have been described and occasionally used for structural assignments. They are indicated in the various chapters along with other physical properties useful for identification purposes. We consider here only the case of unsubstituted thiazole and of some of its typical derivatives.

The ultraviolet absorption spectrum of thiazole was first determined in 1955 in ethanolic solution by Leandri et al. (172), then in 1957 by Sheinker et al. (173), and in 1967 by Coltbourne et al. (174). Albert in 1957 gave the spectrum in aqueous solution at pH 5 and in acidic solution (N HCl) (175). Nonhydroxylic solvents were employed (176, 177), and the vapor-phase spectrum was also determined (123). The results summarized in Table I-15 are homogeneous except for the first data of Leandri (172). Both bands A and B have a red shift of about 3 nm when thiazole is dissolved in hydrocarbon solvents. This red shift of band A increases when the solvent is hydroxylic and, in the case of water, especially when the solution becomes acidic and the extinction coefficient increases simultaneously.

This bathochromic shift is typical of $\pi \rightarrow \pi^*$ transitions. The behavior of the water solution when acidified was attributed by Albert (175) absorption by the thiazolium cation, by analogy with pyridine. However, allowance is made for the very weak basicity of thiazole ($pK_a = 2.52$) compared with that of pyridine ($pK_a = 5.2$), Ellis and Griffiths (176) consider the differences between the spectrum of thiazole in water and in

TABLE I-15. ULTRAVIOLET ABSORPTION SPECTRUM OF THIAZOLE IN THE VAPOUR PHASE AND IN DIFFERENT SOLVENTS

Solvent	Band A		Band B		Ref.
	λ(nm)	log ε	λ(nm)	log ε	
Vapor phase	228	—	206.5	—	123
n-Hexane	231–2	—	209	—	176
Cyclohexane	231–2	—	209–10	—	176
Dichloromethane	234	3.05			177
Methanol	232–3	—	—	—	176
Ethanol	233	3.57	207.5	3.41	174
	232–3	3.52–3.59	—	—	173
	240	3.59	—	—	172
Water	233	3.53	205	3.34	176
Water, pH 13	233	3.54	—	—	176
Water, pH 5	235	3.51	—	—	175
1 N HCl, pH 0	237	3.61	—	—	175
1.8 N HCl	237	3.59	—	—	176

TABLE I-16. ULTRAVIOLET ABSORPTION SPECTRA OF THIAZOLE AND ITS MONOMETHYL DERIVATIVES

	Vapor phase (123)					in CH$_2$Cl$_2$ (177)	
	λ(nm)					λ(nm)	ε
Thiazole	228	206.5	—	—	—	234	1125
2-Methyl	231.6	224.9	211.2	208	205.2	236	4000
4-Methyl	237.4	206	203.2	201.6	198.9	242.5	3400
5-Methyl	234.7	216.6	213	210	—	241.8	3000

hydrochloric acid solutions to be due to hydrogen bonding between the hydronium ion and thiazole being stronger than that between water and thiazole, rather than to protonation in dilute acid, as suggested by Albert. Substitution of a proton by a methyl group (Table I-16) does not fundamentally alter the absorption spectrum of thiazole; it induces a slight bathochromic shift, both in the vapor phase and in dichloromethane solution, the effect decreasing in the order, 4–>5–>2–, with a significant increase in intensity. This behavior is quite similar to that of benzene but differs from that of conjugated dienes for which the absorption intensity is practically unaffected (178). Introduction of a functional group (Table I-17) is characterized by a more pronounced bathochromic effect

TABLE I-17. ULTRAVIOLET ABSORPTION SPECTRA OF SOME TYPICAL DERIVATIVES OF THIAZOLE IN EtOH SOLUTION

	λ(nm)	ε	Ref.
Thiazole	232.3	3900	173
2-Nitro	308	4800	179
2-Me-4-nitro	296	7300	179
5-Nitro	298.5	4800	179
2-Chloro	244.5	5300	180
4-Chloro	245.6	2900	180
2-Bromo	247.6	4900	180
5-Bromo	247.4	4100	180
2-Iodo	253	—	180
2-Methoxy	235	4700	180
2-Ethoxy	237.5	4800	180
5-Ethoxy	255.2	4200	180
2-Dimethylamino	265.3	9200	180

than in the case of methyl substitution, largely because of the extended conjugation including the substituent. With halogens, the bathochromic shift increases with the weight of the halogen atom; just as in the case of benzene, its position has no influence on the red shift but induces a difference in absorption intensities. The largest perturbations are connected with the presence of a nitro group, whereas for alkoxy groups the position of the group is important, but the nature of alkyl substituent does not influence the spectrum.

Quantum chemistry methods allow the prediction of the ultraviolet transitions in good agreement with the experimental values in the case of thiazole and its three methyl derivatives (Table I-18). A very weak absorption has been indicated at 269.5 nm that could correspond to an $n \rightarrow \pi^*$ transition given by calculation at 281.5 nm (133). Ultraviolet absorption spectroscopy has been investigated in connection with steric interactions in the Δ-4-thiazoline-2-thione (**74**) series (181). It was earlier demonstrated by NMR technique that 4-alkyl-3 isopropyl-Δ-4-thiazoline-2-thiones exist in solution as equilibrium mixtures of two conformers (**75** and **76**), the relative populations of which vary with the size of R4 (182): for R4 = tBu the population of rotamer A is 100%, whereas for R4 = Me it is only 28%. Starting from the observed absorption wavelength for

TABLE I-18. COMPARISON OF ULTRAVIOLET ABSORPTION SPECTRA OF THIAZOLE AND ITS METHYL DERIVATIVES WITH CALCULATED VALUES OF TRANSITIONS BY DIFFERENT METHODS

Compounds	Transitions					Methods	Ref.
	Experimental values (nm)						
Thiazole	228	206.5	—	—	—	Gas phase	123
2-Methyl	231.6	224.9	211.2	208	205.2	Gas phase	123
4-Methyl	237.4	206	203.2	201.6	198.9	Gas phase	123
5-Methyl	234.7	216.6	213	210	—	Gas phase	123
	Theoretical values (nm)						
Thiazole	230.8	205.6	159.5	149.9	144.8	PPP CI	123
	243.0	210.1	167.5	151.2	149.3	PPP σ-π CI	133
	247.9	188.9	—	—	—	PPP CI	125
	234.6	217.3	167.6	165.4	—	PPP	127
2-Methyl	243.3	203.9	161.8	159.9	149.8	PPP CI	123
	233.9	199.9	158.9	145.8	140.9	PPP σ-π CI	133
4-Methyl	238.4	208.0	165.7	156.1	148.4	PPP CI	123
	238.4	203.2	158.9	149.3	140.9	PPP σ-π CI	133
5-Methyl	238.8	205.6	162.2	153.6	147.2	PPP CI	123
	243	203.2	158.9	147.6	142.5	PPP σ-π CI	133

exteme cases and making allowance for electronic effects, it was possible to estimate a standard wavelength of absorption for **75** ($R_4 = H$) at 323.8 nm and for **76** ($R_4 = H$) at 320.0 nm (181). The steric compression of the thiocarbonyl chromophore is stronger for **75** than for **76** and results in a red shift of its absorption. This is in agreement with PPP π (183) and CNDO/S (184) calculations that associate the 320-nm absorption of these compounds with an electronic $\pi \rightarrow \pi^*$ transition strongly localized in the $C=S$ π bond. All the results concerning the compounds (**74**) could be rationalized by the assumption that the $C=S$ bond is bulkier in the ground state than in the excited state.

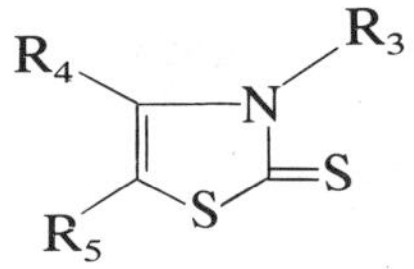

74

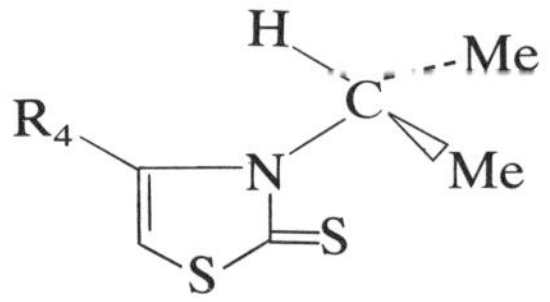

75 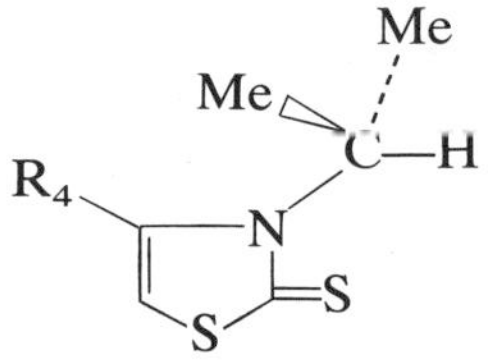   **76**

A. Quaternary Salts

As in the case of pyridine (185), the quaternization of thiazole induces a bathochromic shift of the ultraviolet absorption spectrum: in ethanol the long wavelength maximum at 232.3 nm (3900) for thiazole moves to 240 nm (4200) for 3-methylthiazolium tosylate (186) (Table I-19).

As in the case of the free bases, the substitution of a nuclear hydrogen atom by a methyl group induces a bathochromic shift that decreases in the order of the position substituted: 4–>5–>2– Ferré et al. (187) have proposed a theoretical model based on the PPP (π) method using the fractional core charge approximation that reproduces quite correctly this order of decreasing perturbation.

B. Phosphorescence Spectra

The phosphorescence of thiazole in the 380- to 500-nm region was studied in Ne, Ar, Kr, SF_6, and CH_4 matrices at 4°K (188). Lifetimes

TABLE I-19. ULTRAVIOLET ABSORPTION SPECTRA OF THIAZOLIUM TOSYLATES IN EtOH (186) COMPARED WITH CALCULATED TRANSITION ENERGIES (187)

	λ (nm)	ε	E_{exp}(eV)	E_{calc}(eV)
3-Methyl	240	4200	5.17	5.30
2,3-Dimethyl	243	4800	5.10	5.18
3,4-Dimethyl	252.8	4750	4.90	4.93
3,5-Dimethyl	247	4550	5.03	5.03
2,3,4-Trimethyl	261	6000	4.93	4.90
2,3,5-Trimethyl	252.5	5770	4.90	4.92
3,4,5-Trimethyl	258	4630	4.81	4.75

ranged from 70 msec in Xe to 2 sec in Ne. This emission coincides with the region of the $T_1 \to S_0$ transition predicted from the calculations of Pariser et al. (123). From the absence of fluorescence it is concluded that thiazole, excited to S_1, quenches nonradiatively to T_1 and subsequently emits $T_1 \to S_0$ phosphorescence (188).

3. Photoelectron Spectroscopy

Ultraviolet photoelectron spectroscopy allows the determination of ionization potentials. For thiazole the first experimental measurement using this technique was preformed by Salmona et al. (189) who later studied various alkyl and functional derivatives in the 2-position (190, 191). Substitution of an hydrogen atom by an alkyl group destabilizes the first ionization potential, the perturbation being constant for *iso*-propyl and heavier substituents. Introduction in the 2-position of an amino group strongly destabilizes the first band and only slightly the second.

This important destabilization is yet stronger in the case of 2-dimethylaminothiazole. For these two compounds there is first a complete superposition of the bands corresponding to the second and third ionization potentials. Moreover, a supplementary band appears at 11.35 eV for the first and at 10.30 eV for the second: it corresponds to the ionization of the lone pair of the amino group that is conjugated with the ring π system. In the case of 2-methoxythiazole, the destabilization only affects the first band; the lone pair of oxygen conjugated with the ring π system appears at 11.75 eV. On the other hand, 2-nitrothiazole is uniformly destabilized, with a band at 11.5 eV corresponding to the nitro group. All the assignments could be confirmed with good precision by CNDO/S calculations (Table I-20). Bernardi et al. (131) have studied the halogen derivatives experimentally and theoretically. The experimental assignments have been compared successfully to theoretical results obtained by *ab initio* STO-3G and 4-31G methods (192, 193).

TABLE I-20. IONIZATION POTENTIALS OF VARIOUS 2-SUBSTITUTED THIAZOLES (190)

Compounds	Ionization Potentials (eV)					
	$\pi_3{}^a$	$\pi_2{}^a$	$\sigma_N{}^a$		$\sigma_S{}^a$	$\pi_1{}^a$
Thiazole	9.50	10.25	10.40	—	12.70	13.4
2-Methylthiazole	9.20	10	10.3	—	12.70	13.15
2-Ethylthiazole	9	9.95	10.1	—	—	—
2-i-Propylthiazole	8.95	9.85	10.1	—	—	—
2-i-Butylthiazole	8.95	9.80	10.1	—	—	—
2-t-Butylthiazole	8.95	9.8	10	—	—	—
2,4-Dimethylthiazole	8.71	9.9	10	—	12.35	12.9
2-Bromothiazole	9.4	10.5	10.65	$\sigma_{Br} = 11.24$; $\pi_{Br} = 11.76$	13.1	14
2-Chlorothiazole	9.4	10.6	10.7	$\sigma_{Cl} = 11.9$; $\pi_{Cl} = 12.3$	13.5	14.4
2-Aminothiazole	8.45	10.1	10.1	$\pi_N = 11.35$	12.5	13.6
2-Dimethylaminothiazole	7.8	9.7	9.7	$\pi_N = 10.3$	12.1	13.1
2-Methoxythiazole	8.8	10.15	10.3	$\pi_O = 11.75$	12.8	13.8
2-Nitrothiazole	10.1	11.1	11.1	$\pi_O = 11.5$	13.5	14

[a] Band attribution.

The first band always corresponds to the ionization of a π orbital, as is the second, which is strongly overlapped by the third, a band corresponding to the ionization of a σ orbital mainly centered on the lone pair of the cyclic nitrogen. In the case of 2-chloro- and 4-bromothiazole, both these bands cannot be distinguished. The fourth band, which presents a vibrational fine structure, is assigned to the ionization of the sulphur lone pair. In general, the monohalogen derivatives of thiazole show a band system quite similar to that of thiazole, the first band being destabilized by the halogenation whereas the fine and intense bands of the halogen appear between 11.0 and 12.40 eV. (Table I-21).

TABLE I-21. IONIZATION POTENTIALS OF THIAZOLE AND VARIOUS MONOHALOGENATED THIAZOLES (131)

Compounds	Ionization Potentials (eV)							
	π_3^a	π_2^a	$\pi_2+\sigma_N^a$	σ_N^a	σ_X^a	π_X^a	σ_S^a	π_1^a
Thiazole	9.50	10.24	—	10.48	—	—	17.78	13.5
2-Chlorothiazole	9.37	—	10.67	—	11.93	12.40	13.1	14.5
2-Bromothiazole	9.30	10.52	—	10.62	11.19	11.75	13.06	14.02
4-Bromothiazole	9.23	—	10.64	—	11.02	11.65	12.9	14.1
5-Bromothiazole	9.26	10.35	—	10.45	11.35	12.04	13.3	14.0

[a] Band attribution.

4. Infrared Absorption and Raman Diffusion

Until 1962 the infrared and Raman spectra of thiazole in the liquid state were described by some authors (173, pp. 194–200) with only fragmentary assignments. At that date Chouteau et al. (201) published the first tentative interpretation of the whole infrared spectrum between 4000 and 650 cm^{-1} for thiazole and some alkyl and haloderivatives. They proposed a complete assignment of the normal modes of vibration of the molecule.

The study of the infrared spectrum of thiazole under various physical states (solid, liquid, vapor, in solution) by Sbrana et al. (202) and a similar study, extended to isotopically labeled molecules, by Davidovics et al. (203, 204), gave the symmetry properties of the main vibrations of the thiazole molecule. More recently, the calculation of the normal modes of vibration of the molecule defined a force field for it and confirmed quantitatively the preceeding assignments (205, 206).

A. Numbering, Activity, and Nomenclature of the Vibrations

The planar structure of thiazole (159) implies for the molecule a C_s-type symmetry (Fig. I-8) and means that all the 18 fundamental vibrations are active in infrared and in Raman spectroscopy. Table I-22 lists the predictions made on the basis of this symmetry for thiazole.

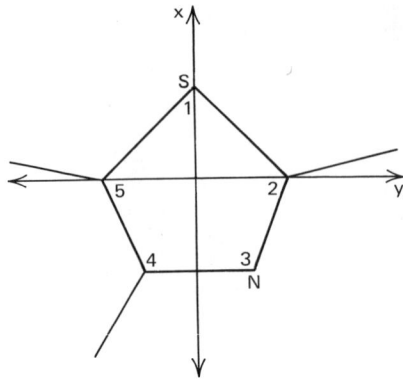

Fig. I-8. Orientation of x and y axes in the plane of the thiazole molecule.

The distinction between in-plane (A' symmetry) and out-of-plane (A'' symmetry) vibrations resulted from the study of the polarization of the diffusion lines and of the rotational fine structure of the vibration-rotation bands in the infrared spectrum of thiazole vapor.

The out-of-plane vibrations of thiazole correspond to C-type vibration-rotation bands and the in-plane vibrations to A, B, or $(A+B)$ hybrid-type bands (Fig. I-9). The Raman diffusion lines of weak intensity were assigned to A''-type oscillations and the more intense and polarized lines to A' vibration modes (Fig. I-10 and Table I-23).

From Table I-24 it appears that A'-type vibrations may, to a first approximation, decompose into six modes of vibration for CH bonds: three for elongation $\nu(CH)$, three for bending $\delta(CH)$, and seven for ring

TABLE I-22. NUMBERING AND ACTIVITY OF THIAZOLE RING VIBRATIONS (C_s SYMMETRY)

Symmetry mode	Symmetry element	Degree of freedom		Molecule	Number of C—H vibrations	Number of skeleton vibrations	Activity of vibrations	
							Raman	Infrared
	σ_z	ω_z						
		3 types	3 types C 1 type S 1 type N					
A'	a	2	2	$T_x T_y R_z$	6	7	P	M
A''	a	1	1	$T_z R_x R_y$	3	2	D	M_z

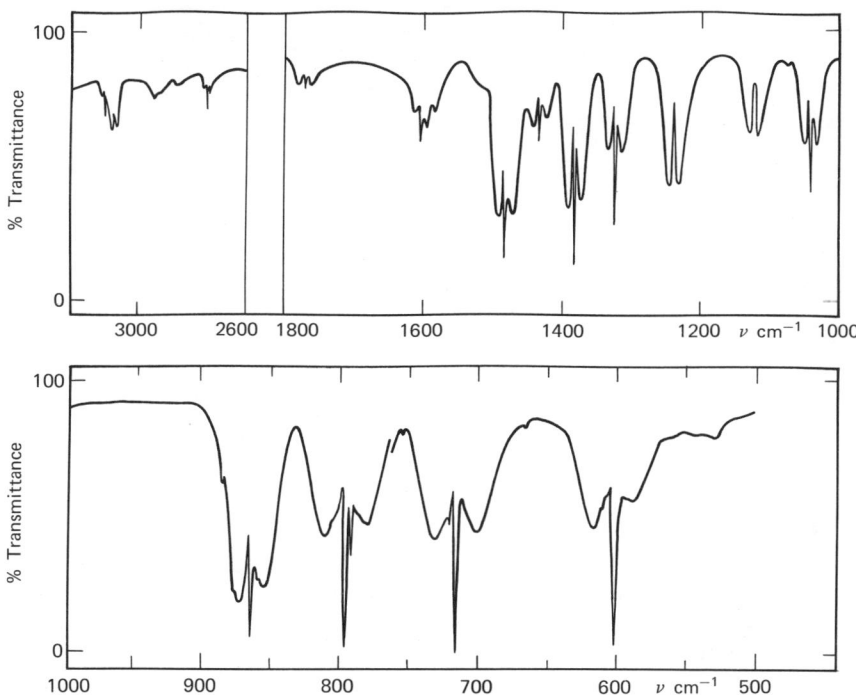

Fig. I-9. Infrared spectrum of thiazole in the vapor phase.

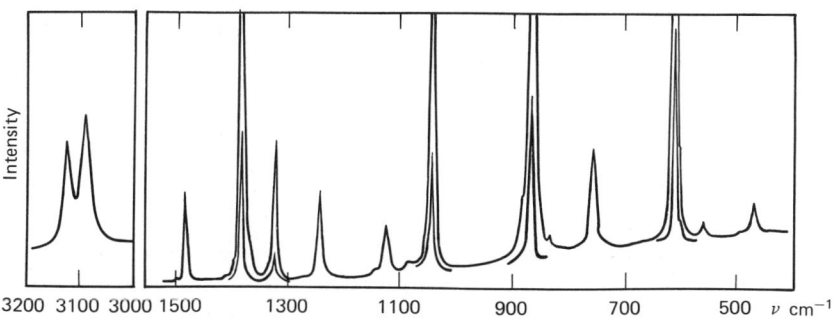

Fig. I-10. Raman spectrum of thiazole in the liquid phase.

oscillations $\omega_i (i = 1$ to 7), whereas A''-type vibrations decompose into three bending modes $\gamma(CH)$ and two ring oscillations Γ_j ($j = 1.2$).

TABLE I-23. FREQUENCIES AND INTENSITIES OF RAMAN SPECTRA[a]

Thiazole[b] (liquid)			2-D thiazole C_2Cl_4 solution (3 M)		Assignments
ν (cm^{-1})	I	ρ	ν (cm^{-1})	I	
3122	f	0.20	—	—	$\nu C_{(5)}H$
—	—	—	—	—	$\nu C_{(4)}H$
3089	f	0.30	—	—	$\nu C_{(2)}H$
1481	f	<0.10	1476	m	ω_1
1380	F	<0.10	1369	F	ω_2
—	—	—	~1312	—	—
1320	f	0.20	1310	m	ω_3
1238	f	<0.10	—	—	δ CH
1123	m	0.70	—	—	δ CH
1042	m	<0.10	~1064	m	δ CH
—	—	—	966	F	δ CD
~882	e	—	—	—	$\gamma C_{(4)}H$
867	F	<0.10	873	m	ω_4
?			789	m	ω_5
757	f	0.70	755	m	ω_6
612	F	<0.10	607	F	ω_7
558	ff	<0.10	—	—	—
470	ff	0.70	—	—	Γ_2

[a] I, intensity; ff, very weak; f, weak; m, medium; F, strong; e, shoulder, ρ depolarization factor.
[b] From a recent Raman study of thiazole on a laser Raman Spectrometer it appears that the actual data differ somewhat from those of Ref. 203.

B. Assignments

a. CH (or CD) BOND OSCILLATIONS (201–204)

VALENCE VIBRATIONS, νCH AND νCD. In the 3100 cm^{-1} region the infrared spectrum of thiazole shows only two absorptions at 3126 and 3092 cm^{-1}, with the same frequencies as the corresponding Raman lines (201–4) (Fig. I-10 and Table I-23). In the vapor-phase spectrum of

II. Physical Properties

TABLE I-24. DEUTEROTHIAZOLE FREQUENCIES AND ASSIGNMENTS

Symmetry mode	D-2 th.	D-4 th.	D-2,5 th.	Assignments
A'	3124	3114	2226	$\nu C_{(5)}$(H or D)
	3087	2305	3090	$\nu C_{(4)}$(H or D)
	2290	3081	2298	$\nu C_{(2)}$(H or D)
	1474	1444	1466	ω_1
	1372	1371	1360	ω_2
	1307	1292	1276	ω_3
	1147	918	1141	δC(H or D)
	1058	1195	741	δC(H or D)
	963	1033	972	δC(H or D)
	871	877	900 or 892	ω_4
	784	842	787	ω_5
	758	732	759	ω_6
	602	609	566	ω_7
A''	880	682	868	$\gamma C_{(4)}$(H or D)
	638	825	638	$\gamma C_{(2)}$(H or D)
	727	696	559	$\gamma C_{(5)}$(H or D)
	587	553	589	Γ_1
	446	451	428	Γ_2

thiazole (Fig. I-9) the highest-frequency band is of A type; the other at 3092 cm^{-1} has a more complex structure (202, 203). In the spectra of 2- and 4-deuterothiazoles there are two maxima at 3120 and 3085 cm^{-1}, whereas in that of 2,5-dideuterothiazole there is only one absorption band at 3090 cm^{-1} (Fig. I-11). From the study of the isotopically labeled molecules and of derivatives it can be concluded that the C–H vibrators are weakly coupled and that their oscillations are practically independant. Moreover, taking into account the disappearance of the high-frequency band for the 5-substituted derivatives and of the lowest-frequency band only for a double substitution, leads to the following order for the νCH vibrations of thiazole: $\nu C_{(5)}H > \nu C_{(2)}H \geqslant \nu C_{(4)}H$ (204). The following modes, $\nu C_{(5)}D$, $\nu C_{(4)}D$ and $\nu C_{(2)}D$, have been assigned to the bands at 2336, 2305, and 2295 cm^{-1} (203, 204), respectively (Table I-24).

BENDING VIBRATIONS, δCH AND δCD. The bands at 1040, 1239, and 1318 cm^{-1} had been proposed for the δCH oscillations (201). The first two assignment were confirmed by the study of deuterated and substituted thiazoles, but the third was modified, being attributed to the band at 1121 cm^{-1} (203). The δCD modes appear at 970, 918, and 740 cm^{-1} for species deuterated at the 2-, 4-, and 5-positions, respectively, (204).

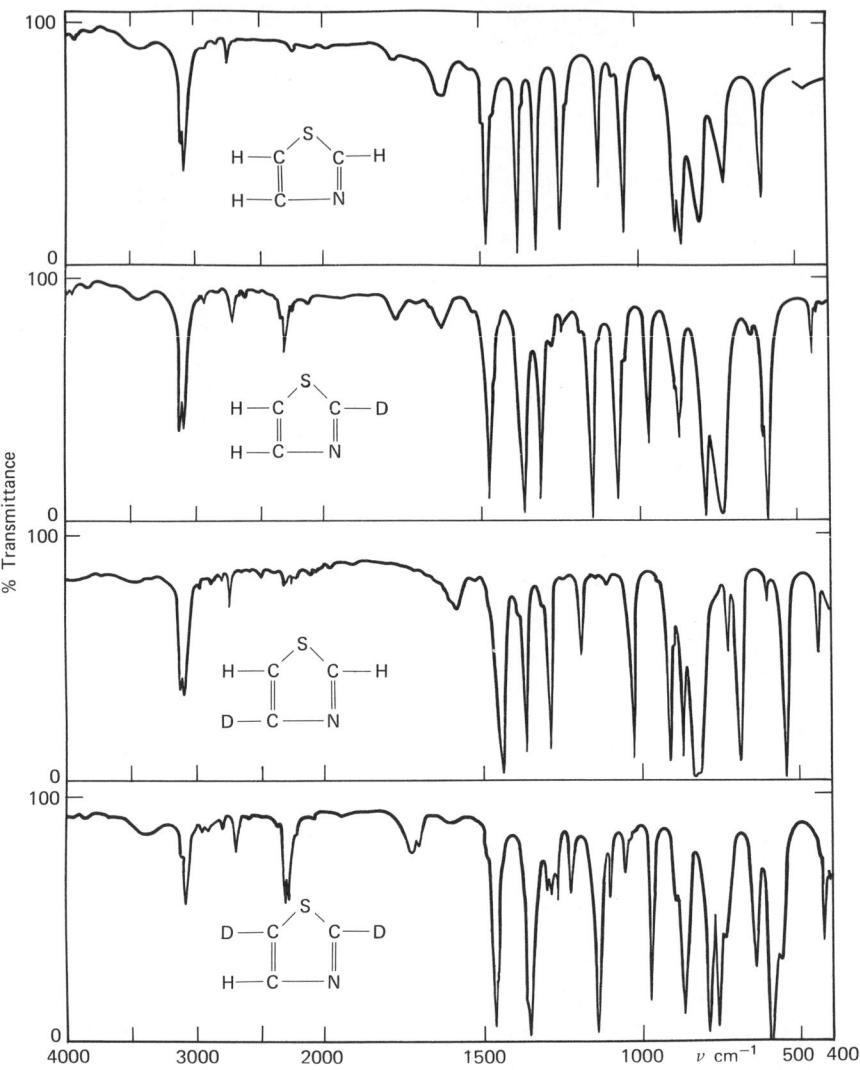

Fig. I-11. Infrared spectra of thiazole and deuterothiazole in the liquid phase.

The observation of the variation of the δCH bands of thiazole with the nature and the position of the substituent has been interpreted as a proof of a fairly strong coupling between the various CH vibrators (203). The couplings are confirmed by the force-field calculation for thiazole that shows that the nature of the 1300–1000 cm^{-1} band is rather complex.

OUT-OF-PLANE VIBRATIONS, γCH AND γCD. In accordance with all the proposed assignments (201–203), the bands at 797 and 716 cm^{-1} correspond to γCH vibrators, which is confirmed by the C-type structure observed for these frequencies in the vapor-phase spectrum of thiazole (Fig. I-9). On the contrary, the assignments proposed for the third γCH mode are contradictory. According to Chouteau et al. (201), this vibration is located at 723 cm^{-1}, whereas Sbrana et al. (202) prefer the band at 849 cm^{-1} and Davidovics et al. (203) the peak at 877 cm^{-1}. This last assignment is the most compatible with the whole set of spectra for the thiazole derivatives (203) and is confirmed by the normal vibration mode calculations (205) (Table I-25). The order of decreasing γCH frequencies, established by the study of isotopic and substituted thiazole derivatives, is (203): $\gamma C_{(4)}H > \gamma C_{(2)}H > \gamma C_{(5)}H$. Both the 2- and 4-positions, which seem equivalent for the νCH modes, are quite different for the γCH out-of-plane vibrations, a fact related to the influence observed for the

TABLE I-25. THIAZOLE FREQUENCIES AND ASSIGNMENTS

Symmetry mode	Experimental frequencies, ν(cm^{-1})	Theoretical frequencies	Theoretical assignment[a] (13)	Experimental assignment (11)
A'	3126	3125	97$\nu C_{(5)}$H	$\nu C_{(5)}$H
		3091	97$\nu C_{(4)}$H	$\nu C_{(4)}$H
	3092	3087	99$\nu C_{(2)}$H	$\nu C_{(2)}$H
	1478	1477	58ν(C=C)+28ν(C–N)	ω_1
	1378	1393	60ν(C=N)+25ν(C=C)	ω_2
	1319	1329	34ν(C–N)+23ν(C=C) 22ν(C=N)	ω_3
	1236	1248	27$\delta C_{(2)}$H+27$\delta C_{(4)}$H +17$\delta C_{(5)}$H	δ CH
	1122	1105	41$\delta C_{(4)}$H+7$\delta C_{(5)}$H +7$\delta C_{(2)}$H	δ CH
	1041	1047	72$\delta C_{(5)}$H+24$\delta C_{(4)}$H	δ CH
	862	875	60δN+38ν(C–S)	ω_4
	811	817	93ν(C–S)	ω_5
	753	749	61ν(C–S)+32δN	ω_6
	610	607	65δN+33ν(C–S)	ω_7
A"	877	877	86$\gamma C_{(4)}$H+11$\gamma C_{(2)}$H	$\gamma C_{(4)}$H
	795	804	68$\gamma C_{(2)}$H+28$\gamma C_{(4)}$H	$\gamma C_{(2)}$H
	718	704	98$\gamma C_{(5)}$H	$\gamma C_{(5)}$H
	602	603	22τ(C=C)+44τ(C–N) +22τ(C=N)	Γ_1
	467	474	46τ(C–S)+12τ(C=C)	Γ_2

[a] δ_N = angular bending of the ring; τ(C–X) = torsion of the C–X bond.

neighbor atoms on the γCH modes, whereas for the γCH the frequency depends only on the C–H bond itself (203). The $\gamma C_{(i)}D$ modes with $i = 2$, 4, and 5 are assigned to the 638, 682, and 540 cm^{-1} bands, respectively (203, 204) (Table I-24).

b. RING VIBRATIONS

A' SYMMETRY VIBRATIONS. Avignon et al. (205) have recently confirmed the earlier assignments of 1967 (202, 203) based on the analysis of the structure of the vibration-rotation bands observed in the vapor-phase infrared spectrum of thiazole and on the depolarization factor of the Raman lines. As in the case of isothiazole (207), the experimental and theoretical assignments (Table I-5), and of oxazole (208) or isoxazole (209), the ring vibration of highest frequency corresponds essentially to the C=C vibrator (Table I-25). The other ring vibrations correspond, in order of decreasing frequencies, to the C=N, C–N, and C–S vibrators as well as the intracyclic angles (Table I-26). Similar results are observed for the isotopic molecules (205).

TABLE I-26. ISOTHIAZOLE (207) FREQUENCIES AND ASSIGNMENTS

Symmetry mode	Experimental frequencies (ν, cm^{-1})	Theoretical frequencies (ν, cm^{-1})	Theoretical assignments (207)
A'	3105	3109	$96\nu C_{(5)}H$
	3086	3090	$96\nu C_{(4)}H$
	3056	3059	$99\nu C_{(3)}H$
	1484	1490	$56\nu(C=C) + 16\nu(C=N)$
	1390	1384	$35\nu(C=N) + 25\nu(C=C) + 21\nu(C-C)$
	1295	1284	$37\nu(C-C) + 17\delta C_{(5)}H$
	1236	1230	$50\delta C_{(3)}H + 20\nu(C=N)$
	1066	1076	$32\delta C_{(5)}H + 42\nu(C-C)$
	1040	1047	$67\delta C_{(4)}H + 30\delta C_{(5)}H$
	869	880	$30\nu(C-S) + 65\delta_N$
	815	823	$47\nu(S-N) + 23\nu(C-S) + 28\delta_N$
	758	734	$55\nu(S-N) + 25\nu(C-S) + 20\delta_N$
	639	636	$83\delta_N + 9\nu(C-S)$
A''	908	906	$64\gamma C_{(3)}H + 27\gamma C_{(4)}H$
	858	873	$28\gamma C_{(4)}H + 24\gamma C_{(3)}H + 24\tau(C-C)$
	728	721	$30\gamma C_{(5)}H + 21\gamma C_{(4)}H + 26\tau(C=C)$
	587	591	$44\tau(C=C) + 26\tau(C=N)$
	474	466	$39\tau(S-N) + 20\tau(C-S) + 13\tau(C=N)$

A" SYMMETRY VIBRATIONS. The first ring vibration of the A" type has been located at 650 cm^{-1}, as a result of the C-type structure observed for that band (Fig. I-9) (202, 203).

The second ring vibration gives rise to a very weak infrared absorption band at 467 cm^{-1} and to a weak and depolarized Raman line at 470 cm^{-1} (202, 203) (Table I-23).

c. OVERTONE AND COMBINATION BANDS

Apart from the band assigned to fundamental vibrations, the infrared spectrum of thiazole shows a certain number of peaks for which the following assignments have been proposed (204): 6116 ($2\nu C_{(5)}H$); ~6072 ($\nu C_{(5)}H + \nu C_{(4)}H$); 6052 ($2\nu C_{(4)}H$) or ($2\nu C_{(2)}H$); 6012 ($\nu C_{(4)}H + 2\omega_1$); ~4595 and 4570 ($\nu CH + \omega_1$); 4495 and 4464 ($\nu CH + \omega_2$); 4435 and 4400 ($\nu CH + \delta_3$); 4357 and 4310 ($\nu CH + \delta CH$); 4233 and 4196 ($\nu CH + \delta CH$); 4146 and 4123 ($\nu CH + \delta CH$); ~3980 and 3950 ($\nu CH + \omega_4$); ~3990 and 3964 ($\nu CH + \gamma C_{(4)}H$); 3876 ($\nu CH + \gamma C_{(2)}H$); 3876 and 3840 ($\nu CH + \omega_6$); 3728 and 3696 ($\nu CH + \omega_7$); 2941 ($2\omega_1$); 2840 ($\omega_1 + \omega_2$); 2749 ($2\omega_2$); 1760 ($2\gamma C_{(4)}H$); ~1670 ($\gamma C_{(4)}H + \gamma C_{(2)}H$); 1602 ($\gamma C_{(4)}H + \gamma C_{(5)}H$); 1591 ($2\gamma C_{(2)}H$; 1516 ($\gamma C_{(2)}H + \gamma C_{(5)}H$; 1436 ($2\gamma C_{(5)}H$); 1158 ($\gamma C_{(5)}H + \Gamma_2$); 1065 ($\Gamma_1 + \Gamma_2$); 928 ($2\Gamma_2$).

d. INFLUENCE OF THE PHYSICAL STATE ON THE VIBRATION SPECTRUM OF THIAZOLE.

The vibration frequencies of C–H bond are noticeably higher for gaseous thiazole than for its dilute solutions in carbon tetrachloride or for liquid samples (Table I-27). The molar extinction coefficient and especially the integrated intensity of the same peaks decrease dramatically with dilution (203). Inversely, the $\gamma(C_{(2)}H)$ and $\gamma(C_{(5)}H)$ frequencies are lower for gaseous thiazole than for its solutions, and still lower than for liquid samples (cf. Table I-27).

A similar frequency shift is observed for their overtones or combination bands (204). It was also established that the proton-donating ability of the thiazole CH groups decreases in the order, $2 > 5 > 4$ (204).

Finally, the $\delta(CH)$ bending frequencies are practically independant of the physical state of the sample as are the nuclear vibration modes (Table I-27).

TABLE I-27. INFRARED FREQUENCIES OF THIAZOLE AS A FUNCTION OF THE PHYSICAL STATE AND ASSIGNMENT

Infrared							
Vapor		Solution		Liquid		Solid	
ν(cm^{-1})	SR	ν(cm^{-1})	ε	ν(cm^{-1})	I	ν(cm^{-1})	Assignments
3143	R						
3134	Q	3126	8	3118	m	3122 m	$\nu C_{(5)}H$
3108	R	3092	10	3083	m	3084 m	$\nu C_{(4)}H$
3093	P					~3077 e	$\nu C_{(2)}H$
				~3062	e	3070 F	
						3067 F	νCH associée
2954	R						
2945	Q	2941	1	2936	ff	—	$2\omega_1$
2930	P						
		2840	1	2837	ff	—	$\omega_1 + \omega_2$
2760	R						
2752	Q	2749	5	2745	f	—	$2\omega_2$
2743	P						
		~2740 e		~2740	e	—	
1780	R						
1771	Q	1760	6	1772	f	—	$2\gamma C_{(4)}H$
1760	P						
		~1670	4	~1680	f	—	$\gamma C_{(4)}H + \gamma C_{(2)}H$
1612	R						
1604	Q	1602	12	1616	f	—	$\gamma C_{(4)}H + \gamma C_{(5)}H$
1595	P						
1585	?	1591	10	—	—	—	$2\gamma C_{(2)}H$
		1516	7				$\gamma C_{(2)}H + \gamma C_{(5)}H$
1492	R						
1484	Q	1478	107	1479	F	1482 F	ω_1
						1464 F	
1473	P						
1444	R						
1434	Q	1436	25	—	—	—	$2\gamma C_{(5)}H$
1424	P						
1392	R						
1383	Q	1378	126	1380	F	1383 m	ω_2
						1378 m	
1374	P						
1334	R						
1325	Q	1319	51	1318	F	1315 FF	ω_3
						~1307 e	
1315	P						
						1248 F	

TABLE I-27 (*Continued*)

Infrared							
Vapor		Solution		Liquid		Solid	
$\nu(\text{cm}^{-1})$	SR	$\nu(\text{cm}^{-1})$	ε	$\nu(\text{cm}^{-1})$	I	$\nu(\text{cm}^{-1})$	Assignments
1246	R	1236	57	1239	F	1244 F	δCH
1234	P					1227 f	
		1158	3				$\gamma C_{(5)}H+\Gamma_2$
1130	R	1122	37	1121	m	1124 F	δCH
1118	P					~1119 e	
		1065	4	1072	f	1074 ff	$\Gamma_1+\Gamma_2$
1052	R	1041	47	1040	F	1033 F	δCH
1043	Q						
1033	P						
		928	4	933	f	—	$2\Gamma_2$
888	Q	877	52	881	m	~888 e	$\gamma C_{(4)}H$
						883 F	
874	R					~868 e	
866	Q	862	223	864	F	861 FF	ω_4
856	P					850 f	
						808 F	ω_5
811	R						
797	Q	795	239	802	F	823 F	$\gamma C_{(2)}H$
793	Q						
780	P						
756	Q	753	10	~756	e	808 F	ω_6
						759 f	
731	R						
721	Q	718	129	726	F	~748 e	$\gamma C_{(5)}H$
716	Q					737 F	
701	P						
617	R						
603	Q	602	131	606	F	616 f	ω_7
583	P					606 F	Γ_1
				467	ff		Γ_2

C. *Thiazole Derivatives*

The infrared and Raman studies of thiazole derivatives are numerous (111, 173, 197–226) though often only fragmentary. The only studies leading to a complete assignment of the observed bands are those of Chouteau and Davidovics et al. (201, 203, 204, 227, 228).

For thiazole derivatives containing polyatomic substituents, such as CH_3, C_2H_5, C_3H_9, NH_2, ND_2, and $N(CH_3)_2$, the assumption was made that these substituents show characteristic group frequencies (204) that make possible the distinction between vibrations essentially originating in the thiazole ring and those of substituents.

Table I-28 lists the mean vibration frequencies characteristic of CH bonds (νCH, δCH, γCH) as a function of the substitution pattern. For the ν(CH) vibrations, the highest frequency peak disappears in the spectra of 5-substituted derivatives, whereas it is unchanged by substitution at the 2- or 4-positions. This band has been assigned to the ν(CH) vibration connected with the CH bond at the 5-position (173).

The peak near 3085 cm^{-1} disappears only for 2,4-disubstituted derivatives. Its frequency is slightly higher for 2- than for 4-substituted compounds. The $\nu C_{(2)}H$ and $\nu C_{(4)}H$ vibrations seem to be nearly equivalent and usually give rise to an unique peak, except in the case of 5-bromo- and 5-isopropylthiazoles in CCl$_4$ solution, where this peak is split into two bands.

The δ(CH) bending vibrations have been located between 1250 and 1000 cm^{-1} and show varying frequencies as a function of the nature and the position of substituents (203). It is possible, however, that the $\delta C_{(2)}H$ mode is located near 1220 cm^{-1} and suffers the weakest influence from 4- or 5-substitution.

The observed order for the γ(CH) bending vibrations in various mono- and disubstituted thiazole derivatives is $\gamma C_{(4)}H > \gamma C_{(2)}H > \gamma C_{(5)}H$. Usually the intensity of $C_{(2)}H$ and $C_{(5)}H$ peaks is very high, whereas that of $C_{(4)}H$ is rather low.

The skeleton vibrations, C_3NSX, C_3NSX_2, C_3NSXY, or C_3NSX_3 (where X or Y is the monoatomic substituent or the atom of the substituent which is bonded to the ring for polyatomic substituents), have been classified into suites, numbered I to X. A suite is a set of absorption bands or diffusion lines assigned, to a first approximation, to a same mode of vibration for the different molecules. Suites I to VIII concern bands assigned to A' symmetry vibrations, while suites IX and X describe bands assigned to A'' symmetry vibrations. For each of these suites, the analysis of the various published works gives the limits of the observed frequencies (Table I-29).

Comparison of the various frequencies for each suite leads to the following remarks:

The frequencies of suite I (related to the ω_1 mode of thiazole) increase under the influence of electron-withdrawing substituents, whatever their positions on the ring: the frequencies increase similarly for suite II, but only when the substituent is in the 2-position.

The frequencies of suite III are usually lower for disubstituted (especially 2,4-dihalo and 2,5-disubstituted thiazoles) and those of the corresponding monosubstituted derivatives.

TABLE I-28. MEAN POSITIONS OF THE C–H VIBRATIONS OF THIAZOLE DERIVATIVES

Position of the substituent	$\nu C_{(5)}H$	$\nu C_{(4)}H$	$\nu C_{(2)}H$	$\delta CH_{(2)}$	$\delta CH_{(4)}$	$\delta CH_{(5)}$	$\gamma C_{(4)}H$	$\gamma C_{(2)}H$	$\gamma C_{(5)}H$
2	3118 ± 8	3087 ± 7	—	—	1164 ± 37	1057 ± 11	874 ± 8	—	702 ± 21
4	3121 ± 11	—	3085 ± 9	1218 ± 20	—	1091 ± 46	—	808 ± 6	722 ± 4
5	—	3089 ± 9	—	1226 ± 8	1109 ± 7	—	851 ± 4	785 ± 2	—
2,4	3120 ± 12	—	—	—	—	1083 ± 55	—	—	707 ± 32
2,5	—	3088 ± 13	—	—	1149 ± 12	—	839 ± 2	—	—
4,5	—	—	3085 ± 7	1215 ± 29	—	—	—	785 ± 3	—

TABLE I-29. LIMITING FREQUENCIES OF THE VARIOUS RING VIBRATIONS OF THIAZOLE DERIVATIVES

Type of substitution		I	II	III	IV	V	V'	VI	VII	VIII	VIII'	IX	X
2	$\nu_M{}^a$	1548	1498	1334	874	1389	—	767	677	567	—	632	513
	ν_m	1474	1365	1318	858	970	—	744	622	335	—	588	462
4	ν_M	1521	1417	1318	883	830	—	932	676	556	—	642	488
	ν_m	1463	1393	1292	872	817	—	899	611	430	—	605	472
5	ν_M	1523	1403	1312	867	1174	—	756	667	554	—	604	481
	ν_m	1477	1383	1294	858	964	—	739	617	336	—	594	466
2,4	ν_M	1541	1517	1324	861	1280	—	959	690	588	550	638	543
	ν_m	1456	1376	1238	838	1014	—	875	634	338	296	595	480
2,5	ν_M	1548	1484	1286	866	1196	1186	756	669	507	—	604	499
	ν_m	1487	1395	1256	859	986	992	727	642	272	—	579	487
4,5	ν_M	1549	1414	1314	832	1156	—	906	700	633	527	641	501
	ν_m	1479	1402	1273	828	1049	—	880	664	503	415	636	493
Triméthyl 2,4,5-thiazole		1555	1488	1304	875	1196	1173	958	674	574	504	632	524

[a] ν_M, maximum frequency; ν_m, minimum frequency.

The frequencies of suite IV are generally near those of the ω_4 mode of thiazole, except for 2,4- and 4,5-disubstituted derivatives for which they are lower.

For 4-monosubstituted derivatives, the frequencies of suite V are near those of the ω_5 mode of thiazole, and for the 2- and 5-ones the frequencies of suite VI are near those of the ω_6 mode of thiazole.

The variable frequencies of suites V and VIII on one side, and VI and VIII on the other, correspond to oscillations resulting from the coupling of the $\nu(C-X)$ vibration with the ω_5 mode in the case of 2- or 5-substituted derivatives and with the ω_6 mode in the case of 4-substituted derivatives. For 2,5-disubstituted thiazoles the ω_6 vibration is only slightly different from that of thiazole itself and the ω_5 oscillation is coupled with both $\nu(C_{(2)}X)$ and $\nu(C_{(5)}X$ or Y) modes, giving rise to three frequencies, two of which are higher and classified in suites V and V', the third, being lower, is assigned to suite VIII.

For 2,4- or 4,5-disubstituted derivatives, there is a double coupling between the ω_5 and $\nu(C_2X)$ or $\nu(C_5X)$ vibration on one hand and oscillations ω_6 and $\nu(C_4X)$ on the other. These interactions induce, for the first one, two frequencies either higher (suite V) or lower (suite VIII), for the second, two other frequencies either higher (suite VI) or lower (suite VIII).

For 2,4,5-trimethylthiazole, the five bands described in suites V, V', and VII on one hand and in suites VI and VIII' on the other result from the coupling of the ω_5, $\nu(C_2x)$, $\nu(C_5X)$ oscillators and of ω_6, $\nu(C_4X)$ oscillators, respectively.

The frequencies responsible for suites IX and X are near the Γ_1 and Γ_2 modes of vibration of thiazole, respectively, and have been assigned to such oscillations.

Suites I to VIII contain infrared frequencies corresponding to vibration-rotation bands of A, B, or (A+B) hybrid types and can thus be assigned to vibrations of A' symmetry; the corresponding Raman lines are generally polarized.

The frequencies classified in suites IX and X belong to depolarized Raman lines and correspond to vibrations-rotation bands of the C type. They can be assigned to oscillations of A'' symmetry.

5. Nuclear Magnetic Resonance

Whatever the derivative considered, the nuclear magnetic resonance spectra of thiazoles are remarkably simple and apparently univoque. The first proton NMR spectrum of thiazole was described by Bak et al. (171). It was followed by a series of works establishing a systematic description of thiazoles, namely the resonance peak assignments, the chemical shifts, and coupling constants as well as the evolution of the spectrum as a function of solvent or temperature (229–236). The signal assignment could be made without ambiguity thanks to the study of various substituted thiazoles, unequivocally synthesized (123), and of 2- and 5-deuteriated thiazole (230). Characteristic proton chemical shifts are given in Table I-30. For protons bonded to sp^2-carbon atoms, these chemical shifts are located at relatively low fields. This deshielding can be correlated to the occurrence of a ring current connected with the "pseudoaromatic" character of thiazole and also to the magnetic anisotropy

TABLE I-30. PROTON CHEMICAL SHIFTS OF THIAZOLE

$\delta H(2)$	$\delta H(4)$	$\delta H(5)$	Solvent	Ref.
−2.31	−1.70	−1.35	—[b]	229
8.68	7.83	7.19	C_6H_{12}	230
10.0	8.45	8.23	CF_3COOH	230
9.15	7.97	7.75	DMSO	234
9.00	7.93	7.65	Acétone	234
8.70	7.84	7.22	C_6H_6	234
8.77	7.86	7.25	CCl_4	235
8.88	7.98	7.41	$CDCl_3$	238
7.72	6.62	6.23	TFA	239
8.55	7.54	7.05	$CDCl_3$	160
	7.86	7.27	CCl_4	224

[a] Chemical shifts are expressed in δ units: ppm of applied magnetic field with internal TMS peak as reference.
[b] In this case H_2O peak is external reference.

of both sulfur and nitrogen heteroatoms. Furthermore, its seems that this latter atom plays a preponderant role in the chemical shifts observed for the protons of benzene (7.30 ppm) on one hand, and those in the α-position of thiophene (7.37 ppm) and pyridine (8.60 ppm) on the other hand. More sophisticated analytical methods have been used to define the characteristics of the ring current: from empirical calculations (237), from experimental data (238), or attempts to determine the "aromaticity" of the ring on the basis of this ring current (126). These studies show the similarity between ring currents in thiazole and thiophene, and more especially, pyrrole (126, 237) and the absence of correlation with resonance energies (238) (Table I-31). In a first analysis no correlation could

TABLE I-31. PROTON CHEMICAL SHIFTS OF THIAZOLE AND VARIOUS MONOSUBSTITUTED THIAZOLES[a]

Thiazole derivatives	H(2)	H(4)	H(5)	Solvent	Ref.
2-Methyl	—	-1.52^b	-1.25^b	—	229
	—	7.44	6.97	CCl_4	235
	—	7.64	7.17	$CDCl_3$	238
	—	7.63	7.36	Acétone	234
	—	7.47	6.92	C_6H_6	234
	—	7.46	7.17	$CDCl_3$	160
	—	7.44	6.97	CCl_4	224
2-Ethyl	—	7.58	7.18	CCl_4	235
	—	7.58	7.17	CCl_4	224
2-n-Propyl	—	7.51	7.04	CCl_4	224
2-i-Propyl	—	7.58	7.09	CCl_4	235
	—	7.58	7.08	CCl_4	224
2-t-Butyl	—	7.58	7.10	CCl_4	235
	—	7.54	7.10	CCl_4	224
2-neo-Pentyl	—	7.61	7.11	CCl_4	224
2-Methoxy	—	7.05	6.62	CCl_4	235
	—	7.16	6.90	Acétone	234
	—	6.16	6.46	C_6H_6	234
2-Ethoxy	—	7.10	6.56	CCl_4	235
2-Carboxy	—	7.80	7.65	CCl_4	235
	—	7.97	7.97	C_6H_6	234
	—	8.10	8.10	DMSO	234
2-Amino	—	6.96	6.60	CCl_4	235
	—	6.93	6.53	DMSO	247
	—	7.14	6.76	TFA	247
	—	7.23	6.85	TFA	248
	—	7.07	6.68	D_2O	248
	—	7.28	6.81	CH_3COOH	248
	—	7.20	6.85	HCOOH	248
2-Nitro	—	8.15	7.93	Acétone	234
	—	8.30	8.07	DMSO	234
2-Chloro	—	7.62	7.62	Acétone	234
	—	7.42	7.08	C_6H_6	248
2-Bromo	—	7.59	7.35	CCl_4	248
	—	7.73	7.14	CH_3COOH	248
	—	7.85	7.10	HCOOH	248
	—	8.23	8.16	TFA	248
	—	8.07	8.07	$H_2SO_4/H_2O(25\%)$	248
	—	7.64	7.64	Acétone	234
	—	7.44	7.11	C_6H_6	234
2-Iodo	—	7.55	7.35	CCl_4	235
2-Hydroxymethyl	—	7.58	7.28	D_2O	234
	—	7.76	7.51	Acétone	234
2-Ethyloxycarbonyl	—	7.82	7.82	DMSO	234
	—	8.01	7.93	Acétone	234
	—	7.89	7.56	C_6H_6	234

TABLE I-31 (*Continued*)

Thiazole derivatives	H(2)	H(4)	H(5)	Solvent	Ref.
2-Dimethylamino	—	7.02	6.37	CCl_4	235
2-Propylamino	—	6.97	6.35	CCl_4	235
2-Diethylamino	—	7.00	6.36	CCl_4	235
2-Dipiperidyl	—	7.02	6.39	CCl_4	235
2-Formyl	—	8.07	7.96	CCl_4	234
	—	7.75	7.32	—[c]	249
	—	8.24	8.24	DMSO	234
	—	7.75	7.32	C_6H_6	234
2-Phenyl	—	7.77	7.16	$CDCl_3$	236
	—	7.64[d]	6.98[d]	C_6H_{12}	236
	—	8.35	8.06	TFA	225
2-*o*-Tolyl	—	7.53	7.14	$CDCl_3$	236
2-Methylthio	—	7.48	7.03	CCl_4	235
2-Ethylthio	—	7.69	7.49	Acétone	234
2-Ethylsulfonyl	—	8.15	8.15	DMSO	234
2-Fluoroformyl	—	8.12	7.95	—[c]	249
2-Thiazolyl	—	7.98	7.77	$CDCl_3$	225
4-Methyl	−2.20[b]	—	−1.07[b]	—	163
	8.64	—	6.87	CCl_4	231
	8.67	—	6.95	D_2O	231
	9.83	—	7.81	TFA	231
	8.50	—	6.74	CCl_4	235
	8.85	—	7.14	Acétone	234
	8.50	—	6.7	C_6H_6	234
	8.65	—	6.88	$CDCl_3$	160
4-Formyl	9.24	—	8.59	DMSO	234
4-Carboxy	9.21	—	8.60	Acétone	234
4-Ethylcarboxy	9.07	—	8.43	Acetone	234
	8.81	—	8.15	C_6H_5	234
4-Phenyl	8.76	—	7.39	$CDCl_3$	236
	8.81[d]	—	7.48[d]	C_6H_{12}	236
	10	—	8.21	TFA	236
4-*p*-Methoxyphenyl	8.76	—	7.27	$CDCl_3$	225
	8.54[d]	—	7.10[d]	C_6H_{12}	225
4-*p*-Nitrophenyl	8.90	—	7.71	$CDCl_3$	225
4-*p*-Bromophenyl	8.97	—	7.38	$CDCl_3$	225
	8.58[d]	—	7.26[d]	C_6H_{12}	225
5-Methyl	8.38	7.39	—	CCl_4	235
	8.44	7.44	—	C_6H_6	234
	8.75	7.57	—	Acetone	234
	8.69	7.48	—	$CDCl_3$	160
5-Chloro	8.57	7.64	—	CCl_4	235
5-Bromo	9.05	7.89	—	Acetone	234
	8.54	7.65	—	C_6H_6	234
5-Phenyl	8.74	8.09	—	$CDCl_3$	225
	8.46	7.57	—	TFA	225

TABLE I-31 (*Continued*)

Thiazole derivatives	H(2)	H(4)	H(5)	Solvent	Ref.
5-Carboxy	9.26	8.48	—	Acetone	234
	9.35	8.58	—	DMSO	234
5-Formyl	9.30	8.65	—	Acetone	234
	9.08	8.53	—	C_6H_6	234
5-Methoxy	8.35	7.23	—	Acetone	234
	8.07	7.08	—	C_6H_6	234
5-Ethylthio	9.08	7.88	—	Acetone	234
	8.70	7.75	—	C_6H_6	234
5-Ethylsulfonyl	9.42	8.45	—	Acetone	160
5-Ethyloxycarbonyl	9.28	8.48	—	Acetone	160

[a] Chemicals shifts are expressed in δ units: ppm of applied magnetic field with internal TMS peak as reference.
[b] In this case, H_2O peak is external reference.
[c] Diethylene glycodiethylether.
[d] Infinite dilution extrapolation.

be found between the electronegativity of the substituent and the variation in chemical shift on the other protons induced by a substitution: a high-field shift is observed with all substituents. For example the introduction in the 4- or 5-position of a chloro or an amino group induces uniformly a shielding of the other protons, consistent with the perturbation of the π-electron distribution (123) and not with any inductive effect at the level of the σ-skeleton. The interpretation of the observed chemical shifts could be attempted using an empirical approach by considering that the observed chemical shift is the sum of particular effects, calculated according to Pople et al. (240).

$$\delta_H = \Delta\delta_{CC} + \Delta\delta_E + \Delta\delta_R + \Delta\delta_A + f(q^*)$$

where $\Delta\delta_{CC}$ is the ring current effect (241, 242), $\Delta\delta_E$ is the lone pair electron effect (243), $\Delta\delta_R$ is the field-reaction effect (243, 244), $\Delta\delta_A$ is the heteroatom anisotropy effect (245, 246), and $f(q^*)$ is the participation of the electronic population of the carbon bearing the considered proton.

Making allowance for those effects gives a good correlation between the chemical shifts and the π- and/or σ-electron density of the carbon atom bearing the proton (133, 236, 237).

More elaborated calculations based on theoretical LCAO models have given interesting results within the CNDO/1 and CNDO/2 approximations (251.) Tables I-32 and I-33).

TABLE I-32. PROTON CHEMICAL SHIFTS OF VARIOUS DISUBSTITUTED THIAZOLES

Thiazole derivatives	δH(2)	δH(4)	δH(5)	Solvent[a]	Ref.
2,4-Dimethyl-	—	—	6.50	CCl_4	235
2,5-Dimethyl	—	7.03	—	CCl_4	235
4,5-Dimethyl	8.26	—	—	CCl_4	235
2-Methyl-4-t-butyl	—	—	6.52	CCl_4	235
2-Ethyl-4-methyl	—	—	6.51	CCl_4	235
2-i-Propyl-4-methyl	—	—	6.52	CCl_4	235
2-t-Butyl-4-methyl	—	—	6.52	CCl_4	235
2-Ethyl-4-t-butyl	—	—	6.55	CCl_4	235
2-i-Propyl-4-t-butyl	—	—	6.52	CCl_4	235
2-t-Butyl-4-t-butyl	—	—	6.53	CCl_4	235
2-Methyl-4-phenyl	—	—	7.91	TFA	225
	—	—	7.12	$CDCl_3$	236
	—	—	7.01	C_6H_{12}	236
2-Ethyl-4-phenyl	—	—	7.84	TFA	225
	—	—	7.15	$CDCl_3$	236
	—	—	7.04	C_6H_{12}	236
2-Propyl-4-phenyl	—	—	7.07	$CDCl_3$	236
	—	—	7.00	C_6H_{12}	236
2-i-Propyl-4-phenyl	—	—	7.14	$CDCl_3$	236
	—	—	7.02	C_6H_{12}	236
2-t-Butyl-4-phenyl	—	—	7.11	$CDCl_3$	236
	—	—	7.01	C_6H_{12}	236
2-Phenyl-4-methyl	—	—	6.71	$CDCl_3$	236
	—	—	6.56	C_6H_{12}	236
	—	—	7.87	TFA	225
2-Phenyl-4-t-butyl	—	—	6.67	$CDCl_3$	236
	—	—	6.63	C_6H_{12}	236
2-Amino-4-phenyl	—	—	6.64	$CDCl_3$	236
2-Chloro-4-phenyl	—	—	7.06	$CDCl_3$	236
2-Hydroxy-4-phenyl	—	—	7.20	$CDCl_3$	236
2-Mercapto-4-phenyl	—	—	6.64	$CDCl_3$	236
2,4-Diphenyl	—	—	7.48	$CDCl_3$	225
2,4-Dichloro	—	—	6.87	CCl_4	235
2,4-Dibromo	—	—	7.07	CCl_4	235
2-Carboxy-4-methyl	—	—	7.48	Acetone	234
2-Methyloxycarbonyl-4-methyl	—	—	7.52	Acetone	234
2-Ethyl-5-ethyl	—	7.13	—	CCl_4	234
2-i-Propyl-5-methyl	—	7.06	—	CCl_4	234
2-t-Butyl-5-ethyl	—	7.14	—	CCl_4	234
2-t-Butyl-5-i-propyl	—	7.16	—	CCl_4	234
2-Phenyl-5-methyl	—	7.39	—	$CDCl_3$	236
	—	7.27	—	C_6H_{12}	234
2-Phenyl-5-ethyl	—	7.34	—	$CDCl_3$	234

TABLE I-32 (*Continued*)

Thiazole derivatives	δH(2)	δH(4)	δH(5)	Solvent[a]	Ref.
2-Phenyl-5-*i*-propyl	—	7.27	—	C_6H_{12}	234
	—	7.45	—	$CDCl_3$	234
	—	7.33	—	C_6H_{12}	234
2,5-Diphenyl	—	7.78	—	$CDCl_3$	225
	—	8.28	—	TFA	225
2-Phenyl-5-bromo	—	7.91	—	DMSO	250
2-Phenyl-5-nitro	—	8.64	—	DMSO	250
2-Amino-5-bromo	—	6.97	—	DMSO	250
	—	6.98	—	Acetone	234
2-Chloro-5-bromo	—	7.36	—	CCl_4	235
2,5-Dibromo	—	7.38	—	CCl_4	235
4-Phenyl-5-methyl	8.48	—	—	$CDCl_3$	236
	8.34	—	—	C_6H_{12}	236
	9.78	—	—	TFA	225
4-Phenyl-5-ethyl	8.55	—	—	$CDCl_3$	236
	8.35	—	—	C_6H_{12}	236

[a] TFA, trifluoroacetic acid; DMSO, dimethylsulfoxide.

TABLE I-33. CHEMICAL SHIFTS OF PROTONS OF THE SUBSTITUENTS IN SUBSTITUTED THIAZOLE (235)

Compounds[a]	Proton in α from the cycle			Proton in β from the cycle		
	2	4	5	2	4	5
2-Methyl	2.68	—	—	—	—	—
4-Methyl	—	2.45	—	—	—	—
5-Methyl	—	—	2.49	—	—	—
2,4-Dimethyl	2.61	2.33	—	—	—	—
2,5-Dimethyl	2.54	—	2.36	—	—	—
4,5-Dimethyl	—	2.35	2.31	—	—	—
2,4,5-Trimethyl	2.51	2.24	2.21	—	—	—
2-Ethyl	3.04	—	—	1.40	—	—
2-*i*-Propyl	3.31	—	—	1.40	—	—
2-*t*-Butyl	—	—	—	1.44	—	—
2-Methyl-4-*t*-butyl	2.63	—	—	—	1.29	—
2-Ethyl-4-methyl	2.91	2.33	—	1.35	—	—
2-*i*-Propyl-4-methyl	3.18	2.34	—	1.34	—	—
2-*t*-Butyl-4-methyl	—	2.35	—	1.38	—	—
2-*i*-Propyl-5-methyl	3.14	—	2.37	1.32	—	—
2-Ethyl-5-ethyl	2.88	—	2.76	1.32	—	1.27
2-Ethyl-4-*t*-butyl	2.90	—	—	1.35	1.29	—
2-*t*-Butyl-5-ethyl	—	—	2.71	1.37	—	1.28

TABLE I-33 (*Continued*)

Compounds[a]	Proton in α from the cycle			Proton in β from the cycle		
	2	4	5	2	4	5
2-*i*-Propyl-4-*t*-butyl	3.31	—	—	1.40	1.29	—
2-*t*-Butyl-5-*i*-propyl	—	—	3.10	1.38	—	1.30
2-*t*-Butyl-4-*t*-butyl	—	—	—	1.40	1.30	—
2-Methoxy	4.03[b]	—	—	—	—	—
2-Ethoxy	4.47[b]	—	—	1.52[c]	—	—
2-Methylthio	2.68[b]	—	—	—	—	—
2-Diallylamino-4-methyl	—	2.18	—	—	—	—

[a] The solvent was CCl_4.
[b] Protons in α position to functional heteroatom.
[c] Protons in β position to functional heteroatom.

TABLE I-34. PROTON CHEMICAL SHIFTS AND COUPLING CONSTANTS J_{HH} OF THIAZOLE IN VARIOUS SOLVENTS AT INFINITE DILUTION (20°C) (235)

Solvent[a]	Concnt.	$\delta H(2)$	$\delta H(4)$	$\delta H(5)$	J_{24}	J_{25}	J_{45}
CCl_4	0	8.733	7.878	7.306	0	1.90	3.10
C_6H_{12}	0	8.635	7.835	7.161	0	1.90	3.20
$CDCl_3$	0	8.890	7.993	7.435	0	1.95	3.15
THF	0	8.991	7.961	7.604	0	1.85	3.15
NO_2Me	0	9.048	8.028	7.640	0	1.90	3.10
C_6D_6	0	8.199	7.678	6.511	0	2.00	3.20
C_6D_5N	0	9.168	8.103	7.562	0	1.95	3.00
DMSO	0	9.125	7.964	7.790	0	1.80	3.10
TEA	0	8.893	7.885	7.476	0	1.90	3.00
MeOH	0	9.050	7.944	7.673	0.60	1.90	3.00
Acetone	0	9.038	7.956	7.667	0	1.92	3.15
DMF	0	9.200	8.033	7.828	0	1.92	2.95
HCOOH	30	9.546	8.225	7.931	0.70	1.55	3.10
AcOH	30	9.179	8.038	7.624	0.50	1.82	3.10
—	100	9.091	8.086	7.496	0	1.90	3.20

[a] THF, tetrahydrofuran; DMSO, dimethylsulfoxide; TEA, triethylamine; DMF, dimethylformamid.

Properties and Reactions of Thiazole

The variation of chemical shifts as a function of dilution could be accounted for only qualitatively (235) because of the large diversity of solute-solvent interactions resulting from the nature and the shape of the solvent molecule (Table I-34).

The proton-proton spin couplings (J_{HH}) have been measured for a large number of thiazole derivatives. The results are shown in Tables I-35 and I-36.

TABLE I-35. COUPLING CONSTANTS J_{HH}(Hz) IN THIAZOLE AND SOME OF ITS DERIVATIVES

Compounds	J_{HH}			Solvent	Ref.
	J_{24}	J_{25}	J_{45}		
Thiazole	0.9	2.2	3.6	CF_3COOH	230
	—	1.9	3.2	C_6H_{12}	230
	—	2.1	3.2	Acetone	234
	—	2.0	3.0	C_6H_6	234
	—	1.95	3.15	DMSO	234
	—	1.8	3.1	$CDCl_3$	160
	—	1.86	—	neat	224
	—	1.8	3.2	neat	253
	0.60	1.90	3.13		113
2-Methyl	—	—	3.40	CCl_4	235
	—	—	3.3	C_6H_6	234
	—	—	3.2	Acetone	234
	—	—	3.2	$CDCl_3$	160
2-Ethyl	—	—	3.2	$CDCl_3$	224
2-i-Propyl	—	—	3.1	$CDCl_3$	224
2-n-Propyl	—	—	3.2	$CDCl_3$	224
2-i-Butyl	—	—	3.2	$CDCl_3$	224
2-t-Butyl	—	—	3.2	$CDCl_3$	224
2-neo-Pentyl	—	—	3.4	$CDCl_3$	224
2-Hydroxymethyl	—	—	3.2	Acetone	234
2-Formyl	—	—	2.8	Acetone	234
	—	1.8	3.1	C_6H_6	234
	—	—	3.1		249
2-Fluoroformyl	—	—	3.3		249
2-Ethyloxycarboxyl	—	—	3.06	C_6H_6	234
	—	—	3.3	Acetone	234
2-Amino	—	—	3.8		248
	—	—	4.5	HCOOH	248
	—	—	4.4	CH_3COOH	248
	—	—	4.6	CF_3COOH	248
	—	—	3.8	in 2-Amino-2-^{13}C-thiazole	113
	—	—	3.27	in 2-Amino-4-^{13}C-thiazole	113
	—	—	3.2	Acetone	234

TABLE I-35 (*Continued*)

Compounds	J_{HH}			Solvent	Ref.
	J_{24}	J_{25}	J_{45}		
2-Nitro	—	—	3.6	Acetone	234
	—	—	3.5	DMSO	234
2-Methoxy	—	—	3.6	Acetone	234
2-Bromo	—	—	3.4	CCl_4	248
	—	—	3.6	CH_3COOH	248
	—	—	3.7	HCOOH	248
	—	—	4.1	CF_3COOH	248
2-Bromo	—	—	3.40	in 2-Bromo-2-^{13}C-thiazole	113
	—	—	3.56		254
2-Methylthio	—	—	3.4	C_6H_6	234
2-Ethylthio	—	—	3.3		234
4-Methyl	—	1.90	—	CCl_4	235
	—	1.80	—	C_6H_6	235
	—	1.60	—		235
4-Formyl	—	1.87	—	Acetone	234
4-Carboxy	—	2.35	—	DMSO	234
4-Ethyloxycarbonyl	—	2.5	—	Acetone	234
5-Methyl	0.50	—	—	CCl_4	234
5-Carboxy	0.5	—	—	DMSO	234
5-Nitro	0.8	—	—	DMSO	234
	0.75	—	—	Acetone	234

TABLE I-36. COUPLING CONSTANTS J_{HH}(Hz) BETWEEN PROTON OF METHYL GROUP AND CYCLIC PROTONS IN METHYLTHIAZOLES IN VARIOUS SOLVENTS (20°C) (235)

Compounds and spin-spin couplings	Solvents							
	$CDCl_3$	THF	NO_2Me	C_6D_6	H_3COH	DMF	C_6H_{12}	CCl_4
2-Methylthiazole								
J Me(2)–H(5)	—	—	—	—	—	—	—	0.55
J H(4)–H(5)	3.35	3.35	3.40	3.40	3.35	3.50	3.35	3.30
4-Methylthiazole								
J H(2)–H(5)	1.90	2.00	1.95	2.00	1.85	1.80	1.95	1.88
J Me(4)–H(5)	1.00	0.93	1.00	1.10	1.10	0.80	0.96	0.50
5-Methylthiazole								
J H(2)–Me(5)	—	—	—	0.80	0.50	0.80	0.50	—
J H(4)–Me(5)	1.30	—	1.30	1.40	1.40	1.40	1.30	1.25
J H(2)–H(4)	—	—	—	0.50	—	—	—	—

J_{HH} spin-coupling results in first-order coupling figures that are easy to interpret. Only the long-range couplings of protons α- to nitrogen are uncertain because of their broad signals. Besides the already mentioned systematic studies concerning signal assignments, where the coupling constants are usually given, three studies are particularly devoted to the determination of homonuclear J_{HH} couplings: Bojesen et al. (113), Jacobsen, et al. (252), and Bildsoe and Schaumburg (251). There is a $^3J_{HH}$ coupling of approximately 3.2 Hz between H(4) and H(5), a $^4J_{HH}$ coupling of approximately 2 Hz between H(2) and H(5), and a very low $^4J_{HH}$ coupling of 0.4 to 0.5 Hz between H(2) and H(4). This last coupling does not usually appear because of the signal broadening due to the nitrogen atom, but could be revealed by certain solvents (235) or by nitrogen decoupling (255). Some attempts to calculate these couplings based on results of theoretical approaches have been performed using the method of finite perturbations or that of state summation (256). The results have not been very convincing (251, 252). Only in the case of $^3J_{HH}$ in fluorinated derivatives of thiazole (253) was it possible to obtain, by CNDO calculations, J_{FH} coupling constants in reasonable agreement with the experimental values (251).

TABLE I-37. CARBON-13 CHEMICAL SHIFTS OF THIAZOLE AND SOME OF ITS DERIVATIVES[a] (257, 258)

Compounds	δC(2)	δC(4)	δC(5)	Carbon of substituents in α position from the cycle		
				2	4	5
Thiazole	153.4	143.7	119.7	—	—	—
2-Methyl	165.2	142.3	119.3	18.1	—	—
2-Ethyl	171.9	142.6	118.4	27.0	—	—
2-i-Propyl	177.0	142.6	117.8	33.5	—	—
2-t-Butyl	180.4	142.5	117.8	38.1	—	—
4-Methyl	153.1	154.2	114.3	—	17.4	—
4-Ethyl	153.1	160.3	113.0	—	25.5	—
2,4-Dimethyl	164.5	152.7	113.1	18.6	16.8	—
2-Methyl-4-t-butyl	164.1	166.4	103.8	19.5	39.3	—
2-Ethyl-4-t-butyl	171.0	166.4	109.0	27.5	35.10	—
2-i-Propyl-4-i-propyl	176.0	163.1	109.4	33.8	31.6	—
2-i-Propyl-4-t-butyl	176.3	166.4	108.7	33.8	35.4	—
2-t-Butyl-4-methyl	179.8	152.4	112.3	37.9	17.6	—
2-t-Butyl-4-t-butyl	179.1	165.9	108.6	37.9	35.0	—
4,5-Dimethyl	143.3	150.2	126.5	—	14.5	10.8
2-Methyl-4-ethyl-5-methyl	160.6	153.8	124.5	18.9	22.7	14.2
2-Methyl-4-i-propyl-5-methyl	160.0	156.9	123.0	18.9	28.0	10.5

[a] All the chemical shifts are expressed in δ units: ppm of applied field and TMS as reference peak.

The chemical shifts of ^{13}C in natural abundance have been measured for thiazole and many derivatives (257, 258). They are given in Tables I-37 and I-38. These chemical shifts are strongly dependent on the nature of the substituent: CNDO/2 calculations have shown (184) that they correlate well with the $(\sigma + \pi)$ net charge of the atom considered. As a consequence, the order of the resonance signals is the same for protons and for carbon atoms.

TABLE I-38. ^{13}CARBON CHEMICAL SHIFTS OF SUBSTITUENTS IN THIAZOLE AND SOME OF ITS DERIVATIVES[a] (257)

Thiazole derivatives	Carbon of substituent in α position of the carbon			Carbon of substituent in β position of the carbon	
	1	4	5	2	4
2-Methyl	18.1	—	—	—	—
2-Ethyl	27.0	—	—	14.4	—
2-i-Propyl	38.5	—	—	23.5	—
2-t-Butyl	38.1	—	—	31.5	—
4-Methyl	—	17.3	—	—	—
4-Ethyl	—	25.5	—	—	14.3
2,4-Dimethyl	18.6	16.8	—	—	—
2-Methyl-4-t-butyl	19.5	39.3	—	—	30.6
2-Ethyl-4-t-butyl	27.5	35.1	—	14.4	30.5
2,4-Di-i-propyl	33.8	31.6	—	23.5	22.8
2-i-Propyl-4-t-butyl	38.8	35.4	—	23.6	30.7
2-t-Butyl-4-methyl	37.9	17.6	—	31.5	—
2-t-Butyl-5-t-butyl	37.9	35.0	—	31.3	30.7
4,5-Dimethyl	—	14.5	10.8	—	—
2-Methyl-4-ethyl-5-methyl	18.9	22.7	14.2	–	10.9
2-Methyl-4-i-propyl-5-methyl	18.9	28.0	10.5	—	22.5

[a] In ppm from TMS.

More difficult to study are the nitrogen chemical shifts: the resonance of ^{15}N in natural abundance has been determined for thiazole, among a lot of other compounds, by Warren and Roberts (259), who propose an empirical rule for the evaluation of these chemical shifts. Their results agree with the experimental determinations of ^{14}N resonance frequencies done by Whitanovsky et al. (128) and later by Nöth et al. (260), as well as with the measurements made by Nagata et al. (261) as part of a study, using the INDOR method, of the chemical shifts of five- and six-membered nitrogen containing heterocycles as a function of the nature (proton accepting or donating) of the solvent.

TABLE I-39. COUPLING CONSTANTS J_{CH} (Hz) IN THIAZOLE AND SOME OF ITS DERIVATIVES

A. $^1J_{C-H}$ Couplings

Compounds	C(2)–H(2)	C(4)–H(4)	C(5)–H(5)	Ref.
Thiazole	210.0	184.2	189.0	168
	210	189	184	262
		182.5	189.0	224
	211.1	186.5	189.1	113
	211.5	184.5	188.6	258
2-Methyl	—	183.8	187.5	168
	—	187.5	183.8	262
	—	183.8	187.5	224
	—	184	188.5	258
2-Ethyl	—	185	188.2	258
	—	184.6	188.5	224
2-i-Propyl	—	182	185.5	258
2-n-Propyl	—	184.5	188.5	224
2-t-Butyl	—	184.2	186.7	258
	—	185.0	188.2	224
2-i-Butyl	—	184.9	188.2	258
2-neo-Pentyl	—	184.8	188.5	258
2-Amino	—	183.2	190.0	168
	—	184.0	191.2	113
2-Dimethylamino	—	182.0	190	168
2-Piperidino	—	184.6	190	168
2-Chloro	—	189.0	189.0	168
2-Bromo	—	191.2	192.2	168
4-Methyl	209	—	187	113
	212.0	—	185.5	168
	212	—	185.5	262
	211.5	—	185.5	258
4-Ethyl	206.5	—	183	258
4-Chloro	214.0	—	194.0	168
5-Methyl	212.5	183.5	—	168
	212.5	183.5	—	262
5-Chloro	214.5	190.0	—	168
5-Bromo	212.0	190.2	—	168
2,4-Dimethyl	—	—	183.0	168
	—	—	188.0	258
2,5-Dimethyl	—	179.5	—	168
	—	183	—	262
4,5-Dimethyl	208.0	—	—	168
	212.0	—	—	258
2-Methyl-4-t-butyl	—	—	184.2	258
2-Ethyl-4-t-butyl	—	—	182.9	258
2-i-Propyl-4-i-propyl	—	—	183.2	258
2-t-Butyl-4-methyl	—	—	185.3	258
2,4-Dichloro	—	—	193.0	168

TABLE I-39 (*Continued*)
A. $^1J_{C-H}$ Couplings

Compounds	C(2)–H(2)	C(4)–H(4)	C(5)–H(5)	Ref.
2-Chloro-5-bromo	—	191.5	—	168
2,4-Dibromo	—	—	194.0	168
2,5-Dibromo	—	192.5	—	168

B. Long-range couplings J_{C-H}

	$^2J_{C-H}$		$^3J_{C-H}$				
Compounds	C(4)–H(5)	C(5)–H(4)	C(2)–H(4)	C(2)–H(5)	C(4)–H(2)	C(5)–H(2)	
Thiazole	14.9	16.4	15.4	6.2	7.2	3.6	113
	15.2	15.8	15.1	6.8	7.3	—	258
2-Amino	6.10	14.26	—	—	—	—	113
2-Bromo	—	—	19.5	8.5	—	—	113
2,4-Dimethyl	6.5	—	—	6.9	—	—	258

The ^{13}C–H spin couplings (J_{CH}) have been dealt with in numerous studies, either by determinations on samples with ^{13}C natural abundance (122, 168, 224, 231, 257, 262, 263) or on samples specifically enriched in the 2-, 4-, or 5-positions (113) (Table I-39). This last work confirmed some earlier measurements and permitted the determination for the first time of $^2J_{CH}$ and $^3J_{CH}$ coupling constants. The $^1J_{CH}$ coupling, between a proton and the carbon atom to which it is bonded, can be calculated (264) with summation rule of Malinovsky (265, 266), which does not distinguish between the 4- and 5-positions, and by use of CNDO/2 molecular wave functions: the numerical values thus obtained are much too low, but their order agrees with experiment. The same is true for $^2J_{CH}$ and $^3J_{CH}$ couplings.

The $^1J_{CH}$ coupling constants are very sensitive to the geometry of the molecule and to the nature of the atoms bonded to the carbon center. Considering the structure H–C$\underset{Y}{\overset{X}{<}}$ (X and Y = CH, O, S, NH, N), Laszlo (262) has established that J_{CH} can be considered as the sum of one term indirectly proportional to the XCY bond angle and of terms accounting for the ionic perturbations induced by the X and Y atoms. Moreover, the

$^1J_{CH}$ coupling varies with eventual self-associations of the H-bonded type produced by solvent and, to a lesser extent, by temperature variations.

Similarly, using another model based on heteroatom increments proposed by Dischler (267), it was possible to calculate a very satisfactory empirical geometry for thiazole (122) with the aid of Shoolery's correlation (268) between $^1J_{CH}$ and C–H bond length.

Concerning ^{15}N-magnetic resonance, Boejesen et al. (113) have measured the following coupling constants on a sample of thiazole enriched in this isotope:

$$^2J_{NH(2)} = -10.56 \text{ Hz}$$
$$^2J_{NH(4)} = -10.6 \text{ Hz}$$
$$^3J_{NH(5)} = -1.97 \text{ Hz}$$

Taking into account the relation, $J_{14NH} = 0.7129 \, J_{15NH}$, these values agree reasonably well with that of 9 Hz evaluated earlier by Kintzinger and Lehn (253) for the ^{14}N–H(2) coupling in a comprehensive study of the width of the resonance lines of thiazole protons. These authors carried out a fundamental study of the mechanism of ^{14}N nuclear quadrupolar relaxation that is responsible for the broadening of the α proton signals resulting from the incomplete destruction of the fine structure induced by the N–H(α) coupling. This relaxation depends on the temperature, solvent, and eventual protonation or quaternization. The line widths of nitrogen or proton resonance have been measured and/or calculated for thiazole (260, 113) and 2-bromothiazole (269, 270). In this last case, Pyper et al. determined the relaxation times T_2 for the protons α to nitrogen:

$$T_{2H(2)} = 0.524 \pm 0.015 \text{ sec}$$
$$T_{2H(4)} = 0.851 \pm 0.070 \text{ sec}$$

whereas the calculated values are:

$$T_{2H(2)} = 0.524 \text{ sec}$$
$$T_{2H(4)} = 0.849 \text{ sec}$$

J_{CC} and J_{CN} have been measured in isotopically enriched molecules, but in this case again the experimental values are in poor accordance with those calculated using the CNDO/2 approximation (Table I-40) (113).

TABLE I-40. COUPLING CONSTANTS $J_{13C-13C}$ AND J_{15N-H} (HZ) IN THIAZOLE (171).

Coupling constants	Experimental values	Calculated values (CNDO/2)
$^2J_{24}$	+0.60	+1.80
$^2J_{25}$	+1.96	+1.34
$^1J_{45}$	+3.13	+2.04
$^2J_{NH(2)}$	−10.56	−4.65
$^2J_{NH(4)}$	−10.6	−4.48
$^3J_{NH(5)}$	−1.97	−1.12

6. Mass Spectrometry

The first mass spectrometric investigation of the thiazole ring was done by Clarke et al. (271). Shortly after, Cooks et al., in a study devoted to bicyclic aromatic systems, demonstrated the influence of the benzo ring in benzothiazole (272). Since this time, many studies have been devoted to the influence of various types of substitution upon fragmentation schemes and rearrangements, in the case of alkylthiazoles by Buttery (273); arylthiazoles by Aune et al. (276), Rix et al. (277), Khnulnitskii et al. (278); functional derivatives by Salmona et al. (279) and Entenmann (280); and thiazoles isotopically labeled with deuterium and ^{13}C by Bojesen et al. (113). More recently, Witzhum et al. have detected the presence of simple derivatives of thiazole in food aromas by mass spectrometry (281).

The first characteristic of the mass spectrum of thiazoles is the large intensity of the molecular peak, resulting from the aromatic character of the ring (271–275). According to Clarke (271), electron impact leads to preferential scission of the S–C(2) and N–C(4) bonds, with retention of the positive charge by the sulfur-containing fragment, which appears as a thiirenium ion. This hypothesis was confirmed by a study of thiazole isotopically labeled with deuterium and ^{13}C (113). This mode of fragmentation leads to the elimination of HCN in the case of thiazole and of R(2)CN in the case of thiazoles bearing a substituent R in the 2-position. The alkyl groups show β-scission relative to the ring double bond, and when the substituent contains at least three carbon atoms, elimination of an olefine accompanied by a McLafferty rearrangement occurs (273, 274, 276). Figure I-12 shows the various possibilities of cleavage. In the case of alkylthiazoles the various schemes of cleavage are summarized in Figs. I-12 and I-13, after Tabacchi (274).

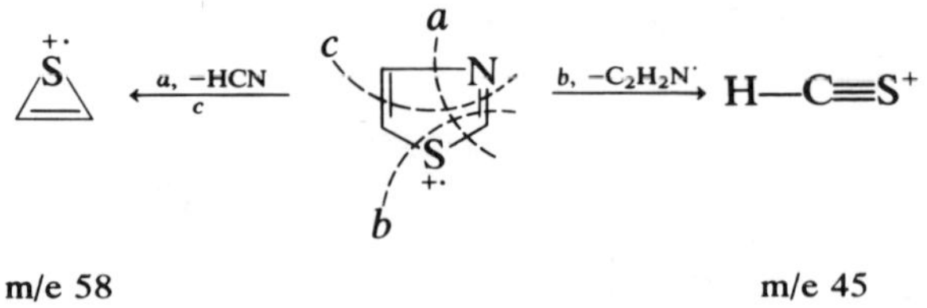

Fig. I-12. Cleavage of thiazoles.

The mass spectra of functional derivatives of thiazole follow the same trends for the molecular peak and the fragmentation pattern. However, the following remarks can be added: 2-amino-5-hydroxythiazole hydrochloride undergoes a preliminary dehydroxylation (280); the nitro group destabilizes the molecule and makes it more fragile (277). On the other hand, the dimethylamino group stabilizes the molecule (279). Finally, it was possible to interpret the differences between the spectra of chloro- and bromothiazoles on the basis of the electronegativity difference of the two halogens (282).

Mass spectrometry can be used to determine ionization potentials by the method of Lossing (283). The values obtained can be compared with those found by photoelectron spectroscopy and those calculated by CNDO/S (134) or *ab initio* (131) methods using the Koopman theorem approximation. The first and second ionization potentials concern a π

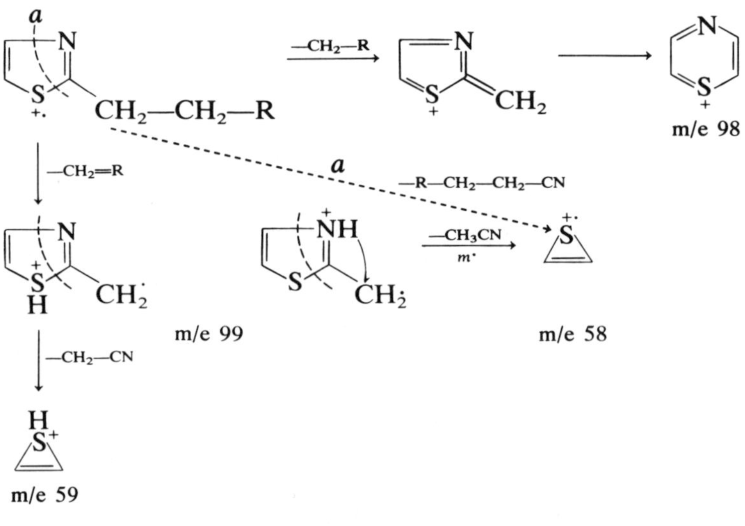

Fig. I-13. Cleavage of 2-alkylthiazole.

II. Physical Properties

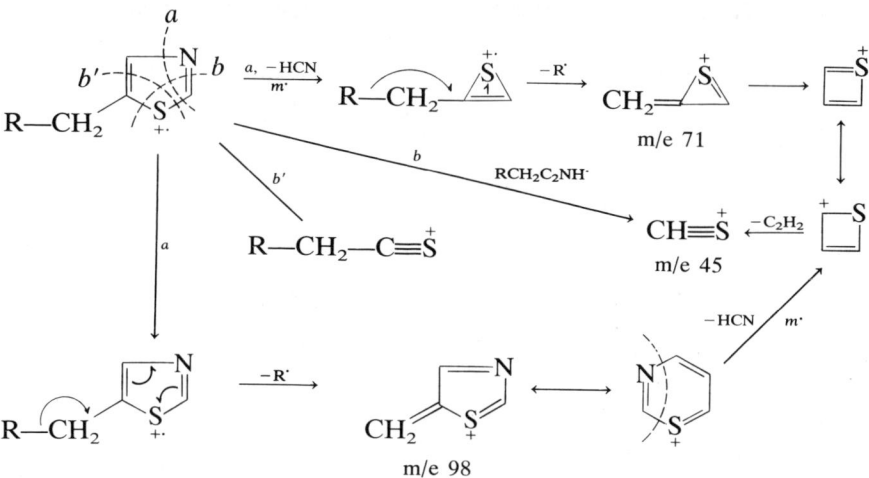

Fig. I-14. Cleavage of 5-alkylthiazole. This scheme is also valid for the cleavage of 4-alkylthiazole.

level, the third a σ_N level. For the subsequent ionization potentials, it is more difficult to make assignments which depend essentially on the nature of the substituents. In mass spectrometry, only the first ionization potential can be determined, but it is possible to measure the apparition potential of the thiirenium ion which is the most important observed fragment. Results are given in Table I-41.

TABLE I-41. IONIZATION POTENTIALS AND APPARITION POTENTIALS (eV) MEASURED FOR VARIOUS THIAZOLE DERIVATIVES (191)

Compounds	Ionization potential	Apparition potential
Thiazole	9.55	12.55
2-Methyl	9.30	12.55
2-Ethyl	9.10	13.15
2-Propyl	8.90	14.45
2-i-Butyl	8.90	14.90
2-t-Butyl	8.95	15.65
2-Methoxy	8.70	12.50
2-Amino	8.50	12.50
2-Chloro	9.40	12.50
2-Bromo	9.50	12.55
2-Nitro	10.20	12.50

7. Electronic Paramagnetic Resonance Spectrometry

The thiazolyl radical has never been directly observed. Torsell (284), however, could indirectly prove its existence by the method of spin trapping in the photolysis of 2-iodothiazole:

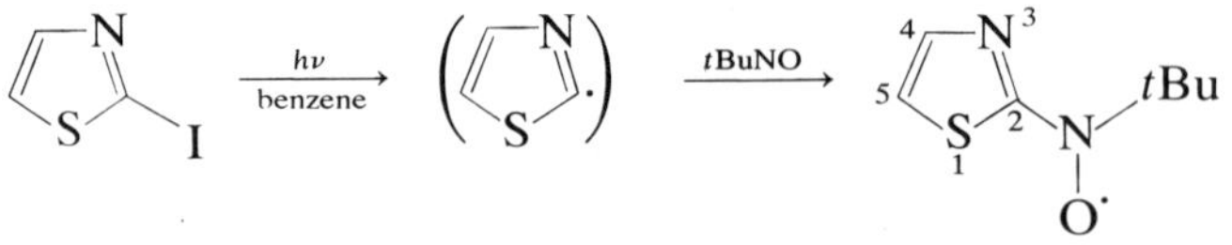

Scheme 46

Coupling constants for this 2-thiazolyl radical are following: $a_{N_0} = 9.90$ gauss, $a_{N_3} = 2.35$ gauss, $a_{H_4} = 0.48$ gauss, $a_{H_5} = 3.15$ gauss.

Anion-radicals derived from the reduction of nitrothiazoles were studied by Tordo et al. (285). The reaction scheme is the following:

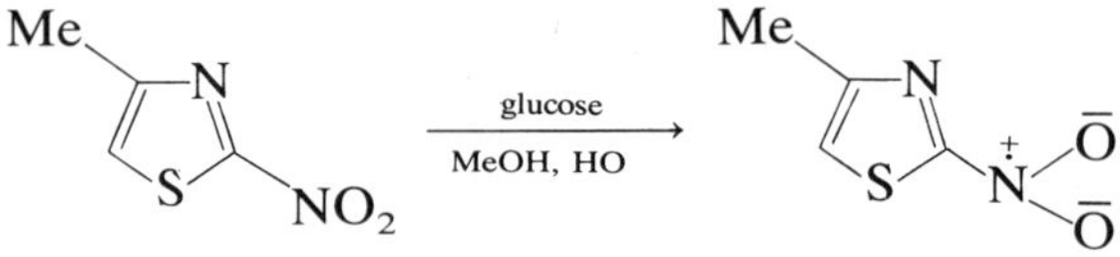

Scheme 47

For 4-nitro-5-alkylthiazoles, the prefered conformations of the alkyl substituents relative to the π-system could be evaluated, and the interactions between alkyl and nitro groups were demonstrated to be negligible.

TABLE I-42. CALCULATED AND EXPERIMENTAL COUPLING CONSTANTS FOR VARIOUS NITROTHIAZOLES (285)

Thiazole derivatives	Thiazole cycle			Substituents		
	a_H		a_N	a_H (alkyl)	a_N (Nitro)	
	exptl	calcd			exptl	calcd
4-Nitro-5-methyl	0.50	0.62	0.50	5.20	13.25	14.30
4-Nitro-5-methyl-2-t-butyl	—	—	0.50	5.70	13.60	—
4-Nitro-2,5-dimethyl	—	—	0.40	0.4(2)	13.20	—
	—	—	—	5.9(5)	—	—
2-Nitro-4-methyl	3.60	3.02	2.66	0.87	9.90	6.70
5-Nitro-4-methyl	4.40	5.25	1.62	6.25	11.50	10.18
4-Nitro-5-ethyl	0.50	—	0.50	4.50	13.20	—
5-Nitro-5-i-propyl	0.50	—	0.50	2.25	13.20	—
5-Nitro-4-methyl-2-t-butyl	—	—	1.50	6.37	12.25	—
5-Nitro-4-ethyl-2-t-butyl	—	—	1.50	3.75	12.25	—
5-Nitro-4-i-propyl-2-t-butyl	—	—	1.50	1.50	12.30	—
5-Nitro-4-t-butyl-2-t-butyl	—	—	1.37	—	12.30	—

Coupling constants, calculated with the aid of McConnel's formula (286) are given in Table I-42 together with the experimental values.

8. Thermodynamic Data

The thermodynamic study of thiazole and of some of its binary mixtures with various solvents has led to the determination of important practical data, but also to the discovery of association phenomena between thiazole and some solvents and of thiazole self-association.

The first binary mixture quantitatively studied was the water–thiazole system, thiazole being a very hygroscopic compound (104). Determining the purity of thiazole sample obtained by distillation, Metzger and Disteldorf (287) observed the existence of a stable azeotropic mixture, the characteristics of which are the following:

Pressure (mm Hg)	Temperature (°K)	Molar fraction of water
695.5	363.15	0.715
750	365.25	0.720

The vapor-liquid equilibrium of the binary mixture is well fitted by Van Laar's equations (228). It was determined from 100 to 760 mm Hg, and the experimental data was correlated by the Antoine equation (289, 290), with P in mm Hg and t in °C:

$$\log_{10} P = 7.14112 - \frac{1424.800}{(t+216.194)}$$

The normal boiling point of pure thiazole is $t_{760} = 118.241 \pm 0.004°C$.

2-Methylthiazole behaves similarly, fitting the Antoine equation (291):

$$\log_{10} P = 7.04109 - \frac{1406.419}{(t+209.257)}$$

The normal boiling point of 2-methylthiazole is $t_{760} = 128.488 \pm 0.005°C$. The purity of various thiazoles was determined cryometrically by Handley et al. (292), who measured the precise melting point of thiazole and its monomethyl derivatives. Meyer et al. (293, 294) extended this study and, from the experimental diagrams of crystallization (temperature/degree of crystallization), obtained the true temperatures of crystallization and molar enthalpies of fusion of ideally pure thiazoles (Table I-43).

TABLE I-43. CRYOSCOPIC DATA FOR VARIOUS THIAZOLES (294)

Thiazole	$T_{f_0}(°C)^a$	ΔH cal · mole$^{-1 b}$
Thiazole	−33.57 ± 0.01	1711
2-Methylthiazole	−24.72 ± 0.02	2576
4-Methylthiazole	−44.13 ± 0.05	2126
5-Methylthiazole	−40.41 ± 0.05	1829
2,4-Dimethylthiazole	−50.31 ± 0.05	691
4,5-Dimethylthiazole	−17.50 ± 0.01	—
2,4-Di-t-butylthiazole	−14.96 ± 0.02	2516
2,4,5-Trimethylthiazole	−32.46 ± 0.02	2152

^{a} True temperature of crystallization of ideally pure sample.
^{b} Molar enthalpy of fusion of ideally pure sample.

The heat capacity of thiazole was determined by adiabatic calorimetry from 5 to 340°K by Goursot and Westrum (295, 296). A glass-type transition occurs between 145 and 175°K. Melting occurs at 239.53°K (−33.62°C) with an enthalpy increment of 2292 cal · mole^{-1} and an entropy increment of 9.57 cal · mole^{-1} · °K^{-1}. Table I-44 summarizes the variations as a function of temperature of the most important thermodynamic properties of thiazole: molar heat capacity C_p, standard entropy $S°$, and Gibbs function $-(G°-H°)/T$.

The variation of Cp for crystalline thiazole between 145 and 175°K reveals a marked inflection that has been attributed to a gain in molecular freedom within the crystal lattice. The heat capacity of the liquid phase varies nearly linearly with temperature to 310°K, at which temperature it rises more rapidly. This thermal behavior, which is not uncommon for nitrogen compounds, has been attributed to weak intermolecular association. The remarkable agreement of the third-law ideal-gas entropy at

TABLE I-44. VARIATION OF SOME THERMODYNAMIC PROPERTIES OF THIAZOLE AS A FUNCTION OF TEMPERATURE (295)

T (°K)	C_p (cal · mole^{-1} · °K^{-1})	$S°$ (Cal · mole^{-1} · °K^{-1})	$-(G°-H°)/T$ (cal · mole^{-1} · °K^{-1})
25	3.301	1.456	0.413
50	7.879	5.324	1.878
100	11.385	12.094	5.389
200	17.18	21.59	11.21
273.15	28.20	38.12	16.06
298.15	28.92	40.62	18.01

298.15°K (66.46 e.u.) with the spectroscopic value calculated from experimental data (66.41 ± 0.009 e.u.) (295, 289) indicates that the crystal is an ordered form at 0°K. Thermodynamic functions of thiazole were also determined by statistical thermodynamics from vibrational spectra (297, 298).

A parallel thermodynamic study of 2-methylthiazole was performed by Goursot and Westrum (299). The stable form was characterized by a temperature of melting of 248.42°K, an enthalpy of melting of 2907 cal · mole^{-1}, and an entropy of melting of 11.70 cal · mole^{-1} · °K^{-1}. A metastable form melting at 246.50°K was also investigated with an entropy of melting of 11.00 cal · mole^{-1} · °K^{-1}. The heat capacity C_p, standard entropy $S°$, and Gibbs function $-(G°-H°)/T$ in cal · mole^{-1} · °K^{-1} for the liquid at 298.15°K are 36.01, 50.64, and 22.50.

The thermal stability of thiazole, pointed out by Johns et al. (145) and measured by Gelus et al. (222, 300), is reported in Table I-45.

The thermal stability can be correlated with the energy of the highest occupied molecular orbital of the molecule (HMO approximation) (300).

The viscosity of thiazole vapor was measured by Nasini (301): $\eta_{99.4}$ = 1.161 10^{-4} poise · dyne^{-1} · sec^{-1} · cm^2 and compared to that of pyridine: $\eta_{98.2}$ = 0.958.

The velocity of sound in liquid thiazole was also measured, and the adiabatic compressibility was determined (302): it was concluded that intermolecular interactions result from the electrical forces, originating in the heteroatoms, between the molecules.

A large number of thermodynamic studies of binary systems were undertaken to find and determine eventual intermolecular associations for thiazole: Meyer et al. (303, 304) discovered eutectic mixtures for the following systems: -thiazole/cyclohexane at −38.4°C, n_T = 0.815; -thiazole/carbon tetrachloride at −60.8°C, n_T = 0.46; -thiazole/benzene at −48.5°C, n_T = 0.70.

The first system is characterized by a partial miscibility of the liquid phases, the second one is instable with incongruent melting points at −54

TABLE I-45. PYROLYSIS TEMPERATURE OF SOME THIAZOLES (222)

Thiazole	T(°C)
Thiazole	530
2,4-Diphenylthiazole	431
2-Phenyl-4-biphenylthiazole	442
4-Phenyl-2-biphenylthiazole	452
2,4-*Bis* (p. biphenyl)thiazole	510

and $-52.8°C$, and the third one is a simple eutectic mixture. The solid/liquid phase diagramm of the binary mixture, thiazole–water, has been plotted (305), and the enthalpy of formation of molecular complexes of thiazole with tetracyanoethylene was measured (306).

Observed deviations from ideality are attributable to thiazole self-association. Such self-association is influenced by steric crowding as indicated by the behavior of methylthiazoles. The constants of self-association have been estimated for benzene solutions of thiazole ($K_{assoc} = 3.2$ at $5.5°C$) and 5-methylthiazole ($K_{assoc} = 7$ at $6.5°C$).

Similarly, molar excess functions have been determined for various thiazole–solvent binary mixtures (Table I-46) (307–310).

For cyclohexane the excess enthalpy (H^E) is positive and large, whereas for solvent with aromatic character it is low and even negative in the case of pyridine.

A parallel study of liquid–vapor equilibrium was reported by Bares et al. for thiazole–CCl_4 and thiazole–C_6H_{12} binary mixtures (311).

The conclusion of all these thermodynamic studies is the existence of thiazole–solvent and thiazole–thiazole associations. The most probable mode of association is of the $n-\pi$ type from the lone pair of the nitrogen of one molecule to the various other atoms of the other. These associations are confirmed by the results of viscosimetric studies on thiazole and binary mixtures of thiazole and CCl_4 or C_6H_{12}. In the case of CCl_4, there is association of two thiazole molecules with one solvent molecule, whereas cyclohexane seems to destroy some thiazole self-associations (aggregates) existing in the pure liquid (312–314). The same conclusions are drawn from the study of the self-diffusion of thiazole (labeled with ^{14}C) in thiazole–cyclohexane solutions (114).

TABLE I-46. PARTIAL MOLAR EXCESS ENTHALPY AT INFINITE DILUTION (H^E) OF THIAZOLE IN VARIOUS SOLVENTS AT 318.15°K

Solvent	H^E(cal · mole^{-1})	Ref.
C_6H_{12}	2380 ± 40	307
	2440.2	309
C_6H_6	230 ± 20	307
	287.1	309
CCl_4	810 ± 20	307
C_6F_6	669.9	309
Thiophene	95.7	309
Pyridine	-71.77	309

II. Physical Properties

9. Magnetic Susceptibility

The molar diamagnetic susceptibility of thiazole and some derivatives was initially determined by the classical Curie-Cheneveau method (5, 315, 316) and later confirmed by a method (317) based on the difference of NMR proton chemical shift of a sample of tetramethylsilane immersed in the liquid to be investigated, according to the shape (cylindrical or spherical) of the sample tube (Table I-47) (318).

TABLE I-47. MOLECULAR DIAMAGNETIC SUSCEPTIBILITIES

Compound	$-\chi_M$ (10^{-6} emu cgs)		
Thiazole	50.55 (315)	50.2 (5)	50.64 (318)
2-Methyl	59.56	59.6	61.07
4-Methyl	—	—	61.33
5-Methyl	—	—	60.97
2,4-Dimethyl	—	—	71.66
4,5-Dimethyl	—	—	71.70
2-Cl	—	63.4	63.93

It is difficult to draw general conclusions from such a small number of values. Nevertheless, it can be noted that, like other five- or six-membered unsaturated rings (thiophene, pyridine) thiazole exhibits a certain aromatic behavior in its magnetic susceptibility.

The aromatic character of thiazole has been deduced from the magnetic susceptibility anisotropy of the molecule (319).

10. Dipole Moment

The charge distribution in the thiazole ring gives rise to a dipole moment. The solvent has practically no influence on the moment, which means that there are no associations of the donor-acceptor type that could lead to an over-estimation of the moment (146). The barycenter of negative charges is situated near the nitrogen atom, whereas that of positive charges is displaced slightly towards the sulfur atom. Calculated and measured dipole moments of thiazole and some of its derivatives are collected in Table I-8. For the calculated values, when the method of calculation is of the all-valence-electrons or *ab initio* type, the total dipole moment is obtained directly, whereas when it is of the π type (HMO-PPP), a σ component obtained from a method like that of Del Re must be added vectorally (136).

11. Refractivity

Refraction index and density were measured for thiazole and some monoalkylthiazoles (Table I-48) (198, 199, 215):

TABLE I-48. REFRACTION INDEX n_D^{25} OF THIAZOLE AND MONOALKYL DERIVATIVES (198, 199, 215)

Position	2	4	5
Thiazole	—	1.5371	—
Methyl	1.5105	1.5221	1.5260
Ethyl	1.5075	1.5144	1.5154
i-Propyl	1.4970	1.5033	1.5062
t-Butyl	1.4909	1.4959	1.4988

The molar refraction deduced for alkyl derivatives, compared to the value obtained by addition, to the experimental molar refraction of thiazole, of the classical (CH_2) increment of Eisenlohr ($R_{CH_2} = 4.618$ cm³), show specific exaltations which are typical for each position of the thiazole ring (Table I-49).

TABLE I-49. SPECIFIC EXALTATION OF MOLAR REFRACTION (IN PERCENTAGE OF THE CALCULLATED VALUE) (198, 199, 215)

Position	2	4	5
Methyl	0.48	1.02	0.25
Ethyl	0.89	1.13	0.32
i-Propyl	0.89	1.19	0.47
t-Butyl	1.53	1.59	0.70

III. CHEMICAL REACTIVITY

1. Theoretical Aspects

Until the end of the forties, when the HMO method was first applied to thiazole, most of the experimental results concerning its chemical reactivity remained of a qualitative nature. Papers devoted to the subject described mostly the principal site of reaction, and when several sites could enter the reaction, the studies indicated only an order of decreasing reactivity. Interpretation of these qualitative results rested essentially on the theory of resonance. The application of the HMO method to thiazole and some derivatives prompted new reactivity studies based on a more quantitative approach. Thus, for example, nucleophilic reactivity of the nitrogen atom was determined by kinetic measurements; electrophilic nitration and free-radical alkylation and arylation were developed in competition experiments. Partial rate factors were determined and interpreted in terms of reactivity indices or Wheland atomic polarization energies. More recently, linear free energy correlations have been applied to some series of substituted thiazole derivatives. More elaborated theoretical treatment, undoubtly interesting for the interpretation of physicochemical properties, have not brought significant progress in the understanding of *chemical* reactivity in the thiazole series (cf. Section I-5).

Most of these theoretical aspects are discussed in the following sections, pointing out the particular suitability of the very dissymmetrical molecular frame of thiazole for quantitative study, on the same species, of a large variety of fundamental organic reactions.

2. Basicity

As early as 1889 Walker (320), using samples of thiazole, 2,4-dimethylthiazole, pyridine, and 2,6-dimethylpyridine obtained from Hantzsch's laboratory, measured the electrical conductivity of their chlorhydrates and compared them with those of salts of other weak bases, especially quinoline and 2-methylquinoline. He observed the following order of decreasing proton affinity (basicity): quinaldine > 2,6-dimethylpyridine > quinoline > pyridine > 2,4-dimethylthiazole > thiazole, and concluded that the replacement of a nuclear H-atom by a methyl group enhanced the basicity of the aza-aromatic substrates.

Potentiometric determinations of the pK_a of thiazole and essentially its alkyl derivatives are summarized in Table I-50. The most reliable values are given by Phan-Tan-Luu et al. (321), who realized a critical study of the classical Henderson method for the determination of pK_a.

In a parallel study Goursot and Wadsö (322) determined calorimetrically the free energies, enthalpies, and entropies of dissociation of the conjugate acids of thiazoles in aqueous media (Table I-51).

TABLE I-50. POTENTIOMETRICALLY DETERMINED pK_a OF THIAZOLE DERIVATIVES

Thiazole	pK_a		
Thiazole	2.53[a]	2.55±0.02[c]	2.518±0.005[d]
2-Amino	5.39[a]	—	—
2-Methyl	—	3.40±0.02	3.43±0.01
4-Methyl	3.07[b]	3.16±0.03	3.150±0.002
5-Methyl	—	3.03±0.02	3.115±0.005
2-Ethyl	—	—	3.372±0.005
4-Ethyl	—	—	3.204±0.005
2-Propyl	—	—	3.353±0.005
2-i-Propyl	—	—	3.283±0.005
2-i-Butyl	—	—	3.377±0.005
2-t-Butyl	—	3.00±0.03	3.150±0.005
4-t-Butyl	—	3.04±0.03	3.056±0.005
2-neo-Pentyl	—	—	3.37±0.01
2,4-Dimethyl	—	3.98	—
2,5-Dimethyl	—	3.91	—
4,5-Dimethyl	—	3.73	—
2,4,5-Trimethyl	—	3.00	—
4-Methyl-5-carbethoxy	1.69[b]	—	—
4-Methyl-5-carboxy	3.51[b]	—	—
4-Methyl-2-carboxy	1.20[b]	—	—

[a] From Ref. 323.
[b] From Ref. 324.
[c] All values in this column from Ref. 322.
[d] All values in this column from Ref. 321.

TABLE I-51. THERMODYNAMICS OF THE AQUEOUS IONIZATION OF THIAZOLIUM IONS (322).

	$BH^+_{aq} + H_2O_{aq} \rightleftharpoons B_{aq} + H_3O^+_{aq}$ at 25°C		
Thiazole	$\Delta G°$ (kcal · mole^{-1})	$\Delta H°$ (kcal · mole^{-1})	$\Delta S°$ (eu)
Thiazole	3.48	2.02	−4.9
2-Methyl	4.64	3.21	−4.8
4-Methyl	4.31	2.99	−4.4
5-Methyl	4.13	2.88	−4.2
2,4-Dimethyl	5.43	4.21	−4.1
2,5-Dimethyl	5.34	3.99	−4.5
4,5-Dimethyl	5.15	3.97	−4.0
2,4,5-Trimethyl	6.24	4.95	−4.3
2-t-Butyl	4.09	3.71	−1.3
4-t-Butyl	4.15	3.86	−1.0

The reaction corresponds to a proton transfer and not to a net formation of ions, and thus the $\Delta S°$ is of minor importance in the whole series, especially for the two t-Bu derivatives. This last effect is believed to be due to a structure-promoting effect of the bulky alkyl groups in the disordered region outside the primary hydration sphere of the thiazolium ion (322).

For the methyl-substituted compounds (322) the increase in $\Delta G_i°$ and $\Delta H_i°$ values relative to the unsubstituted thiazole is interpreted as being mainly due to polar effects. Electron-donating methyl groups are expected to stabilize the thiazolium ion, that is to decrease its acid strength. From Table I-51 it may be seen that there is an increase in $\Delta G_i°$ and $\Delta H_i°$ by about 1 kcal · mole^{-1} for each methyl group. Similar effects have been observed for picolines and lutidines (325).

The order of decreasing basicity of the monomethylthiazoles, $2>4>5$, is in agreement with the order of decreasing interaction of the methyl group with the electronic system of the ring as determined by the HMO method (6). More precisely, for this series the PPP method gave a correlation between the pK_a and the variation in electronic energy (π or $\pi + \sigma$) when protonating the thiazole molecule, allowance being made for the variation in energy of solvation accompanying the protonation of the azaaromatic base (133). The energy of protonation of an organic base may be expressed as follows (326):

$$\Delta E = \Delta E_\sigma + \Delta E_\pi + E_{st} + E_{sol}$$

where ΔE_σ is the energy variation corresponding to the σ frame, ΔE_π is the energy variation corresponding to the π system, ΔE_{st} is the variation of steric interactions, and ΔE_{sol} is the variation of solvation energy. In the case of thiazole the ΔE_{st} term may be neglected in a first approximation (133, 327). The ΔE_{sol} term, the difference between the solvation energies of the thiazolium ion and the thiazole molecule, may be evaluated according to the expression (328, 330):

$$\Delta E_{sol} = -\Delta \left\{ c \sum_i \sum_j \frac{q_i q_j}{2r_{ij}} \left(\frac{1-1}{D} \right) \right\}_{molecule}^{ion}$$

with q_i and q_j being the apparent charges of atoms i and j in the ion and in the molecule separated by the distance r_{ij}, D the dielectric constant of the medium, and c a constant depending on the temperature and the units. ΔE may thus be calculated with the aid of the theoretical models already discussed (Section I.5). It is interesting to note the growing correlation between the experimental value and the calculated one as the theoretical model becomes more sophisticated. Limited to ΔE_π (HMO or

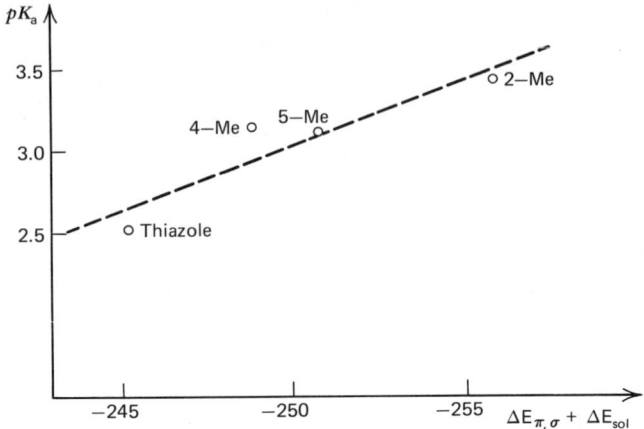

Fig. I-15. Correlation between pK_a and electronic and solvation parameters (in arbitrary units) for thiazole and its three monomethyl derivatives (133).

PPP), this model gives no correlation with pK_a, some correlation appears when the ΔE_{sol} term is included, and finally a fairly good correlation results from the consideration of the term, $\Delta E = \Delta E_{\sigma,\pi} + \Delta E_{sol}$, which is correlated with measured pK_a for thiazole and its three methyl derivatives in Fig. I-15 (133).

3. Additions to the Thiazole System

In contrast to oxazole, thiazole does not undergo the Diels-Alder cycloaddition reaction (331). This behavior can be correlated with the more dienic character of oxazole, relative to thiazole, as shown by quantochemical calculations (184).

In an attempt to synthesize fused aromatic systems of a pentalene-like structure, Boekelheid and Fedoruk (332) submitted the dicyanomethyl ylide of thiazole (**77**) to the addition reaction with dimethyl acetylenedicarboxylate (DMA). They unexpectedly observed the formation of a fused six-membered (**80**) rather than a five-membered-ring (**78**). This ylide (**77**) was readily afforded by the reaction of thiazole (**73**) with tetracyanoethylene oxide and then put into reaction with DMA. The initially formed thiazolopyrrole derivative (**78**) is strongly polarized by the *gem*-dicyano group, and its pyrrole ring is spontaneously cleaved with proton elimination. The ring closure of the intermediate (**79**) leads to the final stable derivative of 5-*H*-thiazolo[3,2-*a*]pyridine (**80**). More recently,

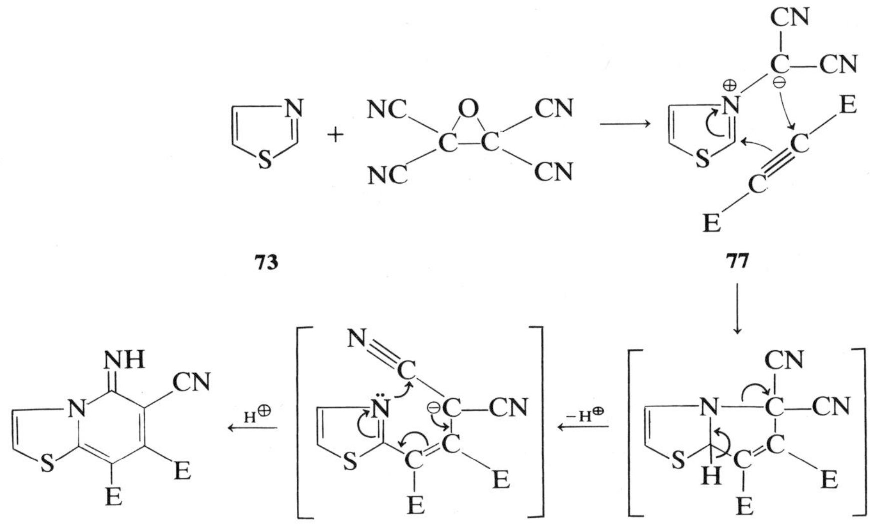

III. Chemical Reactivity

Scheme 48

Potts et al. (333) condensed dipolarophiles (DMA, dibenzoylacetylene, ethyl propiolate) with ylides (**81**) obtained by quaternization of 4-methylthiazole with an α-bromoketone or ester and subsequent deprotonation. In fact the 1:1 molar adduct obtained (**82**) rearranged to a pyrrolothiazine (**83**). One example of this reaction is described Scheme 49.

Like pyridines (334), thiazoles undergo addition reactions with dimethyl acetylenedicarboxylate leading to 2:1 molar adducts, the structure of which has been a matter of controversy (335–339).

Thiazole and DMA (335) react at room temperature in dimethylformamide (DMF) to afford a major product (yield of 20%) forming orange prisms (λ_{max} = 430 nm, log ∈ 3.63). On the basis of the NMR spectrum (12H at δ3.72–3.92 = 4 COOMe; two doublets of an AB system centered at δ6.02 and 6.52 with $J \sim 4.75$ cps; a singlet (1H) at δ8.13), Reid et al. (335) proposed structures **84** or **85** for this adduct (E = COOMe).

The same reaction performed in ether at 0°C (336) gives the same major adduct, but the structure proposed by Acheson et al. corresponds to **86**, although such a structure is hardly compatible with the presence of an isolated low-field proton. Very recently, in a reinvestigation of these cyclo-additions of DMA to azoles (338, 339), Acheson et al. were able to establish the correct structure of the adducts on the base of ^{13}C NMR spectra and X-ray diffraction studies. The adduct of thiazole is represented by formula **87**, and it results from the rearrangement of the

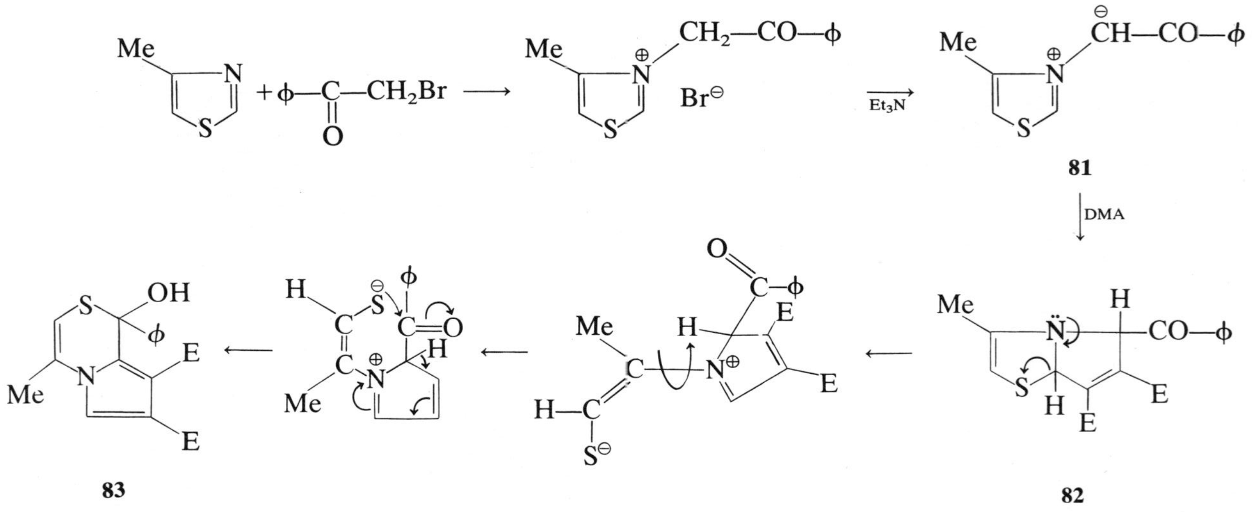
Me
N
S
+ϕ—C—CH2Br
O
Me
N
S
CH2—CO—ϕ
Br
Et3N
Me
N
S
CH—CO—ϕ
81
DMA
Me
N
S
H
CO—ϕ
H
E
E
82
Me
H
C
N
H—C
S
O
C—ϕ
E
E
H
C
S
ϕ
C
O
C
N
H
Me
S
OH
ϕ
E
N
Me
E
83

Scheme 49

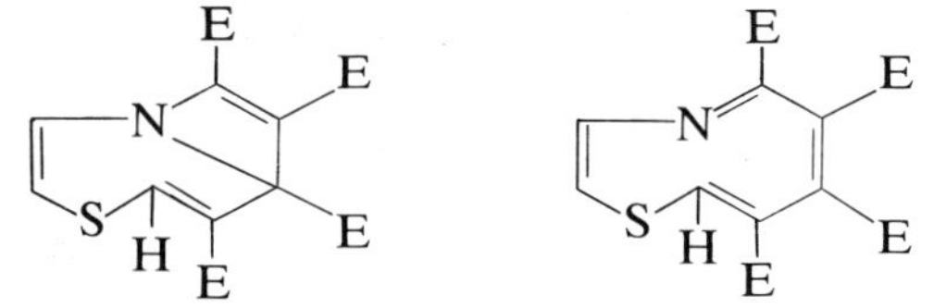

III. Chemical Reactivity

84 **85**

Scheme 50

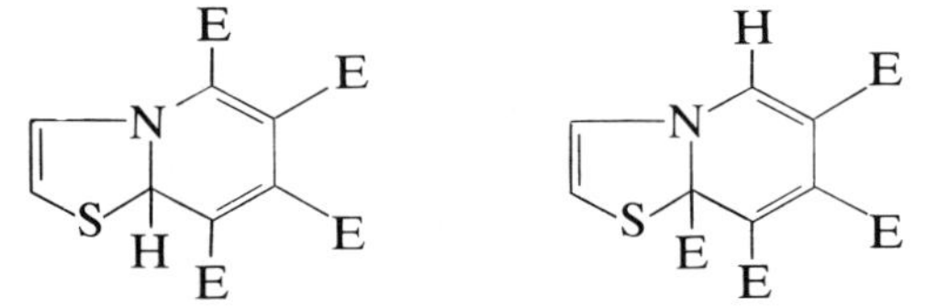

86 **87**

Scheme 51

initially formed adduct (**86**) either via a concerted [1, 5] suprafacial sigmatropic shift or by a nonconcerted pathway proceeding via zwitterion **88** in Scheme 52. When performed in methanol solution the cyclo addition of DMA to thiazole leads to trimethylpyrrolo[2,1-*b*]thiazole-5,6,7-tricarboxylate (**89**) (colorless prisms).

The three monomethylthiazoles and 2,5-dimethylthiazole undergo the same type of cyclo addition with rearrangement when condensed with DMA in DMF (Scheme 54) (335, 339).

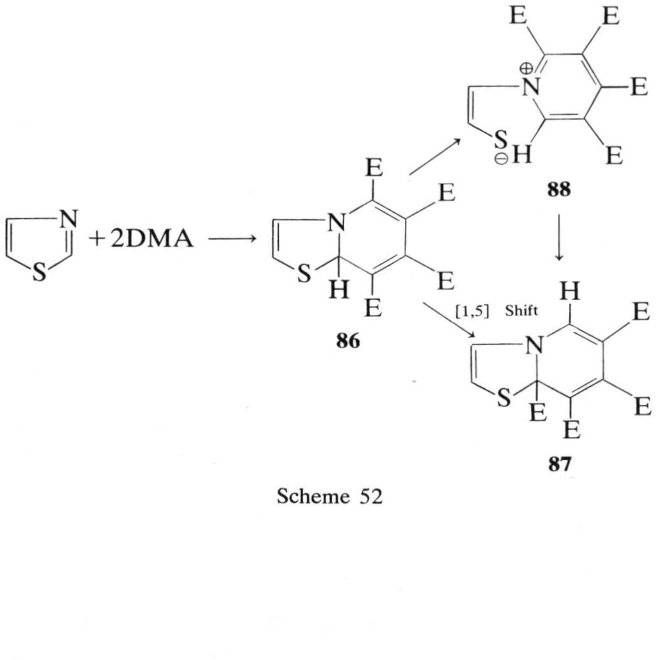

Scheme 52

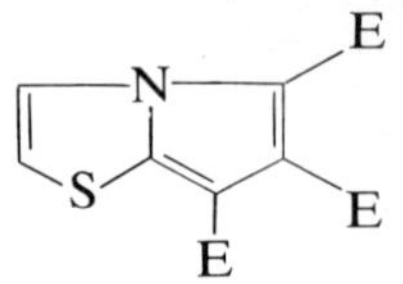

Scheme 53 (structure **89**)

Scheme 54 (structures with **90**)

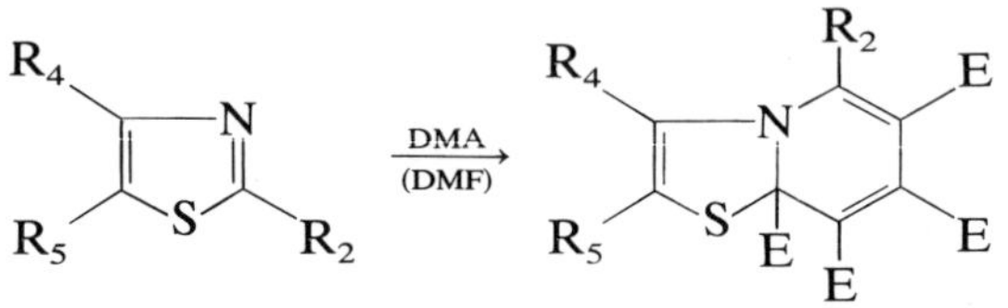

R_2 = Me, H, H, or Me
R_4 = H, Me, H, or H
R_5 = H, H, Me, or Me

A radically different course is followed when the reaction of 2-alkyl-substituted thiazoles is performed in methanol or acetonitrile (335), 2:1 adducts containing seven-membered azepine rings (**91**) are being formed in which two of the original activated hydrogen atoms have altered positions (Scheme 55). A similar azepine adduct (**92**) was obtained by

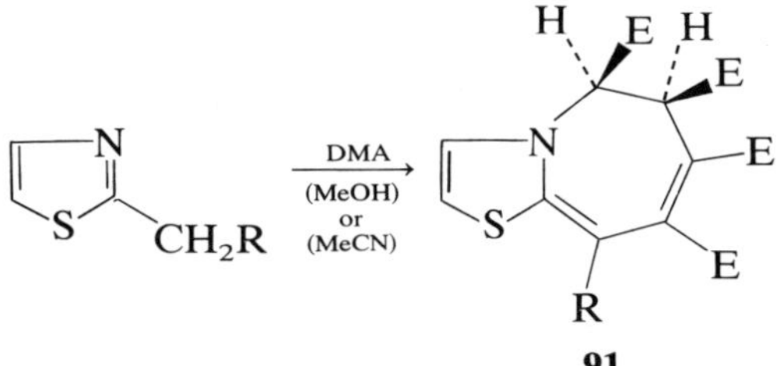

Scheme 55 (structure **91**)

Acheson et al. (336) by the condensation of DMA with 2,4-dimethylthiazole in THF (Scheme 56). As Reid et al. (335) first proposed, the adduct of 2,4-dimethylthiazole with DMA in DMF (**93**) results from the normal cyclo-addition with rearrangement (Scheme 57). The conclusive demonstration of this structure was recently given by Acheson et al. (339)

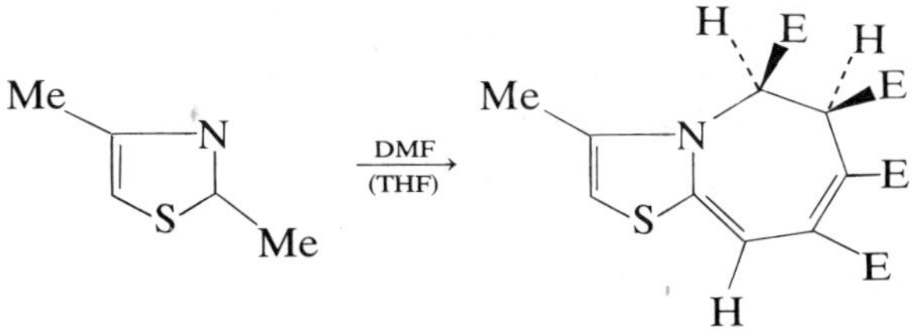

Scheme 56

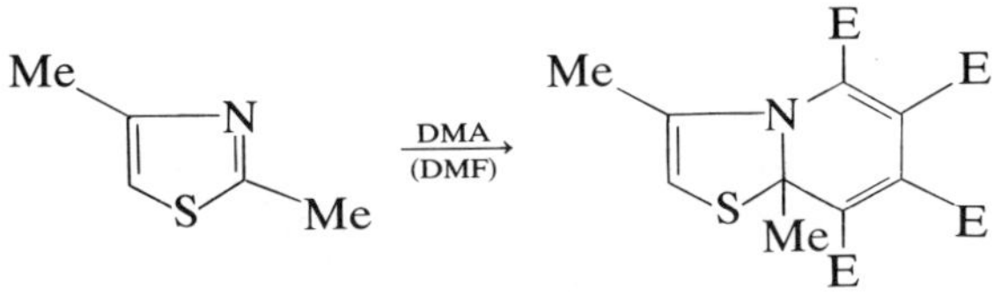

Scheme 57

on the basis of ^{13}C NMR spectroscopy and especially X-ray diffraction studies. The unusual behavior of 2,4-dimethylthiazole toward the rearrangement of the first-formed adduct (**93**) could result from steric interactions between methyl groups when rotation in the intermediate (**88**) is considered, as is shown by Dreiding models.

4. Electrophilic Substitution

Despite its "π excessive character" (340), thiazole, just as pyridine, is resistant to electrophilic substitution. In both cases the ring nitrogen deactivates the heterocyclic nucleus toward electrophilic attack. Moreover, most electrophilic substitutions, which are performed in acidic medium, involve the protonated form of thiazole or some quaternary thiazolium derivatives, whose reactivity toward electrophiles is still lower than that of the free base.

No nitration of thiazole occurs with the classical nitration reagents, even in forcing conditions (341–343). In a study concerning the correlation between the ability of thiazole derivatives to be nitrated and the HNMR chemical shifts of their hydrogen atoms, Dou (239) suggested that only those thiazoles that present chemical shifts lower than 476 Hz can be nitrated. From the lowest field signal of thiazole appearing at 497 Hz one can infer that its nitration is quite unlikely. Thiazole sulfonation occurs only in forcing conditions (341, 344): the action of oleum at 250°C for 3 hr in the presence of mercuric sulfate leads to 65% formation of 5-thiazolesulfonic acid (**94**). Mercuration of thiazole has been carried out by heating it with mercuric acetate in aqueous acetic acid and yields 2,4,5-tri(acetoxymercuri)thiazole (**95**) (345).

$$\underset{\textbf{94}}{\text{HO}_3\text{S} \diagdown\!\!\diagup \text{N} \atop \text{S}} \qquad \underset{\textbf{95}}{\text{AcOHg} \diagdown\!\!\diagup \text{N} \atop \text{AcOHg} \quad \text{S} \quad \text{HgOAc}} \qquad \underset{\textbf{96}}{\text{E}\longrightarrow \diagdown\!\!\diagup \text{N} \atop \text{S} \quad \text{Y}}$$

Thiazole derivatives undergo electrophilic substitution only when the ring is activated by the presence of electron-releasing groups. Among the early literature two schools have most contributed to the knowledge of electrophilic substitution in the thiazole series: the Japanese group of Ochiai and Nagasawa in a series of papers entitled "Polarization of thiazole ring" (341, 347–351) and the Indian group of Ganapathi et al. in a series of papers entitled "Chemistry of thiazoles" (342, 343, 352, 353). The Swiss group of Erlenmeyer and Prijs also developed the study of nitration and sulfonation reactions (344, 354, 355), and more recently a Russian group published results concerning alkylation reactions (356–361). The most significant qualitative results are summarized in Table I-52. Electron-releasing groups in the 2-position promote 5-substitution if that position is free or 4-substitution when it is already substituted (**96**); hydroxy and amino groups are the most effective promotors. Five-substituted thiazoles are nitrated in the 4-position. The 2-position seems to have been nitrated in only one instance, that is, when the 4- and 5-positions were substituted by methyl groups (343). When performed at low temperature (0°C) sulfonation of 2-aminothiazoles affords at first the corresponding 2-sulfamic acid, which, on heating, rearranges to 2-amino-5-sulfonic acid (Scheme 58) (341, 346, 351). Relatively deactivated derivatives like 2-bromo- and 2,5-dibromothiazole undergo sulfonation in oleum at 150 to 200°C in the presence of mercuric sulfate (Scheme 59) (344). Among halogenation reactions only bromination of thiazole derivatives has received attention. Of the monomethylthiazoles only the 2-isomer undergoes bromination (352). When the 5-position is not free, as in 2,5-dimethyl thiazole, no reaction occurs (350). 2-Hydroxy (348, 353) and 2-amino (348, 350) groups strongly activate the 5-position. However, bromination of 2-amino (or acetylamino)-4-(2'-furyl)thiazole (**104**) in mild conditions, occurs successively on the furan

TABLE I-52. QUALITATIVE ORIENTATION OF ELECTROPHILIC SUBSTITUTION IN THE THIAZOLE SERIES

Thiazole	Nitration	Sulfonation	Bromination	Mercuration	Alkylation
2-Substituted					
Me	5 (355)	—	5 (352)	5 (362)	—
CH_2-Ph	Ph (363)				
OH	5 (352)	—	5 (353)	—	—
NH_2, NR_2	5 (351)	5 (351) (344)	5 (350)	—	—
NHAc	5 (342)	—	—	5 (364) 4,5 (365)	—
Cl	failed (342)	—	—	—	—
NO_2	failed (366)	—	—	—	—
4-Substituted					
Me	5 (350, 352–354, 367)	5 (346)	failed (341, 352)	5 (362)	failed (348, 349)
Et, t-Bu	5 (367)	—	—	—	—
Ph	p, Ph (347)	—	—	—	—
5-Substituted					
Me	4 (352, 353, 367)	—	—	4 (362)	—
2,4-Disubstituted					
2,4-diMe	5 (342, 367)	5 (360)	5 (342)	5 (359, 362)	5 (358)
2-Ph-4-Me	—	5 (350)	—	5 (359)	5 (361)
2-OH-4-Me	5 (341, 346)	5 (341, 346)	5 (348)	5 (341)	failed[a] (349) 5[b] (348)
2-NH_2-4-Me	5 (341, 346, 351, 368)	5 (341, 346, 350)	5 (348)	—	5 (356, 357, 360)
2-NHAc-4-Me	5 (342)	—	5 (369)	—	—
2-SH-4-Me	—	—	—	—	failed (348)
2,5-Disubstituted					
2,5-diMe	4 (350, 367)	—	failed (350)	5 (362)	—
2-OH-5-Me	—	4 (350)	—	—	—
2-NH_2-5-Me	failed (350, 351)	4 (350, 351)	failed (350)	—	—
2-Me-5-OEt	—	—	4 (370)	—	—
4,5-Disubstituted					
4,5-diMe	2 (343)	—	—	2 (363)	—
4,5-diPh	p, pH (347)	—	—	—	—

[a] Friedel and Crafts benzoylation.
[b] Gattermann and Reimer Tiemann formylation.

(**105**) and thiazole (**106**) rings (371). Indirect bromination and iodination have been carried out by reaction of bromine or iodine with chloromercuric derivatives (Scheme 61) (345, 359, 362, 364). Acetoxymercuration of thiazole takes place in relatively mild conditions (AcOH aq, $(AcO)_2Hg$, 70–80°C, 30 min). The three positions of the ring are possible sites for acetoxymercuration, but the rate of the reaction decreases in the order, C-5 > C-4 > C-2.

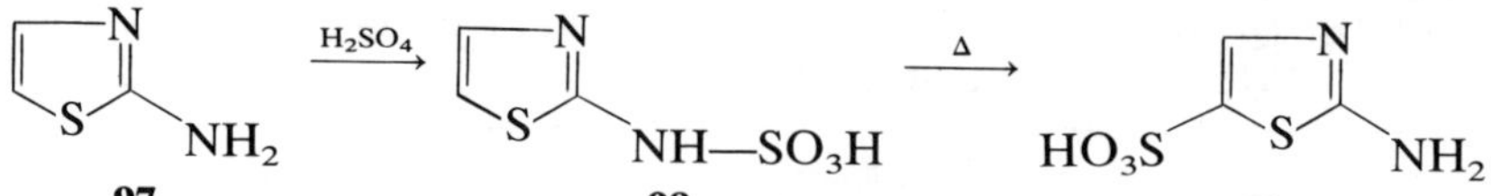

97 **98** **99**

Scheme 58

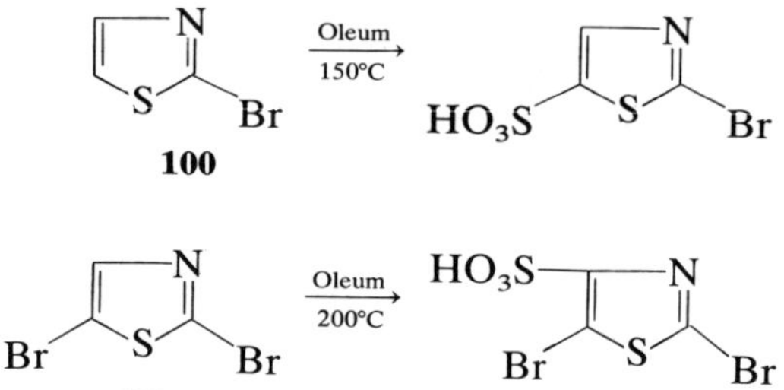

100

102 **103**

Scheme 59

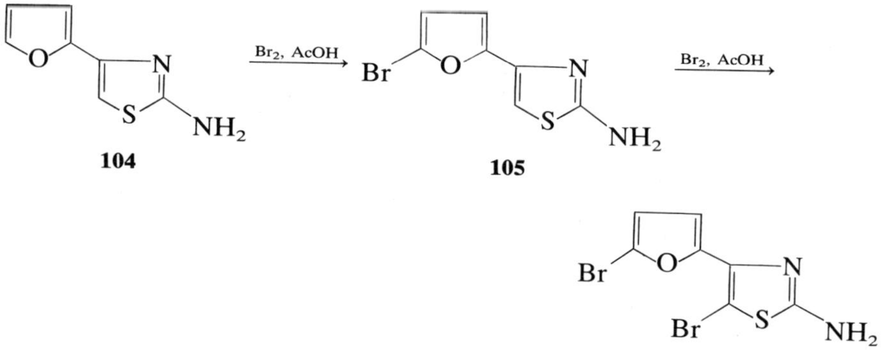

104 **105**

106

Scheme 60

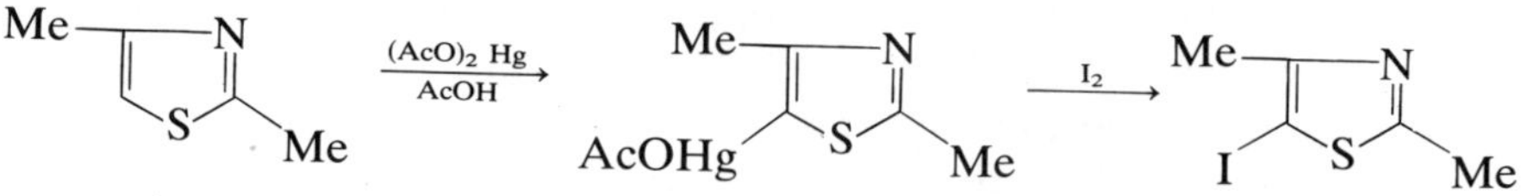

107 **108** **109**

Scheme 61

Only 2-aminothiazole derivatives are reactive enough toward diazonium salts to undergo the diazo-coupling reaction. The azo group fixes exclusively on the 5-position when it is free (Scheme 62) (351).

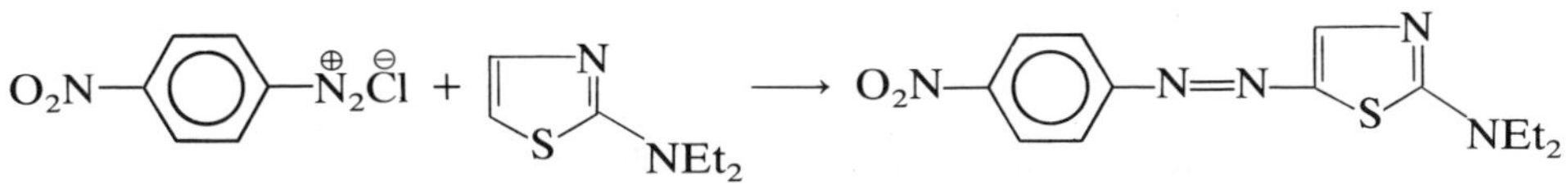

Scheme 62

Under appropriate conditions 2-amino-4-alkylthiazoles are alkylated in the 5-position: 2-acetylamino-4-methylthiazole reacts with dimethylamine and formaldehyde to afford the corresponding Mannich base (**113**) (372). 2-Amino-4-methyl-thiazole is alkylated in the 5-position by heat-

Scheme 63

ing with *t*-butanol in sulfuric acid (**115**) (360). Under similar conditions, 2-phenyl-4-methylthiazole is alkylated by cyclohexanol (361). 2-

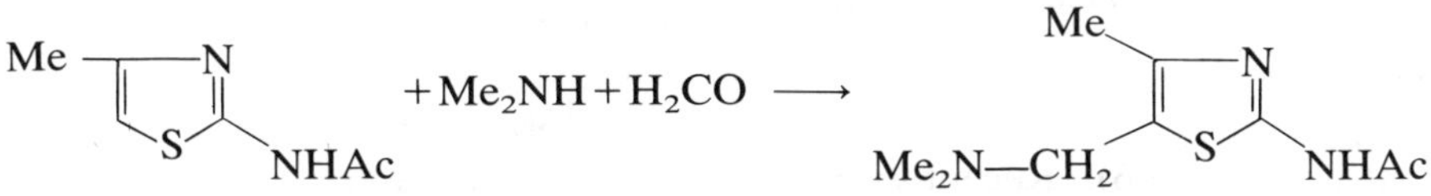

Scheme 64

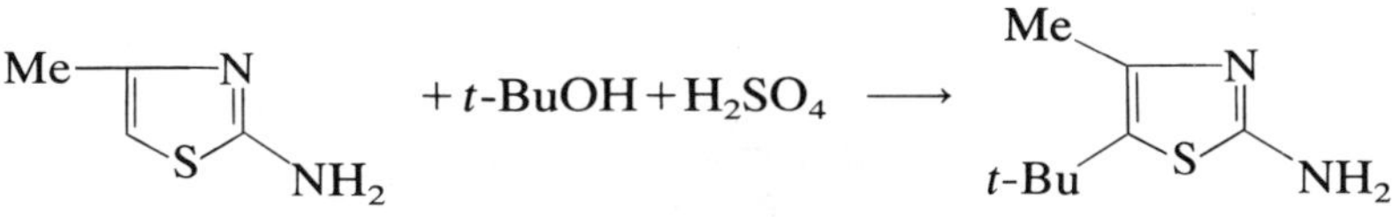

Hydroxy-4-methylthiazole failed to react when submitted to Friedel-Crafts benzoylation conditions (349); on the other hand, it reacted normally in Gattermann and in Reimer-Tiemann formylation reactions, affording the 5-formyl derivative (348). 4-Methylthiazole is insufficiently activated and fails to react under the same conditions. 2,4-Dimethylthiazole undergoes perfluoroalkylation when heated at 200° for 8 hr in a sealed tube with perfluoropropyl iodide and sodium acetate (**116**) (358).

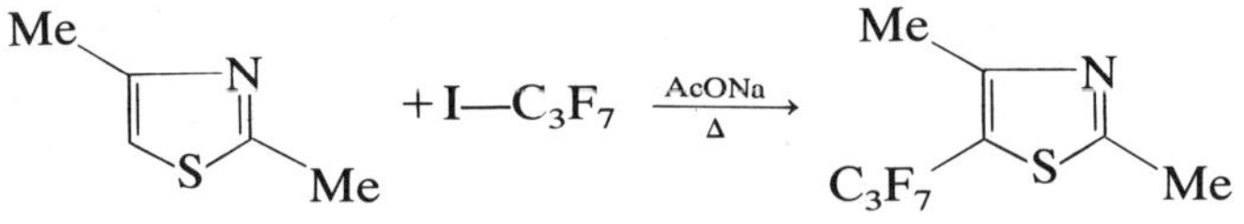

[Scheme 65: compound 107 (2,4-dimethylthiazole) + I–C$_3$F$_7$ →(AcONa, Δ) compound 116 (2-methyl-4-methyl-5-C$_3$F$_7$-thiazole)]

More quantitative results are available for the nitration of alkylthiazoles: Dou et al. (373) determined the reactivity, relative to benzene, of the nitration site of various mono- and dialkylthiazole by competition experiments (Table I-53).

From these results it appears that the 5-position of thiazole is two to three more reactive than the 4-position, that methylation in the 2-position enhances the rate of nitration by a factor of 15 in the 5-position and of 8 in the 4-position, that this last factor is 10 and 14 for 2-Et and 2-t-Bu groups, respectively. Asato (374) and Dou (375) arrived at the same figure for the orientation of the nitration of 2-methyl and 2-propylthiazole; Asato used nitronium fluoroborate and the dinitrogen tetroxide-boron trifluoride complex at room temperature, and Dou used sulfonitric acid at 70°C (Table I-54). About the same proportion of 4- and 5-isomers was obtained in the nitration of 2-methoxythiazole by Friedmann (376). Recently, Katritzky et al. (377) presented the first kinetic studies of electrophilic substitution in thiazoles: the nitration of thiazoles and thiazolones (Table I-55). The reaction was followed spectrophotometrically and performed at different acidities by varying the

TABLE I-53. ORIENTATION AND REACTIVITY RELATIVE TO BENZENE IN THE NITRATION OF ALKYLTHIAZOLES AT 70°C BY SULFONITRIC ACID (373)

Thiazole	Position	Reactivity ($\times 10^3$)
4-Me	5	9
5-Me	4	6
2,4-diMe	5	150
2,5-diMe	4	46
2-Me-5-Et	4	59
2-Et-5-Me	4	63
2-t-Bu-5-Me	4	83

TABLE I-54. ORIENTATION IN THE NITRATION OF 2-ALKYLTHIAZOLES

Thiazole	4-NO_2 (%)	5-NO_2 (%)	Yield (%)	Ref.
2-Me	31	69	86	374
2-Me	23	77	12	375
2-n-Pr	29	71	14	375

H_2SO_4 concentrations. The standard second-order nitration rate constants k_2^0 given in Table I-55 are calculated at $H_0 = 6.60$ from the least-square plots of log k_2 (obsd) against $H_0(T)$ extrapolated to 25°C according to the method described by Katritzky et al. (378) and corrected for minority species where necessary. Such a standardization allows discussion of substrate reactivities under the same conditions. All the alkylthiazoles investigated, and the corresponding quaternary salts, show log k_2^0 values in the range, -6.9 to -7.5, for reaction in both 4- and 5-positions: thus these cations are considerably less reactive than benzene (log $k_2^0 = 0.45$)(377). A 2-methoxy group has a significant rate-enhancing effect on nitration at the 5-position (ca. three log units), and the negatively charged 2-oxido substituent in 2-thiazolones increases the rate by another six log units. Comparison of nitration rates for the polymethyl derivatives shows that the thiazole 5-position is more reactive than the 4-position by a factor of about two. This agrees with the competitive experiments described earlier (375).

TABLE I-55. STANDARD RATE CONSTANTS FOR THIAZOLES AND 2-THIAZOLONES (377)

Compound	pK_a	Species[a]	Position[b]	log k_2^0 [c]
2,4-Dimethylthiazole	3.76	+	5	−6.90
2,5-Dimethylthiazole	3.91	+	4	−7.60
2,3,4-Trimethylthiazolium	—	+	5	−7.52
2,3,5-Trimethylthiazolium	—	+	4	−7.72
5-Ethyl-2-t-butylthiazole	3.48	+	4	−6.96
5-i-Propyl-2-t-butylthiazole	3.74	+	4	−7.43
2-Methoxy-4-methylthiazole	2.22	+	5	−4.62
4-Methylthiazol-2-one	−1.80	0	5	+1.12
3,4-Dimethylthiazol-2-one	−1.70	0	5	+1.82

[a] + = cation, 0 = free base.
[b] Position of nitration.
[c] Calculated at $H_0 = 6.60$ and 25°C.

Hydrogen exchange, in thiazole, especially deuteration, has been quantitatively investigated (379, 380), but the mechanism of the reaction carried out at acidic or neutral pH corresponds to a protonation-deprotonation process (380), different from electrophilic substitution and is discussed in section I.3.E.

Another quantitative approach to the reactivity of thiazole (381) in reactions involving a cationic transition state, though not exactly of the electrophilic substitution type, deserves to be mentioned here because of

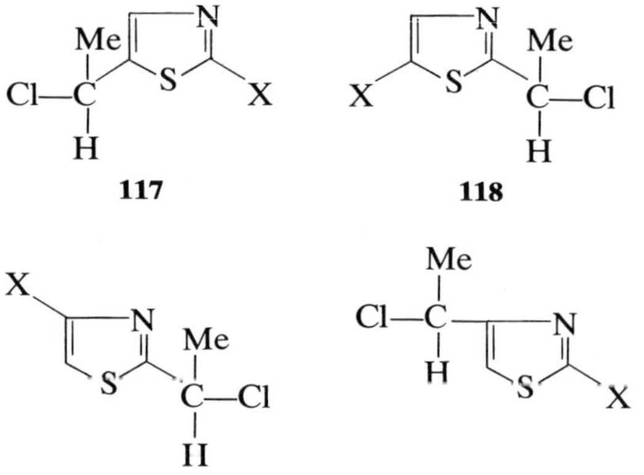

Scheme 66

its general interest. The rates of solvolysis of 2-substituted 1-(5-thiazolyl)ethyl chlorides (**117**) in 80% ethanol have been measured, and log $k^{45°}$ has been correlated with electrophilic substituent constants σ_p^+ of Brown and Okamoto (382) of the X substituent: an excellent correlation was obtained for X = −OMe, −SMe, −Me, −H, −Cl with $\rho = -6.14$. Only the phenyl substituent did not fit the correlation when using the original σ_p^+ value of −0.179 (382): the authors, including the phenyl point in the correlation, deduced a new value of −0.34 for its σ_p^+, supposing that in the compound under study both phenyl and thiazolyl ring were coplanar. The rates of solvolysis (383) of the three isomeric 1-thiazolylethyl chlorides in the same conditions (**121–123**) were measured and compared to that of α-phenylethyl chloride (**124**): the relative rate values are given in Fig. I-16.

The importance of the stabilization by the π system of the heterocycle of the developing positive charge on the α carbon reaction center decreases in the order, 5-thiazolyl > 4-thiazolyl ∼ phenyl > 2-thiazolyl.

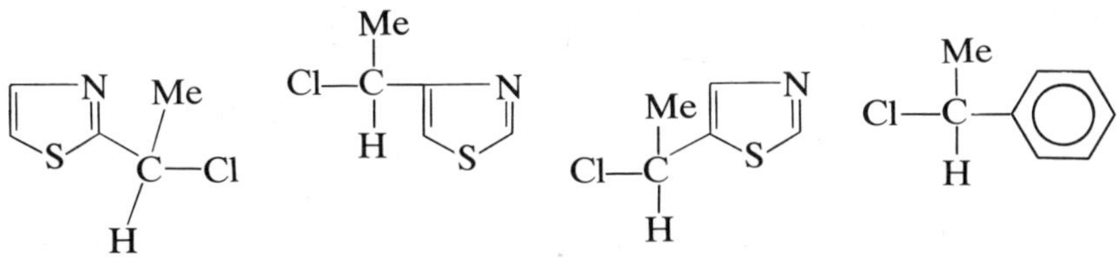

III. Chemical Reactivity

	121	122	123	124
Solvolysis rate:	1	25	174	21

Fig. I-16. Relative rates of solvolysis in 80% ethanol at 45°C (383).

This order compares well with that of the decreasing susceptibility to electrophilic substitution of the three positions of the ring and of benzene. This comparison is justified by the analogy of the transition states of both reactions. The observed order agrees also with that of the calculated π net charge on the site of fixation of the incipient carbocation (133):

$$\frac{C-5(-0.0086)}{C-4(+0.029)} \sim \frac{\text{benzene (0)}}{C-2(+0.100)}$$

Finally, the rates of the same reaction of solvolysis have been measured for 1-(2-X-5-thiazolyl)ethyl chlorides (**117**) 1-(5-X-2-thiazolyl)ethyl chlorides (**118**), 1-(4-X-2-thiazolyl)ethyl chlorides (**119**), and 1-(2-X-4-thiazolyl)ethyl chlorides (**120**) (384, 385). For the first two groups, where substituent X and the carbenium center are conjugated through the ring, good correlations are obtained with the original σ_p^+ constants (382). On the other hand, in the last two cases, where X and the carbenium center are not conjugated (in the sense of resonance theory), the σ_m^+ constant does not give a good correlation. This approach to reactivity is informative on the transmission of substituent effects in the thiazole system. Figure I-17 gives the factors of acceleration for the solvolysis of A = α-chloroethyl when X passes from H to Me. It can be seen that in the thiazole system the transmission of the substituent effects is more favorable between positions 2 and 5 than between positions 2 and 4.

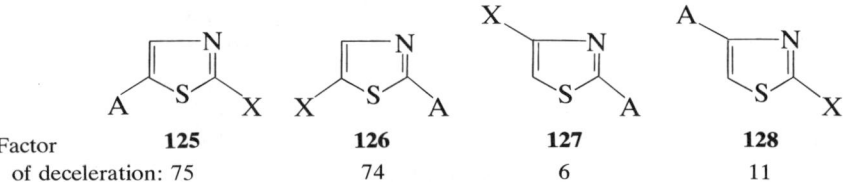

Factor of deceleration:	125	126	127	128
	75	74	6	11

Fig. I-17. Factors of acceleration of the solvolysis of A = CH(Me)Cl, when X passes from H to Me (384).

In the cases of 2-X-5-A (**125**) and 2-A-5-X (**126**) an attempt was made (386) to correlate the experimental ρ values with the calculated changes in the regional charge Δq (387). The correlation also includes 2-A-6-X benzo[b]thiophene and benzofuran, 1-A-4-X benzene, 2-A-5-X pyridine, thiophene, and furan and is reasonably good. The regional charge of an atom is the sum of the net charges on the atom and on any hydrogen atoms bonded to it. Δq refers to the difference in this charge at the substituted site, between the unsubstituted neutral substrate molecule and the cationic transition state.

5. Free Radical Substitution

At elevated temperatures (250–400°C) bromine reacts with thiazole in the vapor phase on pumice to afford 2-bromothiazole when equimolecular quantities of reactants are mixed, and a low yield of a dibromothiazole (the 2,5-isomer) when 2 moles of bromine are used (388–390). This preferential orientation to the 2-position has been interpreted as an indication of the free-radical nature of the reaction (343), a conclusion that is in agreement with the free-valence distribution calculated in the early application of the HMO method to thiazole (Scheme 67) (6, 117).

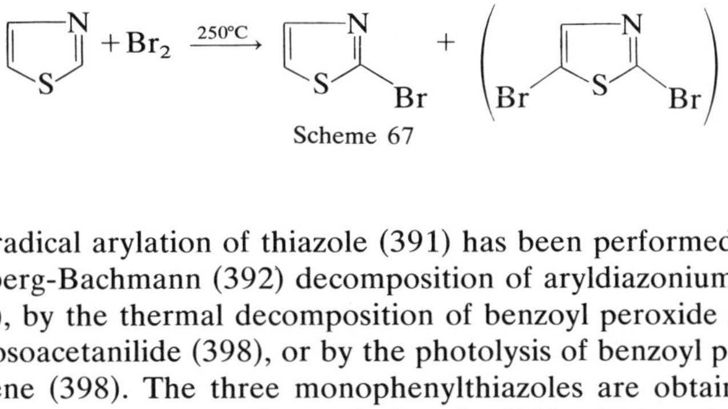

Scheme 67

The free radical arylation of thiazole (391) has been performed either by the Gomberg-Bachmann (392) decomposition of aryldiazonium chlorides (119, 393), by the thermal decomposition of benzoyl peroxide (394–397) or N-nitrosoacetanilide (398), or by the photolysis of benzoyl peroxide or iodobenzene (398). The three monophenylthiazoles are obtained in the practically constant proportions: 2-phenyl, 60%; 5-phenyl, 30%; 4-phenyl, 10%, giving the order, $2>5>4$, of decreasing reactivity of the three positions of thiazole toward phenyl radicals (398). Competition reactions with nitrobenzene (397) gave an estimation of the global reactivity of thiazole relative to benzene of 0.75 with the partial rate factors: $f_2 = 2.2$, $f_5 = 1.9$, $f_4 = 0.5$. When the thermolysis of benzoyl peroxide is performed in acetic acid solution, the substrate in reaction is the conjugate acid of thiazole: the global reactivity is enhanced to 1.25, whereas the isomer percentages and partial rate factors become 2-position, 83%, $f_2 = 6$; 5-position, 13%, $f_5 = 1$; 4-position, 4%, $f_4 = 0.3$ (397). The enhancement of the global reactivity, especially at the 2-position when passing from the thiazole to the thiazolium substrate, is quite similar to that observed in the pyridine series (395) and agrees with the HMO calculations on both substrates of Zahradnik and Koutecki (Fig. I-18) (117). A typical example of the product distribution obtained in a phenylation experiment where 0.02 mole of benzoyl peroxide was decomposed at 110°C in 1 mole of thiazole is given in Table I-56.

These results show that in the phenylation of thiazole with benzoyl peroxide two secondary reactions enter in competition: the attack of thiazole by benzoyloxy radicals, leading to a mixture of thiazolyl benzoates, and the formation of dithiazolyle through attack of thiazole by the thiazolyl radicals resulting from hydrogen abstraction on the substrate and from the dimerization of these radicals. This last reaction is less important than in the case of thiophene but more important than in the case of pyridine (398).

The radical phenylation of a large number of mono- and dialkyl-thiazoles has been investigated (393, 395, 396, 399–405, for a general review cf. 398) and analyzed in terms of partial rate factors. As in other instances the alkyl groups slightly activate the substrate in certain positions toward phenyl radicals, but they also induce some steric hindrance to the approach of the aryl radical from the *ortho* positions (Fig. I-19).

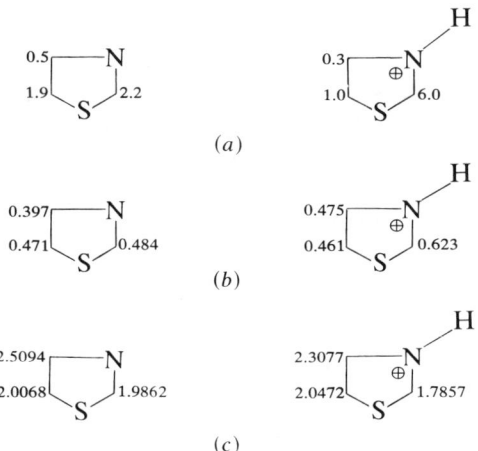

Fig. I-18. (a) Partial rate factors of free radical phenylation, relative to benzene (397). (b) Free valence calculated by HMO method (117). (c) Radical localization energy (in β units) calculated by HMO method (117).

TABLE I-56. TYPICAL PRODUCT DISTRIBUTION OF THE DECOMPOSITION AT 100°C OF BENZOYL PEROXIDE (0.02 MOLE) IN THIAZOLE (1 MOLE) (397)

Product	mmole/mole peroxide
Phenylthiazoles 120	
2-Phenylthiazole	60
4-Phenylthiazole	12
5-Phenylthiazole	48
2,5-Diphenylthiazole	5
2,4,5-Triphenylthiazole	Traces
Diphenyl	5
Dithiazolyls 23	
2,2'-Dithiazolyl	6
2,5'-Dithiazolyl	9
5,5'-Dithiazolyl	3
Other dithiazolyls	5
Thiazolylbenzoates	200
Benzoic acid	1450
Tars	100

Aryl substituents enhance somewhat the reactivity of the ring atoms, the phenyl substituent being arylated in a proportion of 40 to 60% (396, 406, 407).

Methyl free radicals, generated either by thermolysis of lead tetracetate in acetic acid solution (401) or by radical cleavage of dimethylsulfoxide by H_2O_2 and iron (II) salts (408), afford 2- and 5-methylthiazole in the proportion of 86 and 14%, respectively, in agreement with the nucleophilic character of alkyl free radicals and the positive charge of the 2-carbon atom of the thiazole (6).

A similar reaction of benzyl radicals is observed in the thermolysis at 200°C and in the presence of powdered copper of N-benzylthiazolium

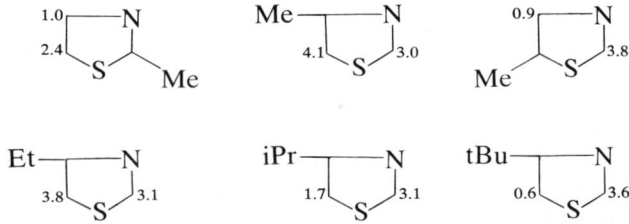

Fig. I-19. Partial rate factors of free radical phenylation relative to benzene (398).

chloride (409): 2- and 5-benzylthiazoles are the main products, which accords with a free-radical process. 2,4-Dimethyl-N-benzylthiazolium chloride subjected to the same treatment afforded not only 2,4-dimethyl-5-benzylthiazole (5%) but mainly the product of benzylation of the 2-methyl side chain, 2-phenethyl-4-methylthiazole (75%), as well as unidentified products (25%) (409).

Cyclohexyl free radicals, generated by photolysis of *t*-butyl peroxide in excess cyclohexane, also possess nucleophilic character (410). Their attack on thiazole in neutral medium leads to an increase of the 2-isomer and a decrease of 5-isomer relative to the phenylation reaction, in agreement with the positive charge of the 2-position and the negative charge of the 5-position (6).

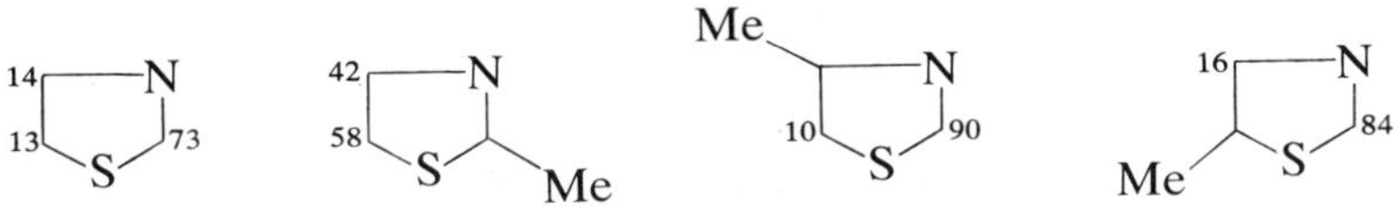

Fig. I-20. Percentage of free-radical cyclohexylation of thiazole and the three monomethyl isomers (411, 412).

The percentage of cyclohexylation is given in Fig. I-20. (411, 412). Hydrogen abstraction from the alkyl side-chain produces, in addition, secondary products resulting from the dimerization of thiazolylalkyl radicals or from their reaction with cyclohexyl radicals (Scheme 68) (411).

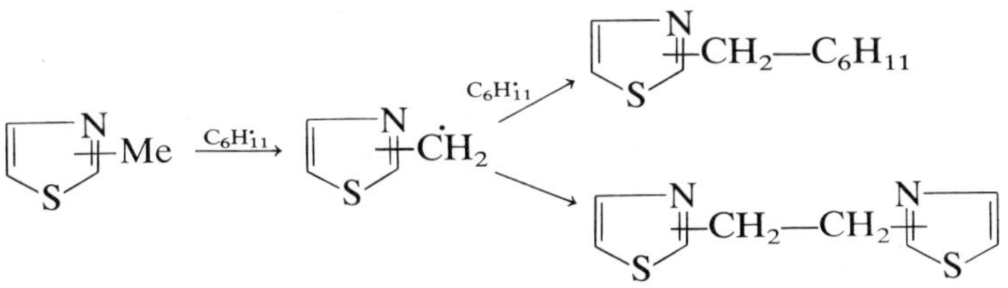

Scheme 68

The (thermal) decomposition of thiazol-2-yldiazonium salts in a variety of solvents at 0°C in presence of alkali generates thiazol-2-yl radicals (413). The same radicals result from the photolysis in the same solvents of 2-iodothiazole (414). Their electrophilic character is shown by their ability to attack preferentially positions of high π-electron density of aromatic substrates in which they are generated (Fig. I-21). The major by-products in the photolysis of 2-iodothiazole were identified as 3-arylisothiazoles, resulting from rearrangement of the initially formed 2-arylthiazole (414). This aspect is discussed in Section III.8.E.

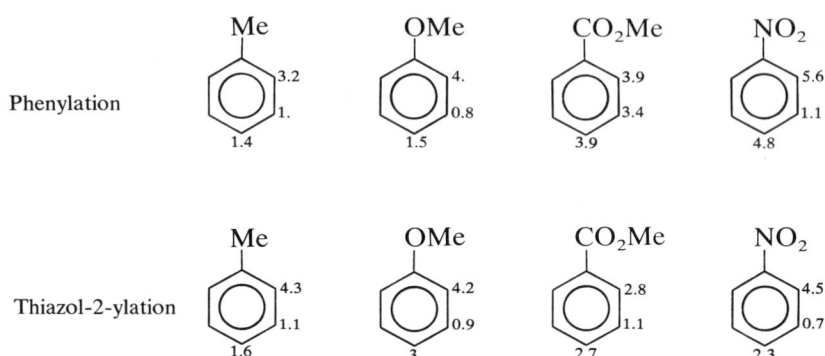

Fig. I-21. Partial rate factors for the phenylation and the thiazol-2-ylation of aromatic substrates (414).

Thiazol-2-yl radicals have also been generated by silver oxide oxidation of thiazol-2-ylhydrazine in various aromatic solvents (Scheme 69). The

$$Th-NH-NH_2 + Ag_2O \longrightarrow Th-N=N-H \longrightarrow T\dot{h} + \dot{H} + N_2$$
Scheme 69

isomer ratios are quite similar to those observed with the preceding methods (415). The three isomeric thiazolecarbonyl peroxides have been used as thiazolyl radical precursors, the free radical being generated by thermolysis of the peroxide in the aromatic substrate (Scheme 70) (415).

$$(Th-CO)_2O \xrightarrow{\Delta} Th-\dot{C}O_2 + T\dot{h} + CO_2$$
Scheme 70

The thermal decomposition of thiazol-2-yl-carbonyl peroxide in benzene, bromobenzene, or cumene affords thiazole together with good yields of 2-arylthiazoles but negligible amounts of esters. Thiazol-4-ylcarbonyl peroxide gives fair yields of 4-arylthiazoles, but the phenyl ester is also a major product in benzene, indicating reactions of both thiazol-4-yl radicals and thiazol-4-carbonyloxy radicals. Thiazole-5-carbonyl peroxide gives no product clearly indicative of thiazol-5-yl radicals or thiazole-5-carbonyloxy radicals. The percentage of *ortho* attack in bromobenzene by thiazol-4-yl radicals is noticeably higher than that by thiazol-2-yl radicals, suggesting that the former are more electrophilic (415).

In agreement with the theory of polarized radicals, the presence of substituents on heteroaromatic free radicals can slightly affect their polarity. Both 4- and 5-substituted thiazol-2-yl radicals have been generated in aromatic solvents by thermal decomposition of the diazoamino derivative resulting from the reaction of isoamyl nitrite on the corresponding 2-aminothiazole (250, 416–418). Introduction in 5-position of electron-withdrawing substituents slightly enhances the electrophilic character of thiazol-2-yl radicals (Table I-57).

TABLE I-57. REACTIVITY OF 5-SUBSTITUTED THIAZOLYL-2-YL FREE RADICALS TO TOLUENE AND NITROBENZENE RELATIVE TO BENZENE (250)

X in 5-X-2-Th	$\frac{\phi_{Me}}{\phi_H}K$	$\frac{\phi_{NO_2}}{\phi_H}K$
H	2.2	1.5
Br	3.2	0.52
SCN	3.25	—
C≡N	3.0	0.35

6. Nucleophilic Substitution

With the exception of the nuclear amination of 4-methylthiazole by sodium amide (341, 346) the main reactions of nucleophiles with thiazole and its simple alkyl or aryl derivatives involve the abstraction of a ring or substituent proton by a strongly basic nucleophile followed by the addition of an electrophile to the intermediate. Nucleophilic substitution of halogens is discussed in Chapter V.

Two types of hydrogen replacement are discussed here: (1) the base-induced hydrogen-deuterium exchange reactions and (2) the hydrogen-metal exchange reactions.

A. *Hydrogen-Deuterium Exchange Reactions*

Hydrogen kinetic acidity of thiazolium ions and thiazole has received much attention since Breslow (419, 420) and Ingraham and Westheimer

(421) discovered that thiamine (**129**), the pyrophosphate of which is the coenzyme factor for a large number of biochemical reactions, catalyzes the nonenzymatic decarboxylation of pyruvic acid via participation of the thiazolium ylide (**130**) resulting from the base-induced abstraction of the

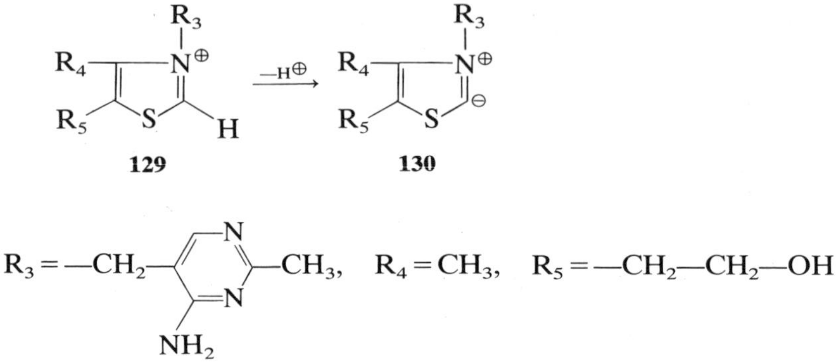

$R_3 = -CH_2-\underset{NH_2}{\overset{N}{\underset{=N}{\bigcirc}}}-CH_3, \quad R_4 = CH_3, \quad R_5 = -CH_2-CH_2-OH$

Scheme 71

2-hydrogen atom at the 2-position. The demonstration was supported by the observation (422) that deuterium exchange takes place very readily at the C-2 position of various thiazolium salts with D_2O and that this exchange occurs in the absence of any basic catalyst. Nuclear magnetic resonance analysis showed a half-time for the disappearance of the H-2 signal of 3-benzyl-4-methylthiazolium bromide in D_2O at 28°C of the order of 20 min. Compared to the conditions necessary to induce exchange for other heterocyclic ylides, this very low value of the half-time prompted the Breslow's remark that "the H at C-2 of thiazolium salts exchanges with D_2O more rapidly than any other active carbon-bound hydrogen so far reported" (422). Further evidence of the unique efficiency of thiamine as catalyst was afforded by studies on model systems, such as acetoin formation (423, 424). The well-known reversible ring cleavage of thiazolium salts in basic medium via their pseudo-bases (**131**) (425) could compete with ylide (**130**) formation, and Breslow was able to predict (426) that there would be an optimal pH below which the concentration of ylide (**130**) is too low for observable catalysis and above which the increasing conversion of the first pseudo-base (**131**) to the open-chain structure (**132**) should prevent catalysis: (Scheme 72). By H-NMR and H/D exchange rate measurements, Haake and Miller (231, 263) found an unusually high acidity of H-2 in oxazolium and thiazolium salts (just as in the oxazole and thiazole bases), which they compared to unusually large $^{13}C-H$ coupling constants at the same

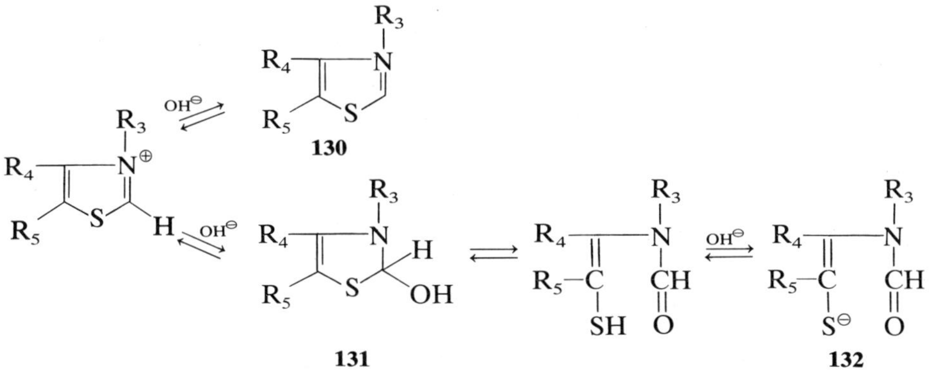

Scheme 72

positions ($J^{13}C_2$–H = 247 Hz for oxazolium and 218 for thiazolium cations). From this comparison a rate factor considerably larger than the observed one (at pH 4 to 5 and 37°C, oxazolium/thiazolium = 40) might well be expected, which would indicate some extra stabilization of the sulfur-containing ylide, attributed to d–σ overlap.

Staab et al. determined the half-time of H/D exchange in CH_3OD at 60°C for oxazole, thiazole, 1-benzylimidazole, and some corresponding quaternary salts and found a decreasing rate of exchange in the order, imidazole > oxazole > thiazole (379). The systematic study of the factors causing the lability of H-2 in thiazolium salts and their quantitative significance was undertaken by Olofson et al. (380, 427–429). Summarizing the available established facts, they ascribed this lability to the combined effect of a number of factors including (1) high s-character of the C_2–H bond as shown by the large ^{13}C–H coupling (analogous with acetylene), (2) an inductive effect of the heterocyclic ring (analogous with HCN acidity), (3) stabilization of the ylide (**130**) by a resonance contribution of the carbene-like structure (**133**), and (4) d–σ overlap of the electron pair of the anion with an empty d-orbital of sulfur (**134**). By a study of the rates of deutero-deprotonation of a series of di- and

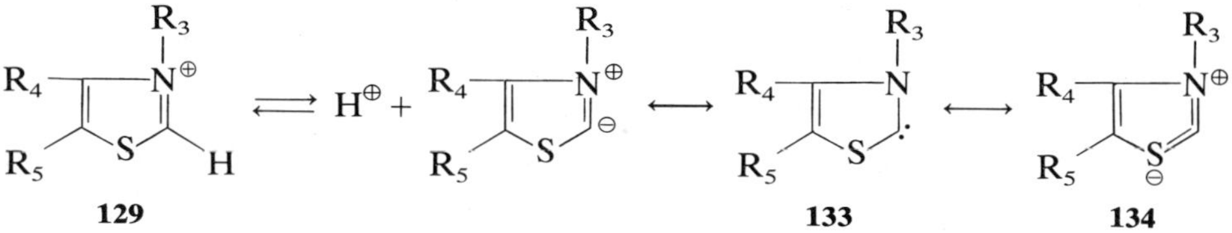

Scheme 73

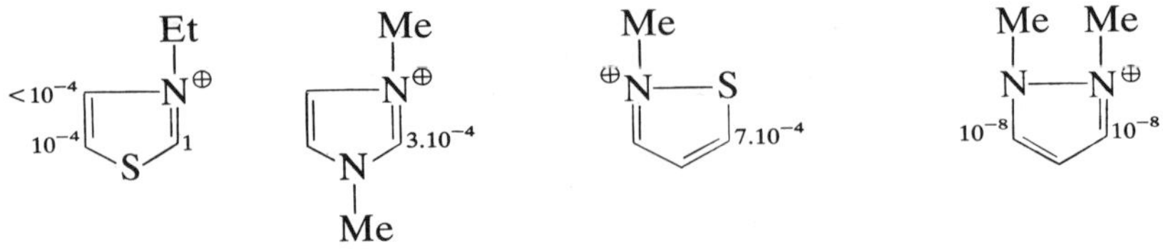

Fig. I-22. Relative rates of H/D exchange at 31°C in basic D_2O (429).

tetraazoles and azolium salts (427) they showed that coulombic and inductive effects are major rate-enhancing factors, especially the positive charge and, to a lesser extent, the presence of an α nitrogen. Then they studied a series of sulfur-nitrogen systems to evaluate the importance of the d-σ overlap in the stabilization of the ylide (130): H-2 in the thiazolium cation (135) is lost 3×10^3 times faster than H-2 in the related 1,3-dimethylimidazolium cation (136), and H-5 in the isothiazolium cation (137) is lost 7×10^4 times faster than H-3 in the corresponding 1,2-dimethylpyrazolium cation (138). From the decrease in electronegativity in replacing N by S, the opposite prediction would be made. The results suggest that another effect, such as d-σ overlap or another special factor, is also operative in sulfur-containing systems (Fig. I-22). To eliminate the stabilization due to the positive charge of the thiazolium cation, they examined the base-induced ionization rates of thiazole (139) and a number of model bases, especially isothiazole (140) and substituted isothiazoles (141 and 142), in which the nitrogen and sulfur atoms should have their greatest influence on different protons (Fig. I-23). Although these bases exchange 10^5 to 10^{10} times more slowly than their related cations, it is evident that there is an extraordinary rate enhancement when a proton is on carbon next to sulfur, suggesting that d-σ overlap is a major factor in stabilizing the corresponding anions. Electron-withdrawing substituents (142) increase the isothiazole (140) exchange rate, whereas electron-releasing substituents (141) slow it down.

Another interesting point is that both H-2 and H-5 in thiazole (139) deprotonate at about the same rate. This observation was also made at

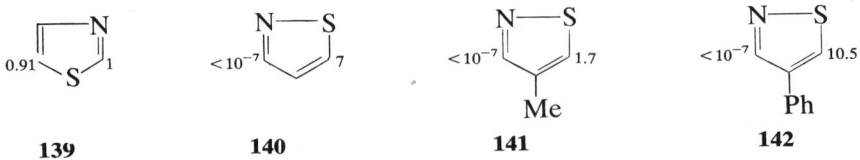

Fig. I-23. Relative rates of H/D exchange in basic MeOD (428).

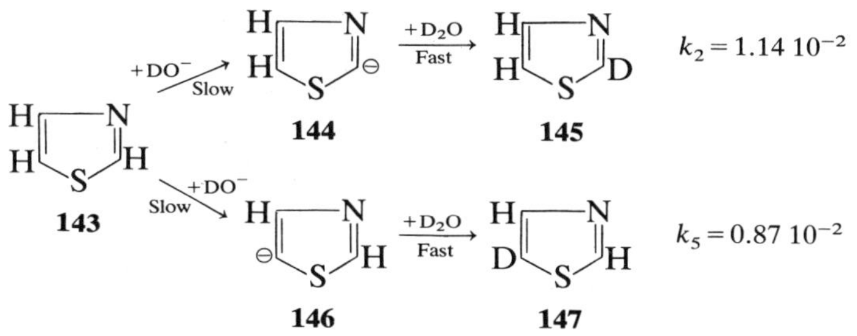

III. Chemical Reactivity

Fig. I-24. Absolute rate of H/D exchange (mole liter^{-1} min^{-1}) in 1.8 N NaOD at 60°C of thiazole (Concentration, 3.2 mole liter^{-1}) (430).

the same time by Derbez, who, in a paper on limited diffusion (430), demonstrated that thiazole exchanges at C-2 and C-5 positions when heated at 60 to 70°C in 1.8 N NaOD in D$_2$O (pD ~ 13). He determined the absolute second-order rates of exchange by H-NMR technique and found the ratio of the rate constants k_2 and k_5 corresponding to both positions to be $k_2/k_5 = 1.31$ (Fig. I-24). This observation afforded further support to the theory of the stabilization value of overlap of the forming anion (**144** and **146**) with an unfilled d orbital on adjacent sulfur and to the suggestion (431) that a significant factor in determining the ease of any C–H ionization process is the change in total vicinal sp^2-electron-pair repulsion in going from a parent heterocycle to the derived carbanion. To rationalize the results obtained with neutral thiazole, Olofson et al. postulated the competition of two exchange mechanisms (380). The first one operates between pD 0 and pD 11, and its two steps describe the ylide (I-**150**) formation by basic H-2 abstraction from the N–D$^+$ conjugate acid of thiazole (**149**) formed in a preequilibrium step and followed by the *trans* deuteration of the ylide (addition-elimination mechanism) (Scheme 74). In such acidic conditions the conjugate acid of thiazole (**149**) exchanges only its H-2 proton to give exclusively 2-deuterothiazole (**151**), a situation quite similar to that observed for the N-ethylthiazolium cation (429).

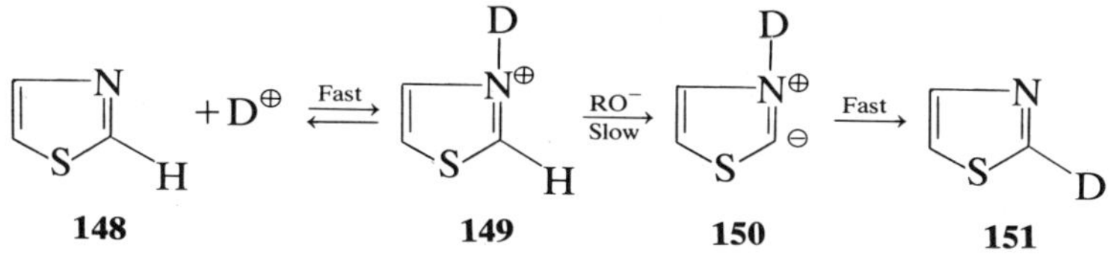

Scheme 74

CH₂—Ph
Me—⟨N⁺⟩
Me—S—T
152

Me—⟨N⁺—CH₂—pyrimidine(Me, NH₂)⟩
HO—CH₂—CH₂—S—T
153

The theoretical curve, deduced from the kinetic expression of the mechanism, fits the experimental points with gratifying exactness, whereas, for $pD > 12$, the simple mechanism reported earlier (428, 430) becomes predominant, and the rate increases very rapidly with pD and becomes first order both in thiazole and deuteroxide concentrations (Fig. I-24). Between pD 12.2 and 13.4 the ratio k_2/k_5 decreases from 12.4 to 1.8.

By protodetritiation of the thiazolium salt (**152**) and of 2-tritiothiamine (**153**) Kemp and O'Brien (432) measured a kinetic isotope effect, k_H/k_T, of 2.7 for (**152**). They evaluated the rate of protonation of the corresponding ylides and found that the enzyme-mediated reaction of thiamine with pyruvate is at least 10^4 times faster than the maximum rate possible with **152**. The scale of this rate ratio establishes the presence within the enzyme of a higher concentration of thiamine ylide than can be realized in water. Thus a major role of the enzyme might be to change the relative thermodynamic stabilities of thiamine and its ylide (432).

A confirmation of the possible role of sulfur in the ylide stabilization by d–σ overlap was afforded by the relatively high rate of decarboxylation of 5-thiazolecarboxylates (433).

Relative equilibrium ion-pair acidities have been determined by Streitwieser and Scannon (434) for thiazole and several heterocyclic compounds by reference to hydrocarbon indicators. The pK_a values for the 2-position, relative to 9-phenylfluorene = 18.49 are given in Fig. I-25. The pK_a of 14 estimated by Haake for 3-methylthiazolium iodide in water (263) shows that the main factor of acidity of thiazolium salts is their positive charge, followed by the aza effect.

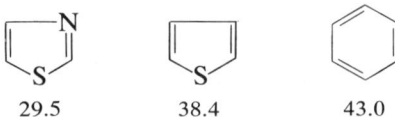

Fig. I-25. pK_a for the 2-position relative to 9-phenylfluorene = 18.49 (434).

More recently, the use of phase-transfer catalysis to promote the deproto-deuteration of thiazole and various alkylthiazoles enabled Spillane and Dou (435) to increase considerably the rate of H/D exchange and afforded the possibility of labeling alkylthiazoles in preparative quantities and at positions otherwise difficult to label.

B. *Hydrogen-Metal Exchange Reactions*

Under the action of strong organometallic bases, thiazole undergoes hydrogen-metal interconversion. Ethylmagnesium bromide reacts at 0°C with thiazole in ether to form an insoluble adduct that upon heating evolves ethane almost quantitatively and affords an etheral solution of thiazol-2-ylmagnesium bromide (**155**) (12). Proof of the structure of this

Scheme 75

Grignard reagent comes from the substitution products it gives with various reactive substrates. When the low-temperature adduct is heated in an autoclave at 90 to 170°C for 3 to 6 hr, it does not rearrange to 2-ethylthiazole (12) as is the case in the pyridine series (436).

Similarly, thiazole reacts at −60°C with phenyllithium affording thiazol-2-yllithium (**156**) (13, 437). As in the case of the Grignard derivative, thiazolyllithium does not rearrange under heating as does the adduct of pyridine and butyllithium (438).

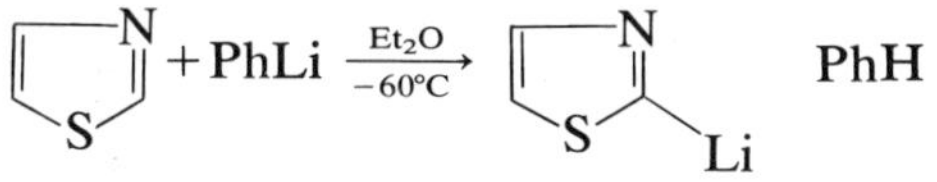

Scheme 76

The most reactive position of thiazole toward Grignard or organolithium reagents is the 2-position. 4,5-Dimethylthiazole (13, 439), 4-ethyl-5-methylthiazole (437), 4-methylthiazole (13, 423, 437), 5-methyl-4-ethyl-thiazole, and 4,5-diethylthiazole (13) readily exchange their hydrogen in the 2-position against lithium or magnesium. The stability of the organolithium derivative increases with alkyl substitution at the 4- and 5-positions: thiazol-2-yllithium decomposes above −60°C, whereas 4-methyl- or 4,5-dimethylthiazol-2-yllithium are stable up to 0°C (437).

The 2-metalated thiazoles react with a variety of electrophilic substrates in a standard way, leading to addition products with aldehydes, ketones, carbon dioxide, epoxides, nitriles, Schiff bases, and to substitution products with alkyl iodides (12, 13, 437, 440).

In an extensive temperature study on the lithiation of 2-methylthiazole (**157**) J. Crousier reported (441) that three lithio salts are formed independently, and not through proton-metal equilibration, at low temperature (−78°C, as shown in Scheme 77). The 4-lithio derivative (**160**) is

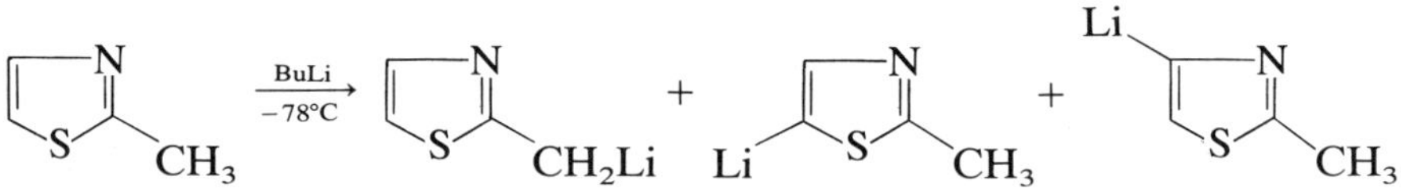

<chemical scheme>
157 → (BuLi, −78°C) → 158 + 159 + 160
</chemical scheme>

Scheme 77

formed only in very minute quantities, observable by the detection of traces of 4-deutero-2-methylthiazole in the mixture obtained by deuterolysis of the lithio salts of **157**. At −100°C the kinetic acidity of the C–5 and 2-methyl protons are quite similar. As the temperature is increased, the 2-lithiomethylderivative (**158**), which is somewhat less stable than the 5-lithio isomer (**159**), decomposes and at 5°C has almost entirely disappeared. Table I-58 collects some data of the quenching by D_2SO_4 of an etheral solution of the lithio salts of 2-methylthiazole at various temperatures (441).

From Table I-58, the practical constancy of the sum of **157** having reacted and the lithio salt being decomposed can be seen.

TABLE I-58. PRODUCT DISTRIBUTION OF THE REACTION AT VARIOUS TEMPERATURES OF 2-METHYLTHIAZOLE (**157**) WITH BUTYLLITHIUM FOLLOWED BY QUENCHING WITH D_2SO_4 (441).

t(°C)	Unreacted **157**	2-CH_2D from **158**	5-D from **159**	4-D from **160**	Decomposed
−100	61	17.5	15.5	1	5
−70	20	12.5	56	1.5	10
+5	10.5	1.5	64	3	21

III. Chemical Reactivity

A similar observation was made more recently in the case of 4-substituted 2-methylthiazoles by A. I. Meyers (442–444), who could identify the product resulting from a temperature increase from −78°C to room temperature of the lithio salt of 2,4-dimethylthiazole (**161**).

2-Lithiomethylthiazole (**162**) adds to unmetalated 2-methylthiazole (**161**), which needs be present in only trace amounts, generating adduct

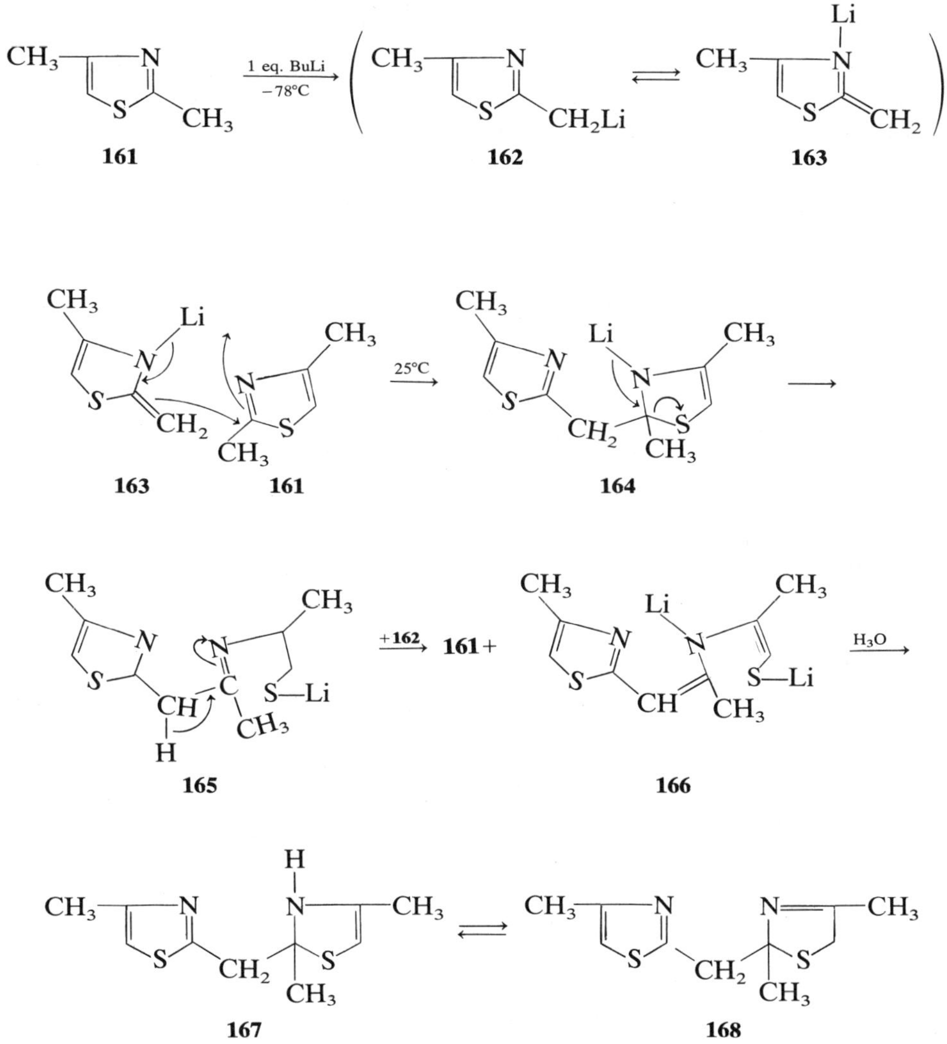

Scheme 78

164. Rearrangement to an open chain imine (**165**) provides an intermediate whose acidity toward lithiomethylthiazole (**162**) is rather pronounced. Proton abstraction by **162** gives the dilithio intermediate (**166**) and regenerates 2-methylthiazole for further reaction. During the final hydrolysis, **166** affords the dimer (**167**) that could be isolated by molecular distillation (433). A proof in favor of this mechanism is that when a large excess of butyllithium is added to (**161**) at −78°C and the solution is allowed to warm to room temperature, the deuterolysis affords only dideuterated thiazole (**170**), with no evidence of any dimeric product. Under these conditions almost complete dianion formation results (**169**), and the concentration of nonmetalated thiazole is nil. (Scheme 79). This dimerization bears some similitude with the formation of 2-methylthiazolium anhydrobase dealt with in Chapter IX. Meyers could confirm the independence of the formation of the benzyl-type (**172**) and the aryl-type

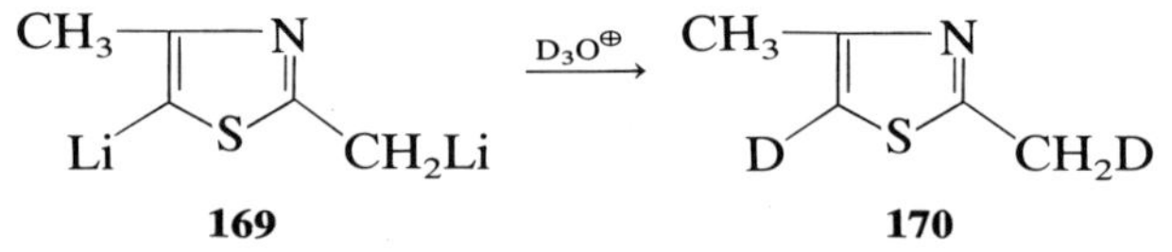

Scheme 79

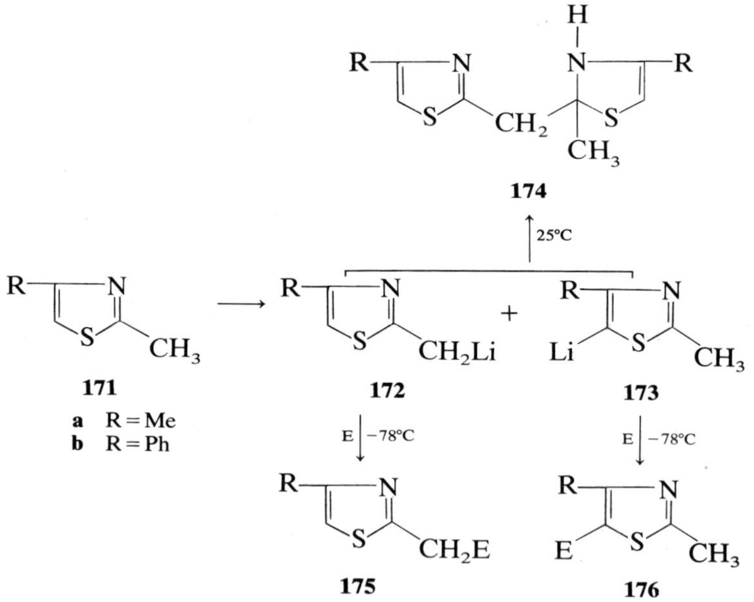

Scheme 80

III. Chemical Reactivity 123

(**173**) lithiated 4-substituted-2-methylthiazoles (**171**) at −78°C (Scheme 80). Crossover experiments at−78 and 25°C using thiazoles bearing different substituents (R = Me, Ph) proved that at low temperature the lithioderivatives (**172** and **173**) do not exchange H/Li and that the product ratios (**175/176**) observed are the result of independent metalation of the 2-methyl and the C–5 positions in a kinetically controlled process (444). At elevated temperatures the thermodynamic acidities prevail and the resonance stabilized benzyl-type anion (Scheme 81) becomes more abundant, so that *in fine* the kinetic lithio derivative is **173**, whereas the thermodynamic derivative is **172**.

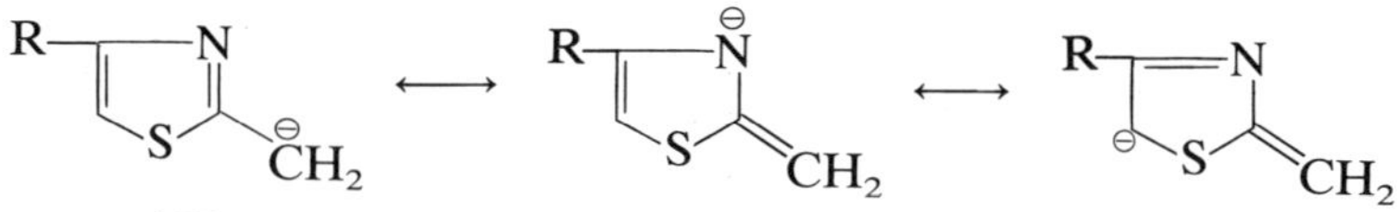

Scheme 81

The electronic influence of the 4-substituent corresponds to a relative increase in the kinetic acidity of the C-5 proton when an electron-withdrawing group (R = Ph) is situated at the 4-position and to a relative increase in the kinetic acidity of the 2-methyl group when an electron-donating group (R = Me) is at the same position (Table I-59).

The steric bulk of the base added to 2-methyl-4-phenylthiazole (**171b**) is another factor of the orientation for the kinetic metalation of 4-substituted 2-methylthiazoles (Table I-60). Even though the C-5 proton

TABLE I-59. METHYLATION AT −78°C OF THE LITHIUM SALT OF 4-R-2-METHYLTHIAZOLE (444)

R	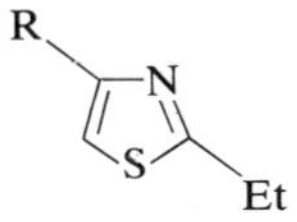 4-R-2-Et-thiazole	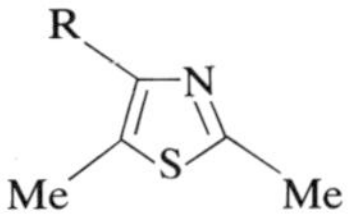 4-R-2-Me-5-Me-thiazole
Me	88	12
Ph	4	91

TABLE I-60. METALATION AND METHYLATION OF **171b** WITH BASES OF INCREASING STERIC BULK (444)

Base	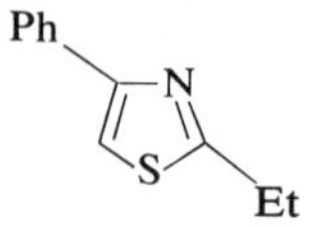 Ph-thiazole-Et (C-5 attack)	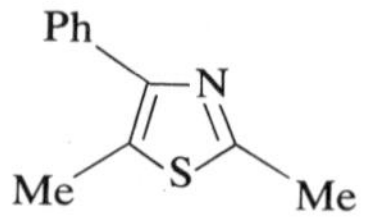 Ph-thiazole-Me, Me (Me attack)
n-BuLi	94.8	5.2
(i-Pr)$_2$NLi	90	10
t-BuLi	34	66

has been shown to be kinetically the more acidic at −78°C, the strongest base employed (t-BuLi), which is also the bulkiest, leads to mainly methyl proton abstraction. Lithium amide has also been utilized for the metalation of 2,4-dimethylthiazole (445).

A primary isotope effect k_H/k_D of 6.4 (extrapolated for 35°C) is observed for the metalation and the methylation of **171b** when the C-5 position is deuterated. This value is in excellent agreement with the primary isotope effect of 6.6 reported for the metalation of thiophene (392) and it confirms that the rate-determining step is the abstraction by the base of the acidic proton.

C. Nucleophilic Amination

There is only one example of substitution of a nuclear hydrogen by a nucleophile: the amination of 4-methylthiazole by NaNH$_2$ described by Ochiai (scheme 82) (341, 346). This reaction is probably similar to the

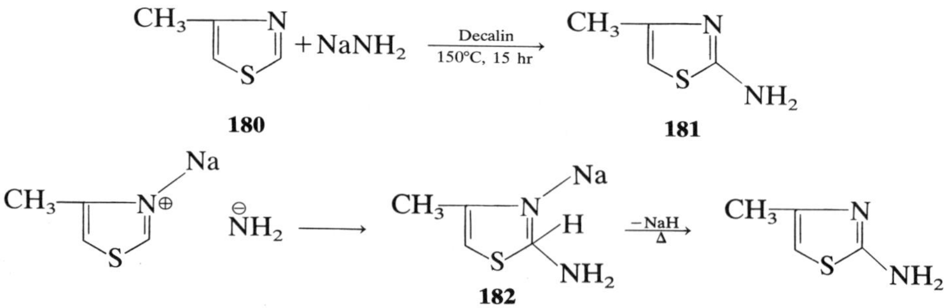

Scheme 82

Tschitschibabin amination of pyridine, the mechanism of which has been established as involving an intermediate σ complex (**182**) (334). Extrapolated to the case of thiazole, this mechanism would give Scheme 83. The orientation to the 2-position is in agreement with the calculated π-charge distribution of the thiazole molecule as well as of the thiazolium ion.

7. Thiazole as an Electron Donor

The most striking analogy between thiazole and pyridine results from the presence in both molecules of a ring nitrogen atom, contributing two σ bonds to the σ framework of the molecule and one $2p_z$ electron to the π system, while a lone pair of electrons remains localized on the nitrogen atom, where it is described by an sp^2-hybrid orbital, the axis of which lies in the σ plane and is directed along the bissector of the C–N–C angle (cf. Section II.1). This lone pair is responsible for the basicity and the nucleophilicity of both molecules.

A. *Quaternary Thiazolium Salts Formation*

In all its reactions the lone pair of thiazole is less reactive than that of pyridine. Table I-61 shows three sets of physicochemical data that illustrate this difference. These are: (1) the thermodynamic basicity, which is three orders of magnitude lower for thiazole than for pyridine; (2) the enthalpy of reaction with BF_3 in nitrobenzene solution, which is 10% lower for thiazole than for pyridine; and (3) the specific rate of quaternization by methyl iodide in acetone at 40°C, which is about 50% lower for

TABLE I-61. PHYSICOCHEMICAL DATA FOR SOME REACTIONS OF THIAZOLE AND PYRIDINE

	Thiazole	Pyridine
pK_a	2.52 (321)	5.27 (321)
$\Delta H\ (BF_3)^a$	30.27 (450)	33.33 (451)
$k_2\ (MeI)^b$	15.8 10^{-6} (452)	35.0 10^{-6} (452)

[a] Enthalpy of reaction with BF_3 in nitrobenzene at 25°C in cal mole^{-1}.
[b] Specific rate of methylation by methyl iodide in acetone at 40°C in liter mole^{-1} sec^{-1}.

thiazole, and which in DMSO at 25°C is of the order of 1/21 of that of pyridine (446).

Although isothiazole ($pK_a = 1.90$) is less basic than thiazole, its rate of quaternization by dinitrophenyl acetate in water at 52°C is approximately 2.5 times higher (447). This deviation from the Brönsted relationship ($\Delta \log k = \beta . \Delta pK$, with β positive) is interpreted as a consequence of the "α effect" of the adjacent sulfur lone pair in isothiazole that is responsible for its higher nucleophilicity (448, 449).

The same situation is observed in the series of alkyl-substituted derivatives. Electron-donating alkyl substituents induce an activating effect on the basicity and the nucleophilicity of the nitrogen lone pair that can be counterbalanced by a deactivating and decelerating effect resulting from the steric interaction of *ortho* substituents. This aspect of the reactivity of thiazole derivatives has been well investigated (198, 215, 446, 452–456) and is discussed in Chapter III.

Because of its dissymmetrical geometry, the thiazole ring affords an unique substrate for the study of the steric implications of $S_N 2$ reactions. In the case of the Menschutkin quaternization reaction, both 2- and 4-positions give the possibility of introducing at the very proximity of the electrophilic carbon center various substituents whose electronic and especially steric effects could be finely adjusted. Thus it was possible to give experimental evidence for the geometrical variations in the transition state of $S_N 2$ reactions induced by changing the leaving group: from a comparative study of quaternization rates of *ortho*-substituted thiazoles and pyridines with various methylating agents (MeI, MeSO$_3$F, MeOTs) it was experimentally established that an increase in the basicity of the leaving group moves the transition state of the $S_N 2$ reaction closer to the products (457).

The properties of thiazolium salts are discussed in Chapter IX.

B. *Thiazole Complexes with Metal Atoms*

All the early literature concerning thiazoles mentions numerous metallic complex-salts formed by addition to the thiazole of the aqueous solution of the metal salt and that could be used for identification purposes. The most usual complexes so obtained are platinum double salts, for example, (4-methylthiazole · HCl)$_2$ · PtCl$_4$ (m.p. decp 204°C) (25), or mercuric chloride derivatives, for example, 2,4-dimethylthiazole · 2 HgCl$_2$ (m.p. decp 176–177°C) (458).

Erlenmeyer and Schmid (459), in the course of a fundamental comparison between pyridine and its isosteric thiazole structures, prepared the first known metal complex of 2,2'-dithiazolyl (**183**) and Fe(II) as a red-yellow coloration appearing in hot aqueous solutions of both constituents but disappearing when cooled. On the other hand, 4,4'-dithiazolyl (**184**) gave no complex in the same conditions (460).

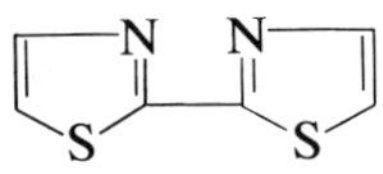

183

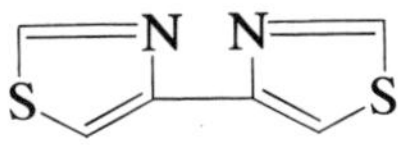

184

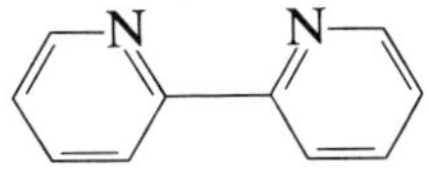

185

Scheme 83

The first identified complexes of unsubstituted thiazole were described by Erlenmeyer and Schmid (461): they were obtained by dissolution in absolute alcohol of both thiazole and an anhydrous cobalt(II) salt (Table I-62). Heating the α-CoCl$_2$·2Th complex in chloroform gives the β isomer, which on standing at room temperature reverses back to the α form. According to Hantzsch (462), these isomers correspond to a *cis-trans* isomerism. Several complexes of 2,2'-(**183**) and 4,4'-dithiazolyl (**184**) were also prepared and found similar to pyridyl analogs (**185**) (Table I-63). Zn(II), Fe(II), Co(II), Ni(II) and Cu(II) chelates of 2,4-*bis*(2-pyridyl)thiazole (**186**) and (2-pyridylamino)-4-(2-pyridyl)thiazole (**187**) have been investigated. The formation constants for species ML^{2+}, ML$_2^{2+}$, and ML$_3^{2+}$ (L = **186** or **187**) have been calculated from data obtained by potentiometric, spectrophotometric, and partition techniques.

TABLE I-62. COMPLEXES OF THIAZOLE (Th) AND PYRIDINE (Py) WITH Co(II) SALTS (461)

Thiazole complex	Color	Pyridine complex	Color
α-CoCl$_2$·2-Th	Red-violet	α-CoCl$_2$·2-Py	Violet
β-CoCl$_2$·2-Th	Blue-violet	β-CoCl$_2$·2-Py	Dark blue
CoCl$_2$·4-Th	Light violet	CoCl$_2$·4-Py	Red-violet
Co(SCN)$_2$·4-Th	Light red	Co(SCN)$_2$·4-Py	Light red

TABLE I-63. COMPLEXES OF 2,2'-DITHIAZOLYL (**183**), 4,4'-DITHIAZOLYL (**184**), AND 2,2'-DIPYRIDYL (**185**) WITH Ni(II) AND Cu(II) SALTS (461)

Complex	Color
Ni · (**184**)$_3$Cl · 6$\frac{1}{2}$H$_2$O	Peach-flower
NiCl$_2$ · (**183**) · 4H$_2$O	Light green
CuSO$_4$ · (**183**) · 2H$_2$O	Light blue
CuSO$_4$ · (**184**)$_2$ · 5H$_2$O	Blue-green
Ni · (**185**)$_3$Cl$_2$ · 6$\frac{1}{2}$H$_2$O	Red
CuSO$_4$ · (**185**) · 2H$_2$O	Light blue

The bonding in the mononuclear metal complexes is considered to involve the two nitrogen atoms of the ligand structure, giving the familiar chelating α-diimine structure $-N=\overset{|}{C}-\overset{|}{C}=N-$. Only in the case of the Fe(II) complexes of **186** is the constant of formation of ML$_3^{2+}$ significantly greater than that of ML$_2^{2+}$, which could be an indication of metal-ligand π-bonding (Scheme 84) (463). A series of simple thiazole complexes has

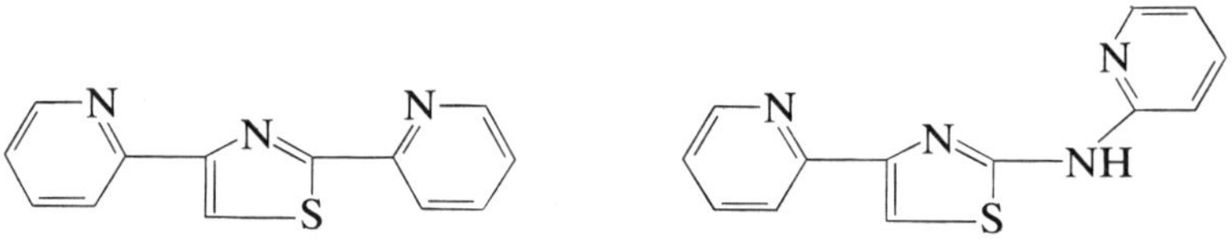

186   **187**

Scheme 84

been investigated: Ni(Th)$_4$Cl$_2$, Co(Th)$_2$Cl$_2$, and Cu(Th)$_2$Cl$_2$ are shown to have essentially octahedral structures (464). In a study of the influence of 4-methyl and 2,4-dialkyl substituents, Hambright et al. (465) confirmed that metal-N coordination, as opposed to metal-S coordination, operates in a large series of 19 complexes of Zn(II), Cu(II), Ni(II), Co(II), and Pt(II) of the general type ML$_2$X$_2$. In a comparison with imidazole they observed that the maximum ligated copper complexes with imidazole are CuCl$_2$L$_4$, whereas pyridine, thiazole, 4-methylthiazole, 2,4-dimethylthiazole, and benzothiazole form predominantly CuCl$_2$L$_2$. This has been explained in terms of σ-donor and π-acceptor properties of the ligand in maintaining an effective electroneutrality of the central ion. In the case

III. Chemical Reactivity

of Pt(II) and Zn(II) derivatives, they noted that the complexes appear to be insensitive to ring substitution, whereas Ni(II) formed mainly NiX_2L_4 complexes with thiazole and NiX_2L_2 complexes with the more crowded 4-methylthiazole and 2,4-dimethyl thiazole. This work was developed by Hughes (466), who discussed the preceding conclusions.

Many other complexes of thiazole with Co(II), Ni(II), Cu(II), Zn(II), and Pd(II), with structures of the types $M(Th)_6X_2$, $M(Th)_4X_2$, $M(Th)_2X_2$, and $M(Th)X_2$, have been prepared, and their spectral characteristics have been determined (467). All these complexes are six-coordinate in the solid state with bridging groups where necessary. The stability of the Co(II), Ni(II), and Zn(II) complexes was determined potentiometrically, and their absorption spectra were recorded in the visible region (468). The far-infrared spectra of the Pd(II) complexes have been investigated (469): the bromo complexes with non-2-substituted ligands showed high metal-bromine frequencies. Rh(III) complexes containing thiazole ligands, trans-$Rh(Th)_4X_2$, have been prepared and studied for their bacteriostatic activity (470). Cu(II) and Co(II) complexes of thiamine chloride hydrochloride and related thiazole $M(Th)_2X_2$ (**188**) and thiazolium derivatives $M(Th)_2X_4$ (**189**) have been prepared, and their electronic and infrared spectra and magnetic moments have been analyzed (471).

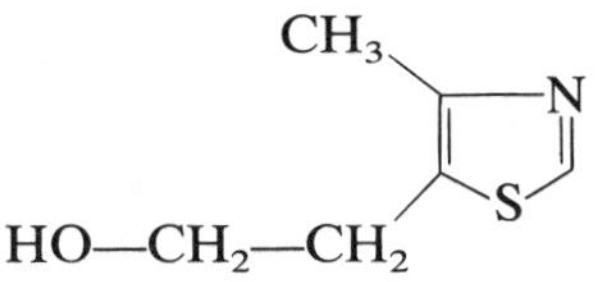

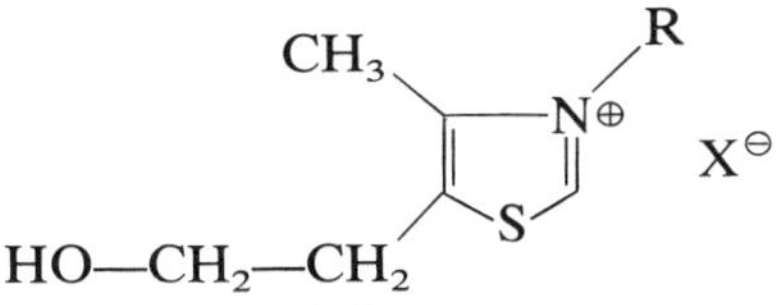

Scheme 85

Experimental confirmation of the metal-nitrogen coordination of thiazole complexes was recently given by Pannell et al. (472), who studied the Cr(O), Mo(O), and W(O) pentacarbonyl complexes of thiazole: $(Th)M(CO)_5$. The infrared spectra are quite similar to those of the pyridine analogs; the H-NMR resonance associated with 2- and 4-protons are sharper and possess fine structure, in contrast to the broad, featureless resonances of free thiazole ligands. This is expected since removal of electron density from nitrogen upon coordination reduces the ^{14}N quadrupole coupling constant that is responsible for the line broadening of the α protons.

The base peak in the mass spectrum of the LM free metal-ligand ion and the fragmentation patterns of this parent ion are of particuliar significance since they illustrate the effect of coordination upon the properties of the thiazole ligand. The free thiazole fragments upon electron impact by two major routes (Scheme 86; also cf. Section II. 6).

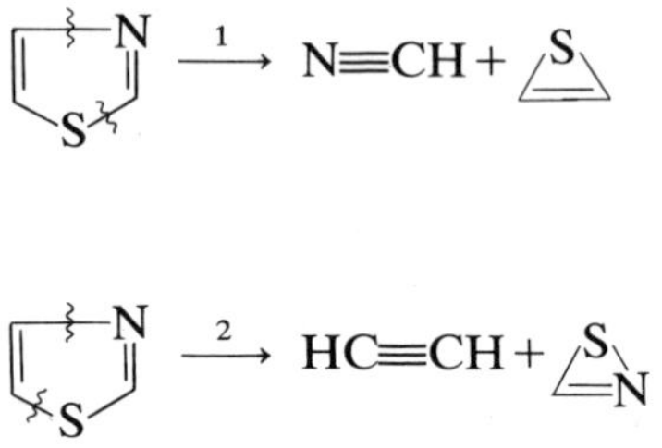

Scheme 86

The fragmentation of $(Th)Cr(CO)_5$ exhibits the two main routes. The loss of HCN with retention of the metal-ligand interaction implies a migration of the metal from nitrogen to sulfur. The presence of the CrS^+ fragment suggests the coordination of the metal *via* the sulfur atom. The ratio of HCN/HC≡CH elimination for $(Th)Cr(CO)_5$ is 1/1; this compares to a ratio of 16/1 for the free thiazole under comparable conditions. The reduction of HCN elimination by N-metal coordination results from the necessity of the process to involve metal migration, whereas acetylene elimination does not imply this migration. In the case of $(Th)W(CO)_5$ only the HCN elimination process is observed: the exclusive migration of tungsten to sulfur under these conditions, as compared to chromium, may be explained on the basis of the relative hardness and softness of the metal and the N and S donor atoms. The larger tungsten atom, being softer, would form thermodynamically more stable bonds to sulfur relative to nitrogen than chromium.

8. Miscellaneous Reactions

A. *Oxidation*

Thiazole is relatively resistant to oxidation, and very few reactions of this type have been recorded. Under photosensitized oxygenation, triphenylthiazole affords various products of ring cleavage depending on the sensitizer and the solvent employed (Scheme 87) (473). A possible mechanism could involve endoperoxide or zwitterionic peroxide.

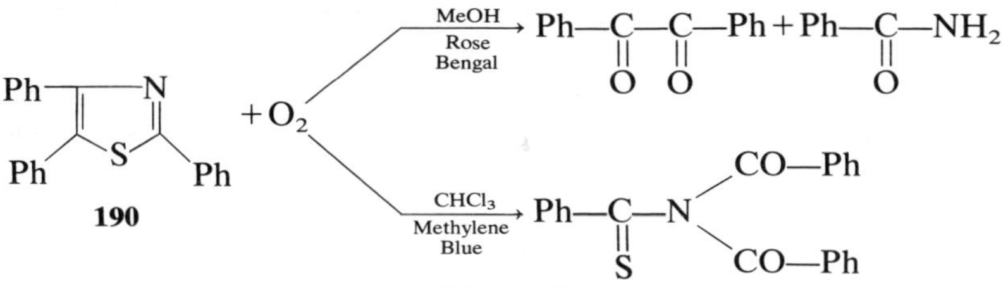

Scheme 87

Thiazole-N-oxides are prepared by the action at low temperature (−10°C) of hydrogen peroxide in acetic acid (474). 4-Methylthiazole and 2,4-dimethylthiazole afforded the corresponding N-oxides with yields of 27 and 58%, respectively (Scheme 88). Thiazole-N-oxides without a methyl group in the 2-position are so unstable that they have a tendency to form 2-hydroxythiazoles and are decomposed by oxidation, whereas a 2-methyl group would prevent such rearrangement (474).

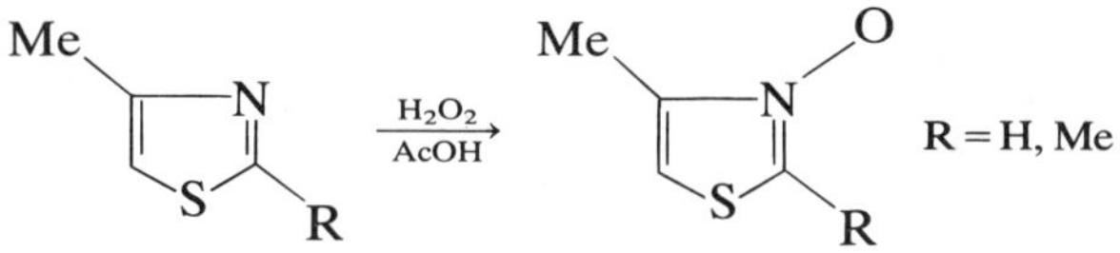

Scheme 88

The antibacterial [(5-nitrofuryl)vinyl]thiazole-N-oxides (**193**) were prepared by oxidizing the corresponding thiazoles with hydrogen peroxide or peracetic acid (Scheme 89) (475).

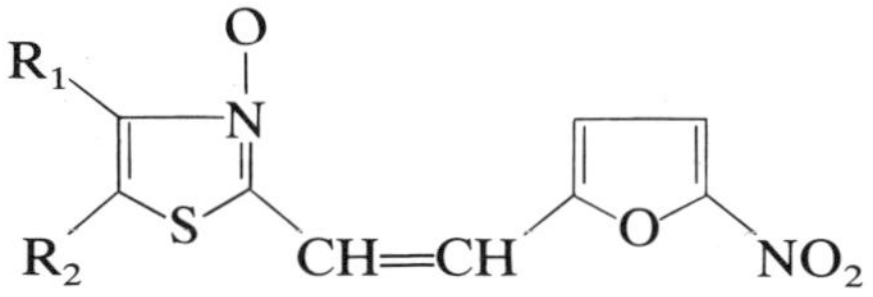

Scheme 89

B. Reduction

Schatzmann, in 1891, tried to prepare 2-thiazolines by hydrogenation of thiazoles and by the action of sodium and ethanol on 2,4-dimethylthiazole, 2-methylthiazole, and 2-methyl-4-phenylthiazole (476). None of these substrates was reduced to thiazoline: the second gave no reaction and the first underwent ring cleavage, leading to a mixture of n-propylmercaptan and ethylamine (Scheme 90). Three years later the same

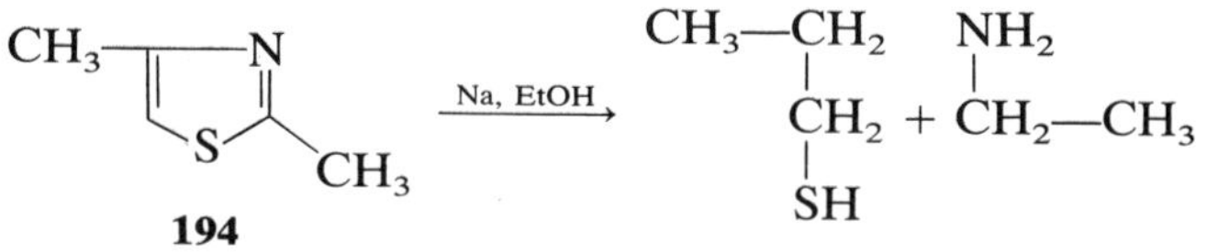

$$\text{194} \xrightarrow{\text{Na, EtOH}} \begin{array}{c} CH_3-CH_2 \\ | \\ CH_2 \\ | \\ SH \end{array} + \begin{array}{c} NH_2 \\ | \\ CH_2-CH_3 \end{array}$$

Scheme 90

reaction carried out in more drastic conditions by Schuftan (477) resulted in complete hydrogenation with desulphurization of the molecule and afforded i-propylethylamine (Scheme 91). This reaction was considered, in those early days of thiazole chemistry, as proof of the position of substituents on the ring.

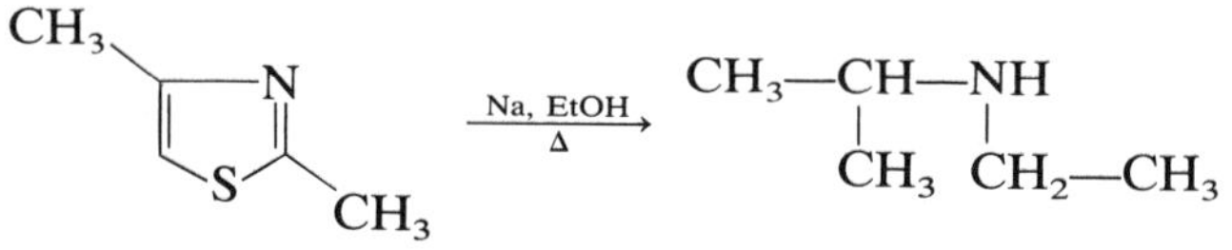

$$\text{194} \xrightarrow[\Delta]{\text{Na, EtOH}} \begin{array}{c} CH_3-CH-NH \\ | \quad\quad\; | \\ CH_3 \;\; CH_2-CH_3 \end{array}$$

Scheme 91

The only reduction investigated more recently on thiazole derivatives concerns the action of sodium borohydride upon thiazolium salts chosen as model molecules for thiamine (478–480).

The alkaline conditions of the reduction with aqueous sodium borohydride leads to competitive reactions of the OH$^-$ nucleophile, but the product usually obtained from a thiazolium salt (**195**) is the corresponding thiazolidine (**196**).

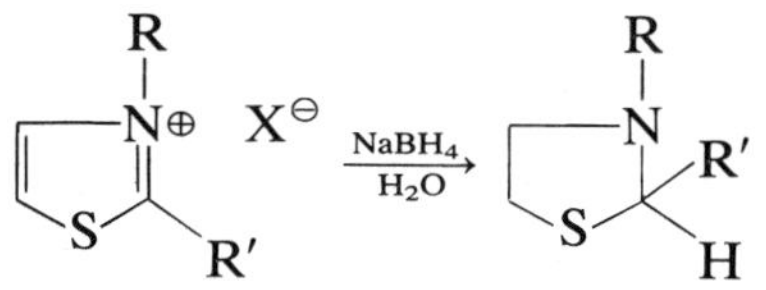

III. Chemical Reactivity

Scheme 92

195 → 196 (NaBH₄, H₂O)

The mechanism and the stereochemistry of the reaction was studied using borodeuteride and/or deuterium oxide (480) and a reaction pathway was suggested (Scheme 93).

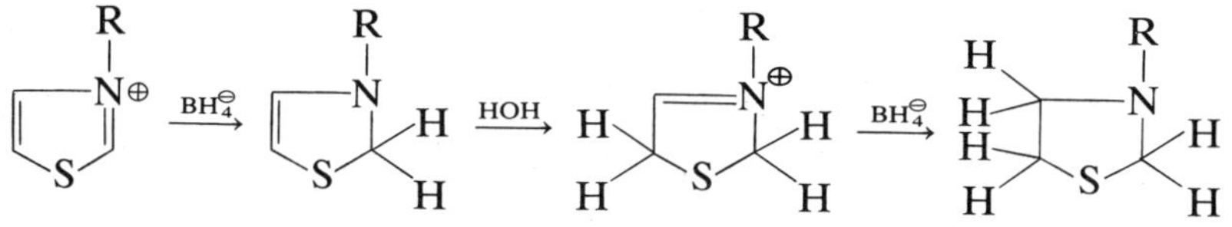

Scheme 93

197 → 198 → 199 → 200

The reduction of a number of thiazolium salts had been shown to yield diastereomeric mixtures of thiazolidines from which it has been possible, in some cases (including that of thiamine), to isolate one pure species (Schemes 94 and 94a).

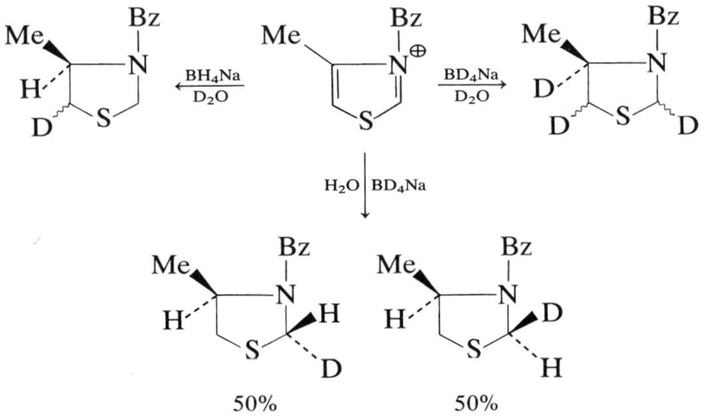

Scheme 94

50%   50%

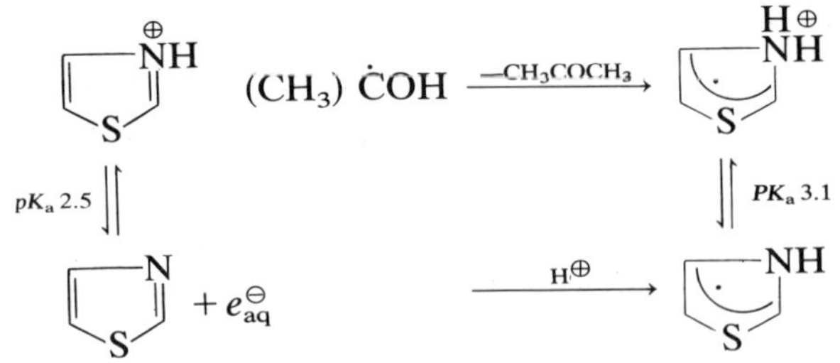

Scheme 94a

C. Ring Cleavage

Active Raney nickel induces desulfurization of many sulfur-containing heterocycles: thiazoles are fairly labile toward this ring cleavage agent. The reaction occurs apparently by two competing mechanisms (481): in the first, favored by alkaline conditions, ring fission occurs before desulfurization, whereas in the second, favored by the use of neutral catalyst, the initial desulfurization is followed by fission of a C–N bond and formation of carbonyl derivatives by hydrolysis (Scheme 95).

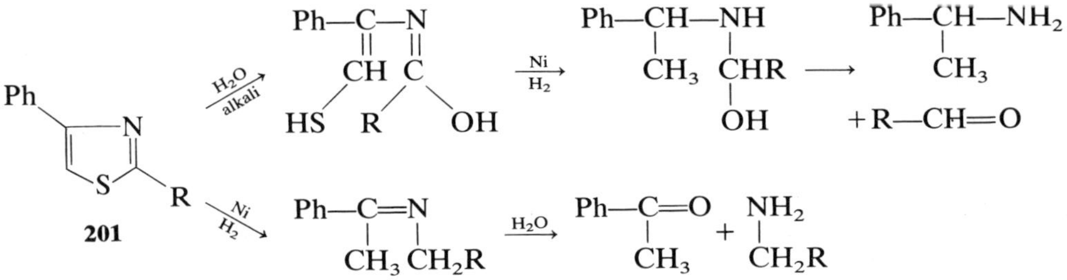

Scheme 95

Because of the easy and versatile synthesis of thiazoles (cf. Chapter II), this reaction could have interesting synthetic applications (481).

2-Alkyl (and aryl) amino-5-nitrothiazoles (482), which are prepared by ready aminolysis of the corresponding 2-bromothiazoles (**202**), give rise to a parallel reaction leading to the cleavage of the thiazole ring, especially under the action of sterically hindered strongly basic secondary amines (483). Ring opening is favored by highly polar solvents such as DMSO. A simple mechanism has been proposed (Scheme 96; also cf. Chapter V). 2,5-Diaminothiazoles (**203**) isomerize readily in alkaline medium to 2-thiolo-5-amino-imidazoles (**204**) (484). A concerted mechanism has been proposed to interpret this rearrangement (cf. Chapter VIII).

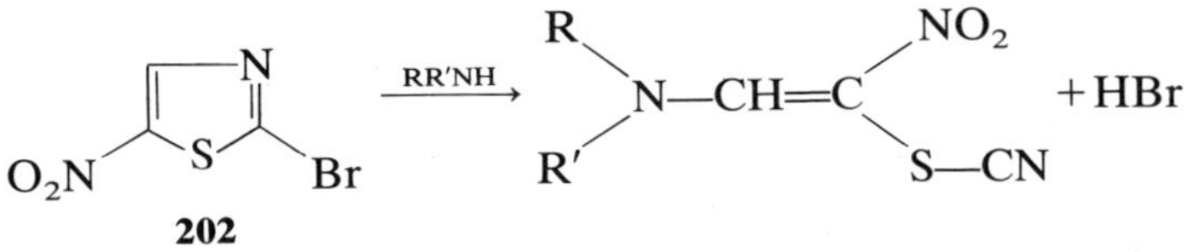

$$\underset{\mathbf{202}}{\text{O}_2\text{N}\overset{\text{N}}{\underset{\text{S}}{\bigsqcup}}\text{Br}} \xrightarrow{\text{RR'NH}} \underset{\text{R'}}{\overset{\text{R}}{\text{N}}}-\text{CH}=\underset{\text{S}-\text{CN}}{\overset{\text{NO}_2}{\text{C}}} + \text{HBr}$$

Scheme 96

$$\underset{\mathbf{203}}{\text{H}_2\text{N}\overset{\text{N}}{\underset{\text{S}}{\bigsqcup}}\text{NHR}} \xrightarrow{\text{HO}^\ominus} \text{H}_2\text{N}\overset{\text{N}}{\underset{\text{S}}{\bigsqcup}}\overset{\ominus}{\text{N}}-\text{R} \longrightarrow \underset{\mathbf{204}}{\text{H}_2\text{N}\overset{\text{N}}{\underset{\text{N}}{\bigsqcup}}\text{SH}}$$

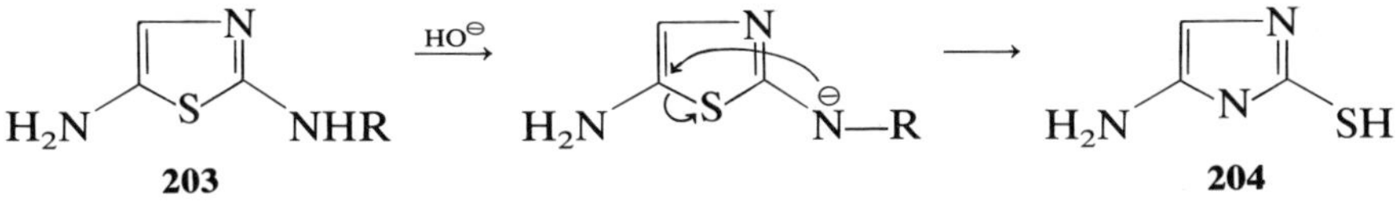

Scheme 97

The one-electron reduction of thiazole in aqueous solution has been studied by the technique of pulse radiolysis and kinetic absorption spectrophotometry (514). The acetone ketyl radical $(\text{CH}_3)_2\dot{\text{C}}\text{OH}$ and the solvated electron e^-_{aq} were used as one-electron reducing agents. The reaction rate constant of e^-_{aq} with thiazole determined at pH 8.0 is $k = 2.1 \times 10^9$ mole^{-1} sec^{-1}, in agreement with 2.5×10^9 mole^{-1} sec^{-1}, the value given by the National Bureau of Standards (513). It is considerably higher than that for thiophene (6.5×10^{-7} mole^{-1} sec^{-1}) (513) and pyrrole (6.0×10^5 mole^{-1} sec^{-1}) (513). The reaction rate constant of acetone ketyl radical with thiazolium ion determined at pH 0.8 is $k = 6.2 = 10^8$ mole^{-1} sec^{-1}. Relatively strong transient absorption spectra are observed from these one-electron reactions: they show λ_{max} (nm) and ε (mole^{-1} cm^{-1}) of 317 and 3.8×10^3 and 340 and 4.1×10^3 for thiazole (pH 5.4 to 13.0) and thiazolium ion (pH 0.8), respectively, with a pK_a of 3.1 for the thiazole radical; Scheme 94a is suggested.

In a study aimed at identifying thiazole derivatives, Travagli (485) obtained the ring cleavage of 2-aminothiazoles (**205**) by their phenylhydrazinolysis in acetic acid.

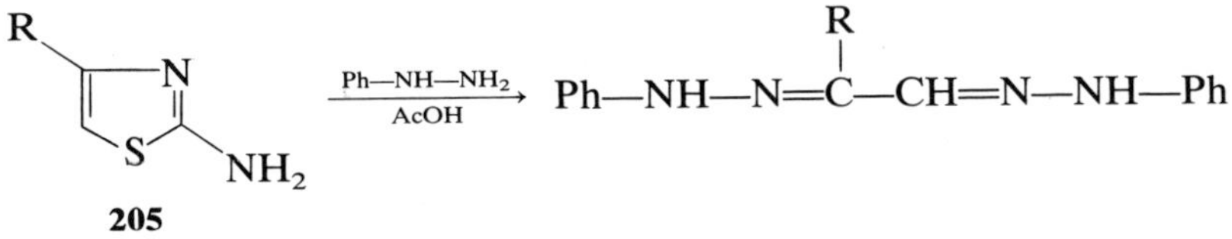

$$\underset{\mathbf{205}}{\overset{\text{R}}{\bigsqcup}\overset{\text{N}}{\underset{\text{S}}{\bigsqcup}}\text{NH}_2} \xrightarrow[\text{AcOH}]{\text{Ph}-\text{NH}-\text{NH}_2} \text{Ph}-\text{NH}-\text{N}=\overset{\overset{\text{R}}{|}}{\text{C}}-\text{CH}=\text{N}-\text{NH}-\text{Ph}$$

Scheme 98

A smoother ring opening of 2-aminothiazoles (and Δ-2-thiazolines) occurs with transition metal salts such as $(PtCl_4)^{2-}$ or $(Pd_2Cl_6)^{2-}$ in basic medium (Scheme 99) (486). Finally, alkaline ring cleavage of thiazolium salts is a classical reaction of this class of compounds and is discussed in Chapter IX. Thiazole ring opening resulting from photosensitized oxidation of thiazoles has been discussed in Section III.7.A.

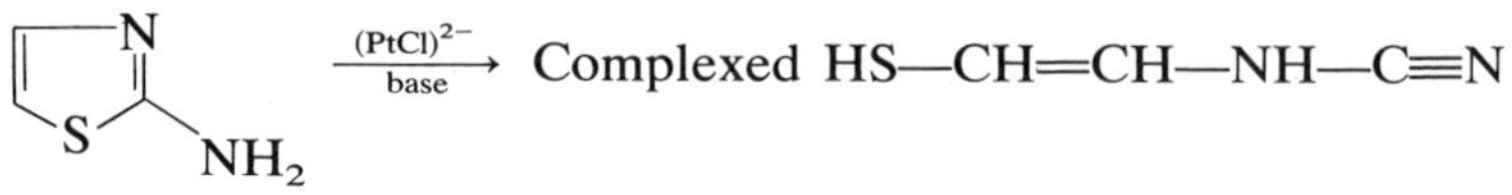

$$\text{thiazole-NH}_2 \xrightarrow[\text{base}]{(PtCl)^{2-}} \text{Complexed HS—CH=CH—NH—C≡N}$$

Scheme 99

D. Photochemically Induced Transformations

In 1969 Lablache-Combier et al. (110) found that in the presence of primary amines isothiazole photoisomerizes to thiazole (Scheme 100).

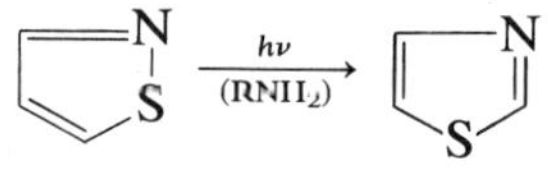

206 $\xrightarrow[(RNH_2)]{h\nu}$ **207**

Scheme 100

In 1970 Kojima and Maeda (487) observed the photoisomerization of 2,5-diphenlythiazole (**208**) to 3,4-diphenylisothiazole (**209**) and 4,5-diphenylthiazole (**210**), and of 2,4-diphenylthiazole (**211**) to 3,5-diphenylisothiazole (**212**) (Scheme 101). Shortly afterward, Ohashi and

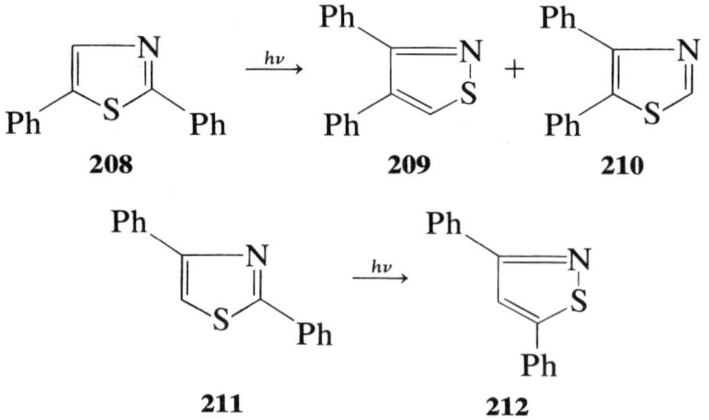

208 $\xrightarrow{h\nu}$ **209** + **210**

211 $\xrightarrow{h\nu}$ **212**

Scheme 101

III. Chemical Reactivity

Yonezawa described the photoisomerization of 3-phenylisothiazole (**213**) to 4-penylthiazole (**214**), of 5-penylisothiazole (**215**) to 3-phenylisothiazole (**213**), and of 3,5-diphenylisothiazole (**212**) to 2,4-diphenylthiazole (**211**) (Scheme 102) (488).

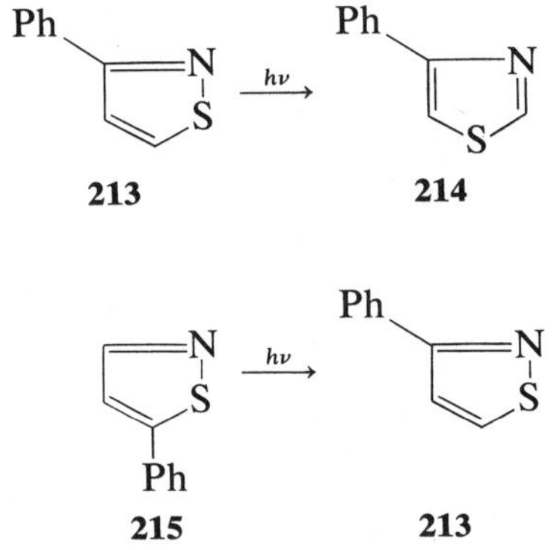

Scheme 102

Vernin et al. independently observed that during the irradiation in benzene of 2-iodothiazole to generate 2-thiazolyl radicals, the resulting 2-phenylthiazole (**216**) isomerized into a mixture of 4-phenylthiazole (**214**) and 3-phenylisothiazole (**213**) (Scheme 103) (489). Marking the 4- and

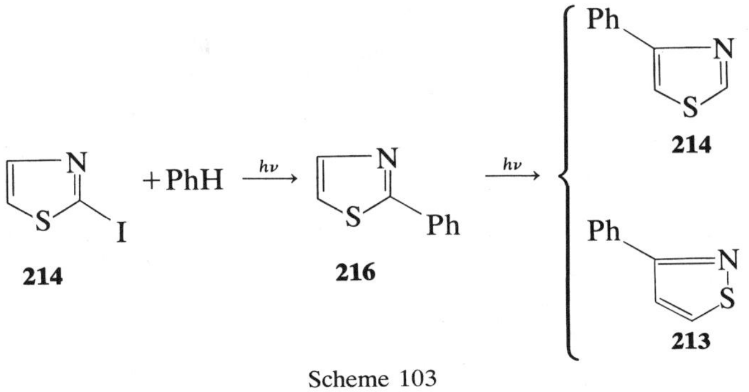

Scheme 103

5-positions of this 2-phenylthiazole, allowed Vernin to investigate the mechanism of this photorearrangement (490, 491). Two possible pathways were suggested (492); the first implies valence-bond isomerization (Scheme 104) (487), Which is in agreement with the evolution of the

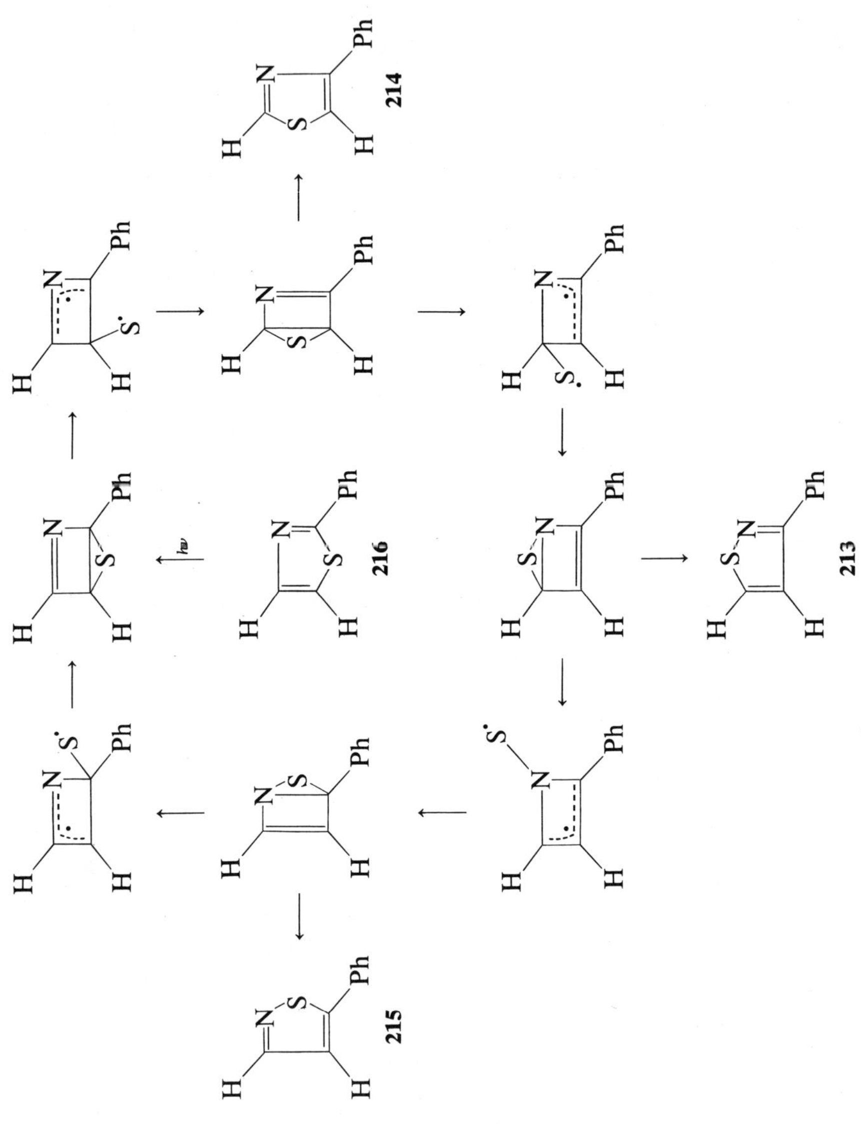

Scheme 104

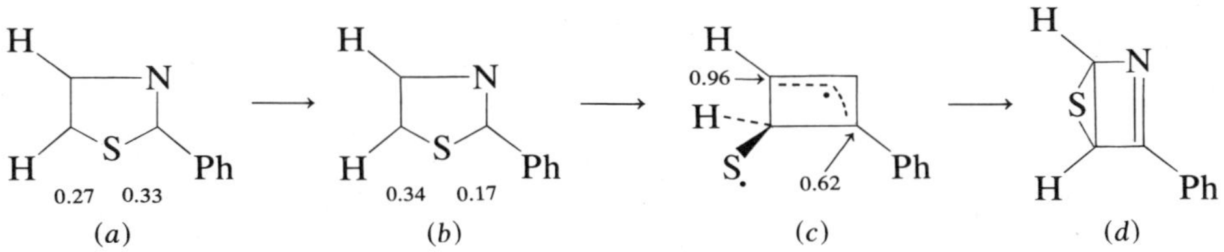

III. Chemical Reactivity

[Structures (a), (b), (c), (d) with bond order values]

(a) 0.27 0.33
(b) 0.34 0.17
(c) 0.96, 0.62
(d)

Fig. I-26. (a) π-Bond order of the C–S bonds in the ground state. (b) π-Bond order of the C–S bonds in the first excited state. (c) Free-valence number of the intermediate diradical. (d) Most probable bicyclic intermediate resulting from the ring closure of the diradical.

π-bond orders (calculated by the PPP method) between the ground state and the first excited state of the 2-phenylthiazole molecule and the free-valence numbers of the intermediate conjugate diradicals (Fig. I-26) (493). The second pathway corresponds to the mechanism proposed by Kellog (494), which implies 180° rotations around the bonds adjacent to the sulfur atom. Details concerning these photorearrangements are developed in Chapter III.

Certain *meso*-ionic derivatives of thiazole (**217**) rearrange into thiazolone (**219**) under irradiation, a thiirenium intermediate (**218**) being postulated (Scheme 105) (495).

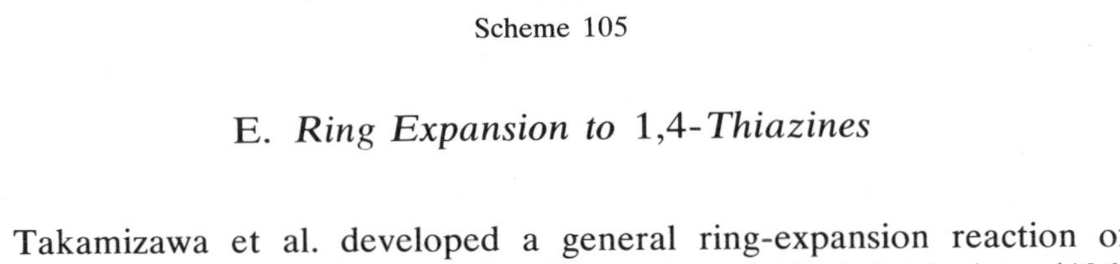

Scheme 105

E. Ring Expansion to 1,4-Thiazines

Takamizawa et al. developed a general ring-expansion reaction of heterocycles that, applied to thiazolium salts, yields 1,4-thiazines (496, 497): thiamine (**220**) reacts with dialkyl acylphosphonates (**221**) to give the tricyclic 1,4-thiazine (**222**) (498), which is easily hydrolyzed to dihydro-1,4-thiazinone (**223**) (499) (Scheme 106). In the case of thiazolium slats containing no functional groups (**224**), 1,4-thiazine derivatives (**226**) were directly obtained in fairly good yields (Scheme 107).

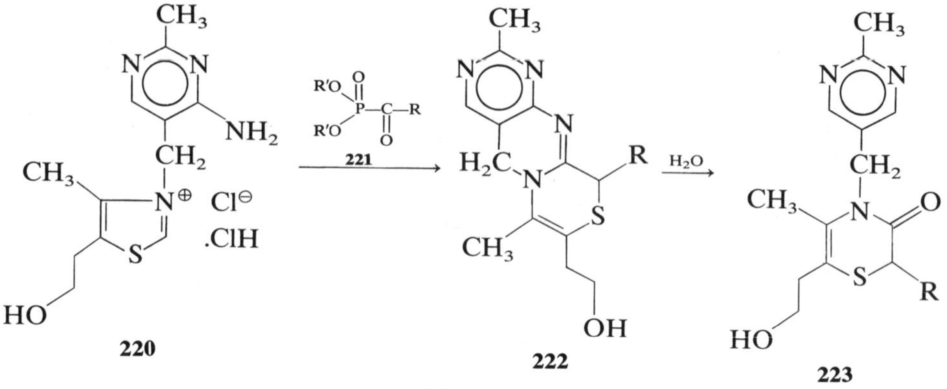

Scheme 106

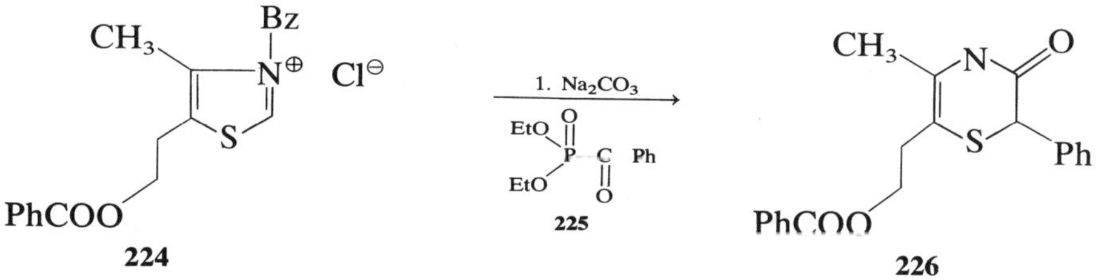

Scheme 107

The 1:1 adduct (**234**) of **227** with **221** was obtained in presence of triethylamine in DMF; it is easily converted to **226** by alkali. These facts reveal that **234** is an intermediate in the ring expansion of thiazolium to 1,4-thiazine. It seems probable that the reaction proceeds by way of Scheme 108. The thiazolium ylide (**228**) produced by treatment of **227** with triethylamine reacts with benzoylphosphonate (**229**) to give the betain (**230**). Nucleophilic attack on the pentavalent phosphorous then occurs to give a cyclic oxyphosphorane (**231**) that rearranges to **232**. Then by a ring-opening–ring-closure process (**233**) rearranges to the thiazine (**235**). Protonation (**232**) yields the intermediate (**234**). This mechanism agrees with the results of the spectroscopic study of the rate of rearrangement of the intermediate (**234**) to the final thiazine (**235**). The same reaction has been applied to yet simpler thiazolium salts (500), viz. 3,4-dimethylthiazolium (**236**) bromide with entirely similar results (Scheme 109). More recently, this interesting ring expansion has been extended with success to analogs of thiamine (501).

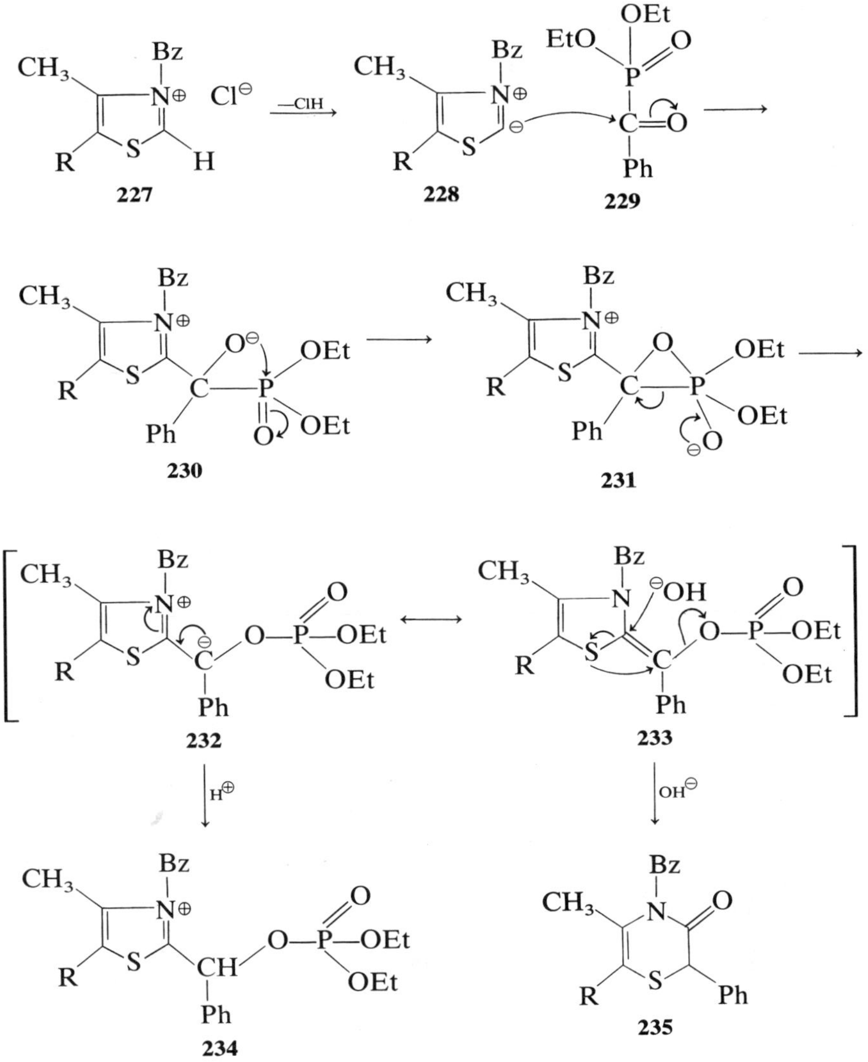

III. Chemical Reactivity

Scheme 108

Other ring-expansion reactions have already been mentioned in regard to addition reactions leading to pyrrolothiazoles (Section III.3), which sometimes rearrange to 1,4-thiazines (333, 497).

Another interesting case is afforded by 2-alkyl-N-phenacyl or N-acetonylthiazolium salt (**239**), which in basic medium gives an intramolecular cyclization product. According to Reid et al. (502), this could

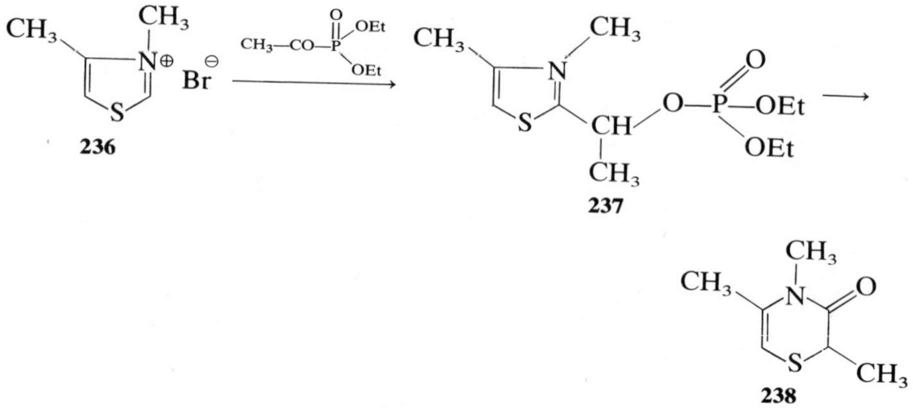

Scheme 109

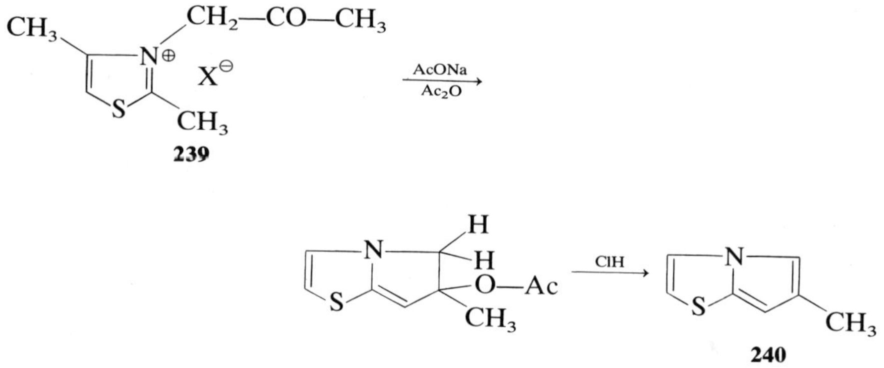

Scheme 110

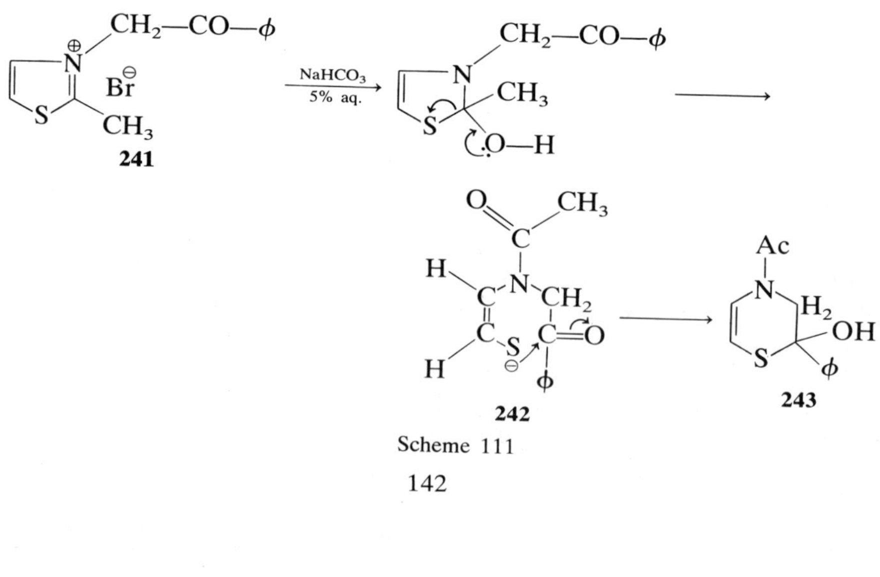

Scheme 111

be a pyrrolo[2,1-*b*]thiazole (**240**) (Scheme 110), but in the opinion of Adam and Wharmby (503) the thiazolium salt (**241**) undergoes ring opening under the action of hydroxyde ions followed by a ring closure of the intermediate vinyl sulfide (**242**) to give a derivative of 4-acyl-2,3-dihydro-2-hydroxy-2-phenyl-4-H-1,4-thiazine (**243**) (Scheme 111).

IV. EFFECTS OF THE THIAZOLE RING ON SUBSTITUENTS

Although these effects are considered in connection with each type of substituent in following chapters, it is of interest to discuss some typical and more quantitative aspects of the influence that the ring exerts on some particular substituents.

The interaction between a substituent and the ring carbon to which it is bonded could be related to some electronic characteristics of the unsubstituted ring and especially to the net charge of its various sites. In that respect the π-net charges diagram discussed in Section I.5 indicates that the electron-withdrawing power of the ring-carbon atoms will decrease in the order, $2>4>5$.

1. Alkylthiazoles

Mills and Smith (504) were the first, in 1922, to develop a systematic study of the reactivity of methyl groups fixed on nitrogen-containing heterocycles. While in alkylpyridines the 2- (or 6) and 4-positions are activated, only the 2-position in thiazole corresponds to an enhanced reactivity of the methyl groups in condensation with aldehydes: 4- and 5-methylthiazoles bear inert methyl groups. Quaternization of the thiazole nitrogen enhances still further the reactivity of the methyl in the 2-position (cf. Chapter IX), but it does not increase the reactivity of a methyl group in the 4-position (504). The authors invoke the possibility for 2- (and 6) methylpyridine and 2-methylthiazole to pass, to some extent, into the reactive enamine form (**245**), while 4-methylthiazole could adopt such a structure only with the participation of an unusual formula such as **247** (Scheme 112).

In 1937, Kondo and Nagasawa confirmed the reactivity of the sole 2-methyl group in the condensation of 2,4-dimethylthiazole with benzaldehyde (505) and in the cyclization to thiazolopyrrole in the reaction with phenacyl bromide (506) (Scheme 113).

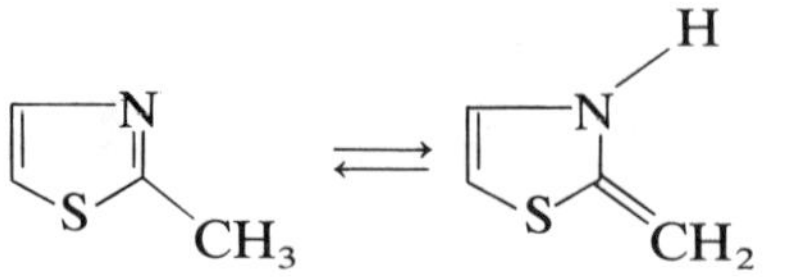

Scheme 112

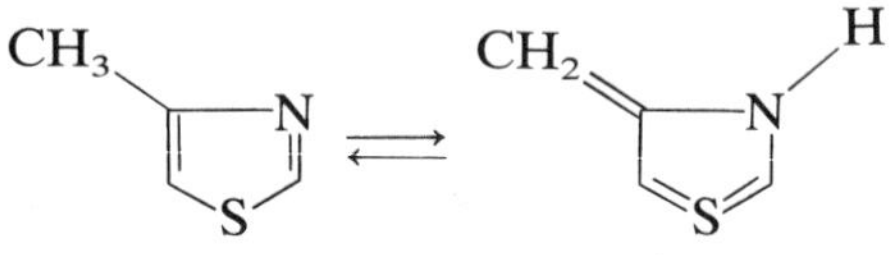

This typical behavior of the very unsymmetrical thiazole ring led to a series of studies from the group of H. Erlenmeyer in Basle: he studied the H/D exchange of 2,4-dimethyl-5-carboxythiazole as well as that of similar methylated nitrogen heterocycles (507). The results are shown in Fig. I-27.

It appears that not only the 2- but also the 4-methyl group of this compound (**252**) exchange with D_2O. The explanation proposed by Erlenmeyer was based on the theory of resonance and concluded that exchange at the 4-position was possible because of the existence of tautomeric molecules such as **247** (507, 508). Pullman and Metzger discussed these conclusions in terms of HMO theory and suggested that the 5-carboxylic group contributes to the lability of the 4-methyl protons (6). In fact, in a pH range of 4 to 11 no exchange takes place at either methyl group of noncarboxylated 2,4-dimethylthiazole (509).

Erlenmeyer et al. also examined the behavior of 4-methyl-, 4,5-dimethyl-, and 5-methylthiazole in the reaction with benzaldehyde at 160°C in the presence of $ZnCl_2$ (439, 510): they were able to show that neither the 4- nor 5-methyl groups reacted under these conditions and that the condensation occurred at the sole unsubstituted 2-position (Scheme 114).

In conclusion, it appears that in neutral or weakly acidic conditions only the methyl in the 2-position shows pseudoacidic behavior. The same conclusion can be drawn from the base-induced hydrogen-metal exchange reactions discussed in Section III.5.B.

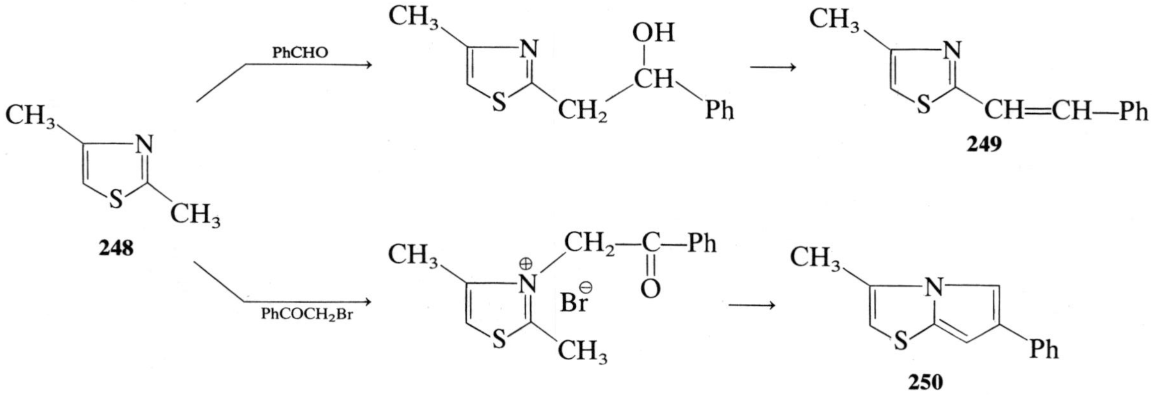

Scheme 113

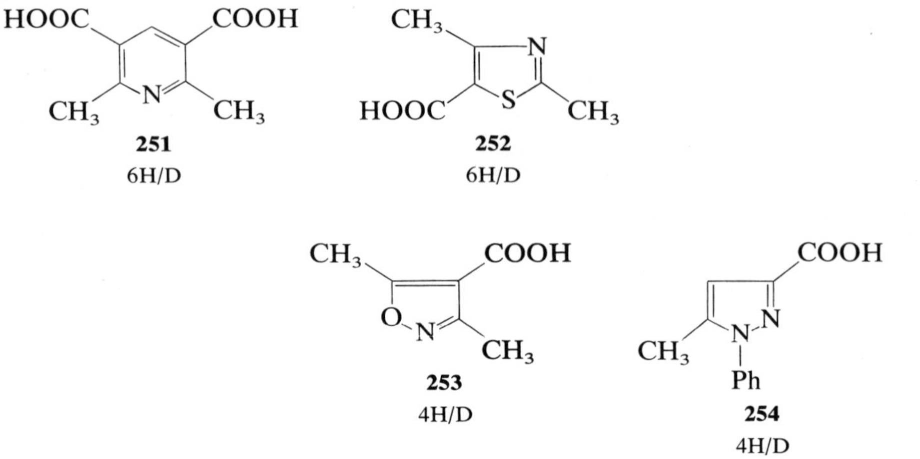

Fig. I-27. Number of hydrogen atoms exchanged against deuterium by dissolution of the sodium salt of the acid in D_2O (507).

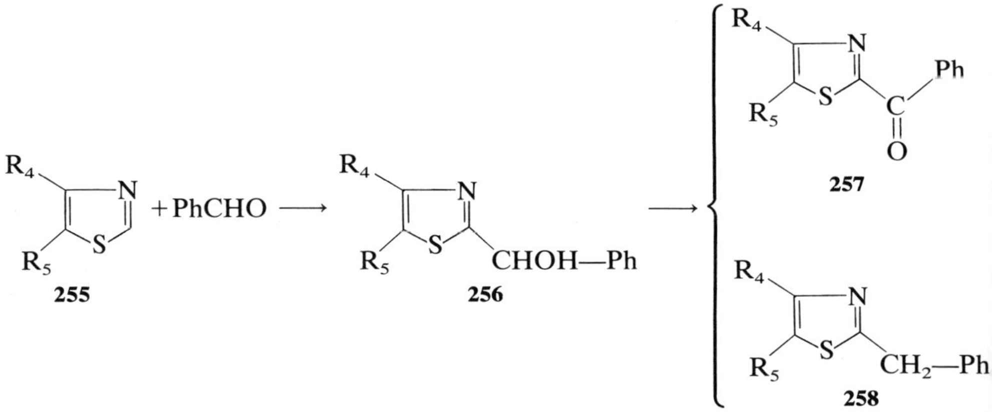

$R_4, R_5 =$ Me, H; Me, Me; H, Me

Scheme 114

2. α-Chloroethylthiazoles

A more quantitative approach to the influence of the thiazole ring on the reactivity of a lateral functional chain was made in a recent study by Noyce and Fike (383), already discussed in Section III.4. The first-order rates of solvolysis for three isomeric 1-thiazolylethyl chlorides were determined in 80% ethanol. The order of relative reactivity observed,

5-thiazolyl (174) >4-thiazolyl (25) >2-thiazolyl (1), reflects well the π net charges calculated by the PPP method (133) for the three positions of the thiazole molecule [5(−0.086)/4(+0.029)/2(+0.100)]. The authors noted that the same calculated charges did not agree with the relative reactivity induced by a phenyl group: such a discrepancy between two different types of substrate is not unexpected because of the crude approximation of the method, but the interesting point is that the right experimental order is reproduced within the same thiazole molecule.

3. Halothiazoles

The activation of halothiazoles toward nucleophilic displacement is discussed in Chapter V: no unique conclusion can be drawn because of the various possible interactions between the halothiazole base and the electrophilic counterpart of the nucleophile.

4. Hydroxy-, Mercapto-, and Aminothiazoles

The tautomerism of these compounds is discussed in Chapters VI and VII.

V. LINEAR FREE-ENERGY RELATIONSHIPS

Only few attempts have been made to apply free-energy relations to thiazole reactivity.

The first was proposed by Imoto and Otsuji (511) and Otsuji et al. (512) and concerned the pK_a of substituted 2-, 4-, and 5-carboxylic acids and the alkaline hydrolysis rate k of their respective ethyl esters (**259, 260,** and **261,** where Y = Et). When Hammett σ_m values were used for

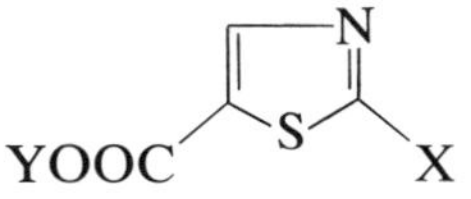

259

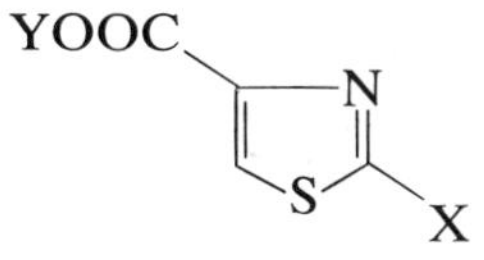

260 Y = H, X = H, Me, Cl, or Br
Y = Et, X = H, Me, Cl, Br, or NH$_2$

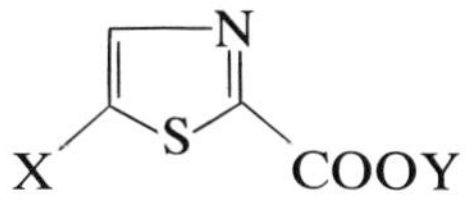

261 Y = H, or Et   X = H or Me

Scheme 115

2,4- and σ_p values for 2,5-disubstituted thiazoles, a linear relation existed between log k and σ (except for the tautomerisable amino compounds) with ρ values of 1.856 for 2-substituted-5-carboxylic acid ethyl esters (1, Y = Et), 1.618 for 5-substituted-2-carboxylic acid ethyl esters (3, Y = Et), and 1.551 for 2-substituted-4-carboxylic acid ethyl esters (2, Y = Et). A linear relationship between pK_a and σ was found, with ρ values of 0.83, 2.35, and 1.34, respectively, for the corresponding acids (**259, 260,** and **261,** where Y = H).

In the second, which belongs to a systematic study of the transmission of substituent effects in heterocyclic systems, Noyce and Forsyth (384–386) showed that for thiazole, as for other simple heterocyclic systems, the rate of solvolysis of substituted hetero-arylethyl chlorides in 80% ethanol could be correlated with σ_p^+ constants of the substituent X only when there is mutual conjugation between X and the reaction center. In the case of thiazole this situation corresponds to 1-(2-X-5-thiazolyl)ethyl chlorides (**262**) and 1-(5-X-2-thiazolyl)ethyl chlorides (**263**).

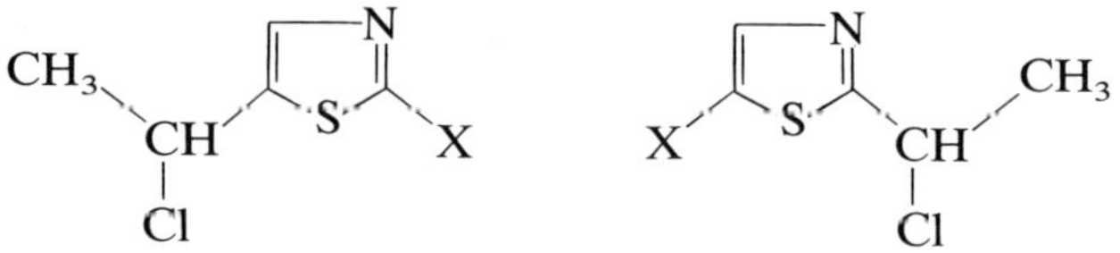

262
X = H, Cl, Me, SMe, or OMe

263
X = H, Me, or SMe

Scheme 116

For solvolysis carried out at 25°C, the correlation gives the following values for ρ: −6.15 for series **262** and −6.68 for series **263**. The ρ values obtained for five other heterocycles and for benzene appear to be related

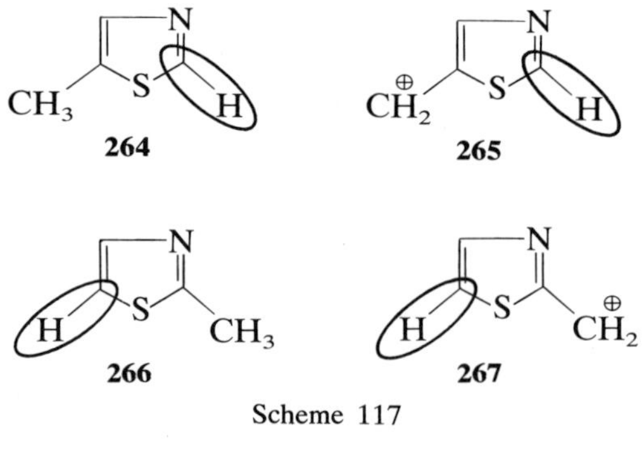

264   **265**

266   **267**

Scheme 117

directly to the amount of transition-state charge Δq delocalized on the site of the substituent attachment. This quantity Δq, calculated in the CNDO/2 approximation, is the difference in regional charge (i.e., the sum of the charges on a carbon atom and on any hydrogen atoms bonded to it) at the substituent site between the unsubstituted neutral parent methylthiazole (**264** and **266**) and the thiazolylmethylene cation (**265** and **267**) taken as models for initial and transition states, respectively.

The relation between ρ values and Δq values is determined by the extent of substituent interaction with the charge developed in the transition state and thus is related to the pattern of charge distribution within the ring itself (386).

VI. REFERENCES

1. A. Hantzsch and H. J. Weber, *Berichte*, **20**, 3118 (1887).
2. A. Hantzsch, *Justus Liebigs Ann. Chem.*, **249**, 1 (1888).
3. T. B. Johnson and E. Gatewood, *J. Amer. Chem. Soc.*, **51**, 1815 (1929); *Chem. Abstr.*, **23**, 3470.
4. J. Metzger, Thesis, University of Nancy, France, 1 (1948).
5. A. Pullman and J. Metzger, *Bull. Soc. Chim. France*, 1021 (1948); *Chem. Abstr.*, **43**, 2511.
6. J. Metzger and A. Pullman, *Bull. Soc. Chim. France*, 1166 (1948).
7. J. Metzger and A. Pullman, *Compt. Rend.*, **226**, 1613 (1948); *Chem. Abstr.*, **42**, 6657.
8. B. Pullman and A. Pullman, "Les Theories Electroniques de la Chimie Organique," Masson, Ed., Paris, France, 1952, p. 623.
9. J. Metzger and B. Koether, *Ann. Univ. Saraviensis*, **1**, 23 (1952); *Chem. Abstr.*, **47**, 10524.
10. J. Metzger and B. Koether, *Ann. Univ. Saraviensis*, **1**, 151 (1952); *Chem. Abstr.*, **47**, 10524.
11. J. Metzger and B. Koether, *Bull. Soc. Chim. France*, **20**, 702 (1953); *Chem. Abstr.*, **48**, 10738.
12. J. Metzger and B. Koether, *Bull. Soc. Chim. France*, **20**, 708 (1953); *Chem. Abstr.*, **48**, 10738.
13. J. Metzger and J. Beraud, *Compt. Rend.*, **242**, 2362 (1956); *Chem. Abstr.*, **51**, 2741.
14. J. Metzger and H. Kuhnen, *Z. Naturforsch.*, **12B**, 27 (1957); *Chem. Abstr.*, **51**, 14686.
15. V. Grignard and R. Rambaud, "Traite de Chimie Organique," Masson, Ed., Paris, France, 1936, Vol. 18, p. 1.
16. A. M. Patterson, L. T. Capell, and D. F. Walker, "The Ring Index," *American Chemical Society*, 2nd., 1960, p. 1.
17. A. W. Hofmann, *Berichte*, **20**, 2262 (1887).
18. J. Tcherniac and C. Norton, *Berichte*, **16**, 345 (1883).
19. J. Tcherniac and R. Hellon, *Berichte*, **16**, 348 (1883).
20. C. Liebermann and A. Lange, *Berichte*, **12**, 1588 (1879).
21. P. Claesson, *Berichte*, **10**, 1346 (1877).
22. M. Nencki, *J. Prakt. Chem.*, **(2)16**, 1 (1877).
23. A. Hantzsch and H. J. Weber, *Berichte*, **20**, 3336 (1887).

24. R. Maly, *Annal. Chem. Pharm.*, **168,** 133 (1873).
25. L. Arapides, *Justus Liebigs Ann. Chem.*, **249,** 7 (1888).
26. R. Dyckerhoff, *Berichte*, **10,** 119 (1877).
27. J. Tcherniac, *Berichte*, **25,** 2607 (1892).
28. J. Tcherniac, *Berichte*, **25,** 2621 (1892).
29. J. Tcherniac, *Berichte*, **25,** 2627 (1892).
30. J. Tcherniac, *Berichte*, **25,** 2629 (1892).
31. A. Hantzsch, *Berichte*, **25,** 3282 (1892).
32. J. Tcherniac, *Berichte*, **25,** 3648 (1892).
33. J. Tcherniac, *J. Chem. Soc.*, **115,** 1071 (1919); *Chem. Abstr.*, **14,** 276.
34. A. Hantzsch, *Berichte*, **60B,** 2537 (1927); *Chem. Abstr.*, **22,** 1158.
35. J. Tcherniac, *Berichte*, **61,** 574 (1928).
36. A. Hantzsch and H. Schwaneberg, *Berichte*, **61B,** 1776 (1928); *Chem. Abstr.*, **23,** 101.
37. W. Heintz, *Annal. Chem. Pharm.*, **136,** 223 (1865).
38. J. Volhard, *Annal. Chem. Pharm.*, **166,** 384 (1873).
39. J. Volhard, *J. Prakt. Chem.*, **(2)9,** 6 (1874).
40. C. Liebermann and M. Völtzkow, *Berichte*, **13,** 276 (1880).
41. M. Völtzkow, *Berichte*, **13,** 1579 (1880).
42. M. Nencki, *Berichte*, **6,** 598 (1873).
43. E. Mulder, *Berichte*, **8,** 1261 (1875).
44. A. Claus, *Berichte*, **9,** 223 (1876).
45. R. Maly, *Berichte*, **10,** 1849 (1877).
46. P. J. Meyer, *Berichte*, **10,** 1965 (1877).
47. A. Lange, *Berichte*, **12,** 595 (1879).
48. R. Andreasch, *Berichte*, **12,** 1385 (1879).
49. O. Wallach, *Berichte*, **11,** 1590 (1878).
50. O. Wallach and H. Bleibtreu, *Berichte*, **12,** 1061 (1879).
51. A. Claus, *Berichte*, **8,** 41 (1875).
52. R. Andreasch, *Berichte*, **13,** 1421 (1880).
53. C. Liebermann and A. Lange, *Justus Liebigs Ann. Chem.*, **207,** 121 (1881).
54. R. Andreasch, *Berichte*, **15,** 324 (1882).
55. R. Andreasch, *Monatsh.*, **2,** 775 (1882).
56. R. Andreasch, *Monatsh.*, **6,** 821 (1885).
57. R. Andreasch, *Monatsh.*, **8,** 407 (1887).
58. F. Evers, *Berichte*, **21,** 962 (1888).
59. A. E. Dixon, *J. Chem. Soc.*, **63,** 815 (1893).
60. R. Trambach, *Justus Liebigs Ann. Chem.*, **280,** 233 (1894).
61. R. Andreasch, *Monatsh.*, **16,** 789 (1895).
62. R. Andreasch, *Monatsh.*, **18,** 56 (1897).
63. H. L. Wheeler and T. B. Johnson, *J. Amer. Chem. Soc.*, **24,** 680 (1902).
64. C. Liebermann and A. Lange, *Justus Liebigs Ann. Chem.*, **207,** 123 (1881).
65. A. Neubert, *Berichte*, **19,** 1822 (1886).
66. H. Goldschmidt and A. Gessner, *Berichte*, **22,** 928 (1889).
67. A. E. Dixon, *J. Chem. Soc.*, **71,** 617 (1897).
68. H. L. Wheeler and T. B. Johnson, *Amer. Chem. J.*, **28,** 121 (1902).
69. A. E. Dixon and J. Taylor, *J. Chem. Soc.*, **101,** 558 (1912).
70. M. Nencki, *Berichte*, **17,** 2277 (1884).
71. J. Ginsburg, and S. Bondzynski, *Berichte*, **19,** 113 (1886).
72. J. Berlinerblau, *Berichte*, **19,** 124 (1886).
73. R. Andreasch, *Monatsh.*, **10,** 73 (1889).

VI. References

74. J. Freydl, *Sitz. Ber. Akad. Wiss. Wien, Math. Naturw. Klasse Abt.* IIB, **98,** 56 (1889).
75. A. Miolati, *Justus Liebigs Ann. Chem.,* **262,** 82 (1891).
76. J. Von Braun, *Berichte,* **35,** 3368 (1902).
77. H. Korner, *Berichte,* **41,** 1901 (1908).
78. R. Andreasch and A. Zipser, *Monatsh.,* **24,** 499 (1903).
79. R. Andreasch and A. Zipser, *Monatsh.,* **25,** 159 (1903).
80. R. Andreasch and A. Zipser, *Monatsh.,* **26,** 1191 (1905).
81. R. Andreasch, *Monatsh.,* **27,** 1211 (1906).
82. A. Wagner, *Monatsh.,* **27,** 1233 (1906).
83. L. Kaluza, *Monatsh.,* **30,** 701 (1909).
84. O. Antulich, *Monatsh.,* **31,** 891 (1910).
85. R. Andreasch, *Monatsh.,* **31,** 785 (1910).
86. H. Nagele, *Monatsh.,* **33,** 941 (1912).
87. R. Andreasch, *Monatsh.,* **38,** 121 (1917); *Chem Abstr.* **12,** 276.
88. B. Holmberg, *J. Prakt. Chem.,* **(2)81,** 451 (1910).
89. B. Holmberg, and B. Psilanderhielm, *J. Prakt. Chem.,* **(2)82,** 440 (1910).
90. W. Will, *Berichte,* **15,** 338 (1882).
91. F. Foerster, *Berichte,* **21,** 1857 (1888).
92. S. Gabriel and W. E. Lauer, *Berichte,* **23,** 87 (1890).
93. P. Hirsch, *Berichte,* **23,** 964 (1890).
94. G. Marchesini, *Gazz. Chim. Ital.,* **22,** 350 (1892).
95. A. Miolati, *Gazz. Chim. Ital.,* **23,** 575 (1893).
96. S. Gabriel and C. Freiherr Von Hirsch, *Berichte,* **29,** 2747 (1896).
97. W. Will, *Berichte,* **14,** 1485 (1881).
98. M. Nencki, and N. Sieber, *J. Prakt. Chem.,* **25,** 72 (1882).
99. M. Steude, *Justus Liebigs Ann. Chem.,* **261,** 22 (1891).
100. W. Will and O. Bielschowski, *Berichte,* **15,** 1309 (1882).
101. R. Andreasch, *Monatsh.,* **4,** 131 (1883).
102. A. Hantzsch and V. Traumann, *Berichte,* **21,** 938 (1888).
103. V. Traumann, *Justus Liebigs Ann. Chem.,* **249,** 31 (1888).
104. G. Popp, *Justus Liebigs Ann. Chem.,* **250,** 273 (1889).
105. A. Hantzsch and L. Arapides, *Berichte,* **21,** 941 (1888).
106. M. Bachstez, *Berichte,* **47,** 3163 (1914).
107. S. Gabriel, *Berichte,* **43,** 134 (1910).
108. S. Gabriel, *Berichte,* **43,** 1283 (1910).
109. H. Schenkel and M. Schenkel, *Helv. Chim. Acta,* **31,** 924 (1948); *Chem. Abstr.,* **42,** 5906.
110. J. P. Catteau, A. Lablache-Combier, and A. Pollet, *J. Chem. Soc. D,* 1018 (1969); *Chem. Abstr.,* **71,** 112850.
111. P. Roussel and J. Metzger, *Bull. Soc. Chim. France,* 2075 (1962); *Chem. Abstr.,* **58,** 13932.
112. J. A. Braun and J. Metzger, *Bull. Soc. Chim. France,* 503 (1967); *Chem. Abstr.,* **67,** 11446.
113. I. N. Bojesen, J. H. Hoeg, J. T. Nielsen, I. B. Petersen, and K. Schaumburg, *Acta Chem. Scand.,* **25,** 2739 (1971); *Chem. Abstr.,* **76,** 24291.
114. M. Meyer, R. Meyer, R. Brun, J. Salvinien, and J. Metzger, *Bull. Soc. Chim. France,* **8,** 3132 (1968).
115. R. F. Hunter, *Proc. Muslim Assoc. Advan. Sci.,* **1,** 1 (1931); *Chem. Abstr.,* **26,** 2978.
116. K. A. Jensen and A. Friediger, *Kgl. Danske Videnskab. Selskab Math. Fys. Medd.,* **20,** 1 (1943); *Chem. Abstr.,* **39,** 2068.

117. R. Zahradnik and J. Koutecki, *Coll. Czech. Chem. Comm.*, **26,** 156 (1961).
118. E. Vincent and J. Metzger, *Bull. Soc. Chim. France*, 2039 (1962); *Chem. Abstr.*, **58,** 9642.
119. J. Vitry and J. Metzger, *Bull. Soc. Chim. France*, 1784 (1963).
120. J. M. Bonnier and M. Gelus, *Rev. Inst. Franc. Petrole*, **22,** 1008 (1967); *chem. Abstr.*, **68,** 2510.
121. J. M. Bonnier, R. Arnaud, and M. Maurey-Mey, *Compt. Rend.*, **C-267,** 10 (1968); *Chem. Abstr.*, **69,** 64067J.
122. E. J. Vincent, R. Phan Tan Luu, and J. Metzger, *Bull. Soc. Chim. France*, **11,** 3530 (1966); *Chem. Abstr.*, **66,** 64913.
123. R. Phan Tan Luu, L. Bouscasse, E. J. Vincent, and J. Metzger, *Bull. Soc. Chim. France*, **9,** 2383 (1967); *Chem. Abstr.*, **68,** 44384.
124. L. Bouscasse, E. J. Vincent, and J. Metzger, *Bull. Soc. Chim. France*, **4,** 1182 (1967); *Chem. Abstr.*, **67,** 85138.
125. C. B. Chowdhury and R. Basu, *J. Indian Chem. Soc.*, **66,** 779 (1969); *Chem. Abstr.*, **71,** 128907.
126. J. Devanneaux and J. F. Labarre, *J. Chim. Phys. Physicochim. Biol.*, **66,** 1780 (1969); *Chem. Abstr.*, **72,** 89603.
127. Y. Zenishi and K. Tsunetoshi, *Theor. Chim. Acta*, **20,** 216 (1971); *Chem. Abstr.*, **75,** 4902.
128. M. Witanowski, L. Stefaniak, H. Januszewski, and Z. Grabowski, *Tetrahedron*, **28,** 637 (1972); *Chem. Abstr.*, **76,** 79007H.
129. E. Corradi, P. Lazzeretti, and F. Taddei, *Mol. Phys.*, **26,** 41 (1973); *Chem. Abstr.*, **79,** 91395.
130. L. Bouscasse, Thesis, University of Marseille, France (1970).
131. F. Bernardi, L. Forlani, P. F. Todesco, F. P. Colonna, and G. Distefano, *J. Electron. Spectrosc.*, **9,** 217 (1976).
132. M. Gelus, P. M. Vay, and G. Berthier, *Theor. Chim. Acta*, **9,** 182, (1967); *Chem. Abstr.*, **68,** 33422.
133. R. Phan Tan Luu, L Bouscasse, E. J. Vincent, and J. Metzger, *Bull. Soc. Chim. France*, **4,** 1149 (1969); *Chem. Abstr.*, **72,** 54567.
134. G. Salmona, Y. Ferre, and E. J. Vincent, *J. Chim. Phys.*, **69,** 1292 (1972).
135. M. Wolfsberg and L. Helmholtz, *J. Chem. Phys.*, **20,** 1837 (1952).
136. G. Del Re, *J. Chem. Soc. B*, 4031 (1958).
137. R. Pariser, *J. Chem. Phys.*, **21,** 568 (1953).
138. R. Pariser and R. G. Parr, *J. Chem. Phys.*, **21,** 466 (1953).
139. R. Pariser and R. G. Parr, *J. Chem. Phys.*, **21,** 767 (1953).
140. H. Kon, *Bull. Chem. Soc. Japan*, **28,** 275 (1955).
141. F. Gallais, D. Voigt, and J. F. LaBarre, *J. Chim. Phys.*, **62,** 761 (1965).
142. K. Nishimoto and L. S. Forster, *Theor. Exp. Chem.*, **4,** 155 (1960).
143. M. Witanowski, L. Stefaniak, H. Januszewski, and G. A. Webb, *Tetrahedron*, **27,** 3129 (1971).
144. K. Nishimoto and N. Mataga, *Z. Phys. Chem.*, **12,** 335 (1957).
145. I. B. Johns, E. A. McElhill, and J. D. Smith, *J. Chem. Eng. Data*, **7,** 277 (1962); *Chem. Abstr.*, **57,** 10981A.
146. J. M. Bonnier and R. Arnaud, *Compt. Rend.*, **C-270,** 885 (1970); *Chem. Abstr.*, **72,** 126203.
147. M. Gelus, Thesis, University of Grenoble, France (1967).
148. J. C. Schug, *Mol. Phys.*, **19,** 121 (1970).
149. P. Lazzeretti and F. Taddei, *Mol. Phys.*, **22,** 941 (1971); *Chem. Abstr.*, **76,** 117691.
150. R. McWeeny, *Phys. Rev.*, **126,** 1028 (1962).

VI. References

151. M. J. S. Dewar and A. J. Haget, *Proc. Roy. Soc.*, **A315,** 443 (1970).
152. M. J. S. Dewar, and A. J. Haget, *Proc. Roy. Soc.*, **A315,** 457 (1970).
153. M. J. S. Dewar and N. Trinajstic, *J. Amer. Chem. Soc.*, **92,** 1453 (1970).
154. N. Bodor, M. Farkas, and N. Trinajstic, *Croat. Chem. Acta*, **43,** 107 (1971); *Chem. Abstr.*, **75,** 98074.
155. L. E. Sutton, "Tables of Interatomic Distances and Configurations of Molecules," The Chemical Society, London, 1958.
156. R. Phan Tan Luu, Thesis, University of Marseille, France (1967).
157. F. Eiden and A. Engelhardt, *Arch. Pharm.*, **300,** 211 (1967); *Chem. Abstr.*, **66,** 104939.
158. B. Bak, D. Christensen, L. Hansen-Nygaard, and J. Rastrup-Andersen, *J. Mol. Spectrosc.* **9,** 222 (1962); *Chem. Abstr.*, **58,** 2029.
159. L. Nygaard, E. Asmussen, J. H. Hoeg, R. C. Maheshwari, C. H. Nielsen, I. B. Petersen, J. Rastrup-Andersen, and C. O. Soerensen, *J. Mol. Structure*, **8,** 225 (1971); *Chem. Abstr.*, **75,** 42785.
160. K. Nishimoto and L. S. Forster, *Theor. Exp. Chem.*, **4,** 155 (1966).
161. G. Streitwieser, "Molecular Orbital Theory for Organic Chemists," Wiley, New York, 1961.
162. K. Inusuka *Bull. Chem. Soc. Japan*, **36,** 1045 (1963).
163. T. Morita, *Bull. Chem. Soc. Japan*, **33,** 1496 (1960).
164. R. S. Mulliken, C. A. Riecke, and R. G. Brown, *J. Amer. Chem. Soc.*, **63,** 41 (1941).
165. C. A. Coulson and V. A. Crawford, *J. Chem. Soc. B*, 2052 (1952).
166. N. Muller, L. Picket, and R. S. Mulliken, *J. Amer. Chem. Soc.*, **76,** 4770 (1954).
167. J. Metzger, *Z. Chem.*, **9,** 99 (1969); *Chem. Abstr.*, **70,** 105567.
168. E. J. Vincent and J. Metzger, *C. R. Acad. Sci. Ser. C*, **261,** 1964 (1965); *Chem. Abstr.*, **63,** 17826.
169. B. Bak, D. Christensen, L. Hansen-Nygaard, and J. Rastrup-Andersen, *J. Mol. Spectrosc.* **7,** 58 (1961).
170. B. Bak, L. Nygaard, E. Pedersen, and J. Rastrup-Andersen, *J. Mol. Spectrosc.* **19,** 283 (1966).
171. B. Bak, L. Hansen-Nygaard, and J. Rastrup-Andersen, *J. Mol. Spectrosc.* **2,** 361 (1958).
172. G. Leandri, A. Mangini, F. Montanavi, and P. Passerini, *Gazz. Chim. Ital.*, **85,** 769 (1955); *Chem. Abstr.*, **50,** 2291.
173. Y. N. Sheinker, V. V. Kushkin, and I. Y. Postovskii, *Zh. Fiz. Khim.*, **31,** 214 (1957); *Chem. Abstr.*, **51,** 17455.
174. N. Colebourne, R. G. Foster, and E. Robson, *J. Chem. Soc. C*, 685 (1967); *Chem. Abstr.*, **66,** 104947.
175. A. Albert, *Chem. Ind.*, 1271 (1957); *Chem. Abstr.*, **52,** 1761.
176. B. Ellis and P. J. F. Griffiths, *Spectrochim. Acta*, **21,** 1881 (1965); *Chem. Abstr.*, **64,** 171.
177. D. Bouin-Roubaud and J. Metzger, *C. R. Acad. Sci., Ser. C*, **281,** 1 (1975).
178. W. F. Forbes, R. Shilton, and A. Balasubramanian, *J. Org. Chem.*, **29,** 3527 (1964).
179. D. Bouin and A. Friedmann, *C. R. Acad. Sci., Ser. C*, **269,** 1343 (1969); *Chem. Abstr.*, **72,** 78116.
180. A. Friedmann, A. Cormons, and J. Metzger, *C. R. Acad. Sci., Ser. C*, **271,** 17 (1970).
181. D. Bouin, C. Roussel, M. Chanon, and J. Metzger, *J. Mol. Structure*, **22,** 389 (1974).
182. C. Roussel, M. Chanon, and J. Metzger, *Tetrahedron Letters*, **21,** 1861 (1971).
183. L. Bouscasse, M. Chanon, R. Phan Tan Luu, J. E. Vincent, and J. Metzger, *Bull. Soc. Chim. France*, 1055 (1972).
184. Y. Ferre, Thesis, University of Marseille, France (1972).

185. Y. Ferre, E. J. Vincent, H. Larive, and J. Metzger, *Bull. Soc. Chim. France*, **7,** 2570 (1971).
186. H. J. M. Dou and J. Metzger, *Bull. Soc. Chim. France*, **10,** 3273 (1966); *Chem. Abstr.*, **67,** 2511.
187. Y. Ferre, E. J. Vincent, H. Larive, and J. Metzger, *Bull. Soc. Chim. France*, **10,** 3862 (1972).
188. L. Williamson and B. Meyer, *Spectrochim. Acta*, **26A,** 331 (1970); *Chem. Abstr.*, **72,** 84606.
189. G. Salmona, R. Faure, and E. J. Vincent, *Compt. Rend.*, **C-280,** 605 (1975); *Chem. Abstr.*, **83,** 8915.
190. G. Salmona, C. Guimon, R. Faure, E. J. Vincent, and G. Pfister-Guillouzo, Unpublished Results, 1977.
191. G. Salmona, Thesis, University of Marseille, France (1977).
192. F. Monichioli and G. Del Re, *J. Chem. Soc. B*, 674 (1969); *Chem. Abstr.*, **71,** 65940n.
193. R. Ditchfield, W. J. Hehre, and J. A. Pople, *J. Chem. Educ.*, **54,** 724 (1971).
194. R. Manzoni Ansidei and G. Travagli, *Gazz. Chim. Ital.*, **71,** 677 (1941); *Chem. Abstr.*, **36,** 7021.
195. A. Taurins, J. G. E. Fenyes, and R. N. Jones, *Can. J. Chem.*, **35,** 423 (1957); *Chem. Abstr.*, **52,** 883.
196. A. R. Katritzky, *Quart. Rev.*, **13,** 353 (1959).
197. M. P. V. Mijovic and J. Walker, *J. Chem. Soc.*, 3381 (1961); *Chem. Abstr.*, **56,** 14252.
198. M. Poite, Thesis, University of Marseille, France (1961).
199. M. Azzaro, Thesis, University of Marseille, France (1962).
200. A. R. Katritzky, "Physical Methods in Heterocyclic Chemistry." Academic, New York, 1963, Vol. 2.
201. J. Chouteau, G. Davidovics, J. Metzger, M. Azzaro, and M. Poite, *Bull. Soc. Chim. France*, 1794 (1962); *Chem. Abstr.*, **58,** 5168.
202. G. Sbrana, E. Castellucci, and M. Ginanneschi, *Spectrochim. Acta*, **A23,** 751 (1967); *Chem. Abstr.*, **66,** 109841.
203. G. Davidovics, C. Garrigou Lagrange, J. Chouteau, and J. Metzger, *Spectrochim. Acta*, **A23,** 1477 (1967); *Chem. Abstr.*, **67,** 37861.
204. G. Davidovics, Thesis, University of Marseille, France (1969).
205. T. Avignon, E. J. Vincent, J. Raymond, and M. Chaillet, *J. Mol Structure*, **21,** 319 (1974); *Chem. Abstr.*, **81,** 36912.
206. D. Y. Movshovich, V. N. Sheinker, A. D. Garnovskii, and O. A. Osipov, *Zh. Org. Khim.*, **11,** 1740 (1975); *Chem. Abstr.*, **83,** 170181.
207. G. Mille, J. Metzger, C. Pouchan, and M. Chaillet, *Spectrochim. Acta*, **31A,** 1115 (1975).
208. G. Mille, C. Pouchan, H. Sauvaitre, and J. Chouteau, *J. Chim. Phys.*, **72,** 37 (1975).
209. C. Pouchan, S. Senez, J. Raymond, and H. Sauvaitre, *J. Chim. Phys.*, **71,** 525 (1974).
210. R. Manzoni ansidei and G. Travagli, *Gazz. Chim. Ital.*, **71,** 680 (1941); *Chem. Abstr.*, **36,** 7021.
211. H. M. Randall, R. G. Fowler, N. Fuson, and J. R. Dangle, "Infrared Determination of organic Structures," Van Nostrand, New York, 1949.
212. J. Weinard, Thesis, University of Sarrebruck, Germany (1952).
213. C. L. Angyal and R. L. Werner, *J. Chem. Soc.*, 2911 (1952); *Chem. Abstr.*, **47,** 7501.
214. P. Roussel, Thesis, University of Marseille, France (1958).
215. M. Carrega, Thesis, University of Marseille, France (1959).
216. Yu N. Scheinker and E. M. Peresleni, *Russ. J. Phys. Chem.*, **36,** 919 (1962).

217. P. Reynaud, M. Robba, and R. C. Moreau, *Bull. Soc. Chim. France*, 1735 (1962); *Chem. Abstr.*, **58,** 6816.
218. P. Bassignana, C. Cogrossi, and M. Gandino, *Spectrochim. Acta*, **19,** 1885 (1963); *Chem. Abstr.*, **59,** 13482.
219. J. Braun, Thesis, University of Marseille, France (1964).
220. C. N. R. Rao and R. Venkataraghavan, *Can. J. Chem.*, **42,** 43 (1964).
221. A. Silberg, E. Hamburg, Z. Frenkel, and L. Cormos, *Rev. Roumaine Chim.*, **9,** 215 (1964); *Chem. Abstr.*, **61,** 14045.
222. R. Arnaud, M. Gelus, J. C. Malet, and J. M. Bonnier, *Bull. Soc. Chim. France*, **9,** 2857 (1966); *Chem. Abstr.*, **66,** 37815.
223. R. Chicheportiche, Thesis, University of Marseille, France (1967).
224. R. Cottet, R. Gallo, and J. Metzger, *Bull. Soc. Chim. France*, **12,** 4499 (1967); *Chem. Abstr.*, **69,** 2894.
225. G. Vernin, J. P. Aune, H. J. M. Dou, and J. Metzger, *Bull. Soc. Chim. France*, 4523 (1967); *Chem. Abstr.*, **69,** 19062t.
226. J. P. Aune, Thesis, University of Marseille, France (1969).
227. J. Chouteau, G. Davidovics, J. Metzger, and A. Bonzom, *Spectrochim. Acta*, **22,** 719 (1966); *Chem. Abstr.*, **64,** 15184.
228. M. Conte, G. Mille, J. Lafon, J. Chouteau, and J. Metzger, *J. Chim. Phys.*, **73,** 569 (1976).
229. A. Taurins and W. G. Schneider, *Can. J. Chem.*, **38,** 1237 (1960); *Chem. Abstr.*, **55,** 3196.
230. B. Bak, J. T. Nielsen, J. Rastrup-Andersen, and M. Schottlaender, *Spectrochim. Acta*, **18,** 741 (1962); *Chem. Abstr.*, **57,** 8089.
231. P. C. Haake and W. B. Miller, *J. Amer. Chem. Soc.*, **85,** 4044 (1963); *Chem. Abstr.*, **60,** 5302.
232. H. A. Staab and A. Mannschreck, *Chem. Ber.*, **98,** 1111 (1965).
233. T. Schaefer and W. G. Schneider, *J. Chem. Phys.*, **32,** 1224 (1960).
234. G. Borgen and S. Gronowitz, *Acta Chem. Scand.*, **20,** 2593 (1966); *Chem. Abstr.*, **66,** 60541.
235. E. J. Vincent, R. Phan Tan Luu, J. Metzger, and J. M. Surzur, *Bull. Soc. Chim. France*, **11,** 3524 (1966); *Chem. Abstr.*, **66,** 64912.
236. J. P. Aune, R. Phan Tan Luu, E. J. Vincent, and J. Metzger, *Bull. Soc. Chim. France*, **7,** 2679 (1972); *Chem. Abstr.*, **78,** 3389.
237. E. J. Vincent, R. Phan Tan Luu, and J. Metzger, *Bull. Soc. Chim. France*, **11,** 3537 (1966); *Chem. Abstr.* **66,** 104623.
238. R. J. Abraham and W. A. Thomas, *J. Chem. Soc. B*, 127 (1966); *Chem. Abstr.* **64,** 9573.
239. H. J. M. Dou and J. Metzger, *Bull. Soc. Chim. France*, 2395 (1966); *Chem. Abstr.* **65,** 18467g.
240. J. A. Pople, W. G. Schneider, and H. J. Bernstein, "High Resolution Magnetic Resonance," McGraw Hill, New York, 1959.
241. R. McWeeny, *Mol. Phys.*, **1,** 311 (1958).
242. A. Veillard, *J. Chim. Phys.*, **59,** 1056 (1962).
243. A. D. Buchingham, *Can. J. Chem.*, **38,** 300 (1960).
244. J. I. Musher, *J. Chem. Phys.*, **37,** 34 (1962).
245. H. M. McConnell, *J. Chem. Phys.*, **27,** 226 (1957).
246. J. A. Pople, *Proc. Roy. Soc.*, **A239,** 541 (1957).
247. R. Dahlbom, T. Ekstrand, S. Gronowitz, and B. Mathiasson, *Acta Chem. Scand.*, **17,** 2518 (1963); *Chem. Abstr.*, **60,** 8011.

248. G. M. Clarke and D. Williams, *J. Chem. Soc.*, 4597 (1965).
249. K. Schaumburg, *Can. J. Chem.*, **49,** 1146 (1971); *Chem. Abstr.*, **74,** 132807.
250. G. Vernin, M. A. Lebreton, H. J. M. Dou, and J. Metzger, *Bull. Soc. Chim. France*, 1085 (1974).
251. H. Bildsoe and K. Schaumburg, *J. Magn. Reson.*, **14,** 223 (1974); *Chem. Abstr.*, **81,** 48913a.
252. J. P. Jacobsen, O. Snerling, E. J. Pedersen, J. T. Nielsen, and K. Schaumburg, *J. Magn. Reson.*, **10,** 130 (1973); *Chem. Abstr.*, **79,** 59730a.
253. J. P. Kintzinger and J. M. Lehn, *Mol. Phys.*, **14,** 133 (1968); *Chem. Abstr.*, **69,** 6961.
254. R. K. Harris, N. C. Pyper, and K. M. Worvill, *J. Magn. Reson.*, **18,** 139 (1975).
255. J. Zajakowsky, J. Berger, and E. J. Vincent, Unpublished Results, 1970.
256. A. D. C. Towl and K. Schaumburg, *Mol. Phys.*, **22,** 49 (1971).
257. E. J. Vincent, R. Phan Tan Luu, and J. Metzger, *C. R. Acad. Sci.,Ser. C.* **270,** 666 (1970); *Chem. Abstr.*, **72,** 116586.
258. R. Garnier, R. Faure, A. Babadjamian, and E. J. Vincent, *Bull. Soc. Chim. France*, **3,** 1040 (1972); *Chem. Abstr.*, **77,** 11964h.
259. J. P. Warren and J. D. Roberts, *J. Phys. Chem.*, **78,** 2507 (1974).
260. H. Noth and B. Wrackmeyer, *Chem. Ber.*, **107,** 3070 (1974); *Chem. Abstr.*, **82,** 56967f.
261. S. Nagata, H. Saito, and Y. Tanaka, *J. Amer. Chem. Soc.*, **95,** 324 (1973); *Chem. Abstr.*, **78,** 83500x.
262. P. Laszlo, *Bull. Soc. Chim. France*, 558 (1966); *Chem. Abstr.*, **65,** 1628g.
263. P. Haake, L. P. Bausher, and W. B. Miller, *J. Amer. Chem. Soc.*, **91,** 1113 (1969); *Chem. Abstr.*, **71,** 12248.
264. K. Tori and T. Nakagawa, *J. Phys. Chem.*, **68,** 3163 (1964).
265. E. R. Malinowsky, *J. Amer. Chem. Soc.*, **83,** 4479 (1961).
266. E. R. Malinowsky, *J. Amer. Chem. Soc.*, **84,** 2649 (1962).
267. B. Dischler, *Z. Naturforsch.*, **19A,** 887 (1964).
268. J. N. Shoolery, *J. Chem. Phys.*, **31,** 1427 (1959).
269. H. R. Kingsley and N. C. Pyper, *Mol. Phys.*, **20,** 467 (1971); *Chem. Abstr.*, **74,** 118099.
270. R. K. Harris and N. C. Pyper, *Mol. Phys.*, **23,** 277 (1972); *Chem. Abstr.*, **77,** 27161e.
271. G. M. Clarke, R. Grigg, and D. H. Williams, *J. Chem. Soc. B*, **4,** 339 (1966); *Chem. Abstr.*, **64,** 17389.
272. R. Graham, I. Cooks, S. Howe, W. Tam, and D. H. Williams, *J. Amer. Chem. Soc.*, **90,** 4064 (1968).
273. R. G. Buttery, L. C. Ling, and R. E. Lundin, *J. Agric. Food Chem.*, **21,** 488 (1973); *Chem. Abstr.*, **79,** 31976.
274. R. Tabacchi, *Helv. Chim. Acta*, **57,** 324 (1974); *Chem. Abstr.*, **81,** 3004.
275. A. Haag and P. Werkhoff, *Org. Magn. Reson.*, **11,** 511 (1976).
276. J. P. Aune and J. Metzger, *Bull. Soc. Chim. France*, **9,** 3536 (1972); *Chem. Abstr.*, **78,** 3421.
277. M. J. Rix and B. R. Webster, *Org. Mass Spectrometry*, **5,** 311 (1971); *Chem. Abstr.*, **75,** 34719.
278. R. A. Khmelnitskii, E. A. Kunina, S. L. Gusinskaya, and V. Y. Telly, *Khim. Geterotsikl. Soedinenii*, **7,** 1372 (1971); *Chem. Abstr.*, **76,** 98611s.
279. A. Friedmann, G. Salmona, G. Curet, R. Phan Tan Luu, and J. Metzger, *Compt. Rend.*, **C-269,** 273 (1969); *Chem. Abstr.* **71,** 123317.
280. G. Entenmann, *Org. Mass Spectrometry*, **10,** 831 (1975).
281. O. G. Vitzthum and P. Werkhoff, *J. Food Sci.*, **39,** 1210 (1974).

VI. References

282. Z. Pelau, D. H. Williams, H. Budzikiewicz, and C. Djerassi, *J. Amer. Chem. Soc.*, **87**, 567 (1965).
283. F. P. Lossing, A. W. Tickner, and W. A. Bryce, *J. Chem. Phys.*, **19**, 1254 (1951).
284. K. Torssell, *Tetrahedron*, **26**, 2759 (1970).
285. P. Tordo, G. Pouzard, H. J. M. Dou, A. Babadjamian, and J. Metzger, *Nouv. J. Chim.*, (1977).
286. C. Heller and H. M. McConnell, *J. Chem. Phys.*, **32**, 1535 (1960).
287. J. Metzger and J. Disteldorf, *J. Chim. Phys.*, **50**, 156 (1953).
288. J. Van Laar, *Zhur. Fiz. Khim.*, **185**, 35 (1929).
289. M. A. Soulie, P. Goursot, A. Peneloux, and J. Metzger, *J. Chim. Phys.*, **66**, 603 (1969); *Chem. Abstr.* **71**, 42994.
290. M. A. Soulie, D. Bares, and J. Metzger, *C. R. Acad. Sci., Ser. C*, **281**, 341 (1975).
291. M. A. Soulie, P. Goursot, A. Peneloux, and J. Metzger, *J. Chim. Phys.*, **66**, 607 (1969); *Chem. Abstr.* **71**, 42995.
292. R. Handley, E. F. G. Herington, M. Azzaro, and J. Metzger, *Bull. Soc. Chim. France*, 1904 (1963).
293. R. Meyer and J. Metzger, *Ann. Fac. Sci. Marseille*, **35**, 33 (1964), *Chem. Abstr.*, **61**, 10053.
294. R. Meyer and J. Metzger, *C.R. Acad. Sci., Ser. C*, **263**, 1333 (1966); *Chem. Abstr.*, **66**, 54922.
295. P. Goursot and E. F. Westrum, Jr., *J. Chem. Eng. Data*, **13**, 471 (1968).
296. P. Goursot, E. F. Westrum, Jr., and J. Metzger, *C.R. Acad. Sci., Ser. C*, **266**, 590 (1968), *Chem. Abstr.*, **69**, 13532z.
297. B. Soptrajanov, *Croat. Chem. Acta*, **39**, 229 (1967), *Chem. Abstr.* **68**, 108663.
298. T. R. Manley and D. A. Williams, *Spectrochim. Acta*, **24A**, 361 (1968), *Chem. Abstr.*, **68**, 99276.
299. P. Goursot and E. F. Westrum, Jr., *J. Chem. Eng. Data*, **13**, 468 (1968).
300. M. Gelus and J. M. Bonnier, *J. Chim. Phys.*, **65**, 253 (1968); *Chem. Abstr.* **69**, 77164.
301. T. M. Lowry and A. G. Nasini, *Proc. Roy. Soc.*, **A123**, 704 (1929); *Chem. Abstr.* **23**, 4114.
302. A. Weissler, *J. Amer. Chem. Soc.*, **71**, 419 (1949); *Chem. Abstr.* **43**, 4532.
303. R. Meyer and J. Metzger, *Bull. Soc. Chim. France*, **5**, 1711 (1967); *Chem. Abstr.* **67**, 57626.
304. R. Meyer, G. Bourrelly, and J. Metzger, *C.R. Acad. Sci., Ser. C*, **267**, 114 (1968), *Chem. Abstr.*, **69**, 70354s.
305. J. C. Rosso, J. Kaloustian, and L. Carbonnel, *C.R. Acad. Sci., Ser. C*, **276**, 455 (1973).
306. J. M. Bonnier and R. Arnaud, *J. Chim. Phys.*, **68**, 1519 (1971); *Chem. Abstr.* **76**, 38105v.
307. D. Bares, P. Goursot, J. Metzger, and A. Peneloux, *Coll. Int. Cnrs*, **156**, 255 (1967); *Chem. Abstr.*, **67**, 85602.
308. D. Bares and J. Metzger, *J. Chim. Phys.*, **10**, 1540 (1973).
309. R. Meyer, M. Meyer, D. Bares, and E. J. Vincent, *Thermochim. Acta*, **11**, 211 (1975).
310. R. Meyer, G. Giusti, M. Meyer, and E. J. Vincent, *Thermochim. Acta*, **13**, 379 (1975).
311. D. Bares, M. Soulie, and J. Metzger, *J. Chim. Phys.*, **10**, 1531 (1973).
312. R. Meyer and J. Metzger, *Bull. Soc. Chim. France*, **12**, 4465 (1967); *Chem. Abstr.*, **68**, 108142.
313. R. Meyer and M. Meyer, *C.R. Acad. Sci., Ser. C*, **266**, 1664 (1968); *Chem. Abstr.*, **69**, 46253n.
314. R. Meyer, M. Meyer, and J. Metzger, *Bull. Soc. Chim. France*, **7**, 2692 (1968); *Chem. Abstr.*, **69**, 89832t.

315. G. B. Bonino and R. Manzoni Ansidei, *Berichte*, **76**, 553 (1943); *Chem. Abstr.* **38**, 1156.
316. J. Metzger, *Bull. Soc. Chim. France*, **D**, 99 (1949).
317. L. N. Mulay and M. Haverbusch, *Rev. Sci. Instr.*, **35**, 756 (1964); *Chem. Abstr.*, **61**, 6549d.
318. E. J. Vincent, R. Phan Tan Luu, J. Metzger, and J. M. Surzur, *Compt. Rend.*, **260**, 6345 (1965); *Chem. Abstr.*, **63**, 6452.
319. M. H. Palmer and R. H. Findlay, *Tetrahedron Lett.*, **3**, 253 (1974).
320. J. Walker, *Z. Phys. Chem.*, **4**, 319 (1889).
321. R. Phan Tan Luu, J. M. Surzur, J. Metzger, J. P. Aune, and C. Dupuy, *Bull. Soc. Chim. France*, **9**, 3274 (1967); *Chem. Abstr.* **68**, 38910.
322. P. Goursot and I. Wadso, *Acta Chem. Scand.*, **20**, 1314 (1966); *Chem. Abstr.* **65**, 17781b.
323. A. Albert, R. Goldacre, and J. Phillips, *J. Chem. Soc.*, 2240 (1948).
324. D. Haake and L. P. Bauscher, *J. Phys. Chem.*, **72**, 2213 (1968).
325. L. Sacconi, P. Paoletti, and M. Ciampolini, *J. Amer. Chem. Soc.*, **82**, 3832 (1960).
326. R. Daudel, *Tetrahedron*, **19**, 351 (1963).
327. J. J. Elliot and S. F. Mason, *J. Chem. Soc. B*, 2352 (1959).
328. N. S. Hush and J. Blackledge, *J. Chem. Phys.*, **23**, 514 (1955).
329. G. J. Hoijtink, E. De Boer, P. Vanmeij, and W. P. Weiland, *Rec. Trav. Chim.*, **75**, 487 (1956).
330. E. S. Amis, "Solvent Effects on Reaction Rates and Mechanisms," Academic, New York, 1966.
331. I. I. Grandberg and A. N. Kost, *Zh. Obshchei Khim.*, **29**, 1099 (1959); *Chem. Abstr.*, **54**, 1500.
332. V. Boekelheide and N. A. Fedoruk, *J. Amer. Chem. Soc.*, **90**, 3830 (1968).
333. K. T. Potts, D. R. Choudhury, and T. R. Westby, *J. Org. Chem.*, **41**, 187 (1976).
334. R. A. Abramovitch and G. M. Singer, "Pyridine and its Derivatives," R. A. Abramovitch, ed., 1st ed., Wiley, New York, 1974, Suppl., Part I, Chap. 1A., p. 55.
335. D. H. Reid, F. S. Skelton, and W. Banthrone, *Tetrahedron Lett.*, 1797 (1964).
336. R. M. Archeson, M. W. Foxton, and G. R. Miller, *J. Chem. Soc.*, 3200 (1965); *Chem. Abstr.*, **63**, 4300.
337. H. Ogura, H. Takayanagi, and K. Furuhata, *Chem. Commun.*, 759 (1974).
338. P. J. Abbott, R. M. Acheson, U. Eisner, D. J. Watkin, and J. R. Carruthers, *Chem. Commun.*, 155 (1975).
339. P. J. Abbott, R. M. Acheson, U. Eisner, D. J. Watkin, and J. P. Carruthers, *J. Chem. Soc., Perkin Trans. 1*, 1269 (1976).
340. A. Albert, "Heterocyclic Chemistry," 2nd ed., Athlone, London, 1968, p. 1.
341. E. Ochiai and H. Nagasawa, *Berichte*, **72B**, 1470 (1939); *Chem. Abstr.*, **33**, 7783.
342. K. Ganapathi and A. Venkataraman, *Proc. Indian Acad. Sci.*, **22A**, 343 (1945); *Chem. Abstr.*, **40**, 4056.
343. K. Ganapathi and K. Kulkarni, *Proc. Indian Acad. Sci.*, **38A**, 45 (1953); *Chem. Abstr.*, **49**, 8256.
344. H. Erlenmeyer and H. Kiefer, *Helv. Chim. Acta*, **28**, 985 (1945); *Chem. Abstr.*, **40**, 1500.
345. G. Travagli, *Gazz. Chim. Ital.*, **85**, 926 (1955); *Chem. Abstr.*, **51**, 13851.
346. E. Ochiai, *J. Pharm. Soc. Japan*, **58**, 1040 (1938); *Chem. Abstr.* **33**, 3791.
347. E. Ochiai, T. Kakuda, I. Nakayma, and G. Masuda, *J. Pharm. Soc. Japan*, **59**, 462 (1939); *Chem. Abstr.*, **34**, 101.
348. H. Nagasawa, *J. Pharm. Soc. Japan*, **59**, 471 (1939); *Chem. Abstr.*, **34**, 102.

349. E. Ochiai, *J. Pharm. Soc. Japan*, **60**, 164 (1940); *Chem. Abstr.*, **34**, 5450.
350. F. Nagasawa, *J. Pharm. Soc. Japan*, **60**, 433 (1940); *Chem. Abstr.*, **35**, 458.
351. E. Ochiai and Y. Kashida, *J. Pharm. Soc. Japan*, **62**, 97 (1942); *Chem. Abstr.*, **45**, 5150.
352. K. Ganapathi and K. D. Kulkarni, *Current Sci.*, **21**, 314 (1952); *Chem. Abstr.*, **48**, 2046.
353. K. Ganapathi and K. D. Kulkarni, *Proc. Indian Acad. Sci.*, **37A**, 758 (1953); *Chem. Abstr.*, **48**, 10006.
354. B. Prijs, J. Ostertag, and H. Erlenmeyer, *Helv. Chim. Acta*, **30**, 2110 (1947); *Chem. Abstr.*, **42**, 1934.
355. H. Babo and B. Prijs, *Helv. Chim. Acta*, **33**, 306 (1950); *Chem. Abstr.*, **44**, 5872.
356. S. I. Burmistrov and V. Krasovskii, *Zh. Obshchei Khim.*, **34**, 685 (1964); *Chem. Abstr.*, **60**, 13235.
357. V. A. Krasovskii and S. I. Burmistrov, *Khim. Geterotsikl. Soedinenii*, **1**, 56 (1969); *Chem. Abstr.*, **70**, 115051.
358. L. M. Yagupol'Skii and A. G. Galushko, *Zh. Obschei Khim.*, **39**, 2087 (1969); *Chem. Abstr.*, **72**, 31676.
359. S. L. Gusinskaya, V. Y. Telly, and T. P. Makagonova, *Khim. Geterotsikl. Soedinenii*, **3**, 345 (1970); *Chem. Abstr.* **73**, 66486.
360. V. A. Krasovskii and S. I. Burmistrov, *Khim. Geterotsikl. Soedinenii*, **7**, 1188 (1971); *Chem. Abstr.*, **76**, 25147.
361. V. Y. Telli and L. M. Tetyukhina, *Nauch. Tr. Tashk. Gos. Univ.*, **462**, 58 (1974); *Chem. Abstr.*, **84**, 4841.
362. G. Travagli and G. Mazzoli, *Studi Urbinati, Fac. Farm.*, **30**, 101 (1956); *Chem. Abstr.*, **54**, 24661.
363. R. Vivaldi, H. J. M. Dou, and J. Metzger, *C.R. Acad. Sci., Ser. C*, **264**, 1652 (1967); *Chem. Abstr.*, **67**, 64280.
364. B. Das, *J. Sci. Ind. Res., India*, **15B**, 613 (1956); *Chem. Abstr.*, **51**, 8725.
365. C. D. Hurd and H. L. Wehrmeister, *J. Amer. Chem. Soc.*, **71**, 4007 (1949); *Chem. Abstr.*, **45**, 155.
366. B. Prijs, J. Ostertag, and H. Erlenmeyer, *Helv. Chim. Acta*, **30**, 1200 (1947); *Chem. Abstr.*, **41**, 7397.
367. A. Friedmann, D. Bouin, and J. Metzger, *Bull. Soc. Chim. France*, **8–9**, 3155 (1970); *Chem. Abstr.*, **74**, 31389.
368. S. Kasman and A. Taurins, *Can. J. Chem.*, **34**, 1261 (1956); *Chem. Abstr.*, **51**, 3567.
369. R. F. Hunter and E. R. Parken, *J. Chem. Soc.*, 1175 (1934).
370. D. Tarbell, H. Hirschler, and R. Carlin, *J. Amer. Chem. Soc.*, **72**, 3138 (1950); *Chem. Abstr.*, **44**, 10703.
371. A. Perkone, N. Saldabols, and S. Hillers, *Khim. Geterotsikl. Soedinenii*, **3**, 498 (1969); *Chem. Abstr.*, **72**, 31670.
372. N. Albertson, *J. Amer. Chem. Soc.*, **70**, 669 (1948); *Chem. Abstr.*, **42**, 3392.
373. H. Dou, A. Friedmann, G. Vernin, and J. Metzger, *C.R. Acad. Sci., Ser. C*, **266**, 714 (1968); *Chem. Abstr.*, **69**, 43182.
374. G. Asato, *J. Org. Chem.*, **33**, 2544 (1968); *Chem. Abstr.* **69**, 27317v.
375. H. J. M. Dou, G. Vernin, and J. Metzger, *J. Heterocyclic Chem.*, **6**, 575 (1969); *Chem. Abstr.*, **71**, 91361.
376. A. Friedmann, *Compt. Rend.*, **C-269**, 1560 (1969); *Chem. Abstr.*, **72**, 78934.
377. A. R. Katritzky, C. Ogretir, H. Tarhan, H. J. M. Dou, and J. Metzger, *J. Chem. Soc. Perkin Trans* 2, 1614 (1975).
378. A. R. Katritzky, B. Terem, E. V. Scriven, S. Clementi, and H. O. Tarhan, *J. Chem. Soc. Perkin Trans* 2, 1600 (1975).

379. H. A. Staab, M. T. Wu, A. Mannschreck, and G. Schwalbach, *Tetrahedron Lett*, **15,** 845 (1964), *Chem. Abstr.*, **60,** 15696.
380. R. A. Coburn, J. M. Landesberg, D. S. Kemp, and R. A. Olofson, *Tetrahedron*, **26,** 685 (1970); *Chem. Abstr.*, **72,** 89488.
381. D. S. Noyce and S. A. Fike, *J. Org. Chem.*, **38,** 2433 (1973); *Chem. Abstr.* **79,** 52486.
382. H. C. Brown and Y. Okamoto, *J. Amer. Chem. Soc.*, **80,** 4979 (1958).
383. D. S. Noyce and S. A. Fike, *J. Org. Chem.*, **38,** 3316 (1973), *Chem. Abstr.*, **79,** 114769.
384. D. S. Noyce and S. A. Fike, *J. Org. Chem.*, **38,** 3318 (1973); *Chem. Abstr.*, **79,** 114768.
385. D. S. Noyce and S. A. Fike, *J. Org. Chem.*, **38,** 3321 (1973); *Chem. Abstr.*, **79,** 114764.
386. D. A. Forsyth and D. S. Noyce, *Tetrahedron Lett.*, 3893 (1972).
387. A. Streitwieser, Jr. and R. G. Jesaitis, "Sigma Molecular Orbital Theory," O. Sinanoglu and K. B. Wiberg, Eds. Yale U.P., New Haven, 1970, p. 197.
388. J. P. Wibaut and H. E. Jensen, *Rec. Trav. Chim.*, **53,** 77 (1934); *Chem. Abstr.*, **28,** 4417.
389. H. E. Jansen and J. P. Wibaut, *Rec. Trav. Chim.*, **56,** 699 (1937); *Chem. Abstr.*, **31,** 6232.
390. J. P. Wibaut, *Berichte*, **72,** 1708 (1939); *Chem. Abstr.*, **33,** 9304.
391. K. C. Bass and P. Nabasing, "Advances in Free Radical Chemistry," G. H. Williams, Ed., 1st ed., Logos, London, 1972, Vol. 4, p. 39.
392. D. A. Shirley and K. R. Barton, *Tetrahedron*, **22,** 515 (1966).
393. J. Vitry, *Compt. Rend.*, **250,** 139 (1960); *Chem. Abstr.*, **55,** 22288.
394. G. Vernin and J. Metzger, *Bull. Soc. Chim. France*, 2504 (1963); *Chem Abstr.*, **60,** 4123.
395. H. J. Dou and B. M. Lynch, *Bull. Soc. Chim. France*, **12,** 3815 (1966); *Chem. Abstr.*, **66,** 104545.
396. G. Vernin, H. Dou, and J. Metzger, *Bull. Soc. Chim. France*, **12,** 4514 (1967); *Chem. Abstr.*, **69,** 10049.
397 H. J. M. Dou, G. Vernin, and J. Metzger, *Tetrahedron Lett.*, **23,** 2223 (1967); *Chem. Abstr.*, **67,** 54069.
398. G. Vernin, H. J. M. Dou, and J. Metzger, *Bull. Soc. Chim. France*, **3,** 1173 (1972).
399. H. J. M. Dou and J. Metzger, *C.R. Acad. Sci., Ser. C*, **262,** 687 (1966); *Chem. Abstr.*, **64,** 19591.
400. J. M. Dou, G. Vernin, and J. Metzger, *C.R. Acad. Sci., Ser. C*, **263,** 1243 (1966); *Chem. Abstr.*, **66,** 54840.
401. H. J. M. Dou, *Bull. Soc. Chim. France*, 1678 (1966); *Chem. Abstr.*, **65,** 8711f.
402. H. J. M. Dou and B. M. Lynch, *Bull. Soc. Chim. France*, **12,** 3820 (1966); *Chem. Abstr.*, **66,** 104546.
403. H. J. M. Dou, G. Vernin, and J. Metzger, *Bull. Soc. Chim. France*, **12,** 4521 (1967); *Chem. Abstr.*, **69,** 2411.
404. G. Vernin and H. J. M. Dou, *C.R. Acad. Sci., Ser. C*, **266,** 822 (1968); *Chem. Abstr.*, **69,** 43233v.
405. G. Vernin, H. J. M. Dou, and J. Metzger, *Bull. Soc. Chim. France*, **8,** 3280 (1968); *Chem. Abstr.*, **70,** 3913.
406. G. Vernin, H. J. Dou, and J. Metzger, *C.R. Acad. Sci., Ser. C*, **263,** 1310 (1966); *Chem. Abstr.*, **66,** 75942.
407. G. Vernin, Thesis, University of Marseille, France (1963).
408. U. Rudquist and K. Torssell, *Acta Chem. Scand.*, **25,** 2183 (1971); *Chem. Abstr.*, **76,** 3146.
409. R. Vivaldi, H. J. M. Dou, and J. Metzger, *C.R. Acad. Sci., Ser. C*, **264,** 1862 (1967); *Chem. Abstr.*, **67,** 90710.

410. J. R. Shelton and C. W. Uzelmeier, *J. Amer. Chem. Soc.*, **88,** 5222 (1966).
411. G. Vernin, H. J. M. Dou, and J. Metzger, *C.R. Acad. Sci., Ser. C,* **272,** 854 (1971); *Chem. Abstr.,* **74,** 125546.
412. M. Baule, G. Vernin, H. J. M. Dou, and J. Metzger, *Bull. Soc. Chim. France,* **6,** 2083 (1971); *Chem. Abstr.* **75,** 76668.
413. G. Vernin, B. Barre, H. Dou, and J. Metzger, *C.R. Acad. Sci., Ser. C,* **268,** 2025 (1969); *Chem. Abstr.* **71,** 61273.
414. G. Vernin, R. Jauffred, C. Ricard, H. J. M. Dou, and J. Metzger, *J. Chem. Soc. Perkin Trans 2,* 1145 (1972); *Chem. Abstr.,* **77,** 74544s.
415. A. L. Lee, D. Mackay, and E. L. Manery, *Can. J. Chem.,* **48,** 3554 (1970); *Chem. Abstr.* **74,** 31707.
416. G. Vernin, H. J. M. Dou, and J. Metzger, *J. Chem. Soc. Perkin Trans 2,* 1093 (1973); *Chem. Abstr.* **79,** 31217.
417. G. Vernin, H. J. M. Dou, J. Metzger, and G. Vernin, *Bull. Soc. Chim. France,* **5–6,** 1079 (1974).
418. G. Vernin, M. Lebreton, H. J. M. Dou, J. Metzger, and G. Vernin, *Tetrahedron,* **30,** 4171 (1974).
419. R. Breslow, *Chem. Ind.,* 28 (1956); *Chem. Abstr.,* **51,** 6802.
420. R. Breslow, *Chem. Ind.,* 893 (1957); *Chem. Abstr.,* **51,** 16484.
421. L. L. Ingraham and F. H. Westheimer, *Chem. Ind.,* 846 (1956); *Chem. Abstr.,* **51,** 2802.
422. R. Breslow, *J. Amer. Chem. Soc.,* **79,** 1762 (1957); *Chem. Abstr.* **51,** 9826.
423. R. Breslow and E. McNelis, *J. Amer. Chem. Soc.,* **81,** 3080 (1959); *Chem. Abstr.,* **54,** 3384.
424. W. Hafferl, R. Lundin, and L. L. Ingraham, *Biochemistry,* **2,** 1298 (1963); *Chem. Abstr.,* **59,** 15537.
425. P. Haake and J. M. Duclos, *Tetrahedron Lett.,* **6,** 461 (1970); *Chem. Abstr.,* **73,** 24565.
426. R. Breslow, *J. Amer. Chem. Soc.,* **80,** 3719 (1958); *Chem. Abstr.,* **53,** 3324.
427. R. A. Olofson, W. R. Thompson, and J. S. Michelman, *J. Amer. Chem. Soc.,* **86,** 1865 (1964).
428. R. A. Olofson, J. M. Landesberg, K. N. Houk, and J. S. Michelman, *J. Amer. Chem. Soc.,* **88,** 4265 (1966); *Chem. Abstr.,* **66,** 10676.
429. R. A. Olofson and J. M. Landesberg, *J. Amer. Chem. Soc.,* **88,** 4263 (1966); *Chem. Abstr.,* **65,** 18439G.
430. M. Derbez, Thesis, University of Marseille, France (1966).
431. R. A. Olofson, R. V. Kendall, A. C. Rochat, J. M. Landesberg, W. R. Thompson, and J. S. Michelman, "Symposium on Properties of Anions," Abstract of the 153rd meeting of Acs, Miami, 1967, p. Q34.
432. D. S. Kemp and J. T. O'Brien, *J. Amer. Chem. Soc.,* **92,** 2554 (1970).
433. P. Haake, L. P. Bausher, and J. P. McNeal, *J. Amer. Chem. Soc.,* **93,** 7045 (1971); *Chem. Abstr.,* **76,** 45573.
434. A. Streitwieser, Jr. and P. J. Scannon, *J. Amer. Chem. Soc.,* **95,** (1973), 6273.
435. W. J. Spillane, H. J. M. Dou, and J. Metzger, *Tetrahedron Lett.,* **26,** 2269 (1976).
436. F. W. Bergstrom and S. H. McAllister, *J. Amer. Chem. Soc.,* **52,** 2845 (1930).
437. J. Beraud and J. Metzger, *Bull. Soc. Chim. France,* 2072 (1962); *Chem. Abstr.,* **58,** 13930.
438. K. Ziegler and H. Zeiser, *Justus Liebigs Ann. Chem.,* **485,** 174 (1931).
439. M. Erne and H. Erlenmeyer, *Helv. Chim. Acta,* **31,** 652 (1948); *Chem. Abstr.,* **42,** 4575.
440. B. D. Compton, *Diss. Abstr. Int. B,* **33,** 2994 (1973).
441. J. Crousier and J. Metzger, *Bull. Soc. Chim. France,* **11,** 4134 (1967); *Chem. Abstr.,* **69,** 10391s.

442. A. I. Meyers and G. N. Knaus, *J. Amer. Chem. Soc.*, **95,** 3408 (1973); *Chem. Abstr.*, **79,** 18789.
443. G. Knaus and A. I. Meyers, *J. Org. Chem.*, **39,** 1189 (1974).
444. G. Knaus and A. I. Meyers, *J. Org. Chem.*, **39,** 1192 (1974).
445. C. Ivanov, V. Dryanska, and I. Arnaudova, *Dokl. Bolg. Akad. Nauk*, **22,** 891 (1969); *Chem. Abstr.*, **72,** 31674.
446. L. W. Deady, *Aust. J. Chem.*, **26,** 1949 (1973).
447. F. Filippini and R. F. Hudson, *Chem. Commun.* **9,** 522 (1972); *Chem. Abstr.*, **77,** 74651Z.
448. W. Adam, A. Grimson, and R. Hoffmann, *J. Amer. Chem. Soc.*, **91,** 2590 (1969).
449. T. Kaufmann and R. Wirthureu, *Angew. Chem.*, **83,** 21 (1971).
450. M. Azzaro, *Bull. Soc. Chim. France*, **9,** 2201 (1964); *Chem. Abstr.* **62,** 4697.
451. H. C. Brown and D. Gintis, *J. Amer. Chem. Soc.*, **78,** 5378 (1956).
452. M. Azzaro and J. Metzger, *Bull. Soc. Chim. France*, **7,** 1575 (1964); *Chem. Abstr.*, **61,** 10554.
453. R. Cottet, R. Gallo, J. Metzger, and J. M. Surzur, *Bull. Soc. Chim. France*, 4502 (1967); *Chem. Abstr.*, **69,** 2397.
454. R. Gallo, Thesis, University of Marseille, France (1972).
455. A. Babadjamian, M. Chanon, R. Gallo, and J. Metzger, *J. Amer. Chem. Soc.*, **95,** 3807 (1973); *Chem. Abstr.*, **79,** 31338.
456. R. Gallo, M. Chanon, H. Lund, and J. Metzger, *Tetrahedron Lett.*, **36,** 3857 (1972); *Chem. Abstr.* **77,** 151308.
457. U. Berg, R. Gallo, J. Metzger, and M. Chanon, *J. Amer. Chem. Soc.*, **98,** 1260 (1976).
458. A. Hantzsch, *Justus Liebigs Ann. Chem.*, **250,** 257 (1889).
459. H. Erlenmeyer and E. H. Schmid, *Helv. Chim. Acta*, **22,** 698 (1939).
460. H. Erlenmeyer and H. Ueberwasser, *Helv. Chim. Acta*, **22,** 938 (1939).
461. H. Erlenmeyer and E. H. Schmid, *Helv. Chim. Acta*, **24,** 869 (1941); *Chem. Abstr.*, **36,** 4433.
462. A. Hantzsch and F. Schlegel, *Z. Anorg. Chem.*, **159,** 273 (1926).
463. W. J. Eilbeck, F. Holmes, T. W. Thomas, and G. Williams, *J. Chem. Soc. A*, 2348 (1968); *Chem. Abstr.*, **69,** 113015.
464. W. J. Eilbeck, F. Holmes, and A. E. Underhill, *J. Chem. Soc. A*, 757 (1967); *Chem. Abstr.*, **67,** 7588.
465. J. A. Weaver, P. T. Hambright, P. T. Albert, E. Kang, and A. Thorpe, *Inorg. Chem.*, **9,** 268 (1970); *Chem. Abstr.*, **72,** 74208.
466. M. N. Hughes and K. J. Rutt, *Inorg. Chem.*, **10,** 414 (1971); *Chem. Abstr.*, **74,** 60322.
467. M. N. Hughes and K. J. Rutt, *J. Chem. Soc. A*, **18,** 3015 (1970); *Chem. Abstr.*, **74,** 18858.
468. B. Lenarcik, J. Kulig, and B. Barszcz, *Roczniki Chem.*, **48,** 2111 (1974); *Chem. Abstr.*, **82,** 160923.
469. M. N. Hughes and K. J. Rutt, *Spectrochim. Acta*, **A27,** 924 (1971); *Chem. Abstr.*, **75,** 55841.
470. A. W. Addison, K. Dawson, R. D. Gillard, B. T. Heaton, and H. Shaw, *J. Chem. Soc. Dalton Trans.*, **5,** 589 (1972).
471. G. Fazakerley and J. C. Russel, *J. Inorg. Nuclear Chem.*, **37,** 2377 (1975); *Chem. Abstr.*, **84,** 53301.
472. K. H. Pannell, C. C. Y. Lee, C. Parkanyi, and R. Redfearn, *Inorg. Chim. Acta*, **12,** 127 (1975).
473. T. Matsuura and I. Saito, *Bull. Chem. Soc. Japan*, **42,** 2973 (1969); *Chem. Abstr.*, **72,** 39061.

474. E. Ochiai and E. Hayashi, *J. Pharm. Soc. Japan*, **67,** 34 (1947); *Chem. Abstr.*, **45,** 9533.
475. Japanese Patent No. 72 42 662; *Chem. Abstr.*, **78,** 72123.
476. P. Schatzmann, *Justus Liebigs Ann. Chem.*, **261,** 1 (1891).
477. A. Schuftan, *Berichte*, **27,** 1009 (1894).
478. G. M. Clarke and P. Sykes, *Chem. Commun.*, **15,** 370 (1965); *Chem. Abstr.*, **63,** 11292.
479. G. M. Clarke and P. Sykes, *J. Chem. Soc. C*, **14,** 1269 (1967); *Chem. Abstr.*, **67,** 72972.
480. G. M. Clarke and P. Sykes, *J. Chem. Soc. C*, **15,** 1411 (1967); *Chem. Abstr.*, **67,** 90285.
481. G. M. Badger and N. Kowanko, *J. Chem. Soc.*, 1652 (1957); *Chem. Abstr.*, **51,** 13849.
482. L. M. Werbel, E. F. Elslager, A. A. Phillips, D. F. Worth, P. J. Islip, and M. C. Neville, *J. Med. Chem.*, **12,** 521 (1969); *Chem. Abstr.*, **71,** 37377.
483. A. O. Ilvespaa, *Helv. Chim. Acta*, **51,** 1723 (1968); *Chem. Abstr.*, **70,** 3911.
484. Y. R. Rao, *Indian J. Chem.*, **7,** 836 (1969).
485. G. Travagli, *Boll. Sci. Fac. Chim. Ind. Bologna*, 161 (1941); *Chem. Abstr.* **37,** 6264.
486. J. Dehand and J. Jordanov, *Chem. Commun.* **18,** 743 (1975).
487. M. Kojima and M. Maeda, *J. Chem. Soc. D*, 386 (1970); *Chem. Abstr.*, **72,** 120787.
488. M. Ohashi, A. Iio, and T. Yonezawa, *J. Chem. Soc. D*, 1148 (1970); *Chem. Abstr.*, **73,** 120547.
489. G. Vernin, H. J. M. Dou, and J. Metzger, *C.R. Acad. Sci.*, *Ser. C*, **271,** 1616 (1970).
490. G. Vernin, J. C. Poite, J. Metzger, J. P. Aune, and H. J. M. Dou, *Bull. Soc. Chim. France*, **3,** 1103 (1971); *Chem. Abstr.*, **75,** 20260.
491. C. Riou, G. Vernin, H. J. M. Dou, and J. Metzger, *Bull. Soc. Chim. France*, **7,** 2673 (1972); *Chem. Abstr.*, **78,** 15316.
492. C. Riou, J. C. Poite, G. Vernin, and J. Metzger, *Tetrahedron*, **30,** 879 (1974).
493. G. Vernin, C. Riou, H. J. M. Dou, L. Bouscasse, J. Metzger, and G. Loridan, *Bull. Soc. Chim. France*, **5,** 1743 (1973); *Chem. Abstr.*, **79,** 91365.
494. R. M. Kellog, *Tetrahedron Lett.*, 1429 (1972).
495. O. Buchardt, J. Domanus, N. Harrit, A. Holm, G. Isaksson, and J. Sandstrom, *Chem. Commun.*, **10,** 376 (1974).
496. A. Takamizawa, Y. Hamashima, Y. Sato, H. Sato, S. Tanaka, H. Ito, and Y. Mori, *J. Org. Chem.*, **31,** 2951 (1966); *Chem. Abstr.*, **65,** 13707h.
497. A. Takamizawa, Y. Hamashima, and M. Sato, *J. Org. Chem.*, **33,** 4038 (1968).
498. A. Takamizawa, Y. Sato, and H. Sato, *Chem. Pharm. Bull. Tokyo*, **15,** 1183 (1967); *Chem. Abstr.* **68,** 114562.
499. A. Takamizawa, Y. Hamashima, Y. Sato, and H. Sato, *Chem. Pharm. Bull. Tokyo*, **15,** 1178 (1967); *Chem. Abstr.*, **68,** 114574.
500. A. Takamizawa, Y. Hamashima, H. Sato, and S. Sakai, *Chem. Pharm. Bull. Japan*, **17,** 1356 (1969); *Chem. Abstr.* **71,** 101758.
501. A. Takamizawa and H. Harada, *Chem. Pharm. Bull.*, **21,** 770 (1973); *Chem. Abstr.* **79,** 66287.
502. B. B. Molloy, D. H. Reid, and F. S. Skelton, *J. Chem. Soc.*, 65 (1965); *Chem. Abstr.*, **62,** 7744.
503. D. J. Adam and M. Wharmby, *Tetrahedron Lett.* **36,** 3063 (1969); *Chem. Abstr.*, **71,** 112873.
504. W. H. Mills and J. L. Smith, *J. Chem. Soc.*, **121,** 2724 (1922); *Chem. Abstr.*, **17,** 1024.
505. H. Kondo and F. Nagasawa, *J. Pharm. Soc. Japan*, **57,** 909 (1937); *Chem. Abstr.*, **32,** 1699.
506. H. Kondo and F. Nagasawa, *J. Pharm. Soc. Japan*, **57,** 1050 (1937); *Chem. Abstr.*, **32,** 3398.

507. H. Erlenmeyer, H. M. Weber, and P. Wiessmer, *Helv. Chim. Acta,* **21,** 1017 (1938); *Chem. Abstr.,* **33,** 603.
508. H. Erlenmeyer and H. M. Weber, *Helv. Chim. Acta,* **21,** 863 (1938); *Chem. Abstr.,* **32,** 7915.
509. H. J. M. Dou, Unpublished Results (1974).
510. H. Erlenmeyer, H. Baumann, and E. Sorkin, *Helv. Chim. Acta,* **31,** 1978 (1948); *Chem. Abstr.,* **43,** 3820.
511. E. Imoto and Y. Otsuji, *Bull. Univ. Osaka,* **6,** 115 (1958); *Chem. Abstr.,* **53,** 3027.
512. Y. Otsuji, T. Kimura, Y. Sugimoto, E. Imoto, Y. Omori, and T. Okawara, *Nippon Kagaku Zasshi,* **80,** 1024 (1959); *Chem. Abstr.,* **55,** 5467.
513. M. Ambar, M. Banbeck, and A. B. Roos, *Nat. Stand. Ref. Data Ser., Nat. Bur. Stand.,* **43,** 1 (1973).
514. P. N. Moorthy and E. Hayon, *J. Org. Chem.,* **42,** 879 (1977).

II

General Synthetic Methods for Thiazole and Thiazolium Salts

GASTON VERNIN

Laboratoire de Chimie Organique, Faculté des Sciences et Techniques de Saint-Jérôme, Marseille, France

```
 I. Introduction. . . . . . . . . . . . . . . . . . . . . . . . . . .  166
    1. Various Types of Thiazole Ring Closure . . . . . . . . . . . .  167
II. Thiazoles from α-Halocarbonyl Compounds and Derivatives. Hantzsch's Synth-
    esis, and Related Condensation (Type I) . . . . . . . . . . . .  169
    1. Reaction with Thioamides . . . . . . . . . . . . . . . . . .  169
       A. Chloroacetaldehyde and Derivatives (Thiazole and its 2-Monosubstituted
          Derivatives     . . . . . . . . . . . . . . . . . . . . .  169
       B. Higher α-Haloaldehydes (5- and 2,5-Disubstituted Thiazoles) . . . . .  172
       C. α-Haloketones and Derivatives . . . . . . . . . . . . . .  175
       D. Hantzsch's Synthesis Mechanism . . . . . . . . . . . . . .  209
    2. Reaction with N-Substituted Thioamides (Thiazolium Salts) . . . . . . . .  211
    3. Reaction with Thiourea . . . . . . . . . . . . . . . . . .  213
       A. α-Halocarbonyl Compounds and Derivatives: 2-Aminothiazoles . . . .  213
       B. Mechanism    . . . . . . . . . . . . . . . . . . . . . .  232
    4. Reaction with N-Substituted Thioureas . . . . . . . . . . . . .  232
       A. N-Monosubstituted Thioureas . . . . . . . . . . . . . . .  232
       B. Dithiobiuret   . . . . . . . . . . . . . . . . . . . . .  244
       C  N,N-Disubstituted Thioureas . . . . . . . . . . . . . . .  244
       D. Thiosemicarbazides and Thiosemicarbazones . . . . . . . . . . . . .  249
    5. Reaction with Salts and Esters of Thiocarbamic Acid: 2-Hydroxythiazoles
       and Derivatives   . . . . . . . . . . . . . . . . . . . . .  258
    6. Reaction with Salts and Esters of Dithiocarbamic Acid: 2-Mercaptothiazole
       Derivatives and 2-Thiazolyl Sulfides . . . . . . . . . . . . .  260
       A. 2-Mercaptothiazole Derivatives. . . . . . . . . . . . . . .  260
       B. 2-Thiazolyl Sulfides . . . . . . . . . . . . . . . . . .  266
       C. Mechanism     . . . . . . . . . . . . . . . . . . . . .  269
```

III. Thiazoles from Rearrangement of the α-Thiocyanatoketones (Type Ib). . . . 271
   1. Acid or Alkaline Hydrolysis 271
   2. Action of Dry Hydrogen Halides 273
   3. Action of Labile Sulfur. 276
   4. Action of Labile Nitrogen 278
IV. Thiazoles from Acylaminocarbonyl Compounds and Phosphorus Pentasulfide and Related Condensations (Gabriel's Synthesis (Type III). 278
V. Thiazoles from α-Aminonitriles (Cook–Heilbron's Synthesis) (Type II) 284
   1. Salts and Esters of Dithioacids: 5-Aminothiazole Derivatives and Related Condensations. 284
   2. Carbon Disulfide: 2-Mercapto-5-aminothiazole Derivatives 286
   3. Carbon Oxysulfide: 2-Hydroxy-5-aminothiazole Derivatives. 288
   4. Isothiocyanates: 2,5-Diaminothiazole Derivatives 289
VI. Thiazoles from Nitriles and α-Mercaptoketones or Acids: 2,4-Disubstituted and 4-Hydroxythiazole Derivatives 291
   1. α-Mercaptoketones 291
   2. α-Mercaptoacids: 4-Hydroxythiazole Derivatives 293
   3. 2,4-Diaminothiazoles from α-Halonitriles and Thiourea 296
VII. Miscellaneous Reactions. 297
   1. 2-Aminothiazole Derivatives 297
   2. 4-Aminothiazole Derivatives 301
   3. 5-Hydroxythiazole Derivatives. 303
   4. 4,5-Dihalogenothiazoles and 2,4,5-Trihalogenothiazoles 304
   5. 4-Tosylthiazoles 305
   6. Alkylthiazoles 305
   7. Cyanothiazoles. 306
   8. Miscellaneous 306
VIII. Thiazoles from Other Heterocyclic Compounds 308
   1. Thiazoles from Δ-3-Thiazoles 308
   2. Thiazoles from Oxazoles 309
   3. Thiazoles from Isothiazoles 310
IX. References. 310

I. INTRODUCTION

The first syntheses of the thiazolic ring were made at the end of the nineteenth century when the initial research was carried out by scientists such as Hantzsch, Hubacher, Traumann, Miolatti, Tcherniac, and Gabriel.

Together with the derivatives of pyridine, the thiazoles soon constituted an important part of heterocyclic chemistry, as much from the point of view of the initial research as from the practical aspect. Their biological and pharmaceutical interest is in fact important as they appear in the composition of certain vitamins such as vitamin B_1 (thiamine) and in the penicillins. Reduced thiazoles serve in the study of polypeptides and proteins and occur as structural units in compounds of biological significance, for example, firefly luciferins and in antibiotics bacitracin A and

I. Introduction

thiostrepton. Equally some derivatives of the 2-aminothiazoles are used as fungicides, pesticides, and bacteriocides, other possess mitodepressive and mitostatic properties (698), and a large range of 2-amino (and hydrazino) 5-nitrothiazoles (nitridazole) are devoid of schistosomicidal activity (727).

Certain Schiff bases derived from 2-amino-5-phenylthiazole and their reduction products show diuretic properties (661). Others such as rhodanines are used as intermediates in the synthesis of amino acids, peptides, and purines. In industry, several mercaptothiazole derivatives serve to accelerate the vulcanization of rubber, and alkyl- and acyl-thiazoles are known to be interesting flavoring agents (711, 785). Finally, derivatives of thiazoles are also to be found in certain natural products: a new amino acid incorporating the thiazole ring has been recently isolated from the fungus *Xerocomus substomentosus* (662).

In this chapter we intend to outline the general methods by which the thiazolic ring is synthetized from open-chain compounds. The conversion of one thiazole compound to another is not discussed here, but in appropriate later chapters. Thus the conversion of thiazole carboxylic acids, halogeno-, amino-, hydroxy-, and mercaptothiazoles, to the corresponding unsubstituted thiazoles is treated in Chapters IV through VII, respectively.

These subjects have been reviewed up the year 1975 in various works. Prij's Card Index (363) of thiazole compounds provides swift access to information on individual compounds. Syntheses of thiazoles have been carefully reviewed by Wiley et al. (361), and in 1957 the subject was dealt with in an excellent survey by Sprague and Land (448).

This list was usefully supplemented in 1970, 1973, and 1975 by the publications of Kurzer (699), and a number of books on penicillin contain much information on the reduced thiazole system (312, 540).

Asinger and Offermanns (607) have reviewed the chemistry of Δ-3 thiazolines, and Ohta and Kato's (663) comprehensive survey on sydnones includes a section of mesionic thiazoles (cf. Chapter VIII).

In our study, literature coverage has been extended to 1976. If important omissions have occurred, I ask the indulgence of the reader.

1. Various Types of Thiazole Ring Closure

Several methods for the synthesis of thiazole compounds are available, which can be classified into the partial structures illustrated in Scheme 1. The first of these structures (**Ia**) is by far the most useful and versatile of all the thiazole syntheses. By a judicious choice of reactants it allows alkyl, aryl, aralkyl, or heterocyclic substituents to be placed in any one of the 2-, 3-, 4-, or 5-positions of the ring. This method, better known by the name of the German chemist Hantzsch who originated it in 1887, involves the condensation of a compound bearing the two heteroatoms on the same carbon with a compound bearing one halogen and one carbonyl function on two neighboring carbons. A great variety of compounds may serve as nucleophilic reagent in this reaction, such as thioamide, thiourea, ammonium thiocarbamate or dithiocarbamate, and derivatives. The two carbon fragments may be α-halocarbonyl compounds or their functional derivatives. This synthetic method has been the subject of thiazole research for a hundred years and is one of the classic reaction of thiazole chemistry. Its importance is clearly reflected by its continued wide use. A variant of this method **Ib** was introduced by Tcherniac who cyclized α-thiocyanatoketones (obtained from α-haloketones and metal thiocyanates) to thiazoles under the action of labile hydrogen compounds.

This reaction leads to thiazoles variously substituted in the 2-position, depending on the experimental conditions. Although this method has been studied rather extensively, it has limited scope and frequently fails to give thiazoles.

The thiazole ring can be obtained directly by other methods, but they have limited application. An example is the synthesis of Cook and Heilbron using α-aminonitriles or α-aminoamides and carbon disulfide (or thioacid derivatives) as reactants of type **II**.

The reaction of phosphorus pentasulfide with α-acylamino carbonyl compounds of type **IIIa** also yields thiazoles. Even more commonly, a mercaptoketone is condensed with a nitrile of type **IVa** or α-mercaptoacids or their esters with Schiff bases. This ring closure is limited to the thiazolidines. In the **Va** ring-closure type, β-mercaptoalkylamines serve as the principal starting materials, and ethylformate is the reactant that supplies the carbon at the 2-position of the ring. These syntheses constitute the most important route for the preparation of many thiazolidines and 2-thiazolines. In the **Vb** type of synthesis, one of the reactant supplies only the carbon at the 5-position of the resultant thiazole. Then in these latter years new modern synthetic methods of thiazole ring have been developed (see Section 7; also Refs. 515, 758, 807, 812, 822).

Ia	Ib	II
Hantzsch (1887)	Tcherniac (1919)	Cook and Heilbronn (1947)

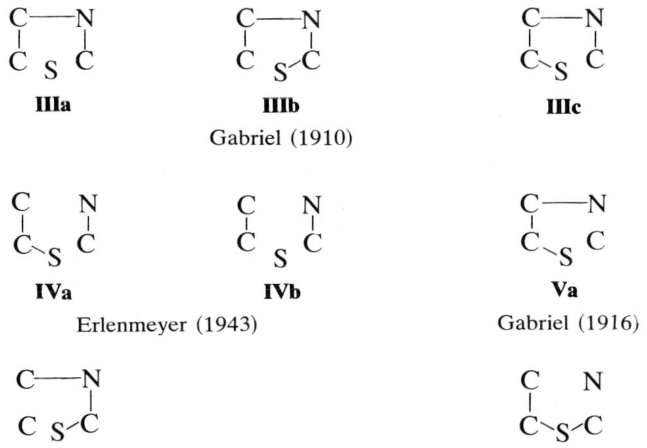

II. Thiazoles from α-Halocarbonyl Compounds and Derivatives 169

IIIa **IIIb** **IIIc**

Gabriel (1910)

IVa **IVb** **Va**

Erlenmeyer (1943) Gabriel (1916)

Vb **VI**

Hartke and Seib (1971) Dubs (1974)

Scheme 1. Various types of ring closure for the thiazoles, thiazolines, and thiazolidines.

II. THIAZOLES FROM α-HALOCARBONYL COMPOUNDS AND DERIVATIVES. HANTZSCH'S SYNTHESIS, AND RELATED CONDENSATION (TYPE I)

The cyclization of α-halocarbonyl compounds is carried out with a great variety of reactants including thioamides, thioureas, their mono- or disubstituted derivatives, and salts and esters of monothiocarbamic acid, leading to variously substituted thiazoles.

1. Reaction with Thioamides

A. *Chloroacetaldehyde and Derivatives* (*Thiazole and its 2-Monosubstituted Derivatives*) (*Table II-1*)

Thiazole itself (**3**), $R_1 = H$, can be obtained by condensing chloroacetaldehyde (**1**) and thioformamide (**2**), $R_1 = H$ (39, 127).

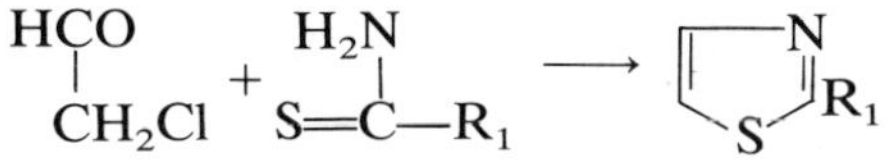

Scheme 2

TABLE II-1. 2-SUBSTITUTED THIAZOLES

$$R_1 \underset{S}{\overset{N}{\underset{\|}{\bigg|}}} \overset{H}{\underset{H}{\bigg|}}$$

R_1	Conditions[a]	Yield (%)	Ref.
H	Cl, P_2S_5, dioxane, or $ClCH_2CH(Cl)OMe$, H_2O, 40–50°C	21–25	39, 127, 455
Me	$ClCH_2CH(Cl)OEt$, heat		4, 10, 22, 175
	$BrCH_2CH(Br)OEt$, P_2S_5, $MgCO_3$, heat 40–60°C; Cl or Br, P_2S_5, C_6H_6, $(Ac)_2O$, heat	39 30–49	578 4, 10, 15, 22, 285, 578
Et	$BrCH_2CH(Br)OEt$, P_2S_5, heat, 40–60°C	50	578
	Br, P_2S_5, dioxane, $MgCO_3$, heat 40–60°C	26–50	285, 578
iso-Pr	$BrCH_2CH(Br)OEt$ or Cl, P_2S_5, dioxane $MgCO_3$, heat 40–60°C	39–62	285, 578
n-Pr	Cl, P_2S_5, heat	49	578
iso-Bu	$BrCH_2CH(Br)OEt$ or Cl, P_2S_5, dioxane, $MgCO_3$ heat 40–60°C and then 80–100°C	45	578, 711
tert-Bu	$BrCH_2CH(Br)OEt$ or Cl, P_2S_5, dioxane	56	578
Neopentyl	$BrCH_2CH(Br)OEt$, P_2S_5, dioxane, $MgCO_3$ heat 40–50°C and then 80–100°C	20	578
$CH_2(CH_2)_9$-$CH=CH_2$	Cl, heat	63	195
$HC(CH_3)CO_2Ph$	Cl or Br, C_6H_6, reflux 2 hr	38–84	
Bz	Cl or Br, heat	10–60	75, 595
CH_2CH_2Ph	Cl or Br, P_2S_5, dioxane, heat		595
CHCHPh	Cl, abs. alcohol, reflux	44	285
Cyclohexyl	Br, P_2S_5, dioxane, heat	4–10	649
Ph	$ClCH_2(Cl)OEt$, AcONa, heat,	—	15
	Br, abs. alcohol, piperidine (drops),	72	264
	Cl, abs. alcohol, $MgCO_3$, reflux 10 hr	82	641
o-MeC_6H_4	$BrCH_2CH(Br)OEt$, abs. alcohol, $MgCO_3$, reflux 5 hr	58	641
m-MeC_6H_4	$BrCH_2CH(Br)OEt$, abs. alcohol, $MgCO_3$, reflux 5 hr	—	738
p-MeC_6H_4	$BrCH_2CH(Br)OEt$, abs. alcohol, $MgCO_3$, reflux 5 hr	—	336
o-BrC_6H_4	Cl in excess, reflux 10–12 hr	—	654
m-BrC_6H_4	Cl in excess, reflux 10–12 hr	—	646
p-BrC_6H_4	Cl in excess, reflux 10–12 hr	—	336, 646, 738
p-ClC_6H_4	Cl in excess, reflux 10–12 hr	—	141, 738
p-$MeOC_6H_4$	$BrCH_2CH(OEt)_2$	—	75, 221
p-$EtOC_6H_4$	$ClCH_2CH(Cl)OEt$	—	101
m- and p-AcC_6H_4	$BrCH_2CH(OEt)_2$, alcohol, reflux	—	569
3,4-$(CH_2O_2)C_6H_3$	$BrCH_2CH(OEt)_2$	—	75
3,4-$(HO)_2C_6H_3$	$BrCH_2CH(OEt)_2$	—	75, 101
$CONHNH_2$	Cl, 30 min at 100°C	30	444

[a] Cl or Br designates the corresponding α-chloro or α-bromo-aldehyde or ketone; P_2S_5 designates thioamides formed in situ from amides.

This reaction is explosive and proceeds in low yield ($\simeq 21\%$) because of the instability of the thioformamide that is destroyed as soon as it is cyclized with **1** (113, 491). The thioformamide is better prepared directly in the reaction mixture by condensing phosphorus pentasulfide and formamide at room temperature, in dioxane solution, according to reaction 1 (491, 492),

$$HCONH_2 + 1/5 P_2S_5 \rightarrow HCSNH_2 \qquad (1)$$

but secondary reactions can occur:

$$HCONH_2 + P_2O_5 \rightarrow HCN + 2HPO_3 \qquad (2)$$
$$HCSNH_2 \rightarrow HCN + H_2S \qquad (3)$$

Another difficulty in this reaction lies in the preparation of pure chloroacetaldehyde. The low yield observed is due to simultaneous formation of by-products (polyhalogenation). So vinylchloride was used as a starting material for this synthesis (449). A simpler method is to react chlorine with vinylchloride in aqueous solution and then to dehydrate the semihydrated chloroacetaldehyde by distillation through a column of calcium chloride heated to 70 to 90°C (451).

When chloroacetaldehyde is condensed with higher thioamides prepared from amides and phosphorus pentasulfide according to Schwarz's method (222), 2-substituted thiazoles are obtained (4, 10, 22, 175).

For example, with thioacetamide prepared in situ in dioxane solution at 45°C, Cottet and Metzger (578) prepared the 4-methylthiazole (**3**), $R_1 =$ Me, in 39% yield, while Erlenmeyer et al. (285) obtained a similar result in benzene and acetic anhydride.

Cyclohexyl, benzyl, and phenethylthioamides give a low yield (4 to 10%) of the corresponding 2-substituted thiazoles (595, 649).

With arylthioamides except for some nitrothiobenzamides (101), yields are usually higher than those obtained above, due to the increased stability of these amides under acidic conditions (**3**), $R_1 =$ Ph, yield 70 to 82% (264, 285, 336, 483, 578, 641). In this case, cyclizations are carried out several hours to reflux, in absolute alcohol, in the presence of melted sodium acetate and few drops of piperidine.

Aromatic thioamides can be prepared as described in the literature by different ways, either by $S \rightarrow O$ exchange between the corresponding benzamides and phosphorus pentasulfide in pyridine solution in the presence of triethylamine (65, 646) as strong base, or by action of H_2S on the appropriate nitrile with pyridine and triethylamine solvents using the method of Fairfull et al. (34, 374, 503). In this reaction, thioacetamide in acidic medium can also be used as a H_2S generator with dimethylformamide as the solvent (485).

It is also possible to start from chloroacetaldehyde derivatives such as 1,2-dihalogeno ethyl acetate: yields can reach 90% (356). These compounds can be easily obtained by addition of halogen to the double bond of vinylacetate at 0 to 10°C.

Aliphatic thioamides cyclized with α,β-dibromoether in the presence of $MgCO_3$ as base, yield the corresponding 2-alkylthiazoles in the 20 to 60% range (578).

α-Bromoacetal condensed with thiobenzamide (513) also give 2-phenylthiazole. But thiazole is very often prepared by indirect methods. Among them, deamination of 2-aminothiazole (49), subsequent reduction of the diazo compound using hypophosphorus acid, or the reduction of 2-halogenothiazoles (mainly the 2-chloro derivatives) by acetic acid and zinc according to the method of Mac Lean and Muir (175) are most commonly used. Another method by which thiazole can be indirectly synthetized is by oxidation of 2-mercaptothiazole (77% yield) (135, 203).

Therefore, the synthesis of the unsubstituted heterocycle is usually more difficult to realize than that of its derivatives.

B. Higher α-Haloaldehydes (5- and 2,5-Disubstituted Thiazoles, Table II-2.).

The cyclization of α-haloaldehydes (4) with thioformamide has been reported (4, 10, 12, 22, 506); the products are 5-alkyl- or arylthiazoles (5), $R_1 = H$, $R_2 = $ alkyl or aryl (Table II-2).

TABLE II-2. 5-ALKYL (OR ARYL) THIAZOLES AND 2,5-ALKYLARYL THIAZOLES

R_1	R_2	Conditions[a]	Yield (%)	Ref.
H	Me	Cl, P_2S_5, dioxane, heat	8–30	245, 364, 455
		Br, P_2S_5, dioxane, $MgCO_3$, heat	30–60	4, 10, 22, 219, 286, 492, 506, 512
H	Et	Br, P_2S_5, dioxane, $MgCO_3$, heat 20 mn at 40–50°C and then 5 hr at 80–100°C	37–50	492, 512
H	CH_2OH	Cl, heat	36	587
H	$(CH_2)_2OH$	Cl, heat	65	587
H	iso-Pr	Br, heat	30–40	492, 512
H	tert-Bu	Br, heat	7–20	492, 512
H	CO_2Me	Cl, heat	—	359
H	Ph	Cl, abs. alcohol, heat or $R_2CH(Br)CH(Br)OEt$	10–15	75, 101, 189, 503
		Br, P_2S_5, dioxane, $MgCO_3$, heat.		127, 519, 641

TABLE II-2 (*continued*)

R_1	R_2	Conditions[a]	Yield (%)	Ref
Me	Me	Cl, or	—	15
		Br, P_2S_5, dioxane, $MgCO_3$, heat	44	175, 285, 492, 512
Et	Me	Br, P_2S_5, dioxane, $MgCO_3$, heat	61	492, 512
iso-Pr	Me	Br, P_2S_5, dioxane, $MgCO_3$, heat	74.5	492, 512
tert-Bu	Me	Br, P_2S_5, dioxane, $MgCO_3$, heat	74	492, 512
Me	Et	Br, P_2S_5, dioxane, $MgCO_3$, heat	37	492, 512
Et	Et	Br, P_2S_5, dioxane, $MgCO_3$, heat	42	492, 512
iso-Pr	Et	Br, P_2S_5, dioxane, $MgCO_3$, heat	72.5	492, 512
tert-Bu	Et	Br, P_2S_5, dioxane, $MgCO_3$, heat	68.5	492, 512
Me	iso-Pr	Br, P_2S_5, dioxane, $MgCO_3$, heat	35.5	492, 512
Et	iso-Pr	Br, P_2S_5, dioxane, $MgCO_3$, heat	35	492, 512
iso-Pr	iso-Pr	Br, P_2S_5, dioxane, $MgCO_3$, heat	40	492, 512
tert-Bu	iso-Pr	Br, P_2S_5, dioxane, $MgCO_3$, heat	30.5	492, 512
Me	tert-Bu	Br, P_2S_5, dioxane, $MgCO_3$, heat	13	492, 512
Me	CO_2Et	Cl or Br, heat	—	145
Me	Ph	Br, abs. alcohol, heat	—	450
Me	p-HOC_6H_4	Br	11	450
Bz	Me	Br	40	294
CH=CH–Ph	Ph	Br, abs. alcohol, heat	11	374
Ph	Me	Cl, abs. alcohol, $MgCO_3$ 2 hr at 40–80°C and then reflux 5 hr	82	641
Ph	Et	Br, abs. alcohol, $MgCO_3$, 2 hr at 40–80°C and then reflux 5 hr	54	641
Ph	iso-Pr	Br, abs. alcohol, $MgCO_3$, 2 hr at 40–80°C and then reflux 5 hr	45	641
Ph	CO_2Et	Cl,	37	196
Ph	Ph	Br, abs. alcohol, $MgCO_3$, heat	50	519
p-FC_6H_4	Ph	Br, abs. alcohol, heat	62	457
o-ClC_6H_4	Ph	Br, abs. alcohol, heat	12	457
p-ClC_6H_4	Ph	Br, abs. alcohol, heat	56	457
m-IC_6H_4	Ph	Br, abs. alcohol, heat	38	457
p-$O_2NC_6H_4$	Ph	Br, abs. alcohol, heat	13	457
p-Biphenylyl	Ph	Br, abs. alcohol, heat	59	457
Ph	α-naphthyl	Br, abs. alcohol, heat	28	457
α-Naphthyl	Ph	Br, abs. alcohol, heat	54	457
α-Thienyl	Ph	Br, abs. alcohol, heat	23	457
β-Pyridyl	Ph	Br, abs. alcohol, heat	62	457
γ-Pyridyl	Ph	Br, abs. alcohol, heat	21	457
CO_2Me	CO_2Me	Cl, water bath, 1.5 hr	35	383
CO_2Et	Me	Cl or Br or $BrCH_2CH(Br)OEt$	—	201, 208, 210, 236, 245

[a] See Table II-1.

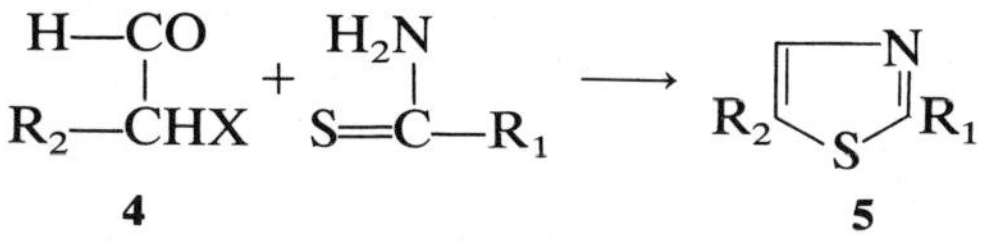

$$\underset{4}{\overset{H-CO}{\underset{R_2-CHX}{|}}} + \underset{\underset{R_1}{|}}{\overset{H_2N}{\underset{S=C}{|}}} \longrightarrow \underset{5}{R_2 \underset{S}{\overset{N}{\diagdown}} R_1}$$

Scheme 3

Thioformamide is prepared in situ at 25 to 30°C, as described previously, and in the presence of magnesium carbonate (492, 512, 578). The mixture is then mildly heated on a water bath, and when temperature reaches 70°C, α-haloaldehyde is added in small quantities. At the end of this addition the reaction mixture is stirred for 2 hr at 100°C. Thiazoles were isolated in the usual manner by a double steam distillation.

The yields are on the whole fairly low and do not exceed 50 to 60% for the 5-methylthiazole (**5**), $R_1 = H$, $R_2 = Me$ (492, 512).

The use of α-bromoaldehydes, more reactive than the α-chloro derivatives, gives better results (492, 512). They have permitted the yields of the cyclization to be increased from 8 to 60% in the latter case. With the higher aldehydes the yields decrease. Thus for 5-*t*-butylthiazole it is not higher than 7 to 20% (492, 512). On the other hand, α-bromoaldehydes are particularly difficult to obtain.

α-Haloaldehydes can be prepared by Guinot and Tabuteau's method (456) by the halogenation of aliphatic aldehydes in strong aqueous solution of hydrochloric acid using methylene chloride as solvent. The yields are in the order of 50 to 60%. But a secondary reaction occurs, resulting in the trimerization of the aldehyde in acidic medium (492, 512), due to the formation of a carbocation that reacts with the surrounding molecules of aldehyde:

$$R_2CH_2\overset{+}{C}HOH \xrightarrow{2R_2CH_2CHO} (R_2CH_2CHO)_3 + H^+ \qquad (4)$$

The trimers obtained in this manner are very stable, but it is impossible to synthesize thiazoles from this form of α-bromoaldehyde, hence the advantage of Yanowskaya and Terent'ev's method (365) that consists of brominating the aldehydes at −5°C in ether or methylene chloride using dioxane dibromide as an halogenating agent (yields ranged from 60 to 75%) (219). This latter is prepared by Favorskii's method (37). Moreover, in this way the formation of polybromide derivatives can be avoided. The solution is neutralized by aqueous sodium carbonate (492, 512). In some cases, when the α-bromoaldehyde is purified by distillation, it polymerizes, especially in the case of phenylacetaldehyde; hence the reaction is carried out in dioxane solution. However the 5-phenylthiazole is itself obtained with a yield not higher than 15% (188, 189, 641).

Another method of aldehyde bromination, apart from Riehl's established method (432) from bromine at 20°C, is to use trimethylphenylammonium bromide in tetrahydrofuran solution, prepared by Vorlander and Siebert's method (50). However, the yield of 5-phenylthiazole using this method with thioformamide dissolved in dioxane is only 8% (513).

The reaction of α-bromo (or α-chloro) aldehydes with higher

thioamides was run under the same conditions as for thioformamide and gave 2,5-dialkyl (492, 512), diaryl (457), or arylalkylthiazoles (432), with yields rarely exceeding 60%. With aliphatic thioamides, formation takes place at a higher temperature: 40–50°C.

In the series of 2,5-dialkylthiazoles prepared by Poite and Metzger (492, 512), the yields decrease from 5-methyl-2-alkylthiazoles (44 to 75%) to 5-t-butyl-2-alkylthiazoles (0 to 3%).

When thioamides such as thiobenzamide are used directly, neither dioxane nor magnesium carbonate is necessary. Instead absolute alcohol with fused sodium acetate in the presence of piperidine is used (457).

Owing to the instability of α-halogenoaldehydes it is occasionally preferable to use more stable derivatives, such as enol acetate prepared according to Bedoukian's method (204) and α-bromoacetals (4, 8, 10, 16, 22, 67, 101, 426). An advantage is said to be in the yield; however, this appears to be slight. The derivatives react in the same sense as the aldehydes themselves, that is, the acetal group as the more polarized reacts first and enters the C-4 position. It is likely that the condensation and cyclization occur by direct displacement of alkoxide ions. Ethyl-α,β-dihalogeno ethers (159, 164, 177, 248) have also been used in place of the free aldehydes in condensation with thioamides.

Syntheses of α,β-dihalogenoethers can be achieved in various ways: the classical method (37), wherein a current of dry gaseous hydrochloric acid, is made to react in an equimolar mixture of ethanol and aldehyde at 20°C first to form the monochloroether (50% yield) and then by the action of bromine, the dibromoether (80 to 90% yield) can be used. The second and simpler method is the direct bromination of ethylvinylether in a chloroformic or dioxane solution if the product is used directly without purification.

C. α-Haloketones and Derivatives

a. CONDENSATION WITH THIOFORMAMIDE (4-MONO-SUBSTITUTED AND 4,5-DISUBSTITUTED THIAZOLES, TABLE II-3)

The only method that yields the 2-unsubstituted thiazole derivatives directly involves the condensation of α-haloketones with thioformamide. As in the case of previously reported α-haloaldehydes, yields are better when more reactive bromoketones are used instead of α-chloroketones. Cyclization can be achieved by adding ketones dissolved in dioxane in small quantities to the thioformamide formed in situ at below 40°C. The temperature is kept below 70°C during the addition, and then the

TABLE II-3. 4- AND 4,5-DISUBSTITUTED THIAZOLES

$$\begin{array}{c} N \underset{S}{\overset{R_1}{=}} R_2 \\ H \end{array}$$

R_1	R_2	Conditions[a]	Yield (%)	Ref.
Me	H	Cl or Br, C_6H_6 at room temperature and then heated (reflux) with C_6H_6 or without solvent	32–44	127, 156, 221, 324, 426
		Br, P_2S_5, dioxane, <40°C, and then 1 hr in water bath at 80–100°C	67–73	366, 455
Et	H	Br, without solvent, heat	48	385
		Br, P_2S_5, dioxane	67–78	455
1-(4-thiazolyl)2-phtalimido-propane	H	Cl,	—	347
N-phtalimidomethyl	H	Br, abs. alcohol at (°C and then kept overnight at room temperature	75	337
N-phtalimidoethyl	H	Br, abs. alcohol at (°C and then kept overnight at room temperature	97	337
$CH_2CH_2CO_2Me$	H	Br, heat	—	557
CH_2CH_2Cl	H	Br, heat	49	153b
CH_2CH_2OH	H	Br, heat	49	557
$CH_2CH(OH)CH_3$	H	Br, heat	—	156
CH_2CO_2Et	H	Br, heat	—	119, 247, 273
CH_2Cl	H	$(CH_2Cl)_2CO$, heat	—	119
iso-Pr	H	Br, P_2S_5, dioxane, heat	51–67	87
tert-Bu	H	Br, P_2S_5, dioxane, heat	63–87	87
C(=NOH)Me	H	Br, heat	15–50	152
C(=NOH)Ph	H	Br, heat	—	152
COMe	H	Br, abs. alcohol at <20°C, and then 12 hr at low temperature	—	711
CO_2Ph	H	Br, heat	79–85	152
Ph	H	Cl or Br, P_2S_5, ether, heat	40–87	127, 264, 432, 472, 518

p-MeC$_6$H$_4$	H	Br, ether, heat	—	264
p-ClC$_6$H$_4$	H	Br, ether, heat	—	264
o-BrC$_6$H$_4$	H	Br, ether, heat	—	646
m-BrC$_6$H$_4$	H	Br, ether, heat	—	646
p-BrC$_6$H$_4$	H	Br, ether, heat	—	336, 646
2-(4-Me)imidazolyl	H	Br, heat	—	153b
4-(HOCH$_2$CH$_2$)$_2$NC$_6$H$_4$	H	Cl, abs. EtOH	—	721
Me	Me	Cl, heat	40	156, 221, 364, 426
		Br, P$_2$S$_5$, dioxane	67–72	220
Me	Et	Cl, P$_2$S$_5$ at 10°C let stand overnight, and then heat 2 hr; Br, P$_2$S$_5$, dioxane	23–29	227, 364
			69	426, 711
	−(CH$_2$)$_3$−	Cl, heat	—	262
	−(CH$_2$)$_4$−	Cl, heat	—	188
Me	(CH$_2$)$_3$Br	Br, heat	—	304
Me	(CH$_2$)$_3$OH	Br, heat	48	227
Me	(CH$_2$)$_2$Br	Br, heat	—	89, 161, 229, 304
Me	(CH$_2$)$_2$OH	Cl or Br, C$_6$H$_6$, reflux 1–2 hr	26–62	95, 106, 127, 130
		Br, picoline heat 10 hr at 60 to 80°C	76	131, 190, 357, 464
Me	(CH$_2$)$_2$OEt	Cl, heat	—	587
Me	CH$_2$CH(OH)CH$_3$	Cl or Br, P$_2$S$_5$, C$_6$H$_6$, reflux 2 hr	24–40	227, 587
Me	CH$_2$CO$_2$Me	Cl, P$_2$S$_5$, heat	16	230
Me	CH$_2$CO$_2$Et	Cl or Br, heat	71	105, 160, 230
Me	CH$_2$Ph	Cl or Br, EtOH, reflux	47	472
Me	CH$_2$Br	Br, P$_2$S$_5$, heat	29	230, 304, 587
Me	CH$_2$OH	Cl or Br, C$_6$H$_6$, heat	27–28	230, 304, 587
Me	(CH$_2$)$_2$OH	Br, C$_6$H$_6$, reflux 1–2 hr	62	464
Me	iso-Pr	Br, P$_2$S$_5$, dioxane	51	426
Me	CH(OH)CH$_3$	Br, heat	—	190
Me	COMe	Cl, heat, P$_2$S$_5$ first at room temperature, then under reflux	55	180, 227
Me	CO$_2$Et	Cl or Br, P$_2$S$_5$, C$_6$H$_6$, heat	50–58	92, 118, 132, 220

TABLE II-3 (continued)

R_1	R_2	Conditions[a]	Yield (%)	Ref.
Me	$CO_2(CH_2)_2Cl$	Br, heat	25	247
Me	Ph	Br, P_2S_5, heat	55	149, 413, 542
Me	$(O=)P(Et)_2$	Br, heat	—	372
CH_2CO_2Et	CO_2Et	Cl, heat	35	298
Et	Me	Br, heat	6	156, 426
Et	Et	Br, H_2O, heat	30	377
Pr	Et	Cl, ether, steam bath	35	377
CH_2Ph	Me	Br, EtOH, reflux	—	472
CH_2Ph	Et	Br, EtOH, reflux	—	472
iso-Pr	Me	Br, P_2S_5, dioxane	51	426
$CONHNH_2$	Me	Cl, cooling, and then heat 30 mn at 100°C	67	445
CO_2Et	CO_2Et	Br, H_2O, 24 hr at 5°C	—	110, 118, 383
Ph	CO_2Et	Br, heat	—	118, 252

[a] See Table II-1.

reaction mixture is maintained at 100°C for several hours. In some cases where α-haloketones are stable, the treatment of the mixture, without any solvent, yielded good results (221, 385), but the reaction can be rather violent.

Depending on the starting ketone (**6**), the reaction leads to the formation of 4-monosubstituted or 4,5-disubstituted thiazoles (**7**) as shown in Scheme 4.

$$R_1-CO \atop R_2-CHX \quad + \quad {H_2N \atop S=C-H} \quad \longrightarrow \quad {R_1 \diagup\!\!\!\!\diagdown N \atop R_2 \diagdown\!\!\!\!\diagup S}$$

　　　　　6　　　　　　　　　　　　　　　**7**

Scheme 4

Initially, 4-methylthiazole (**7**), $R_1 =$ Me, $R_2 =$ H, was obtained in low yield (<40%) from chloroacetone (102, 127). Kurkjy and Brown (364) using bromoacetone with dioxane as solvent increased the yield to 73%, whereas it only yields 39% in benzene (426). Kurkjy and Brown's method was extended later to 4-alkylthiazoles by Metzger and Carrega (455) with 75 to 90% yields. 4-Vinyl- and 4-isopropenylthiazoles (629), 4-adamanthylthiazole and derivatives (705), 4-carboranylmethylthiazole (706), 4,5-dialkylthiazoles (156, 220, 426, 703, 810), 4,5-alkylarylthiazoles (149, 189, 413), and some chiral thiazoles (783) syntheses have been also reported.

The 4-arylthiazoles are obtained from phenacylbromide and its substituted derivatives (in 40 to 87% yields) (264, 335, 646), and 4-phenyl-5-methylthiazole (129, 641) and 4,5-diphenylthiazole (519) from α-bromopropiophenone and desylchloride, respectively.

4-methyl-5-(β-hydroxyethyl)thiazole, an intermediate in the synthesis of vitamin B_1, has been prepared by several authors (95, 106, 127, 130, 131, 190, 464, 707). Either α-chloro-α-acetyl-γ-butyrolactone or 4-acetyl-4-chloro-1-pentanol are used as the starting α-chloroketone. The best yield (76%) was obtained using picoline as solvent (357, 464).

5-(-2-Chloroethyl)-4-methylthiazole and analogs have been also produced by this method (735).

In a similar manner, the condensation of diethyl (bromooxoalkyl)phosphonates (**8**) with thioformamide gave the corresponding thiazole (**9**) (Scheme 5).

$$(EtO)_2 P(=O)CH(Br)Ac + HCSNH_2 \quad \longrightarrow \quad {Me \diagup\!\!\!\!\diagdown N \atop (EtO)_2(O=)P \diagdown\!\!\!\!\diagup S}$$

　　　　　　8　　　　　　　　　　　　　　　　　**9**

Scheme 5

BROMOMETHYLKETONE SYNTHESIS. The preparation of nonsymmetrical ketones monobrominated on the less substituted carbon presents a difficult problem. We know that in alkaline medium bromine reacts on the less substituted carbon but gives polybrominated products (306). On the other hand, in acidic medium bromination takes place chiefly on the more substituted carbon (426). In a recent study on the solvent effect, it has been shown that the use of methanol at room temperature instead of carbon tetrachloride, gives bromomethylketones in good yields (683). These brominations can equally be achieved using sodium hypobromite (80% yield) (493) in dioxane dibromide (365) as bromination agent. But the only method that gives bromomethylketones in quantitative yields is that of Catch and Elliot (306), which uses the reaction of acylhalides with 2 moles of diazomethane in either ethereal or methylene chloride solution. Diazomethane is prepared by the method of Lutz (246), that is, the action of potassium hydroxide on nitrosomethylurea at 0 to 5°C. Acetophenone and its substituted derivatives are usually prepared according to the method of Gilman and Blatt (329) by direct bromination in ethereal or acetic acid solution at room temperature.

b. CONDENSATION WITH HIGHER THIOAMIDES (2,4-DISUBSTITUTED AND 2,4,5-TRISUBSTITUTED THIAZOLES)

The reaction of a thioamide with α-halocarbonyl compounds has been applied extensively, and many thiazoles (**10**) with alkyl, aryl, aralkyl, or heteroaryl functional groups at the three 2-, 4-, or 5-positions have been reported (Scheme 6).

In this section we successively examine each class of these compounds.

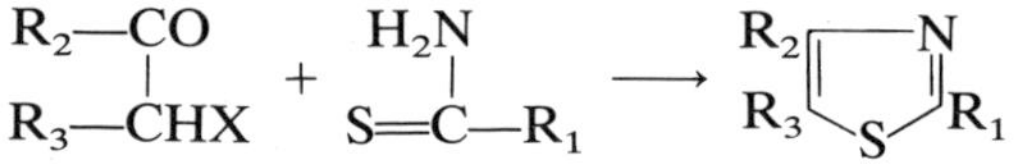

$$\begin{array}{c} R_2\text{—CO} \\ | \\ R_3\text{—CHX} \end{array} + \begin{array}{c} H_2N \\ | \\ S\text{=}C\text{—}R_1 \end{array} \longrightarrow \begin{array}{c} R_2 \diagup N \\ R_3 \diagdown S \diagup R_1 \end{array}$$

6   **10**

Scheme 6

2,4-DIALKYLTHIAZOLES AND DERIVATIVES (TABLE II-4). The condensation of α-haloalkylketones (**6**), $R_3 = H$, with alkylthioamides or their derivatives gave the 2,4-dialkylthiazoles. For example, 2,4-dimethylthiazole (**10**), $R_1 = R_2 = Me$, $R_3 = H$, and 2-ethyl-4-methylthiazole (**10**), $R_1 = Et$, $R_2 = Me$, $R_3 = H$, were prepared for the first time in 1890 by Hubacher (15) and Roublef (16) from chloroacetone and thioacetamide, but the yields were low because of the instability on the thioamide in acidic medium.

TABLE II-4. 2,4-DIALKYLTHIAZOLES AND DERIVATIVES

$\underset{R_1}{\overset{N}{\underset{S}{\bigvee}}}R_2$

R_1	R_2	Conditions[a]	Yields (%)	Ref.
Me	Me	Cl or Br, dioxane or alcohol heat	—	4, 10, 22, 102, 127, 220, 222, 364, 426
Me	Et	Br, heat	—	16, 493
Me	$(CH_2)NMe_2$	Br, alcohol reflux	—	392
Me	$(CH_2)_2N$-piperidyl	Br, alcohol reflux	—	392
Me	$(CH_2)_2N$-phthalimido	Br, alcohol reflux	—	337
Me	$(CH_2)_2N$-phthalimido	Br, alcohol reflux	—	337
Me	CH_2Cl	Cl, Me_2CO, heat	—	87, 98
Me	tert-Bu	Br, P_2S_5, dioxane, heat	—	739
Et	Me	Cl or Br, heat	77	15
Et	tert-Bu	Br, P_2S_5, dioxane, heat	86	739
Pr	Me	Cl, alcohol heat	69	281
$(CH_2)HC(Me)_2$	Me	Br, alcohol heat	—	128

TABLE II.4 (*continued*)

R_1	R_2	Conditions[a]	Yields (%)	Ref.
$(CH_2)_{10}NHCOPh$	Me	Cl, alcohol heat	—	281
N-Phtalimidopropyl	Me	Cl, heat	—	281
$(CH_2)_2NHCOPh$	Me	Cl, alcohol reflux	—	281, 337
$(CH_2)_2NHCOPh$	$(CH_2)_2$N-phthalimido	Cl, alcohol reflux	71	337
$(CH_2)_2NHCOPh$	Me	Cl, alcohol heat	—	274
$(CH_2)NHCOPh$	$(CH_2)_2$N-phthalimido	Br, abs. alcohol at 0°C, and then kept overnight at room temperature	70	337
CH_2Ph	Me	Cl, alcohol heat	79	297, 595
CH_2Ph	$(CH_2)_2NMe_2$	Br, alcohol reflux	64	392
CH_2Ph	$(CH_2)_2$N-piperidinyl	Br, alcohol reflux	47	392
CH_2Ph	COEt	Br, alcohol reflux	—	332

Structure		R	Conditions	Yield	Ref.
(phthalimide-CH$_2$N structure)	CH$_2$Cl			32	190
		iso-Pr	Cl, abs. alcohol reflux		
		tert-Bu	Br, P$_2$S$_5$, dioxane, heat	80	739
		Me	Cl, heat	—	306
HC(Me)NHCOPh		Me	Br, alcohol reflux 3 hr, then hydrolysis	—	569
HC(Me)OH		Me	Cl, heat	—	281
HC(Me)NHSO$_2$C$_6$H$_4$-NHCOPh-p-					
HC(Ph)$_2$		Me	Cl, CHCl$_3$, reflux 24 hr	—	612
HC(Ph)$_2$		Et	Cl, CHCl$_3$, reflux 24 hr	—	612
HC(Ph)$_2$		Pr	Br, CHCl$_3$, reflux 24 hr	—	612
HC(p-MeOC$_6$H$_4$)$_2$		Me	Cl, alcohol reflux 5 hr	—	624
HC(p-MeOC$_6$H$_4$)$_2$		Et	Cl, alcohol reflux 5 hr	—	624
HC(p-MeOC$_6$H$_4$)$_2$		Pr	Cl, alcohol reflux 5 hr	—	624
HC(p-MeOC$_6$H$_4$)$_2$		tert-Bu	Cl, alcohol reflux 5 hr	—	624
HC(Ph)OH		Me	Cl, alcohol reflux 5 hr	22–91	75, 297
HC(Ph)OH		CH$_2$Cl	Cl, Me$_2$CO, reflux	60	75
(CHOAc)$_3$CH$_2$OAc (D-glucose)		Me	Cl, alcohol reflux	27	660
		tert-Bu	Br, P$_2$S$_5$, dioxane, heat	55	739
		tert-Bu	Br, P$_2$S$_5$, dioxane, heat	62	739
C(Me)$_2$NHAc		3,4-(OH)$_2$C$_6$H$_3$	Br, heat	—	73
CH=CHAr		Me	Br, alcohol reflux 2 hr	—	576
Cyclohexyl		Me	Cl, P$_2$S$_5$, dioxane	—	649

^{a} See Table II-1.

As described in the case of 4,5-dialkylthiazole synthesis, yields can be increased by preparing thioamides directly in the reaction mixture. In this way Schwarz (222) obtained 2,4-dimethylthiazole in 45% yield, in benzene as solvent and in the presence of an excess of chloroacetone. With the more reactive bromoacetone, Kurkjy and Brown (364) and Beraud and Metzger (426), on the other hand, increased the yield to 70 to 76% using the same solvent and an excess of amide. But it seems that the best solvent for this cyclization is dioxane, as used by Kurkjy and Brown (364). Using this solvent with magnesium carbonate added to reduce its acidity, they prepared 2,4-dialkylthiazoles in good yields (55 to 80%) (739).

2-Arylvinylthiazoles (**12**) were prepared from **11** and bromomethylketones (Scheme 7) (576), with R_1 = aryl or heteroaryl groups (576) and R_2 = Ph (80 to 95% yield) and Me (50 to 60% yield).

By condensing α-haloketones with diacetylaminothioacetamide (**13**), Pyl et al. (533) obtained the corresponding 2-aminomethylthiazoles (**15**) after hydrolysis (Scheme 8).

Other 2-acetyl and benzoyl aminoalkylthiazole derivatives have been prepared similarly (73, 80, 281, 337).

4-Dimethylaminoethylthiazoles (**17**) were prepared from thioamides and 1-bromo-4-dimethylamino-2-butanone (**16**), with R_1 = Me, CH_2Ph, $(CH_2)NHCOPh$, and $(Ac)_2NCH_2$ (Scheme 9) (337, 392).

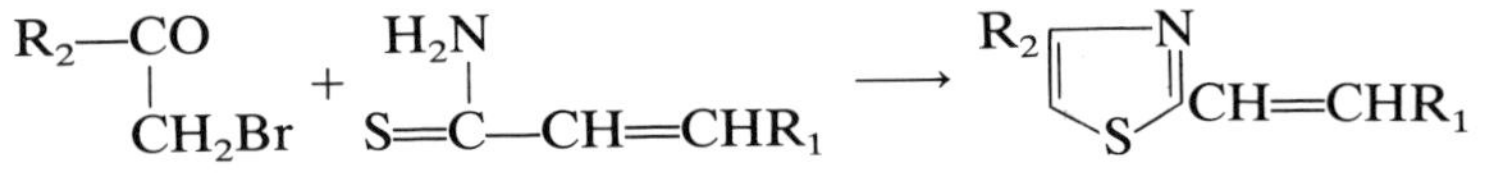

Scheme 7

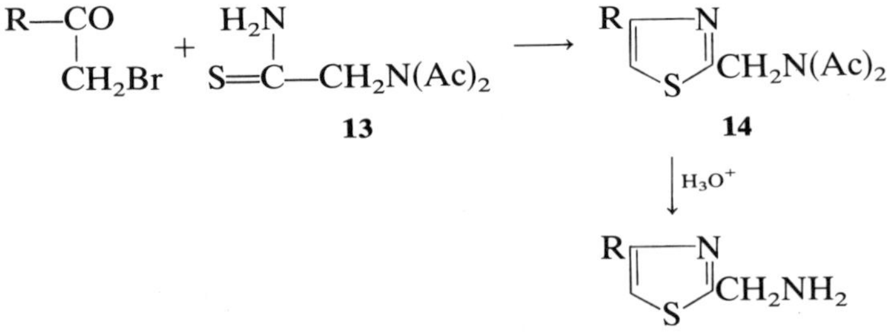

Scheme 8

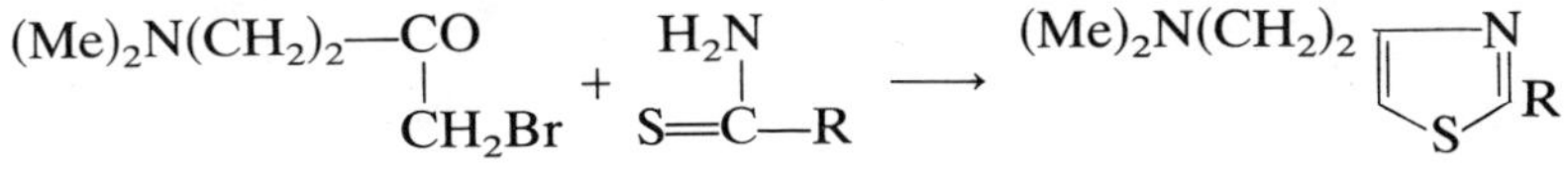

II. Thiazoles from α-Halocarbonyl Compounds and Derivatives

$$(Me)_2N(CH_2)_2-\underset{\underset{16}{CH_2Br}}{\overset{CO}{|}} + \underset{S=\underset{|}{C}-R}{H_2N} \longrightarrow (Me)_2N(CH_2)_2\underset{17}{\underset{S}{\left[\begin{array}{c}N\\\end{array}\right]}R}$$

Scheme 9

The 4-piperidinoethyl and 4-phtalimidoethyl derivatives have also been prepared by treating α,α-diarylthioacetamide (**18**) with α-haloketones either in an inert solvent such as chloroform or in alcohol (624): 2-benzhydrylthiazole derivatives (**19**) were obtained (Scheme 10).

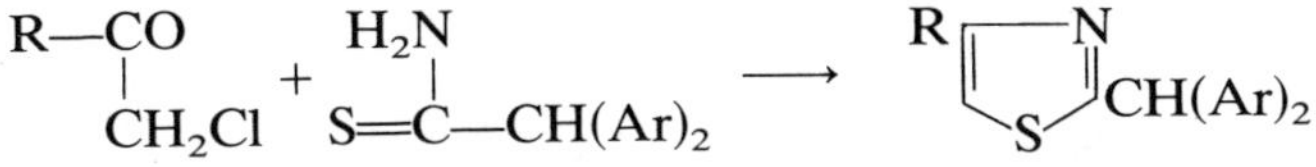

$$\underset{\underset{CH_2Cl}{|}}{R-CO} + \underset{\underset{18}{S=\underset{|}{C}-CH(Ar)_2}}{H_2N} \longrightarrow \underset{19}{\underset{S}{R\left[\begin{array}{c}N\\\end{array}\right]CH(Ar)_2}}$$

Scheme 10

sym-Dichloroacetone (**20**) condensed with alkylthioamides (Table II-4) or arylthioamides (Table II-5) (362, 630, 633, 638, 651) in acetone at room temperature first gave the insoluble 2-thiazoline hydrochloride salt (**21**), which is converted almost quantitatively by refluxing for 30 min in alcohol (or acetone) to the corresponding 2-substituted 4-chloromethylthiazoles (**22**), with $R_1 = Me$ (88, 98), alkyl (351, 584), HC(Ph)OH (75), *N*-phtalimidobutyl (190, 454) (Scheme 11). Yields ranged from 50 to 60%, and with $R_1 = $ aryl (618, 638, 651) yields reached 80 to 90%).

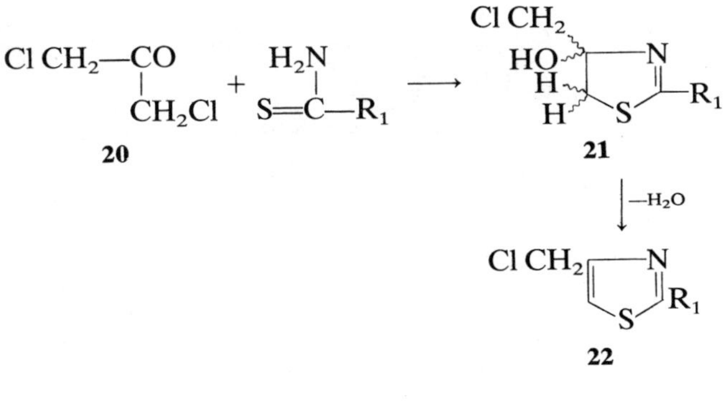

$$\underset{\underset{CH_2Cl}{|}}{Cl\,CH_2-CO} + \underset{S=\underset{|}{C}-R_1}{H_2N} \longrightarrow \underset{21}{\overset{Cl\,CH_2}{\underset{H}{\overset{HO}{\underset{H}{\left[\begin{array}{c}N\\S\end{array}\right]}}-R_1}}}$$

$$\downarrow -H_2O$$

$$\underset{22}{\underset{S}{Cl\,CH_2\left[\begin{array}{c}N\\\end{array}\right]R_1}}$$

Scheme 11

TABLE II-5. 2,4-ALKYLARYL AND ARYLALKYLTHIAZOLES

$$\underset{R_1}{\overset{N}{\underset{S}{\bigvee}}}R_2$$

R_1	R_2	Conditions[a]	Yield (%)	Ref.
Me	Ph	Br, P_2S_5, dioxane 20 min at 45°C then heat 100°C 6 hr; or Br, alcohol reflux 1–2 hr	66–85	6, 54, 101, 157, 472, 563, 641
Me	o-$HCO_2C_6H_4$	Br, alcohol heat	—	272
Me	p-MeC_6H_4	Br, alcohol reflux 1–2 hr	—	652
Me	p-ClC_6H_4	Br, alcohol reflux 1–2 hr	—	652
Me	p-BrC_6H_4	Br, alcohol reflux 1–2 hr	—	652
Me	p-biphenyl	Br, alcohol reflux 1–2 hr	—	652
Me	p-$O_2NC_6H_4$	Br, alcohol reflux	91	720
Me	p-$PhOC_6H_4$	Br, alcohol reflux	—	652
Me	p-$(HOCH_2CH_2)_2NC_6H_4$	Cl, alcohol reflux	—	721
Et	Ph	Br, P_2S_5, dioxane, heat	67	641
Et	p-$H_2NC_6H_4$	Cl, alcohol reflux 1 hr	—	580
Et	p-$O_2NC_6H_4$	Br, alcohol reflux 1 hr	—	580
n-Pr	Ph	Br, P_2S_5, dioxane, heat	62	641
$(CH_2)_3N\begin{smallmatrix}O\\O\end{smallmatrix}$	3,4-$(OH)_2C_6H_3$	Br, alcohol reflux	—	80
$(CH_2)_2NHCOPh$	Ph	Cl or Br, alcohol reflux	—	43, 337
$CH_2HC(Me)NHMe$	3,4-$(OH)_2C_6H_3$	Br, alcohol reflux	—	281
CH_2NHAc	3,4-$(OH)_2C_6H_3$	Cl, alcohol reflux	—	73
$CH_2N(Me)Ac$	3,4-$(OH)_2C_6H_3$	Cl, alcohol reflux	—	73
CH_2CO_2Ph	p-$O_2NC_6H_4$	Br, alcohol reflux	84	409
CH_2Ph	p-$H_2NC_6H_4$	Cl, abs. alcohol reflux 1 hr	—	580

p-MeOC₆H₄CH₂	p-H₂NC₆H₄		Cl, abs. alcohol reflux 1 hr	—	580
CH₂OPh	p-H₂NC₆H₄		Cl, abs. alcohol reflux 1 hr	—	580
o- and p-MeC₆H₄OCH₂	p-H₂NC₆H₄		Cl, abs. alcohol reflux 1 hr	—	580
iso-Pr	Ph		Br, P₂S₅, dioxane, heat	60	641
HC(CH₂NHMe)OH	3,4-(OH)₂C₆H₃		Cl, alcohol reflux	—	281
(CHOAc)₃CH₂OAc	Ph		Br, alcohol reflux	—	660
(CHOAc)₃CH₂OAc	Ph		Br, alcohol reflux	23	660
CH=CHAr	Ph		Br, alcohol reflux	—	285, 576
HC(PhCO₂)COPh	Ph		Br, alcohol reflux	—	419
HC(PhCO₂)COPh	p-MeC₆H₄		Br, alcohol reflux	—	419
HC(PhCO₂)COPh	p-ClC₆H₄		Br, alcohol reflux	—	419
HC(PhCO₂)COPh	p-MeOC₆H₄		Br, alcohol reflux	—	419
HC(PhCO₂)COPh	p-O₂NC₆H₄		Br, alcohol reflux	—	419
HC(PhCO₂)COPh	2,5-Cl₂C₆H₃		Br, alcohol reflux	—	419
HC(PhCO₂)Ph	3,5-Cl₂C₆H₃		Br, alcohol reflux	—	419
tert-Bu	Ph		Br, P₂S₅, dioxane, heat	60	641
Ph	Me		Cl, abs. alcohol reflux 6 hr	80	641
Ph	(CH₂)₂NMe₂		Br, alcohol reflux	49	392
Ph	(CH₂)₂N⟨⟩ (piperidine)		Br, alcohol reflux	42	392
Ph	CH₂Cl		Cl, Me₂CO overnight at room temperature, then alcohol or Me₂CO reflux 4 hr	—	638
Ph	tert-Bu		Cl, abs. alcohol reflux 6 hr	68	641
m-MeC₆H₄	CH₂Cl		Cl, Me₂CO, reflux 4 hr	85	651
p-MeC₆H₄	CH₂Cl		Cl, Me₂CO, reflux 4 hr	90	651
p-MeOC₆H₄	CH₂Cl		Cl, Me₂CO, reflux 4 hr	80	651
p-ClC₆H₄	CH₂Cl		Cl, Me₂CO, reflux 4 hr	—	651
o-AcNHC₆H₄	Me		Br, alcohol reflux	—	239
m-AcNHC₆H₄	Me		Br, alcohol reflux	—	239
α-naphthyl	CH₂Cl		Cl, Me₂CO, reflux 4 hr	—	651
β-naphthyl	CH₂Cl		Cl, Me₂CO, reflux 4 hr	—	651

[a] See Table II-1.

These products are used as starting material for the preparation of 2-substituted thiazol-4-ylacetic acids. α-Benzoyloxythiopropionamide and α-benzoyloxy-α-benzoylthioacetamide condensed with an equimolar amount of an α-haloketone in alcoholic solution yield the following compounds: (409, 419, 569): **24**, $R_1 = CH_3$, PhCO, $R_2 = $ Me or Ph, and $R_3 = $ H (Scheme 12), which after hydrolysis give the corresponding 2-(α-hydroxyalkyl)thiazoles (**25**).

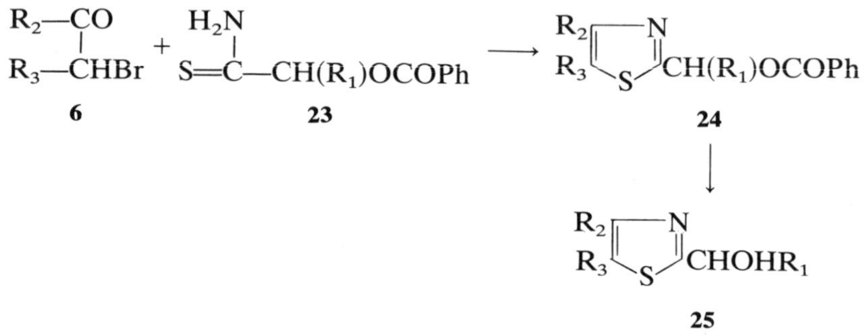

$$R_2\text{—CO} \quad H_2N$$
$$R_3\text{—CHBr} \quad + \quad S=C\text{—CH}(R_1)\text{OCOPh} \quad \longrightarrow \quad \underset{S}{\underset{R_3}{R_2}}\diagdown N \diagup \text{CH}(R_1)\text{OCOPh}$$
$$\quad 6 \qquad\qquad\qquad 23 \qquad\qquad\qquad\qquad\qquad 24$$

$$\downarrow$$

$$\underset{S}{\underset{R_3}{R_2}}\diagdown N \diagup \text{CHOHR}_1$$
$$25$$

Scheme 12

In a similar way, DL-2-(α-hydroxyalkyl)- and 2-(α-alkoxycarbonyl)-4-methyl-5-(β-hydroxyethyl)thiazoles were synthetized from the corresponding thioamides and 4-hydroxy-3-bromo-2-pentanone (615).

The cyclization of pentaacetyl-D-gluconic thioamide with chloroacetone and of pentaacetyl-D-galactonic acid thioamide with phenacyl bromide give the corresponding 4-substituted 2-(D-galactopentaacetoxypentyl)-thiazoles (**27**) (660) but in low yield (23 to 27%) (Scheme 13). The products may be deacetylated in the usual way. These compounds are interesting from a pharmacological point of view.

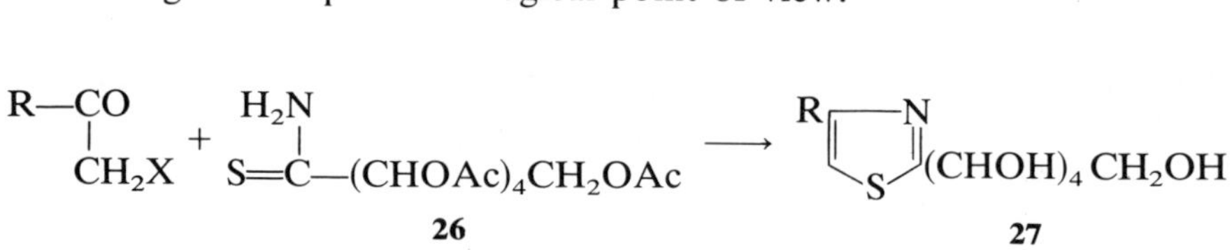

$$R\text{—CO} \quad H_2N$$
$$CH_2X \quad + \quad S=C\text{—(CHOAc)}_4CH_2OAc \quad \longrightarrow \quad \underset{S}{R}\diagdown N \diagup \text{(CHOH)}_4\,CH_2OH$$
$$\quad\qquad\qquad\qquad 26 \qquad\qquad\qquad\qquad\qquad 27$$

Scheme 13

2,4-ARALKYL OR ALKYLARYLTHIAZOLES AND 2,4,5-TRISUBSTITUTED THIAZOLES (TABLES II-5 AND II-6). Alkylthioamides and their substituted derivatives have been condensed with ω-bromoacetophenones (**28**) to

TABLE II-6. 2,4,5-TRISUBSTITUTED THIAZOLES

$$R_1 \underset{S}{\overset{N}{\underset{\|}{\bigvee}}} \genfrac{}{}{0pt}{}{R_2}{R_3}$$

R_1	R_2	R_3	Conditions[a]	Yield (%)	Ref.
Me	Me	Me	Cl, or Br, P_2S_5, dioxane or C_6H_6 (anhydrous) at 40°, and then heat;	42–70	4, 10, 16, 22
Me	Me	Et	Br, P_2S_5, C_6H_6 (anhydrous), heat.	26.5–76	426, 711
Me	Me	$(CH_2)_2OH$	Br, alcohol reflux	—	612b
Me	Me	CH_2Ph	1-Cl, 1,2-epoxide, alcohol reflux 17–20 hr	60	472, 504
Me	Me	$2,5\text{-Me}_2C_6H_3CH_2$	1-Cl, 1,2-epoxide, alcohol reflux 17 hr	73	504
Me	Me	$2,4,6\text{-Me}_3C_6H_2CH_2$	1-Cl, 1,2-epoxide, alcohol reflux 17 hr	43	504
Me	Me	COMe	Cl, P_2S_5, C_6H_6, alcohol heat	85	220, 711
Me	Me	Ph	Br, alcohol reflux	—	47, 285
Me	Me	α-Naphthyl	1-Cl, 1,2-epoxide, alcohol reflux 24 hr	32	504
Me	Et	Ph	Br, alcohol reflux	—	472
Me	iso-Pr	Me	Br, P_2S_5, C_6H_6	15	426
Me	CH_2Ph	Me	Br, alcohol reflux	—	472
Me	CH_2Ph	Et	Br, alcohol reflux	50	472
Me	Ph	Me	Br, alcohol reflux	—	54, 472, 618
Me	Ph	Et	Br, alcohol reflux	50	472
Me	Ph	Ph	Br, alcohol reflux	—	563
Me	Ph	$p\text{-ClC}_6H_4$	Br, alcohol reflux	—	563
Me	Ph	$p\text{-O}_2NC_6H_4$	Br, alcohol reflux	—	563
Me	Ph	N(R)R'	Cl, tetrahydrofuran at 0°C	63–100	754
Me	$p\text{-ClC}_6H_4$	Ph	Cl, tetrahydrofuran at 0°C	—	563
Me	$2,4\text{-(MeO)}_2C_6H_3$	$p\text{-O}_2NC_6H_4$	Cl, tetrahydrofuran at 0°C	—	563
CH_2Ph	Me	Me	Br, P_2S_5, dioxane, heat	—	595
CH_2Ph	Me	Me	Br, P_2S_5, dioxane, heat	—	595

TABLE II-6. (continued)

R_1	R_2	R_3	Conditions[a]	Yield (%)	Ref.
HC(p-MeOC$_6$H$_4$)$_2$	Me	Me	Cl, CHCl$_3$, reflux 24 hr	—	624
HC(p-MeOC$_6$H$_4$)$_2$	Et	Me	Cl, CHCl$_3$, reflux 24 hr	—	624
HC(p-MeOC$_6$H$_4$)$_2$	—(CH$_2$)$_4$—		Cl, CHCl$_3$, reflux 24 hr	—	624
HC(p-MeOC$_6$H$_4$)$_2$	Ph	Me	Cl, CHCl$_3$, reflux 24 hr	—	624
CH$_2$OAr	Me	CONHPh	Cl, MeOH, reflux	—	724
CH$_2$OAr	Me	CO$_2$Et	Cl, MeOH, reflux	—	724
HC(Me)OAr	Me	CONHPh	Cl, MeOH, reflux	—	724
HC(Me)OAr	Me	CO$_2$Et	Cl, MeOH, reflux	—	724
HC(p-MeOC$_6$H$_4$)$_2$	Ph	Ph	Cl, CHCl$_3$, reflux 24 hr	—	624
HC(Me)CO$_2$Ph	Me	Me	Cl, C$_6$H$_6$, reflux 2 hr	70	352
HC(Me)OH	Me	Me	Br, alcohol reflux 3 hr, and hydrolysis	—	569
HC(Me)OH	Me	(CH$_2$)$_2$OH	Br, alcohol reflux 3 hr	—	615
HC(Me)OH	Et	Me	Br, alcohol reflux 3 hr	—	569
Cyclohexyl	Me	Me	Br, alcohol reflux 3 hr	—	649
Ph	Me	Me	Cl, alcohol, Na$_2$CO$_3$, reflux	64	101, 515
Ph	o-MeC$_6$H$_4$	Me	Br, abs. alcohol 2 hr, steam bath	85	589
Ph	m-MeC$_6$H$_4$	Me	Br, abs. alcohol 2 hr, steam bath	85	589
Ph	p-MeC$_6$H$_4$	Me	Br, abs. alcohol 2 hr, steam bath	85	589
Ph	o-O$_2$NC$_6$H$_4$	Me	Br, abs. alcohol 2 hr, steam bath	45	589
Ph	m-O$_2$NC$_6$H$_4$	Me	Br, abs. alcohol 2 hr, steam bath	82	589
Ph	p-O$_2$NC$_6$H$_4$	Me	Br, abs. alcohol 2 hr, steam bath	85	589
Ph	Ph	Ph	Cl, alcohol reflux	—	15, 519

[a] See Table II-1.

give 2-alkyl-4-arythiazoles (**29**) in fairly good yields (6, 80, 275, 285, 337, 419, 172, 563, 576, 580, 634, 641, 652, 733, 767):

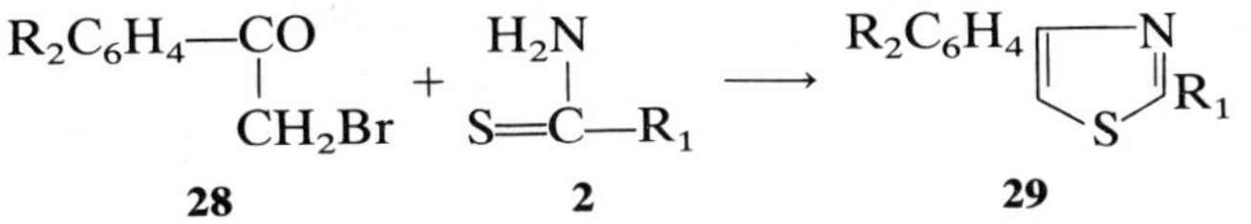

$$R_2C_6H_4-\underset{\underset{\textbf{28}}{CH_2Br}}{CO} \quad + \quad \underset{\underset{\textbf{2}}{S=C-R_1}}{H_2N} \quad \longrightarrow \quad \underset{\textbf{29}}{R_2C_6H_4\underset{S}{\overset{N}{\diagdown\diagup}}R_1}$$

Scheme 14

Results are summarized in Table II.5.

The reaction can be carried out in two steps (641). First, equimolar amounts of amide and phosphorus pentasulfide are mixed under stirring in dioxane, the temperature being kept below 45°C. After 20 minutes, the α-halocarbonyl compounds (in dioxane solution) are added in small portions. At the end of the addition the temperature reaches 80 to 100°C, and the reaction mixture is kept at this temperature for another hour.

4-Alkyl-2-arylthiazoles and 4,5-disubstituted-2-(p-aminophenyl thia-zoles were similarly prepared from arylamides and α-halomethyl-ketones in alcoholic (239, 392, 641, 792) or acetonic solution (638, 651).

3-Iminothiobutyramide (**30**), containing four nucleophilic centers (only two of which might react with two electrophilic sites in phenacylbromide), undergoes the Hantzsch reaction preferentially, yielding the enamine (**31**) in dry dioxane or (4-phenylthiazol-2-yl)acetone (**32**) in isopropanol. Other enamines are obtainable from the ketone (**32**) by standard methods (626) (Scheme 15).

Scheme 15

1-Chloro-1,2-epoxydes (**33**) condensed with thioamides lead to 2,4,5-trisubstituted thiazoles (Scheme 16 and Table II-6) (504).

For example, by refluxing thioacetamide with 1-phenyl-2-chloro-2,3-epoxybutane (**33**), R_2 = Me, R_3 = CH_2Ph, in alcohol solution for 17 hr, 2,4-dimethyl 5-benzylthiazole was obtained (**10**), R_1 = R_2 = Me, R_3 = CH_2Ph. This product was found identical to that obtained from 4-phenyl-3-chloro-2-butanone.

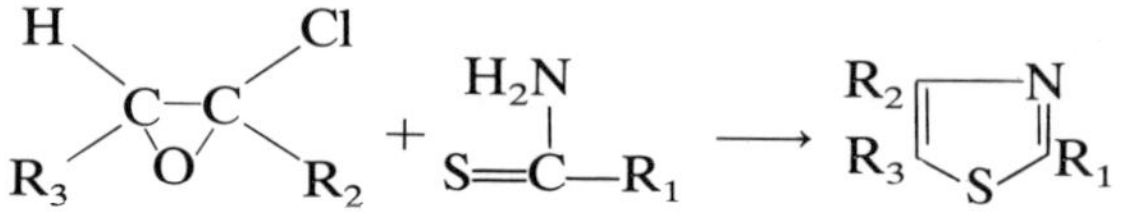

Scheme 16

2,4-Diarylthiazoles and Polycyclic Compounds with Thiazole Ring. 2,4-Diarylthiazoles (**34**) (Table II-7) were prepared in 1950 by Erlenmeyer et al. (327) from thiobenzamide and ω-bromoacetophenones in alcoholic solution, with yields of the order of 50%, whereas the 2,4-diphenylthiazole (**34**), R_1 = R_2 = H was obtained in 80–85% yield (Scheme 17) (15, 327, 519); others have prepared **34** with R_1 and R_2 = H, p-Me, p-Cl, p-Br (327), p-NH_2 (580), p-NO_2 (589), 3,4-$(OH)_2$ (73, 75, 78), m- and p-Ac (569).

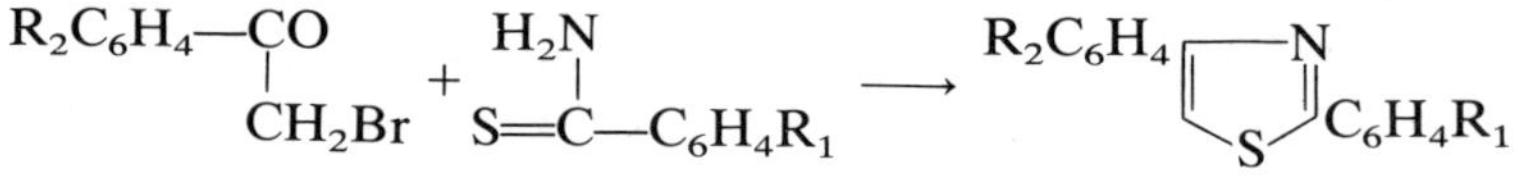

Scheme 17

TABLE II-7. 2,4-DIARYLTHIAZOLES

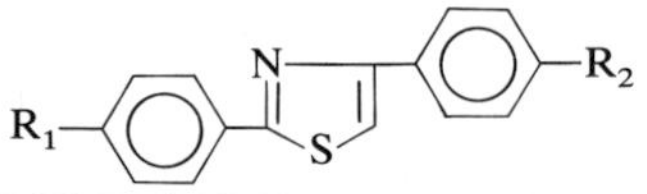

R_1	R_2	Conditions[a]	Yield (%)	Ref.
H	H	Br, abs. alcohol reflux	80–85	15, 327, 519
H	Me	Br, abs. alcohol reflux	54	327
H	Cl	Br, abs. alcohol reflux	47	327
H	Br	Br, abs. alcohol reflux	41	327
H	NH_2	Cl, abs. alcohol reflux	—	580
H	NO_2	Br, abs. alcohol reflux	80	589
H	Ph	Br, abs. alcohol reflux	49	574
H	3,4-$(OH)_2C_6H_3$	Cl or Br, abs. alcohol reflux	—	73, 78
Me	H	Br, abs. alcohol reflux	51.5	327
Cl	H	Br, abs. alcohol reflux	53	327

TABLE II-7. (continued)

R_1	R_2	Conditions[a]	Yield (%)	Ref.
Br	H	Br, abs. alcohol reflux	49	327
R	H	Cl, abs. alcohol reflux	—	580
m-Ac	H	Br, abs. alcohol reflux	—	569
p-Ac	H	Br, abs. alcohol reflux	—	569
Ph	H	Br, abs. alcohol reflux	50	569
Me	Me	Br, abs. alcohol reflux	51.5	327
Me	Cl	Br, abs. alcohol reflux	46	327
Me	Br	Br, abs. alcohol reflux	45	327
Cl	Me	Br, abs. alcohol reflux	51.5	327
Cl	Cl	Br, abs. alcohol reflux	54	327
Cl	Br	Br, abs. alcohol reflux	55	327
Br	Me	Br, abs. alcohol reflux	48.5	327
Br	Cl	Br, abs. alcohol reflux	50	327
Br	Br	Br, abs. alcohol reflux	52	327
MeO	3,4-(OH)$_2$	Br, abs. alcohol reflux	—	75
3,4-(OH)$_2$	3,4-(OH)$_2$	Br, abs. alcohol reflux	80	75
3,4-(CH$_2$O$_2$)	3,4-(OH)$_2$	Br, abs. alcohol reflux	90	75
m-Ac	Cl	Br, abs. alcohol reflux	—	569
p-Ac	Cl	Br, abs. alcohol reflux	—	569
m-Ac	NO$_2$	Br, abs. alcohol reflux	—	569
p-Ac	NO$_2$	Br, abs. alcohol reflux	—	569
Ph	Ph	Br, abs. alcohol reflux	68.5	574

[a] See Table II-1.

Several 2-phenyl-4-(substituted)arylthiazoles and Schiff's bases derived therefrom have been also reported (732).

Thiobenzamide reacts also with α- or β-naphthylbromomethylketone to give 2-phenyl-4-α- or β-naphthylthiazoles (54).

A series of *meta*- and *para*-bis (2-thiazolyl) benzenes and of *meta*- and *para*-bis(4-thiazolyl)benzenes of general formula **35** and **36** was prepared in higher yields (60–90%) from the appropriate bis-(haloacetyl)benzenes with a suitable thioamide (Scheme 18a and Table II-8) (573, 574).

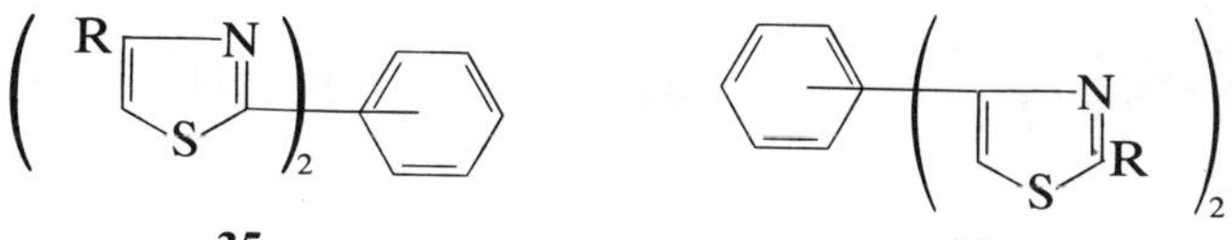

35   **36**

Scheme 18a

TABLE II-8. META- AND PARA-BIS (2-THIAZOLYL) BENZENES (35) AND META- AND PARA-BIS(4-THIAZOLYL) BENZENES (36)

		35		36	
		Yield (%)			
R	35a[a]	35b	36a	36b	Ref.
Me	—	44	88	—	573, 574
Ph	—	64.4	—	59.5	574
p-Biphenylyl	—	65	—	44	576
p-O$_2$NC$_6$H$_4$	70	—	—	—	592

[a] a and b, m- and p-phenylene compounds, respectively.

Similarly 2-thiazolylphenacylbromide (37) condenses with thiobenzamides in alcohol to give 38, R = H, Ph, p-ClC$_6$H$_4$, p-O$_2$NC$_6$H$_4$ (Scheme 18b) (569).

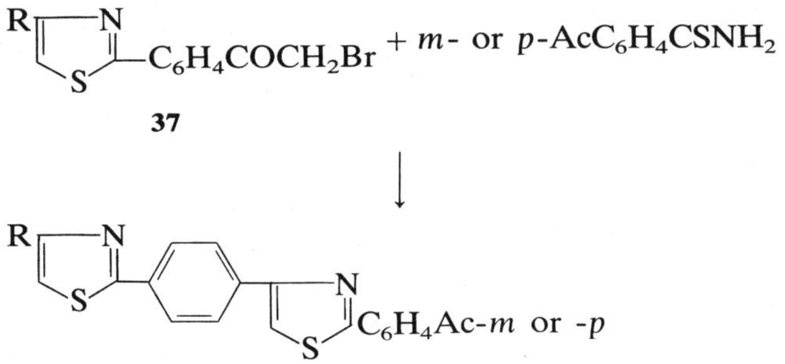

Scheme 18b

The bis α-chloroketone (39), X = O or S, condensed with thioacetamide in alcohol gave the corresponding thiazole (40) in 85% yield (Scheme 19) (410).

$(p\text{-ClCH}_2\text{COC}_6\text{H}_4\text{—})_2\text{X} + \text{MeCSNH}_2 \longrightarrow$

Scheme 19

2,2'-Oxydi-p-phenylene-bis(4-phenyl and biphenyl-4-yl)thiazoles were also prepared in 62 to 63% yield (574).

The following 2,2'-R,R-disubstituted 4,4'-bithiazoles (**41**) were obtained by refluxing dibromobiacetyl and thioamides for 20 min to 3 hr in alcohol (Scheme 20) (508):

$$(BrCH_2CO)_2 + RCSNH_2 \longrightarrow$$

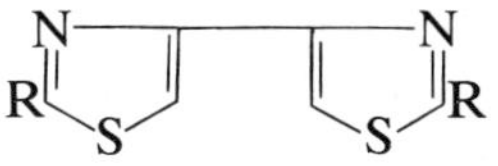

41

Scheme 20

41 (R)	Yield (%)
Ph	70
CH_2CO_2Ph	70
$HC(Me)CO_2Ph$	60
$HC(Ph)CO_2Ph$	20

Bromomethyl-1-adamantyl ketones were condensed with thioamides of carboxylic or carbonic acids to give the corresponding thiazoles (613).

N,N'-Diacetylethylenediamine-N,N'-dithioacetamide (**42**) reacts with ω-bromoacetophenone and its p-substituted derivatives to give the expected N,N'-diacetyl-N,N'-bis(4-phenyl-2-thiazolylmethylethylene diamine) (**43**) (Scheme 21); with R = H, Me, Cl, Br, MeO, yields ranged from 65 to 80% (482).

2. $(RC_6H_4COCH_2Br) + (H_2NC(=S)CH_2N(Ac)CH_2-)_2$

42

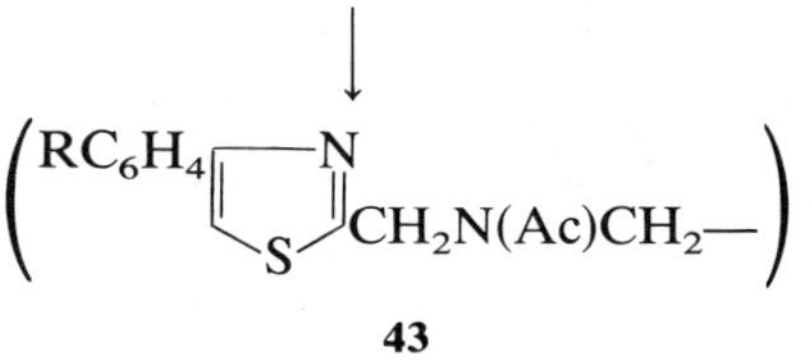

↓

$$\left(RC_6H_4 \underset{S}{\overset{N}{\diagup\!\!\!\diagdown}} CH_2N(Ac)CH_2- \right)$$

43

Scheme 21

THIAZOLES WITH HETEROCYCLIC SUBSTITUENTS. Thiazoles with heterocyclic substituents in the 2- or 4-position have been synthesized (Table II-9). Thus thioacetamide (or its α-substituted derivatives) react with bromomethyl heteroarylketones under reflux in alcohol to give the corresponding 2-methyl-4-heteroarylthiazoles: heteroaryl groups in the 4-position were 2'-thienyl (213, 692); α-pyrrolyl and 3-methyl derivatives

TABLE II-9. THIAZOLES WITH HETEROCYCLIC SUBSTITUENTS (10)

R_1	R_2	R_3	Conditions[a]	Yield (%)	Ref.
Me	α-Thienyl	H	Br, alcohol reflux	—	213, 692
Me	α-Pyrrolyl	H	Br, alcohol reflux	—	272
Me	3-Me-α-Pyrrolyl	H	Br, alcohol reflux	—	153b
Me	α-Furyl	H	Br, alcohol reflux	87	272, 692
Me	5-Aryl-2-furyl	H	Br, alcohol reflux	65–70	755
HC(Me)OH	4-Ph-2-thiazolyl	H	Br, alcohol reflux 5 hr	60	569
HC(Me)OH	4-(p-$O_2NC_6H_4$)-2-thiazolyl	H	Br, alcohol reflux 5 hr	—	569
HC(p-MeOC_6H_4)$_2$	α-, β-, or γ-pyridyl	H	Cl, alcohol reflux 5 hr	—	624
α-thienyl	p-BrC_6H_4	H	Br, alcohol reflux 5 hr	60	589
α-furyl	H	H	Cl, reflux	—	758
α-, β-, or γ-pyridyl	Me	H	Cl, alcohol reflux	—	238
3-pyridyl	Me, Et, CO_2Et	$CONH_2$, $CON(Et)_2$	Cl, reflux	—	759
3-Pyridyl	CONHMe, Ph	$CO_2(CH_2)_2$OMe CON(Et)Ph	Cl, reflux	—	759
2-, 3-, 4-, 5-, or 6-Quinolyl	Me	H	Cl, alcohol reflux	80–100	271
4-(p-BrC_6H_4)-2-thiazolyl	p-BrC_6H_4	H	Br, alcohol reflux 5 hr	73	590
![imidazolidinone structure]—C_6H_4R-p	p-RC_6H_4	—	Br, alcohol reflux	80–96	567
![piperidinyl structure]	p-RC_6H_4	—	Br, alcohol reflux	90–95	482
![piperidinyl structure]	Me	—	Cl, alcohol reflux	19	482

[a] See Table II-1.

II. Thiazoles from α-Halocarbonyl Compounds and Derivatives 197

(272); 2'-furyl (272, 692); 2'-, 3'-, or 4'-pyridyl (624); 4'-phenyl-2'-thiazolyl (569); 2'-quinolyl (643); 3'-indolyl (676); and thieno-[2,3d] (644).

In the reverse reaction, thioheteroaryl amides reacted under reflux in alcohol with haloketones or aldehydes to give the corresponding 2-heteroarylthiazole derivatives (238, 271, 482, 550, 751, 765, 776, 781). 2,2'-Bithiazoles (4,4'-disubstituted) have been obtained in 80 to 90% yield by cyclocondensation of 1 mole rubeanic acid with 2 moles of α-bromoketones in polyphosphoric acid at 95 to 135°C (780). Some multiheteroaryl substituted thiazoles have been also reported (704).

Piperazine-N,N'-dithioacetamide (**44**) condensed with ω-bromoacetophenones (482), give the corresponding thiazoles (**45**) Scheme 22), with R = H, Br, MeO; yield is 90 to 95% (482).

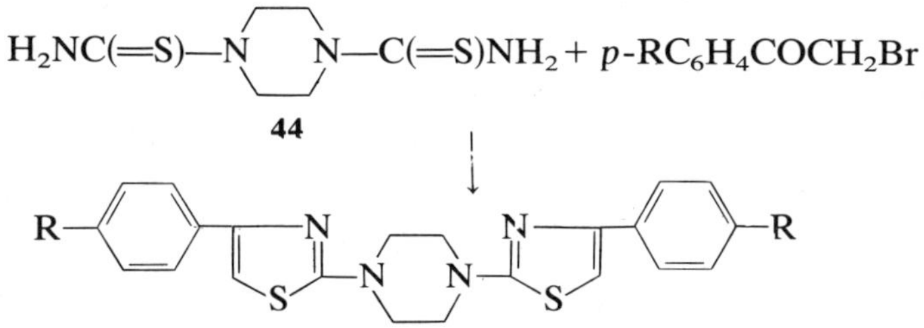

$$H_2NC(=S)—N\underset{\underset{44}{}}{\overset{}{\frown}\!\!\!\!\!\smile}N—C(=S)NH_2 + p\text{-}RC_6H_4COCH_2Br$$

$$R\text{-}C_6H_4\text{-thiazole-}N\text{-piperazine-}N\text{-thiazole-}C_6H_4\text{-}R$$

45

Scheme 22

N-acetylsarcosinethioamide reacts in a similar way (482).

1-(2-Thiazolyl)-2-imidazolones (**47**) were synthesized in alcoholic solution from 1-thiocarbamoyl-2-imidazolones (**46**) and α-bromoketones in

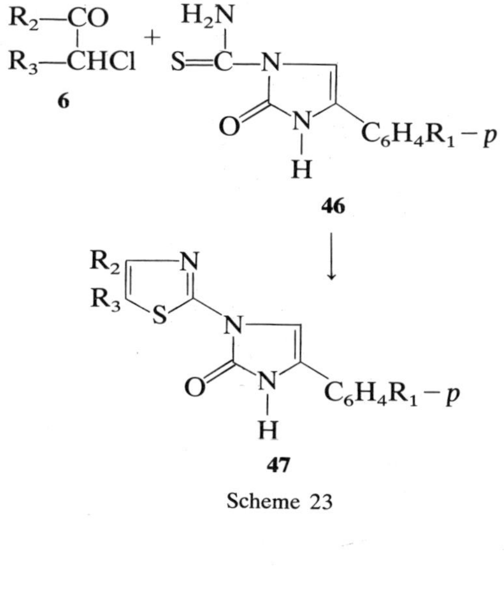

$$\underset{6}{\overset{R_2—CO}{\underset{R_3—CHCl}{}}} + \underset{46}{\overset{H_2N}{\underset{S=C—N}{}}} \cdots C_6H_4R_1\text{-}p$$

↓

47 — R_2,R_3-thiazole linked to imidazolone-$C_6H_4R_1$-p

Scheme 23

good yields (70 to 95%) (Scheme 23) (567), with R_1 = H, Me, Br, MeO; R_2 = H, Me, Br, MeO, NO_2; and R_3 = H or Ph.

An unusual reaction was reported in which 2-(4-pyridyl)-4-carboxyethylthiazole (**50**) prepared from thioisonicotinamide (**49**) and ethylbromopyruvate (**48**) was converted to **51**, which upon treatment with KNCS in the presence of $NaHCO_3$ gave **52** (618). This was again cyclized with **48** to yield 2-(4-pyridyl)-4-(4-ethoxycarbonyl-2-thiazolyl hydrazinocarbonyl)thiazole (**53**) (Scheme 24).

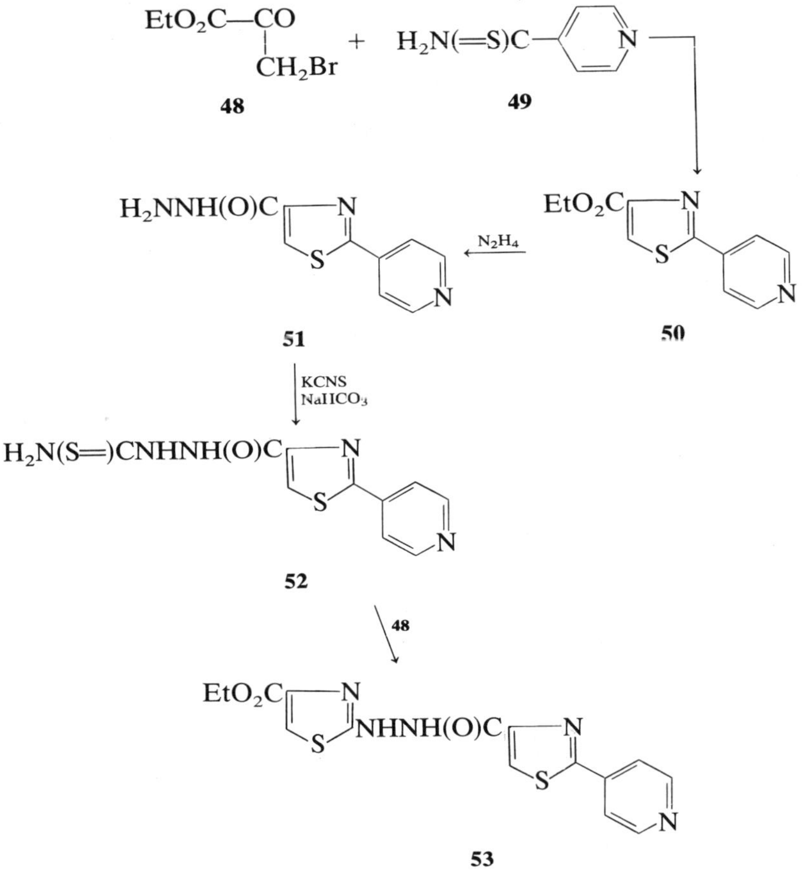

Scheme 24

By refluxing **54** and thioformamide for 16 hr in aqueous acetone solution, the expected compound **55** was obtained (Scheme 25). **55** in acetic medium, after treatment with hydrochloric acid and zinc, gives 2-(4-thiazolyl)benzimidazole derivatives **56** (554, 593).

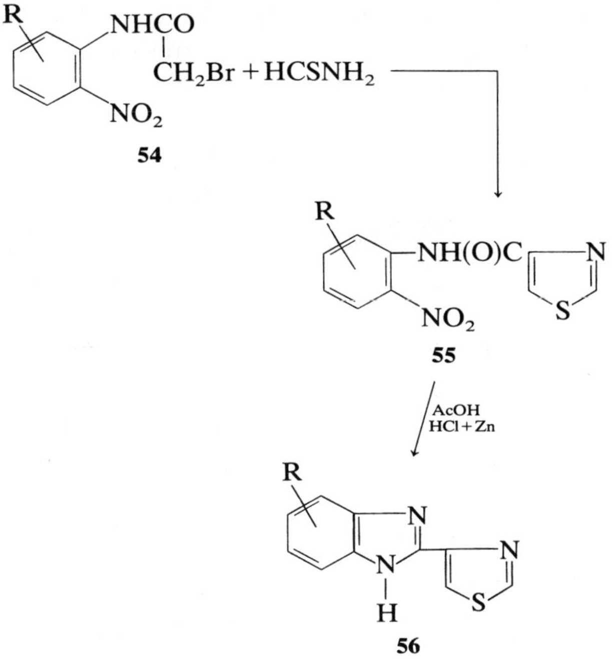

II. Thiazoles from α-Halocarbonyl Compounds and Derivatives

Scheme 25

2-Benzothiazolylbromomethyl ketone (**57**) reacts with thioacetamide to give 2-methyl-4-(2-benzothiazolyl)thiazole (**58**) in 56% yield (468) (Scheme 26). Phenanthro[3,4d]thiazoles were also prepared from 3-phenanthrenylamine (726).

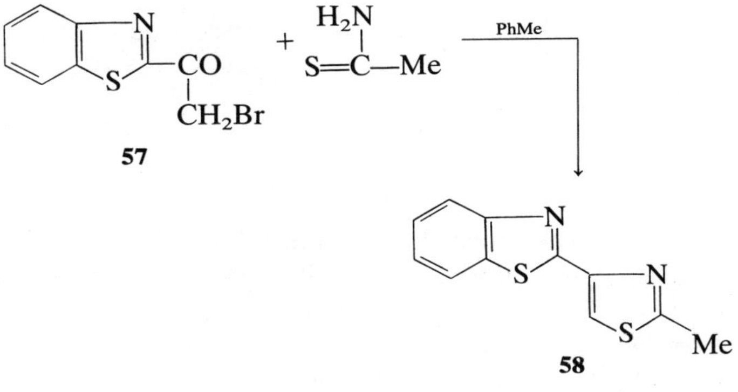

Scheme 26

Benzocycloheptathiazoles (**60**) were prepared by the reaction of 6-bromo-1-benzosuberone (**59**) with the corresponding thioamides, by refluxing for 4 hr in alcoholic solution (Scheme 27); with R = Me, yield is 35%.

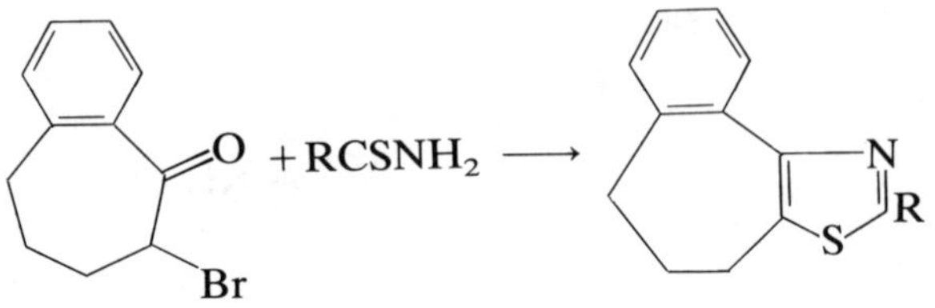

Scheme 27

Heating 5-amino-4-mercapto pyrimidines (**61**) with formic acid affords the corresponding thiazolo[5,4d]pyrimidines (**62**) (Scheme 28) (357, 382, 411, 431).

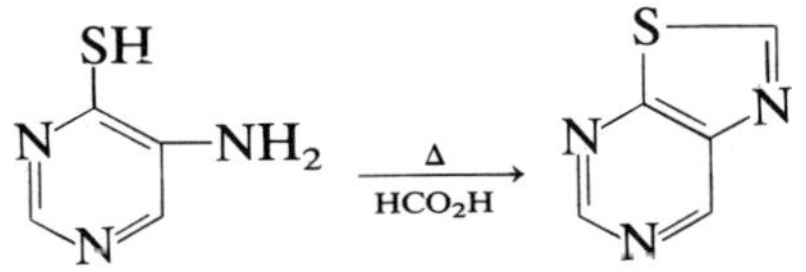

Scheme 28

In a similar way, 6-amino-1,3-dimethyluraciles (**63**) undergo easy conversion to the corresponding thiazolopyrimidines (**64**) upon treatment with thionyl chloride in pyridine solution (except with R = CF$_3$, where SO$_2$Cl$_2$ is more effective in the absence of pyridine) (Scheme 29) (654), with R = H, CO$_2$H, CO$_2$Et, Ph, or CF$_3$.

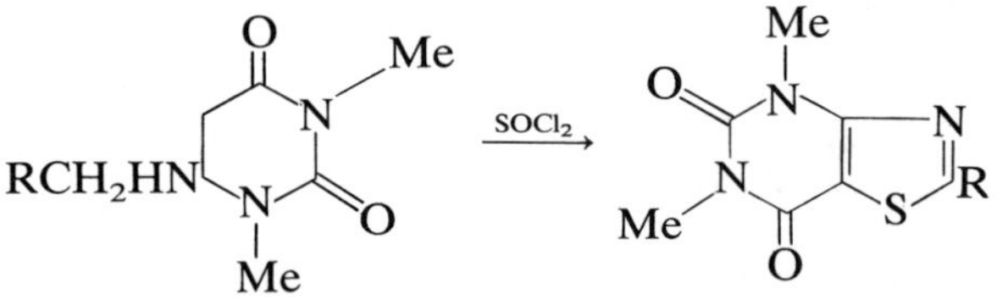

Scheme 29

Syntheses of some thiazolo[5,4d]thiazoles (685), 2-methylthieno-[3,2d]thiazoles (644), and 5-thiazolylbenzimidazoles (670) were also reported.

EXTENSION OF THE HANTZSCH'S SYNTHESIS TO THIAZOLE CARBOXYLIC AND THIAZOLE ACETIC ACIDS (TABLE II-10). Mono-, di-, and tricarboxylic acids are among the most easily prepared thiazole derivatives.

TABLE II-10. THIAZOLE CARBOXYLIC, THIAZOLE ALKANOIC ACIDS, AND THEIR DERIVATIVES (10)

R_1	R_2	R_3	Conditions[a]	Yield (%)	Ref.
Me	Me	CO_2Et	Br, heat	—	4, 10, 22, 58
Me	$CH_2CO_2(CH_2)_2Cl$	H	Cl or Br, C_6H_6, heat	82	4, 10, 22, 122, 220, 242
Me	CO_2Et	H	Cl, heat	50	247
Me	CO_2Et	COPh	Br, heat	60–75	19, 220, 221, 247, 273, 295
Me	CO_2Et	CO_2Et	Cl, heat	—	717
Et	CO_2Et	CO_2Et	Cl or Br	—	103, 145, 639
CH_2OAc	CH_2OAc	H	Cl, heat 2 hr	69	577, 639
CH_2CO_2Me	Ph	H	Cl, heat	—	753
CH_2CO_2Et	Me	H	Br, EtOH, 48 hr at 5°C	75(HBr)	469
CH_2NHAc	CO_2Et	p-$O_2NC_6H_4$	Cl, alcohol	—	295
CH_2NHAc	p-$O_2NC_6H_4$	CO_2Et	Cl, heat	57	489
$CH_2NHCOPh$	Me	CO_2Et	Br, MeOH, heat	91	489
$CH_2NHCOPh$	CH_2CO_2Et	H	Cl, heat	—	281
$CH_2NHCOPh$	CO_2Et	H	Cl, heat	—	281
$CH_2NHCOPh$	CO_2Et	p-$O_2NC_6H_4$	Br, heat	—	281
CH_2NHCO_2Et	Ph	H	—	60	489
$CH_2NHCOPh$	p-$O_2NC_6H_4$	CO_2Et	Br, heat	—	43
$CH_2NHCOPh$	Ph	H	Br, MeOH, heat	93	489
$CH_2NHCONH_2$	p-$O_2NC_6H_4$	CO_2Et	Br, heat	—	43
Phtaloylaminomethyl	CO_2Me	H	Br, heat	80	489
$HC(R)NH_2$	CH_2CO_2Et	H	Br, heat	—	486
C_6H_{11}	CH_2CO_2Et	H	Br, heat	—	640
Me	CH_2CO_2Et	H	Br, heat	75	4, 10, 22, 414
Me	$HC(Me)CO_2Et$	H	Br, heat	—	4, 10, 22
$CON=CMe_2$	Ph	H	Br, cool, then heat to 100°C for a few minutes	80	445
CO_2Et	Me	H	Cl, heat	—	201, 209
CO_2Et	Me	Me	Cl, heat	80	294
CO_2Et	Me	$(CH_2)_2CO_2Me$	Cl, heat	—	201, 210
CO_2Et	Me	$(CH_2)_2OH$	Br, heat	—	201, 210

TABLE II-10. (continued)

R_1	R_2	R_3	Conditions[a]	Yield (%)	Ref.
CO_2Et	Me	CO_2Et	Cl, heat	43	242
CO_2Et	Me	CO_2Ph	Cl, heat	43	242
CO_2Et	CO_2Et	H	Br, heat	47	298
CO_2Et	CO_2Et	CO_2Et	Cl, heat	50	298
Ph	Me	CO_2Et	Cl, heat	—	15
Ph	CH_2CO_2Et	H	Cl, heat	—	640
Ph	HC(Me)CO_2Et	H	Cl, heat	—	640
Ph	CO_2Et	H	Br, heat	—	208, 603
Ph	CO_2Et	CO_2Et	Cl, heat	83	110
Ph	CO_2Et	Ph	Br, heat	—	299
Ph	Ph	$(CH_2)_2CO_2H$	Br, iPrOH, Na_2CO_3 0.5 hr at 60–70°C	—	657
Ph	Ph	CH_2CO_2H	Br, iPrOH, Na_2CO_3 0.5 hr at 60–70°C	—	657
Ph	Ph	HC(Me)CO_2H	Br, iPrOH, Na_2CO_3 0.5 hr at 60–70°C	—	657
CH_2CO_2Ph	Ph	H	Cl, heat	—	80
CH_2Ph	CH_2CO_2H	H	Br, heat	—	273
HC(Me)CO_2Ph	Me	H	Br, heat	75	352
HC(Me)CO_2Ph	Ph	H	Br, heat	80	80, 352
HC(Me)CO_2Ph	3,4-$(OH)_2C_6H_3$	H	Br, heat	83	80
HC(Me)CO_2Ph	3,4-$(OMe)_2C_6H_3$	H	Br, heat	76	83
$C(Me)_2CO_2Ph$	Me	H	Cl, C_6H_6, reflux 2 hr	—	352
2,5-$(Me)_2C_6H_3$	CH_2CO_2Et	H	Br, heat	—	275
[structure: OH, CH₂CH=CH₂ on benzene ring with OH and methyl]	CH_2CO_2Et	H	Br, heat	—	275

γ-pyridyl	CO$_2$Et	H	Br, heat	—	618
α-furyl	CO$_2$Et	CO$_2$Et	Cl, heat	—	577
α-thienyl	CO$_2$Et	CO$_2$Et	Cl, heat	—	577
β-thienyl	CO$_2$Et	CO$_2$Et	Cl, heat	—	577
γ-pyridyl	Me	CH$_2$CO$_2$Et	Br, EtOH, 10 hr at 50–60°C	—	695
γ-pyridyl	CH$_2$CO$_2$Et	H	Br, EtOH	—	695
4-ClC$_6$H$_4$	CH$_2$CO$_2$Et	H	Br, EtOH	—	640
4-MeOC$_6$H$_4$	CH$_2$CO$_2$Et	H	Br, EtOH	—	640

[a] See Table II-1.

Ethyl-4-thiazole carboxylate and derivatives (**65**) prepared by treating a suitably substituted ethylbromopyruvate (**48**) with a thioamide (**2**), followed by hydrolysis gave the corresponding 4-carboxythiazoles (**66**) (Scheme 30), which can be also prepared in good yields from α-bromopyruvic acid (250, 787). These compounds are convertible into 4-hydrazides, azides, and other functional derivatives in the usual manner. By a suitable choice of starting material, the 2-substituent (R_1) may be hydrogen, alkyl, or aminomethyl and derivatives (19, 220, 221, 247, 273, 281, 295, 486, 489, 795), phenyl (208, 299, 603), aryl (786), heteroaryl (557), or esters (298); and the 5-substituent (R_2) may be hydrogen (208, 220, 221, 247, 281, 298, 486, 603), aryl (299, 489), ester (103, 110, 145, 298, 577, 639), benzoyl (728), or acetyl (786).

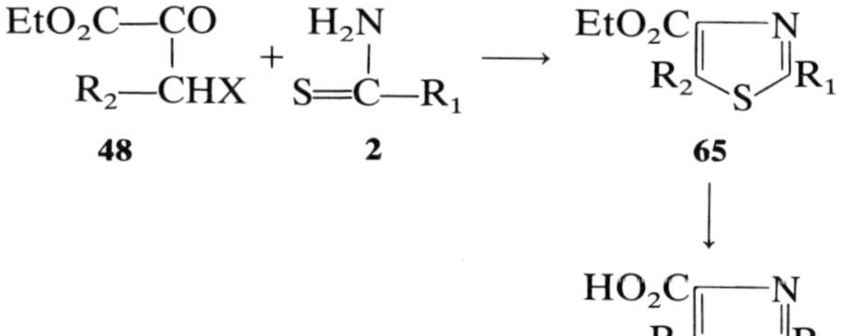

Scheme 30

Analogously, 1-bromo-3-oximinobutan-2-one yields the 4-oximinoethyl compound (**67**), while chloromalonic dialdehyde gives rise to the 5-formylthiazole (**68**) (Scheme 31) (618).

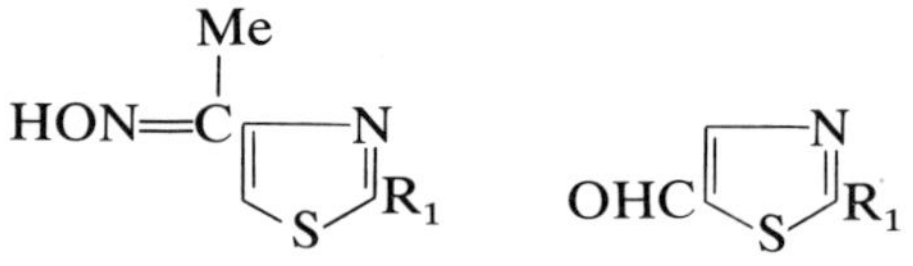

Scheme 31

Similarly, ethyl (or methyl) α-formyl chloroacetate (**69**), $R_2 = H$, and its substituted derivatives, condensed with thioformamide or higher thioamides give 5-ethyl- or 5-methyl-thiazole carboxylates (**70**) in good yields (Scheme 32) (4, 10, 15, 22, 58, 242, 281, 489, 810, 815, 819), with $R_1 = H$, CH_2NHAc, $CH_2NHCOPh$; $R_2 = H$, Me, $p\text{-}O_2NC_6H_4$; and $R_1 = $ Me, $R_2 = $ Ph.

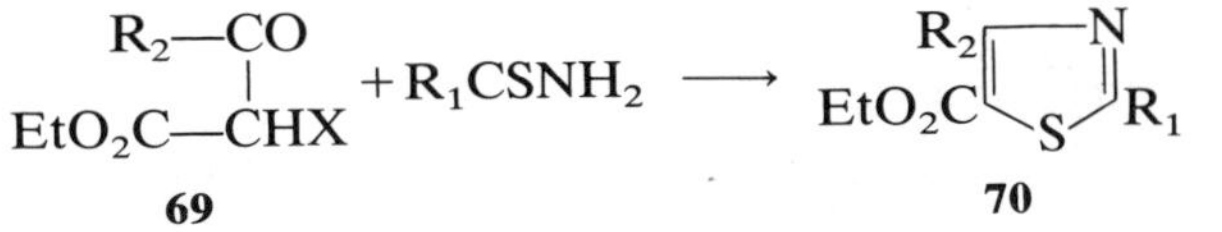

$$\underset{69}{\underset{EtO_2C-CHX}{R_2-CO}} + R_1CSNH_2 \longrightarrow \underset{70}{\underset{EtO_2C\diagdown_S\diagup R_1}{R_2\frown N}}$$

Scheme 32

5-Thiazole carboxylic acid (**70**), $R_1 = R_2 = H$, can be also obtained from decarboxylation of 2,5-thiazole dicarboxylic acids.

3-Arylamino-2-chloroprop-2- enoic esters (**72**) obtained from 2-chloroaceto acetic ester (**71**) and arylamines, react with thiourea to yield substituted 2-aminothiazoles (**73**), probably by initial nucleophilic substitution of the chloro atom of **72**, followed by cyclization with loss of aniline (Scheme 33) (729).

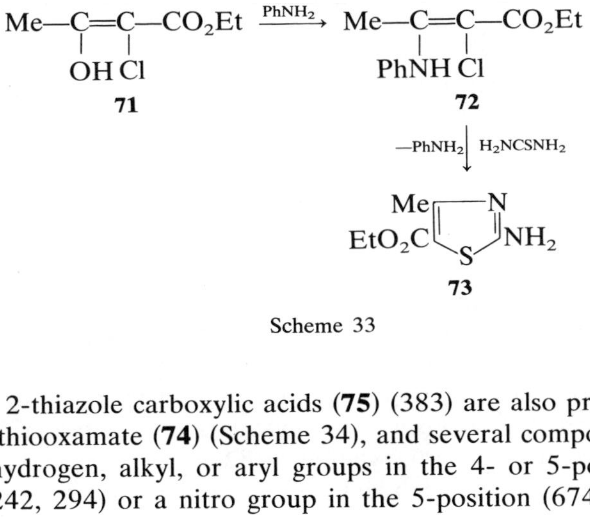

$$\underset{71}{\underset{OH\ Cl}{Me-C=C-CO_2Et}} \xrightarrow{PhNH_2} \underset{72}{\underset{PhNH\ Cl}{Me-C=C-CO_2Et}}$$

$$\downarrow \, {}_{-PhNH_2} \, {}_{H_2NCSNH_2}$$

$$\underset{73}{\underset{EtO_2C\diagdown_S\diagup NH_2}{Me\frown N}}$$

Scheme 33

Esters of 2-thiazole carboxylic acids (**75**) (383) are also prepared from ethyl monothiooxamate (**74**) (Scheme 34), and several compounds of this type with hydrogen, alkyl, or aryl groups in the 4- or 5-position (201, 209, 210, 242, 294) or a nitro group in the 5-position (674) have been reported.

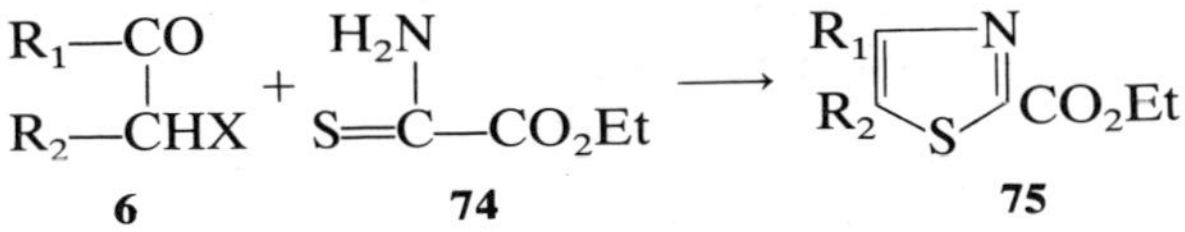

$$\underset{6}{\underset{R_2-CHX}{R_1-CO}} + \underset{74}{\underset{S=C-CO_2Et}{H_2N}} \longrightarrow \underset{75}{\underset{R_2\diagdown_S\diagup CO_2Et}{R_1\frown N}}$$

Scheme 34

Ethyl-4,5-thiazole dicarboxylates (**77**), $R_1 = H$, Me, Et, Ph, or heteroaryl, were prepared from diethyl-α-chloro-β-ketosuccinate (**76**) and thioamides in boiling ethanol (Scheme 35) (103, 110, 145, 298, 577, 639).

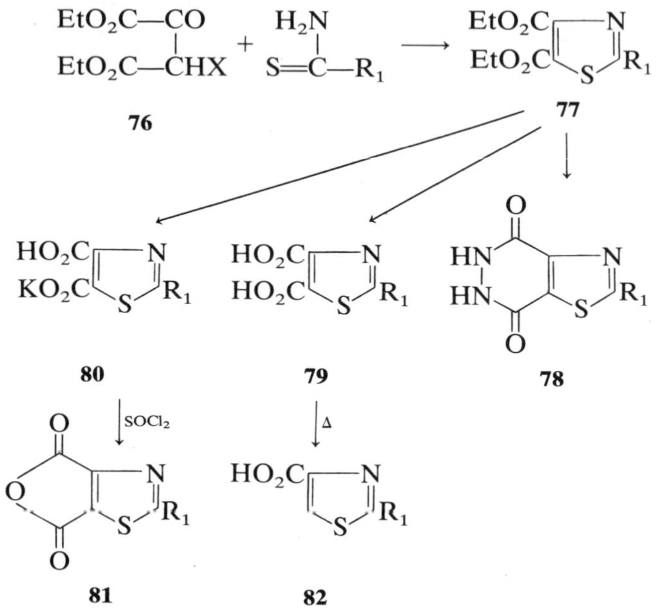

Scheme 35

Some of these compounds are used as potential intermediates for the preparation of 4,7-dioxo-4,5,6,7-tetrahydrothiazolo[4,5d]pyridazines (**78**). The diesters (**77**) are hydrolyzed under appropriate conditions to free acids (**79**), whose monopotassium salts (**80**) yield the cyclic anhydrides (**81**) under the influence of thionylchloride. Pyrolysis of **79**, $R_1 = \alpha$-thienyl, results in its decarboxylation to **82**.

Thiazole carboxylic acid hydrazides were prepared in a similar way (444, 445). Thus by refluxing thioacetamide or thiobenzamide with γ-bromoaceto acetic ester arylhydrazones (**83**) for several hours in alcohol the 4-carboxythiazole derivatives (**84**) listed in Table II-11 were obtained (Scheme 36) (656). This reaction is presumed to proceed via dehydration of the intermediate, thiazoline-S-oxide.

Similarly, **85**, in which $R_1 = H$, p-Me, p-Cl, p-MeO, 2,4-dichloro, and 3,4-dichloro, was synthesized (Scheme 37) (576).

II. Thiazoles from α-Halocarbonyl Compounds and Derivatives

TABLE II-11. 4-CARBOXYTHIAZOLES DERIVATIVES[a,b]

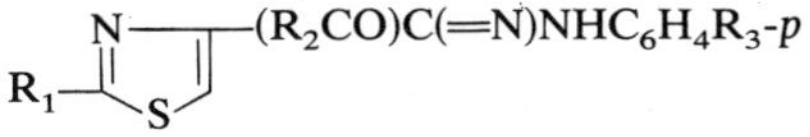

R_1—thiazole—$(R_2CO)C(=N)NHC_6H_4R_3$-p

R_1	R_2	R_3	Yield (%)
Me	OEt	OMe	74
Me	OEt	Me	60
Me	NHNH$_2$	OMe	100
Me	NHNHCSNHPh	OMe	70
Ph	OEt	OMe	53

[a] Conditions: Br, alcohol reflux for several hours.
[b] Ref. 656.

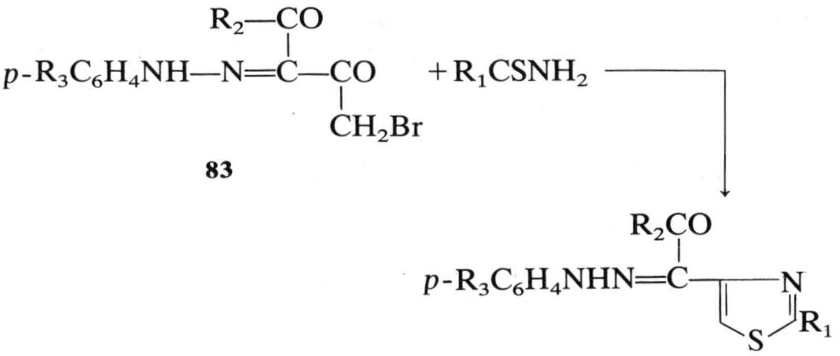

$p\text{-}R_3C_6H_4NH\text{—}N=C(COR_2)(CH_2Br)\text{—}CO$ + R_1CSNH_2 ⟶

83

$p\text{-}R_3C_6H_4NHN=C(R_2CO)\text{—thiazole—}R_1$

84

Scheme 36

$p\text{-}R_2C_6H_4\text{—}C(=O)\text{—}C(=NOH)\text{—thiazole—}C_6H_4R_1$

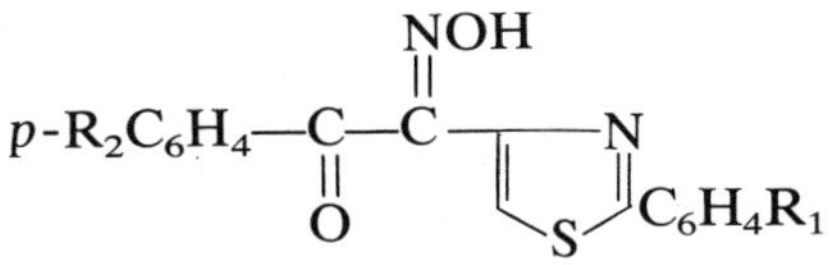

85

Scheme 37

Thiazole acetic acids and their homologs can also be prepared by cyclization procedures: 4-thiazole alkanoic acids and their salts were prepared by treating a thioamide with a γ-chloro- or γ-bromoacetoacetic or their α-alkyl derivatives (4, 10, 16, 22, 273, 275, 281, 640, 647, 695).

For example, (2-phenyl-4-thiazolyl) acetate (**87**), R = H, was obtained in good yield from thiobenzamide and ethyl-γ-bromo acetoacetate (**86**) as in Scheme 38 (640).

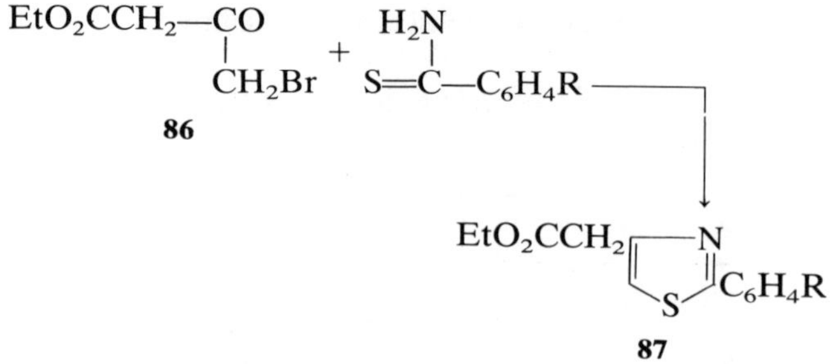

EtO_2CCH_2—CO
 |
 CH_2Br
 86

+

H_2N
 |
 S=C—C_6H_4R

→

EtO_2CCH_2 $\underset{S}{\underset{|}{\diagdown}}\overset{N}{\underset{}{\diagup}}$ C_6H_4R
87

Scheme 38

Similarly, 5-thiazole alkanoic acids and their salts are obtained from thioamides and β-halo γ-keto acids (695). Thus thioarylamides condensed with 3-aroyl-3-bromopropionic acid (**88**) in isopropanolic solution in the presence of Na_2CO_3 give first 4-hydroxy-2-aryl-Δ-2-thiazoline-5-acetic acid intermediates (**89**), which were dehydrated in toluene with catalytic amounts of p-toluene sulfonic acid to 2,4-diaryl-5-thiazole acetic acid (**90**) (Scheme 39) (657), with R = H or Me; Ar = Ph, o-, m- or p-tolyl, o-, m-, or p-ClC_6H_4, o-, m-, or p-$MeOC_6H_4$, p-$CF_3C_6H_4$, α-thienyl, α-naphthyl (657).

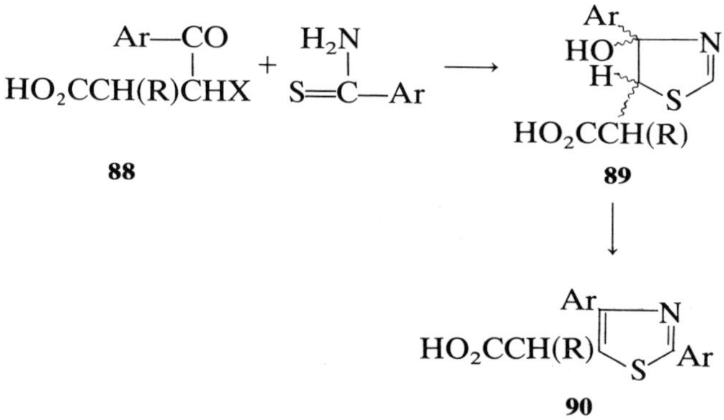

Ar—CO
 |
 $HO_2CCH(R)CHX$
 88

+

H_2N
 |
 S=C—Ar

→

Ar, HO, H N
 ⟩—⟨
 S
 $HO_2CCH(R)$
 89

↓

Ar $\underset{S}{\underset{|}{\diagdown}}\overset{N}{\underset{}{\diagup}}$ Ar
$HO_2CCH(R)$
90

Scheme 39

Thiazole-2-acetic acid derivatives (**92**) are prepared from ethyl thiomalonamate (**91**), R_1 = H, alkyl, phenyl (195, 295, 319), or monothiomalonamide as the thioamide reactant (Scheme 40) (80, 83, 279, 295, 352, 469).

II. Thiazoles from α-Halocarbonyl Compounds and Derivatives

When ethyl-γ-bromoacetoacetate is the carbonyl reactant, ethylthiazole-2,4-diacetate is produced (**92**) ($R_1 = R_3 = H$, $R_2 = CH_2CO_2Et$).

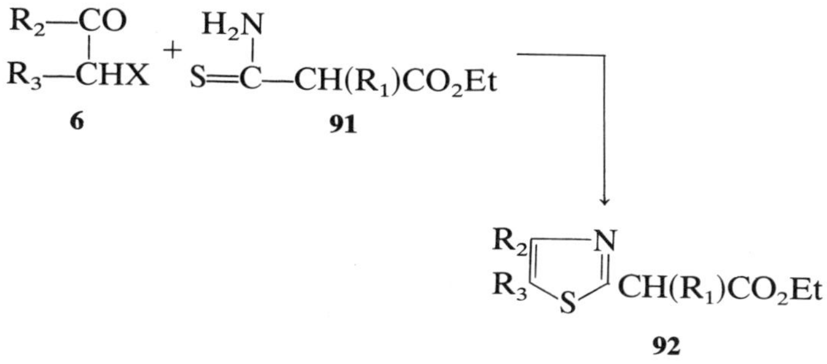

$$\begin{array}{c} R_2-CO \\ | \\ R_3-CHX \end{array} + \begin{array}{c} H_2N \\ | \\ S=C-CH(R_1)CO_2Et \end{array} \longrightarrow$$

6   **91**

$$\underset{\textbf{92}}{\begin{array}{c} R_2 \diagup N \\ R_3 \diagdown_S \diagup CH(R_1)CO_2Et \end{array}}$$

Scheme 40

D. Hantzsch's Synthesis Mechanism

The mechanism of the Hantzsch's synthesis was studied at a very early stage by several authors. The intermediates were generally assumed to be open-chain α-thioketones, but in a series of papers by Murav'eva and Schukina (470, 490) the isolation of hydroxythiazolines from the reaction between α-haloketones and a variety of thioureas was reported.

The use of a reagent bearing a basic center or the addition of a base to the reaction mixture was recognized as necessary to prevent the acid-catalyzed elimination of the elements of water from the intermediates. Since the publication of this work, a number of similar intermediates have been isolated from thioamides and α-halogeno carbonyl compounds (608, 609, 619, 739, 754, 801), and as a result of kinetic studies, the exact mechanism of this reaction has been well established (739, 821).

In this type of synthesis, the ring closure can be presented as occurring stepwise, as shown in Scheme 41. Undoubtedly the first step in this reaction is the formation of an acyclic intermediate (**93**) by nucleophilic attack of the sulfur by one of its lone pairs, on the halogen-carrying carbon to form the C–S–C bond, this attack occurs with Walden inversion.

In several cases, this intermediate has been successfully isolated at a relatively low temperature, particularly with the following substituents on **98**: $R_1 = Me$ and $R_2 = CH_2CO_2C_2H_5$, and $R_1 = Me$ or Ph and $R_2 = CH_2Cl$ (19, 99, 145, 339).

The next step is attack at the carbon carrying the carbonyl function by the lone pair of nitrogen atom, giving rise to a new cyclic intermediate

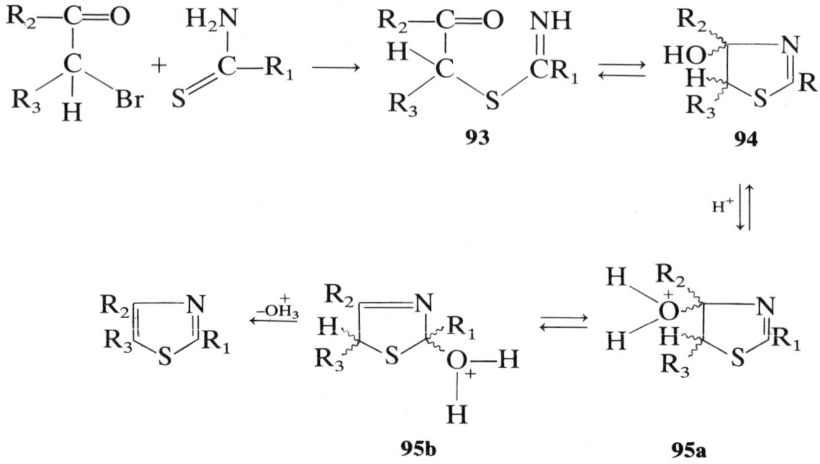

Scheme 41

(**94**) whose structure has been well established in the following cases: $R_1 = R_2 = Ar$, $R_3 = CHRCO_2H$ (657); $R_1 = Ar$, $R_2 = CH_2Cl$, $R_3 = H$ (618, 638, 651); $R_1 = Me$, $R_2 = Ph$, $R_3 = N(R, R')$ (761); $R_1 = Ar$, $R_2 = Ac$, $R_3 = H$ (763); $R_1 = Ar$, $R_2 = H$ or CH_2Cl, $R_3 = Ac$ or CO_2H (763); $R_1 = Ph$, Me, NHCOMe, NH_2, $R_2 = CHBr_2$, $R_3 = H$ (625).

These intermediates upon treatment by acid lead to thiazole by loss of $H_3\overset{+}{O}$ from **95a** or **95b**.

It is interesting to note that in each of these cases only one intermediate, either the α-thioketone (**93**) or the hydroxythiazoline (**94**), was isolated from the reaction mixture, thus suggesting that hydroxythiazolines are in equilibrium with the corresponding α-thioketones. Similar conclusions are implicit in Baganz and Ruger's production of 4-formylthiazoles by this synthesis (625).

The rates of the reaction of phenacyl bromides with thiobenzamides were followed by Okamiya (549) using conductivity measurements. By application of the Hammett correlation, a value of ρ close to 0.48 was found for the reaction of $ArCOCH_2Br$ and $PhCSNH_2$, indicating that sulfur is attacking as a nucleophile. On the other hand, reactions between $PhCOCH_2Br$ and $ArCSNH_2$ gave $\rho = 0.93$. The results indicate that the reaction rate k at 30°C can be expressed by the equation: $\log k = 0.176 + 0.48\sigma_1 - 0.93 \sigma_2$, where σ_1 and σ_2 are the substituent constants of $ArCOCH_2Br$ and $ArCSNH_2$, respectively. The rate constants at 40°C can be obtained by substituting 0.176 in the equation by 0.474. The activation energy is 12 to 16 kcal/mole.

2. Reaction with N-Substituted Thioamides (Thiazolium Salts)

Thiazolium salts can be obtained successfully by a modification of the Hantzsch's thiazole synthesis. This method is particularly valuable for those thiazolium compounds in which the substituent on the ring nitrogen cannot be introduced by direct alkylation, for example, aryl or heteroaryl thiazolium salts (Scheme 42).

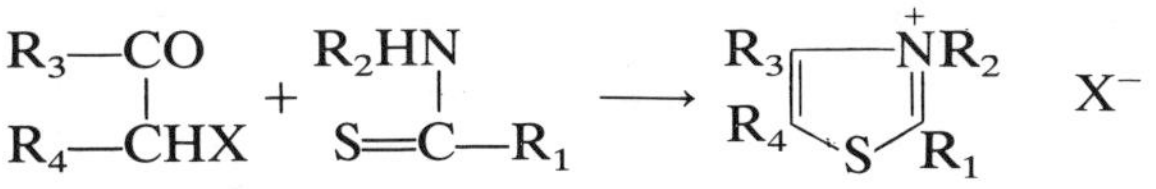

$$\begin{array}{c} R_3-CO \\ | \\ R_4-CHX \end{array} + \begin{array}{c} R_2HN \\ | \\ S=C-R_1 \end{array} \longrightarrow \begin{array}{c} R_3 \\ \\ R_4 \end{array} \underset{S}{\overset{+}{\underset{}{N}R_2}} \overset{}{\underset{R_1}{}} \quad X^-$$

6   **96**   **97**

Scheme 42

N-Monosubstituted thioamides (**96**) have been cyclized with α-halocarbonyl compounds to give thiazolium salts (**97**) in excellent yields (89, 99, 102, 305, 722).

For example, when an N-methylthioacetamide (**96**), $R_1 = R_2 = $ Me, was condensed with chloroacetone, a 2,3,4-trimethylthiazolium chloride was obtained in quantitative yield. The reaction is usually run in aqueous or alcoholic solution at room temperature. At low temperature, with N-phenylthioacetamide (**96**), $R_1 = $ Me, $R_2 = $ Ph and chloroacetone, an acyclic intermediate (**98**) was isolated and characterized (Scheme 43). It was easily converted to 2,4-dimethyl-3-phenylthiazolium chloride (**97**), $R_1 = R_3 = $ Me, $R_2 = $ Ph, by heating (99, 102, 145).

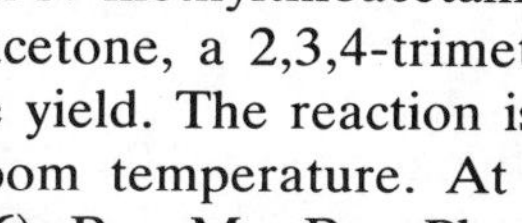

$$\begin{array}{cc} R_3C=O & NHR_2 \\ | & | \\ R_4(H)C & CHR_1 \\ \diagdown S \diagup \end{array}$$

98

Scheme 43

The reaction of N-methyl-(p-dimethylamino)thiobenzamide (**99**) with a number of α-haloketones and α-bromoheptaldehyde gave stable 4-hydroxythiazolinium salts (**100**), which could be subsequently dehydrated by methanolic hydrogen chloride to the thiazolium salts (**101**), (Scheme 44) (622).

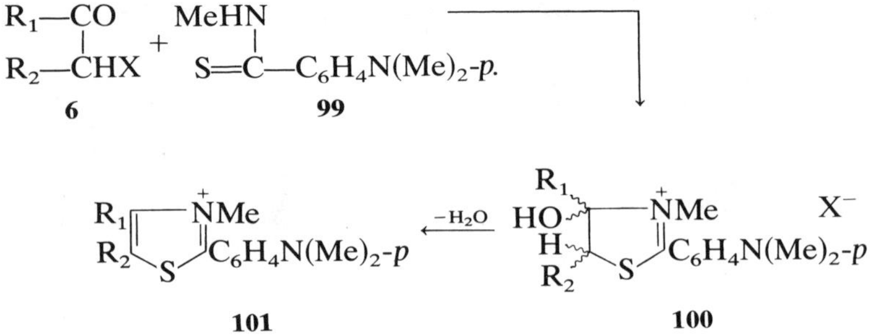

R_1—CO   MeHN
  | + |
R_2—CHX   S=C—$C_6H_4N(Me)_2$-p.
  6   **99**

R_1⌐——⌐$\overset{+}{N}$Me R_1⌐——⌐$\overset{+}{N}$Me
  | | ←—H_2O HO—| | X^-
R_2⌊__⌋$C_6H_4N(Me)_2$-p   H⋯| |$C_6H_4N(Me)_2$-p
     S R_2 S
     101   **100**

Scheme 44

R_1	R_2
Me	H
Me	n-Butyl
n-Pentyl	H
Et	Me
p-BrC$_6$H$_4$	H
H	n-Pentyl

In a similar way, the thioamide, EtNHC(=S)SMe, condensed with α-haloketones gives the corresponding 2-thiomethyl-3-ethylthiazolium salts (722).

When the intermediates were also substituted in the 5-position both possible diastereoisomeric forms were detected by NMR spectroscopy.

Thiazolium salts with alkyl (103, 722), arylalkyl (116), aryl (305), or heteroaryl (96) substituents on the nitrogen have been also prepared by this procedure. As in the thiazole series, N-substituted thioamides can be formed directly in the reaction mixture from phosphorus pentasulfide and N-substituted amides (127). These methods are important in the synthesis of thiamine **102** (vitamin B_1) (Scheme 45).

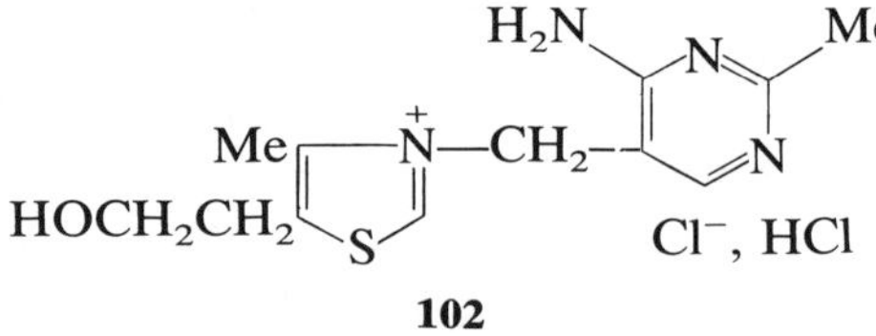

102

Scheme 45

II. Thiazoles from α-Halocarbonyl Compounds and Derivatives

3. Reaction with Thiourea

A. *α-Halocarbonyl Compounds and Derivatives*: *2-Aminothiazoles*

Of all the methods described for the synthesis of thiazole compounds, the most efficient involves the condensation of equimolar parts of thiourea (**103**) and α-haloketones or aldehydes to yield the corresponding 2-aminothiazoles (**104a**) or their 2-imino-Δ-4-thiazoline tautomers (**104b**) with no by-products (Method A, Scheme 46).

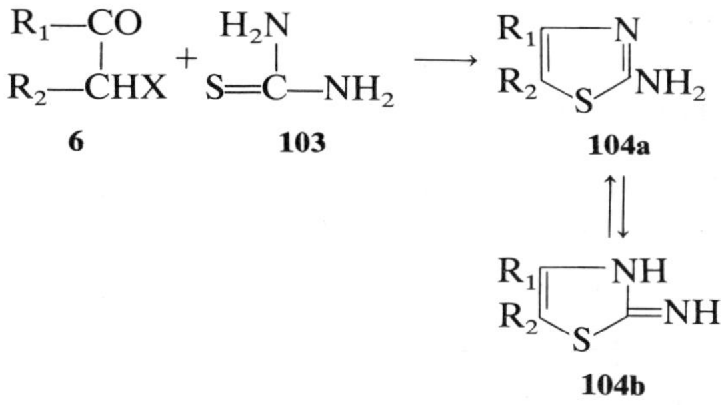

$$\begin{array}{cc} R_1-CO & H_2N \\ | & | \\ R_2-CHX & S=C-NH_2 \\ \mathbf{6} & \mathbf{103} \end{array} \longrightarrow \begin{array}{c} R_1 \diagup\!\!\!\diagdown N \\ R_2 \diagdown_S \diagup NH_2 \\ \mathbf{104a} \end{array}$$

$$\updownarrow$$

$$\begin{array}{c} R_1 \diagup\!\!\!\diagdown NH \\ R_2 \diagdown_S \diagup\!\!=\!\!NH \\ \mathbf{104b} \end{array}$$

Scheme 46

This method was initially proposed by Popp (11) and Traumann (8). The antimicrobic and fungicidal properties of this class of compounds (477, 535) give them great commercial importance, which explains the large amount of work done in this field, 2-aminothiazoles being used as starting materials for obtaining pharmacologically more active substances. This reaction occurs more readily than the reaction with thioamides as starting materials; it can be carried out in aqueous or alcoholic solution on a water bath, even in a distinctly acidic medium, an advantage not shared by the thioamides, which are often unstable in acid.

The 2-aminothiazoles are extracted with ether after alkalinization of the reaction mixture. The yields are almost theoretical with α-haloketones and lower with α-haloaldehydes.

A new method of synthesizing 2-aminothiazoles from thiourea and ketones has been developed by Dodson (225, 261, 280). It was improved by King and Lavacek (328) and later taken up by several other workers (328, 779, 798) (Method B).

This latter method consists in treating 2 moles of thiourea and 1 mole of ketone, having a methylene group adjacent to the carbonyl, with 1 mole of iodine overnight on a steam bath. Unreacted products are then extracted with ether after alkalinization of the reaction mixture.

This simplified method gives 2-aminothiazole in good yield (50 to 70%) (311, 330). Other reactants can replace iodine, for example, chlorine, bromine, sulfuryl chloride, chlorosulfonic acid, or sulfur monochloride also give good results.

In many cases, the α-haloketone does not appear to be an intermediate in this reaction, since reagents such as sulfur trioxide, sulfuric, or 60% nitric acid lead to 2-aminothiazole but with lower yields (11 to 43%). Formamidine disulfide $[-S-C(=NH)NH_2]_2$, a product of the oxidation of thiourea, seems to be the intermediate in this reaction, since upon treatment with ketones, it gives 2-aminothiazole (604). However, the true mechanism of this reaction has not yet been completely elucidated.

Iodomercuriketones have been also used as starting material for the synthesis of 2-aminothiazoles (354) (Method C).

Another method used less often for these syntheses consists in condensing ketones with cyanamid and sulfur (527) (Method D).

2-Aminothiazole itself (**104a**), $R_1 = R_2 = H$ (an intermediate in the preparation of sulfathiazole), was synthetized as early as 1888 from α,β-dihalogenoethylethers, which give chloroacetaldehyde in acidic medium (8). Many other derivatives have now been used for this synthesis, and they are listed in Table II-12. All give excellent yields.

Labeled 2-aminothiazole (an intermediate in the synthesis of an antiparasitic drug) has been obtained from ^{14}C-thiourea (666).

By choosing the appropriate halomethylketone, the substituents in the

TABLE II-12. 2-AMINOTHIAZOLES FROM THIOUREA AND α-HALOACET-ALDEHYDE AND DERIVATIVES

Starting material and conditions	Yield (%)	Ref.
$ClCH_2CHO$, water, heat	70–80	205, 267, 449
$(ClCH_2CHO)_2$, water + 2 or 3 drops of HCl	92	539
$(BrCH_2CHO)_3$	65	163
$ClCH_2CH(OMe)_2$	75–92	206, 232
$ClCH_2CH(OEt)_2$	60–90	205, 206, 508
$ClCH_2CHClOEt$	60–100	8, 16, 67, 438
$ClCH_2CHClOMe$	90	356
$ClCH_2CHClOBu$	80	205
iso-$ClCH_2CHClOPr$	72	205
$(ClCH_2CHCl)_2O$	100	248
$(BrCH_2CHBr)_2O$	86	248
$ClCH_2CHClOAc$, abs. alcohol for 30 min, water bath	50–80	159, 164, 177, 248
$BrCH_2CHBrOAc$	50–80	164
$ClCH_2CHClOC(=O)Pr$	—	164
$ClCH_2CHO$, in 10% aqueous sol, 30 min at room temperature and then 1 hr at 70°C	97–98.5[a] (HCl)	697

[a] 2-Aminothiazole isolated as hydrochloride

II. Thiazoles from α-Halocarbonyl Compounds and Derivatives

4-position, that is, R_1, can be alkyl (3, 8, 67, 164, 181, 192, 225, 261, 280, 308, 353, 429) or their substituted derivatives such as aminoalkyl (249, 254, 267, 516, 526), chloroalkyl (84), hydroxyalkyl (249), benzyl (407, 421), β-arylethyl (708), alkylsulfones (278), α-carboxyalkyl (33, 158, 182, 214), or methyl or (p-tolyl)arylazo (734).

4-Acetyl-2-aminothiazole (**104a**), R_1 = COMe, R_2 = H) (277, 371), and its oxime (**104a**), R_1 = [C(=NOH)] Me, R_2 = H) (152, 575), were obtained by refluxing the α-diketone (XCH$_2$COCOMe) or its oxime in alcohol with thiourea.

Similarly, by condensing γ-bromoacetoacetic esters with thiourea, 4-carboxy-2-aminothiazoles (**104a**), R_1 = CO$_2$H, R_2 = H, were obtained after hydrolysis (183, 185, 221, 441). These compounds can be decarboxylated by heating (158).

The α-halogenated acids or their esters (**105**) also react with thiourea to give 2-amino-4-hydroxythiazoles (**106a**) or their 2-amino-4-thiazolone (**106b**) (1, 247, 254, 530) or 2-imino-4-oxathiazolidine (**106c**) tautomers (Scheme 47).

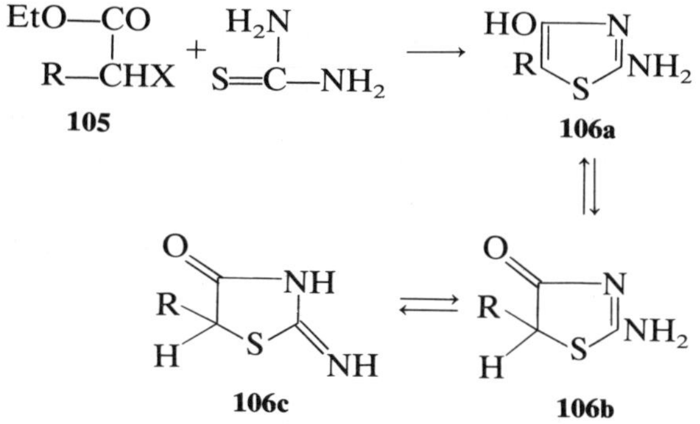

Scheme 47

Thus with ethyl α-chloroacetate (**105**), R = H, X = Cl, Robba and Moreau (530) obtained the 2-amino-4-hydroxythiazole (**106**), R = H, known as pseudothiohydantoine.

In the course of this reaction it is possible to isolate an acyclic intermediate (**107**) (Scheme 48).

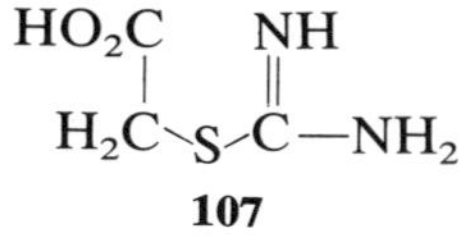

Scheme 48

2,4-Diaminothiazole hydrochloride (**110**), R = H, is readily prepared (80 to 86% yield) by the reaction of thiourea with chloroacetonitrile (**108**), R = H, in warm alcoholic solution (Scheme 49) (127, 220, 254).

The reaction probably proceeds through the acyclic intermediate (**109**), which can be isolated when the reaction is carried out in cold acetone and can be then cyclized by heating.

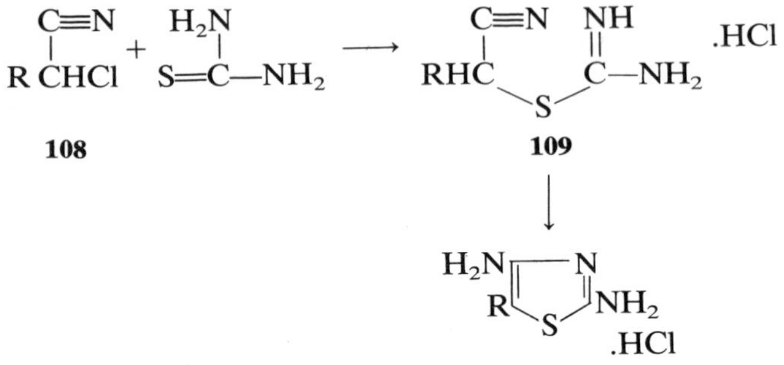

Scheme 49

Several derivatives **110**, R = Me, Ph, have been prepared in a similar way from the corresponding nitriles **108**.

4-Aryl-2-aminothiazoles (**104a**), $R_1 = C_6H_4R$, $R_2 = H$, in which R = alkyl (328, 420, 427, 460, 480, 496, 519), carbonyl (272), phenyl (341, 463), hydroxy and ether (341, 420, 427, 463, 496, 605), halogeno (310, 328, 420, 519, 606, 672), amino (328, 427), and nitro (225, 254, 261, 280, 328, 435, 505, 519), mainly in the *para* position, have been synthetized by either Method A or B in good yields (769, 770).

In the same way, some polysubstituted derivatives (78, 91, 254, 303, 435, 478), 4-α- or β-naphthyl (436, 496), 4-heteroaryl-2-aminothiazoles (153, 272, 422, 538, 569, 677, 680, 692, 735, 736, 752, 793), 2-aminothiazole sulfonyl derivatives (671, 798), 2-amino-4-(4-alkyl-selenophenyl)thiazoles (631), and 4-substituted 2-aminothiazoles (691) have been also reported (Table II-13).

Bis (*meta* or *para* haloacetyl) benzenes $(ClCH_2CO)_2C_6H_4$ condensed with thiourea yield the corresponding *meta* or *para* bis(4-thiazolyl-2-amino)phenylene (573, 574), analogous to **36** with $R = NH_2$.

By refluxing in alcohol solution *m*- or *p*-2-thiazolylphenacyl bromide (**37**) with thiourea, compound **111**, in which R = H, Ph, p-ClC_6H_4, and p-$O_2NC_6H_4$, was obtained (569).

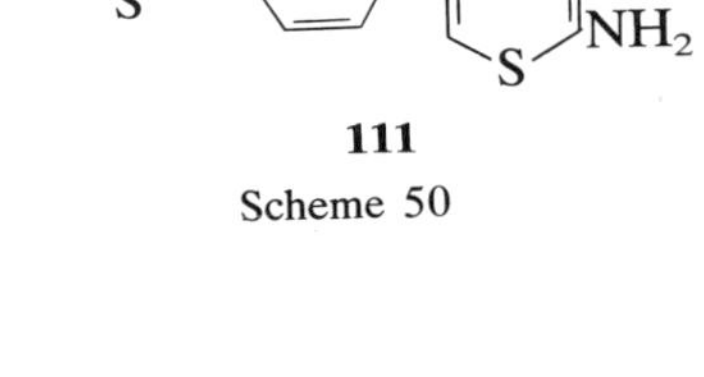

111

Scheme 50

TABLE II-13. 4-ALKYL, 4-ARYL, OR 4-HETEROARYL 2-AMINOTHIAZOLES

$$R_1 \underset{S}{\overset{N}{\rightleftarrows}} NH_2$$

R_1	Conditions[a]	Yield (%)	Ref.
Me	A(Cl), heat, water bath	70–75	3, 8, 64, 192, 308, 429
Me	A(Br), heat, water bath	36	225, 261
Me	A(CH$_3$CHClCHClCO$_2$Me)	—	164
Me	B(I$_2$)	77	225, 261
Et	A(Cl), heat, water bath	64	377
n-Pr	A(Cl), heat	—	353
n-Bu	A(Cl), heat	—	181, 353
n-Am	A(Cl), heat	—	181
n-C$_7$H$_{15}$	A(Cl), heat	—	181
n-C$_9$H$_{19}$	A(Cl), heat	—	181
n-C$_{11}$H$_{23}$	A(Cl), heat	—	181
n-C$_{13}$H$_{27}$	A(Cl), heat	—	181
n-C$_{15}$H$_{31}$	A(Cl), heat	—	181
iso-Bu	A(Cl), steam bath	—	353
(CH$_2$)$_5$NH$_2$	A(Br), alcohol reflux	50	516
(CH$_2$)$_3$N⟨(CO)$_2$C$_6$H$_4$⟩	A(Br), alcohol reflux, 5 hr	43	526
(CH$_2$)$_2$NH$_2$	A(Br), alcohol reflux, 5 hr	—	254
(CH$_2$)$_2$N(Me)$_2$	A(Br), alcohol reflux	81	267
(CH$_2$)$_2$N(Et)$_2$	A(Br), alcohol reflux	69	249

TABLE II-13. (continued)

R_1	Conditions[a]	Yield (%)	Ref.
(phthalimide) (CH$_2$)$_2$N	A(Br), PrOH, reflux 5 hr	32	526
(CH$_2$)$_2$N(piperidine)	A(Br), alcohol reflux	83	267
(CH$_2$)$_2$N<	A(Br), alcohol reflux	52	267
(CH$_2$)$_2$Cl	A(Cl), alcohol reflux	—	84
p-(CH$_2$)$_2$C$_6$H$_4$SO$_2$Me	A(Cl), alcohol reflux	—	296
CH$_2$—CHC$_6$H$_5$ \\ CH$_2$	A(Br)	94	288
CH$_2$CO$_2$Et	A(Cl or Br)	86–100	2, 4, 10, 22, 33, 212
CH$_2$NH$_2$	A(Br), alcohol reflux	75	516
CH$_2$N(Me)$_2$	A(Br), alcohol reflux	76	249
CH$_2$N(piperidine)	A(Br), alcohol reflux		249
CH$_2$OH	A(Cl or Br)	56	249
CH$_2$OBu	A(Cl), 2 hr at 100°C	—	401
CH$_2$Ph	A(Cl), alcohol reflux 3 hr	—	407, 421
CH$_2$SC(=NH)NH$_2$	A(Cl), alcohol reflux	84	249
p-CH$_2$SO$_2$C$_6$H$_4$O$_2$N	A(Br)	—	278
iso-Pr	A(Cl), steambath	77.5	353, 429
CH$_2$—CHC$_6$H$_4$O$_2$N-p \\ CH$_2$	A(Br)	99.5	288

HC(CH$_2$COOH)SC(=NH)NH$_2$	A(Br)	—	32
HC(CO$_2$H)CH$_3$	A(Br)	100	33
HC(CO$_2$Et)Et	A(Br)	42	158
HC(CO$_2$Et)CH$_2$CO$_2$Et	A(Br)	45	214
HC(CO$_2$H)(CH$_2$)$_3$CH$_3$	A(Br)	—	182
HC(CO$_2$Et)(CH$_2$)$_3$CH$_3$	A(Br)	33	158
HC(CO$_2$Et)(CH$_2$)$_5$CH$_3$	A(Br)	45	158
HC(CO$_2$Et)(CH$_2$)$_2$CO$_2$Et	A(Br)	64	158, 214
HC(CO$_2$Et)(CH$_2$)$_3$CO$_2$Et	A(Br)	51	214
p-HC=CHC$_6$H$_4$OMe	A(Br)	60	420
HC=CH — [furan with NO$_2$]	A(Br), alcohol reflux	65–70	655
C(Me)=CH — [furan with NO$_2$]	A(Br), alcohol reflux	83	655
C(Cl)=CH — [furan with NO$_2$]	A(Br), alcohol reflux	89	655
C(CO$_2$Et)$_2$Me	A(Br)	—	21
C(=NOH)Me	A(Cl or Br)	—	152, 575
C(=O)Me	A(Cl), alcohol reflux	—	277, 371
CO$_2$H	A(Cl or Br), alcohol reflux	56	2, 250
CO$_2$Me	A(Cl or Br), alcohol reflux	60–80	221, 441
CO$_2$Et	A(Br), alcohol reflux	66	183, 185
NH$_2$	A(ClCH$_2$COCN)	80–86	127, 221, 254
OH	A(Cl)	60	254, 530

TABLE II-13. (continued)

R_1	Conditions[a]	Yield (%)	Ref.
Ph	A(Cl or Br) water or alcohol steambath	50–90	8, 11, 18, 67, 115, 221, 341, 366, 460, 519
Ph	$B(Br_2)$, abs. alcohol	—	601
Ph	C	85	341, 354
p-MeC$_6$H$_4$	A(Cl or Br)	85	328, 427, 519
	$B(I_2)$	85	420, 480
p-EtC$_6$H$_4$	A(Cl) or $B(I_2)$	—	427, 460, 480
p-iso-PrC$_6$H$_4$	B	—	460, 480
p-BuC$_6$H$_4$	B	—	460, 480
p-AmC$_6$H$_4$	B	—	460, 480
p-hexyl C$_6$H$_4$	B	—	460, 480
p-octyl C$_6$H$_4$	B	—	460, 480
p-nonyl C$_6$H$_4$	B	—	460, 480
p-decyl C$_6$H$_4$	B	—	460, 480
p-undecyl C$_6$H$_4$	B	—	460, 480
p-dodecyl C$_6$H$_4$	B	—	460, 480
p-cyclopentyl C$_6$H$_4$	A(Br) alcohol steam bath	—	496
o-HO$_2$CC$_6$H$_4$	A(Br)	—	272
p-biphenylyl	A(Br) or $B(Br_2)$	—	341, 463
p-HOC$_6$H$_4$	A or B	59.6	463, 496, 605
o-HOC$_6$H$_4$	A or B	—	463, 496
p-MeOC$_6$H$_4$	A or B, 4 hr, water bath	49–80	341, 420, 605
p-EtOC$_6$H$_4$	A or B, 4 hr, water bath	79.5–80	341, 420, 605
p-PrOC$_6$H$_4$	A or B, 4 hr, water bath	47.0	605
p-iso-PrOC$_6$H$_4$	A or B, 4 hr, water bath	59.8	605
p-BuOC$_6$H$_4$	A or B, 4 hr, water bath	58.4	605
p-iso-BuOC$_6$H$_4$	A or B, 4 hr, water bath	46.3	605
p-ClC$_6$H$_4$	A(Br)	89	328, 519

p-BrC$_6$H$_4$	B(I$_2$) 4 hr, water bath	61–70	420, 606
	A(Br)	93	310, 328, 519
	B(I$_2$) 4 hr, water bath	54	606
p-IC$_6$H$_4$	A(Br)	97	328
p-NH$_2$C$_6$H$_4$	A(Br or Cl)	—	328, 427
o-NO$_2$C$_6$H$_4$	A(Br)	—	505
m-NO$_2$C$_6$H$_4$	B(Br$_2$)	75	225, 261, 280
	B(Br$_2$)	95	225, 261, 280
	B(I$_2$)	52	225, 261, 280
p-NO$_2$C$_6$H$_4$	A(Br)	—	254, 328, 435, 519
3,4-(HO)$_2$C$_6$H$_3$	A(Cl)	—	78, 91
3,4-(MeO)$_2$C$_6$H$_3$	A(Cl)	—	303
3,4-(EtO)$_2$C$_6$H$_3$	A(Cl)	—	303
2,5-(Cl)$_2$C$_6$H$_3$	A(Br)	—	254, 435
2-OH, 5-MeC$_6$H$_3$	B	46	478
2-OH, 4-MeC$_6$H$_3$	B	42	478
2-OH, 4-EtC$_6$H$_3$	B	52	478
2-OH, 3-ClC$_6$H$_3$	B	39	478
2-OH, 3,5-Cl$_2$C$_6$H$_2$	B	25	478
2-OH, 3-Cl, 4-MeC$_6$H$_2$	B	48	478
![N-S thiazole]-C$_6$H$_4$-m (H$_2$N-thiazole-C$_6$H$_4$-m)	A(Cl or Br), alc. reflux	88–91	573, 574
![N-S thiazole]-C$_6$H$_4$-p (H$_2$N-thiazole-C$_6$H$_4$-p)	A(Cl or Br), HCONH$_2$, 2 hr at room t°C	—	573, 574
![N-S thiazole]-C$_6$H$_4$-X-C$_6$H$_4$-p (H$_2$N-thiazole-C$_6$H$_4$-X-C$_6$H$_4$-p)	A(Cl or Br), alcohol 1.5 hr at 60°C, with X=O or S	85–90	410

TABLE II-13. (continued)

R_1	Conditions[a]	Yield (%)	Ref.
α-Naphthyl	A(Br)	—	457
β-Naphthyl	A(Br)	—	457
2-OH, 1-naphthyl	A(Br)	—	463
7-MeO, 2-naphthyl	A(Br), alcohol steam bath	—	496
α-furyl	A(Br), alcohol or B(Br$_2$), 6 hr steam bath	73–74	272, 537, 677
5-aryl, 2-furyl	A(Br), alcohol	65–90	755
(1,3-dioxolan-2-yl)	A(Cl), alcohol	—	422
(pyrrolyl with R, R$_1$=H or Me; NH)	A(Cl or Br) with R=H or Me	—	153b, 272
(thiazolyl with R$_1$, R$_2$)	A(Br), alcohol reflux	—	569
α-Pyridyl	A(Br), alcohol	90.8	680
β-Pyridyl	A(Br), alcohol	90.6	680
γ-Pyridyl	A(Br), alcohol	90.0	680
α-Quinolyl	A(Br), alcohol	61.0	680
γ-Quinolyl	A(Br), alcohol	87.4	680

A(Br), alcohol	—	538
A(Br), Me$_2$CO	—	723

[a] 2-Aminothiazoles were synthetized from thiourea by three methods: Method A, from α-haloketones or aldehydes designated as (Cl) or (Br); Method B, from ketones and iodine (I$_2$) or bromine (Br$_2$); Method C, from iodomercuriketones. Method D consists in condensing ketones with cyanamid and sulfur.

Similarly, the ether and thioether (**112**), with X = O or S, were prepared in 80 to 90% yield by condensing **39** with thiourea (410).

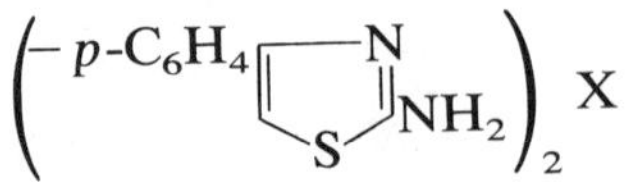

$$\left(-p\text{-}C_6H_4 \underset{S}{\overset{N}{\underset{\|}{\bigsqcup}}} NH_2 \right)_2 X$$

112

Scheme 51

2-Bromoacetylthiazole (**113**) condensed with thiourea gave the corresponding 2-aminothiazoles (**114**) in which R_1 = H, Me, Et, Ph, and p-$O_2NC_6H_4$, and R_2 = H and Me (Scheme 52) (569).

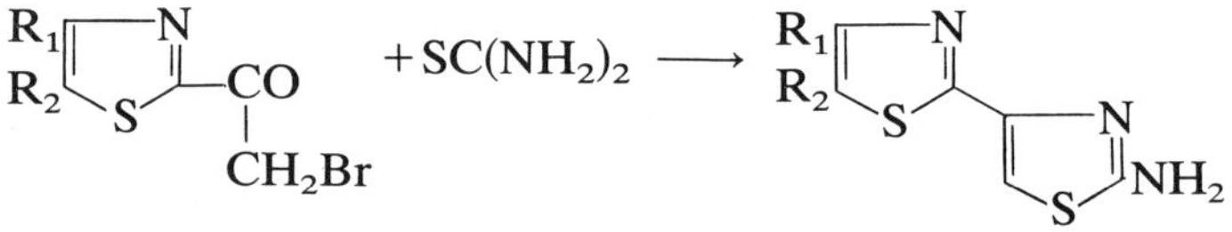

$$R_1 \underset{S}{\overset{N}{\bigsqcup}} -\underset{\underset{CH_2Br}{|}}{C}O \quad + SC(NH_2)_2 \longrightarrow R_1 \underset{S}{\overset{N}{\bigsqcup}} \underset{S}{\overset{N}{\bigsqcup}} NH_2$$

113 **114**

Scheme 52

Some 5-substituted 2-aminothiazoles with alkyl (15, 173, 175, 224, 366, 396), ester (173, 184, 220), aryl (115, 265, 396, 414), α-naphthyl (463), sulfur and sulfones derivatives (329, 373), isonitrosomethyl (772), and chloro groups were synthetized from the corresponding α-haloaldehyde and thiourea in lower yield (Table II-14).

Condensation of thiourea with ketones of the general formula $R_1COCHXR_2$ gave the 4,5-disubstituted 2-aminothiazoles (Table II-14) with R_1 and R_2 = alkyl (96, 102, 165, 170, 172, 173, 258, 265, 267, 353, 354, 377, 426, 443, 527, 591, 604); R_1 = Me and R_2 = benzyl (335), aryl (341, 463, 505), α-naphthyl (463), or a functional group such as alkanoic acid or ester (32, 112, 139, 276, 302), alkanoic alcohols (162, 184, 229, 520), esters (4, 10, 13, 22, 33, 221, 225, 247, 261, 280, 301, 531), sulfide (355), O = P(OEt)$_2$ (372), or anilide (773); R_1 = CO$_2$R and R_2 = Me (616), Ph (299), or CO$_2$Et (16, 208, 250); R_2 = CO$_2$R and R_1 = CH$_2$OH (635), CH$_2$CO$_2$Et (250), and so forth (172, 509, 681); R_1 = Ph or aryl and R_2 = Me (225, 261, 280, 335, 341, 463, 585), CH$_2$CO$_2$R (231), Ph (341, 463), CH$_2$–CH$_2$–CO$_2$Et (794), CH$_2$–CH$_2$N< (800), or carboxyanilide (771).

TABLE II-14. 5-SUBSTITUTED AND 4,5-DISUBSTITUTED 2-AMINOTHIAZOLES

$$\begin{array}{c} R_1 \diagdown \!\!\!\!\!\!\! \diagup N \\ R_2 \diagup \!\!\!\!\!\!\! \diagdown S \diagdown \!\!\!\! NH_2 \end{array}$$

R_1	R_2	Conditions[a]	Yield (%)	Ref.
H	Me	A(Cl or Br), water, heat	55	15, 175, 224
H	Et	A(Cl or Br), water, heat	61	173, 366, 396
H	n-Pr	A(Br)	—	173
H	n-Bu	A(Cl), water	—	173
H	n-Pentyl	A(Cl), water	—	173, 396
H	CO_2Me	A(Cl), steam bath	—	220
H	CO_2Et	A(Cl or Br), steam bath	39–60	173, 184, 198, 220
H	Ph	A(Br)	50–60	172, 265, 414
H	p-$NH_2C_6H_4$	$A(CH_2ClCHCl)_2O$ and phenylthio-semicarbazide	—	173
H	α-naphthyl	B	—	463
H	p-$SC_6H_4O_2N$	A(Cl or Br), alcohol heat or $B(Br_2)$, AcOH, 20 min at 90°C	30	340, 373
H	p-$SO_2C_6H_4Me$	$B(Br_2)$, AcOH, 40 min at 100°C	—	373
H	p-$SO_2C_6H_4O_2N$	$B(Br_2)$, AcOH, 40 min at 100°C	—	373
H	Cl	A, $ClCH_2CHClOEt$	—	257, 277
Me	Me	A, (Cl or Br), alcohol steam bath	28–60	157, 265, 354
Me	Et	A(Cl)	50	173, 353
Me		$D(S+H_2NCN)$	53	527
Me	Pr	A(Cl)	50	173, 353
		A(Cl)	40.6	276
—$(CH_2)_3$—		$D(S+H_2NCN)$	73	527
—$(CH_2)_4$—		A(Cl)	47.3–49.6	170, 172, 258, 604
		$D(S+H_2NCN)$	85	527
—$(CH_2)_5$—		A(Cl)	—	170
—$(CH_2)_6$—		A(Cl)	—	170
—$(CH_2)_3CH(CH_2NMe_2)$—		A(Br)	91	267

TABLE II-14. (continued)

R_1	R_2	Conditions[a]	Yield (%)	Ref.
—(CH$_2$)$_3$CH(CH$_2$N⟨piperidine⟩)—		A(Br)	64	267
Me	Bu	A(Cl)	50	173, 353
Me	Am	A(Cl)	50	173, 353
Me	Isoamyl	A(Cl)	50	173
Me	R	A(Cl)	—	591
Me	Hexyl	A(Cl)	50	173
Me	Decyl	A(Cl) 60% alcohol, 20 hr	—	443
Me	(CH$_2$)$_2$CO$_2$H	A(Br)	—	302
Me	(CH$_2$)$_2$CO$_2$Me	A(Cl)	—	139
Me	(CH$_2$)$_2$OH	A(Cl or Br) or halolactone	79–91	15, 140, 184, 229, 282, 291, 587, 760
Me	CH$_2$—CH—CH$_2$ (epoxide)	A(Cl)	—	162
Me	CH$_2$CO$_2$H	A(Br)	56.7	32, 276
Me	CH$_2$CO$_2$Et	A(Cl)	—	112
Me	CH$_2$Ph	A(Cl)	—	335
Me	iso-Pr	A(Br)	98	353, 426
Me	CO$_2$Me	A(Cl) or C	—	221
Me	CO$_2$Et	A(Cl or Br), steam bath B(I$_2$ or Br$_2$)	60–100 63	4, 10, 13, 22, 33 225, 261, 301, 708
Me	CO$_2$(CH$_2$)$_2$NEt$_2$	A(Br)	20	247
Me	CO$_2$(CH$_2$)$_2$Cl	A(Br)	95	247
Me	C≡N	A(Br), H$_2$O, 80°C water bath	50	380
Me	Ph	A(Br), alcohol reflux	—	341
Me	p-MeOC$_6$H$_4$	A(Br), alcohol reflux	—	341

Me		p-NO$_2$C$_6$H$_4$	A(Br), alcohol reflux	—	505
Me		α-Naphthyl	B(I$_2$)	—	463
Me		OBu	A(Br), alcohol reflux 1 hr	—	401
Me		P(=O)(EtO)$_2'$	A(Br), alcohol reflux	33.6	372
Me		Me—⟨N═⟩—NH$_2$ (thiazole) S	A(Br), alcohol, H$_2$O, reflux	—	355
Et		Me	A(Br)	92	165, 426
Et		Et	A(Br), water bath	90	377
Pr		Et	A(Cl), steam bath	50	377
	–CH$_2$C(Me)$_2$CH$_2$C(=O)–		A(Cl or Br), alcohol reflux 2 hr	84	509
CH$_2$CO$_2$Et		CO$_2$Et	A(Cl)	84	250
CH$_2$SC(=NH)NH$_2$		CH$_2$CO$_2$H	A(Br)	—	172
CH$_2$OH		CO$_2$Me	A(Br)	74.5	635
CH$_2$OH		CO$_2$Et	A(Br)	64.3	635
CH$_2$OH		CO$_2$Pr	A(Br)	60.2	635
CO$_2$Et		CO$_2$Et	A(Cl), alcohol	84	46, 208, 250
CO$_2$Et		Ph	A(Br)	—	299
CO$_2$Bu		Me	A(Br), AcOH	45	586
	(CH$_2$)$_3$ (o-tolyl)		A(Br)	—	681
Ph		Me	B(I$_2$, Br$_2$ or Cl$_2$)	69–94	225, 261, 280, 335
Ph		Et	B(I$_2$)	—	328, 335
Ph		CH$_2$CO$_2$H	A(Br)	90	231
Ph		CH$_2$CO$_2$Me	A(Br)	87	231
Ph		CH$_2$N⟨piperidine⟩	A(Cl)	—	267
Ph		HC(Me)CO$_2$H	A(Br)	83	231
Ph		CO$_2$Et	A(Cl or Br)	90	61, 272, 583

TABLE II-14. (continued)

R_1	R_2	Conditions[a]	Yield (%)	Ref.
o-MeC$_6$H$_4$	Me	A(Br), alcohol reflux	75	585
m-MeC$_6$H$_4$	Me	A(Br), alcohol reflux	100	585
p-MeC$_6$H$_4$	Me	A(Br), alcohol reflux	100	585
p-MeC$_6$H$_4$	CH$_2$CO$_2$H	A(Br), alcohol reflux	90	231
p-HOC$_6$H$_4$	Me	B(I$_2$)	—	463
p-MeOC$_6$H$_4$	Me	B(I$_2$)	—	341
p-ClC$_6$H$_4$	Me	B(I$_2$)	55	478
o-NO$_2$C$_6$H$_4$	Me	A(Br), alcohol	—	585
m-NO$_2$C$_6$H$_4$	Me	A(Br), alcohol	95	585
p-NO$_2$C$_6$H$_4$	Me	A(Br), alcohol	90	585
Ph	Ph	B(I$_2$)	—	341
Ph	p-NO$_2$C$_6$H$_4$	A(Br)	—	505
Ph	p-NH$_2$C$_6$H$_4$	A(Cl)	—	67
p-HOC$_6$H$_4$	Ph	B(I$_2$)	—	463
o-HOC$_6$H$_4$	Ph	B(I$_2$)	—	463
p-NO$_2$C$_6$H$_4$	p-NH$_2$C$_6$H$_4$	A(Br)	—	505
Ph	α-Naphthyl	B(I$_2$)	—	463
Ph	p-OC$_6$H$_4$I	A(Br)	—	80
α-Naphthyl	Me	B(I$_2$)	—	463
α-Naphthyl	CH$_2$CO$_2$H	A(Br), alcohol reflux	90	231
α-Naphthyl	Ph	B(I$_2$)	—	463
α-Naphthyl	α-Naphthyl	B(I$_2$)	—	463
β-Naphthyl	CH$_2$CO$_2$H	A(Br), alcohol reflux	90	231
α-Thienyl	CH$_2$CO$_2$H	A(Br)	90	231
![N-S ring]—Me	p-NH$_2$C$_6$H$_4$	A(Br), alcohol reflux	—	569

Me$\underset{S}{\overset{N}{=}}$R	p-NH$_2$C$_6$H$_4$	A(Br), alcohol		
			—	569
OH	Me	B(I$_2$)	—	396
β-Pyridyl	Me	B(I$_2$)	—	680
β-Pyridyl	Ph	B(I$_2$)	—	680
β-Pyridyl	β-Pyridyl	B(I$_2$)	—	680
β-Pyridyl	γ-Pyridyl	B(I$_2$)	—	680
α-Pyridyl	α-Pyridyl	B(I$_2$), 10 hr at 85–95°C	52.1	680
γ-Pyridyl	α-Pyridyl	B(Br$_2$), 2 hr at 55–60°C in HBr solution	71.8	680
α-Pyridyl	γ-Pyridyl	B(Br$_2$), 2 hr at 55–60°C	60.0	680
γ-Pyridyl	γ-Pyridyl	B(Br$_2$), 2 hr at 55–60°C	72.6	680
β-Quinolyl	Me	B(I$_2$)	—	680
β-Quinolyl	Ph	B(I$_2$)	—	680
β-Quinolyl	γ-Pyridyl	B(I$_2$)	—	680

[a] See Table II-13.

4-Aryl-2-amino-5-(p-aminophenyl)thiazoles of the general formula **116** were prepared by condensing phenylthiosemicarbazides (**115**) with either ω-bromoacetophenone by refluxing in alcohol for 2 hr (Method A) or with acetophenones and iodine on a steam bath for 8 hr (Method B) Scheme 53 (517), with R = para Me, MeO, Cl, Br, I, NO_2, NH_2, Ph; ortho Me, NO_2; or meta Br, I, NO_2. Yields ranged from 55 to 90% from Method A and 40 to 70% from Method B.

In a similar way, 4-naphthyl (517), 4-(4-methyl-2-thiazolyl) (**117**) (517), and 4-(4-methyl-5-thiazolyl) thiazoles (**118**) were prepared from the corresponding bromomethylketones (Scheme 54) (569).

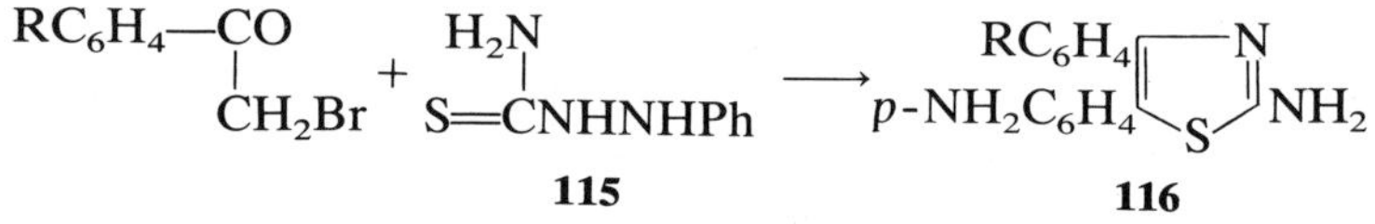

$$RC_6H_4-CO \atop CH_2Br \quad + \quad {H_2N \atop S=CNHNHPh} \quad \longrightarrow \quad p\text{-}NH_2C_6H_4\underset{S}{\overset{RC_6H_4\text{—}N}{\diagdown}}NH_2$$

115   116

Scheme 53

117   118

R = Me, Ph and OH

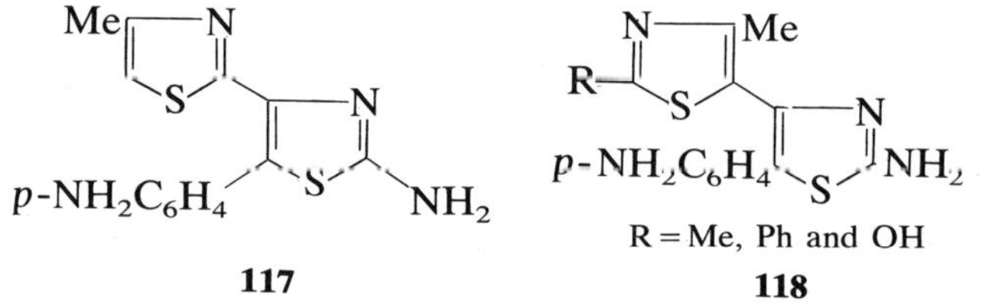

Scheme 54

3-Bromo-4-piperidone hydrochloride condensed with thiourea 3 days at 20°C give the 2-amino-4,5,6,7-tetrahydrothiazolo[5,4c]pyridine (**119**) (Scheme 55) (648).

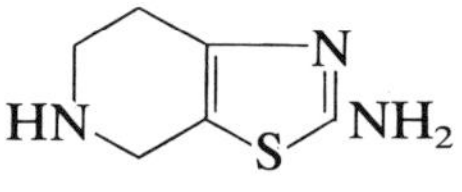

119

Scheme 55

Compounds other than α-halocarbonyls and thiourea can lead to 2-aminothiazoles. Thus 3-halogenoalkynes (**120**) condensed with thiourea in absolute alcohol give 2-aminothiazoles (Scheme 56a) (497), with R_1 = Me, Et, n-Bu, H_2CPh, and $H_2CC_6H_4R$-p; R_2 = H, Ph, C_6H_4R, and α-naphthyl. Yields ranged from 20 to 80% (497).

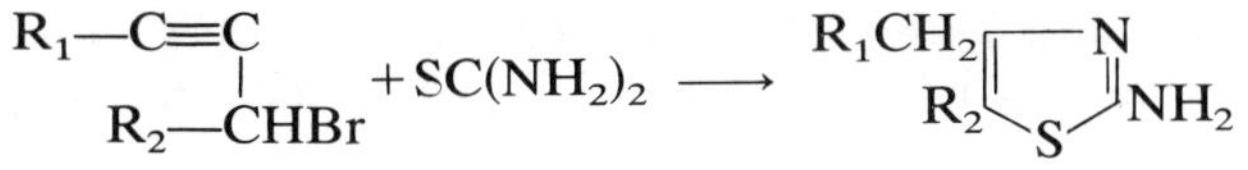

II. Thiazoles from α-Halocarbonyl Compounds and Derivatives

$$R_1-C\equiv C\atop R_2-CHBr \quad +SC(NH_2)_2 \longrightarrow \underset{120}{\overset{R_1CH_2\underset{S}{\overset{N}{\diagdown}}NH_2}{R_2}}$$

Scheme 56a

Similarly, 2,4-dichloro-2,3-epoxyalkanoates react with thiourea in methanol to yield 2-amino-4-methoxycarbonyl-5-(1-methoxyalkyl)-thiazoles (Scheme 56b) (755).

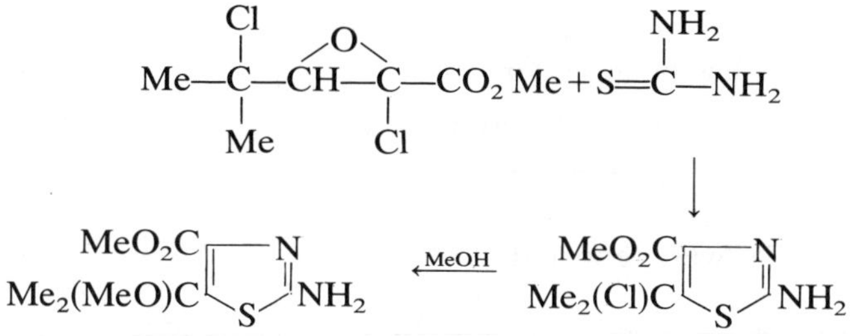

Scheme 56b

A variation of Hantzsch's synthesis, using thioureas in conjunction with α-diazoketones in place of α-halogenoketones, has proved to be generally applicable. In this manner, King and Miller (310) obtained 2-amino-4-phenylthiazole in 67% yield. A wide range of 4-substituted 2-alkyl (or aryl) aminothiazoles and 2-arylimino-3,4-diarylthiazolines have been prepared by Hampel and Muller (627, 665, 666).

The reaction involving thiourea may occur in the steps shown in Scheme 57.

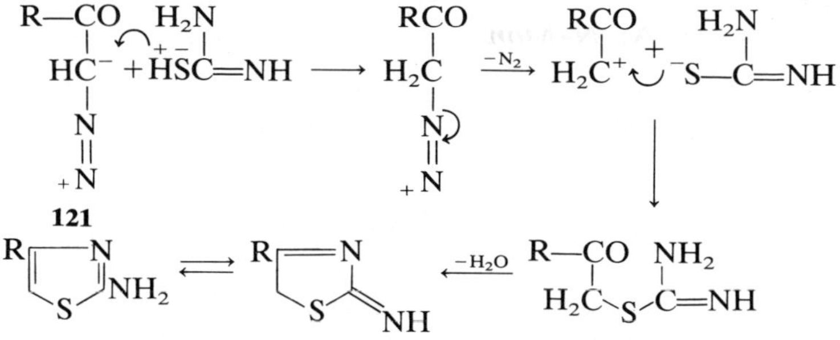

Scheme 57

The conversion of 2-aminothiazole derivatives into those unsubstituted in the 2-position was tried by the two available methods involving:

1. The Sandmeyer reaction, the substitution of an amino group by an halogen atom that is subsequently substituted after reduction by hydrogen atom.
2. The direct replacement of the diazonium group by hydrogen. These methods are discussed further in Chapter VI.

B. Mechanism

2-Aminothiazole formation as with alkyl- or arylthiazoles from thioamides can be divided into two parts: reaction between α-halocarbonyl compounds and thiourea form first an acyclic isothiourea; this intermediate has been seldom isolated because of the greater reactivity of thiourea over thioamide. However, in the case where an α-halogeno acid or ester is the reactant acyclic pseudothiohydantoin has been isolated (1, 126, 247). The second step of the reaction is the cyclization of this intermediate. The rate of the reaction of thiourea with phenacylbromide was reported (505). The reaction rate in the first reaction, followed by conductivity methods, was found to be second order and the activation energy was 10 to 11 kcal/mole. The reaction rate of the second reaction was measured by spectrometry. The rate constant has a linear relationship with the Hammett σ values as shown by studies with m- or p-substituted phenacylbromide, and ρ was found to be 0.774. The reaction with o-$O_2NC_6H_4COCH_2Br$ is slow due to the steric effect. The yields of thiazoles as measured by conductimetry and by spectrometry agree in the case of p-$O_2NC_6H_4COCH_2Br$ and disagree in the case of p-$O_2NC_6H_4CHBrCOPh$ and $PhCHClCOPh$, indicating that thiazole formation is a two-step reaction.

The cyclization rate is faster at higher temperatures, and the yield of thiazole attains a maximum after several days.

4. Reaction with N-Substituted Thioureas

A. N-Monosubstituted Thioureas

The cyclization of N-substituted thioureas (**123**) with halocarbonyl compounds gives 2-monosubstituted aminothiazoles (**124**) (Scheme 58) (768).

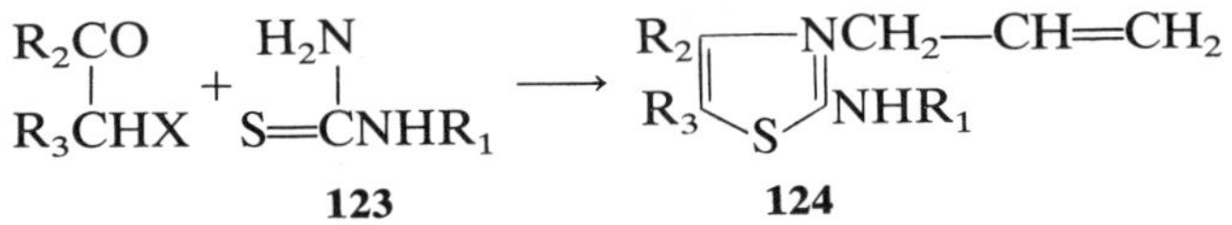

II. Thiazoles from α-Halocarbonyl Compounds and Derivatives

$$R_2CO \atop R_3CHX \ + \ {H_2N \atop S=CNHR_1} \ \longrightarrow \ \underset{R_3}{\overset{R_2}{\vphantom{|}}}\!\!\overset{\displaystyle NCH_2-CH=CH_2}{\underset{\displaystyle S}{\bigcirc}}\!\!NHR_1$$

                123   **124**

Scheme 58

This reaction has been widely studied, and a great variety of compounds (**124**), with R_1 = methyl (8, 36, 62, 68, 78, 178, 616, 621), benzyl (600), phthalimidomethyl (650), alkyl (555, 583, 796), acetyl (78, 582, 655), phenyl (8, 23, 36, 39, 255, 338, 398, 423, 508, 510, 583, 586, 601, 650, 800), aryl (737, 793), α- and β-naphthyl (399, 400, 423, 599, 604, 617, 621, 653), β-pyridyl (680), β-quinolyl (680), heteroaryl (810), or NHCONHMe (799), were prepared in fairly good yields by either Method A or B (Table II-15).

Allylthiourea condensed with α-halocarbonyl compounds gives the corresponding 2-allylaminothiazoles (53, 91, 262, 557, 565, 586, 588, 621, 655, 800) (Table II-16).

When the carbonyl reactant is ethylchloroacetate, a 2-imino-4-thiazolone isomer (**125**) is formed (Scheme 59) (85).

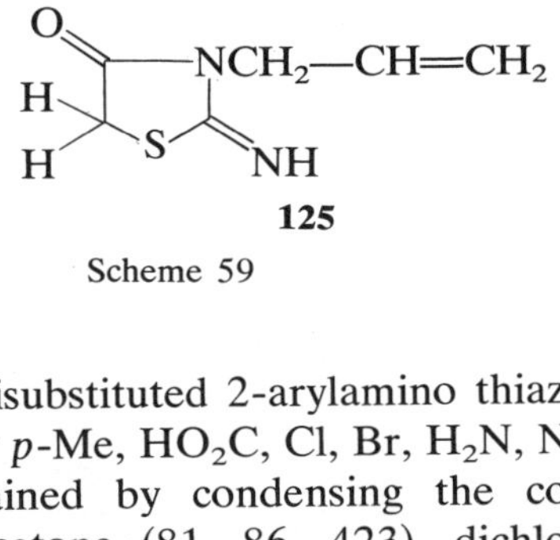

125

Scheme 59

Various 4-, 5-, or 4,5-disubstituted 2-arylamino thiazoles (**124**), R_1 = C_6H_4R with R = o-, m-, or p-Me, HO_2C, Cl, Br, H_2N, NHAc, NR_2, OH, OR, or O_2N, were obtained by condensing the corresponding N-arylthiourea with chloroacetone (81, 86, 423), dichloroacetone (510, 618), phenacylchloride or its p-substituted methyl, t-butyl, n-dodecyl or undecyl (653), or 2-chlorocyclohexanone (653) (Method A) or with 2-butanone (423), acetophenone or its p-substituted derivatives (399, 439), ethyl acetate (400), ethyl acetyl propionate (621), α- or β-unsaturated ketones (691), benzylidene acetone, furfurylidene acetone, and mesityl oxide in the presence of Br_2 or I_2 as condensing agent (Method B) (Table II-17).

2-Sulfonamido, 2-phthalimidomethyl (650) and 2-(N-4-maleylsulphanilamido)thiazoles (710), and N-thiazolylphosphoramido thiazoles were prepared similarly (790).

TABLE II-15. N-SUBSTITUTED 2-AMINOTHIAZOLES DERIVATIVES FROM N-SUBSTITUTED THIOUREAS AND α-HALOCARBONYL COMPOUNDS

$$R_1HN-\overset{N}{\underset{S}{\diagdown}}\overset{R_2}{\underset{R_3}{<}}$$

R_1	R_2	R_3	Conditions[a]	Yield (%)	Ref.
Me	H	H	A, CH_2ClCHO or $ClCH_2CHClOEt$	—	8, 23, 36, 68
Me	Me	H	A, (Cl)	—	8, 62
Me	Me	CH_2CO_2Et	A, (Br), alcohol reflux 9 hr, or Cl or dioxane	—	621
Me	Me	CO_2Et	A, (Br), alcohol, 2 hr at 100°C	—	61, 600, 752
Me	Me	COMe	A, (Cl)	—	696
Me	Ph	H	A, (Br), alcohol reflux	—	8
Me	3,4-$(OH)_2C_6H_3$	H	A(Cl), alcohol reflux	—	78
$C_{12}H_{25}$	Me	H	A(Cl), alcohol reflux 3 hr	—	555
CH_2Ph	Me	H	A(Br), alcohol reflux	—	600
CH_2Ph	Me	CO_2Et	A(Br), alcohol reflux	—	600
Phtalimidomethyl	H (or CO_2Et)	H (or allyl)	A(Br)	—	650
R	3,4-$(OH)_2C_6H_3$ 2,3,4-$(HO)_3C_6H_2$	H	A(Cl), alcohol reflux	—	616
iso-Pr	Ph	COPh	A(Br), alcohol reflux	—	583
COMe	CH_2Cl	H	A(Cl), Me_2CO	—	582
COMe	3,4-$(HO)_2C_6H_3$	H	A(Cl)	—	78
COPh	Me	Me	C	—	354
COPh	Et	H	C	—	354
COMe	⟨furan-NO_2⟩	H	A(Br), alcohol reflux	—	655
Ph	H	H	A($ClCH_2CHClOEt$)	—	8, 23, 650
Ph	Me	H	A(Cl)	—	8, 36, 255, 398

Ph	CH$_2$Cl	H	A(Cl), Me$_2$CO, 24 hr at room temperature	—	510
Ph	Me	Me	A(Br)	55.8	604
Ph	—(CH$_2$)$_4$—		B(I$_2$) steam bath 24 hr	65	423
Ph			A(Cl), alcohol reflux	24–40	599, 653
Ph	Me	CH$_2$CO$_2$Et	A(Br), alcohol reflux 9 hr	—	621
Ph	Me	CO$_2$Et	A(Br)	25	604
Ph			B(I$_2$) steam bath 24 hr	67	400
Ph	CO$_2$Et	H (or allyl)	A(Br), alcohol reflux	—	653
Ph	CO$_2$Bu	Me	A(Br), HOAc, 1 hr water bath	—	586
Ph	Ph	H	A(Br) or B(Br$_2$), C$_6$H$_6$	50.9	601, 604
Ph	Ph	Ph	A(Br), alcohol reflux	—	215
Ph	p-MeOC$_6$H$_4$	p-MeOC$_6$H$_4$	A(Br), alcohol reflux 3 hr	—	646
R	Ph	COPh	A(Br), alcohol reflux 9 hr	—	583, 696
Ph		H	A(BrCH$_2$CO)$_2$, alcohol, 5.6 hr at room temperature	80	508
Ph	⟨N/S–NHPh thiazole⟩	H	A(ClCH$_2$CN), 1 hr reflux, then 3 hr at room temperature	—	338
α-Naphthyl	H	p-C$_6$H$_4$Ph	A(Cl), alcohol reflux	—	617
α-Naphthyl	Me	H	A(Cl), alcohol reflux	—	653
α-Naphthyl	n-undecyl	H	A(Cl), alcohol reflux	—	653
α-Naphthyl	tert-Bu	H	A(Cl), alcohol reflux	—	653
α-Naphthyl	Me	Me	B(I$_2$) water bath 24 hr	65	423
α-Naphthyl	—(CH$_2$)$_4$—		A(Cl), alcohol reflux	—	599, 653
α-Naphthyl	Me	CH$_2$CO$_2$Et	A(Br), alcohol, 9 hr	—	621
α-Naphthyl	Me	CO$_2$Et	B(I$_2$) water bath 8 hr	60	402
α-Naphthyl	Ph	H	B(I$_2$)	60	399, 599
α-Naphthyl	p-tert-BuC$_6$H$_4$	H	B(I$_2$)	—	599
β-Naphthyl	Me	Me	B(I$_2$) water bath 24 hr	65	423
β-Naphthyl	Me	CH$_2$CO$_2$Et	A(Br), alcohol reflux 9 hr	—	621

TABLE II-15. (continued)

R_1	R_2	R_3	Conditions[a]	Yield (%)	Ref.
β-Naphthyl	Me	CO_2Et	$B(I_2)$	62	400
β-Naphthyl	Ph	H	$B(I_2)$	58	399
β-Naphthyl	Ph	Ph	A(Cl), alcohol reflux	—	617
β-Pyridyl	Me	H	A(Br), alcohol	—	680
β-Pyridyl	Ph	H	A(Br), alcohol	—	680
β-Pyridyl	β-Pyridyl	H	A(Br), alcohol	—	680
β-Pyridyl	γ-Pyridyl	H	A(Br), alcohol	—	680
β-Quinolyl	Me	H	A(Br), alcohol	—	680
β-Quinolyl	Ph	H	A(Br), alcohol	—	680
β-Quinolyl	γ-Pyridyl	H	A(Br), alcohol	—	680
o-MeOC$_6$H$_4$	Ph	H	A(Cl), EtOH	—	645
o-HOC$_6$H$_4$	Ph	H	A(Cl), EtOH	—	645
o-H$_2$NC$_6$H$_4$	Ph	H	A(Cl), EtOH	—	645
p-H$_2$NC$_6$H$_4$	Ph	H	A(Cl), EtOH	—	645
m-HOC$_6$H$_4$	Ph	H	A(Cl), EtOH	—	645
m-H$_2$NC$_6$H$_4$	Ph	H	A(Cl), EtOH	—	645
p-HOC$_6$H$_4$	Ph	H	A(Cl), EtOH	—	645
3,5-di-tert-butyl-4-HOC$_6$H$_2$	Ph	H	A(Cl), EtOH	—	645
5-(1,1,3,3-Tetramethylbutyl)-2-hydroxyphenyl	Ph	H	A(Cl), EtOH	—	645
p-Et$_2$NC$_6$H$_4$	Ph	H	A(Cl), EtOH	—	645
p-AcNHC$_6$H$_4$	Ph	H	A(Cl), EtOH	—	645
p-Lauramidophenyl	Ph	H	A(Cl), EtOH	—	645
p-Lauramidophenyl	p-tert-BuC$_6$H$_4$	H	A(Cl), EtOH	—	645
p-Stearamidophenyl	Ph	H	A(Cl), EtOH	—	645
p-MeO$_2$CHNC$_6$H$_4$SO$_2$	H	H	[ClCH$_2$CH(OH)]$_2$O, heat at 50–55°C, then 1 hr at 90°C	95	698

[a] See Table II-13.

TABLE II.16. 2-ALLYLAMINOTHIAZOLES FROM ALLYLTHIOUREA AND α-HALOCARBONYL COMPOUNDS

$$CH_2=CH-CH_2NH \underset{S}{\overset{N}{\rightthreetimes}} \overset{R_1}{\underset{R_2}{\lessgtr}}$$

R_1	R_2	Conditions[a]	Yield (%)	Ref.
H	Me	A(Br), alcohol reflux	—	588
Me	H	A(Cl), alcohol reflux	—	258
Me	Me	A(Cl), alcohol reflux	—	258
—(CH$_2$)$_3$—		A(Cl), alcohol reflux	—	258
—(CH$_2$)$_4$—		A(Cl), alcohol reflux	—	262
Me	Ac	A(Cl), alcohol reflux	—	588
Me	CH$_2$CO$_2$Et	A(Br), alcohol reflux 9 hr	—	621
Me	CO$_2$Et	A(Cl or Br), alcohol reflux 1 hr	—	565, 588
CH$_2$Cl	H	A(Cl), alcohol reflux 1 hr	45–50	557
(C=O)Me	—	A(Br), alcohol reflux	73	557
C(=NOH)Me	H	A(Br), alcohol reflux ½ hr	83	557
CO$_2$Et	H	A(Br), alcohol reflux ½ hr	40–55	557
CO$_2$Bu	Me	A(Br), alcohol reflux	—	586
Ph	H	A(Cl), alcohol	—	45
p-BrC$_6$H$_4$	H	A(Br), alcohol reflux 2 hr	91.5	557
p-NO$_2$C$_6$H$_4$	H	A(Br), alcohol	85	557
3,4-(OH)$_2$C$_6$H$_3$	H	A(Cl)	—	91
![NHCH$_2$=CHCH$_2$ thiazole]	H	A(Br), alcohol reflux 1 hr	86	506, 557
—C(R)=CH—(furan)—NO$_2$	H	A(Br), alcohol reflux, with R = H, Me, or Cl	80–90	655

[a] See Table II-13.

TABLE II-17. 2-ARYLAMINOTHIAZOLE DERIVATIVES

$$R_1C_6H_4NH\underset{S}{\overset{N}{\diagdown}}\underset{}{\overset{R_2}{\diagup}}R_3$$

R_1	R_2	R_3	Conditions[a]	Yield (%)	Ref.
o-Me	Me	Me	B(I_2), steam bath 24 hr	51	423
p-Me	Me	H	A(Cl)	—	81
p-Me	Me	Me	B(I_2), steam bath 24 hr	50	423
o-Me	Me	CH_2CO_2Et	A(Br), alcohol reflux	—	604
m-Me	Me	CH_2CO_2Et	B(I_2)	—	621
p-Me	Me	CH_2CO_2Et	B(I_2)	—	621
o-Me	Me	CO_2Et	B(I_2), water bath 8 hr	54	400
p-Me	Me	CO_2Et	B(I_2), water bath 8 hr	—	400
p-Me	CH_2Cl	H	A(Cl), Me_2CO, 24 hr at room temperature	15	604
			A(Er), alcohol reflux	—	510
o-Me	Ph	H	B(I_2), water bath 20 hr	88	399
m-Me	Ph	H	B(I_2), water bath 20 hr	—	399
p-Me	Ph	H	B(I_2), water bath 20 hr	80	399
p-Me	Ph	Ph	A(Br), alcohol	50.7	45, 584
2,4-Me_2	Me	H	A(Br), alcohol reflux	—	45
o-HO_2C	Me	CO_2Et	A(Cl)	—	86
o-HO_2C	Me	Me	B(I_2), water bath 24 hr	—	400
o-HO_2C	Ph	H	B(I_2), water bath 24 hr	—	423
m-HO_2C	Me	Me	B(I_2)	45	399
m-HO_2C	Me	CO_2Et	B(I_2)	44	423
p-HO_2C	Me	Me	B(I_2)	—	400
p-HO_2C	Me	CO_2Et	B(I_2)	43	423
p-HO_2C	Ph	H	B(I_2)	—	400
p-HO_2C	Ph	H	B(I_2)	50	399
m-HO_2C	Ph	H	B(I_2)	50	423

o-Cl	Me	Me	B(I$_2$)	50	423
o-Cl	Me	CO$_2$Et	B(I$_2$)	—	400
o-Cl	Ph	H	B(I$_2$)	25	399
m-Cl	Me	Me	B(I$_2$)	56	423
m-Cl	Me	CH$_2$CO$_2$Et	B(I$_2$)	—	621
m-Cl	Me	CO$_2$Et	B(I$_2$)	—	400
m-Cl	Ph	H	B(I$_2$)	45	399
p-Cl	Me	Me	B(I$_2$)	53	423
p-Cl	Me	H	A(Cl)	33	86
p-Cl	Me	CH$_2$CO$_2$Et	B(I$_2$)	—	621
p-Cl	Me	CO$_2$Et	B(I$_2$)	—	400
p-Cl	CH$_2$Cl	H	A(Cl), Me$_2$CO 24 hr at room temperature	90	618
p-Br	Me	H	A(Cl)	—	86
p-Br	Me	CH$_2$CO$_2$Et	B(I$_2$)	—	621
p-Br	CH$_2$Cl	H	A(Cl), Me$_2$CO, 24 hr at room temperature	—	510
o-Br	Ph	H	B(I$_2$), water bath 20 hr	72	439
o-Br	p-BrC$_6$H$_4$	H	B(I$_2$), water bath 20 hr	65	439
m-Br	Ph	H	B(I$_2$), water bath 20 hr	70	439
m-Br	p-BrC$_6$H$_4$	H	B(I$_2$), water bath 20 hr	81	439
p-Br	Ph	H	B(I$_2$), water bath 20 hr	75	439
p-Br	p-BrC$_6$H$_4$	H	B(I$_2$)	72	439
2,4-Br$_2$	Me	H	A(Cl)	—	86
p-I	Me	H	A(Cl)	—	86
o-H$_2$N	Ph or p-tert-BuC$_6$H$_4$	H	A(Cl), C$_6$H$_6$ and alcohol reflux	—	599, 653
m-H$_2$N	Ph or p-tert-BuC$_6$H$_4$	H	A(Cl), C$_6$H$_6$ and alcohol reflux	—	599, 653
p-H$_2$N	Ph or p-tert-BuC$_6$H$_4$	H	A(Cl), C$_6$H$_6$ and alcohol reflux	—	599, 653
m-AcNH	Ph or p-tert-BuC$_6$H$_4$	H	A(Cl), C$_6$H$_6$ and alcohol reflux	—	599, 653
p-AcNH	H or CO$_2$Et	H or allyl	A(Cl), C$_6$H$_6$ and alcohol reflux	—	650
p-AcNH	Ph or p-tert-BuC$_6$H$_4$	H	A(Cl), C$_6$H$_6$ and alcohol reflux	—	599, 653
p-Et$_2$N	Ph or p-tert-BuC$_6$H$_4$	H	A(Cl), C$_6$H$_6$ and alcohol reflux	—	599, 653
p-Laurylamido	Ph or p-tert-BuC$_6$H$_4$	H	A(Cl), C$_6$H$_6$ and alcohol reflux	—	599, 653

TABLE II-17 (continued)

R_1	R_2	R_3	Conditions[a]	Yield (%)	Ref.
p-Stearylamido	Ph or p-tert-BuC$_6$H$_4$	H	A(Cl), C$_6$H$_6$ and alcohol reflux	—	599, 653
o-OH	Me	tert-Bu	A(Cl)	—	599, 653
o-OH	H	n-Pentyl	A(Cl)	—	599, 653
o-OH	Ph or p-tert-BuC$_6$H$_4$	H	A(Cl)	—	599, 653
m-OH	Ph or p-tert-BuC$_6$H$_4$	H	A(Br)	—	599, 653
p-OH	H	n-Pentyl	A(Br)	—	599, 653
p-OH	Me	H	A(Cl)	—	86
p-OH	Ph or p-tert-BuC$_6$H$_4$	H	A(Cl)	—	599, 653
o-MeO	Me	CH$_2$CO$_2$Et	B(I$_2$)	—	621
p-MeO	Me	CH$_2$CO$_2$Et	B(I$_2$)	—	621
o-MeO	Ph or p-tert-BuC$_6$H$_4$	H	A(Cl), C$_6$H$_6$ and alcohol	—	599, 653
p-PhO	H	H	A(ClCH$_2$CHClOEt), 30 min in water	—	410
p-PhO	Me	H	A(Cl), water	80	423
p-EtO	Me	H	A(Cl), water	—	86
p-NH$_2$O$_2$SC$_6$H$_4$	H	H	A, CH$_2$ClCHClOEt	—	147
p-NH$_2$O$_2$SC$_6$H$_4$	Me	H	A(Cl)	92	148
p-NH$_2$O$_2$SC$_6$H$_4$	—(CH$_2$)$_4$—		A(Cl)	—	302
p-NH$_2$O$_2$SC$_6$H$_4$	Ph	H	A(Br)	—	147
p-NH$_2$O$_2$SC$_6$H$_4$	Me	(CH$_2$)$_2$OH	A(Cl)	25	302
p-NH$_2$O$_2$SC$_6$H$_4$	Me	CH$_2$CO$_2$Et	A(Br)	—	147
p-NH$_2$O$_2$SC$_6$H$_4$	Me	CO$_2$Et	A(Br)	—	147
p-MeCONHO$_2$SC$_6$H$_4$	Me	H	A(Cl)	—	186, 187, 302
p-MeCONHO$_2$SC$_6$H$_4$	—(CH$_2$)$_4$—		A(Br)	—	302
p-MeCONHO$_2$SC$_6$H$_4$	Ph	H	A(Br)	—	302
p-MeCONHO$_2$SC$_6$H$_4$	Me	(CH$_2$)$_2$OH	A(Cl)	—	181
CH$_3$CONHC$_6$H$_4$SO$_2$	CH$_2$Cl	H	A(Cl)	—	254
p-NH$_2$C$_6$H$_4$SO$_2$	Me	H	A(Cl)	—	302
p-NH$_2$C$_6$H$_4$SO$_2$	Ph	H	A(Cl)	—	302

3,5-di-tert-Bu-4-OH	Ph or p-tert-BuC$_6$H$_4$	H	A(Cl)	—	599, 653
5-(1,1,3,3-tetraMe)Bu	Ph or p-tert-BuC$_6$H$_4$	H	A(Cl)	—	599, 653
2-OH-5-octyl	Ph or p-tert-BuC$_6$H$_4$	H	A(Cl)	—	599, 653
m-O$_2$N	Me	Me	B(I$_2$), water bath 24 hr	55	423
m-O$_2$N	Me	CO$_2$Et	B(I$_2$)	—	400
p-O$_2$N	Me	H	A(Cl)	—	86
p-O$_2$N	Me	Me	B(I$_2$)	43	423
p-O$_2$N	Me	CO$_2$Et	B(I$_2$)	—	400
R	–(CH$_2$)$_4$–		A(Cl)	—	553, 653

[a] See Table II-13.

2-Arylamino thiazoles unsubstituted in the 4- or 5-position were obtained from α,β-dihalogenoethers and arylthioureas (8, 23, 147).

N-monosubstituted thioureas also react with α-haloacids or esters to give stable compounds of type **126**, in which R is aryl or acyl. When R is alkyl such as an allyl group (85), isomer **125** is formed (Scheme 60).

$$\underset{\text{CH}_2\text{Cl}}{\overset{\text{EtOCO}}{|}} + \underset{\text{S}=\text{CNHR}}{\overset{\text{H}_2\text{N}}{|}} \longrightarrow \underset{\mathbf{126}}{\text{[thiazolinone-NHR]}}$$

Scheme 60

Steroid compounds (**127**) have been obtained (524) by condensing several iodo analogs of Reichstein's compound with N-alkyl (or aryl) thioureas in acetonic solution (Scheme 61), with X = H$_2$ or O, R = Et, Pr, iPr, n-Bu, s-Bu, Ph, p-MeOC$_6$H$_4$, p-EtOC$_6$H$_4$, p-MeC$_6$H$_4$, m-MeC$_6$H$_4$, m-ClC$_6$H$_4$, 3,5-Cl$_2$C$_6$H$_3$, and so forth.

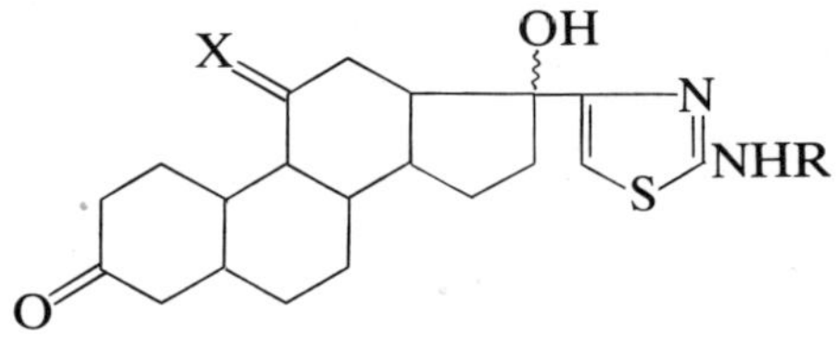

127

Scheme 61

Ortho and *para* phenylene dithioureas (**128**) treated with α-halocarbonyl compounds afford NN′-*ortho*- and *para*-phenylene-bis(2-amino-4-R-thiazole) (**129**) (Scheme 62) (653).

o- or p-C$_6$H$_4$(—NH(S=)CNH$_2$)$_2$ + RCOCH$_2$X ⟶
128

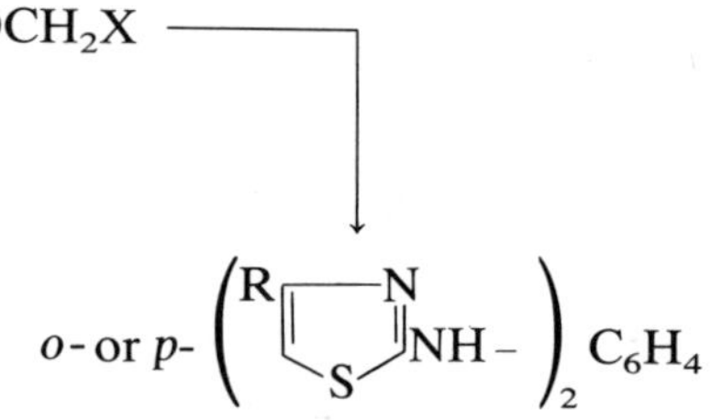

o- or p- $\left(\underset{\text{S}}{\overset{\text{R}\underset{}{\frown}\text{N}}{\bigcup}}\text{NH}- \right)_2$ C$_6$H$_4$

129

Scheme 62

II. Thiazoles from α-Halocarbonyl Compounds and Derivatives 243

Compound **129**, with R = Ph, p-MeC$_6$H$_4$, p-MeOC$_6$H$_4$, p-EtOC$_6$H$_4$, and 2-furyl, was obtained in 70 to 80% yield when p-phenylene bis-thiourea was heated with iodine and p-AcC$_6$H$_4$R$_1$ according to Method B (741).

Similarly, p-biphenylene-bis(2-thiourea) (**130**) in methyl cellosolve gave N,N'-p-biphenylene-bis(2-amino-4-R-thiazole) (**131**), in which R = t-Bu, Ph, n-dodecylphenyl, and n-undecylphenyl (Scheme 63) (553, 599).

p-(C$_6$H$_4$NHCSNH$_2$)$_2$ + RCOCH$_2$X ⟶
   130

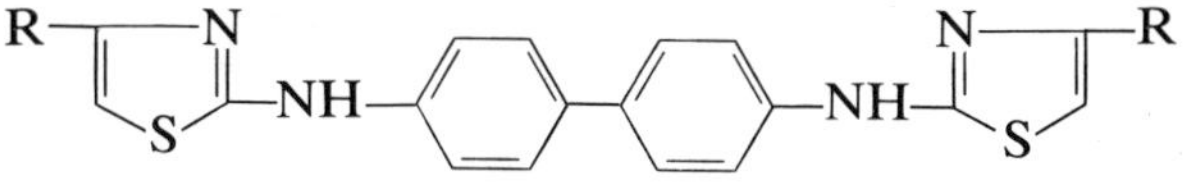

[structure of **131**]

Scheme 63

Some 2-arylamino arsenic derivatives of thiazole have been also prepared (598) (Table II-18).

TABLE II-18. 2-ARYLAMINO ARSENIC DE-
RIVATIVES OF THIAZOLES[a]

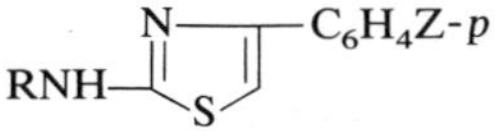

RNH—[thiazole]—C$_6$H$_4$Z-p (I) p-ZC$_6$H$_4$NH—[thiazole]—R (II)

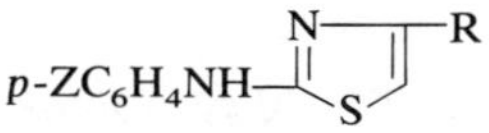

R(I)	Ia	Ib	Ic
H	80	70	65
Ph	75	70	55
p-O$_2$NC$_6$H$_4$	68	70	60
p-ClC$_6$H$_4$	70	70	60
p-MeC$_6$H$_4$	72	70	55
p-MeOC$_6$H$_4$	70	70	63
p-EtOC$_6$H$_4$	70	70	61
m-O$_2$NC$_6$H$_4$	65	70	58
m-ClC$_6$H$_4$	67	70	60
α-Pyridyl	60	70	55
2-Phenyl-4-thiazolyl	62	70	51
2-Benzothiazolyl	60	70	60
α-Naphthyl	70	70	60
β-Naphthyl	70	70	60

[a] With Z = p-AsO$_3$H$_2$ (Ia), AsCl$_2$ (Ib), AsO (Ic) 598).
[b] Compound **II** was also prepared. Yields ranged from 60 to 75% (598).

B. Dithiobiuret

Compound **132** condensed with 1 or 2 moles of aliphatic or aromatic α-haloketones in acetonic or alcoholic solution yielded either the corresponding 2-thiazolythiourea (**133**) (559, 753, 797) or *sym*-substituted bis(2-thiazolyl)amine (**134**) (Scheme 64 and Table II-19) (430, 553, 653).

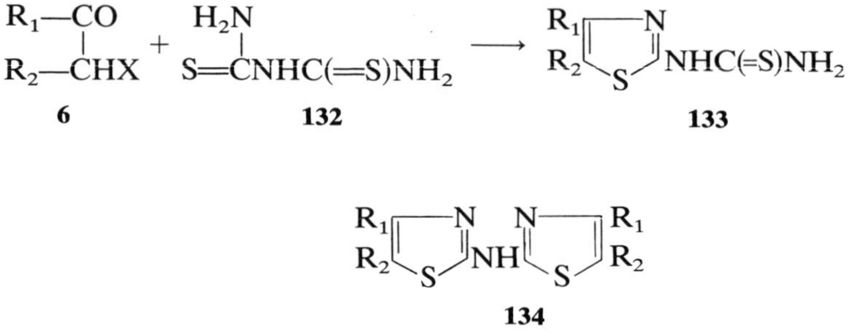

$$R_1-CO \atop R_2-CHX \quad + \quad {H_2N \atop S=CNHC(=S)NH_2} \longrightarrow R_1 {\underset{S}{\diagup}}^N \diagdown NHC(=S)NH_2$$
$$\textbf{6} \qquad\qquad \textbf{132} \qquad\qquad\qquad \textbf{133}$$

$$R_1 {\underset{S}{\diagup}}^N \diagdown NH {\underset{S}{\diagup}}^N \diagdown R_1 \atop R_2$$
$$\textbf{134}$$

Scheme 64

C. N,N-Disubstituted Thioureas

The *N,N*-disubstituted thioureas (**135**) condensed with α-halocarbonyl compounds give 2-disubstituted aminothiazoles (**136**) but in lower yields (30 to 70%) (Scheme 65 and Table II-20) (518). For example, *N,N*-dialkylthioureas condensed with chloroacetaldehyde or dibromoether lead to *N,N*-dialkyl-2-aminothiazoles in **136**, $R_1 = R_2 =$ methyl (342, 404, 436, 637), ethyl (343, 436), *n*-propyl (518), *n*-butyl (518), allyl (518), and benzyl (26, 29). When chloroacetone and dichloroacetone are the carbonyl reactants the corresponding 4-methyl (518) and 4-chloromethyl derivatives (572) were obtained.

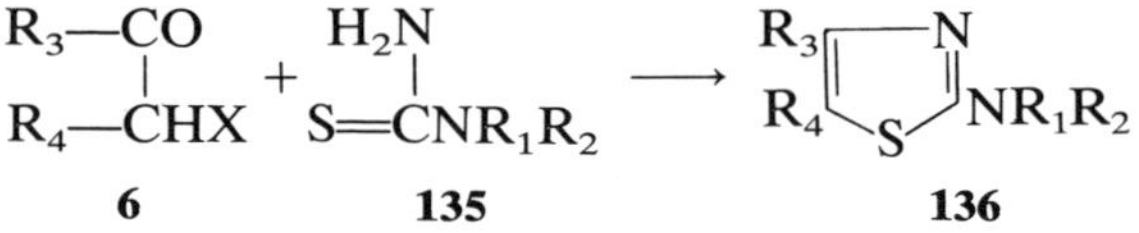

$$R_3-CO \atop R_4-CHX \quad + \quad {H_2N \atop S=CNR_1R_2} \longrightarrow R_3 {\underset{S}{\diagup}}^N \diagdown NR_1R_2$$
$$\textbf{6} \qquad\qquad \textbf{135} \qquad\qquad \textbf{136}$$

Scheme 65

TABLE II-19. DITHIAZOLYLAMINES FROM DITHIOBIURET AND α-HALOCARBONYL COMPOUNDS

$$\text{H}_2\text{N(S=)CHN} \underset{\mathbf{133}}{\overset{\displaystyle \begin{array}{c}\text{N}\!\!=\!\!\overset{\displaystyle R_1}{\underset{\displaystyle }{}}\\ \end{array}}{\left\langle_{\text{S}}^{R_2}\right.}} \qquad \text{NH}\underset{\mathbf{134}}{\left(\!\!\begin{array}{c}\text{N}\!\!=\!\!\overset{\displaystyle R_1}{\underset{\displaystyle }{}}\\ \underset{\text{S}}{}\!\!R_2\end{array}\right)_2}$$

			Yield		
R_1	R_2	Conditions[a]	133	134	Ref.
H	H	A(ClCH$_2$CHCl)$_2$O, alcohol, water bath	29	64	430
Me	H	A(Cl), alcohol, water bath	51.5	70	430
Me	Me	A(Br), alcohol	—	78	430
$C_{11}H_{23}$	H	A(Cl), alcohol	—	—	553
undecyl	H	A(Cl), alcohol, 55°C then 3 hr at 75°C under N_2	—	—	653
tert-Bu	H	A(Cl), C_6H_6 + alcohol	—	—	553
Ph	H	A(Br), alcohol	—	93	430
p-MeC$_6$H$_4$	H	A(Br), Me$_2$CO, reflux 1 hr	67	11	559
p-n-dodecyl C$_6$H$_4$	H	A(Br), Me$_2$CO, reflux 1 hr	74	7	559
p-C$_{18}$H$_{37}$C$_6$H$_4$	H	A(Cl), alcohol	—	—	653
p-tert-BuC$_6$H$_4$	H	A(Cl), alcohol	—	—	653
p-MeOC$_6$H$_4$	H	A(Cl), alcohol	—	—	653
p-ClC$_6$H$_4$	H	A(Br), Me$_2$CO, reflux 1 hr	80	6	559
p-BrC$_6$H$_4$	H	A(Br), Me$_2$CO, reflux 1 hr	61	8	559
p-BrC$_6$H$_4$	H	A(Br), Me$_2$CO, reflux 1 hr	74	9	559
p-BrC$_6$H$_4$	Me	A(Br), Me$_2$CO, reflux 1 hr	85	14	559
m-O$_2$NC$_6$H$_4$	H	A(Br), alcohol	—	85	430
p-O$_2$NC$_6$H$_4$	H	A(Br), Me$_2$CO, reflux 1 hr	58	18	559
Ph	Ph	A(Br), alcohol	—	80	430
		B(Br), Me$_2$CO, reflux 1 hr	40	20	559

[a] See Table II-13.

TABLE II-20. N,N-DISUBSTITUTED 2-AMINOTHIAZOLES FROM α-HALOCARBONYL COMPOUNDS AND N,N DISUBSTITUTED THIOUREAS

$$\begin{array}{c} R_1 \\ \diagdown \\ N \\ \diagup \\ R_2 \end{array} \begin{array}{c} N \!=\! \overset{R_3}{\underset{S}{\diagdown}} \\ \overset{}{\underset{}{\diagup}} R_4 \end{array}$$

R_1	R_2	R_3	R_4	Conditions[a]	Yield (%)	Ref.
Me	Me	H	H	A(Cl)	72–75	342, 404
					—[b]	436, 637
Me	Me	Me	H	A(Cl)	—	223
Me	Me	CH_2Cl	H	A(Cl), Me_2CO	50–67	572
Me	Me	CO_2Me	H	A(Cl)	70	637
Me	Ph	Ph	COPh	A(Br)	90	583
Et	Et	H	H	A(Cl)	—	343, 436
Et	Et	Me	H	A(Cl) or (Br)	85–94	129
Et	Et	CH_2Cl	H	A(Cl), Me_2CO	weak	572
Et	Ph	Ph	COPh	A(Br)	83	583
Me	Ph	Me	H	A(Cl)	—	86
Me	p-MeC_6H_4	Me	H	A(Cl)	—	518
n-C_3H_7	n-C_3H_7	H	H	A(Cl) or $BrCH_2CHBrOEt$	58–68	518
n-C_3H_7	n-C_3H_7	Me	H	A(Cl)	62	518
n-C_4H_9	n-C_4H_9	H	H	A(Cl)	45–50	518
n-C_4H_9	n-C_4H_9	Me	H	A(Cl)	77–80	518
CH_2=CH–CH_2	CH_2=CHCH$_2$	Me	H	A(Cl)	39	518
CH_2=CH–CH_2	CH_2=CHCH$_2$	Me	H	A(Cl)	60	518
CH_2Ph	CH_2Ph	Me	H	A(Cl)	—	—
CH_2Ph	CH_2Ph	Ph	H	A(Br)	—	26, 29
CH_2Ph	$(CH_2)_2NMe_2$	H	H	A(ClCH$_2$CH(OEt)$_2$)+HCl, 1 hr at 100°C	80	380
CH_2Ph	p-MeC_6H_4	Ph	H	A(Br)	—	460
p-$MeOC_6H_4CH_2$	$(CH_2)_2NMe_2$	H	H	A(ClCH$_2$CH(OEt)$_2$)+HCl, 1 hr at 100°C	—	380

3,4-(MeO)₂C₆H₃CH₂	(CH₂)₂NMe₂	H	A(ClCH₂CH(OEt)₂)+HCl, 1 hr at 100°C	—	380
p-ClC₆H₄CH₂	(CH₂)₂NMe₂	H	A(ClCH₂CH(OEt)₂)+HCl, 1 hr at 100°C	—	380
p-MeOC₆H₄CH₂	(CH₂)₃NMe₂	H	A(ClCH₂CH(OEt)₂)+HCl, 1 hr at 100°C	—	380
p-MeC₆H₄CH₂	(CH₂)₂NMe₂	H		—	380
p-MeSC₆H₄CH₂	(CH₂)₂NMe₂	H		—	380
2,5-MeS(Br)C₆H₃CH₂	CH₂NMe₂	H	A(Br), Me-cellosolve	—	380
Ph	Ph	H	A(Br)	—	653
Ph	Ph	COPh	A(Br), Me-cellosolve	92	583
Ph	p-tert-BuC₆H₄	H	A(Br), Me-cellosolve	—	653
Ph	p-n-dodecyl C₆H₄	H	A(Br), Me-cellosolve	—	653
Ph	p-n-undecyl C₆H₄	H		—	653
p-H₂NC₆H₄	Me	H	A(Cl)	—	653
p-OHC₆H₄	Ph	H	A(Cl)	—	653
2-OH-5-octyl C₆H₃	lauroyl	H	A(Cl)	—	653
COMe	Ph	COPh	A(Br)	90	583
COMe	tert-Bu	H	C(iodomercuriketone)	—	354

[a] See Table II-13.
[b] Isolated as hydrochloride

The synthesis of these disubstituted thioureas takes place in three steps. First the alkyl bromide is prepared by the action of hydrobromic acid on the corresponding alcohol (518). Then the dialkylcyanamide is obtained by treatment at 25°C with calcium cyanamide. The yields are of the order of 30 to 60%. Thioureas are obtained in a third step from the cyanamide by reaction at 40°C with H_2S in the presence of pyridine. Yields ranged from 57 to 90% (518).

With two different N,N-substituents in **135**, variously N,N-disubstituted (**136**) are obtained. Thus **136**, in which R_1 = Me (86, 583), Et (583), benzyl (380, 460, 653), acyl (354, 583), or phenyl (653) and R_2 = Ph (86, 583), aryl (460, 653), or dimethylaminoethyl (380), were prepared.

The symmetrical disubstituted thioureas such as **137** do not give a thiazolic ring (Scheme 66), but give compounds of type **138** or **139**, which are derived from the tautomer imino form of the 2-aminothiazole (86).

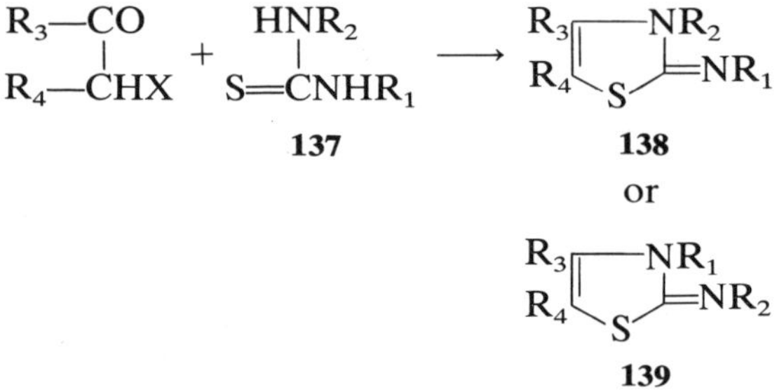

$$R_3\text{—CO} \atop R_4\text{—CHX} \quad + \quad {HNR_2 \atop S=CNHR_1} \quad \longrightarrow \quad {R_3\text{—}\!\!\!-\!\!\!\text{NR}_2 \atop R_4\diagdown_S\diagup\!\!=\!\!\text{NR}_1}$$

137   **138**

or

$${R_3\text{—}\!\!\!-\!\!\!\text{NR}_1 \atop R_4\diagdown_S\diagup\!\!=\!\!\text{NR}_2}$$

139

Scheme 66

With R_1 different from R_2 two isomeric compounds (**138** and **139**) are possible, depending on the direction of ring closure (86). However, only one form is generally obtained. Finally, the trisubstituted thioureas such as N,N,N'-trimethylthiourea react with chloroacetone to give a thiazolium salt, in a reaction identical to that of the N-monosubstituted thioamides (Scheme 67).

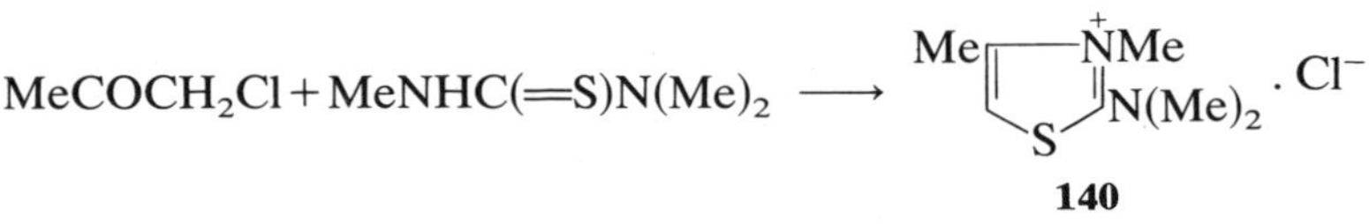

$$\text{MeCOCH}_2\text{Cl} + \text{MeNHC}(=S)\text{N}(\text{Me})_2 \longrightarrow {\text{Me}\text{—}\!\!\!-\!\!\!\overset{+}{\text{N}}\text{Me} \atop \diagdown_S\diagup\!\!\text{N}(\text{Me})_2} \cdot \text{Cl}^-$$

140

Scheme 67

II. Thiazoles from α-Halocarbonyl Compounds and Derivatives

D. *Thiosemicarbazides and Thiosemicarbazones*

Thiosemicarbazides (**141**), $R_1 = H$, yield the 2-hydrazinothiazoles (Scheme 68). Compound **142**, $R_1 = H$, on reaction with α-halocarbonyl

$$R_2COCXHR_3 + R_1NHNHCSNH_2 \longrightarrow \underset{\textbf{142}}{\underset{R_3}{R_2}\underset{S}{\diagdown}\hspace{-0.5em}\diagup\hspace{-0.5em}\underset{}{N}}NHNHR_1$$

141

Scheme 68

compounds and several derivatives of this type, has been reported with $R_1 = R_3 = H$ and $R_2 = Me$ (72, 379, 614, 678), Ph (558), or CO_2Et (467).

With 7-bromobenzo[*a*]cyclohept-2-ene-1-one in alcohol solution, the 9,10-dihydro-8H-benzo-4,5-cyclohepta-[*d*]-thiazol-2-ylhydrazine (**143**) was obtained in 46% yield (Scheme 69) (679, 688, 689).

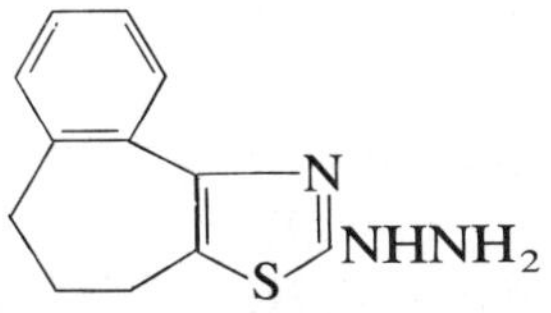

143

Scheme 69

But the reaction with aliphatic α-halocarbonyl compounds is usually complex, and a variety of compounds can be formed depending on the reactants and the reaction conditions. With chloroacetone in neutral medium (alcohol) the acyclic intermediate (**144**) analogous to those obtained with thiourea and thioamides was isolated (Scheme 70).

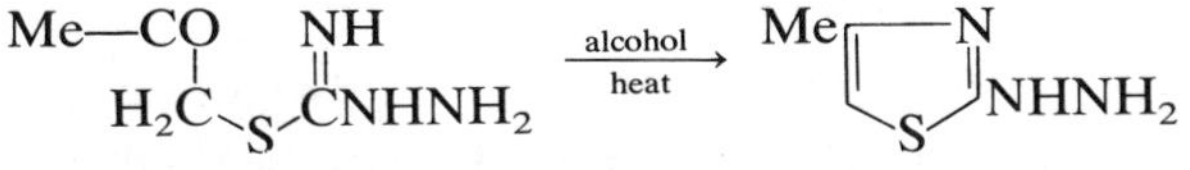

$$\underset{\textbf{144}}{Me-\underset{|}{\overset{CO}{C}}-\underset{S}{\diagdown}\underset{}{\overset{NH}{\underset{\|}{C}}}NHNH_2} \xrightarrow[\text{heat}]{\text{alcohol}} \underset{S}{\overset{Me}{\diagdown}}\underset{}{\diagup}\hspace{-0.5em}\overset{N}{\diagdown}NHNH_2$$

Scheme 70

Under acidic conditions, a thiosemicarbazone intermediate (**145**) has been isolated, this can be cyclized in alcohol either into the corresponding 2-hydrazinothiazole (**142**) in the presence of benzaldehyde or into 1,3,4-thiadiazine (**146**) in the absence of benzaldehyde (Scheme 71) (375, 397, 408).

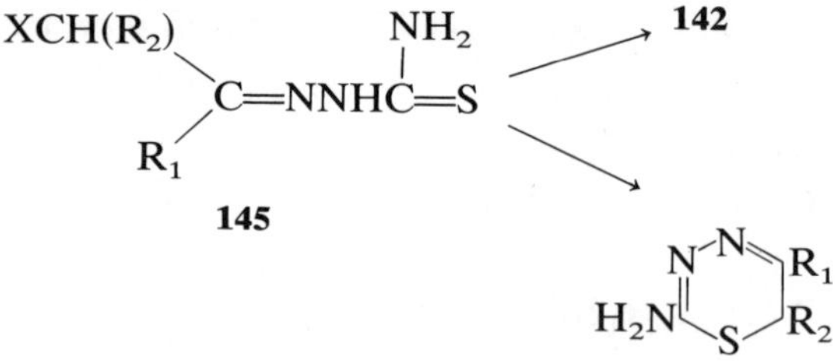

$$\text{XCH}(R_2) \diagdown \atop R_1 \diagup C=NNHC \diagup^{NH_2}_{=S} \quad\rightarrow\quad 142$$

145

$$\underset{146}{H_2N \underset{S}{\overset{N-N}{\diagdown\diagup}} \overset{R_1}{\underset{R_2}{}}}$$

Scheme 71

Similarly substituted thiosemicarbazides yield substituted 2-hydrazinothiazoles (**142**), R = alkyl, allyl (36), acyl (59, 60, 67, 69, 500, 545), amido (391), aroyl (525), aryl (71, 247, 396, 466), α-naphthyl (59, 60, 67, 69, 71, 140), or α-quinolyl (149) (Tables II-21 and II-22).

In some cases two products have been isolated. Thus the condensation of phenylthiosemicarbazide (**141**), R_1 = Ph with dichloroether (396) or chloroacetone, gives two isomers, **142**, R_1 = Ph, R_2 = H or Me, R_3 = H, and **147**, R_1 = H or Me, that are the result of benzidine-type rearrangement (Scheme 72).

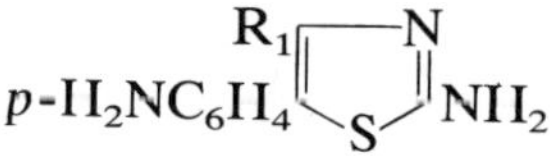

$$p\text{-H}_2NC_6H_4 \underset{S}{\overset{R_1\quad N}{\diagdown\diagup}} NH_2$$

147

Scheme 72

When the 5-position is substituted no rearrangement occurs. Phenacylbromide and its substituted derivatives seem to give only products of type **147**, in which R_1 = aryl (517).

Several sulfonamides (**142**), $R_1 = p\text{-H}_2NSO_2C_6H_4$, have been obtained by condensing the corresponding thiosemicarbazide (**141**), $R_1 = p\text{-H}_2NSO_2C_6H_4$, with α-halocarbonyl compounds (396, 466) in alcohol solution.

In the case of dichloroether (396) an acyclic intermediate **148** was isolated (Scheme 73). This intermediate was cyclized by heating to the expected **141**, $R_1 = p\text{-H}_2NSO_2C_6H_4$, $R_1 = R_2 = H$.

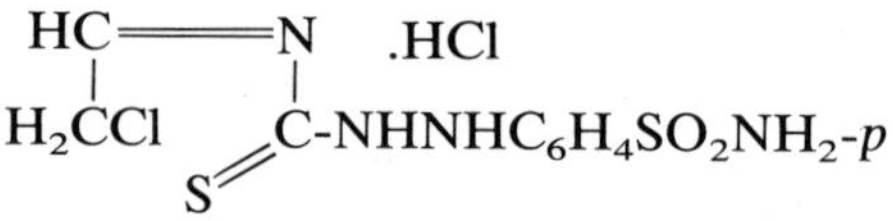

$$\underset{H_2CCl}{\overset{HC=N}{|}} \underset{S}{\diagup} \overset{.HCl}{C\text{-NHNHC}_6H_4SO_2NH_2\text{-}p}$$

148

Scheme 73

TABLE II-21. 2-HYDRAZINOTHIAZOLES FROM THIOSEMICARBAZIDES AND α-HALOCARBONYL COMPOUNDS

$$R_1NHNH\underset{S}{\overset{N}{\diagdown}}\!\!\!=\!\!\!\underset{}{\overset{R_2}{\diagup}}\!\!\!R_3$$

R_1	R_2	R_3	Conditions[a]	Yield (%)	Ref.
H	Me	H	A(Cl), CHCl$_3$, reflux 5–6 hr	20–31 (HCl)[b]	72, 379, 558
H	Ph	H	A(Cl)	—	558
H	CO$_2$Et	Me	A(Br)	—	467
CH$_2$=CHCH$_2$	Me	H	A(Cl)	—	36
Ac	H	H	A, BrCH$_2$CHBrOEt, alcohol	30	500
Ac	Ph	H	A(Br), Alcohol	—	50, 60, 67, 69, 500
Ac	p-MeC$_6$H$_4$	H	A(Cl), alcohol	79	545
Ac	p-BrC$_6$H$_4$	H	A(Cl), alcohol	79	545
Ac	p-MeOC$_6$H$_4$	H	A(Cl), alcohol	66	545
(O=)CNH$_2$	H	H	A(ClCH$_2$CHCl)$_2$O	70	391
(O=)CNH$_2$	Me	H	A(Cl), alcohol	75	391
(O=)CNH$_2$	Ph	H	A(Br), 60% alcohol	80	391
(O=)CNH$_2$	Me	CO$_2$Et	A(Br), 50% alcohol	70	391
(O=)CNH$_2$	Ph	Ph	A(Cl), 60% alcohol	70	391
Ph	Me	H	A(Cl)	—	255
Ph	Ph	H	A(Cl)	—	59, 60, 67, 69
o-MeC$_6$H$_4$	Me	H	A(Cl)	—	295
o-MeC$_6$H$_4$	Ph	H	A(Br)	—	295
o-MeC$_6$H$_4$	p-MeC$_6$H$_4$	H	A(Br)	—	295
m-MeC$_6$H$_4$	Me	H	A(Cl)	—	295
m-MeC$_6$H$_4$	Ph	H	A(Br)	—	295
m-MeC$_6$H$_4$	p-MeC$_6$H$_4$	H	A(Br)	—	295
p-MeC$_6$H$_4$	Me	H	A(Cl)	—	295

TABLE II-21. (continued)

R_1	R_2	R_3	Conditions[a]	Yield (%)	Ref.
p-MeC$_6$H$_4$	Ph	H	A(Br)	—	295
p-MeC$_6$H$_4$	p-MeC$_6$H$_4$	H	A(Br)	—	295
m-O$_2$NC$_6$H$_4$	Me	H	A(Cl)	—	71
m-O$_2$NC$_6$H$_4$	Me	H	A(Cl)	—	71
m-O$_2$NC$_6$H$_4$	p-MeC$_6$H$_4$	H	A(Br)	—	71
p-O$_2$NC$_6$H$_4$	Me	H	A(Cl)	—	71
p-H$_2$NSO$_2$C$_6$H$_4$	H	H	A, ClCH$_2$CF=(OH)$_2$, alcohol	65–67	396, 466
p-H$_2$NSO$_2$C$_6$H$_4$	Me	H	A(Cl), alcohol	60	466
p-H$_2$NSO$_2$C$_6$H$_4$	H	Me	A(Br), alcohol	40	466
p-H$_2$NSO$_2$C$_6$H$_4$	Me	Me	A(Br), alcohol	50	466
p-H$_2$NSO$_2$C$_6$H$_4$	Me	CO$_2$Et	A(Cl), alcohol	50	466
α-Naphthyl	Me	H	A(Cl), alcohol	—	71
α-Naphthyl	Ph	H	A(Cl), alcohol	—	50, 60, 67, 69, 71, 149
α-Naphthyl	p-MeC$_6$H$_4$	H	A(Cl), alcohol	—	71
α-Quinolyl	Ph	H	A(Cl)	20	149, 367

[a] See Table II-13.
[b] Isolated as hydrochloride

TABLE II-22. 1-AROYL-2-(2-THIAZOLYL) HYDRAZINES FROM AROYLTHIOSEMICARBAZIDES[a]

$$\underset{R_2}{\overset{R_1}{\bigvee}}\!\!\!\underset{S}{\overset{N}{\bigvee}}\!\!-\!\text{NHNHCOC}_6\text{H}_4\text{R-}p \quad (\text{HCl})$$

R	R_1	R_2	Yield
H	H	H	91
p-HO	H	H	75
p-MeO	H	H	90
p-EtO	H	H	96
p-PrO	H	H	98
p-iso-Pr	H	H	94
p-BuO	H	H	96
p-iso-Bu	H	H	56
p-Bu	H	H	91
p-AmO	H	H	88
p-(n-C$_6$H$_{13}$)	H	H	93
p-PhCH$_2$	H	H	96
4-Pyridyl	H	H	75
p-O$_2$N	H	H	80
H	Me	H	96
p-HO	Me	H	99
p-MeO	Me	H	101
p-EtO	Me	H	96
p-PrO	Me	H	91
p-iso-Pr	Me	H	91
p-PhCH$_2$O	Pr	H	96
4-Pvridyl	Pr	H	65
p-O$_2$N	Pr	H	86
H	Bu	H	96
p-HO	Bu	H	96
p-MeO	Bu	H	96
p-EtO	Bu	H	96
p-PrO	Bu	H	96
p-BuO	Bu	H	96
p-AmO	Bu	H	96
p-PhCH$_2$O	Bu	H	96
H	Am	H	96
p-HO	Am	H	91
p-MeO	Am	H	96
p-EtO	Am	H	96
p-PrO	Am	H	96
p-iso-Pr	Am	H	96
p-BuO	Am	H	96
p-sec.BuO	Am	H	96
p-AmO	Am	H	96
p-(n-C$_6$H$_{13}$O)	Am	H	96
p-PhCH$_2$O	Am	H	96
4-Pyridyl	Am	H	80

TABLE II-22. (continued)

R	R_1	R_2	Yield
p-O$_2$N	Am	H	91
H	–(CH$_2$)$_4$–		91
p-HO	–(CH$_2$)$_4$–		75
p-MeO	–(CH$_2$)$_4$–		96
p-EtO	–(CH$_2$)$_4$–		91
p-PrO	–(CH$_2$)$_4$–		91
p-BuO	Me	H	96
p-iso-BuO	Me	H	65
p-sec-BuO	Me	H	91
p-AmO	Me	H	96
p-(n-C$_6$H$_{13}$O)	Me	H	96
p-PhCH$_2$O	Me	H	96
4-Pyridyl	Me	H	60
p-O$_2$N	Me	H	91
Ph	Et	H	96
p-HO	Et	H	96
p-MeO	Et	H	96
p-EtO	Et	H	96
p-PrO	Et	H	96
p-iso-Pr	Et	H	96
p-BuO	Et	H	96
p-iso-BuO	Et	H	96
p-sec-BuO	Et	H	96
p-AmO	Et	H	96
p-(n-C$_6$H$_{13}$O)	Et	H	91
p-PhCH$_2$O	Et	H	96
4-Pyridyl	Et	H	80
p-O$_2$N	Et	H	86
H	Pr	H	96
p-HO	Pr	H	96
p-MeO	Pr	H	96
p-EtO	Pr	H	96
p-PrO	Pr	H	96
p-iso-Pr	Pr	H	91
p-Bu	Pr	H	96
p-AmO	Pr	H	96
p-(n-C$_6$H$_{13}$O)	Pr	H	96
p-iso-Pr	–(CH$_2$)$_4$–		96
p-BuO	–(CH$_2$)$_4$–		96
p-AmO	–(CH$_2$)$_4$–		96
p-(n-C$_6$H$_{13}$O)	–(CH$_2$)$_4$–		91
p-PhCH$_2$O	–(CH$_2$)$_4$–		96
4-Pyridyl	–(CH$_2$)$_4$–		75
p-O$_2$N	–(CH$_2$)$_4$–		80

[a] Conditions: A(Cl), EtOH, reflux 0.5 to 2 hr (479).

II. Thiazoles from α-Halocarbonyl Compounds and Derivatives 255

Condensation of thiosemicarbazides of nicotinic and isonicotinic acids (528) and their hydrochlorides, with α-halocarbonyl compounds gave the corresponding hydrazinylthiazoles (**149**), in which R_1 = methyl, phenyl, or aryl and R_2 = H (Scheme 74).

149

Scheme 74

1-(2-Thiazolyl)semicarbazide (**150**) (391) was similarly obtained from the N-amidothiosemicarbazide in aqueous solution (Scheme 75). Compound **150** can be cyclized with another mole of α-halocarbonyl compound in the presence of phosphorus pentasulfide to give the corresponding 2,2′-hydrazo-bis-thiazole (**151**). Compounds of this type have been synthetized from bis thiourea (**152**) (284, 321, 322, 339, 375) and α-halocarbonyl compounds (Scheme 76).

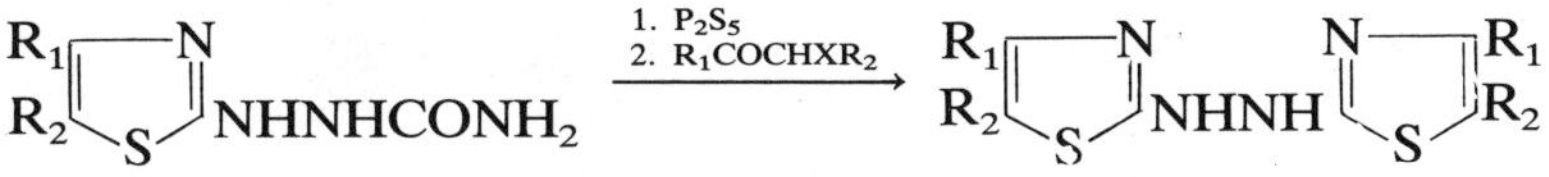

150 → **151**

Scheme 75

$(-NHCSNH_2)_2 + R_1COCHXR_2 \longrightarrow$ **151**
152

Scheme 76

Disubstituted thiosemicarbazides, such as **153**, gave disubstituted 2-hydrazinothiazole derivatives (**154**), in which R_1 = Me and Ph and R_2 = H, Me, and phenyl. Yields ranged from 60 to 90% (Scheme 77) (425).

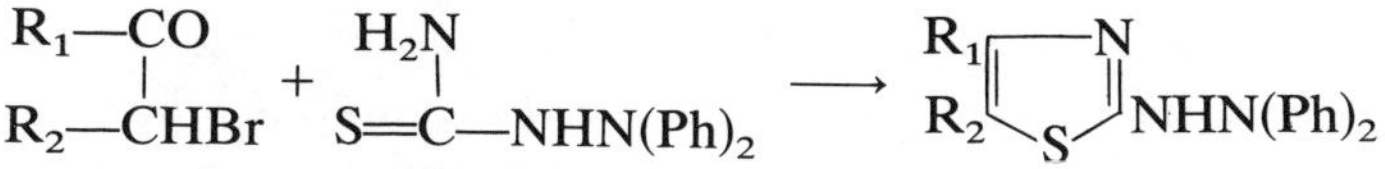

6   **153**   **154**

Scheme 77

Thiosemicarbazones of the general formula **155** react either with α-halocarbonyl compounds (59, 60, 67, 69, 71, 107, 743, 744, 788) or with 1-chloro-1,2-epoxides (504) to yield the corresponding 2-hydrazinothiazoles (**156**) (Scheme 78 and Tables II-22 and II-23).

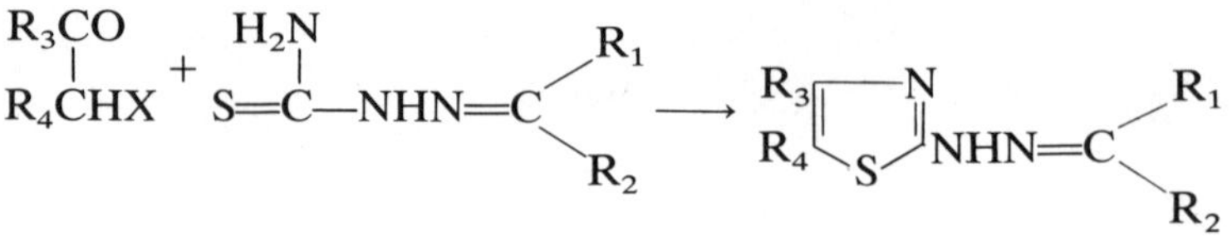

$$R_3CO\text{—}R_4CHX + H_2N\text{—}S=C\text{—}NHN=C(R_1)(R_2) \longrightarrow R_3\text{—}R_4\text{—}[S,N\text{-thiazole}]\text{—}NHN=C(R_1)(R_2)$$

6　　　　　155　　　　　　　　　156

Scheme 78

The interaction of *sym*-dichloroacetone with thiosemicarbazones similarly yields 2-arylidenehydrazino-4-chloromethylthiazoles, which are successively acetylated and brominated (Scheme 79) (636).

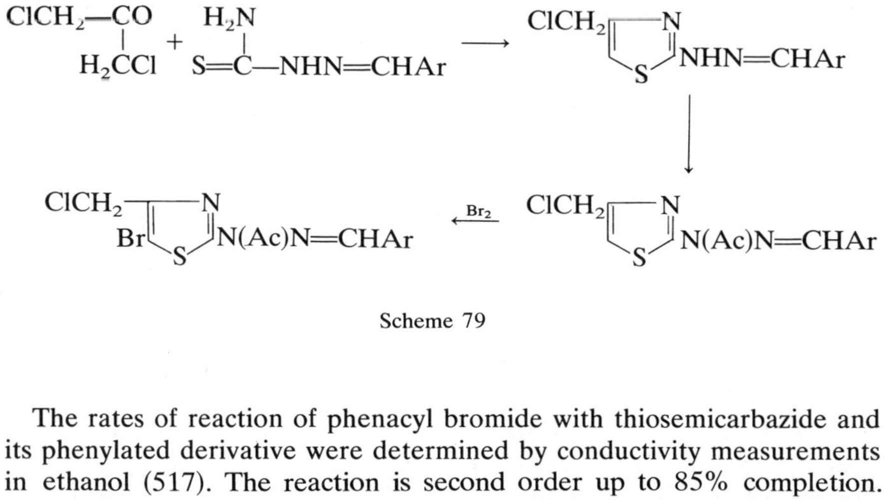

$$ClCH_2\text{—}CO\text{—}CH_2Cl + H_2N\text{—}S=C\text{—}NHN=CHAr \longrightarrow ClCH_2\text{-thiazole-}NHN=CHAr$$

$$\downarrow$$

$$ClCH_2\text{-thiazole(Br)-}N(Ac)N=CHAr \xleftarrow{Br_2} ClCH_2\text{-thiazole-}N(Ac)N=CHAr$$

Scheme 79

The rates of reaction of phenacyl bromide with thiosemicarbazide and its phenylated derivative were determined by conductivity measurements in ethanol (517). The reaction is second order up to 85% completion. The activation energies are 10.5 to 11.3 kcal/mole with the phenyl thiosemicarbazide and 8.5 to 9.3 kcal/mole for the unsubstituted derivatives.

The reaction with the oxime of phenacyl bromide is much slower than that of the parent ketone itself. Hammett's correlation applicated to these reactions gave $\rho = 0.63$ and 0.74 for **141**, $R_1 = Ph$ and $R_1 = H$, respectively (517).

TABLE II-23. HYDRAZINOTHIAZOLE DERIVATIVES FROM THIOSEMICARBAZONES AND α-HALOCARBONYL COMPOUNDS

$$R_2 \underset{S}{\overset{N}{\underset{}{\bigsqcup}}} NHN{=}(R_1)$$
$$R_3$$

R_1	R_2	R_3	Conditions[a]	Yield (%)	Ref.
$(Me)_2C$	H	H	A(Cl)	—	107
$(Me)_2C$	H	H	A(Cl)	—	107
$(Me)_2C$	Me	H	A(Cl)	—	107
$(Me)_2C$	Me	CH_2Ph	A(epoxide), $CHCl_3$, reflux 2 hr	70	504
$(Me)_2C$	Me	$2,5\text{-Me-}2(C_6H_3CH_2)$	A (epoxide), $CHCl_3$, reflux 2 hr	57.9	504
$(Me)_2C$	Ph	H	A(Br)	—	59, 60, 67, 69
$(Me)_2C$	Me	$2,4,6\text{-Me}_3C_6H_2CH_2$	A(epoxide), $CHCl_3$, reflux 5 hr	51.6	504
PhCH	Me	H	A(Cl)	—	59, 60, 67, 69
PhC(Me)	Me	H	A(Cl)	—	59, 60, 67, 67, 69
ArC(Me)	Ar	H	A(Br), MeOH, reflux	60–70	744
$(Me)_2C$	Me	H	A(Cl)	—	71
	Ph	H	A(Br)	—	71

(additional rows with structural drawings)

R_1	R_2	R_3	Conditions[a]	Yield (%)	Ref.
(cyclohexadienone-NO₂ structure)	Ph	H	A(Br)	—	71
(cyclohexadienone structure with N-Me, S, Ph thiazoline)	Me	H	A(Cl)	—	272
RC_6H_4CH (R = o-HO, p-MeO, 3,4-MeO (HO) o-Cl, p-Cl)	(coumarin-3-yl)	H	A(Br)	—	749
RC_6H_4CH (R = H, OH, Cl, N(Me)$_2$)	CH_2CO_2Et	H	A(Br)	—	743
RCH	(coumarin-3-yl)	H	A(Br)	—	788

[a] See Table II-13.

5. Reaction with Salts and Esters of Thiocarbamic Acid: 2-Hydroxythiazoles and Derivatives

This method, initiated by Marchesini in 1893 (26, 29), consists in the condensation of an α-halocarbonyl compound with ammonium thiocarbamate (**157**), $R_1 = NH_4$ or its esters (**157**), $R_1 =$ alkyl (Scheme 80).

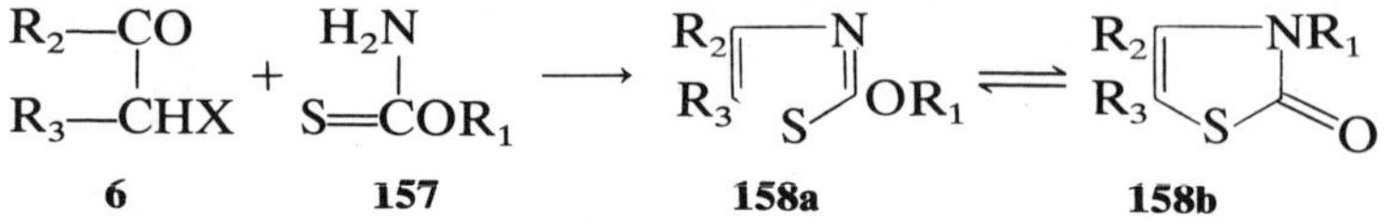

$$\underset{\mathbf{6}}{\overset{R_2-CO}{\underset{R_3-CHX}{|}}} + \underset{\mathbf{157}}{\overset{H_2N}{\underset{S=COR_1}{|}}} \longrightarrow \underset{\mathbf{158a}}{R_3\underset{S}{\overset{R_2\frown N}{\diagdown}}OR_1} \rightleftharpoons \underset{\mathbf{158b}}{R_3\underset{S}{\overset{R_2\frown NR_1}{\diagdown}}O}$$

Scheme 80

The reaction is carried out at low temperature in aqueous medium and then allowed to stand overnight (221). Ammonium thiocarbamate is prepared from a cold saturated solution of ammonium thiocyanate, which is gradually added to dilute sulfuric acid at 25°C. The liberated carbonyl sulfide is passed into a saturated solution of alcoholic ammonia at about 10°C (221). The fairly low yield indicates that the reaction has not been greatly developed.

Thus the condensation of dichloroether or chloroacetone fails to give the parent compound, 2-hydroxythiazole (**158a**), $R_1 = R_2 = R_3 = H$ (221). However, 2-hydroxythiazole can be obtained in 12% yield from chloroacetaldehyde (386). The condensation of ammonium thiocarbamate with α-chloroketones gives the corresponding 2-hydroxy derivatives in 25 to 70% yields (76, 221, 304, 412) (Table II-24). These compounds condensed with $ClP(S)(OEt)_2$ give the corresponding 2-thiazolylthiophosphates (791).

On the other hand, treatment of ethyl α-chloroacetylacetone with **157**, $R_1 = NH_4$, furnished a product (m.p. 175°C) in very good yield, whose properties agree with the structure of ethyl-2-hydroxy-4-methylthiazole-5-carboxylate (**158**), $R_1 = H$, $R_2 = Me$, $R_3 = CO_2Et$) (Scheme 81) (221).

$$\underset{}{\overset{Me-CO}{\underset{MeCO-CHCl}{|}}} + \underset{}{\overset{H_2N}{\underset{S=CONH_4}{|}}} \longrightarrow \underset{\mathbf{159}}{MeCO\underset{S}{\overset{Me\frown N}{\diagdown}}OH}$$

Scheme 81

In some cases **158a**, $R_1 = NH_4$, $R_2 = R_3 = Me$ and $R_2 = CH_2Cl$, $R_3 = H$, the low yield obtained is due to the production of an oily by-product that has not been identified (221).

TABLE II-24. 2-HYDROXYTHIAZOLE DERIVATIVES FROM α-HALO-CARBONYL COMPOUNDS AND SALTS AND ESTERS OF THIOCARBAMIC ACID (158a)

R_1	R_2	R_3	Conditions[a]	Yield (%)	Ref.
H	H	H	Cl, water	12	386
H	Me	H	Cl, water, cool, then stand overnight	90	221
H	CH_2Cl	H	Cl, water, cool, then stand overnight	25	221
H	CH_2SCN	H	Cl, water, cool, then stand overnight	85	221
H	Ph	H	$ClCH_2COOH$, water	—	89, 287, 324
H	Me	Me	Cl, water, cool, then stand overnight	25	221
H	Me	$(CH_2)_2Br$	Br, water	—	304
H	Me	COMe	Cl, water	55	221
H	Me	CO_2Et	Cl, water	57	221
H	Ph	CO_2Et	Br, water	—	76
H	Ph	Ph	Cl, heat, 30 min	55	392
H	Ph	Ph	Cl, heat, 30 min	55	392
Me	H	H	Dichlorether	—	132
Me	Me	H	Cl, AcONa	—	68
Et	H	H	Dichlorether	—	77
Et	Me	H	Cl	—	68
Et	Me	CH_2CO_2Et	Br	—	77
Et	H	H	Dichlorether	—	77
Et	Ph	H	Br	—	15

[a] Cl or Br designates the corresponding α-chloro- or α-bromo-aldehyde or ketone.

2-Hydroxythiazoles give 2-chlorothiazole derivatives almost quantitatively upon treatment with phosphorus oxychloride (221, 229, 428). This constitutes a convenient synthesis method for these compounds when the conversion of 2-aminothiazoles to 2-chlorothiazole derivatives fails. Esters of thiocarbamic acid or thiourethanes also react with α-halocarbonyl compounds to give the corresponding 2-alkoxythiazoles (50, 68, 209, 272).

For example, p-bromophenacylaryl (or alkyl) sulfides (**160**) with **157**, R_1 = Et, yield the 5-(aryl or alkyl)-thiothiazolyl-2-ones (**161**) (Scheme 82) (487).

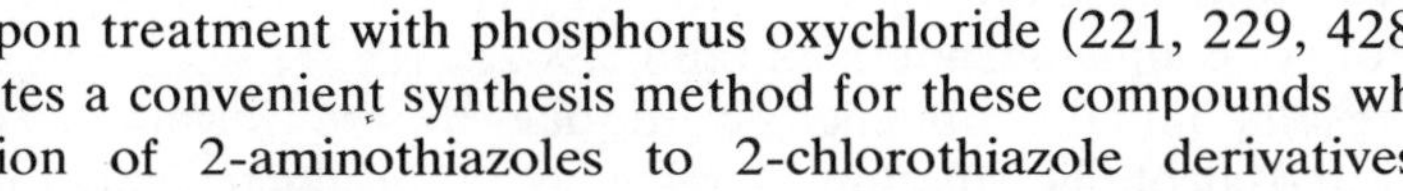

Scheme 82

However, 2-alkoxythiazoles (711) are usually prepared by refluxing 2-halogenothiazoles and a metallic alcoholate in the corresponding anhydrous alcohol for several hours (see Chapter V).

6. Reaction with Salts and Esters of Dithiocarbamic Acid: 2-Mercaptothiazole Derivatives and 2-Thiazolyl Sulfides

A. 2-Mercaptothiazole Derivatives

Initiated by Miolati in 1893 (27), taken up later by Levi (82), and successively by many other authors, the reaction of ammonium dithiocarbamate (**162**) with α-halocarbonyl compounds gives 2-mercaptothiazole derivatives with yields varying greatly according to experimental conditions (Table II-25).

These compounds are commercially important as accelerators in the vulcanization of rubber (Scheme 83).

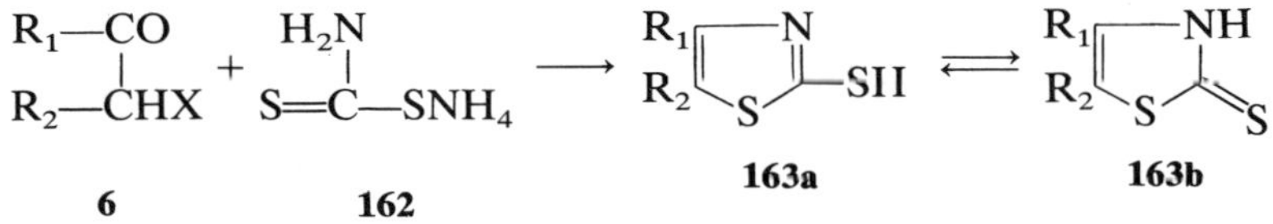

$$R_1-CO \atop R_2-CHX \quad + \quad {H_2N \atop S=C-SNH_4} \quad \longrightarrow \quad R_1 \underset{S}{\overset{N}{\longmapsto}} SH \quad \rightleftarrows \quad R_1 \underset{S}{\overset{NH}{\longmapsto}} S$$

    6   **162**   **163a**   **163b**

Scheme 83

Similar to their 2-hydroxy analogs, these compounds exhibit properties (cf. Chapter VII) that are characteristic of each of the two possible tautomeric forms: the thiol (**163a**) and the thione or 2-thioxo-Δ-4-thiazoline (**163b**).

2-Mercaptothiazole (**163a**), $R_1 = R_2 = H$, is prepared from chloroacetaldehyde or α-,β-dihalogenoethers (268, 296, 521, 597).

The results initially obtained were due to the formation in both aqueous and alcoholic solution of resinous by-products. This formation results from the decomposition of the ammonium dithiocarbamate, or from the self-condensation of chloroacetaldehyde or the formation of intermediate products.

A recent study instigated by Kolosova (521) and taken up and proved by Chanon and Metzger (597) increased the yield of this reaction to 70% in the case of chloroacetaldehyde and to 50% in the case of dibromoether, whereas other authors have been either unable to obtain the product in this way or in very low yield (11%) (316).

Various compounds were obtained from the corresponding halomethylketones. Yields are generally better than in the case of α-chloroaldehydes

TABLE II-25. 2-MERCAPTOTHIAZOLE DERIVATIVES FROM AMMONIUM DITHIOCARBAMATE AND α-HALOCARBONYL COMPOUNDS (**163a**)

R_1	R_2	Conditions[a]	Yield (%)	Ref.
H	H	ClCH$_2$CHO, water or alcohol+water	39–50	233, 289, 498
		ClCH$_2$CHO, water, pH 5.3	70	521, 530
		BrCH$_2$CHBrOEt, water	11–50	233, 597
H	Me	MeCHBrCHBrOEt, water, HCl	10–50	521, 597
Me	H	Cl or Br, water, ether or alcohol, several hours at room temperature	78–85	27, 55, 82, 150, 156, 290, 314, 332, 345, 597
Et	H	Cl or Br, water or alcohol	29–97	345, 346
n-Pr	H	Cl, water or alcohol 60°C 2 hr	82	290, 314, 332, 345
		R$_1$C≡CCH$_2$Br, alcohol	59.5	497
n-Bu	H	Cl, water, or alcohol 60°C 2 hr	78	315, 551
n-Pentyl	H	Cl, abs. alcohol several hours at room temperature	81	270
n-Pentyl	H	R$_1$C≡CCH$_2$Br, alcohol	47	497
n-Hexyl	H	Cl, abs. alcohol several hours at room temperature	78	270
n-Undecyl	H	Cl, abs. alcohol several hours at room temperature	65	270
n-Dodecyl	H	Cl, abs. alcohol several hours at room temperature	45	270

TABLE II-25. (continued)

R_1	R_2	Conditions[a]	Yield (%)	Ref.
n-Tridecyl	H	Cl, abs. alcohol	58	270
n-Tetradecyl	H	Cl, abs. alcohol	52	270
n-Pentadecyl	H	Cl, abs. alcohol	55	270
CH_2CH_2Ph	H	$R_1C \equiv CCH_2Br$, alcohol	—	497
CH_2CO_2Et	H	Cl, water	95	388
Ph	H	Cl, water	95	156
		Br, ether or alcohol	30–55	27, 82, 199
		Cl, iso-PrOH	91	200
$p\text{-EtC}_6H_4$	H	Cl, alcohol	40	270
$p\text{-tert-BuC}_6H_4$	H	Cl, alcohol	72	270
p-biphenylyl	H	Cl, alcohol	82	270
CO_2Et	H	Br, abs. alcohol, 18 hr at room temperature, then 2 hr at 70–80°C	66	488
CON(H)Ph	Me	Cl, Me_2CO	85	751
α-Thienyl	H	Cl, alcohol	86	292
5-Cl-2-thienyl	H	Cl, alcohol	91	292
2-SH-4-thiazolyl	H	$(BrCH_2CO)_2$, alcohol reflux 2 hr	60	508
1-Me-2-benzimidazolyl	H	Cl, water, 3 hr at 10°C, 1 hr at 30°C	—	602
Me	Me	Cl, alcohol	47–53	135, 156, 316
		Br, abs. MeOH, 12 hr at room temperature	55–85	316, 711
Me	Et	Cl, abs. MeOH, 12 hr at room temperature	100	711
—$(CH_2)_3$—		Cl	—	262
—$(CH_2)_4$—		Cl or Br, alcohol	70	188, 453

Me	(CH$_2$)$_2$CO$_2$Me	Cl, alcohol or Br, ether	71	135, 150, 229
Me	(CH$_2$)$_2$COMe	Br, ether	—	200, 229
Me	(CH$_2$)$_2$OH	Br, ether	—	229
		Chloracetone, water	40	229
Me	(CH$_2$)$_2$Cl	Cl, ether	—	135
Me	COMe	Cl, abs. MeOH, 12 hr at room temperature	46–74	387, 711
Me	CO$_2$Me	Cl, water	73.4	387
Me	CO$_2$Et	Cl, water or abs. MeOH, 12 hr at room temperature, then heat 1 hr on steam bath	50–75	27, 82, 156, 199, 387
Me	CO$_2$Bu	Cl, heptane	56.6	313
Me	CONHPh	Cl, alcohol	98	243
Me	p-CONHC$_6$H$_4$Me	Cl, alcohol	—	243
Me	o-CONHC$_6$H$_4$OMe	Cl, alcohol	—	243
Me	m-CONHC$_6$H$_4$OH	Cl, alcohol	—	243
Me	CONH (α-naphthyl)	Cl, alcohol	—	243
Me	SPh	Cl, alcohol, 2 hr at 60°C	55	561
Me	p-MeC$_6$H$_4$S	Cl, alcohol, 2 hr at 60°C	27	561
Me	p-O$_2$NC$_6$H$_4$S	Cl, alcohol, 2 hr at 60°C	41	561
Et	Ph	R$_1$C≡CCHBrR$_2$, alcohol	76	497
n.C$_5$H$_{11}$	Et	R$_1$C≡CCHBrR$_2$, alcohol	62	497
(CH$_2$)$_2$Ph	Ph	R$_1$C≡CCHBrR$_2$, alcohol	76	497
2-Mercapto-9, 10-dihydro-8H-benzo[4,5]-cycloheptal[d]-thiazole	—	Br, alcohol	—	689

[a] Cl or Br designates the corresponding α-chloro- or α-bromo-aldehyde or ketone.

or dihalogenoethers. Alkylchloromethylketones are conveniently prepared in excellent yield from the corresponding acid chlorides and ethereal diazomethane followed by treatment with dry hydrogen chloride.

These condensations are carried out by slow addition at room temperature of the halomethylketone to the dithiocarbamate, the latter in 10 to 50% molar excess to prevent the formation of 2-ketonylthiothiazoles (270, 291).

With chloroacetone in aqueous solution, yields range from 78 to 85% (27, 82, 199, 313, 597), while in ether or ethanol they are lower (156, 229, 290, 314, 332, 345).

The more reactive bromacetone gives not only 2-mercapto-4-methylthiazole but also its substitution products. The higher homologs, as far as C_{15}, are obtained in reasonably good yield in absolute ethanol (150, 156, 234, 316, 530). The best result (85%) was obtained by working in aqueous solution with the 3-bromobutan-2-one (597).

Ethyl-α-chloroacetoacetate gives 5-carbethoxy-4-methyl-2-thiazole thiol (387), while 3-chloro-2,4-pentanedione affords the 2-mercapto-4-methyl-5-thiazolylmethylketone in good yield (74%) (387).

Similarly, ethylbromopyruvate yields ethyl-2-mercapto-4-thiazole carboxylate (488).

1-Alkylsulfonyl-3-bromo-2-propanones (**164**) react in the cold with ammonium dithiocarbamate (in 10% excess in aqueous solution) to give initially the 4-hydroxythiazolidine-2-thione intermediates (**165**), which on heating dehydrate to give the corresponding Δ-4 thiazoline-2-thiones (**166**) (522). Yields are good at least for the first terms of the series. Thus for $n = 1, 2, 3, 4, 5$, and 6, yields are 100, 91, 89, 74, 62, and 24%, respectively (Scheme 84).

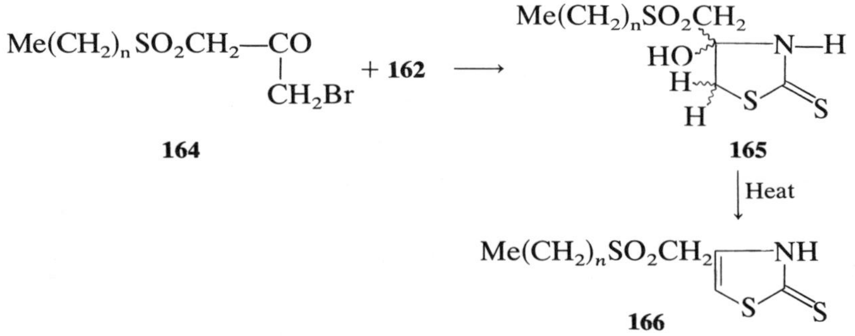

$$Me(CH_2)_n SO_2 CH_2 - CO \underset{CH_2Br}{|} + 162 \longrightarrow$$

164

165

| Heat

166

Scheme 84

Alkylsulfonylketones (**164**) were prepared from α-chloroketones and sodium alkylsulfinates (RSO_2Na).

α-Bromoalkynes also give 2-mercaptothiazoles by condensation with ammonium dithiocarbamate in alcoholic solution: they are not isolated in

II. Thiazoles from α-Halocarbonyl Compounds and Derivatives

their free form but treated directly in the solution mixture with methyliodide in order to obtain their methylated derivatives, which are easier to separate (497). The mechanisms of this reaction (Scheme 85) have been discussed (534); with R_1 = Me, Et, Bu, CH_2Ph and R_2 = H, Me, Et, Ph, yields ranged from 47 to 76% (534).

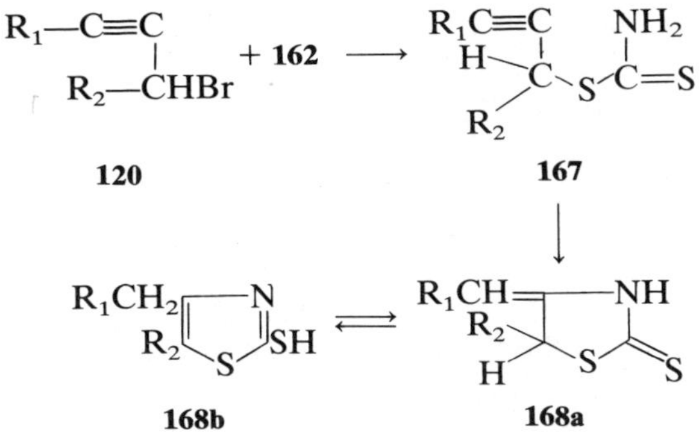

Scheme 85

Aromatic chloro- or bromomethylketones react with **162** to give 4-aryl-2-mercaptothiazoles (27, 82, 156, 199, 200, 519).

Alkylphenacyl chlorides are synthetized from alkylbenzenes using chloroacetyl chloride as catalyst (270, 291).

2-Mercaptothiazoles with heterocyclic substituents in the 4-position have also been prepared (292, 508, 602). For example, N-alkyl-2-benzimidazolyl chloromethylketones (**169**) give the corresponding 2-mercaptothiazole derivatives (**170**) with R_1 = H, Me, CH_2Ph, Ac, COPh and R_2 = H, Me, 5,6-dichloro (Scheme 86) (602).

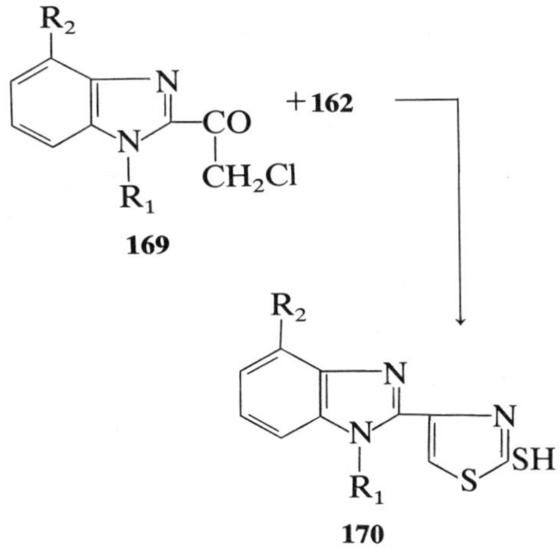

Scheme 86

B. 2-Thiazolyl Sulfides

Substitution of an ester of dithiocarbamic acid such as alkyl or benzyl ester (**171**), or their salts leads directly to 2-substituted mercaptothiazoles (**172**) (Scheme 87) (272, 461). Some of these compounds have an antifungal activity (561).

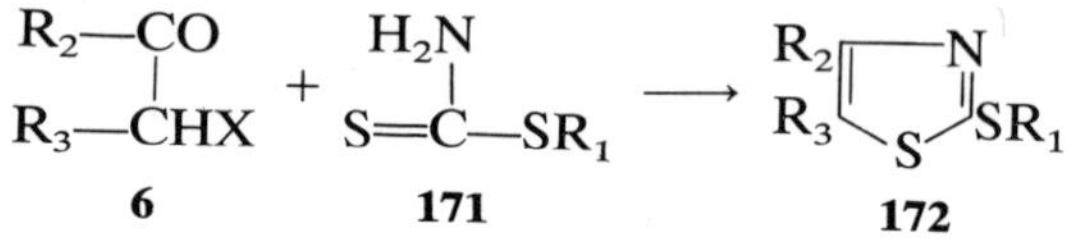

$$R_2-CO \atop R_3-CHX \quad + \quad {H_2N \atop S=C-SR_1} \quad \longrightarrow \quad R_2 \underset{S}{\overset{N}{\diagup\!\!\!\!\diagdown}} SR_1$$

$$\text{6} \qquad\qquad \text{171} \qquad\qquad \text{172}$$

Scheme 87

By this general method, compound **172**, in which R_1 = alkyl (231, 521, 561, 597), allyl (488), acetonyl (156), phenacyl (199, 292, 519), benzyl (272, 561), aryl (561), or carboxyalkyl (561), R_2 = methyl (156, 199, 231, 561, 597), aryl (231, 270, 291, 561), or heteroaryl (231, 272, 292), and R_3 = hydrogen, acetic acid (231), sulfur (561), or ester (199), have been prepared (Table II-26).

Sulfides (**172**) in which R_1 = alkyl can be obtained also by direct alkylation of the 2-mercaptothiazoles either in alcaline medium (156, 597) or by phase-transfer catalysis in better yield (824).

With an excess of halocarbonyl reactant or a more reactive ketone like bromoketone, compounds of type **173** may result through reaction of the 2-mercaptothiazole (**163a**) with the excess of bromoketone (Scheme 88) (156, 199, 270, 291, 292, 519). Thus when R_1 = phenyl and R_2 = hydrogen, **173** was obtained in 76% yield (292) in aqueous solution and in 20 to 40% in alcoholic solution (292, 519).

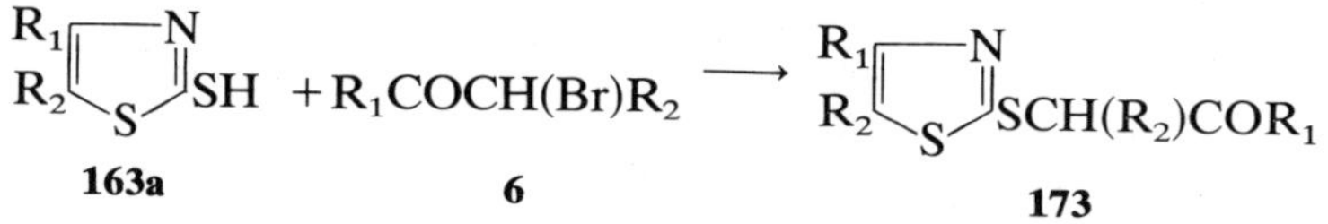

$$R_1 \underset{S}{\overset{N}{\diagup\!\!\!\!\diagdown}} SH \quad + R_1COCH(Br)R_2 \quad \longrightarrow \quad R_1 \underset{S}{\overset{N}{\diagup\!\!\!\!\diagdown}} SCH(R_2)COR_1$$

$$\text{163a} \qquad\qquad \text{6} \qquad\qquad\qquad \text{173}$$

Scheme 88

With the less reactive phenacylchloride, 4-phenyl-2-mercaptothiazole was obtained in quantitative yield (289, 292) either in alcoholic or aqueous solution.

The synthesis of sulfides (**172**) can be carried out by condensing 2-mercaptothiazoles and the required α-bromo compounds in basic solution at room temperature (488).

TABLE II-26. 2-THIAZOLYL SULFIDES

$$\begin{array}{c} R_1S \diagdown \underset{S}{\overset{N}{\diagup}} \hspace{-4pt} =\hspace{-4pt} \underset{}{\overset{R_2}{\diagdown}} R_3 \end{array}$$

R_1	R_2	R_3	Conditions[a]	Yield (%)	Ref.
Me	H	H	Cl, water	85	597
Me	H	Me	BrCH$_2$CHBrOEt	40–54	521, 597
Me	Me	H	MeCHBrCHBrOEt	—	597
Me	Me	Me	Br	21	318
Me	Ph	CH$_2$COOH	Br or Cl	96	156, 597
Me	p-MeC$_6$H$_4$	CH$_2$COOH	Br	39	231
Me	Ph	H	Br	38	231
Me	α-Furyl	H	Br	—	231
Me	α-Naphthyl	CH$_2$COOH	Br	70	272
Me	α-Naphthyl	CH$_2$COOH	Br	—	231
Me	β-Naphthyl	CH$_2$COOH	Br	—	231
Et	Ph	H	Br	—	231
Et	Me	SEt	Cl, alcohol, 60°C 2 hr	72	561
Et	Ph	CH$_2$COOH	Br	23	231
Pr	Ph	H	Cl, alcohol, 60°C 2 hr	60.5	561
Hexyl	Me	H	Cl, alcohol, 60°C 2 hr	44.4	561
Octadecyl	Me	H	Cl, alcohol, 60°C 2 hr	80.6	561
CH$_2$CH=CH$_2$	CO$_2$Et	H	Br	57.6	488
CH$_2$CCl=CH$_2$	CO$_2$Et	H	Br	53.3	488
CH$_2$C=CH	CO$_2$Et	H	Br	61.6	488
CH$_2$COCH$_3$	Me	H	Cl, alcohol	—	156
CH$_2$COPh	Ph	H	Br, water or alcohol	60–76	199, 292, 519
p-CH$_2$COC$_6$H$_4$Br	p-BrC$_6$H$_4$	H	Br, alcohol	94	270
CH$_2$Ph	Ph	H	Br, alcohol	—	272
2-OH, 5-O$_2$NC$_6$H$_3$CH$_2$	Ph	H	Cl, alcohol	88	561
4-OH, 5-O$_2$NC$_6$H$_3$CH$_2$	Ph	H	Cl, alcohol	53	646

TABLE II-26. (continued)

R_1	R_2	R_3	Conditions[a]	Yield (%)	Ref.
$CH_2(O=)C\!-\!\langle{}_S\rangle\!-\!Cl$	[chlorothienyl]		Cl, alcohol	21	292
HC(COMe)CO$_2$R	Me	CO$_2$R	Cl	—	199
CO$_2$Me	Me	H	Cl, alcohol, 60°C 2 hr	32	561
CO$_2$Pr	Me	H	Cl, alcohol, 60°C 2 hr	32.6	561
CO$_2$Bu	Me	H	Cl, alcohol, 60°C 2 hr	30.8	561
CO$_2$Am	Me	H	Cl, alcohol, 60°C 2 hr	41	561
CO$_2$-Hexyl	Me	H	Cl, alcohol, 60°C 2 hr	52.2	561
CO$_2$-Octyl	Me	H	Cl, alcohol, 60°C 2 hr	79	561
2,5-Me$_2$C$_6$H$_3$	Ph	H	Cl, alcohol, 60°C 2 hr	91	561
2,4-O$_2$NC$_6$H$_3$	Me	H	Cl, alcohol, 60°C 2 hr	80.6	561
2,4-O$_2$NC$_6$H$_3$	Ph	H	Cl, alcohol, 60°C 2 hr	79.5	561

[a] Cl or Br designates the corresponding α-chloro-, or α-bromo-aldehyde or ketone.

II. Thiazoles from α-Halocarbonyl Compounds and Derivatives

The N-substituted derivatives (**175**) are also obtained from salts of N-monosubstituted dithiocarbamic acid (**174**) (Scheme 89) (35, 348).

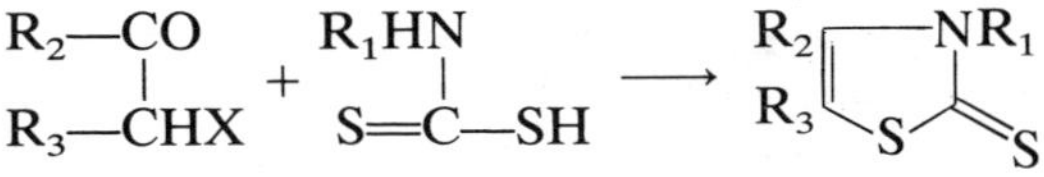

$$\underset{\textbf{6}}{\begin{array}{c} R_2\text{—CO} \\ | \\ R_3\text{—CHX} \end{array}} + \underset{\textbf{174}}{\begin{array}{c} R_1\text{HN} \\ | \\ S\!=\!C\text{—SH} \end{array}} \longrightarrow \underset{\textbf{175}}{\begin{array}{c} R_2 \\ R_3 \end{array}\!\!\diagdown\!\!\underset{S}{\bigcirc}\!\!\diagup\!\! S}\text{—NR}_1$$

Scheme 89

C. Mechanism

Probably the mechanism of this reaction is similar to that proposed for the Hantzsch's synthesis and discussed in Section II.1.D.

The first step of the reaction involves the formation of the S–C bond with the elimination of a molecule of ammonium salt. The fact that it has been possible to isolate the acyclic intermediate (**176**), R = Me or Ph, would confirm this hypothesis, particularly when the reaction is carried out for a short time in the cold in ethereal solution (27, 82). These intermediates (**176**) can be cyclized quantitatively on standing or on being treated by hydrochloric acid. However, no evidence has been advanced concerning their structures.

Humphlett and Lamon (522) have recently studied the intermediary compounds of this reaction and have shown with the help of infrared and ultraviolet spectroscopy that **176** was not present in the reaction mixture (Scheme 90); instead, a compound containing an hydroxyl radical and not a carbonyl function was present (Scheme 91).

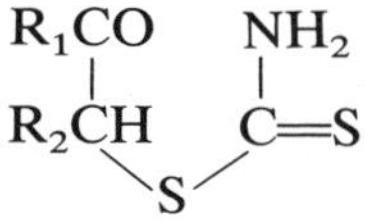

$$\begin{array}{c} R_1\text{CO} \quad NH_2 \\ | \qquad\quad | \\ R_2\text{CH} \quad C\!=\!S \\ \diagdown S \diagup \end{array}$$

176

Scheme 90

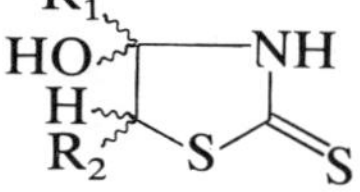

177

Scheme 91

Chloroacetone, phenacylbromide, α-bromoisobutyrophenone, 3-bromo-3-methyl-2-butanone, 1-alkylsulfonyl-3-bromo-2-propanone, and ethyl-γ-chloroacetoacetate give with ammonium dithiocarbamate the corresponding 4-hydroxythiazolidine-2-thiones (**177**), which have a characteristic absorption between 273 and 279 nm. Dehydration by heating with dilute HCl can be followed by ultraviolet spectroscopy because the products formed (**175**) absorb at 315 to 340 nm.

This result was confirmed more recently by Chanon and Metzger (597, 611), who investigated the mechanism of the Δ_4-thiazoline-2-thione formation from a dithiocarbamate and α-halocarbonyl compounds. During this study, it was possible by precise control of the pH to isolate an acyclic intermediate (**178**) analogous to **176**, which is in equilibrium with its cyclic isomer (**179**) (Scheme 92).

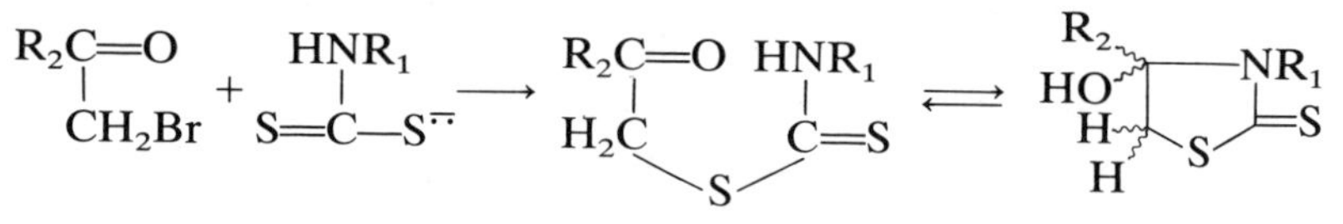

Scheme 92

On the other hand, the large influence of steric effects on the 3-position has been demonstrated. Thus with **175**, $R_1 = t$-Bu and $R_2 = H$, dealkylation at nitrogen was observed upon dehydration of the intermediate (**179**) (Scheme 93) (712, 713, 714).

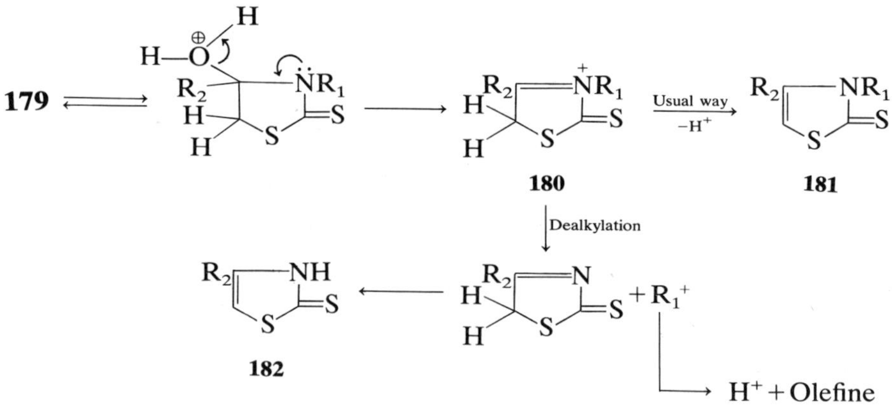

Scheme 93

III. Thiazoles from Rearrangement of the α-Thiocyanatoketones   271

In the cationic intermediate (**180**), when the steric interaction between R_1 and R_2 becomes too great, R_1 is eliminated as an olefin leaving the nitrogen-unsubstituted Δ-4 thiazoline-2-thione (**182**).

N-alkylthiazoline-2-thione (**181**) is produced by the normal mechanism.

III. THIAZOLES FROM REARRANGEMENT OF THE α-THIOCYANATOKETONES (TYPE Ib)

Taking into account the experimental conditions, a fairly large variety of thiazoles, variously substituted at the 2-position can be obtained from α-thiocyanatoketones. This method, more widely known as Tcherniac's synthesis, is a variation of the first synthesis group.

The α-thiocyanatoketones are easily obtainable from α-halocarbonyl compounds and metal thiocyanates (sodium, potassium, barium, or lead thiocyanate) (416, 484, 519, 659) in an alcoholic solution. Yields ranged from 80 to 95%. They are very sensitive substances that isomerize when reacted upon by acids, bases, or labile hydrogen and sulfur compounds.

1. Acid or Alkaline Hydrolysis

The cyclization of α-thiocyanatoketones (**183**) in aqueous acid, concentrated sulfuric acid in acetic acid and water, or alkaline solution leads to 2-hydroxythiazoles after dilution in water.

These reactions can be carried out in several hours at room temperature or by heating for 1 or 2 hr on a steam bath (Scheme 94) (24, 110, 136, 220, 229, 304, 369, 392, 416, 428, 484, 519, 762).

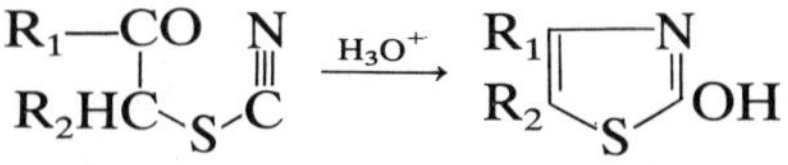

$$R_1-CO \quad N \xrightarrow{H_3O^+} \quad R_1 \diagup\!\!\!\!\diagdown N$$
$$R_2HC\diagdown_S\diagup C \qquad\qquad R_2 \diagdown_S\diagup OH$$
$$\textbf{183} \qquad\qquad\qquad \textbf{184}$$

Scheme 94

The yields with the lower alkyls rarely exceed 50%. For example, with $R_1 = R_2 =$ Me, the yield is 20% in an aqueous acidic solution (219) or 40% in an acetic solution acidified with sulfuric acid (369).

2-Hydroxy-4-methylthiazole has been prepared in 68% yield through the reaction of barium thiocyanate with chloroacetone (70).

Compounds of type **184** were prepared similarly with $R_1 =$ methyl and $R_2 =$ alkanoic esters (136, 229, 304) or carboxyethyl (4, 10, 13, 22, 220) (Table II-27). Dichloroacetone reacts with ammonium thiocyanate, but

TABLE II-27. 2-HYDROXYTHIAZOLE DERIVATIVES BY CYCLIZATION OF α-THIOCYANATO KETONES IN ACIDIC MEDIA

$$\underset{HO}{\overset{N}{\diagdown}}\!\!=\!\!\!\underset{S}{\overset{R_1}{\diagup}}\!\!R_2$$

R_1	R_2	Conditions	Yield (%)	Ref.
Me	H	NaHCO$_3$, water	42–68	24, 70, 110
CH$_2$SCN	H	HOAc, H$_2$SO$_4$, water	50	220
Ph	H	HOAc, H$_2$SO$_4$	87	416, 519
p-MeC$_6$H$_4$	H	HOAc, H$_2$SO$_4$	90	416, 519
p-ClC$_6$H$_4$	H	HOAc, H$_2$SO$_4$	85	416, 473
p-BrC$_6$H$_4$	H	HOAc, H$_2$SO$_4$	80	416, 473
p-MeOC$_6$H$_4$	H	HOAc, H$_2$SO$_4$	84	416, 473
2,4-(HO)$_2$C$_6$H$_3$	H	HOAc, H$_2$SO$_4$, water, steam bath 1 hr	92.5	87, 484
3,6-Me(OH)C$_6$H$_3$	H	HOAc, H$_2$SO$_4$	70	484
4,6-Me(OH)C$_6$H$_3$	H	HOAc, H$_2$SO$_4$	85	484
2,3,4-(HO)$_3$C$_6$H$_2$	H	HOAc, H$_2$SO$_4$, water, steam bath 1 hr	40	484
Me	Me	HOAc, H$_2$SO$_4$	40	369
Me	(CH$_2$)$_2$COMe	HOAc, H$_2$SO$_4$	—	229
Me	(CH$_2$)$_2$CO$_2$Et	HOAc, H$_2$SO$_4$	—	136, 304
Me	(CH$_2$)$_2$CO$_2$Me	HOAc, H$_2$SO$_4$	—	109, 136, 229, 304
Me	(CH$_2$)$_2$CO$_2$Ph	HOAc, H$_2$SO$_4$	—	109, 304
Me	(CH$_2$)$_2$OH	HOAc, H$_2$SO$_4$	—	109, 304
Me	(CH$_2$)$_2$Br	H$_2$SO$_4$	—	109, 304
Me	CH$_2$CO$_2$Et	H$_2$SO$_4$	—	109, 304
Me	CO$_2$Et	HOAc, H$_2$SO$_4$	20	4, 10, 13, 22, 220
Et	(CH$_2$)$_2$CO$_2$Ph	H$_2$SO$_4$	—	304
Ph	Ph	HOAc, H$_2$SO$_4$, water, reflux 2 hr	85	392, 420

III. Thiazoles from Rearrangement of the α-Thiocyanatoketones

the reaction proceeds further giving 2-hydroxy-4-thiocyanatomethylthiazole (220).

With α-thiocyanatoacetophenones, 4-aryl-2-hydroxythiazoles can be obtained in 80 to 90% yields in an acetic acid solution with the addition of dilute sulfuric acid (87, 392, 416, 428, 484, 519).

α-Thiocyanoheteroarylketones were cyclized by acid to give 2-hydroxy derivatives in 60 to 90% yields (**184**), R_1 = 5-aryl-2-furyl and R_2 = H. The mechanism of this reaction can be explained by the preliminary acid-catalyzed hydration of the nitrilic triple bond. Ring closure is then achieved when the nitrogen attacks the carbon carrying the carbonyl function. Finally a molecule of water is eliminated.

This hypothesis is based on the fact that Arapides (9) was able to demonstrate the formation of the acyclic compound (**185**), which is an intermediate in the conversion of α-thiocyanatoketone to 2-hydroxythiazole, but an intermediate such as **186** seems to be more probable (Scheme 95).

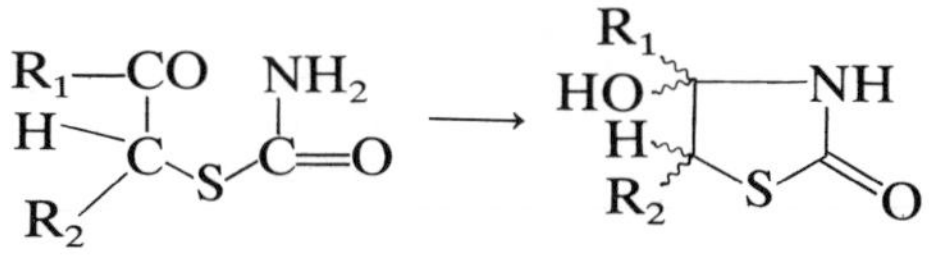

Scheme 95

The poor results obtained in some cases were attributed to the formation of resins and by-products as a result of ring closure by another route leading to the formation of 2-imino-3-oxa-4-thiolenes (**187**) (Scheme 96). (7, **221**, **294**).

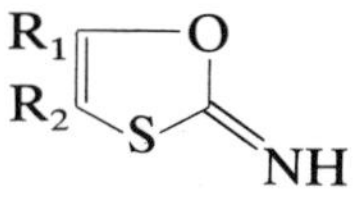

Scheme 96

2. Action of Dry Hydrogen Halides

Treatment of α-thiocyanatoketones at low temperature with dry hydrogen chloride in ether solution gives satisfactory yields of 2-chlorothiazole derivatives (**188**). The use of phosphorus pentachloride leads to the same results, but in this case chlorination can also occur at the 5-position (Scheme 97) (18, 68).

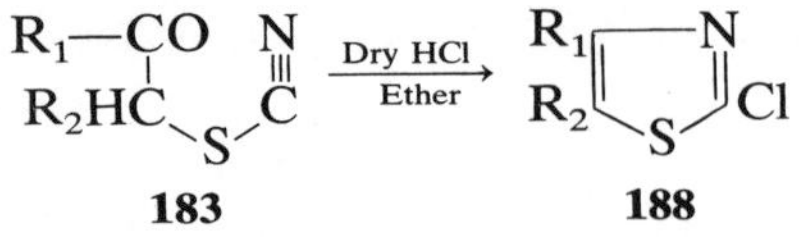

$$R_1-CO\ N \atop R_2HC\diagdown_S\diagup C \quad \xrightarrow[\text{Ether}]{\text{Dry HCl}} \quad R_1\diagup^N\diagdown \atop R_2\diagdown_S\diagup Cl$$
    183   188

Scheme 97

This method has been applied to the synthesis of 4-methyl (659), 4-aryl (416, 519, 659), 4,5-dimethyl (137, 220, 221), 4,5-dialkyl (229, 681), 4-methyl-5-(β-acetoxyethyl) (229), 4-methyl-5-(β-carbethoxyethyl) (229), 4-aryl-5-bromo (579), and 2-chlorothiazoles from the corresponding α-thiocyanatoketones (Table II-28).

TABLE II-28. 2-CHLOROTHIAZOLE DERIVATIVES FROM α-THIOCYANATO-KETONES AND DRY HYDROCHLORIC ACID IN ETHEREAL SOLUTION

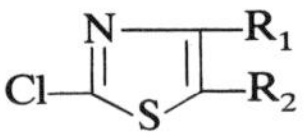

$$Cl\diagup^N\diagdown_{S}\diagup R_1 \atop R_2$$

R_1	R_2	Conditions	Yield (%)	Ref.
Me	H	Ether, dry HCl at 0°C	53	659
Ph	H	Ether, dry HCl at 0°C	85	416, 519, 659
p-MeC$_6$H$_4$	H		86	416, 519
p-ClC$_6$H$_4$	H	Ether, dry HCl at 0°C	84	416, 519, 659
p-BrC$_6$H$_4$	H	Ether, dry HCl at 0°C	80	416, 519
p-MeOC$_6$H$_4$	H	Ether, dry HCl at 0°C	85	416, 519
3,4-Cl$_2$C$_6$H$_3$	H	Ether, dry HCl at 0°C	90	484
4,6-Me(OH)C$_6$H$_3$	H	Ether, dry HCl at 0°C	90	484
Me	Me	Ether, dry HCl at 0°C	52–66	220
Me	(CH$_2$)$_2$COMe	Ether, HCl	—	220
Me	(CH$_2$)$_2$CO$_2$Me	Ether, HCl	—	229
–(CH$_2$)$_4$–		dry HCl 53 hr at room temperature	—	659
–(CH$_2$)$_5$–		dry HCl 53 hr at room temperature	—	659
Ph	Br	Ether, dry HCl at 0°C	67.6	579
p-MeC$_6$H$_4$	Br	Ether, dry HCl at 0°C	66	579
p-ClC$_6$H$_4$	Br	Ether, dry HCl at 0°C	70.9	579
p-BrC$_6$H$_4$	Br	Ether, dry HCl at 0°C	66.7	579
p-MeOC$_6$H$_4$	Br	Ether, dry HCl at 0°C	69.8	579
Ph	Ph	Ether, dry HCl at 0°C	—	

These compounds can also be obtained from the corresponding 2-hydroxythiazoles and phosphorylchloride (68, 109, 137, 221, 229, 428, 519), but the method is restricted by the availability of the hydroxy compounds, which are discussed in another section (Chapter VII).

The same general procedure was applied satisfactorily to the synthesis of 2-bromothiazoles using hydrogen bromide below 0°C (465, 507) (Table II-29).

III. Thiazoles from Rearrangement of the α-Thiocyanatoketones

TABLE II-29. 2-BROMOTHIAZOLE DERIVATIVES FROM α-THIOCYANATOKETONES AND DRY HYDROBROMIC ACID

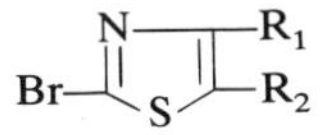

$$\text{Br}\underset{S}{\overset{N}{\underset{\|}{\bigwedge}}}\genfrac{}{}{0pt}{}{R_1}{R_2}$$

R_1	R_2	Yield	Ref.
Et	H	59	507
Ph	H	62	475
p-MeC$_6$H$_4$	H	80	475
p-EtC$_6$H$_4$	H	61	475
p-ClC$_6$H$_4$	H	70.8	475
p-BrC$_6$H$_4$	H	86.6	475
p-MeOC$_6$H$_4$	H	50.6	475
p-EtOC$_6$H$_4$	H	60.8	475
p-AcNHC$_6$H$_4$	H	46	507
2,5-(MeO)$_2$C$_6$H$_3$	H	56	507
2,4-(OH)$_2$C$_6$H$_3$	H	60.1	475
3,4-Me(OH)C$_6$H$_3$	H	75	507
2,5-Me(OH)C$_6$H$_3$	H	56	507
3,4-(MeO)$_2$C$_6$H$_3$	H	66	507
H$_2$N	H	—	507
Me	Me	65	507
p-MeC$_6$H$_4$	Me	74	507
p-EtC$_6$H$_4$	Me	64	507
p-ClC$_6$H$_4$	Me	70	507

A new synthesis of 2-bromo-4-aminothiazole was reported from α-cyanoalkylthiocyanates and hydrogen bromide (**189**), $R_1 = NH_2$, $R_2 = H$ (Scheme 98) (523). Hydrogen chloride is unsatisfactory as a cyclizing agent, whereas hydrogen iodide causes further reduction and leads directly to derivatives of the previously inaccessible 4-aminothiazoles (see Section VII. 2).

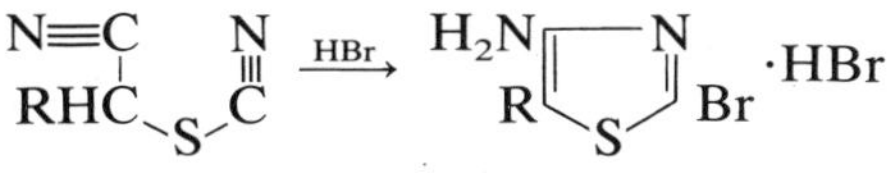

$$\underset{\textbf{189}}{\overset{N\equiv C}{\underset{RHC\diagdown S \diagup C}{\overset{|}{}\overset{}{}\overset{N}{\underset{\|}{}}}}} \xrightarrow{HBr} \underset{\textbf{190}}{\overset{H_2N}{\underset{R\diagdown S \diagup Br}{\overset{}{}\overset{N}{}}}} \cdot HBr$$

Scheme 98

The starting α-cyanoalkylthiocyanates (**189**), in which R = Ph, o-ClC$_6$H$_4$, p-C$_6$H$_4$, 2,4-Cl$_2$C$_6$H$_3$, o-MeOC$_6$H$_4$, o-NO$_2$C$_6$H$_4$, and p-AcNHC$_6$H$_4$, were prepared from α-chloroalkyl and α-(p-toluenesulfonyloxy)alkylcyanides.

If anhydrous dimethylformamide is used as solvent, the reaction is complete within 30 min at room temperature, and the products are generally obtained in high yield.

2-Bromo-4-aminothiazoles (**190**) were usually prepared either by passing anhydrous hydrogen bromide through a dry ethereal solution of **189** at 0°C for $1\frac{1}{2}$ hr, or by dissolving α-cyanoalkylthiocyanate in three times its weight of glacial acetic acid and adding this solution under stirring to a solution of acetic acid containing three equivalents of hydrogen bromide at 0°C. In this case, after adding an excess of acetic anhydride compounds **190** were isolated as N-acetyl derivatives. Yields ranged from 20 to 94%.

3. Action of Labile Sulfur

Thioacids (**191**) react with α-thiocyanatoacetophenone to produce 2-mercapto-4-phenylthiazole (**192**) (Scheme 99). With thioacetic acid **192** is obtained directly while with thiobenzoic acid, an acyclic intermediate (**193**) has been isolated that is cyclized by heating with dilute acid (Scheme 100) (143).

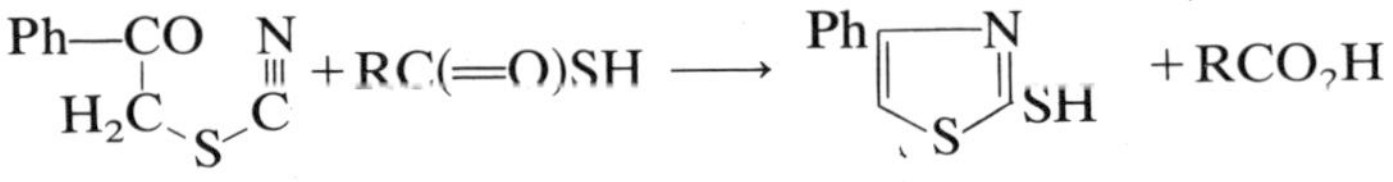

$$\underset{\textbf{191}}{\text{Ph—CO}\atop\text{H}_2\text{C}-\text{S}-\text{C}\equiv\text{N}} + \text{RC}(=\text{O})\text{SH} \longrightarrow \underset{\textbf{192}}{\left[\begin{array}{c}\text{Ph}\\ \\ \text{S}\end{array}\right]\text{SH}} + \text{RCO}_2\text{H}$$

Scheme 99

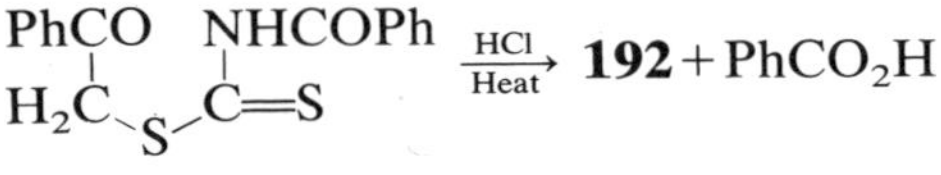

$$\underset{\textbf{193}}{\text{PhCO}\quad\text{NHCOPh}\atop\text{H}_2\text{C}-\text{S}-\text{C}=\text{S}} \xrightarrow[\text{Heat}]{\text{HCl}} \textbf{192} + \text{PhCO}_2\text{H}$$

Scheme 100

Other sulfur compounds such as thiourea, ammonium dithiocarbamate, or hydrogen sulfide also lead to 2-mercaptothiazoles. Thus thiourea has been used in the syntheses of 4,5-dimethyl (369) and 4-aryl-2-mercaptothiazoles (Table II-30) (519). The reactions were carried out by condensing the α-thiocyanatoketones with thiourea in alcohol and water acidified with hydrochloric acid. By this procedure, 4-aryl-2-mercaptothiazoles were obtained in yields of 40 to 80% with bis-(4-aryl-2-thiazolyl) sulfides as by-products (519). These latter products (**194**) have also been observed as a result of the action of thiourea on 2-chloro-4-arylthiazole under the same experimental conditions. They can be separated from 2-mercaptothiazoles because of their different degrees of solubility in sodium hydroxide solution at 5%. In this medium bis-(4-phenyl-2-thiazolyl)sulfide is

TABLE II-30. 2-MERCAPTOTHIAZOLE DERIVATIVES FROM α-THIOCYANATOKETONES AND LABILE SULFUR[a]

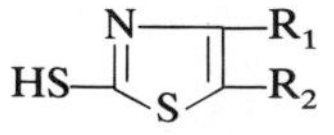

$$\text{HS}-\underset{S}{\overset{N}{\|}}\underset{}{\overset{}{\diagup}}\begin{smallmatrix}R_1\\R_2\end{smallmatrix}$$

R_1	R_2	Yield (%)	Ref.
Me	H	—	369
Ph	H	79	519
p-MeC$_6$H$_4$	H	85	519
p-ClC$_6$H$_4$	H	60	519
p-BrC$_6$H$_4$	H	39	519
p-MeOC$_6$H$_4$	H	75	519
Me	Me	—	369

[a] Reactions were carried out by condensing α-thiocyanatoketones with thiourea in alcohol and water acidified with hydrochloric acid.

insoluble, whereas 2-mercapto-4-phenylthiazole (**192**) dissolves and is precipitated by the addition of dilute hydrochloric acid.

The behavior of the 2-mercapto-4-arylthiazoles in this reaction would seem to be analogous to that of the 2-mercaptobenzothiazoles (137). It appears that monosulfide compound (**195**) cannot be obtained from 2-chloro- and 2-mercapto-4-phenylthiazoles (given the difficulty of preparing it in this way) but rather by the action of the 2-mercapto-4-phenylthiazole on the intermediary, 4-phenyl-2-isothiazolyl isothiouronium chloride (**194**), as in Scheme 101.

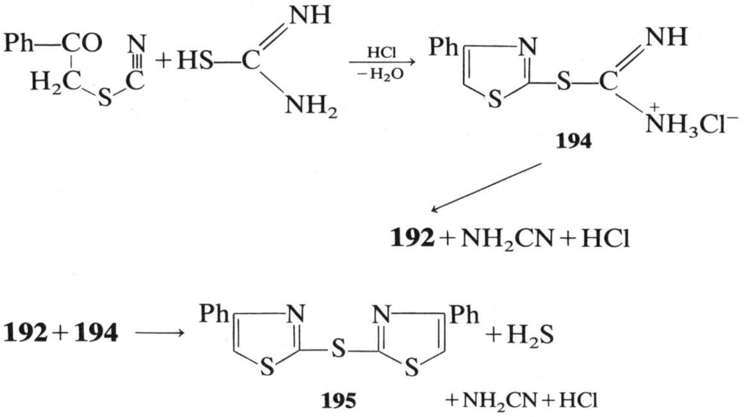

Scheme 101

The percentage of these various compounds during the reaction depends upon the experimental conditions.

4. Action of Labile Nitrogen

α-Thiocyanatoketones also react with ammonium chloride or amine hydrochloride to give 2-aminothiazoles or their N-substituted derivatives.

For example, the action of α-thiocyanatoacetone on ammonia in ether solution gives 4-methyl-2-aminothiazole but in very low yield (137). Methylamine in ether at 0°C gives in a first step S-acetonyl-N-methylisothiourea (**196**) in 80% yield (Scheme 102) (137). The cyclization of this intermediate occurs either after a prolonged rest at room temperature, either by fusion or by heating with dilute hydrochloric acid to afford the 4-methyl-2-methylaminothiazole (**197**).

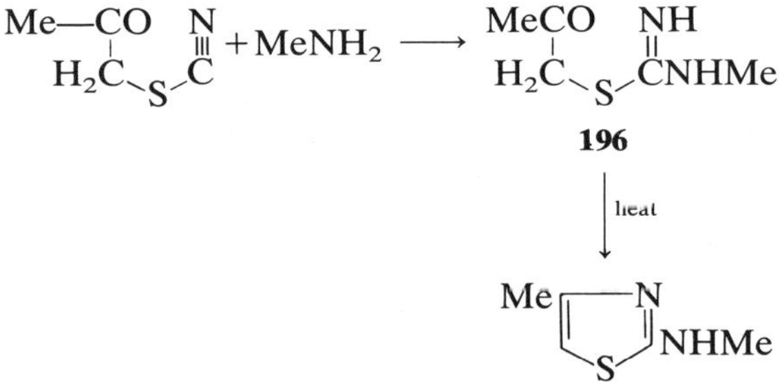

$$\begin{array}{c} Me-CO \quad N \\ | \quad \quad \; ||| \\ H_2C_{\diagdown S} \diagup C \end{array} + MeNH_2 \longrightarrow \begin{array}{c} MeCO \quad NH \\ | \quad \quad \; || \\ H_2C_{\diagdown S} \diagup CNHMe \end{array}$$

196

↓ heat

Me—⌐─N
 ‖   ‖ NHMe
 ⌐S─┘

Scheme 102   **197**

No reaction occurs with dimethylamine.

α-Thiocyanatoacetophenones also react with ammonium chloride to give 2-amino-4-arylthiazoles (23). But this method has limited scope.

3-Thiocyanato-2-aminothiophenes give by ring closure the corresponding 2-aminothieno[2,3d]thiazoles (764).

IV. THIAZOLES FROM ACYLAMINOCARBONYL COMPOUNDS AND PHOSPHORUS PENTASULFIDE AND RELATED CONDENSATION (GABRIEL'S SYNTHESIS) (TYPE III)

This reaction was first described by Gabriel in 1910 (40), when he warmed an acylaminoketone (**197a**) with an equimolecular amount of phosphorus pentasulfide. The reaction (Scheme 103) is similar to the preparation of other five-membered oxygen- and sulfur-containing rings from 1,4-dicarbonyl compounds.

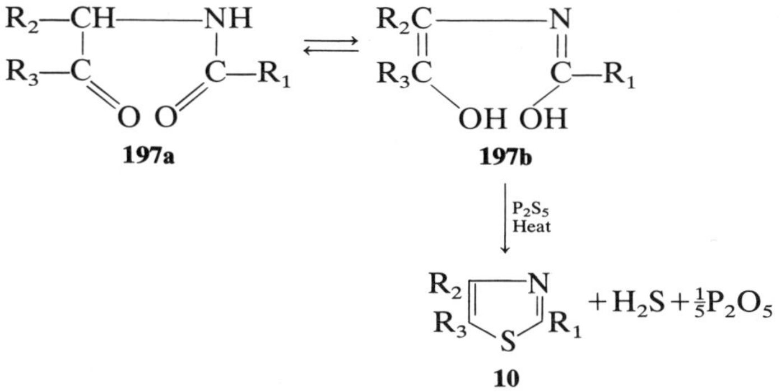

IV. Thiazoles from Acylaminocarbonyl Compounds

$$\underset{\textbf{197a}}{\underset{R_3-C}{\overset{R_2-CH\text{——}NH}{|}}\underset{\overset{\|}{O}\ \overset{\|}{O}}{\diagdown\diagup}\underset{}{C-R_1}} \rightleftarrows \underset{\textbf{197b}}{\underset{R_3C}{\overset{R_2C\text{——}N}{\|}}\underset{\overset{|}{OH}\ \overset{|}{OH}}{\diagdown\diagup}\underset{}{\overset{\|}{C}-R_1}}$$

$$\downarrow \text{P}_2\text{S}_5 \text{ Heat}$$

$$\underset{\textbf{10}}{\underset{R_3}{\overset{R_2}{}}\underset{S}{\boxed{}}\overset{N}{}R_1} + H_2S + \tfrac{1}{5}P_2O_5$$

Scheme 103

The method has not been studied extensively and is restricted to the preparation of alkyl-, aryl-, or alkoxy-substituted thiazoles mostly in 2-, 5-, or 2,5-positions. Yields ranged from 45 to 80%. Sometimes this method gives good results when the usual Hantzsch's synthesis fails. There has been very little speculation about the mechanism of this reaction.

The conditions under which this reaction occurs are comparatively mild, and for this reason it has been assumed that dehydration takes place through the enol form without rearrangement to give the 2,5-disubstituted derivatives as indicated in Scheme 103.

Thiazole itself was obtained in a 62% yield (47) from formylaminoacetal (Scheme 104).

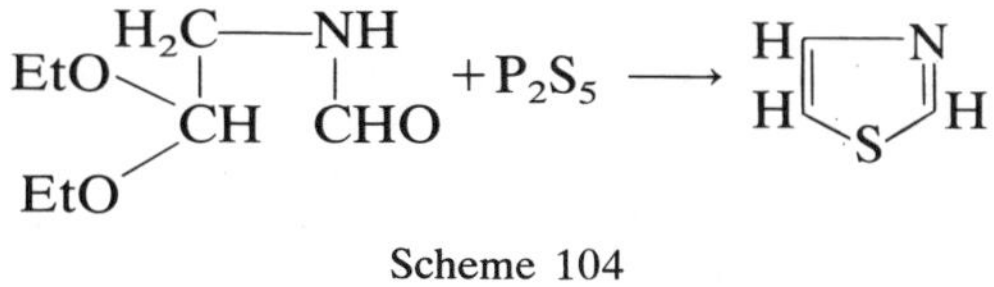

$$\underset{EtO}{\overset{EtO}{\diagdown}}\underset{CH}{\overset{H_2C\text{——}NH}{|}}\underset{CHO}{|} + P_2S_5 \longrightarrow \underset{H}{\overset{H}{}}\underset{S}{\boxed{}}\overset{N}{}H$$

Scheme 104

5-Arylthiazoles were prepared either from ω-formaminoacetophenones and phosphorus pentasulfide in 70% yield (47, 641) or by treating thioformaminoketones with concentrated sulfuric acid in water (344). Thioformaminoacetophenone itself was obtained by the action of potassium dithioformate on aminoacetophenone (251).

The Gabriel's synthesis is also applicable when a polysubstituted thiazole is required (381, 550). Thus 2,4,5-trisubstituted thiazoles are obtained by treating the corresponding α-acylaminoketones with phosphorus pentasulfide for a few minutes at 100°C (550) or at higher temperature for heavier substituents (381). (Table II-31).

TABLE II-31. THIAZOLES FROM α-ACYLAMINOKETONES AND PHOSPHORUS PENTASULFIDE (Gabriel's synthesis)

$$\underset{R_1}{\overset{N}{\diagdown}}\underset{S}{\overset{R_2}{\diagup}}R_3$$

R_1	R_2	R_3	Conditions[a]	Yield (%)	Ref.
H	H	H	P_2S_5, heat	62	47
H	H	Ph	HCSNHCH$_2$COPh,	90	351
			H$_2$SO$_4$, H$_2$O, 30 min		
H	H	p-ClC$_6$H$_4$	HCSNHCH$_2$COPh,	90	351
			H$_2$SO$_4$, H$_2$O, 30 min	—	351
H	H	p-O$_2$NC$_6$H$_4$	HCSNHCH$_2$COPh,		
			H$_2$SO$_4$, H$_2$O, 30 min	—	351
Me	H	Me	P_2S_5, heat	—	351
Me	H	Ph	P_2S_5, heat	—	153a, 168
Me	H	p-MeC$_6$H$_4$	P_2S_5, heat	—	40
CH$_2$OEt	H	Me	P_2S_5, heat, pyridine, reflux 45 min	49	46
CH$_2$OBu	H	Ph	P_2S_5, heat, pyridine, reflux 45 min	50	409
CH$_2$OPh	H	Ph	P_2S_5, heat, pyridine, reflux 45 min	50	409
o-MeOC$_6$H$_4$OCH$_2$	H	Ph	P_2S_5, pyridine, reflux 45 min	48	409
Me	Me	Me	P_2S_5, heat	67	461
Me	Me	Ph	P_2S_5, heat	50	47
Me	C$_6$H$_6$	Me	P_2S_5, heat	65	488
Me	p-MeOCC$_6$H$_4$CH$_2$	Me	P_2S_5, 100°C	35	550
Me	Et	Ph	P_2S_5, heat	85	488
Me	Pr	Me	P_2S_5, 100°C 10–30 min	45	550
Me	iso-Bu	Me	P_2S_5, 100°C 10–30 min	25	550
Me	tert-Bu	Me	P_2S_5, 100°C	62	461
Me	C$_{10}$H$_{21}$	Me	P_2S_5, 100°C	59	550

Me	C$_{14}$H$_{29}$	Me	P$_2$S$_5$, 100°C	42	550
Me	C$_{16}$H$_{33}$	Me	P$_2$S$_5$, 100°C	53	550
Me	Pr	Et	P$_2$S$_5$, 100°C	33	550
Et	Pr	Et	P$_2$S$_5$, 100°C	50	550
Et	iso-Bu	Et	P$_2$S$_5$, 100°C	38	550
C$_5$H$_{11}$	iso-Bu	C$_5$H$_{11}$	P$_2$S$_5$, 100°C	60	550
C$_7$H$_{15}$	iso-Bu	C$_7$H$_{15}$	P$_2$S$_5$, 100°C	41	550
C$_{11}$H$_{23}$	Pr	C$_{11}$H$_{23}$	P$_2$S$_5$, 100°C	32	550
C$_{17}$H$_{35}$	iso-Bu	C$_{17}$H$_{35}$	P$_2$S$_5$, 100°C	32	550
Ph	H	H	P$_2$S$_5$, 100°C	70	47, 641
Ph	H	Me	P$_2$S$_5$, 90°C oil bath	45	46, 562
Ph	H	Ph	P$_2$S$_5$, heat	—	40
Ph	H	p-MeC$_6$H$_4$	P$_2$S$_5$, heat	—	46
Ph	H	p-biphenyl	P$_2$S$_5$, 200–300°C	—	381
α-furyl	H	p-biphenyl	P$_2$S$_5$, 200–300°C	—	381

[a] Thiazoles were usually obtained by treating the corresponding α-acylaminoketones with phosphorus pentasulfide for a few minutes at 100°C

For example, N-propynylbenzamides lead to 2-phenyl-5-methylthiazole (**10**), R_1 = phenyl, R_2 = H, R_3 = methyl (562).

The synthesis of α-aminoketones can be achieved using α-halogenated ketones as starting material. These latter are converted into the hexamethylene tetraminium salts by the method of Mannich and Hahn (42). This reaction proceeds in two steps:

$$R_2COCHXR_2 + (CH_2)_6N_4 \rightarrow R_1COCHR_2N_4^+(CH_2)_6Cl^- \quad (5)$$
$$\mathbf{198}$$

The addition product is then hydrolyzed with alcohol and hydrochloric acid according to eq. 6.

$$\mathbf{198} + 3HCl + 12C_2H_5OH \rightarrow R_1COCHR_2\overset{+}{N}H_3Cl^- \quad (6)$$

Another procedure for obtaining α-aminoketones is by reduction of α-nitrosoketones in the presence of the required carboxylic acid. Acylaminoketones are prepared either by reacting acids with the chlorhydrate of α-aminoketones according to the method of Pictet and Gauss (41) or by the action of acid anhydrides upon α-amino acids (550).

The Gabriel's synthesis has been further extended to alkoxythiazoles. Thus 2-alkoxy (or aryloxymethyl) 5-methyl (or phenyl) thiazoles (**200**) were prepared by refluxing the corresponding acylaminoketone (199) in dry pryridine in the presence of P_2S_5 (Scheme 105) (409).

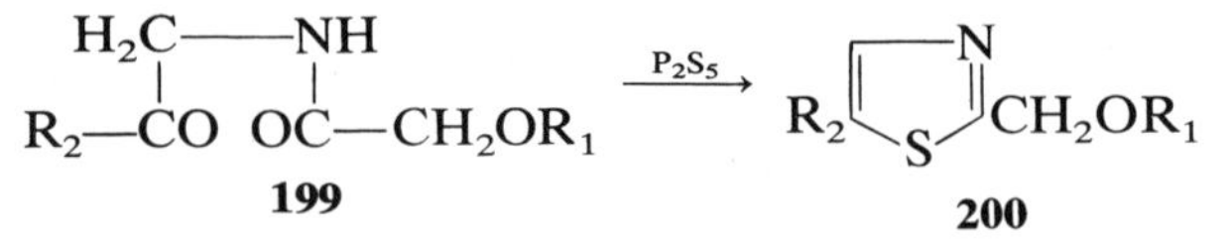

$$\underset{\mathbf{199}}{\underset{R_2-CO\ OC-CH_2OR_1}{H_2C-NH}} \xrightarrow{P_2S_5} \underset{\mathbf{200}}{R_2\underset{S}{\overset{N}{\diagup}}CH_2OR_1}$$

Scheme 105

5-Alkoxythiazole derivatives (**202**) are formed in a similar manner by the action of phosphorus pentasulfide on α-acylamino esters (**201**) (64, 334, 711).

The reactions are carried out by warming the ester in an inert solvent such as chloroform or benzene with P_2S_5 on the steam bath for 8 to 10 hr (Scheme 106). Products are isolated by ether extraction of the aqueous alkali-treated reaction mixture.

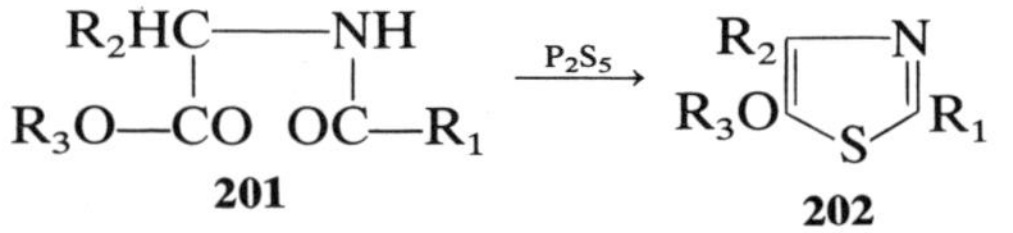

$$\underset{\mathbf{201}}{\underset{R_3O-CO\ OC-R_1}{R_2HC-NH}} \xrightarrow{P_2S_5} \underset{\mathbf{202}}{\underset{R_3O}{R_2}\underset{S}{\overset{N}{\diagup}}R_1}$$

Scheme 106

N-benzoyl derivatives of α-aminoacids give 2-phenyl-5-alkoxythiazoles (64). A list of the thiazoles prepared by this method is given in Table II-32 (711).

TABLE II-32. 5-ALCOXYTHIAZOLE DERIVATIVES FROM α-ACYLAMINOESTERS AND PHOSPHORUS PENTASULFIDE[a]

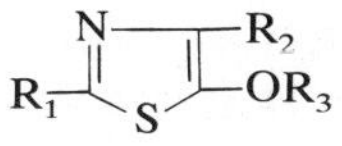

$$\underset{R_1}{\overset{N}{\|}}\underset{S}{\overset{R_2}{\|}}OR_3$$

R_1	R_2	R_3	Yield (%)	Ref.
H	Me	Me	22	334
H	H	Me	—	711
H	iso-Bu	Me	38	711
H	iso-Bu	Et	50	711
Me	H	Me	—	711
Me	H	Et	65	64
Me	H	n-Bu	—	711
Me	Me	Me	—	711
Me	iso-Pr	Pr	—	711
Me	iso-Bu	Me	58	711
Me	iso-Bu	Et	54	711
OEt	iso-Bu	Et	—	711
Ph	iso-Bu	Et	—	64
Ph	Me	Et	85	64

[a] Reactions were carried out by refluxing 24 hr the corresponding α-acylaminoesters in $CHCl_3$ in the presence of P_2S_5.

The preparation of 2-phenylthiazole from thiobenzamidoacetal (**203**) (readily prepared from aminoacetal) has also been reported (Scheme 107) (440). Compound **203** loses 1 mole of ethanol under mild dehydrating conditions to give 5-ethoxy-2-phenyl-Δ-2-thiazoline (**204**). This then loses a second mole of ethanol under more vigorous conditions to give 2-phenylthiazole (**205**).

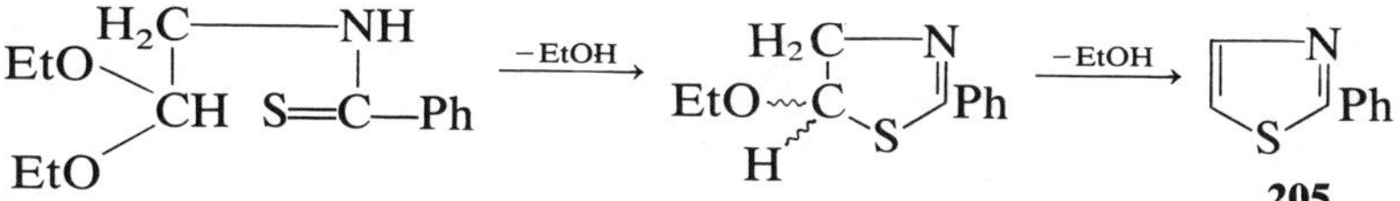

Scheme 107

To the previous reaction we related the dehydration of ethyl-2-acetamido-2-thiocarbamoylacetate. This latter compound, heated in the presence of $POCl_3$ 20 min at 80°C and then 35 min at 95°C, gives 5-amino-4-carboxyethyl-2-methylthiazole in 77.5% yield. In a similar manner, 2-formaminothioacetamide gives the 5-aminothiazole (718).

V. THIAZOLES FROM α-AMINONITRILES (COOK–HEILBRON'S SYNTHESIS) (TYPE II)

This type of synthesis, which was investigated by Cook and Heilbron (323) and Takahashi (393, 394) between 1947 and 1953, gives 5-aminothiazoles variously substituted in the 2-position by reacting an aminonitrile with salts and esters of dithioacids, carbon disulfide, carbon oxysulfide, and isothiocyanates under exceptionally mild conditions.

1. Salts and Esters of Dithioacids: 5-Aminothiazole Derivatives and Related Condensations

By condensing the salts or the esters of either dithioformic (**207**) or dithiophenacetic acids with α-aminonitriles (**206**) 5-aminothiazoles (**209**), in which R_1 = hydrogen, benzyl and R_2 = phenyl, carbethoxy, or carbophenoxy, were obtained in fairly good yields (Scheme 108) (271). These reactions were carried out in aqueous ethereal solution at room temperature. Acyclic thioamides as intermediates in this reaction have been isolated in some cases (**208**).

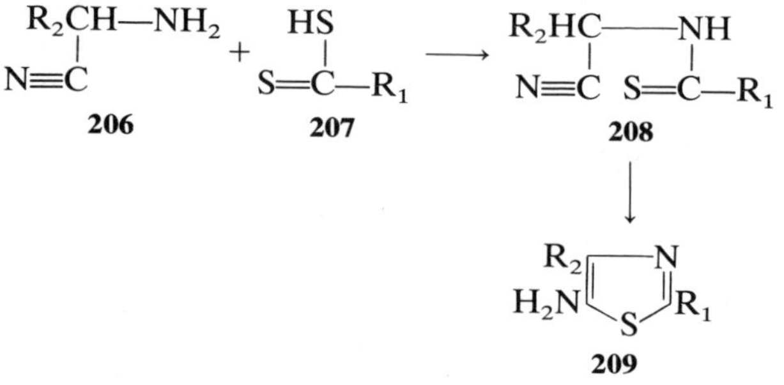

Scheme 108

V. Thiazoles from α-Aminonitriles

Similarly, potassium thiophenylacetate gives 2-benzyl-5-aminothiazole in 38% yield (**209**), R_1 = benzyl, R_2 = hydrogen (68). Another method of synthesis of 5-aminothiazole from ethylformate is shown in Scheme 109. The resulting nitriles (**210**) treated with hydrogen sulfide were converted to the corresponding thioamides in the usual manner. These thioamides (**211**), R_1 = H, afford the corresponding 5-acetamidothiazoles unsubstituted in the 2-position after refluxing with acetic anhydride (**212**), R_1 = H.

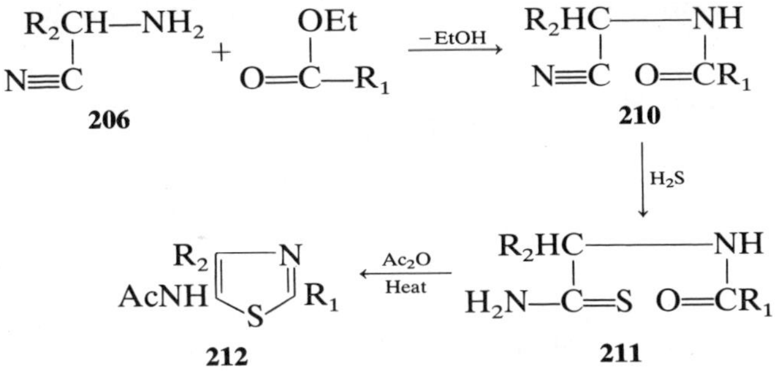

$$R_2CH-NH_2 \atop N\equiv C \quad + \quad {OEt \atop O=C-R_1} \quad \xrightarrow{-EtOH} \quad {R_2HC-NH \atop N\equiv C \ \ O=CR_1}$$

206   **210**

$$\downarrow H_2S$$

$$\underset{\mathbf{212}}{\underset{S}{AcNH \diagdown \diagup R_1}^{R_2 \diagup N}} \quad \xleftarrow{Ac_2O \atop Heat} \quad \underset{\mathbf{211}}{R_2HC-NH \atop H_2N-C=S \ \ O=CR_1}$$

Scheme 109

Similarly, 2-methyl-5-acetamidothiazoles were obtained from the corresponding nitriles, NCCH(R_2)NHCOMe, in which R_2 = hydrogen, or amido, methylamido, and dimethylamido groups (551, 571) (Table II-33).

5-Aminothiazole derivatives (**209**), R_1 = Me or Ph and R_2 = H, $CONH_2$, CO_2Et, were also prepared by ring closure of thioamides such as $R_1CONHCH(R_2)C(=S)NH_2$ with polyphosphoric acid (2 hr at 120°C) (718).

TABLE II-33. 5-ACETYLAMINOTHIAZOLE DERIVATIVES[a]

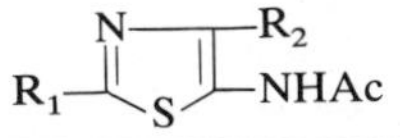

$$R_1 {-}\!\!\!{\diagdown \atop S} \!\!\!{\diagup \atop }\!\!{-NHAc \atop -R_2}^{N}$$

R_1	R_2	Yield (%)	Ref.
H	H	28	551
H	$CONH_2$	53–56	551, 571
H	CONHMe	50–77	551, 571
H	$CON(Me)_2$	40	551
Me	H	23–30	551, 571
Me	$CONH_2$	44	551, 571
Me	CONHME	47	551, 571
Me	$CONMe_2$	25	551

[a] Thiazoles were obtained from the corresponding thioamides [$R_1CONHCH(R_2)C(=S)NH_2$] by refluxing them 30 min in acetic anhydride.

2. Carbon Disulfide: 2-Mercapto-5-aminothiazole Derivatives

Carbon disulfide readily reacts with α-aminonitriles giving 2-mercapto-5-aminothiazoles (213), (271, 293) which can be converted to 5-aminothiazoles unsubstituted in the 2-position (Scheme 110 and Table II-34a). If this reaction is carried out in the presence of benzyl chloride in phosphorus tribromide, a 2-S-substituted thiazole derivative (214) is obtained in quantitative yield (Scheme 111), with R = hydrogen or phenyl (68, 304).

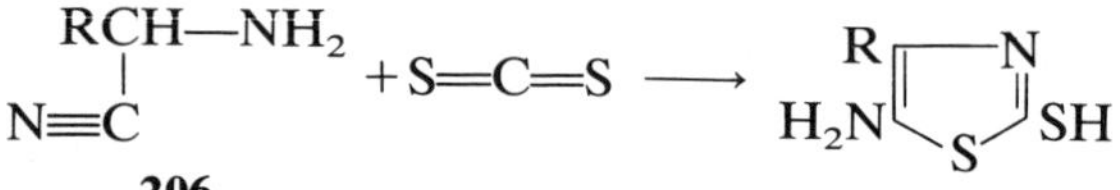

$$\underset{206}{\underset{N\equiv C}{\overset{RCH-NH_2}{|}}} + S=C=S \longrightarrow \underset{213}{\underset{H_2N}{\overset{R}{\diagdown}}\underset{S}{\diagup}SH}$$

Scheme 110

TABLE II-34a. 2-MERCAPTO-5-AMINOTHIAZOLE DERIVATIVES FROM α-AMINONITRILES AND CARBON DISULFIDE

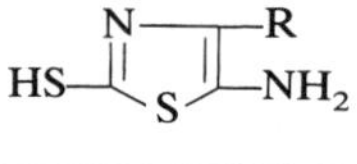

$$HS\underset{S}{\overset{NR}{\diagdown\diagup}}NH_2$$

R	Conditions	Yield (%)	Ref.
iso-Pr	CHCl$_3$, reflux 30 min	55	358
CO$_2$Et	Ether	—	271
H	NH$_2$CH$_2$CN, AcOEt, 0°C, P$_2$O$_5$, CS$_2$	60–82	42, 393
n-Hexyl	Petroleum ether	—	293
n-Heptyl	Petroleum ether	—	293
(CH$_2$)$_4$CO$_2$Et	Petroleum ether	—	293
Ph	Water	95	271
HC(C$_2$H$_5$)(CH$_2$)$_3$CH$_3$	Petroleum ether	—	293

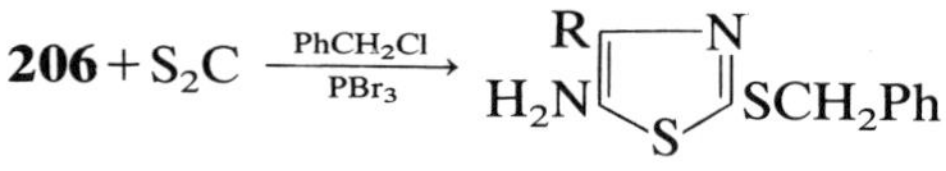

$$206 + S_2C \xrightarrow[PBr_3]{PhCH_2Cl} \underset{214}{\underset{H_2N}{\overset{R}{\diagdown}}\underset{S}{\diagup}SCH_2Ph}$$

Scheme 111

When benzaldehyde or its substituted derivatives are added to carbon disulfide and α-aminonitrile, the corresponding 2-mercapto-5-(p-R-benzylideneamino)thiazoles (215), R = hydrogen atom or a propenyl or phenyl group and Ar = aryl, are obtained (Scheme 112) (393, 442, 694). Yields ranged from 40 to 60% (Table II-34b).

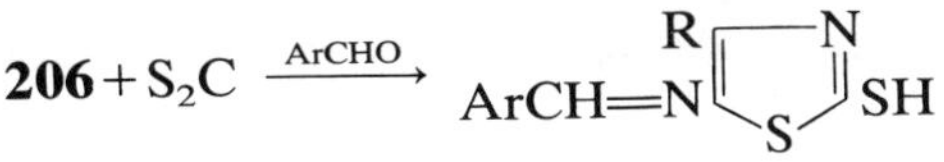

V. Thiazoles from α-Aminonitriles

$$206 + S_2C \xrightarrow{ArCHO} ArCH=N-\underset{S}{\overset{R\underset{}{}N}{[}}\rangle-SH$$
215

Scheme 112

TABLE II-34b. 2-MERCAPTO-5-AMINOTHIAZOLE DERIVATIVES

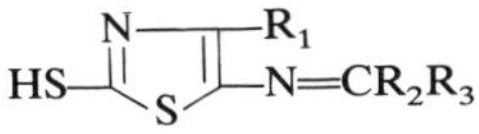

$$HS-\underset{S}{\overset{N\underset{}{}R_1}{[}}\rangle-N=CR_2R_3$$

R_1	R_2	R_3	Conditions [a]	Yield	Ref.
H	H	H	AcOEt, 1 hr at 10°C,	40	393
H	o-(OH)C$_6$H$_4$	H	AcOEt,	—	393
H	p-(OH)C$_6$H$_4$	H	AcOEt,	—	393
H	2,4-(Cl$_2$)C$_6$H$_3$	H	Et$_2$O, 1 hr at 8–10°C	—	694
HC=CHMe	p-(O$_2$N)C$_6$H$_4$	H	AcOEt, 3.5–5 hr	—	393
HC=CHMe	o-(OH)C$_6$H$_4$	H	AcOEt, 5 hr	—	393
HC=CHMe	p-ClC$_6$H$_4$	H	AcOEt, 4 hr	—	393
Ph	α-furyl	H	MeOH, reflux	91.3	442
Ph	p-Me$_2$NC$_6$H$_4$	H	MeOH, reflux 1–1.5 hr	80.2	442
Ph	p-MeOC$_6$H$_4$	H	MeOH, reflux 1–1.5 hr	95.3	442
Ph	3,4-(MeO)(OH)C$_6$H$_3$	H	MeOH, reflux 1–1.5 hr	82	442
Ph	3,4-(CH$_2$O$_2$)C$_6$H$_3$	H	MeOH, reflux 1–1.5 hr	96.6	442
H	Me	Me	Acetone	70	304
Ph	Me	Me	Acetone	—	68
Ph	Ph	H	Acetone	—	68
(CH$_2$)$_5$CH$_3$	Me	H	Acetone	—	293

[a] Reactions were carried out by condensing α-aminonitriles [NCCH(R$_1$)NH$_2$] with carbon disulfide in the presence of aldehydes (R$_2$CHO) or ketones (R$_2$COR$_3$) in these solvents.

Acetone reacts similarly to give the corresponding 2-mercapto-5-isopropylideneaminothiazoles (68, 293, 304).

Methylaminoacetonitrile (**216**) reacts with carbon disulfide in the presence of acetic anhydride with ethyl acetate as solvent to give 2-thio-3-methyl-Δ-4-thiazoline in 74% yield (Scheme 113a) (326). If the reaction is carried out using benzaldehyde in place of acetic anhydride, the corresponding 5-benzylideneamino derivative of **217** is obtained in 70% yield.

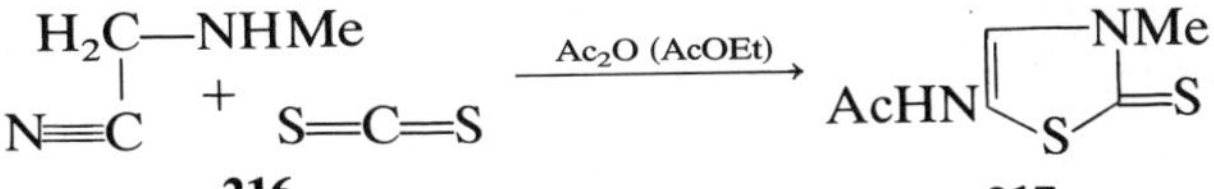

$$\underset{216}{\overset{H_2C-NHMe}{\underset{N\equiv C}{|}} + S=C=S} \xrightarrow{Ac_2O\ (AcOEt)} \underset{217}{AcHN-\underset{S}{\overset{NMe}{[}}\rangle=S}$$

Scheme 113a

Another example of condensation with carbon disulfide is shown in (Scheme 113b) (774).

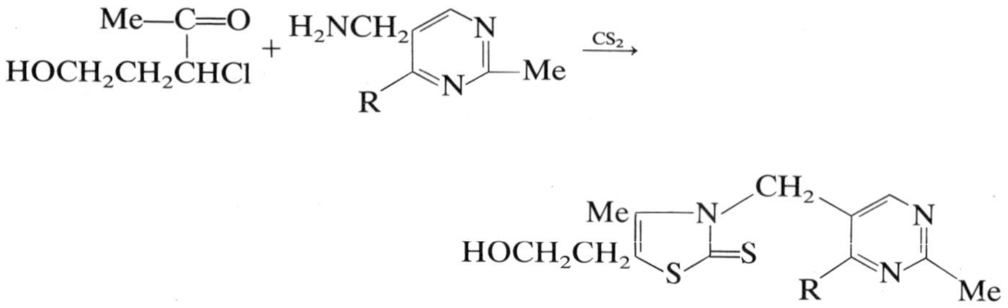

Scheme 113b

3. Carbon Oxysulfide: 2-Hydroxy-5-aminothiazole Derivatives

By condensing carbon oxysulfide with α-aminonitriles the corresponding 2-hydroxy-5-aminothiazoles can be obtained. In the presence of benzaldehyde or its substituted derivatives the reaction leads to 5-benzylideneaminothiazole derivatives (**218**) in good yields (Scheme 114 and Table II.35) (393, 442). However, the reaction fails with α-amino acetonitrile (**206**), R = H (317). The 2-alkoxy analogs (**220**), R = Me, Et, Pr, Bu, vinyl, were similarly obtained from **219** and benzylideneamino acetonitrile (Scheme 115a) (393).

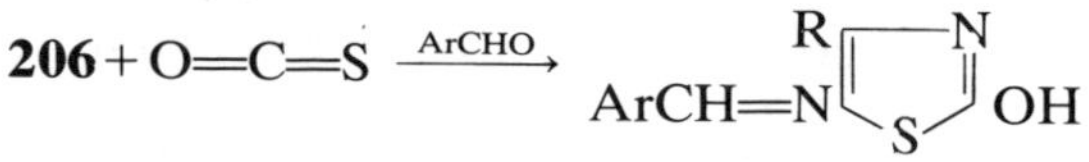

$$206 + O\!=\!C\!=\!S \xrightarrow{\text{ArCHO}} \mathbf{218}\ (\text{ArCH}\!=\!\text{N, R, S, OH})$$

Scheme 114

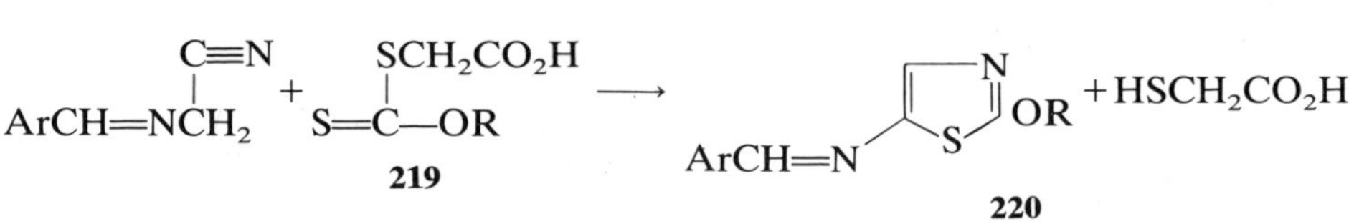

Scheme 115a

V. Thiazoles from α-Aminonitriles

TABLE II-35. 2-HYDROXY-5-AMINOTHIAZOLES DERIVATIVES FROM α-AMINONITRILE AND CARBON OXYSULFIDE[a]

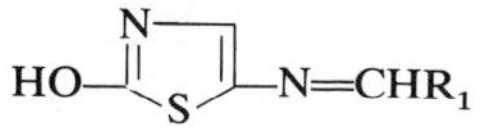

$$HO-\underset{S}{\overset{N}{\underset{\|}{C}}}-N=CHR_1$$

R_1	Yield (%)	Ref.
Ph	—	393
p-ClC$_6$H$_4$	—	393
o-EtOC$_6$H$_4$	—	393
p-AcNHC$_6$H$_4$	—	393
p-Me$_2$NC$_6$H$_4$	—	393
o-O$_2$NC$_6$H$_4$	60.6	442
m-O$_2$NC$_6$H$_4$	841.1	442
p-O$_2$NC$_6$H$_4$	75.9	442
3,4-(MeO)$_2$C$_6$H$_3$	88.4	442
3,4-(CH$_2$O$_2$)C$_6$H$_3$	75	442
2,4-Cl$_2$C$_6$H$_3$	57.4	442
α-Naphthyl	92.7	442
α-Furyl	73	442
β-Pyridyl	45.4	442

[a] Reactions were carried out by condensing the α-aminoacetonitrile with carbon oxysulfide in the presence of benzaldehyde or its substituted derivatives, in alcohol as solvent.

2-Thiazolones in which R_1 = H, alkyl, or phenyl and R_3 = CO$_2$R, CONHR, or CN (**220b**) have been obtained by Grobe and Heitzer (Scheme 115b) (775).

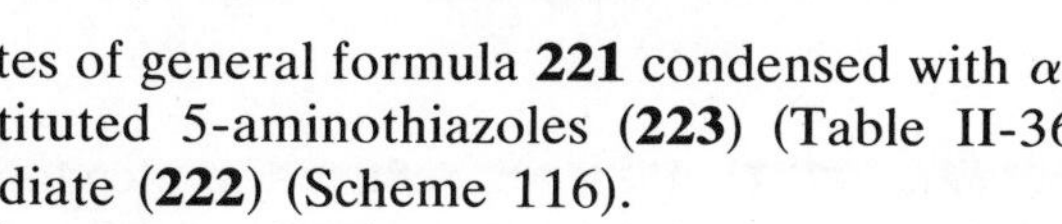

$$\underset{R_3CH}{\overset{R_2C-NHR_1}{\|}} + \underset{ClSC=O}{\overset{Cl}{|}} \longrightarrow \underset{R_3}{\overset{R_2}{}}\underset{S}{\overset{}{}}\overset{NR_1}{\underset{=O}{}}$$

220b

Scheme 115b

4. Isothiocyanates: 2,5-Diaminothiazole Derivatives

Isothiocyanates of general formula **221** condensed with α-aminonitriles lead to 2-substituted 5-aminothiazoles (**223**) (Table II-36) through an acyclic intermediate (**222**) (Scheme 116).

Methyl (299), phenyl (299), acetyl (394), benzoyl (299, 394), and carbethoxy (292) isothiocyanates (**221**), R_1 = Me, Ph, Ac, PhCO, CO$_2$Et, have been successfully used. In some cases, a 2,5-disubstituted aminothiazole (**224**) was isolated as by-product (292).

TABLE II-36. 2,5-DIAMINOTHIAZOLE DERIVATIVES FROM α-AMINO-NITRILES AND ISOTHIOCYANATES.

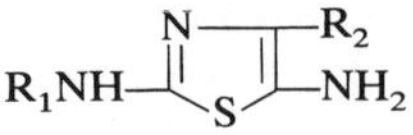

$$R_1NH-\underset{S}{\overset{N}{\underset{\|}{\bigsqcup}}}-NH_2$$
with R_2 at the 4-position

R_1	R_2	Conditions	Yield (%)	Ref.
Me	H	H$_2$NCH$_2$CN, MeNCS	—	299
Me	CO$_2$Et	H$_2$NCH(CO$_2$Et)CN, MeNCS	—	299
Ph	H	H$_2$NCH$_2$CN, PhNCS	—	
Ph	Ph	H$_2$NCHPhCN, PhNCS	—	271
Ph	CO$_2$Et	H$_2$NCH(CO$_2$Et), PhNCS	—	299
COMe	H	H$_2$NCH$_2$CN, MeCONCS, Et$_2$O, AcOEt 10% alcohol, HCl	—	394
COPh	H	H$_2$NCH$_2$CN, PhCONCS	81	299
COPh	Ph	H$_2$NCHPhCN, PhCONCS	—	299
COPh	CH=CHPh	dl-MeCH=CH(CN)NH, PhCONCS, ether, 2 days at 0°C	—	394
COPh	CO$_2$Et	H$_2$NCH(CO$_2$Et), PhCONCS	—	292
CO$_2$Et	H	H$_2$NCH$_2$CN, EtCO$_2$NCS	—	292
CO$_2$Et	Ph	H$_2$NCHPhCN, EtCO$_2$NCS	—	292
CO$_2$Et	CO$_2$Et	H$_2$NCH(CO$_2$Et)CN, EtCO$_2$NCS	—	292
CO$_2$Ph	Me	H$_2$NCHMeCN, PhCO$_2$NCS	—	292

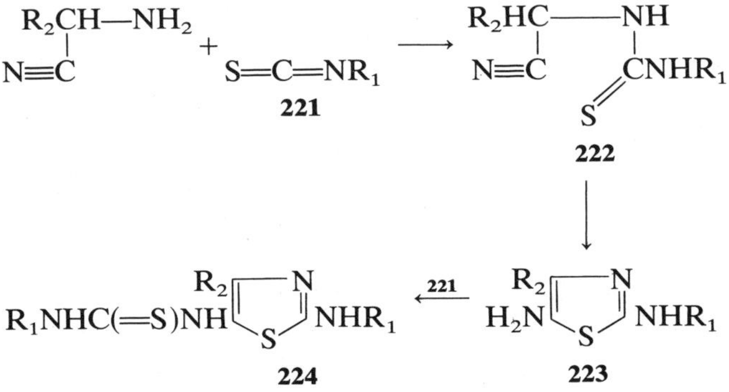

Scheme 116

Acylisothiocyanates (**221**), R_1 = Ac, refluxed in alcoholic solution with α-aminoacetonitrile, in the presence of benzaldehyde or its substituted derivatives, afford the corresponding 5-benzylideneaminothiazole derivatives (**225**) in which R_1 = hydrogen or propenyl and R_2 = aryl or styryl (Scheme 117 and Table II-37) (394).

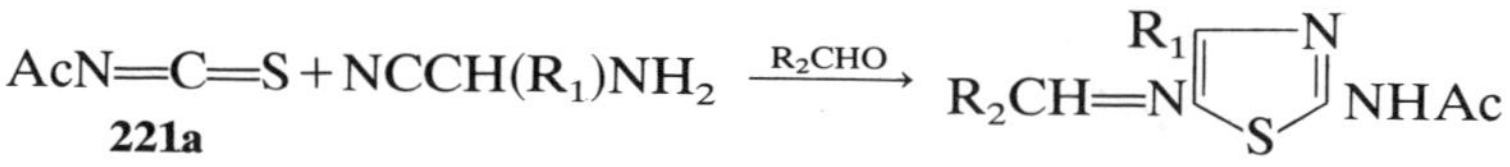

$$\text{AcN}=\text{C}=\text{S} + \text{NCCH}(R_1)\text{NH}_2 \xrightarrow{R_2\text{CHO}} R_2\text{CH}=N\underset{S}{\overset{R_1\quad N}{\underset{}{\boxed{}}}}\text{NHAc}$$
221a   **225**

Scheme 117

TABLE II-37. 2,5-DISUBSTITUTED AMINOTHIAZOLES DERIVATIVES FROM ACYLISOTHIOCYANATES[a]

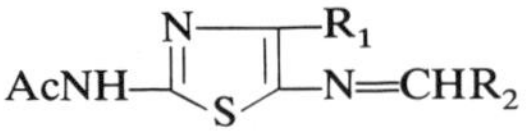

$$\text{AcNH}-\underset{S}{\overset{N\quad R_1}{\boxed{}}}-N=\text{CHR}_2$$

R_1	R_2	Conditions
H	Ph	NH_2CH_2CN, PhCOH, alcohol, ether
H	o-OHC$_6$H$_4$	NH_2CH_2CN, ArCOH, alcohol, ether
H	p-OHC$_6$H$_4$	NH_2CH_2CN, ArCOH, alcohol, ether
H	CH=CH–Ph	NH_2CH_2CN, PhCH=CHOH, alcohol, ether
H	p-Me$_2$NC$_6$H$_4$	NH_2CH_2CN, ArCOH, alcohol, ether
CH=CH–CH$_3$	o-OHC$_6$H$_4$	$NH_2CH(CN)CH=CH$–CH$_3$, ArCOH
CH=CH–CH$_3$	CH=CHPh	$NH_2CH(CN)CH=CH$–CH$_3$, PhCH=CHCOH

[a] Ref. 394

VI. THIAZOLES FROM NITRILES AND α-MERCAPTO-KETONES OR ACIDS: 2,4-DISUBSTITUTED AND 4-HYDROXYTHIAZOLE DERIVATIVES

Besides α-halocarbonyl compounds, α-mercaptoketones and acids are also used for the preparation of thiazoles from nitriles and aldehydes oximes.

1. α-Mercaptoketones

Miyatake and Yashikawa have prepared several 2,4-disubstituted thiazoles in fairly low yield (16 to 40%) by the action of α-mercaptoketones (**226**) on nitriles (**227**) (Scheme 118 and Table II-38). The reaction was carried out in benzene solution at 0°C by passing a current of dry hydrogen chloride through the mixture. After 3 hr the mixture was filtered and washed with benzene. When the resins had been removed and the remaining solution alkalinized, the product was extracted.

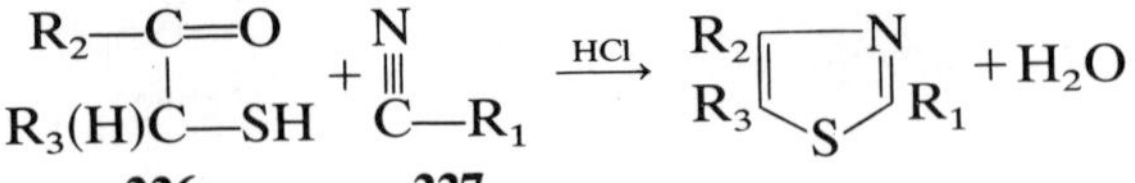

General Synthetic Methods for Thiazole and Thiazolium Salts

$$R_2-C=O \atop R_3(H)C-SH \quad + \quad {N \atop \underset{C-R_1}{\|||}} \quad \xrightarrow{HCl} \quad {R_2 \diagup\!\!\!\!\diagdown N \atop R_3 \diagdown_S\diagup R_1} \quad + H_2O$$

$$\textbf{226} \qquad\qquad \textbf{227} \qquad\qquad\qquad \textbf{10}$$

Scheme 118

TABLE II-38. THIAZOLES FROM α-MERCAPTOKETONES AND NITRILES

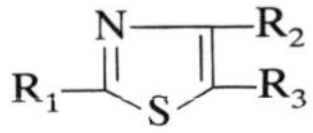

$$R_1-\underset{S}{\overset{N}{\|}}\!\!-\!\!{R_2 \atop R_3}$$

R_1	R_2	R_3	Conditions[a]	Yield (%)	Ref.
H	Me	H	Dry HCl	35–40	446
H	Me	CH_2CH_2OH	Dry HCl	—	446
Me	Me	H	Dry HCl	—	446
Me	Ph	H	Dry HCl	15	446
Me	Me	CH_2CH_2OH	From $H_2NN=CHMe$	60	446, 529
Me	OH	H	Dry HCl	—	446
Ph	Me	H	Dry HCl	18	446
CH_2Ph	Me	H	Dry HCl	27	446
H_2N	$-(CH_2)_4-$		From $H_2NC\equiv N$, MeOH 50–60°C for 20 min	80	527

[a] α-Mercaptoketones were condensed with nitriles under a current of dry hydrogen chloride in benzene solution at 0°C.

The following 2,4,5-trisubstituted thiazoles (**10**) have been prepared in good yields refluxing α-mercaptoketones (**226**), R_3 = Me, and aldehyde oximes, R_1 = –CH=NOH, (719) or nitriles (809) 2 hr at 100°C.

R_1	R_2	R_3	Ref.
Me	Et	Me	809
$PhCH_2$	Me	Me	719
$PhCH_2$	Et	Me	809
$PhCH_2$	$HOCH_2CH_2$	Me	719
$PhCH_2$	$PhCO_2CH_2CH_2$	Me	719
Ph	Et	Me	809
Ac	Me	Me	719
Ac	$HOCH_2$	Me	719
$4\text{-}ClC_6H_4CO$	$HOCH_2CH_2$	Me	719
α-Furoyl	$HOCH_2CH_2$	Me	719

VI. Thiazoles from Nitriles and α-Mercaptoketones or Acids

Asinger and Thiel (473) use an aldehyde and ammonia instead of nitrile. Thus the reaction of mercaptoacetone with benzaldehyde and ammonia gives 4-methyl-2-phenylthiazole in 80% yield and 4-methyl-2-phenyl-Δ-4-thiazoline as the main by-product.

3-Acetyl-3-mercaptopropanol reacts with acetaldehyde in the presence of hydrazine hydrate to yield 2,4-dimethyl-5-(β-hydroxyethyl)thiazole (**10**), $R_1 = R_2 =$ Me, $R_3 = CH_2CH_2OH$ (556).

Cyanamide (**227**), $R_1 = NH_2$, condensed with α-mercaptoketones gives the corresponding 2-aminothiazoles (527).

Ohta (344) prepared 2,4-dimethylthiazole (**10**), $R_1 = R_2 =$ Me, $R_3 =$ H, in fairly low yield by condensing α-mercaptoacetone with acetamide in the presence of anhydrous zinc chloride.

The cyclization of α-mercaptoketones with ammonium thiocyanate leads to the corresponding 2-mercaptothiazoles (144). For example, 2-mercapto-3-pentanone in ethereal solution with sulfuric acid gives 4-ethyl-5-methyl-2-mercaptothiazole (**10**), $R_1 =$ SH, $R_2 =$ Et, $R_3 =$ Me, when allowed to stand for 3 hr without heating with ammonium thiocyanate.

With alkylisothiocyanates (RNCS) the corresponding 2-alkylmercaptothiazoles are obtained (**10**), $R_1 =$ SR, $R_2 =$ Et, $R_3 =$ H, Me (809).

2. α-Mercaptoacids: 4-Hydroxythiazole Derivatives

α-Mercaptoacids (**228**) condense with nitriles (**227**) to afford 4-hydroxythiazoles (**230a**) and 4-thiazolone tautomers (**230b**). This reaction is carried out in ethereal or alcoholic solution saturated with dry hydrogen chloride for several hours at 0°C (Scheme 119).

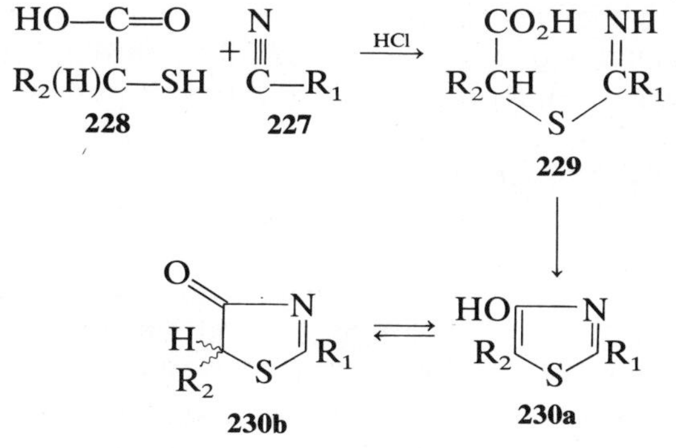

Scheme 119

In some cases, the acyclic intermediates (**229**) of **230**, R_1 = aryl, R_2 = hydrogen, were isolated (424). They undergo cyclization to thiazoles upon heating in toluene. Some 2-aryl-4-hydroxythiazoles prepared by this procedure are listed in Table II-39.

TABLE II-39. 4-HYDROXYTHIAZOLE DERIVATIVES FROM NITRILES AND α-MERCAPTO ACIDS

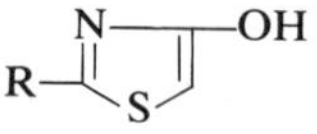

$$R-\underset{S}{\overset{N}{\|}}-OH$$

R	Conditions	Yield (%)	Ref.
Me	Et_2O saturated with HCl	—	542
CH_2CO_2Et	Et_2O saturated with HCl	—	542
Ph	Et_2O saturated with HCl, 3 days at 0°C, then reflux 3 hr in MeC_6H_5	54	424, 542
p-$MeOC_6H_4$	Et_2O saturated with HCl, 3 days at 0°C, then reflux 3 hr in MeC_6H_5	73	424
p-$O_2NC_6H_4$	Et_2O saturated with HCl, 3 days at 0°C then reflux 3 hr in MeC_6H_5	52	424
m-$O_2NC_6H_4$	Et_2O saturated with HCl, 3 days at 0°C, then reflux 3 hr in MeC_6H_5	60	424
p-$CF_3C_6H_4$	Et_2O saturated with HCl, 3 days at 0°C, then reflux 3 hr in MeC_6H_5	78	424
Ph	$(ClCH_2CO)_2S$ or $ClCH_2COSH$	—	542

Ethylcyanoacetate (**227**), $R_1 = CH_2CO_2Et$, also reacts with ethylthioglycolate to afford the corresponding ethyl-2-thiazolylacetate (**230**), $R_1 = CH_2CO_2Et$, $R_2 = H$, after cyclization of the acyclic intermediate in the presence of sodium acetate (542).

The action of thioamides (**1**) on α-haloacids or esters gives the same products (**230**) in moderate yields (20 to 30%) through intermediates analogous to **229**. In **230**, R_1 is phenyl, benzyl, or β-pyridyl, R_2 is methyl or hydrogen (287, 324).

α-Cyano-α-acetylthioacetamide reacts with ethyl-α-bromoacetate to give the corresponding 4-hydroxy-2-(α-cyano-α-acetyl)methylthiazole (804).

In some cases, a product of self-condensation of **230** was obtained (424). For example, refluxing chloroacetic acid with thiobenzamide (**1**), R_1 = Ph, gave a mixture of (29%) (**230**), R_1 = Ph, R_2 = H, and (12%) (**231**), R_1 = Ph. With a slight excess of thiobenzamide **230**, R_1 = phenyl, R_2 = H, is obtained in 55.7% yield (424). This latter refluxed in toluene for 3 hr gave **231** in 84% yield (Scheme 120a) (424).

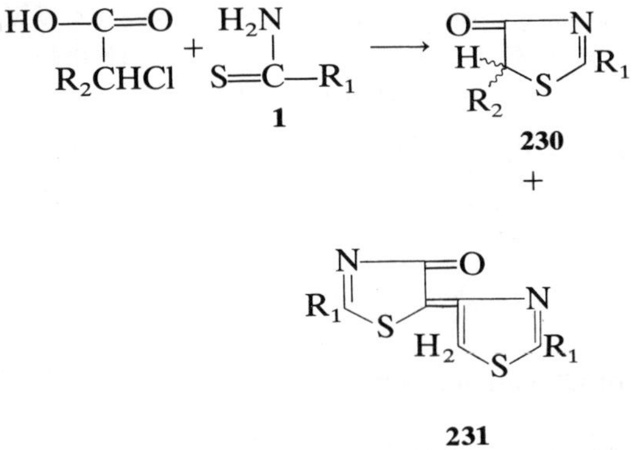

VI. Thiazoles from Nitriles and α-Mercaptoketones or Acids

$$\begin{array}{c}\text{HO—C=O} \quad \text{H}_2\text{N} \\ | \quad\quad + \quad | \\ \text{R}_2\text{CHCl} \quad \text{S=C—R}_1 \\ \textbf{1}\end{array} \longrightarrow \begin{array}{c}\text{O=}\!\!\diagup\!\!\text{—N} \\ \text{H}\diagdown\!\!\diagdown\!\!\diagup\text{R}_1 \\ \text{R}_2 \quad \text{S} \\ \textbf{230}\end{array}$$

+

$$\begin{array}{c}\text{N———O} \\ \text{R}_1\diagdown\!\!\diagup\quad\diagdown\!\!—\text{N} \\ \quad\text{S}\quad\text{H}_2\diagdown\!\!\diagup\text{R}_1 \\ \quad\quad\quad\text{S} \\ \textbf{231}\end{array}$$

Scheme 120a

In a similar reaction, bromosuccinic acid and thiobenzamide in ethylacetate yielded an acyclic intermediate (**229**), $R_1 = Ph$, $R_2 = CH_2CO_2H$, which by heating in water cyclizes to the thiazole (**230**), $R_1 = Ph$ and $R_2 = CH_2CO_2H$. (260).

Oxiranes condensed in dioxane solution at room temperature with thioamides led to 4-hydroxy-5-arylthiazoles in 36 to 55% yield (804), $R_1 = H$, Me, Ph and $R_2 = H$, Cl, MeO, NO_2 (Scheme 120b).

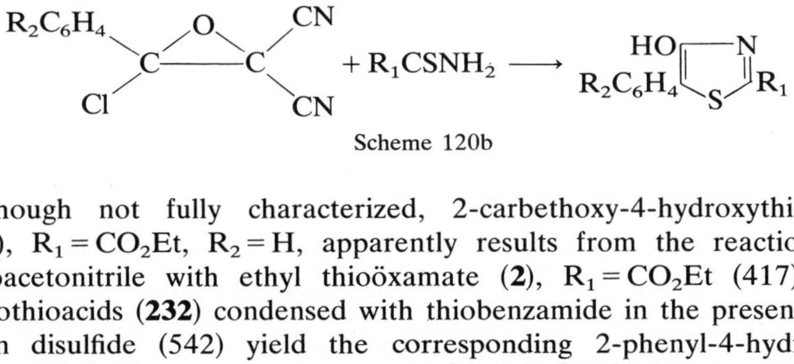

$$\begin{array}{c}\text{R}_2\text{C}_6\text{H}_4\diagdown\quad\text{O}\quad\diagup\text{CN} \\ \quad\quad\text{C——C} \\ \diagup\quad\quad\quad\diagdown\text{CN} \\ \text{Cl}\end{array} + \text{R}_1\text{CSNH}_2 \longrightarrow \begin{array}{c}\text{HO——N} \\ \text{R}_2\text{C}_6\text{H}_4\diagdown\!\!\diagup\text{R}_1 \\ \quad\quad\text{S}\end{array}$$

Scheme 120b

Although not fully characterized, 2-carbethoxy-4-hydroxythiazole (**230a**), $R_1 = CO_2Et$, $R_2 = H$, apparently results from the reaction of chloroacetonitrile with ethyl thioöxamate (**2**), $R_1 = CO_2Et$ (417). α-Chlorothioacids (**232**) condensed with thiobenzamide in the presence of carbon disulfide (542) yield the corresponding 2-phenyl-4-hydroxythiazole (**234**). The same product was obtained from **233** (Scheme 121).

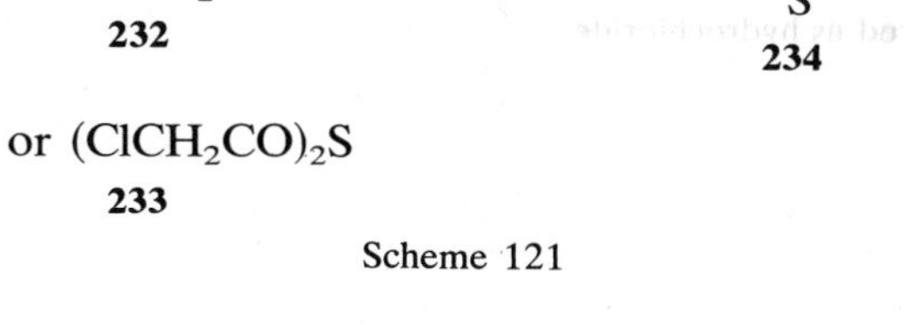

$$\begin{array}{c}\text{HS—CO} \quad \text{H}_2\text{N} \\ | \quad\quad + \quad | \\ \text{CH}_2\text{Cl} \quad \text{S=C—Ph} \\ \textbf{232}\end{array} \xrightarrow{\text{CS}_2} \begin{array}{c}\text{HO——N} \\ \diagdown\!\!\diagup\text{Ph} \\ \text{S} \\ \textbf{234}\end{array}$$

or $(ClCH_2CO)_2S$
233

Scheme 121

It has also been found that p-nitrophenylthiourea (**235**) reacts with chloroacetic acid by boiling in alcohol for 2 hr to afford 2-p-nitrophenylimino-4-thiazolidone and its derivatives (**236**) (Scheme 122) (434).

$$ClCH_2CO_2H + p\text{-}O_2NC_6H_4NHCSNH_2 \xrightarrow{\text{alc.}} \underset{\textbf{236}}{\text{O=}\underset{S}{\bigsqcup}\text{NH}\,=\!NC_6H_4NO_2\text{-}p}$$
$$\textbf{235}$$

Scheme 122

3. 2,4-Diaminothiazoles from α-Halonitriles and Thiourea

α-Halonitrile (**237**) can replace α-halogenocarbonyl compounds in the Hantzsch's synthesis (338, 418, 544). Thus the reaction of thiourea with an α-halonitrile in boiling alcohol gives 2,4-diaminothiazole hydrochloride (**239**), R = alkyl or aryl according to Scheme 123 (Table II-40).

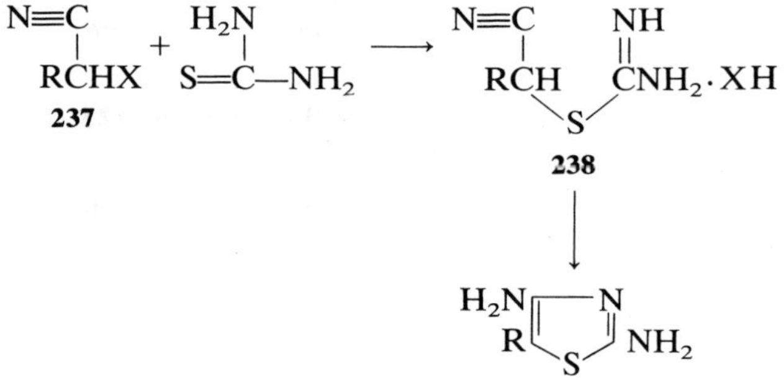

$$\underset{\textbf{237}}{\underset{RCHX}{N\!\equiv\!C}} + \underset{S=C-NH_2}{H_2N} \longrightarrow \underset{\textbf{238}}{\underset{RCH\diagdown_S\diagup CNH_2 \cdot XH}{N\!\equiv\!C \quad NH}}$$

$$\downarrow$$

$$\underset{\textbf{239}}{\underset{R}{H_2N}\!\!\diagdown\!\!\underset{S}{\bigsqcup}\!\!\diagup\!NH_2}$$

Scheme 123

TABLE II-40. 2,4-DIAMINOTHIAZOLE DERIVATIVES FROM THIOUREA AND α-HALOGENONITRILES

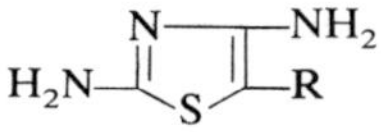

$$H_2N\!-\!\!\underset{S}{\bigsqcup}\!-\!R \text{ with } NH_2 \text{ at 4-position}$$

R	Conditions	Yield (%)[a]	Ref.
H	Cl, alcohol, heat 4 hr	85–95 (HCl)	217, 338
Me	Cl, alcohol, 10 weeks at 40°C	51 (HCl)	338
Ph	Cl. or Br, Me$_2$CO, alcohol 3 to 6 days at room temperature or 6 hr at 100°C	80–83 (HCl or HBr)	338
o-ClC$_6$H$_4$	o-ClC$_6$H$_4$CHO, PhSO$_2$Cl, NaCN	(as benzene sulfonate)	447
p-ClC$_6$H$_4$	Br, alcohol, heat	60 (HCl)	418
p-FC$_6$H$_4$	p-FC$_6$H$_4$CHO, PhSO$_2$Cl, NaCN	(as benzene sulfonate)	447

[a] Of products isolated as hydrochloride or benzene sulfonate.

The reaction probably proceeds through the *S*-cyanomethyl isothiouronium (**238**) salt, which can be isolated when the reaction is carried out in cold acetone. Cyclization takes place upon heating.

The 5-aryl derivatives have been prepared in a similar way from substituted phenylbromoacetonitrile (**239**), R = phenyl, X = Br.

2,6-Diamino-4,4-dimethyl-1,3,5-thiadiazine hydrobromide was isolated as by-product (418). Benzene sulfonates of cyanohydrin prepared from sodium cyanide and an halobenzoaldehyde, when treated with thiourea or its derivatives, afford 2,4-diamino-5-(*p*-halogenophenyl)-thiazole benzene sulfonates (447). Similarly, cyanoamido thiocarbamates obtained from cyanamide and isothiocyanates yield substituted 2,4-diaminothiazoles (598).

VII. MISCELLANEOUS REACTIONS

A few thiazole syntheses that have not been discussed in preceding sections are included here.

1. 2-Aminothiazole Derivatives

2-Aminothiazole derivatives (**243**) can be prepared by treatment of enamines of type **240** with sulfur and cyanamide at room temperature in ethanol (701); yields range from 30 to 70%, and no catalyst is required. Initial formation of the thiolated intermediate (**241**) is probably followed by addition of cyanamide, yielding **242** (Scheme 124).

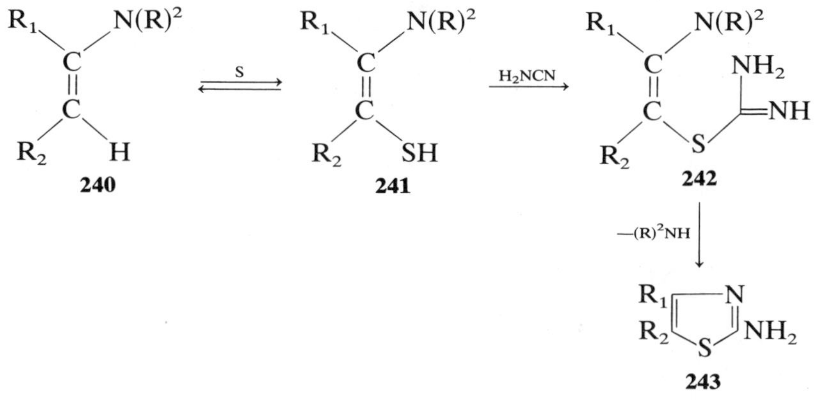

Scheme 124

2-Aminothiazoles (**243**), R_1 = Me, Et, Pr, Ph, or PhCH$_2$ and R_2 = Me, Et, iso-Pr, Ph, or CO$_2$Et, can also be prepared by condensing aminothiocyanogen (**244**) with a variety of ketones in ether (Scheme 125a) (608, 690). An enamine intermediate (**245a**), R_1 = Me and R_2 = CO$_2$Et, was isolated.

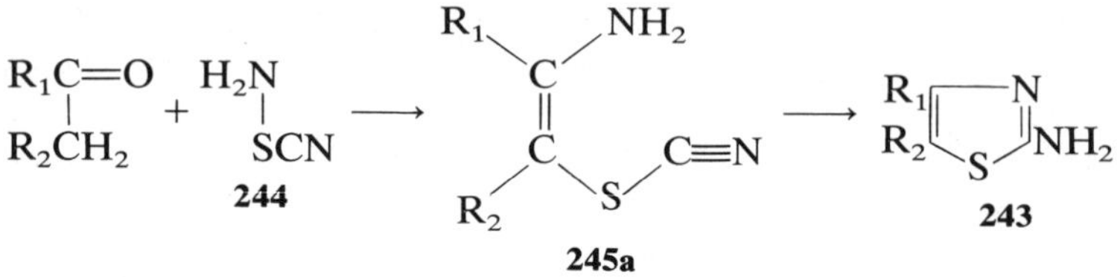

Scheme 125a

If the condensation is done with β-aminocrotonic ester or (2-aminopent-2-en-4-one)enamine, intermediates **245b** are also obtained; then they are cyclized either to 2-aminothiazoles (**243b**) under the influence of alkalis or to Δ-4-thiazol-2-ones by acids (Scheme 125b) (728).

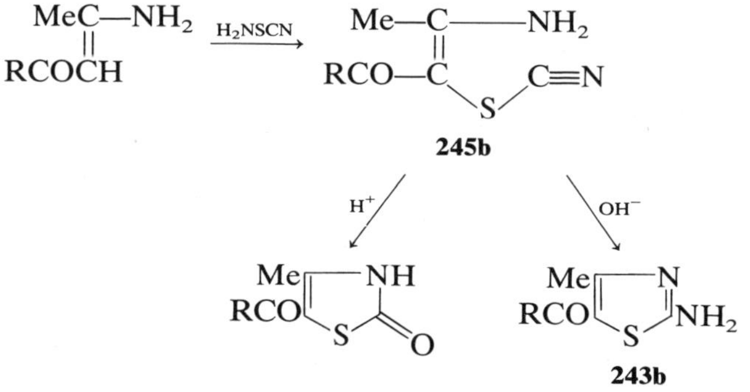

Scheme 125b

N-Methyl-Δ-4-thiazolines (**247**) are similarly accessible from N-methylaminothiocyanogen (**246**) (Scheme 126).

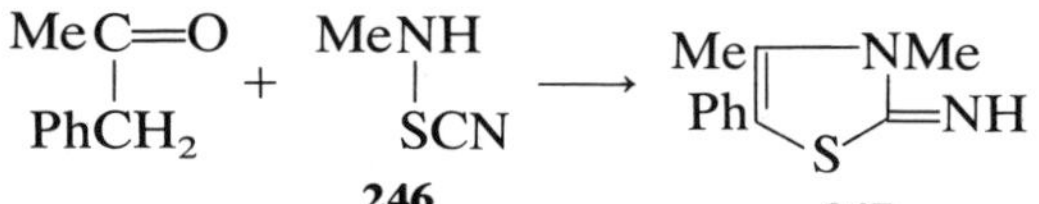

Scheme 126

VII. Miscellaneous Reactions

Another approach to 2-aminothiazole derivatives was recently developed by Zbiral and Hengstberger (667, 700); thus the condensation of β-acylvinylphosphonium salts (**248**) with thiourea affords the thiazolylmethylphosphonium salt (**249**) via an acyclic intermediate analogous to the Hantzsch's synthesis. Final alkaline hydrolysis of **249** furnishes the 2-aminothiazoles (**250**) (Scheme 127) (700).

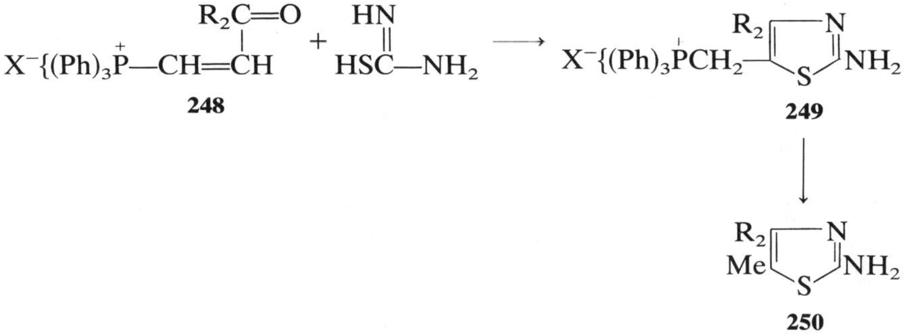

$$X^-\{(Ph)_3\overset{+}{P}-CH=CH\underset{248}{|}\overset{R_2C=O}{} + \underset{HSC-NH_2}{\overset{HN}{\|}} \longrightarrow X^-\{(Ph)_3\overset{+}{P}CH_2\underset{249}{\overset{R_2\diagup\!\!\!\!-N}{\diagdown_S\diagdown NH_2}}$$

$$\underset{250}{\overset{R_2\diagup\!\!\!\!-N}{Me\diagdown_S\diagdown NH_2}}$$

Scheme 127

In a similar way the use of the 2-methyl-3-isothiocyanato-4-thiocyanato-hept-3-ene (**252**) prepared from thiocyanogen and the oxoalkylene phosphorane (**251**) yields the 2-anilino-4-propyl-5-isopropylthiazole (**253**) by condensation with aniline (Scheme 128).

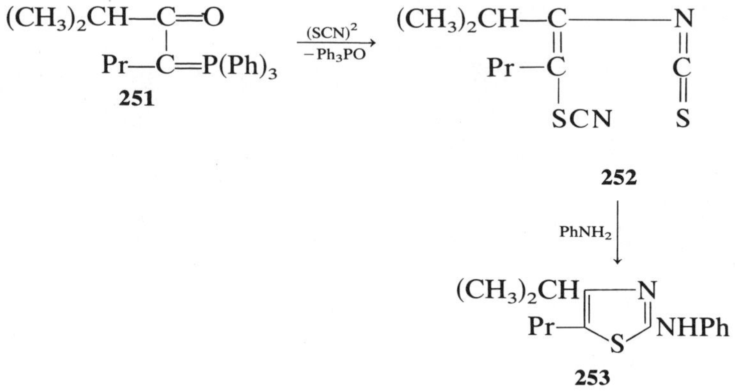

Scheme 128

The action of ammonia on *N*-(aryl-1,3-oxathiol-2-ylidine) tertiary iminium salts (**254**) yields linear intermediates (**255**) that cyclize to 2-amino-4-phenyl thiazoles (**256**) on crystallization from acetic acid (Scheme 129) (730).

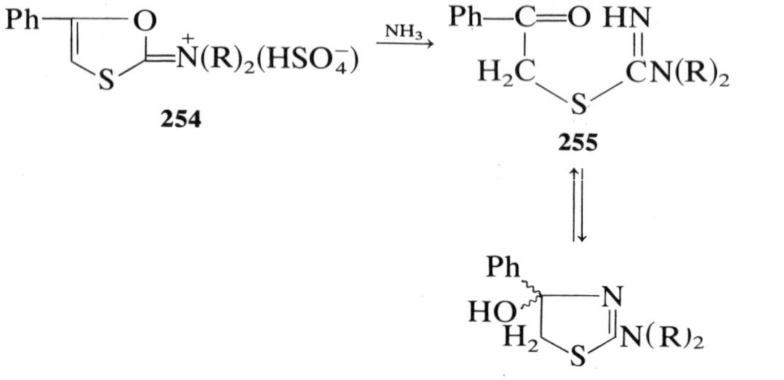

N(R)$_2$ = Piperidino, Morpholino, or dimethylamino

Scheme 129

Isothiocyanates (R$_1$NCS) react with 1,4-diamino-2-butynes to give 2-amino-5-β-aminoethylidene-Δ-2-thiazolines, which can be isomerized into 2-amino-5-β-aminoethylthiazoles with R$_1$, R$_2$, R$_3$ = alkyl (Scheme 130) (789).

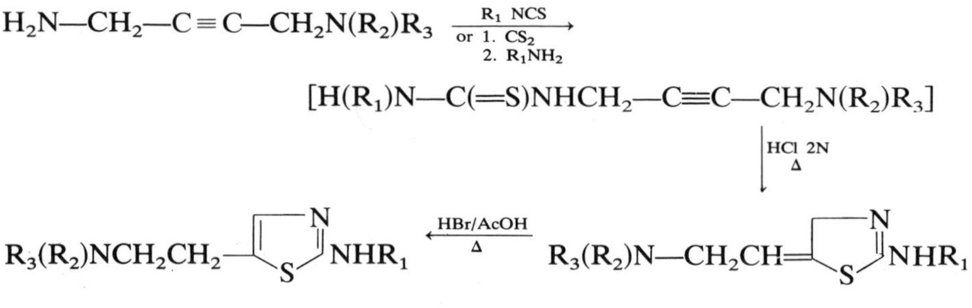

$$H_2N-CH_2-C\equiv C-CH_2N(R_2)R_3 \xrightarrow[\text{2. } R_1NH_2]{\text{R}_1 \text{ NCS or 1. CS}_2}$$

$$[H(R_1)N-C(=S)NHCH_2-C\equiv C-CH_2N(R_2)R_3]$$

$$\downarrow \text{HCl 2N, } \Delta$$

R$_3$(R$_2$)NCH$_2$CH$_2$–[thiazole]–NHR$_1$ ←(HBr/AcOH, Δ)— R$_3$(R$_2$)N—CH$_2$CH=[thiazoline]–NHR$_1$

Scheme 130

These compounds are obtained by reaction of carbon disulfide on diamines followed by condensation of ammonia or primary or secondary amines on 1-isothiocyanato-4-amino-2-butynes.

2-N-Substituted aminothiazoles, with R = H, Me, CHMe$_2$, R$_1$ = H, Me, R$_2$ = H, Me, CHMe$_2$, Ph, and R$_1$, R$_2$ = CH$_2$CH$_2$, have been recently obtained by condensing thioureas with 1,2-dichloroethylisothiocyanate (Scheme 131) (805).

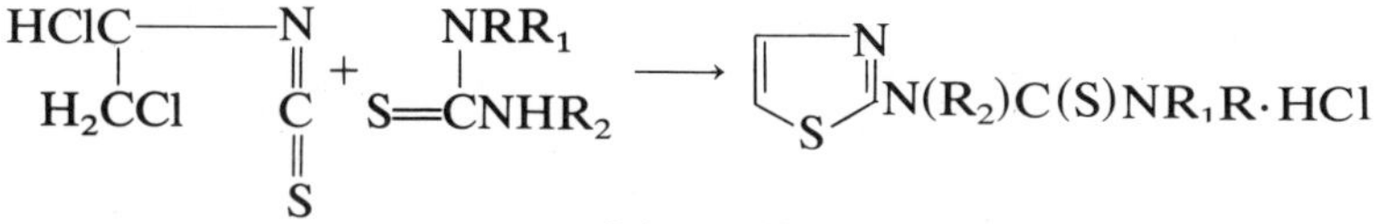

Scheme 131

2. 4-Aminothiazole Derivatives

Very few 4-aminothiazoles have been synthetized directly. The reaction of α-halonitriles with thioamides generally fails and only extensive decomposition results. However, the benzene sulfonic ester of mandelonitrile reacts with thiobenzamide to give 2,5-diphenyl-4-aminothiazole (**257**), R$_1$ = R$_2$ = Ph, in 37% yield (Scheme 132) (417) Similarly, α-cyano-α-acetylthioacetamide condensed with α-chloroacetonitrile give **257**, R$_1$ = CH(CN)CH$_3$ and R$_2$ = H (804).

On the other hand, 4-amino-5-carbethoxy-2-methylthiothiazole (**260**) was obtained by condensing methyl-N-cyanoiminodithiocarbamic ester (**259**) with ethyl-α-mercaptoacetate (**258**) in foramidine in the presence of triethylamine (Scheme 133) (564).

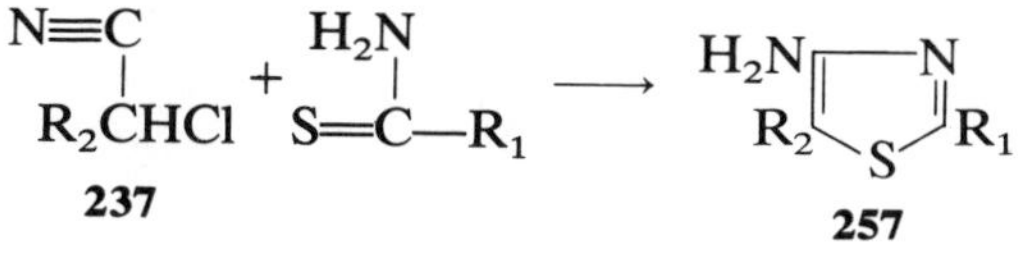

Scheme 132

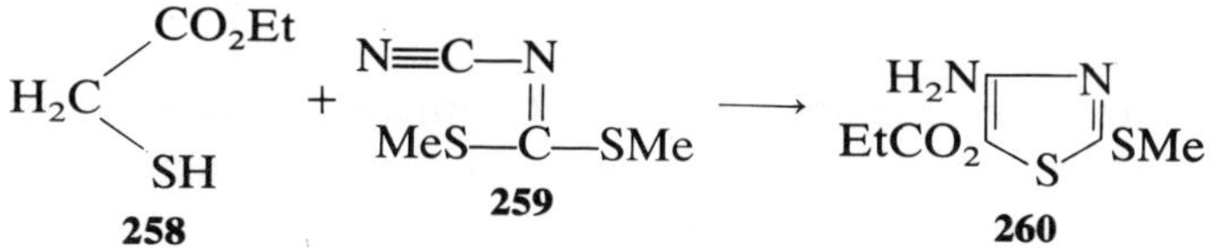

Scheme 133

A similar reaction from disulfide (**261**) and α-chloroalkanoic acids is reported by Dahlbom (Scheme 134) (392).

$$(NCN=CHS-)_2 + RCHClCO_2H \longrightarrow \underset{\textbf{262}}{\underset{S}{H_2N}\!\!\!\diagdown\!\!\!\diagup\!\!\!N\!\!-\!\!R}$$

261

Scheme 134

Several 4-amino-2,5-disubstituted thiazoles (**257**) have been obtained recently (702, 756, 776, 814, 820) by a ring cyclization reaction of halogeno compounds with cyanamide derivatives (**263**) according to the general Scheme 135.

R_1	R_2	Ref.
Ph	Ph	417
CH(CN)CH$_3$	H	804
Me	C(=S)OEt	776
Me	CN	702
SMe	alkyl	756
SMe	CH$_3$CO	814
SR	ClCH$_2$CO	820

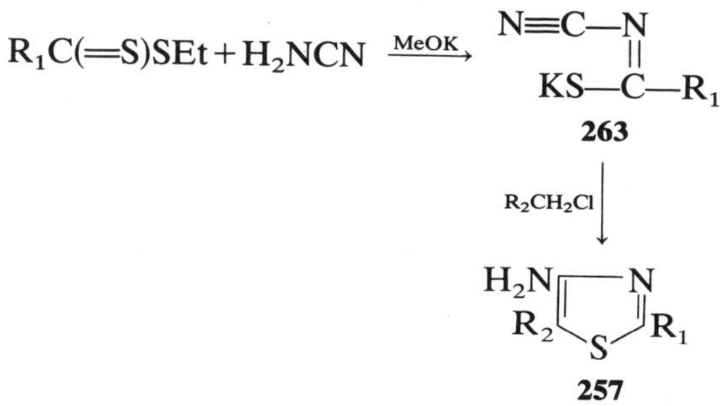

$$R_1C(=S)SEt + H_2NCN \xrightarrow{MeOK} \underset{\textbf{263}}{\overset{N\equiv C-N}{\underset{KS-C-R_1}{\|}}}$$

$\downarrow R_2CH_2Cl$

$$\underset{\textbf{257}}{\underset{S}{H_2N}\!\!\!\diagdown\!\!\!\diagup\!\!\!N}\!\!\!\underset{R_2}{}\!\!\!\underset{S}{}\!\!\!R_1$$

Scheme 135

For example, the condensation of the O-ethyl-α-chlorothioacetate, $R_2 =$ C(=S)OEt, with the potassium salt of thioacetylcyanamide (**263**), $R_1 =$ Me, produces the substituted 4-aminothiazole (**257**), in 79% yield (776). 2-Substituted 4-amino-5-cyanothiazoles (**257**), $R_1 =$ Me and $R_2 =$ CN, have been similarly obtained.

VII. Miscellaneous Reactions 303

Reaction of 1,3-oxathiolium salts (**264**) with cyanamide in the presence of sodium ethoxide produces also substituted 4-aminothiazoles (**265**) (Scheme 136) (777).

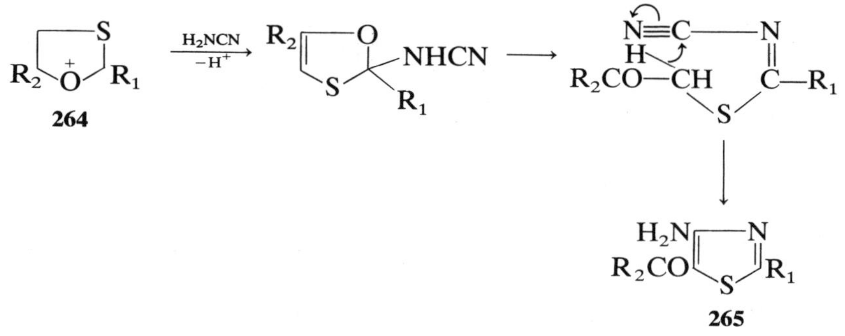

Scheme 136

3. 5-Hydroxythiazole Derivatives

2-Substituted 5-hydroxythiazoles (**267b**), R_1 = alkylmercapto, acylamino, and *sec*-amino, are prepared by cyclization of N-thioacylamino acids (**266**) with phosphorus tribromide or acetic anhydride (Scheme 137) (317, 350). When the cyclization of **266**, R_2 = H, is carried out with acetic anhydride in the presence of benzaldehyde (317, 325) or ethylformate (317), the benzylidene (**268**), R_2 = Ph, R_1 = SR or CH_2Ph, or 4-ethoxymethylene (**268**), R_1 = SR and R_2 = OEt, derivative is obtained directly (Scheme 138).

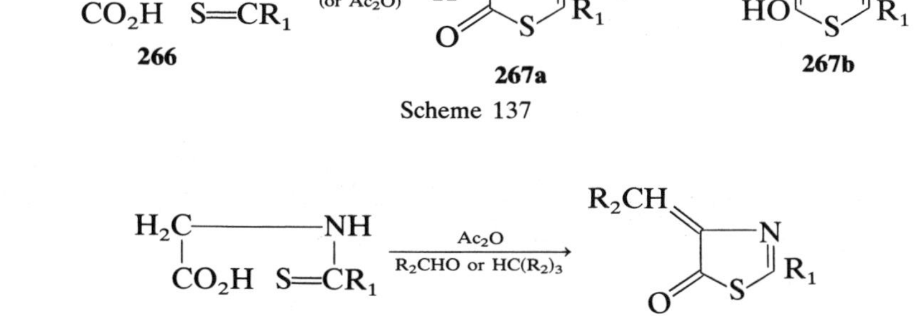

Scheme 137

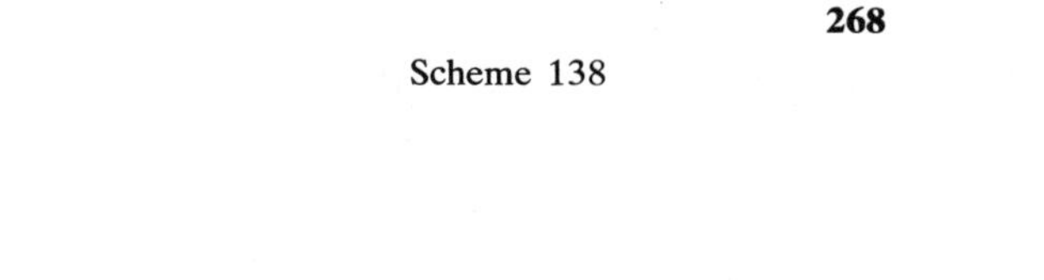

Scheme 138

The same products can be also obtained from **267** and benzaldehyde. This behavior indicates the presence of an active methylene group and supports the thiazolone structure (**267a**). Alkyl or aryl ethers of **267** are prepared by two different procedures (Scheme 139).

$$R_2O \underset{S}{\overset{N}{\diagdown}} R_1$$
269

Scheme 139

Alkylethers (**269**), R_2 = alkyl, are obtained by the action of phosphorus pentasulfide on alkyl esters of α-acylamino acids (64, 334, 711) by means of the Gabriel's synthesis (Section II.4), while aryl ethers (**269**), R_2 = aryl, are formed in the Hantzsch's synthesis from thioamides and ω-bromo-ω-aryloxyacetophenones (376).

4. 4,5-Dihalogenothiazoles and 2,4,5-Trihalogenothiazoles

An unusual reaction has been reported by Reynaud et al. (515) in which all the atoms are in the starting molecule.

4,5-Dihalogenothiazoles (**271**) thus can be prepared from sodium acetylaminomethane sulfonate (**270**) and thionyl halide, but this reaction proceeds in low yield. With X = Cl, the yield is 20%, while with X = Br the yield decreases to 2% (Scheme 140).

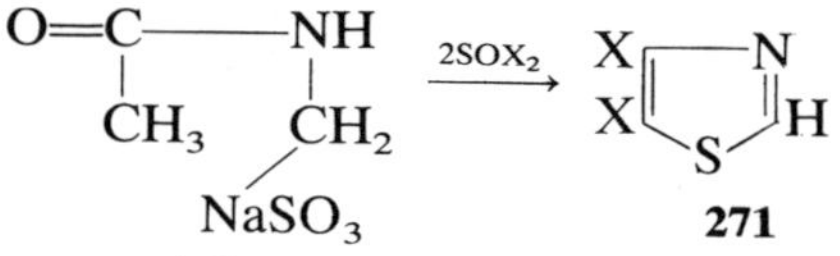

$$\underset{\underset{270}{NaSO_3}}{\overset{O=C-NH}{\underset{CH_3 \quad CH_2}{|\quad\quad|}}} \xrightarrow{2SOX_2} \underset{271}{X \underset{S}{\overset{N}{\diagdown}} H}$$

Scheme 140

2,4,5-Trichlorothiazole was obtained in 73.5% yield by condensing $ClCH_2CH(Cl)N=CCl_2$ with S_2Cl_2 or SCl_2 (Scheme 141) (822).

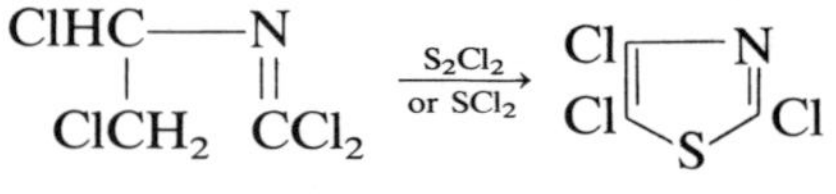

$$\underset{ClCH_2 \quad CCl_2}{\overset{ClHC-N}{\underset{|\quad\quad||}{}}} \xrightarrow[\text{or } SCl_2]{S_2Cl_2} \underset{Cl}{\overset{Cl}{\diagdown}} \underset{S}{\overset{N}{\diagdown}} Cl$$

Scheme 141

5. 4-Tosylthiazoles

Recently Oldenziel and Van Leusen (750) have reported a new synthesis of 4-tosylthiazoles (**275**), R = p-MeOC$_6$H$_4$, p-MeC$_6$H$_4$, Ph, p-ClC$_6$H$_4$, and 2-furyl, from tosylisocyanide (**272**) and carboxymethyldithioates (**273**) (Scheme 142); yields ranged from 50 to 80%.

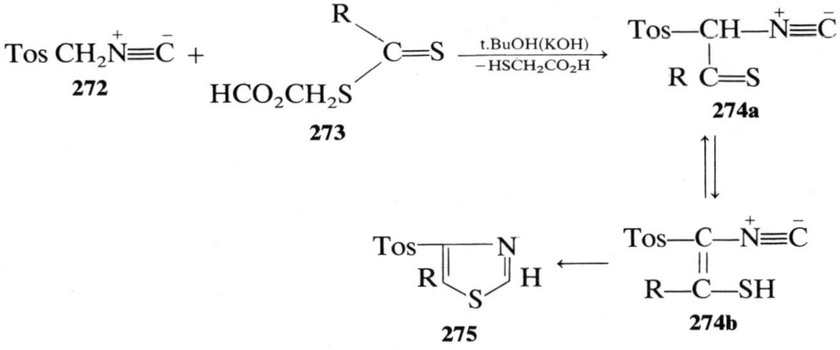

Scheme 142

6. Alkylthiazoles

An interesting new synthesis of 2,4,5-trisubstituted and 2,4-disubstituted alkylthiazoles has been reported by Dubs (807). Heating the easily obtainable derivatives (**276**) with ammonium acetate in acetic acid gives thiazoles (**10**), R_1, R_2 = alkyl and R_3 = H or alkyl in good yields (Scheme 143).

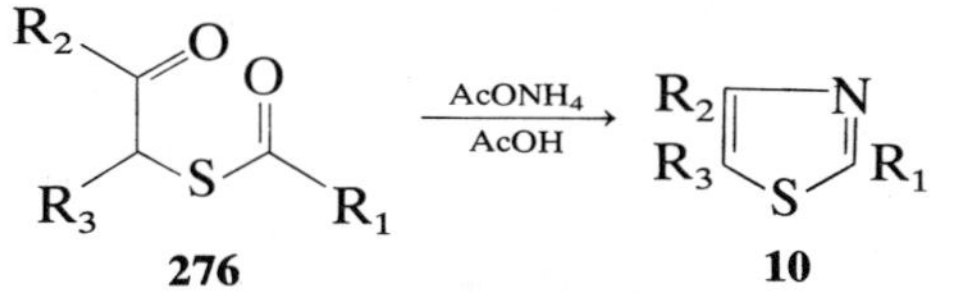

Scheme 143

2-Methyl and 2,4-dimethylthiazole were prepared by the vapor-phase reaction (450 to 500°C) of sulfur with diethylamine and diisopropylamine, respectively (816); yields were 66 and 59%.

7. Cyanothiazoles

A new synthesis of 4-cyanothiazole derivatives via α-metallated isocyanides and thionoesters have been reported by Hartman and Weinstock (Scheme 144) (812), with R_2 = H, Me, n-C_3H_7, Ph, $PhCH_2$. 4-Cyanothiazole (in 50% yield) was also formed by condensing thioformamide with $Cl_2C=C(NH_2)CN$ (803).

4-Hydroxy-5-cyanothiazoles (817) were prepared by cyclizing EtO_2-CNHCSR (R = OEt, OBu, SEt, SCH_2Ar, OCH_2Ar) with α-chloroacetonitrile (Scheme 145).

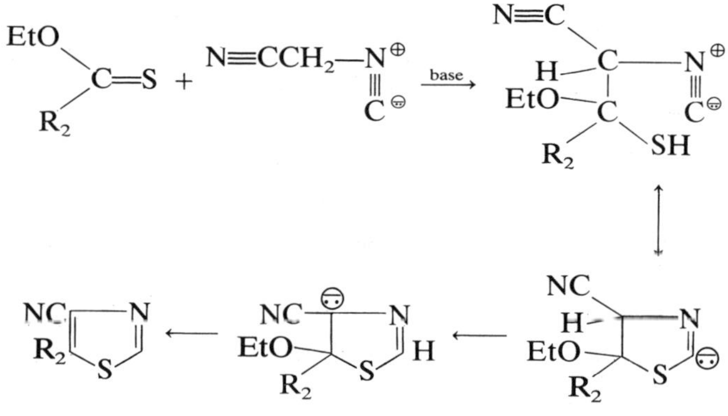

Scheme 144

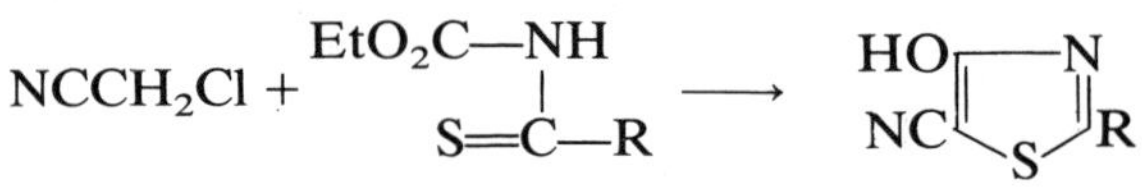

Scheme 145

8. Miscellaneous

The condensation of thioacetic acid with amino acids under drastic conditions provides a useful new synthesis of thiazoles (Scheme 146) (668, 669). Instead of the amino acid, N-acyl (**279**) or N-thioacylamino acids (**278**) are used.

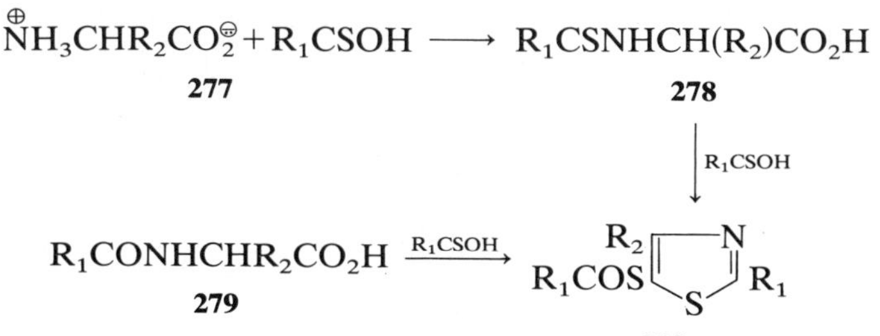

$\overset{\oplus}{N}H_3CHR_2CO_2^\ominus + R_1CSOH \longrightarrow R_1CSNHCH(R_2)CO_2H$
277   **278**

$R_1CONHCHR_2CO_2H \xrightarrow{R_1CSOH}$ [thiazole with R_1COS, R_2, R_1]
279   **280**

Scheme 146

Ethylisocyanoacetate with carbon disulfide leads to the 1,3-thiazole in 49% yield (Scheme 147) (778).

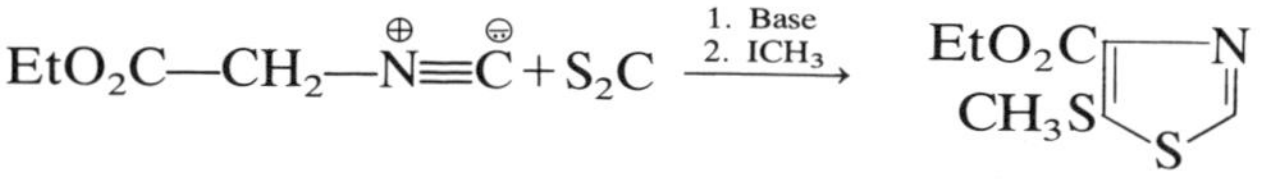

$EtO_2C-CH_2-\overset{\oplus}{N}\equiv\overset{\ominus}{C} + S_2C \xrightarrow[\text{2. ICH}_3]{\text{1. Base}}$ [thiazole: EtO_2C, CH_3S]

Scheme 147

Cycloadditions of benzonitril-4-nitrobenzylide to a variety of carbon to heteroatom multiple bonds including methyl dithiobenzoate and dimethyl trithiocarbonate (Scheme 148), X = MeS or Ph, have been examined in detail by Huisgen et al. (757).

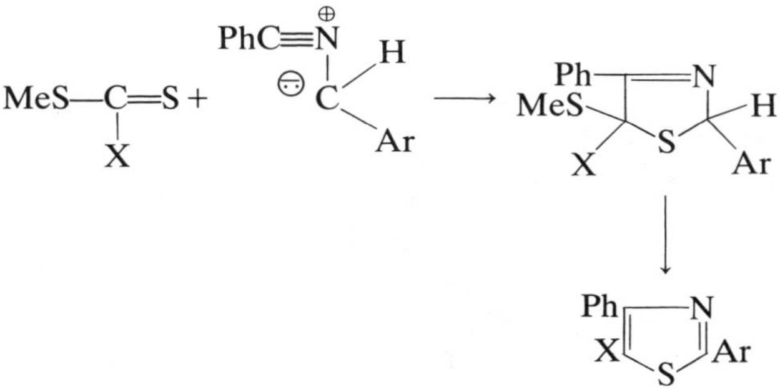

[Scheme 148 reaction: MeS—C(=S)—X + PhC≡N⁺—C⁻(H)(Ar) → thiazoline (Ph, MeS, X, H, Ar) → thiazole (Ph, X, Ar)]

Scheme 148

Variously trisubstituted thiazoles were prepared by Ried and Kaiser by condensing PhNHCPh=NCSR with BrCH$_2$R$_1$ (Scheme 149), R = OEt, morpholino and R$_1$ = 4-NO$_2$C$_6$H$_4$, Ac, Bz, CO$_2$Et, CN, 4-PhC$_6$H$_4$CO, NO$_2$, 4-NO$_2$C$_6$H$_4$CO, 4-BrC$_6$H$_4$CO, COCH$_2$Br; yields ranged from 45 to 92% (811).

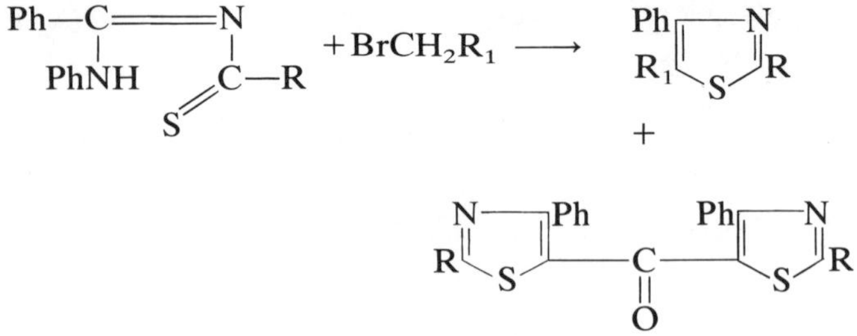

Scheme 149

VIII. THIAZOLES FROM OTHER HETEROCYCLIC COMPOUNDS

Closing this chapter on thiazole synthetic methods, I would like to point out that it is possible to obtain thiazoles from other heterocyclic compounds.

1. Thiazoles from Δ-3-Thiazolines

During their researches on thiazolines, Asinger and Thiel (437, 452, 518) showed that Δ-3-thiazolines (**281**) can be dehydrogenated to thiazoles in good yields (Scheme 150).

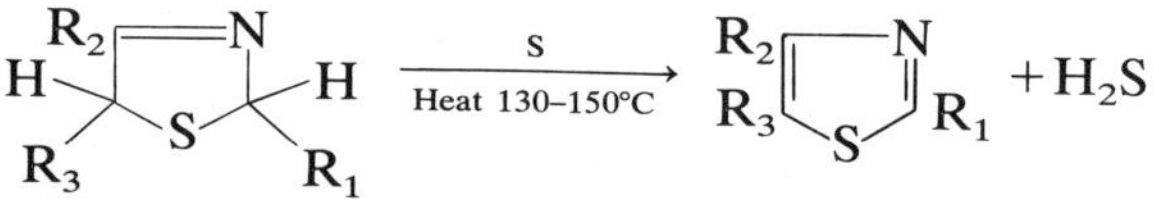

Scheme 150

A satisfactory dehydrogenation agent is sulfur. The reaction is usually complete in 3 hr at 130 to 150°C. Ferric chloride in absolute alcohol at 60 to 70°C and potassium ferrocyanide have been also used as deshydrogenating agents (453). Δ-3-Thiazolines are themselves produced by bubbling a current of ammonia through a benzenic or alcoholic solution of α-mercaptoketones or aldehydes. Thus these authors obtained the following compounds **10** from the corresponding Δ-3-thiazolines: R_1 = Me, Et, or Ph and R_2 = R_3 = H, 55% yield; R_1 = Ph, R_2 = H and R_3 = Ph, 75% yield; R_1 = R_2 = Me, R_3 = CH$_2$Ph, 50% yield; R_1 = Et, R_2 = tert-Bu and R_3 = H, 65% yield; R_1 = H, R_2 = Me, R_3 = Ph, 90% yield (453).

Another reaction has been reported in which action of sulfur and ammonia on acetophenone affords 2,5-diphenylthiazole (511).

An efficient synthesis of 2-substituted thiazoles from 2,5-dihydrothiazoles was reported recently (784). Δ-3-Thiazolines were obtained from the mercaptoacetaldehyde dimer, 2,5-dihydroxy-1,4-dithione (**282**), ammonia, and aldehydes according to Asinger's method (Scheme 151) (437). These are easily transformed to the corresponding thiazoles in good yields refluxing them in benzene 30 to 60 min, in the presence of an equimolar amount of quinones such as chloranil. With R = iso-Pr, iso-Bu, n-hexyl, CH_2CH_2SMe, or $CH_2CH_2CH=C(Me)_2$, yields ranged from 67 to 81% for **3**.

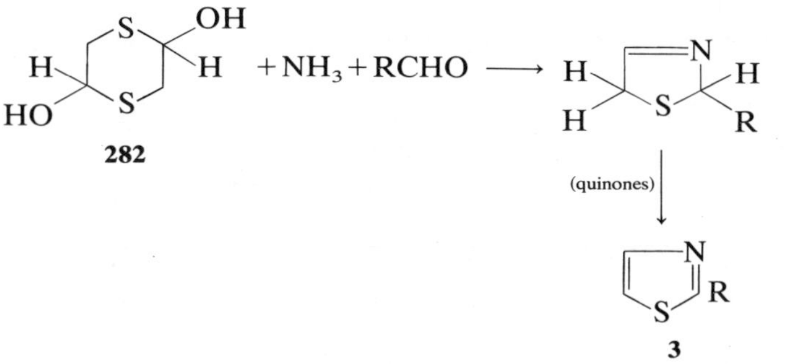

Scheme 151

2. Thiazoles from Oxazoles

Oxazoles (**283**) heated with phosphorus pentasulfide can be converted to thiazoles. In this manner, Keyer and Brooker obtained 2-methyl-5-phenylthiazoles (**10**), R_1 = Me, R_2 = H, R_3 = Ph in 85% yield (Scheme 152) (389).

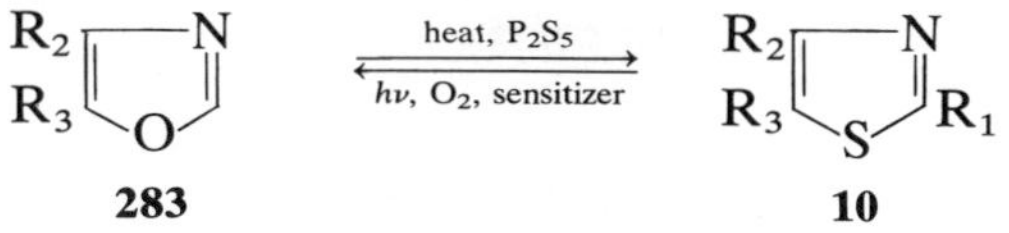

Scheme 152

With short periods of irradiation (with high-pressure mercury lamps) under oxygen in chloroform containing methylene blue as a sensitizer, variously substituted 2-arylthiazoles are converted in the corresponding 2-aryloxazoles (823).

3. Thiazoles from Isothiazoles

It has been recently found that upon irradiation isothiazoles can be converted to thiazole and isothiazole isomers among other products (Scheme 153).

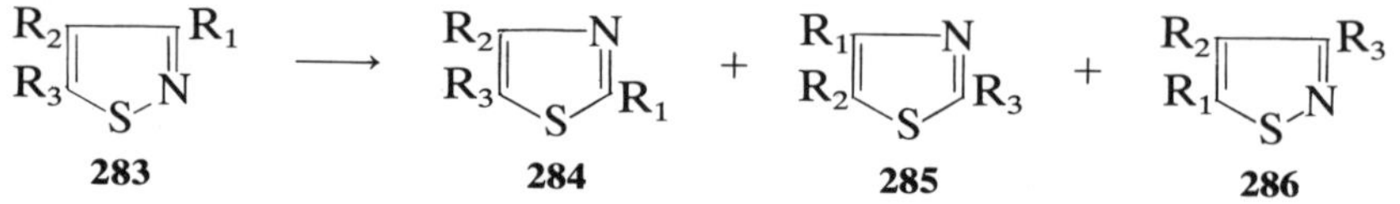

$$\underset{\textbf{283}}{R_3 \diagdown_S \diagup^N \diagdown R_1^{R_2}} \longrightarrow \underset{\textbf{284}}{R_3 \diagdown_S \diagup^N \diagdown R_1^{R_2}} + \underset{\textbf{285}}{R_2 \diagdown_S \diagup^N \diagdown R_3^{R_1}} + \underset{\textbf{286}}{R_1 \diagdown_S \diagdown^N \diagup R_2^{R_3}}$$

Scheme 153

Isothiazole itself (**283**), $R_1 = R_2 = R_3 = H$, is converted to thiazole in 7% yield, in propylamine as solvent using a low-pressure mercury lamp (642).

When R_1, R_2, and R_3 are methyl or alkyl groups, yields are low, while the yields with phenyl groups are higher. Thus 3-phenylisothiazole (**283**), $R_1 = Ph$, $R_2 = R_3 = H$, gave 4-phenylthiazole (**285**), $R_1 = Ph$, $R_2 = R_3 = H$ and diphenyl-3,5-isothiazole gave 2,4-diphenylthiazole (**285**), $R_1 = R_3 = Ph$, $R_2 = H$, in 12 and 48% yields, respectively (684). But these reactions are of little interest given the difficulty in preparing isothiazoles themselves.

On the other hand, 2-arylthiazoles are easily isomerized to 3-arylisothiazoles in 40% yield upon irradiation with a high-pressure mercury lamp, in benzene solution in the presence of iodine (738). A valence bond isomerization was proposed among several alternatives to account for these results.

A computer assisted approach of heterocyclic structures (825, 826) predicted a new method of synthesis for the thiazole (826). This method was reported soon after by P. Dubs (807). The same program proposed two other methods not yet experimentally checked (827).

IX. REFERENCES

1. C. Liebermann and A. Lange, *Berichte*, **12**, 1588 (1879).
   A. Lange, *Berichte*, **12**, 595 (1879).
2. M. Nencki and N. Sieber, *J. Prakt. Chem.*, **25**, 72 (1882).
3. J. Tcherniac and C. Norton, *Berichte*, **16**, 345 (1883).
   J. Tcherniac and R. Hellon, *Berichte*, **16**, 348 (1883).
4. A. Hantzsch and H. J. Weber, *Berichte*, **20**, 3118 (1887).
   A. Hantzsch and H. J. Weber, *Berichte*, **20**, 3122 (1887).
   A. Hantzsch and H. J. Weber, *Berichte*, **20**, 3129 (1887).

IX. References

5. A. Hantzsch and V. Traumann, *Berichte*, **21,** 938 (1888).
6. A. Hantzsch, *Berichte*, **21,** 942 (1888).
7. A. Hantzsch, *Justus Liebigs Ann. Chem.*, **249,** 1 (1888).
8. V. Traumann, *Justus Liebigs Ann. Chem.*, **249,** 31 (1888).
9. L. Arapides, *Justus Liebigs Ann. Chem.*, **249,** 27 (1888).
10. A. Hantzsch, *Justus Liebigs Ann. Chem.*, **250,** 257 (1889).
11. G. Popp, *Justus Liebigs Ann. Chem.*, **250,** 273 (1889).
12. G. Popp, *Justus Liebigs Ann. Chem.*, **250,** 943 (1889).
13. A. Zurcher, *Justus Liebigs Ann. Chem.*, **250,** 281 (1889).
14. A. Hantzsch, *Berichte*, **23,** 2339 (1890).
15. K. Hubacher, *Justus Liebigs Ann. Chem.*, **259,** 228 (1890).
16. T. Roubleff, *Justus Liebigs Ann. Chem.*, **259,** 253 (1890).
17. M. Wohmann, *Justus Liebigs Ann. Chem.*, **259,** 277 (1890).
18. P. Schatzmann, *Justus Liebigs Ann. Chem.*, **261,** 1 (1891).
19. M. Steude, *Justus Liebigs Ann. Chem.*, **261,** 22 (1891).
20. A. Miolati, *Justus Liebigs Ann. Chem.*, **262,** 82 (1891).
21. A. Hantzsch and H. Schiffer, *Berichte*, **25,** 728 (1892).
22. A. Hantzsch, *Berichte*, **25,** 3282 (1892).
23. E. Naf, *Justus Liebigs Ann. Chem.*, **265,** 108 (1892).
24. J. Tcherniac, *Berichte*, **25,** 2067 (1892).
25. G. Marchesini, *Gazz. Chim. Ital.*, **22,** 350 (1892).
26. G. Marchesini, *Gazz. Chim. Ital.*, **23,** 437 (1893).
27. A. Miolati, *Gazz. Chim. Ital.*, **23,** 575 (1893).
28. A. Hantzsch and J. Epprecht, *Justus Liebigs Ann. Chem.*, **278,** 61 (1893).
29. G. Marchesini, *Gazz. Chim. Ital.*, **24,** 65 (1894).
30. A. Schuftan, *Berichte*, **27,** 1009 (1894).
31. C. Boettinger, *Arch. Pharm.*, **232,** 349 (1894).
32. M. Conrad and L. Schmidt, *Justus Liebigs Ann. Chem.*, **285,** 203 (1895).
33. M. Conrad and L. Schmidt, *Berichte*, **29,** 1043 (1896).
34. F. Saulmann, *Chem. Ber.*, **33,** 2634 (1900).
35. J. Von Braun, *Berichte*, **35,** 3368 (1902).
36. G. Young and S. I. Crooks, *J. Chem. Soc.*, **89,** 59 (1906).
37. H. E. Favorski, *Chem. Zentralbl.*, **1,** 25 (1907).
38. R. Von. Walther, *J. Prakt. Chem.*, **75,** 187 (1907).
39. R. Willstatter and T. Wirth, *Berichte*, **42,** 1908 (1909).
40. S. Gabriel, *Berichte*, **43,** 1283 (1910).
    S. Gabriel, *Berichte*, **43,** 134 (1910).
41. A. Pictet and A. Gauss, *Chem. Ber.*, **43,** 2387 (1910).
42. C. Mannich and R. M. Hahn, *Chem. Ber.*, **44,** 1545 (1911).
43. T. B. Johnson and G. Burnham, *Amer. Chem. J.*, **47,** 232 (1912).
44. T. B. Johnson and G. Burnham, *Amer. Chem. J.*, **47,** 232 (1912).
45. R. Von Walther and H. Roch, *J. Prakt. Chem.*, **87,** 27 (1913).
46. K. Rudenburg, *Chem. Ber.*, **46,** 3555 (1913).
47. M. Bachstez, *Berichte*, **47,** 3163 (1914).
48. S. Gabriel and M. Bachstez, *Berichte*, **47,** 3169 (1914).
49. R. Willstatter, *Chem. Ber.*, **42,** 1909 (1918).
50. D. Vorlander and E. Siebert, *Chem. Ber.*, **52,** 283 (1919).
51. A. Wohl, *Berichte*, **52,** 51 (1919).
52. J. Tcherniac, *J. Chem. Soc.*, **115,** 1071 (1919); *Chem. Abstr.*, **14,** 276.
53. A. Wohl and K. Jaschinowski, *Berichte*, **54,** 476 (1921).

54. W. H. Mills and J. L. Smith, *J. Chem. Soc.*, **121,** 2724 (1922); *Chem. Abstr.*, **17,** 1024.
55. G. Bruni and E. Romani, *Atti Accad. Naz. Lincei, Rend. Classe Sci. Fis. Mat. Nat.*, **31,** 86 (1922); *Chem. Abstr.*, **16,** 4093.
56. J. L. Smith, *J. Chem. Soc.*, **123,** 2288 (1923); *Chem. Abstr.*, **18,** 264.
57. M. T. Bogert and M. Chertcoff, *J. Amer. Chem. Soc.*, **46,** 2864 (1924); *Chem. Abstr.*, **19,** 513.
58. J. L. Smith, E. H. Flack, and A. R. Inggs, *S. African J. Sci.*, **21,** 227 (1924); *Chem. Abstr.*, **19,** 1706.
59. P. K. Bose, *Quart. J. Indian Chem. Soc.*, **1,** 51 (1924); *Chem. Abstr.*, **19,** 831.
60. P. K. Bose, *Quart. J. Indian Chem. Soc.*, 95 (1925); *Chem. Abstr.*, **20,** 416.
61. E. L. Hirst, A. K. Macbeth, and D. Traill, *Proc. Roy. Irish. Acad.*, **37B,** 47 (1925); *Chem. Abstr.*, **19,** 2931.
62. R. Burtles, F. L. Pyman, and J. Roylance, *J. Chem. Soc.*, **127,** 581 (1925); *Chem. Abstr.*, **19,** 1709.
63. V. K. Nimkar and F. L. Pyman, *J. Chem. Soc.*, 2746 (1925); *Chem. Abstr.*, **20,** 415.
64. E. Miyamichi, *J. Pharm. Soc. Japan*, **528,** 103 (1926); *Chem. Abstr.*, **20,** 2679.
65. K. Kindler and E. Trev, *Annalen*, **450,** 813 (1926).
66. R. M. Hann, *J. Wash. Acad. Sci.*, 31 (1926); *Chem. Abstr.*, **20,** 1080.
    R. M. Hann and K. S. Markley, *J. Wash. Acad. Sci.*, 169 (1926); *Chem. Abstr.*, **20,** 1980.
67. P. K. Bose, *Quart. J. Indian Chem. Soc.*, **4,** 331 (1927); *Chem. Abstr.*, **22,** 1158.
68. A. Hantzsch, *Berichte*, **60B,** 2537 (1927); *Chem. Abstr.*, **22,** 1158.
69. P. K. Bose and B. K. Sen, *J. Indian Chem. Soc.*, **5,** 643 (1928); *Chem. Abstr.*, **23,** 1410.
70. A. Hantzsch and H. Schwaneberg, *Berichte*, **61B,** 1776 (1928); *Chem. Abstr.*, **23,** 101.
71. B. J. Das-Gupta and P. K. Bose, *J. Indian Chem. Soc.*, **6,** 495 (1929); *Chem. Abstr.*, **24,** 1095.
72. J. L. B. Smith and R. H. Sapiro, *Trans. Roy. Soc. S. Africa*, **18,** 229 (1929); *Chem. Abstr.*, **24,** 2130.
73. T. B. Johnson and E. Gatewood, *J. Amer. Chem. Soc.*, **51,** 1815 (1929); *Chem. Abstr.*, **23,** 3470.
74. P. K. Bose and B. K. Nandi, *J. Indian Chem. Soc.*, **7,** 733 (1930); *Chem. Abstr.*, **25,** 1532.
75. C. M. Suter and T. B. Johnson, *J. Amer. Chem. Soc.*, **52,** 1585 (1930); *Chem. Abstr.*, **24,** 2460.
76. C. M. Suter and T. B. Johnson, *Rec. Trav. Chim.*, **49,** 1066 (1930); *Chem. Abstr.*, **25,** 952.
77. H. Schwaneberg, *Dissertation Liepzig* (1930); *Chem. Abstr.*, **25,** 3664.
78. U.S. Patent No. 1 743 083; *Chem. Abstr.*, **24,** 1126.
79. W. S. Hinegardner and T. B. Johnson, *J. Amer. Chem. Soc.*, **52,** 3724 (1930); *Chem. Abstr.*, **24,** 5038.
    W. S. Hinegardner and T. B. Johnson, *J. Amer. Chem. Soc.*, **52,** 4139 (1930); *Chem. Abstr.*, **24,** 5751.
80. J. F. Olin and T. B. Johnson, *J. Amer. Chem. Soc.*, **53,** 1470 (1931); *Chem. Abstr.*, **25,** 2722.
    J. F. Olin and T. B. Johnson, *J. Amer. Chem. Soc.*, **53,** 1475 (1931); *Chem. Abstr.*, **25,** 2722.
81. G. M. Dyson, R. F. Hunter, J. W. T. Jones, and E. R. Styles, *J. Indian Chem. Soc.*, **8,** 147 (1931); *Chem. Abstr.*, **25,** 4880.
82. T. G. Levi, *Gazz. Chim. Ital.*, **61,** 719 (1931); *Chem. Abstr.*, **26,** 1602.

IX. References

83. D. W. Maccorquodale and T. B. Johnson, *Rec. Trav. Chim.*, **51,** 483 (1932); *Chem. Abstr.*, **26,** 3794.
84. R. H. Carroll and G. B. Smith, *J. Amer. Chem. Soc.*, **55,** 370 (1933); *Chem. Abstr.*, **27,** 955.
85. F. B. Dains and F. Eberly, *J. Amer. Chem. Soc.*, **55,** 3859 (1933); *Chem. Abstr.*, **27,** 5075.
86. R. F. Hunter and E. R. Parken, *J. Chem. Soc.*, 347 (1934).
87. F. E. Hooper and T. B. Johnson, *J. Amer. Chem. Soc.*, **56,** 470 (1934).
88. J. P. Wetherill and R. M. Hann, *J. Amer. Chem. Soc.*, **56,** 970 (1934); *Chem. Abstr.*, **28,** 3072.
89. H. T. Clarke and S. Gurin, *J. Amer. Chem. Soc.*, **57,** 1876 (1935).
90. P. L. Julian and B. M. Sturgis, *J. Amer. Chem. Soc.*, **57,** 1126 (1935); *Chem. Abstr.*, **29,** 5087.
91. Z. Horii, *J. Pharm. Soc. Japan*, **55,** 21 (1935); *Chem. Abstr.*, **29,** 3338.
92. M. L. Tomlinson, *J. Chem. Soc.*, 1030 (1935); *Chem. Abstr.*, **29,** 6596.
93. U.S.Patent No. 1 970 656; *Chem. Abstr.*, **28,** 6250.
94. J. P. Wetherill and R. M. Hann, *J. Amer. Chem. Soc.*, **57,** 1752 (1935); *Chem. Abstr.*, **29,** 7328.
95. E. R. Buchman, *J. Amer. Chem. Soc.*, **58,** 1803 (1936); *Chem. Abstr.*, **30,** 7572.
96. A. R. Todd, F. Bergel, and A. Jacob, *J. Chem. Soc.*, 1555 (1936); *Chem. Abstr.*, **31,** 401.
97. A. R. Todd, F. Bergel, H. F. Fraenkel-Conrat, and A. Jacob, *J. Chem. Soc.*, 1601 (1936); *Chem. Abstr.*, **31,** 690.
98. U.S. Patent No. 2 014 498; *Chem. Abstr.*, **29,** 7344.
99. A. R. Todd, F. Bergel, and Karimullah, *Berichte*, **69B,** 217 (1936); *Chem. Abstr.*, **30,** 3431.
100. A. R. Todd, F. Bergel, and Karimullah, *J. Chem. Soc.*, 1957 (1936); *Chem. Abstr.*, **31,** 401.
101. B. S. Friedman, M. Sparks, and R. Adams, *J. Amer. Chem. Soc.*, **59,** 2262 (1937); *Chem. Abstr.*, **32,** 556.
102. Karimullah, *J. Chem. Soc.*, 961 (1937); *Chem. Abstr.*, **31,** 6231.
103. A. R. Todd, F. Bergel, Karimullah, and R. Keller, *J. Chem. Soc.*, **217,** 361 (1937).
104. British Patent No. 456 751; *Chem. Abstr.*, **31,** 2232.
105. L. R. Cerecedo and J. G. Tolpin, *J. Amer. Chem. Soc.*, **59,** 1660 (1937); *Chem. Abstr.*, **31,** 7873.
106. French Patent No. 803 495; *Chem. Abstr.*, **31,** 2616.
107. J. MacLean and J. W. Forsyth, *J. Chem. Soc.*, 556 (1937); *Chem. Abstr.*, **31,** 4316.
108. A. R. Todd and F. Bergel, *J. Chem. Soc.*, 1504 (1937); *Chem. Abstr.*, **32,** 159.
109. H. Andersag and K. Westphal, *Berichte*, **70,** 2035 (1937); *Chem. Abstr.*, **32,** 2127.
110. H. Erlenmeyer, A. Epprecht, and H. Meyenburg, *Helv. Chim. Acta*, **20,** 514 (1937); *Chem. Abstr.*, **31,** 6230.
111. H. Erlenmeyer and H. Meyenburg, *Helv. Chim. Acta*, **20,** 204 (1937); *Chem. Abstr.*, **31,** 3916.
112. British Patent No. 472 459; *Chem. Abstr.*, **32,** 1408.
113. T. Kazanskii and N. Gushnev, *Bull. Acad. Sci., U.S.S.R.*, **65,** 1061 (1938).
114. H. Erlenmeyer and A. Kleiber, *Helv. Chim. Acta*, **21,** 111 (1938); *Chem. Abstr.*, **32,** 3350.
115. U. P. Basu and S. J. Dasgupta, *J. Indian Chem. Soc.*, **15,** 160 (1938); *Chem. Abstr.*, **32,** 7040.
116. J. Goetze, *Berichte*, **71B,** 2289 (1938); *Chem. Abstr.*, **33,** 978.

117. T. Imai and K. Makino, *Z. Phys. Chem. Frankfurt,* **252,** 1014 (1938).
118. Swiss Patent Nos. 192 069 and 192 849; *Chem. Abstr.,* **32,** 4285.
119. British Patent No. 471 416.
120. German Patent No. 658 353; *Chem. Abstr.,* **32,** 4728.
121. Swiss Patent Nos. 192 069 and 192 849; *Chem. Abstr.,* **32,** 4285.
122. German Patent No. 664 789; *Chem. Abstr.,* **33,** 177.
123. French Patent No. 824 519.
124. H. Erlenmeyer, *Helv. Chim. Acta,* **21,** 1013 (1938); *Chem. Abstr.,* **33,** 752.
125. T. Schinzel and G. Benoit, *Bull. Soc. Chim. France,* **501** (1939); *Chem. Abstr.,* **33,** 5807.
126. W. J. Doran and H. A. Shonle, *J. Org. Chem.,* **3,** 193 (1938).
127. U.S. Patent No. 2 160 867; *Chem. Abstr.,* **33,** 7320.
128. German Patent No. 670 131; *Chem. Abstr.,* **33,** 2909.
129. E. Ochiai, T. Kakuda, I. Nakayama, and G. Masuda, *J. Pharm. Soc. Japan,* **59,** 462 (1939); *Chem. Abstr.,* **34,** 101.
130. German Patent No. 673 174; *Chem. Abstr.,* **33,** 4271.
131. A. G. Pesina, *J. Gen. Chem. U.S.S.R.,* **9,** 804 (1939); *Chem. Abstr.,* **34,** 425.
132. E. R. Buchman and E. M. Richardson, *J. Amer. Chem. Soc.,* **61,** 891 (1939); *Chem. Abstr.,* **33,** 4242.
133. H. Erlenmeyer and H. Ueberwasser, *Helv. Chim. Acta,* **22,** 938 (1939).
134. German Patent No. 664 789; *Chem. Abstr.,* **33,** 177.
135. German Patent No. 678 153; *Chem. Abstr.,* **33,** 7819.
136. U.S. Patent No. 2 139 570; *Chem. Abstr.,* **33,** 2287.
137. G. W. Watt, *J. Org. Chem.,* **4,** 436 (1939).
138. E. Ochiai and H. Nagasawa, *J. Pharm. Soc. Japan,* **59,** 43 (1939).
139. German Patent No. 675 617.
140. Dutch Patent No. 41 419.
141. H. Erlenmeyer and E. H. Schmid, *Helv. Chim. Acta,* **22,** 698 (1939).
142. J. R. Byers and J. B. Dickey, *Org. Synth.,* 10 (1939); *Chem. Abstr.,* **33,** 5395.
143. F. B. Dains and O. A. Krober, *J. Amer. Chem. Soc.,* 1830 (1939); *Chem. Abstr.,* **33,** 6840.
144. F. C. Schmelkes, *Science,* 113 (1939); *Chem. Abstr.,* **33,** 7858.
145. A. Schoberl and M. Stock, *Chem. Ber.,* **73B,** 1240 (1940); *Chem. Abstr.,* **35,** 3636.
146. U.S. Patent No. 2 230 962; *Chem. Abstr.,* **35,** 3270.
147. K. Ganapathi, *Proc. Indian Acad. Sci.,* **12A,** 274 (1940); *Chem. Abstr.,* **35,** 1772.
148. J. Walker, *J. Chem. Soc.,* 1304 (1940); *Chem. Abstr.,* **35,** 113.
149. B. K. Nandi, *J. Indian Chem. Soc.,* **17,** 449 (1940); *Chem. Abstr.,* **35,** 2146.
150. U.S. Patent No. 2 186 419; *Chem. Abstr.,* **34,** 3537.
151. A. Schoberl and M. Stock, *Chem. Ber.,* **73B,** 1240 (1940); *Chem. Abstr.,* **35,** 3636.
152. H. Erlenmeyer and H. Ueberwasser, *Helv. Chim. Acta,* **23,** 197 (1940).
153. F. Nagasawa, *J. Pharm. Soc. Japan,* **60,** 433 (1940); *Chem. Abstr.,* **35,** 458. E. Ochiai, Y. Tamamusi and H. Nagasawa, *Berichte,* **73B,** 28 (1940); *Chem. Abstr.,* **34,** 2373.
154. Swiss Patent No. 204 688.
155. J. R. Stevens and G. A. Stein, *J. Amer. Chem. Soc.,* **62,** 1045 (1940); *Chem. Abstr.,* **34,** 6270.
156. E. R. Buchman, A. O. Reims, and H. Sargent, *J. Org. Chem.,* **6,** 764 (1941); *Chem. Abstr.,* **36,** 1606.
157. K. A. Jensen and T. Thorsteinsson, *Dansk Tidsskr. Farm.,* **15,** 41 (1941); *Chem. Abstr.,* **35,** 5109.

IX. References

158. W. M. Ziegler, *J. Amer. Chem. Soc.*, **63**, 2946 (1941); *Chem. Abstr.*, **36**, 470.
159. U.S. Patent No. 2 242 237; *Chem. Abstr.*, **35**, 5518.
160. U.S. Patent No. 2 209 092; *Chem. Abstr.*, **35**, 141.
161. German Patent No. 702 831; *Chem. Abstr.*, **36**, 784.
162. H. Beyer, *Chem. Ber.*, **74B**, 1100 (1941); *Chem. Abstr.*, **36**, 4819.
163. U.S. Patent No. 2 230 962; *Chem. Abstr.*, **35**, 3270.
164. W. G. Christiansen, *J. Amer. Chem. Soc.*, **63**, 632 (1941); *Chem. Abstr.*, **35**, 2144.
165. J. M. Sprague and L. W. Kissinger, *J. Amer. Chem. Soc.*, **63**, 578 (1941); *Chem. Abstr.*, **35**, 2144.
166. U.S. Patent No. 2 252 921; *Chem. Abstr.*, **35**, 7660.
167. U.S. Patent No. 2 209 244; *Chem. Abstr.*, **35**, 282.
168. R. Manzoni Ansidei and G. Travagli, *Gazz. Chim. Ital.*, **71**, 680 (1941); *Chem. Abstr.*, **36**, 7021.
169. H. J. Backer and J. De Jonge, *Rec. Trav. Chim.*, **60**, 495 (1941); *Chem. Abstr.*, **36**, 5793.
170. H. Erlenmeyer and W. Schoenauer, *Helv. Chim. Acta*, **24**, 172 (1941); *Chem. Abstr.*, **36**, 5171.
171. J. R. Hatcher, *J. Amer. Chem. Soc.*, **69**, 465 (1941).
172. U. P. Basu and S. J. Das-Gupta, *J. Indian Chem. Soc.*, **18**, 167 (1941).
173. K. Ganapathi, M. V. Shirsat, and C. V. Deliwala, *Proc. Indian Acad. Sci.*, **14A**, 630 (1941); *Chem. Abstr.*, **36**, 4102.
174. M. Kofler and L. Sternbach, *Helv. Chim. Acta*, **24**, 1014 (1941); *Chem. Abstr.*, **36**, 2556.
175. J. McLean and G. D. Muir, *J. Chem. Soc.*, 383 (1942); *Chem. Abstr.*, **36**, 5815.
176. British Patent No. 540 032; *Chem. Abstr.*, **36**, 4138.
177. H. Morren and R. Dupont, *J. Pharm. Belg.*, **1**, 126 (1942); *Chem. Abstr.*, **38**, 3284.
178. E. Ochiai and Y. Kashida, *J. Pharm. Soc. Japan*, **62**, 97 (1942); *Chem. Abstr.*, **45**, 5150.
179. E. Ochiai and Y. Kashida, *J. Pharm. Soc. Japan*, **62**, 31 (1942).
180. P. Baumgarten, A. Dornow, K. Gutschmidt, and H. Krehl, *Berichte*, **75B**, 442 (1942); *Chem. Abstr.*, **37**, 3091.
181. K. A. Jensen, P. Falkenberg, T. Thorsteinsson, and M. Lauridsen, *Dansk Tidsskr. Farm.*, **16**, 141 (1942); *Chem. Abstr.*, **38**, 3264.
    K. A. Jensen and A. Kjaer, *Dansk Tidsskr. Farm.*, **16**, 110 (1942); *Chem. Abstr.*, **38**, 2326.
182. K. Ganapathi, C. V. Deliwala, and M. V. Shirsat, *Proc. Indian Acad. Sci.*, **16A**, 126 (1942); *Chem. Abstr.*, **37**, 1404.
183. H. Erlenmeyer and C. J. Morel, *Helv. Chim. Acta*, **36**, 4102 (1942).
184. H. J. Backer and J. De Jonge, *Rec. Trav. Chim.*, **61**, 463 (1942); *Chem. Abstr.*, **38**, 2327.
185. H. Erlenmeyer and C. J. Morel, *Helv. Chim. Acta*, **25**, 1073 (1942); *Chem. Abstr.*, **37**, 1702.
186. French Patent No. 867 318.
187. Swiss Patent No. 216 270.
    Swiss Patent No. 216 994; *Chem. Abstr.*, **42**, 6381.
188. H. Erlenmeyer and M. Simon, *Helv. Chim. Acta*, **25**, 362 (1942); *Chem. Abstr.*, **36**, 5816.
189. H. Erlenmeyer and M. Simon, *Helv. Chim. Acta*, **25**, 528 (1942); *Chem. Abstr.*, **36**, 6539.
190. Y. F. Chi and S. Y. Tshin, *J. Amer. Chem. Soc.*, **64**, 90 (1942); *Chem. Abstr.*, **36**, 1322.

191. Y. M. Slobodin and E. E. Hel'ms, *Compte Rend. Acad. Sci. U.S.S.R.* (*In English*), **39,** 145 (1943); *Chem. Abstr.*, **38,** 1239.
192. J. R. Byers and J. B. Dickey, *Org. Synth.*, **31,** 31 (1943).
193. K. A. Jensen, P. Falkenberg, and T. Thorsteinsson, *Dansk Tidsskr. Farm.*, 141 (1939).
194. H. Coates, A. H. Cook, I. H. Heilbron, and F. B. Lewis, *J. Chem. Soc.*, 419 (1943); *Chem. Abstr.*, **38,** 106.
195. F. Brody and M. T. Bogert, *J. Amer. Chem. Soc.*, **65,** 1080 (1943); *Chem. Abstr.*, **37,** 4733.
196. E. H. Huntress and K. Pfister, *J. Amer. Chem. Soc.*, **65,** 1667 (1943); *Chem. Abstr.*, **37,** 6263.
197. Y. M. Slobodin, M. S. Zigel and M. V. Yanishevskaya, *J. Appl. Chem. U.S.S.R.*, **16,** 280 (1943); *Chem. Abstr.*, **39,** 702.
198. O. Dann, *Chem. Ber.*, **76,** 419 (1943); *Chem. Abstr.*, **37,** 6260.
199. I. Ubaldini and A. Fiorenza, *Gazz. Chim. Ital.*, **73,** 169 (1943); *Chem. Abstr.*, **38,** 5827.
200. A. I. Gravin, *J. Appl. Chem. U.S.S.R.*, **16,** 105 (1943); *Chem. Abstr.*, **38,** 1239.
201. British Patent No. 546 994; *Chem. Abstr.*, **37,** 5556.
202. German Patent No. 729 853; *Chem. Abstr.*, **38,** 382.
203. P. Karrer and M. C. Sanz, *Helv. Chim. Acta*, **26,** 1778 (1943); *Chem. Abstr.*, **38,** 4599.
204. P. Z. Bedoukian, *J. Amer. Chem. Soc.*, **66,** 1325 (1944).
205. I. Ya. Postovskii, V. I. Khmelevskii, and N. P. Bednyagina, *J. Appl. Chem. U.S.S.R.*, **17,** 65 (1944); *Chem. Abstr.*, **39,** 1410.
206. U.S. Patent No. 2 330 223; *Chem. Abstr.*, **38,** 1250.
207. British Patent No. 549 846; *Chem. Abstr.*, **38,** 1078.
208. H. Erlenmeyer, P. Buchmann, and H. Schenkel, *Helv. Chim. Acta*, **27,** 1432 (1944); *Chem. Abstr.*, **39,** 3281.
209. H. Schenkel, E. Marbet, and H. Erlenmeyer, *Helv. Chim. Acta*, **27,** 1437 (1944); *Chem. Abstr.*, **39,** 3281.
210. U.S. Patent No. 2 341 687; *Chem. Abstr.*, **38,** 4385.
211. H. Erlenmeyer, P. Buchmann, and H. Schenkel, *Helv. Chim. Acta*, **27,** 1432 (1944); *Chem. Abstr.*, **39,** 3281.
212. Y. Garreau, *Compt. Rend.*, **218,** 597 (1944); *Chem. Abstr.*, **40,** 2445.
213. British Patent No. 537 691.
214. W. M. Ziegler, *J. Amer. Chem. Soc.*, **66,** 744 (1944); *Chem. Abstr.*, **38,** 3648.
215. G. Carrara and G. Bonacci, *Chim. Ind.* **26,** 75 (1944).
216. H. Erlenmeyer and G. Bischoff, *Helv. Chim. Acta*, **27,** 412 (1944); *Chem. Abstr.*, **38,** 6275.
217. German Patent No. 729 853; *Chem. Abstr.*, **38,** 382.
218. H. Lehr and H. Erlenmeyer, *Helv. Chim. Acta*, **27,** 489 (1944); *Chem. Abstr.*, **38,** 6275.
    H. Erlenmeyer, W. Buchler, and H. Lehr, *Helv. Chim. Acta*, **27,** 969 (1944); *Chem. Abstr.*, **39,** 1641.
219. R. E. Kent and S. M. McElmain, *Org. Synth.*, 25 (1945).
220. K. Ganapathi and A. Venkataraman, *Proc. Indian Acad. Sci.*, **22A,** 343 (1945); *Chem. Abstr.*, **40,** 4056.
221. K. Ganapathi and A. Venkataraman, *Proc. Indian Acad. Sci.*, **22A,** 359 (1945); *Chem. Abstr.*, **40,** 4059.
222. G. Schwarz, *Org. Synth.*, **25,** 35 (1945); *Chem. Abstr.*, **40,** 330.
223. U.S. Patent No. 2 389 126; *Chem. Abstr.*, **40,** 991.
224. J. McLean and G. D. Muir, *J. Chem. Soc.*, 383 (1942); *Chem. Abstr.*, **36,** 5815.

IX. References

225. R. M. Dodson and L. C. King, *J. Amer. Chem. Soc.*, **67,** 2242 (1945); *Chem. Abstr.*, **40,** 1162.
226. A. Favorskii and M. N. Shchukina, *Zh. Obshchei Khim.*, **15,** 385 (1945).
227. E. R. Buchman and E. M. Richardson, *J. Amer. Chem. Soc.*, **67,** 395 (1945); *Chem. Abstr.*, **39,** 1871.
228. L. C. Leitch, B. E. Baker, and L. Brickman, *Can. J. Res.* **23B,** 139 (1945); *Chem. Abstr.*, **40,** 556.
229. E. M. Gibbs and F. A. Robinson, *J. Chem. Soc.*, 925 (1945); *Chem. Abstr.*, **40,** 1830.
230. E. R. Buchman and H. Sargent, *J. Amer. Chem. Soc.*, **67,** 400 (1945); *Chem. Abstr.*, **39,** 1872.
231. E. B. Knott, *J. Chem. Soc.*, 455 (1945); *Chem. Abstr.*, **39,** 4869.
232. British Patent No. 558 956; *Chem. Abstr.*, **39,** 4632.
233. H. J. Backer and J. A. Buisman, *Rec. Trav. Chim.*, **64,** 102 (1945); *Chem. Abstr.*, **40,** 3414.
234. J. F. Bunnet and D. S. Tarbell, *J. Amer. Chem. Soc.*, **67,** 1944 (1945); *Chem. Abstr.*, **40,** 830.
235. British Patent No. 559 106; *Chem. Abstr.*, **39,** 4436.
236. H. Erlenmeyer, R. Marbet, and H. Schenkel, *Helv. Chim. Acta*, **28,** 924 (1945); *Chem. Abstr.*, **40,** 869.
237. H. Erlenmeyer and C. J. Morel, *Helv. Chim. Acta*, **28,** 362 (1945); *Chem. Abstr.*, **40,** 1501.
238. P. Karrer and J. Schukri, *Helv. Chim. Acta*, **28,** 820 (1945); *Chem. Abstr.*, **40,** 1501.
239. A. H. Cook, I. H. Heilbron, and K. J. Reed, *J. Chem. Soc.*, 182 (1945); *Chem. Abstr.*, **39,** 2978.
240. H. Lehr, W. Guex, and H. Erlenmeyer, *Helv. Chim. Acta*, **28,** 1281 (1945); *Chem. Abstr.*, **40,** 2146.
241. S. C. De and P. K. Datta, *Sci. Culture*, **11,** 150 (1945); *Chem. Abstr.*, **40,** 1804.
242. W. R. Boon, *J. Chem. Soc.*, 601 (1945); *Chem. Abstr.*, **40,** 573.
243. U.S. Patent No. 2 402 066; *Chem. Abstr.*, **40,** 4908.
244. U.S. Patent No. 2 396 894; *Chem. Abstr.*, **40,** 5598.
245. H. Erlenmeyer and P. Schmidt, *Helv. Chim. Acta*, **29,** 1957 (1946); *Chem. Abstr.*, **41,** 1667.
246. R. E. Lutz, *J. Amer. Chem. Soc.*, **68,** 1818 (1946).
247. E. R. H. Jones, F. A. Robinson, and M. N. Strachan, *J. Chem. Soc.*, 91 (1946); *Chem. Abstr.*, **40,** 3442.
248. U.S. Patent No. 2 387 212; *Chem. Abstr.*, **40,** 1179.
249. J. M. Sprague, A. H. Land, and C. Ziegler, *J. Amer. Chem. Soc.*, **68,** 2155 (1946); *Chem. Abstr.*, **41,** 960.
250. J. M. Sprague, R. M. Lincoln, and C. Ziegler, *J. Amer. Chem. Soc.*, **68,** 266 (1946); *Chem. Abstr.*, **40,** 2126.
251. J. A. Buisman, Thesis, 1946.
252. H. Erlenmeyer and R. Marbet, *Helv. Chim. Acta*, **29,** 1946 (1946); *Chem. Abstr.*, **41,** 1668.
   H. Erlenmeyer and R. Marbet, *Helv. Chim. Acta*, **29,** 1233 (1946).
253. M. Erne, F. Ramirez, and A. Burger, *J. Amer. Chem. Soc.*, **68,** 266 (1946).
254. A. Land, C. Ziegler, and J. Sprague, *J. Org. Chem.*, **11,** 617 (1946); *Chem. Abstr.*, **41,** 412.
255. C. D. Hurd and N. Kharasch, *J. Amer. Chem. Soc.*, **68,** 653 (1946); *Chem. Abstr.*, **40,** 3442.
256. P. K. Das Gupta and P. Gupta, *J. Indian Chem. Soc.*, **23,** 241 (1946).

257. J. P. English, J. H. Clark, J. W. Clapp, D. Seeger, and R. H. Ebel, *J. Amer. Chem. Soc.*, **68**, 453 (1946); *Chem. Abstr.*, **40**, 2808.
258. N. F. Kicherova and K. A. Kocheshkov, *J. Gen. Chem. USSR.*, **16**, 1701 (1946); *Chem. Abstr.*, **41**, 6242.
259. H. Erlenmeyer and M. Erne, *Helv. Chim. Acta*, **29**, 275 (1946); *Chem. Abstr.*, **40**, 2829.
260. B. Holmberg, *Ark. Kemi Mineral. Geol.*, **24A**, 10 (1946); *Chem. Abstr.*, **44**, 4423.
261. R. M. Dodson and L. C. King, *J. Amer. Chem. Soc.*, **68**, 871 (1946); *Chem. Abstr.*, **40**, 4070.
262. H. Erlenmeyer and G. Bischoff, *Helv. Chim. Acta*, **29**, 280 (1946); *Chem. Abstr.*, **40**, 2447.
263. U.S. Patent No. 2 404 416; *Chem. Abstr.*, **40**, 7238.
264. H. Erlenmeyer, C. Becker, E. Sorkin, H. Bloch, and E. Suter, *Helv. Chim. Acta*, **30**, 2058 (1947); *Chem. Abstr.*, **42**, 2253.
265. B. Prijs, J. Ostertag, and H. Erlenmeyer, *Helv. Chim. Acta*, **30**, 2110 (1947); *Chem. Abstr.*, **42**, 1934.
266. U.S.S.R. Patent No. 66 044; *Chem. Abstr.*, **41**, 1713.
267. A. Land, C. Ziegler, and J. Sprague, *J. Amer. Chem. Soc.*, **69**, 125 (1947); *Chem. Abstr.*, **41**, 2038.
268. M. J. Baker, *J. Org. Chem.*, **12**, 163 (1947).
269. U.S. Patent No. 2 426 397; *Chem. Abstr.*, **42**, 224.
270. J. Ritter and H. Sokol, *J. Amer. Chem. Soc.*, **69**, 2069 (1947).
271. A. H. Cook, I. H. Heilbron, and A. L. Levy, *J. Chem. Soc.*, 1598 (1947).
272. E. B. Knott, *J. Chem. Soc.*, 1656 (1947); *Chem. Abstr.*, **42**, 2969.
273. A. Burger and G. E. Ullyot, *J. Org. Chem.*, **12**, 342 (1947); *Chem. Abstr.*, **41**, 4489.
274. R. Schneider, *Ann. Pharm. Franc.*, **5**, 112 (1947); *Chem. Abstr.*, **42**, 323.
275. E. Seebeck, *Helv. Chim. Acta*, **30**, 149 (1947).
276. H. Gregory and L. F. Wiggins, *J. Chem. Soc.*, 1400 (1947); *Chem. Abstr.*, **42**, 1262.
277. E. Pedley, *J. Chem. Soc.*, 431 (1947); *Chem. Abstr.*, **41**, 5509.
278. E. Hoggarth, *J. Chem. Soc.*, 110 (1947); *Chem. Abstr.*, **41**, 3458.
279. M. Ohta, R. Sudo, and K. Kato, *J. Pharm. Soc. Japan*, **68**, 40 (1948); *Chem. Abstr.*, **44**, 3974.
280. L. C. King and I. Ryden, *J. Amer. Chem. Soc.*, **69**, 1813 (1947).
281. A. Goldberg and W. Kelly, *J. Chem. Soc.*, 1372 (1947); *Chem. Abstr.*, **42**, 1263.
282. U.S. Patent No. 2 433 388; *Chem. Abstr.*, **42**, 2990.
283. J. Hatcher, *J. Amer. Chem. Soc.*, **69**, 465 (1947); *Chem. Abstr.*, **41**, 3458.
284. D. Markees, M. Kellerhals, and H. Erlenmeyer, *Helv. Chim. Acta*, 304 (1947); *Chem. Abstr.*, **41**, 3096.
285. H. Erlenmeyer, O. Weber, P. Schmidt, G. Kung, C. Zinsstag, and B. Prijs, *Helv. Chim. Acta*, **31**, 1142 (1948); *Chem. Abstr.*, **42**, 7291.
286. J. R. Catch, D. F. Elliott, D. H. Hey, and E. R. H. Jones, *J. Chem. Soc.*, **1**, 278 (1948).
287. P. Chabrier and S. Renard, *Compt. Rend.*, **226**, 582 (1948); *Chem. Abstr.*, **42**, 5453.
288. D. G. Markees and A. Burger, *J. Amer. Chem. Soc.*, **70**, 3329 (1948); *Chem. Abstr.*, **43**, 635.
289. R. Mathes and A. Beber, *J. Amer. Chem. Soc.*, **70**, 1451 (1948); *Chem. Abstr.*, **42**, 5453.
290. U.S. Patent No. 2 445 722; *Chem. Abstr.*, **43**, 1602.
291. J. Ritter and H. Sokol, *J. Amer. Chem. Soc.*, **70**, 3419 (1948); *Chem. Abstr.*, **43**, 633.
292. W. S. Emerson and T. M. Patrick, Jr., *J. Org. Chem.*, **13**, 722 (1948); *Chem. Abstr.*, **43**, 5393.

IX. References

293. A. H. Cook, I. H. Heilbron, and E. Stern, *J. Chem. Soc.*, 2031 (1948); *Chem. Abstr.*, **43**, 3409.
294. M. Erne and H. Erlenmeyer, *Helv. Chim. Acta*, **31**, 652 (1948); *Chem. Abstr.*, **42**, 4575.
295. R. G. Jones, Q. F. Soper, O. K. Behrens, and J. W. Corse, *J. Amer. Chem. Soc.*, **70**, 2843 (1948); *Chem. Abstr.*, **43**, 3362.
296. H. Forrest, A. Fuller, and J. Walker, *J. Chem. Soc.*, 1501 (1948); *Chem. Abstr.*, **43**, 1739.
297. H. Erlenmeyer, H. Baumann, and E. Sorkin, *Helv. Chim. Acta*, **31**, 1342 (1948); *Chem. Abstr.*, **43**, 3820.
298. H. Erlenmeyer, J. Junod, W. Guex, and M. Erne, *Helv. Chim. Acta*, **31**, 1342 (1948); *Chem. Abstr.*, **43**, 5021.
299. A. H. Cook, J. Downer, and I. H. Heilbron, *J. Chem. Soc.*, 2028 (1948); *Chem. Abstr.*, **43**, 3408.
300. C. N. Capp, A. H. Cook, J. D. Downer, and I. H. Heilbron, *J. Chem. Soc.*, 1340 (1948).
301. A. T. Blomquist and F. H. Baldwin, *J. Amer. Chem. Soc.*, **70**, 29 (1948).
302. N. Albertson, *J. Amer. Chem. Soc.*, **70**, 669 (1948); *Chem. Abstr.*, **42**, 3392.
303. L. Szekeres, *Gazz. Chim. Ital.*, **78**, 681 (1948); *Chem. Abstr.*, **43**, 2967.
304. A. H. Cook, I. H. Heilbron, and A. Levy, *J. Chem. Soc.*, 201 (1948); *Chem. Abstr.*, **42**, 8798.
305. A. I. Kiprianov, F. I. Asnina, and I. K. Ushenko, *J. Gen. Chem. U.S.S.R.*, **18**, 165 (1948); *Chem. Abstr.*, **42**, 7293.
306. J. R. Catch, D. F. Elliott, D. H. Hey, and E. R. H. Jones, *J. Chem. Soc.*, 272 (1948).
307. H. Erlenmeyer and J. Ostertag, *Helv. Chim. Acta*, **31**, 26 (1948); *Chem. Abstr.*, **42**, 4166.
308. J. Monche, *An. Quim.*, 1299 (1948); *Chem. Abstr.*, **44**, 2981.
309. C. D. Hurd and H. L. Wehrmeister, *J. Amer. Chem. Soc.*, **71**, 4007 (1949); *Chem. Abstr.*, **45**, 155.
310. L. King and F. Miller, *J. Amer. Chem. Soc.*, **71**, 367 (1949); *Chem. Abstr.*, **43**, 2992.
311. S. Schulman, *J. Org. Chem.*, **14**, 382 (1949).
312. T. Clarke, J. R. Johnson, and Sir R. Robinson, "Chemistry of Penicillin," Princeton U.P., Princeton, N.J., 1949.
313. A. Beber and R. Mathes, *Ind. Eng. Chem.*, **41**, 2637 (1949); *Chem. Abstr.*, **44**, 3292.
314. U.S. Patent No. 2 459 020.
315. E. Carr, G. Smith, and G. Alliger, *J. Org. Chem.*, **14**, 921 (1949); *Chem. Abstr.*, **44**, 3976.
316. F. Stewart and R. Mathes, *J. Org. Chem.*, **14**, 1111 (1949); *Chem. Abstr.*, **44**, 3975.
317. A. H. Cook, I. H. Heilbron, and G. Hunter, *J. Chem. Soc.*, 1443 (1949); *Chem. Abstr.*, **44**, 1965.
318. J. D. Kendall and H. G. Suggate, *J. Chem. Soc.*, 1503 (1949); *Chem. Abstr.*, **44**, 466.
319. A. H. Cook, D. J. Brown, and I. H. Heilbron, *J. Chem. Soc.*, **5**, 111 (1949).
320. J. Batty and B. Weedon, *J. Chem. Soc.*, 786 (1949); *Chem. Abstr.*, **44**, 540.
321. G. Fodor, *Acta Univ. Szegediensis Chem. Phys.*, **21**, 167 (1949); *Chem. Abstr.*, **44**, 6414.
322. H. Beyer, H. Schulte, and G. Henseke, *Chem. Ber.*, **82**, 143 (1949); *Chem. Abstr.*, **44**, 2511.
323. I. H. Heilbron, *J. Chem. Soc.*, 2099 (1949).
324. P. Chabrier, S. Renard, and K. Smarzewska, *Bull. Soc. Chim. France*, 237 (1949); *Chem. Abstr.*, **44**, 5347.
325. J. W. Cornforth, "Chemistry of Penicillin," Princeton U.P. Princeton, N.J., 1949, p. 848.

326. A. H. Cook and S. Cox, *J. Chem. Soc.*, 2337 (1949); *Chem. Abstr.*, **44**, 1968.
327. J. Eckenstein, E. Broglie, E. Sorkin, and H. Erlenmeyer, *Helv. Chim. Acta*, **33**, 1353 (1950).
328. L. King and R. Hlavacek, *J. Amer. Chem. Soc.*, **72**, 3722 (1950); *Chem. Abstr.*, **45**, 2934.
329. H. Gilman and A. H. Blatt, *Org. Synth.*, **1**, 109 (1950).
    H. Gilman and A. H. Blatt, *Org. Synth.*, **1**, 127 (1950).
    H. Gilman and A. H. Blatt, *Org. Synth.*, **1**, 180 (1950).
330. L. H. Conover and D. S. Tarbell, *J. Amer. Chem. Soc.*, **72**, 5221 (1950); *Chem. Abstr.*, **45**, 5148.
331. U.S. Patent No. 2 514 181.
332. R. Mory and H. Schenkel, *Helv. Chim. Acta*, **33**, 405 (1950); *Chem. Abstr.*, **44**, 5873.
333. H. Erlenmeyer and C. J. Morel, *Helv. Chim. Acta*, **34**, 143 (1950).
334. D. Tarbell, H. Hirschler, and R. Carlin, *J. Amer. Chem. Soc.*, **72**, 3138 (1950); *Chem. Abstr.*, **44**, 10703.
335. M. Kopp, *Bull. Soc. Chim. France*, **17**, 582 (1950); *Chem. Abstr.*, **48**, 1340.
336. H. Erlenmeyer, J. Eckenstein, E. Sorkin, and H. Meyer, *Helv. Chim. Acta*, **33**, 1271 (1950).
337. R. G. Jones, E. C. Kornfeld, and K. C. MacLaughlin, *J. Amer. Chem. Soc.*, **72**, 4526 (1950); *Chem. Abstr.*, **45**, 3383.
338. W. Davies, J. MacLaren, and L. Wilkinson, *J. Chem. Soc. A*, 3491 (1950); *Chem. Abstr.*, **45**, 7110.
339. H. Beyer and W. Lassig, *Chem. Ber.*, **84**, 463 (1951); *Chem. Abstr.*, **46**, 5581.
    H. Beyer and A. Kreutzberger Reese, *Chem. Ber.*, **84**, 518 (1951), *Chem. Abstr.*, **46**, 5582.
340. M. Izumi, M. Kimiti, and M. Yokoo, *J. Pharm. Soc. Japan*, **71**, 540 (1951).
341. Y. Tajika, Y. Nitta, J. Yomoda, and H. Oya, *J. Pharm. Soc. Japan*, **71**, 709 (1951); *Chem. Abstr.*, **46**, 2050.
342. T. E. Young and E. D. Amstutz, *J. Amer. Chem. Soc.*, **73**, 4773 (1951).
343. U.S. Patent No. 2 547 677; *Chem. Abstr.*, **45**, 6351.
344. M. Ohta, *J. Pharm. Soc. Japan*, **71**, 869 (1951); *Chem. Abstr.*, **46**, 4002.
345. U.S. Patent No. 2 560 020.
346. F. O. Stewart and R. A. Mathes, *Ind. Eng. Chem.*, **43**, 1569 (1951).
347. M. Erne, F. Ramirez, and A. Burger, *Helv. Chim. Acta*, **34**, 143 (1951); *Chem. Abstr.*, **45**, 7565.
348. P. Sykes and A. R. Todd, *J. Chem. Soc.*, 534 (1951); *Chem. Abstr.*, **46**, 2078F.
349. U.S. Patent No. 2 532 573; *Chem. Abstr.*, **45**, 3424.
350. P. Aubert, E. Knott, and L. A. Williams, *J. Chem. Soc.*, 2185 (1951).
351. W. T. Caldwell and S. M. Fox, *J. Amer. Chem. Soc.*, **73**, 2935 (1951); *Chem. Abstr.*, **46**, 2538.
352. D. Schoene, *J. Amer. Chem. Soc.*, **73**, 1970 (1951); *Chem. Abstr.*, **46**, 972.
353. H. Cardwell and A. Kilner, *J. Chem. Soc.*, 2430 (1951); *Chem. Abstr.*, **46**, 7999.
354. A. Morton and H. Penner, *J. Amer. Chem. Soc.*, **73**, 3300 (1951); *Chem. Abstr.*, **46**, 2995.
355. M. Ohta and K. Sato, *J. Pharm. Soc. Japan*, **71**, 9 (1951); *Chem. Abstr.*, **45**, 7111.
356. Japanese Patent No. 177 625; *Chem. Abstr.*, **45**, 7153.
357. Japanese Patent No. 178 630; *Chem. Abstr.*, **46**, 1593.
358. A. Davis and A. Levy, *J. Chem. Soc.*, 2419 (1951); *Chem. Abstr.*, **46**, 8097.
359. Y. Garreau, *Compt. Rend.*, **232**, 847 (1951); *Chem. Abstr.*, **45**, 8006.
360. S. J. Childress and R. L. McKee, *J. Amer. Chem. Soc.*, **73**, 3862 (1951).
361. R. Wiley, D. C. England, and L. C. Behr, *Org. Reactions*, **6**, 367 (1951).

362. A. Silberg, I. Simiti, and H. Mantsch, *Chem. Ber.*, **94,** 2887 (1961); *Chem. Abstr.*, **56,** 7295.
363. B. Prijs, "Kartothek der Thiazol Verbindungen," Karger, Basel, 1952.
364. R. P. Kurkjy and E. V. Brown, *J. Amer. Chem. Soc.*, **74,** 5778 (1952); *Chem. Abstr.*, **49,** 1013.
365. L. A. Yanovskaya, and A. P. Terent'sev, *Zh. Obshchei Khim.*, **22,** 1921 (1952).
366. R. McKee and J. Thayer, *J. Org. Chem.*, **17,** 1494 (1952); *Chem. Abstr.*, **47,** 9319.
367. French Patent No. 959 908; *Chem. Abstr.*, **46,** 5619.
368. I. Kaye and C. Parris, *J. Amer. Chem. Soc.*, **74,** 2271 (1952); *Chem. Abstr.*, **48,** 5856.
369. J. Gregory and R. Mathes, *J. Amer. Chem. Soc.*, **74,** 1719 (1952); *Chem. Abstr.*, **48,** 164.
370. K. Okada, *Ann. Rep. Takamine Lab.*, **4,** 40 (1952); *Chem. Abstr.*, **49,** 11665.
371. French Patent Nos. 1 006 623 and 1 006 624; *Chem. Abstr.*, **51,** 12148.
372. N. D. Dawson and A. Burger, *J. Amer. Chem. Soc.*, **74,** 5312 (1952).
373. M. Izumi, I. Aiko, M. Yokoo, and H. Kimoto, *J. Pharm. Soc. Japan*, **72,** 21 (1952); *Chem. Abstr.*, **46,** 11182.
374. A. E. S. Fairfull, J. L. Lowe, and D. A. Peak, *J. Chem. Soc.*, 742 (1952).
375. H. Beyer, H. Hohen, and W. Laessig, *Chem. Ber.*, **85,** 1122 (1952); *Chem. Abstr.*, **47,** 11183.
376. E. B. Knott, *J. Chem. Soc.*, 4099 (1952); *Chem. Abstr.*, **47,** 6799F.
377. J. Metzger and B. Koether, *Ann. Univ. Saraviensis*, **1,** 151 (1952); *Chem. Abstr.*, **47,** 10524.
378. K. Murata, *Bull. Chem. Soc. Japan*, **25,** 16 (1952); *Chem. Abstr.*, **48,** 4521.
379. G. Fodor and G. Wilheim, *Acta Chim. Acad. Sci. Hung.*, **2,** 189 (1952); *Chem. Abstr.*, **48,** 3346.
380. U.S. Patent No. 2 639 285; *Chem. Abstr.*, **48,** 8265.
381. F. N. Hayes, L. C. King, and D. E. Peterson, *J. Amer. Chem. Soc.*, **74,** 1106 (1952); *Chem. Abstr.*, **47,** 12355.
382. F. L. Rose, *J. Chem. Soc.*, 3448 (1952).
383. M. Erne, *Helv. Chim. Acta*, **36,** 138 (1953); *Chem. Abstr.*, **48,** 2688.
384. B. Koether, Thesis, University of Sarrebruck, F.R. of Germany 1953; *Chem. Abstr.*, **47,** 10524.
385. J. Metzger and B. Koether, *Bull. Soc. Chim. France*, **20,** 702 (1953); *Chem. Abstr.*, **48,** 10738.
386. K. Ganapathi and K. Kulkarni, *Proc. Indian Acad. Sci.*, **37,** 58 (1953); *Chem. Abstr.*, **49,** 8256.
387. J. D'Amico, *J. Amer. Chem. Soc.*, **75,** 102 (1953); *Chem. Abstr.*, **47,** 12361.
388. E. Creed, J. D'Amico, and M. W. Harman, *Chem. Eng.*, **46,** 808 (1953).
389. U.S. Patent No. 2 652 396.
390. German Patent No. 869 490.
391. H. Beyer and W. Schindler, *Chem. Ber.*, **86,** 1410 (1953); *Chem. Abstr.*, **49,** 1710.
392. R. Dahlbom, *Acta Chem. Scand.*, **7,** 374 (1953); *Chem. Abstr.*, **48,** 2687.
    R. Dahlbom, *Acta Chem. Scand.*, **7,** 885 (1953); *Chem. Abstr.*, **49,** 4626.
393. T. Takahashi and S. Nishigaki, *J. Pharm. Soc. Japan*, 1071 (1953); *Chem. Abstr.*, **48,** 12081.
394. T. Takahashi, S. Nishigaki, and S. Sakamoto, *J. Pharm. Soc. Japan*, **73,** 1076 (1953); *Chem. Abstr.*, **48,** 12082.
395. A. Krattiger, *Bull. Soc. Chim. France*, 222 (1953); *Chem. Abstr.*, **48,** 1270.
396. H. Beyer and W. Laessig, *Chem. Ber.*, **86,** 1342 (1953); *Chem. Abstr.*, **49,** 1709.
397. H. Beyer, W. Laessig, and G. Rulhig, *Chem. Ber.*, **86,** 764 (1953); *Chem. Abstr.*, **48,** 10736.

398. Y. Garreau, *Compt. Rend.*, **236,** 1575 (1953); *Chem. Abstr.*, **48,** 3346.
399. G. Mahapatra and M. K. Rout, *J. Indian Chem. Soc.*, **30,** 398 (1953); *Chem. Abstr.*, **49,** 1011.
400. M. K. Rout and H. Pujari, *J. Amer. Chem. Soc.*, **75,** 4057 (1953); *Chem. Abstr.*, **48,** 11393.
401. Y. Yamamoto, *J. Pharm. Soc. Japan*, **73,** 938 (1953); *Chem. Abstr.*, **48,** 10739.
402. S. Ban, *J. Pharm. Soc. Japan*, **73,** 533 (1953); *Chem. Abstr.*, **48,** 9361.
403. S. Akiyoshi and K. Okuno, *J. Amer. Chem. Soc.*, **76,** 693 (1954); *Chem. Abstr.*, **49,** 3946.
404. Y. Garreau, *Bull. Soc. Chim. France*, 564 (1954); *Chem. Abstr.*, **48,** 10740.
    Y. Garreau, *Bull. Soc. Chim. France*, 1048 (1954); *Chem. Abstr.*, **49,** 12445.
405. C. Rodriquez and F. J. Aparicio, *An. Real. Soc. Espan. Fis. Quim. Madrid*, **50B,** 705 (1954).
406. H. Beyer and U. Schultz, *Chem. Ber.*, **87,** 78 (1951).
407. A. C. B. Smith, W. Wilson, and R. Woodger, *Chem. Ind.*, 309 (1954).
408. H. Beyer, W. Laessig, E. Bulka, and D. Behrens, *Chem. Ber.*, **87,** 1392 (1954); *Chem. Abstr.*, **49,** 15868.
    H. Beyer, W. Laessig, and E. Bulka, *Chem. Ber.*, **87,** 1385 (1954); *Chem. Abstr.*, **49,** 15868.
409. S. G. Fridman, *Zh. Obshchei Khim.*, **24,** 1059 (1954); *Chem. Abstr.*, **49,** 8923.
410. M. Tomita, H. Kumaoka, and M. Takase, *J. Pharm. Soc. Japan*, **74,** 850 (1954); *Chem. Abstr.*, **49,** 9623.
411. G. P. Hager and C. Kaiser, *J. Amer. Pharm. Assoc.*, **44,** 193 (1954).
412. G. Vasiliu and L. Gertler, *Rev. Univ. C. I. Parhon SI politeh. Bucuresti, Ser. Stiint. Nat.*, **4,** 97 (1955); *Chem. Abstr.*, **52,** 1994.
413. W. Wilson and R. Woodger, *J. Chem. Soc.*, 2943 (1955); *Chem. Abstr.*, **50,** 1782.
414. C. D. Hurd and H. L. Wehrmeister, *J. Amer. Chem. Soc.*, **77,** 663 (1955).
415. K. Miyatake and T. Yoshikawa, *J. Pharm. Soc. Japan*, **75,** 1054 (1955); *Chem. Abstr.*, **50,** 5633.
416. D. Bariana, H. Sachdev, and K. Narang, *J. Indian Chem. Soc.*, **32,** 427 (1955); *Chem. Abstr.*, **50,** 10711.
417. E. C. Taylor, Jr., J. A. Anderson, and G. A. Berchtold, *J. Amer. Chem. Soc.*, **77,** 5444 (1955); *Chem. Abstr.*, **50,** 6434.
418. B. Chase and J. Walker, *J. Chem. Soc.*, 4443 (1955); *Chem. Abstr.*, **50,** 10712.
419. S. Fridman, *Zh. Obshchei Khim.*, 970 (1955); *Chem. Abstr.*, **50,** 3410.
420. B. Das and M. K. Rout, *J. Indian Chem. Soc.*, **32,** 663 (1955); *Chem. Abstr.*, **50,** 12026.
421. K. Kaji, H. Nagashima, N. Ninoi, and T. Hanada, *J. Pharm. Soc. Japan*, **75,** 438 (1955); *Chem. Abstr.*, **50,** 2548.
422. M. Astle and J. Pierce, *J. Org. Chem.*, **20,** 178 (1955); *Chem. Abstr.*, **49,** 13969.
423. N. Kishore Das, G. N. Mahapatra, and M. K. Rout, *J. Indian Chem. Soc.*, **32,** 55 (1955); *Chem. Abstr.*, **49,** 11626.
424. F. Stepanov and Z. Moiseeva, *Zh. Obshchei Khim.*, **25,** 1170 (1955); *Chem. Abstr.*, **50,** 3409.
425. H. Beyer, C. Kroger, and M. Zander, *Chem. Ber.*, **88,** 1233 (1955); *Chem. Abstr.*, **50,** 6433.
426. J. Metzger and J. Beraud, *Compt. Rend.*, **242,** 2362 (1956); *Chem. Abstr.*, **51,** 2741.
427. G. N. Mahapatra, *J. Indian Chem. Soc.*, **33,** 527 (1956); *Chem. Abstr.*, **52,** 9079.
428. H. Beyer and G. Ruhlig, *Chem. Ber.*, **89,** 107 (1956); *Chem. Abstr.*, **50,** 15510.
429. S. Kasman and A. Taurins, *Can. J. Chem.*, **34,** 1261 (1956); *Chem. Abstr.*, **51,** 3567.
430. H. Beyer and G. Berg, *Chem. Ber.*, **89,** 1602 (1956); *Chem. Abstr.*, **51,** 5757.
431. G. B. Elion, W. H. Lange, and G. W. Hitchings, *J. Amer. Chem. Soc.*, **78,** 2858 (1956); *Chem. Abstr.*, **50,** 14771.

IX. References

432. J. J. Riehl, *Compt. Rend.*, **245,** 1321 (1957); *Chem. Abstr.*, **52,** 7286.
433. B. Das and M. K. Rout, *J. Indian Chem. Soc.*, **34,** 505 (1957); *Chem. Abstr.*, **52,** 2621.
434. M. K. Rout, *J. Indian Chem. Soc.*, **33,** 690 (1956); *Chem. Abstr.*, **51,** 14685.
435. J. B. Dickey, E. B. Towne, M. S. Bloom, W. H. Moore, and H. M. Hill, *J. Org. Chem.*, 187 (1959); *Chem. Abstr.*, **53,** 18004.
     J. B. Dickey and E. B. Towne, *J. Org. Chem.*, **20,** 499 (1957).
436. Y. N. Sheinker, V. V. Kushkin, and I. Y. Postovskii, *Zh. Fiz. Khim.*, **31,** 214 (1957); *Chem. Abstr.*, **51,** 17455.
437. F. Asinger, M. Thiel, and E. Pallas, *Annalen*, **602,** 37 (1957); *Chem. Abstr.*, **51,** 12074.
     F. Asinger, M. Thiel, and G. Esser, *Annalen*, 33, (1957); *Chem. Abstr.*, **52,** 9141.
438. A. I. Vogel, "Practical Organic Chemistry," Longmans, London, 1957, p. 840.
439. G. N. Mahapatra, *J. Amer. Chem. Soc.*, 988 (1957); *Chem. Abstr.*, **51,** 9589.
440. A. Lawson and C. E. Searle, *J. Chem. Soc.*, 1556 (1957); *Chem. Abstr.*, **51,** 12078.
441. J. Ratusky and F. Sorm, *Chem. Listy*, **51,** 1091 (1957); *Chem. Abstr.*, **51,** 13843.
442. T. Takahashi and S. Nogawa, *Yakugaku Zasshi*, **77,** 458 (1957); *Chem. Abstr.*, **51,** 14686.
443. R. L. Huang, *J. Chem. Soc.*, 2528 (1957); *Chem. Abstr.*, **51,** 15451.
444. British Patent No. 768 481; *Chem. Abstr.*, **51,** 15590.
445. British Patent No. 768 482; *Chem. Abstr.*, **51,** 15590.
446. Japanese patent No. 792656, *Chem. Abstr.* **52,** 14698.
447. U.S. Patent No. 2 776 978; *Chem. Abstr.*, **51,** 12148.
448. J. M. Sprague and A. H. Land, "Heterocyclic Compounds," R. C. Elderfield, Ed., Wiley, New York, 1957, Vol. 5, p. 484.
449. B. G. Yasnitskii, *Materialy Po Obmenu Peredovym Opyrom I Nauch. Dostizhen. V. Khim. Farm. Prom.*, 63 (1957); *Chem. Abstr.*, **54,** 1499.
450. E. D. Sych, *Ukrain. Khim. Zhur.*, **25,** 767 (1958); *Chem. Abstr.*, **54,** 13143.
451. P. Roussel, Thesis, University of Marseille, France, 1958.
452. M. Thiel, F. Asinger, and M. Fedtke, *Justus Liebigs Ann. Chem.*, **615,** 77 (1958).
453. F. Asinger, M. Thiel, G. Peschel, and K. H. Meinicke, *Annalen*, **619,** 145 (1958); *Chem. Abstr.*, **53,** 17105.
454. Io Fon Chi and Shi Yun Zin, *Zh. Obshchei Khim.*, **28,** 1492 (1958); *Chem. Abstr.*, **53,** 1312.
455. M. Carrega and J. Metzger, *Bull. Soc. Chim. France*, 583 (1958).
     M. Carrega, Thesis, University of Marseille, France, 1959.
456. H. Guinot and J. Tabuteau, *Compt. Rend.*, **231,** 234 (1959).
457. V. N. Kerr, F. N. Hayes, D. G. Ott, and E. Hansbury, *J. Org. Chem.*, **24,** 1861 (1959); *Chem. Abstr.*, **54,** 9891.
458. E. D. Sych, *Ukrain. Khim. Zhur.*, **25,** 344 (1959); *Chem. Abstr.*, **54,** 5619.
459. U.S. Patent No. 2 900 299; *Chem. Abstr.*, **54,** 2361.
460. F. Ueda, *Yakugaku Zasshi*, **79,** 1248 (1959); *Chem. Abstr.*, **54,** 4542.
461. V. N. Kerr, F. N. Hayes, D. G. Ott, and E. Hansbury, *J. Org. Chem.*, **24,** 1861 (1959); *Chem. Abstr.*, **54,** 9891.
462. F. Asinger, M. Thiel, and H. G. Hauthal, *Annalen*, **634,** 131 (1959); *Chem. Abstr.*, **55,** 2616.
463. J. Okamiya, *Nippon Kagaku Zasshi*, **80,** 903 (1959); *Chem. Abstr.*, **55,** 5471.
464. I. M. Lisnganskii and E. S. Zhdanovich, *Tr. Vsesoyuz. Nauch. Issledovatel. Vitamin. Inst.*, **6,** 2830 (1959); *Chem. Abstr.*, **55,** 19903.
465. A. Nederlof, *Bull. Soc. Chim. Belges*, **68,** 148 (1959); *Chem. Abstr.*, **53,** 21891.
466. H. Beyer, G. Henseke, E. Bulka, H. Drews, and E. Muller, *Chem. Ber.*, **92,** 1105 (1959); *Chem. Abstr.*, **53,** 17100.

467. E. Bulka, H. Beyer, and G. Brandenburg, *Chem. Ber.*, **92,** 1447 (1959); *Chem. Abstr.*, **53,** 21891.
468. V. M. Zubarovskii, *Zh. Obshchei Khim.*, **29,** 2018 (1959); *Chem. Abstr.*, **54,** 8792.
469. C. Scherschener, *Pr Montevideo*, **9–10,** 45 (1959); *Chem. Abstr.*, **58,** 5655.
470. K. M. Murav'eva and M. N. Shchukina, *Dokl. Akad. Nauk SSSR*, **126,** 1274 (1959); *Chem. Abstr.*, **54,** 498.
471. J. P. Aune, Thesis, University of Marseille, France, 1960.
472. H. Bredereck, R. Gompper, and F. Reich, *Chem. Ber.*, **93,** 1389 (1960); *Chem. Abstr.*, **54,** 21105.

    H. Bredereck, R. Gompper, and F. Reich, *Chem. Ber.*, **93,** 723 (1960); *Chem. Abstr.*, **54,** 15360.
473. F. Asinger, M. Thiel, W. Dathe, O. Hampel, E. Mittag, E. Plaschil, and C. Schroder, *Annalen*, **639,** 146 (1960); *Chem. Abstr.*, **56,** 2435.
474. K. Ruehlmann, A. Grosalski, and U. Schraepler, *J. Prakt. Chem.*, **11,** 54 (1960); *Chem. Abstr.*, **56,** 465.
475. N. K. Ralhan, G. S. Sandhu, H. S. Sachdev, and K. S. Narang, *J. Indian Chem. Soc.*, **37,** 773 (1960).
476. F. Asinger, M. Thiel, W. Dathe, O. Hampel, E. Mittag, E. Plaschil, and C. Schroder, *Annalen*, **639,** 146 (1960); *Chem. Abstr.*, **56,** 2435.
477. A. Berlin and V. Bronovitskaia, *Zh. Obshchei Khim.*, **13,** 324 (1960).
478. A. B. Sen, and A. K. Roy, *J. Indian Chem. Soc.*, **37,** 427 (1960); *Chem. Abstr.*, **55,** 1582.
479. U.S. Patent No. 2 937 183; *Chem. Abstr.*, **55,** 3615.
480. Japanese Patent No. 727460; *Chem. Abstr.*, **55,** 6499.

    T. Ueda, S. Kato, S. Toyoshima, and F. Ueda, *J. Chem. Soc. Japan*, 7274 (1960); *Chem. Abstr.*, **55,** 6499.
481. M. Thiel, F. Asinger, W. Schafer, and H. G. Hauthal, *Annalen*, **638,** 174 (1960); *Chem. Abstr.*, **55,** 15464.
482. H. A. Braun, H. Kuehne, and B. Prijs, *Helv. Chim. Acta*, **43,** 659 (1960); *Chem. Abstr.*, **55,** 25917.
483. S. Huenig, H. Balli, and W. Brenninger, *Chem. Ber.*, **93,** 1518 (1960); *Chem. Abstr.*, **54,** 22578.
484. H. S. Sachdev, K. S. Dhami, and K. S. Narang, *J. Sci. Ind. Res.*, India, **19C,** 9 (1960); *Chem. Abstr.*, **54,** 24661.
485. E. L. Taylor and J. A. Zoltewicz, *Helv. Chim. Acta*, **82,** 2656 (1960).
486. G. W. Kenner, R. C. Sheppard, and C. E. Stehr, *Tetrahedron Lett.*, **1,** 23 (1960); *Chem. Abstr.*, **54,** 13104.
487. W. Groebel, *Chem. Ber.*, **93,** 896 (1960); *Chem. Abstr.*, **54,** 18480.
488. J. J. D'Amico and T. W. Bartram, *J. Org. Chem.*, **25,** 1336 (1960); *Chem. Abstr.*, **54,** 24662.
489. A. Balog and A. Benko, *Studia Univ. Babes Bolyai, Ser. Chem.*, **1,** 155 (1960); *Chem. Abstr.*, **58,** 4534.
490. K. M. Murav'eva and M. N. Shchukina, *Zh. Obshchei Khim.*, **30,** 2327 (1960); *Chem. Abstr.*, **55,** 9376.

    K. M. Murav'eva and M. N. Shchukina, *Zh. Obshchei Khim.*, **30,** 2334 (1960); *Chem. Abstr.*, **55,** 9376.

    K. M. Murav'eva and M. N. Shchukina, *Zh. Obshchei Khim.*, **30,** 2344 (1960); *Chem. Abstr.*, **55,** 9377.
491. R. Didier and J. Metzger, *Compt. Rend.*, **252,** 1619 (1961); *Chem. Abstr.*, **55,** 16516.

IX. References

492. M. Poite, Thesis, University of Marseille, France, 1961.
493. A. Kergomard, *Bull. Soc. Chim. France*, 2361 (1961).
494. T. Takahashi and M. Hayami, *Yakugaku Zasshi*, **81**, 1419 (1961); *Chem. Abstr.*, **56**, 11713.
495. A. Kirrmann, *Bull. Soc. Chim. France*, 657 (1961).
496. E. Fournier and L. Petit, *Compt. Rend.*, **252**, 291 (1961); *Chem. Abstr.*, **55**, 12389.
497. British Patent No. 951 885; *chem. Abstr.*, **61**, 5657.
498. Japanese Patent No. 25 676; *Chem. Abstr.*, **60**, 5509 (1963).
499. F. Asinger, L. Schroeder, and S. Hoffmann, *Annalen*, **648**, 83 (1961); *Chem. Abstr.*, **56**, 7297.
500. T. Uno and S. Akihama, *Yakugaku Zasshi*, **81**, 585 (1961); *Chem. Abstr.*, **55**, 19904.
501. F. Asinger, M. Thiel, and K. Gewald, *Annalen*, **639**, 133 (1961); *Chem. Abstr.*, **55**, 23495.
502. N. Gill, N. K. Ralhan, H. S. Sachdev, and K. S. Narang, *J. Org. Chem.*, **26**, 968 (1961); *Chem. Abstr.*, **55**, 23552.
503. R. N. Hurd and G. De La Mater, *Chem. Rev.*, **61**, 45 (1961).
504. A. A. Durgaryan, S. A. Titanyan, and R. A. Kazaryan, *Izvest. Akad. Nauk Armyam. S.S.R., Khim. Nauki*, **14**, 165 (1961); *Chem. Abstr.*, **56**, 4741.
505. J. Okamiya, *Nippon Kagaku Zasshi*, **82**, 87 (1961); *Chem. Abstr.*, **56**, 10122.
506. J. E. Mulvaney and C. S. Marvel, *J. Org. Chem.*, **26**, 95 (1961); *Chem. Abstr.*, **55**, 19902.
507. J. S. Bhalla, N. K. Ralhan, and K. S. Narang, *J. Sci. Ind. Res., India*, 291 (1962); *Chem. Abstr.*, **57**, 9836.
508. T. Uno, S. Akihama, and K. Asakawa, *Yakugaku Zasshi*, **82**, 257 (1962); *Chem. Abstr.*, **58**, 3410.
509. I. Gaile, E. Gudriniece, and G. Vanags, *Doklady Akad. Nauk SSSR*, 817 (1962); *Chem. Abstr.*, **58**, 6816.
510. I. Simiti, M. Farkas, and S. Silberg, *Berichte*, **95**, 2672 (1962); *Chem. Abstr.*, **58**, 9043.
511. F. Asinger, M. Thiel, P. Puechel, F. Haaf, and W. Schaefer, *Annalen*, **660**, 85 (1962); *Chem. Abstr.*, **58**, 9058.
512. M. Poite and J. Metzger, *Bull. Soc. Chim. France*, 2078 (1962); *Chem. Abstr.*, **58**, 11340.
513. A. Cormons, Thesis, University of Marseille, France, 1962.
514. H. Erlenmeyer and M. Simon, *Helv. Chim. Acta*, **25**, 528 (1942); *Chem. Abstr.*, **36**, 6539.
515. P. Reynaud, M. Robba, and R. C. Moreau, *Bull. Soc. Chim. France*, 1735 (1962); *Chem. Abstr.*, **58**, 6816.
516. J. Jonas, M. Brokl, and J. Borkovec, *Spisy Priorodovedecke Fak. Univ. Brne*, **429**, 43 (1962); *Chem. Abstr.*, **58**, 4534.
517. J. Okamiya, *Nippon Kagaku Zasshi*, **83**, 209 (1962); *Chem. Abstr.*, **59**, 3905.
518. A. Bonzom and J. Metzger, *Bull. Soc. Chim. France*, **11**, 2582 (1963); *Chem. Abstr.*, **60**, 5474.

    A. Bonzom, Thesis, University of Marseille, France, 1963.
519. G. Vernin and J. Metzger, *Bull. Soc. Chim. France*, 2498 (1963); *Chem. Abstr.*, **60**, 4122.

    G. Vernin, Thesis, University of Marseille, 1963.
520. German Patent No. 1 159 451; *Chem. Abstr.*, **60**, 9282.
521. M. O. Kolosova, *Zh. Priklad. Khim.*, **36**, 931 (1963); *Chem. Abstr.*, **59**, 6380.
522. W. J. Humphlett and R. W. Lamon, *J. Amer. Chem. Soc.*, **85**, 2144 (1963).
523. F. Johnson and W. A. Nasutavicus, *J. Org. Chem.*, **28**, 1877 (1963); *Chem. Abstr.*, **59**, 3905.

524. Japan Patent Nos. 18 690 and 18 691; *Chem. Abstr.* **60,** 4217 (1963).
525. G. A. Cortiguera, *Rev. Real Akad. Cienc. Exact. Fis. Nat. Madrid,* **57,** 293 (1963); *Chem. Abstr.,* **60,** 9259.
526. J. Jonas, *Spisy Priorodovedecke Fak. Univ. Brne,* **447,** 405 (1963); *Chem. Abstr.* **60,** 15852.
527. K. Gewald, H. Boettcher, and R. Mayer, *J. Prakt. Chem.,* **23,** 298 (1964); *Chem. Abstr.,* **60,** 15853.
528. S. Bilinski, J. Matysik, and T. Uban, *Ann. Univ. M. Curie Sklodowska, Sect. AA,* **19,** 49 (1964); *Chem. Abstr.,* **60,** 46364.
529. Japanese Patent No. 10 413; *Chem. Abstr.,* **63,** 5656 (1965).
530. M. Robba and R. C. Moreau, *Ann. Pharm. Franc.,* **22,** 14 (1964).
531. A. L. Mndzhoyan, A. A. Aroyan, M. A. Kaldrikyan, T. R. Ovsepyan, and R. H. Arshakyan, *Izvest. Akad. Nauk Armyan. S.S.R., Khim. Nauki,* **17,** 204 (1964); *Chem. Abstr.,* **61,** 8298.
532. M. Lora-Tamayo, C. Sunkel, and R. Madronero, *Bull. Soc. Chim. France,* **2,** 251 (1964); *Chem. Abstr.,* **61,** 7005.
533. T. Pyl, H. D. Dinse, and O. Sietz, *Justus Liebigs Ann. Chem.,* **676,** 141 (1964).
534. I. Iwai and T. Hiraoka, *Chem. Pharm. Bull.,* **12,** 813 (1964); *Chem. Abstr.,* **61,** 9487.
535. A. B. Sen and S. S. Chatter, *J. Indian Chem. Soc.,* **41,** 6485 (1964).
536. J. B. Hendrickson, R. Rees, and J. F. Templeton, *J. Amer. Chem. Soc.,* **86,** 107 (1964); *Chem. Abstr.,* **60,** 6804.
537. N. Saldabols and A. Medne, *Zh. Obshchei Khim.,* **34,** 980 (1964); *Chem. Abstr.,* **60,** 15852.
538. L. Giammanco, *Atti Accad. Sci., Lettere Artl. Palermo,* **123,** 139 (1964); *Chem. Abstr.,* **62,** 10425.
539. B. G. Yasnitskii and E. B. Dol'berg, *Metody Polucheniya Khim. Reak. Prep.,* **11,** 22 (1964); *Chem. Abstr.,* **64,** 17568.
540. F. P. Doyle and J. H. C. Nayler, "Penicillin and Related Structures," in "Advances in Drug Research," N. J. Harper and B. Simmonds, Eds., Academic, New York, 1964, Vol. 1, p. 2.
541. A. Babadjamian, Thesis, University of Marseille, France, 1965.
542. H. Behringer and D. Weber, *Annalen,* **682,** 196 (1965); *Chem. Abstr.,* **62,** 14649.
543. German Patent No. 1 187 620.
544. Netherlands Appl. No. 6 508 138; *Chem. Abstr.,* **64,** 17611A.
545. E. Bulka and H. D. Dinse, *Z. Chem.,* **5,** 376 (1965); *Chem. Abstr.,* **64,** 3514.
546. E. Bulka, H. G. Rohde, and H. Beyer, *Chem. Ber.,* **98,** 259 (1965); *Chem. Abstr.,* **62,** 10426.
547. W. R. Shernan and A. Von Esch, *J. Med. Chem.,* **8,** 25 (1965); *Chem. Abstr.,* **62,** 5269.
548. J. M. Craven and T. M. Fischer, *J. Polymer Sci. Part B, Polymer Letters,* **3,** 35 (1965); *Chem. Abstr.,* **62,** 6630.
549. J. Okamiya, *Nippon Kagaku Zasshi,* **86,** 315 (1965); *Chem. Abstr.,* **63,** 4123.
550. N. Gerencevic, A. Castek, and M. Prostenik, *Bull. Sci., Conseil Acad. R.S.F., Yougoslavie,* **10,** 34 (1965); *Chem. Abstr.,* **63,** 16326.
551. M. Sekiya and Y. Osaki, *Chem. Pharm. Bull. Tokyo,* **13,** 1319 (1965); *Chem. Abstr.,* **64,** 12674.
552. E. Haruki, S. Izumita, and E. Imoto, *Nippon Kagaku Zasshi,* **86,** 942 (1965); *Chem. Abstr.,* **65,** 13688A.
553. U.S. Patent No. 3 201 409; *Chem. Abstr.,* **63,** 18098.
554. French Patent No. 1 413 905; *Chem. Abstr.,* **64,** 6658H.

555. Y. Yamamoto, H. Nakamura, R. Yoda, N. Kaneko, M. Mikawa, and S. Mizutani, *Kyoritsu Yakka Daigaku Kenkyu Nempo*, **10**, 56 (1956); *Chem. Abstr.*, **66**, 94939.
556. *Netherlands* Appl. 6 410 554; *Chem. Abstr.*, **63**, 9923.
557. F. Gagiu, G. Csavassy, and G. Valau, *Bull. Soc. Chim. France*, **2**, 686 (1966); *Chem. Abstr.*, **64**, 17568.
558. S. Bilinski and L. Bielak, *Ann. Univ. M. Curie Sklodowska, Sect. D*, **21**, 263 (1966); *Chem. Abstr.*, **69**, 9526.
559. H. Beyer and K. Pommerening, *Chem. Ber.*, **99**, 2931 (1966); *Chem. Abstr.*, **65**, 18571F.
560. S. Avramovici, E. Apachitei, and I. Zugravescu, *Anal. Sti., Univ. Al. I. Cuza Iasi. Sect. Ic*, **12**, 73 (1966); *Chem. Abstr.*, **67**, 3019.
561. Y. Usui and C. Matsumura, *Yakugaku Zasshi*, **86**, 87 (1966); *Chem. Abstr.*, **64**, 12658.
562. K. E. Schulte, J. Reisch, and M. Sommer, *Arch. Pharm.*, **299**, 107 (1966); *Chem. Abstr.*, **64**, 14190.
563. P. C. Rath, P. K. Jesthi, P. K. Mishra, and M. K. Rout, *Indian J. Chem.*, **4**, 24 (1966); *Chem. Abstr.*, **64**, 17752.
564. R. Gompper, M. Gaeng, and F. Saygin, *Tetrahedron Lett.*, **17**, 1885 (1966); *Chem. Abstr.*, **65**, 713D.
565. F. Gagiu and A. L. Mavrodin, *Farm. Bucharest*, **14**, 21 (1966); *Chem. Abstr.*, **65**, 2243F.
566. R. G. Dubenko, *Zh. Org. Khim.*, **2**, 485 (1966); *Chem. Abstr.*, **65**, 8890D.
    R. G. Dubenko, *Zh. Org. Khim.*, **2**, 465 (1966); *Chem. Abstr.*, **65**, 8800.
567. H. Beyer and H. Schilling, *Chem. Ber.*, **99**, 2118 (1966); *Chem. Abstr.*, **65**, 8892B.
568. V. S. Bhagwat and K. S. Nargund, *J. Indian Chem. Soc.*, **43**, 323 (1966); *Chem. Abstr.*, **65**, 8890H.
569. J. Okamiya, *Nippon Kagaku Zasshi*, **87**, 594 (1966); *Chem. Abstr.*, **65**, 15362B.
570. C. Hennart, *Bull. Soc. Chim. France*, 2093 (1966); *Chem. Abstr.*, **65**, 18573A.
571. Japanese Patent No. 4785; *Chem. Abstr.*, **65**, 721E (1966).
572. A. Silberg, M. Ruse, and A. Bodor, *Stud. Univ. Babes Bolyai, Ser. Chem.*, 57 (1966); *Chem. Abstr.*, **66**, 2507.
573. A. De Rooker and P. De Radzitky, *Bull. Soc. Chim. Belges*, **75**, 641 (1966); *Chem. Abstr.*, **66**, 37814.
574. R. Arnaud, M. Gelus, J. C. Malet, and J. M. Bonnier, *Bull. Soc. Chim. France*, **9**, 2857 (1966); *Chem. Abstr.*, **66**, 37815.
575. M. Masaki, M. Sugiyama, S. Tayama, and M. Ohta, *Bull. Chem. Soc. Japan*, **39**, 2745 (1966); *Chem. Abstr.*, **66**, 46360.
576. G. Pappalardo, B. Tornetta, and G. Scapini, *Farm. Pavia Ed. Sci.*, **21**, 740 (1966); *Chem. Abstr.*, **66**, 46363.
577. M. Robba and Y. Le Guen, *Compt. Rend.*, 1385 (1966); *Chem. Abstr.*, **66**, 94981.
578. R. Cottet, R. Gallo, and J. Metzger, *Bull. Soc. Chim. France*, **12**, 4499 (1967); *Chem. Abstr.*, **69**, 2894T.
579. G. M. Sharma, B. Parshad, and K. S. Narang, *Indian J. Chem.*, **5**, 586 (1967).
580. S. S. Sabnis, *Indian J. Chem.*, **5**, 619 (1967).
581. M. Peretyazhko and P. Pel'kis, *Khim. Geterosikl. Soedinenii*, **3**, 471 (1967); *Chem. Abstr.*, **68**, 2837.
582. A. Silberg and Z. Frenkel, *Bull. Soc. Chim. France*, **6**, 2235 (1967); *Chem. Abstr.*, **68**, 12889.
583. K. Ziemelis, E. Gudriniece, and E. K. Matseevskaya, *Latv. Psr Zinat. Akad. Vestis, Kim. Ser.*, **4**, 445 (1967); *Chem. Abstr.*, **68**, 68920.

E. Gudriniece, E. K. Matseyevskaya, and K. Ziemelis, *Latv. Psr Zinat. Akad. Vestis, Kim. Ser.*, **5,** 559 (1967); *Chem. Abstr.*, **68,** 39527.
584. M. O. Kolosova, *Khim. Farm. Zh.*, **6,** 27 (1967); *Chem. Abstr.*, **68,** 49493.
585. A. E. Silberg, Z. Frenkel, I. Szotyori, and V. Axente, *Farm. Bucharest*, **15,** 275 (1967); *Chem. Abstr.*, **68,** 49498.
586. A. Ya. Strakov and D. Brutane, *Latv. Psr Zinat. Akad. Vestis*, **5,** 591 (1967); *Chem. Abstr.*, **68,** 49499.
587. U. H. Lindberg, J. Pedersen, and B. Ulff, *Acta Pharm. Suecica*, **4,** 269 (1967); *Chem. Abstr.*, **68,** 105069.
588. F. Gagiu, G. Csavassy, and G. Todor, *Arch. Pharm. Weinheim*, **300,** 964 (1967); *Chem. Abstr.*, **68,** 105070.
589. A. Silberg, Z. Frenkel, L. Szotyori, and E. Stanciu, *Rev. Roumaine Chim.*, **12,** 905 (1967).
590. J. Sheckri and S. Alazawe, *J. Indian Chem. Soc.*, **44,** 800 (1967); *Chem. Abstr.*, **68,** 114489.
591. V. A. Krasovskii and S. I. Burmistrov, *Isobret., Prom. Obratztsy, Tovarnye Znaki*, **44,** 26 (1967); *Chem. Abstr.*, **68,** 105187.
U.S.S.R. Patent No. 199 895; *Chem. Abstr.*, **68,** 105187.
592. U.S. Patent No. 3 335 149; *Chem. Abstr.*, **67,** 73603S.
593. French Patent No. 88 794; *Chem. Abstr.*, **67,** 82213.
594. G. Vernin, J. P. Aune, H. J. M. Dou, and J. Metzger, *Bull. Soc. Chim. France*, **12,** 4521 (1967); *Chem. Abstr.*, **69,** 19062.
595. R. Vivaldi, H. J. M. Dou, and J. Metzger, *Compt. Rend.*, **264,** 1652 (1967); *Chem. Abstr.*, **67,** 64280.
596. G. Vernin, J. P. Aune, H. J. M. Dou, and J. Metzger, *Bull. Soc. Chim. France*, **12,** 4521 (1967); *Chem. Abstr.*, **69,** 19062.
597. P. Bastianelli, M. Chanon, and J. Metzger, *Bull. Soc. Chim. France*, **6,** 1948 (1967); *Chem. Abstr.*, **67,** 73544.
M. Chanon, Thesis, University of Marseille, France, 1967.
598. K. Gewald, P. Blauschmidt, and R. Mayer, *J. Prakt. Chem.*, **35,** 97 (1967); *Chem. Abstr.*, **66,** 85721.
599. U.S. Patent No. 3 299 087; *Chem. Abstr.*, **66,** 85783T.
600. A. K. Bhattacharya, *J. Indian Chem. Soc.*, **44,** 57 (1967); *Chem. Abstr.*, **67,** 43723.
601. B. C. Dash and G. N. Mahapatra, *Indian J. Chem.*, **5,** 40 (1967); *Chem. Abstr.*, **67,** 73559.
602. French Patent No. 88 775; *Chem. Abstr.*, **67,** 7360W.
603. A. Benko, J. Zsako, and P. Nagy, *Chem. Ber.*, **100,** 2178 (1967); *Chem. Abstr.*, **67,** 81735.
604. A. K. Bhattacharya, *Indian J. Chem.*, **5,** 62 (1967); *Chem. Abstr.*, **67,** 82179.
605. A. L. Mndzhoyan, M. A. Kaldryakan, R. Melikogandzhanyan, and A. A. Aroyan, *Armyan. Khim. Zhur.*, **20,** 51 (1967); *Chem. Abstr.*, **67,** 100049.
606. A. S. Azaryan, R. G. Melik-Ogandzhanyan, M. A. Kaldrikyan, and A. A. Aroyan, *Armyan. Khim. Zhur.*, **20,** 135 (1967); *Chem. Abstr.*, **67,** 100051.
607. F. Asinger and H. Offermanns, *Angew. Chem. Internat. Edn.*, **79,** 953 (1967).
608. A. Takamizawa, K. Hirai, T. Ishiba, and Y. Matsumoto, *Chem. Pharm. Bull.*, **15,** 731 (1967).
609. B. M. Regan, F. T. Galysh, and R. N. Morris, *J. Med. Chem.*, **10,** 649 (1967); *Chem. Abstr.*, **67,** 73272.
610. E. Schmitz, R. Ohme, and S. Schramm, *Justus Liebigs Ann. Chem.*, **702,** 131 (1967).
611. M. Chanon and J. Metzger, *Bull. Soc. Chim. France*, **7,** 2842 (1968).

IX. References

612. U.S. Patent No. 3 391 151; *Chem. Abstr.*, **69,** 67368.
     U.S. Patent No. 3 379 700; *Chem. Abstr.*, **69,** 11257.
613. U.S.S.R. Patent No. 216 004; *Chem. Abstr.*, **69,** 67369.
614. V. S. Bhagwat and K. S. Nargund, *J. Indian Chem. Soc.*, **45,** 270 (1968); *Chem. Abstr.*, **69,** 106602.
615. J. Kiss, R. D'Souza, and H. Spiegelberg, *Helv. Chim. Acta*, **51,** 325 (1968); *Chem. Abstr.*, **68,** 78184.
616. J. M. Singh, *Can. J. Chem.*, **46,** 1168 (1968); *Chem. Abstr.*, **68,** 105066.
617. French Patent No. 1 526 370; *Chem. Abstr.*, **71,** 30468.
618. A. Benko and A. Levente, *Justus Liebigs Ann. Chem.*, **717,** 148 (1968); *Chem. Abstr.*, **70,** 37693.
619. A. Babadjamian, M. Chanon, and J. Metzger, *Compt. Rend.*, **267,** 918 (1968); *Chem. Abstr.*, **70,** 47344.
620. A. Babadjamian and J. Metzger, *Bull. Soc. Chim. France*, **12,** 4878 (1968); *Chem. Abstr.*, **70,** 68234.
621. G. Westphal and H. Wasicki, *Z. Chem.*, **8,** 337 (1968); *Chem. Abstr.*, **70,** 11612.
622. R. S. Egan, J. Tadanier, D. L. Garmaise, and A. P. Gaunce, *J. Org. Chem.*, **33,** 4422 (1968); *Chem. Abstr.*, **70,** 47352.
623. I. I. Chizhevskaya, M. I. Zavadskaya, and N. N. Khovratovich, *Khim. Geterotsikl. Soedinenii*, **6,** 1008 (1968); *Chem. Abstr.*, **70,** 68257.
624. British Patent No. 1 112 128; *Chem. Abstr.*, **69,** 77258.
625. H. Baganz and J. Rueger, *Chem. Ber.*, **101,** 3872 (1968); *Chem. Abstr.*, **70,** 19965.
626. T. R. Govindachari, S. Rajappa, and K. Nagarajan, *Helv. Chim. Acta*, **51,** 2102 (1968); *Chem. Abstr.*, **70,** 28861.
627. W. Hampel and I. Muller, *J. Prakt. Chem.*, **38,** 320 (1968); *Chem. Abstr.*, **70,** 37690.
628. G. C. Barrett and J. R. Chapman, *Chem. Commun.* **6,** 335 (1968); *Chem. Abstr.*, **69,** 10704.
629. C. L. Schilling and J. E. Mulvaney, *Macromolecules*, **1,** 445 (1968); *Chem. Abstr.*, **69,** 107146.
630. A. Silberg and Z. Frenkel, *Rev. Roumaine Chim.*, **13,** 1251 (1068); *Chem. Abstr.*, **70,** 96689.
631. E. Hannig, and H. Ziebandt, *Pharmazie*, **23,** 552 (1968); *Chem. Abstr.*, **70,** 37374.
632. D. L. Garmaise, C. H. Chambers, and R. C. Macrae, *J. Med. Chem.*, **11,** 1205 (1968); *Chem. Abstr.*, **70,** 3909.
633. I. Simiti and M. Farkas, *Bull. Soc. Chim. France*, **9,** 3862 (1968); *Chem. Abstr.*, **70,** 3914.
634. A. Boucherle, G. Carraz, M. I. Hicter, H. Beriel, and M. Trivin, *Chim. Ther.*, **3,** 360 (1968); *Chem. Abstr.*, **70,** 106424.
635. E. Gudriniece and V. Kuzmina, *Latv. Psr Zinat. Akad. Vestis, Kim. Ser.*, **4,** 476 (1969); *Chem. Abstr.*, **72,** 55312.
636. I. Simiti and M. M. Coman, *Bull. Soc. Chim. France*, **9,** 3276 (1969); *Chem. Abstr.*, **72,** 21635.
637. A. Friedmann and J. Metzger, *Compt. Rend.*, **269,** 1000 (1969); *Chem. Abstr.*, **72,** 43551.
638. K. Brown and R. A. Newberry, *Tetrahedron Lett.* **32,** 2797 (1969); *Chem. Abstr.*, **71,** 91367.
639. M. Robba and Y. Le Guen, *Bull. Soc. Chim. France*, **5,** 1762 (1969); *Chem. Abstr.*, **71,** 81255.
640. French Patent No. 1 561 433; *Chem. Abstr.*, **72,** 43654.
641. J. P. Aune, Thesis, University of Marseille, France, 1969.

642. J. P. Catteau, A. Lablache-Combier, and A. Pollet, *J. Chem. Soc. D*, 1018 (1969); *Chem. Abstr.*, **71**, 112850.
643. G. Kempter, H. Schaefer, and G. Sarodnick, *Z. Chem.*, **9**, 186 (1969); *Chem. Abstr.*, **71**, 30388.
644. U.S.S.R. Patent No. 242 902; *Chem. Abstr.*, **71**, 70591.
645. U.S. Patent No. 3 467 666; *Chem. Abstr.*, **72**, 3480.
646. U.S. Patent No. 3 458 526; *Chem. Abstr.*, **72**, 31777.
647. German Patent No. 1 917 432; *Chem. Abstr.*, **72**, 43655.
648. French Addn. 94 123; *Chem. Abstr.*, **72**, 100685.
649. R. Vivaldi, H. J. M. Dou, G. Vernin, and J. Metzger, *Bull. Soc. Chim. France*, **11**, 4014 (19); *Chem. Abstr.*, **72**, 55313.
650. F. Gagiu, G. Csavassy, and E. Bebesel, *J. Prakt. Chem.*, **311**, 168 (1969); *Chem. Abstr.*, **70**, 87643.
651. British Patent No. 1 137 529; *Chem. Abstr.*, **71**, 30469.
652. South African Patent No. 68 05975; *Chem. Abstr.*, **72**, 3479.
653. U. S. Patent No. 3 467 666; *Chem. Abstr.*, **72**, 3480.
654. I. M. Goldman, *J. Org. Chem.*, **34**, 3285 (1969); *Chem. Abstr.*, **72**, 12658.
655. N. Saldabols and V. V. Krylova, *Khim. Geterotsikl. Soedinenii*, **3**, 555 (1969); *Chem. Abstr.*, **72**, 21633.
656. M. O. Lozinskii, S. N. Kukota, P. S. Pel'kis, and T. N. Kudrya, *Khim. Geterotsikl. Soedinenii*, **4**, 757 (1969); *Chem. Abstr.*, **72**, 43547.
657. German Patent No. 1 917 432; *Chem. Abstr.*, **72**, 43655.
658. Japanese Patent No. 69 32 406; *Chem. Abstr.*, **72**, 79030.
659. British Patent No. 1 171 524; *Chem. Abstr.* **72**, 90519.
660. R. Bognar, I. Farkas, L. Szilagy, M. Menyhart, E. Nemes, and F. Szabo, *Acta Chim. Acad. Sci. Hung.*, **62**, 179 (1969); *Chem. Abstr.*, **72**, 90801.
661. P. Bessin, O. Tetu, and M. Selim, *Chim. Ther.*, **4**, 220 (1969); *Chem. Abstr.*, **71**, 101762.
662. J. Jadot, J. Casimir, and R. Warin, *Bull. Soc. Chim. Belges*, **78**, 299 (1969).
663. M. Ohta and H. Kato, "Nonbenzenoid Aromatics," J. P. Snyder, Ed., Academic, New York, 1969, Vol. 1, p. 117.
664. W. Hampel, *Z. Chem.*, **9**, 61 (1969).
665. W. Hampel and I. Mueller, *J. Prakt. Chem.*, **311**, 684 (1969); *Chem. Abstr.*, **71**, 91363.
666. J. W. Faigle and H. Keberle, *J. Labelled Compounds*, **5**, 173 (1969); *Chem. Abstr.*, **71**, 70531.
667. E. Zbiral and H. Hengstberger, *Justus Liebigs Ann. Chem.*, **721**, 121 (1969); *Chem. Abstr.*, **70**, 96863.
668. G. C. Barrett, A. R. Khokhar, and J. R. Chapman, *Chem. Commun.*, **14**, 818 (1969); *Chem. Abstr.*, **71**, 81251.
669. G. C. Barrett and A. R. Khokhar, *J. Chromatogr.*, **39**, 47 (1969).
670. J. M. Singh, *J. Med. Chem.*, **12**, 553 (1969).
671. R. D. Desai, G. S. Saharia, and H. S. Sodhi, *J. Indian Chem. Soc.*, **46**, 115 (1969); *Chem. Abstr.*, **71**, 12872.
672. S. Giri and A. Mahmood, *J. Indian Chem. Soc.*, **46**, 441 (1969).
673. B. S. Kulkarni, B. S. Fernandez, M. R. Patel, R. A. Bellare, and C. V. Deliwala, *J. Pharm. Sci.*, **58**, 852 (1969).
674. D. W. Henry, *J. Med. Chem.*, **12**, 303 (1969); *Chem. Abstr.*, **70**, 96684.
675. W. Hepworth, B. B. Newbould, D. S. Platt, and G. J. Stagey, *Nature*, **221**, 582 (1969).
676. Y. I. Smushkevich, T. M. Babueva, and N. N. Suvorov, *Kim. Geterotsikl. Soedinenii*, **1**, 91 (1969); *Chem. Abstr.*, **70**, 115049.

677. A. Perkone, N. Saldabols, and S. Hillers, *Khim. Geterotsikl. Soedinenii*, **3**, 498 (1969); *Chem. Abstr.*, **72**, 31670.
678. A. A. Aroyan and N. S. Bol'shakova, *Armyan. Khim. Zhur.*, **22**, 601 (1969); *Chem. Abstr.*, **71**, 112588.
679. Japanese Patent No. 70 14 068; *Chem. Abstr.*, **73**, 45502.
680. A. Taurins and A. Blaga, *J. Heterocyclic Chem.*, **7**, 1137 (1970).
681. Japanese Patent No. 70 14 068; *Chem. Abstr.*, **73**, 45502.
682. A. L. Lee, D. MacKay, and E. L. Manery, *Can. J. Chem.*, **48**, 3554 (1970); *Chem. Abstr.*, **74**, 31707.
683. M. Gaudry and A. Marguet, *Tetrahedron*, **26**, 5635 (1970).
684. M. Ohashi, A. Iio, and T. Yonezawa, *J. Chem. Soc. D*, 1148 (1970); *Chem. Abstr.*, **73**, 120547.
685. D. A. Thomas, *J. Heterocyclic Chem.*, **7**, 457 (1970); *Chem. Abstr.*, **72**, 121420.
686. A. K. Pinagrahi, P. L. Nayak, and M. K. Rout, *J. Inst. Chem., Calcutta*, 17 (1970); *Chem. Abstr.*, **72**, 121423.
687. Y. Tamura, T. Miyamoto, K. Shimooka, and T. Masui, *Chem. Ind. London*, 1470 (1970).
688. K. Yamane, K. Fujimori, and R. Oguchi, *Nippon Kagaku Zasshi*, **91**, 395 (1970); *Chem. Abstr.*, **73**, 45396.
689. K. Yamane, K. Fujimori, and M. Hirabagashi, *Yakugaku Zasshi*, **90**, 1363 (1970); *Chem. Abstr.*, **74**, 31705.
690. E. Schmitz and H. Striegler, *J. Prakt. Chem.*, **312**, 359 (1970); *Chem. Abstr.*, **73**, 77118.
691. H. Tripathy, B. C. Dash, and G. N. Mahapatra, *Indian J. Chem.*, **8**, 586 (1970); *Chem. Abstr.*, **73**, 87827.
692. S. P. Misra and P. K. Jesthi, *Current Sci.*, **39**, 417 (1970); *Chem. Abstr.*, **73**, 120551.
693. V. Barkane and E. Gudriniece, *Latv. Psr Zinat. Akad. Vestis, Kim. Ser.*, **6**, 695 (1970); *Chem. Abstr.*, **74**, 53619.
694. Japanese Patent No. 70 37 771; *Chem. Abstr.*, **74**, 87953.
695. German Patent No. 2 023 425; *Chem. Abstr.*, **74**, 87964.
696. S. Rajappa and B. G. Advani, *Indian J. Chem.*, **8**, 1145 (1970); *Chem. Abstr.*, **74**, 125532.
697. B. G. Yanitskii and E. B. Dol'berg, *Metody Polucheniya Khim. Reak. Prep.* **21**, 70 (1970); *Chem. Abstr.*, **76**, 72444.
    B. G. Yanitskii and E. R. Dol'berg, *Metody Polucheniya Khim. Reak. Prep.*, **21**, 79 (1970); *Chem. Abstr.*, **76**, 85741.
698. F. Gagiu, T. Suciu, O. Henegaru, and Z. Gyorfi, *Arch. Pharm. Weinheim*, **303**, 102 (1970); *Chem. Abstr.*, **72**, 129466.
699. F. Kurzer, "Organic Compounds of Sulphur, Selenium and Tellurium" D. H. Reid, Ed., The Chemical Society, London, 1970, Vol. 1, p. 378.
    F. Kurzer, "Organic Compounds of Sulphur, Selenium and Tellurium" D. H. Reid, Ed., The Chemical Society, London, 1973, Vol. 2, p. 587.
    F. Kurzer, "Organic Compounds of Sulphur, Selenium and Tellurium" D. H. Reid, Ed., The Chemical Society, London, 1975, Vol. 3, p. 566.
700. E. Zbiral, *Tetrahedron Lett.*, 5107 (1970).
701. K. Gewald, H. Spies, and R. Mayer, *J. Prakt. Chem.*, **312**, 776 (1970); *Chem. Abstr.*, **74**, 141622.
702. K. Hartke and B. Seib, *Arch. Pharm. Weinheim*, **303**, 625 (1970); *Chem. Abstr.*, **73**, 98859.
703. U. H. Lindberg, G. Bexell, J. Pedersen, and S. Ross, *Acta Pharm. Suecica*, **7**, 423 (1970); *Chem. Abstr.*, **73**, 118624.

704. G. Kempter, H. Schafer, and G. Sarodnick, *Z. Chem.*, **10,** 402 (1970).
705. F. N. Stepanov and S. D. Isaev, *Zh. Org. Khim.*, **6,** 1189 (1970); *Chem. Abstr.*, **73,** 66488.
706. L. I. Zakharkin, A. V. Grebennikov, and A. I. L'vov, *Izvest. Akad. Nauk SSSR Ser. Khim.*, **1,** 106 (1970); *Chem. Abstr.*, **72,** 111544.
707. I. A. Rubstov and B. I. Shapira, *Khim. Farm. Zh.*, **4,** 49 (1970); *Chem. Abstr.*, **73,** 56015.
708. B. Das and G. Mahapatra, *J. Indian Chem. Soc.*, **47,** 98 (1970); *Chem. Abstr.*, **72,** 100579.
709. F. Gagiu, C. Draghici, E. Banu, G. Csavassy, G. Vrejoiu, and M. Theodorescu, *Chim. Ther.*, **5,** 194 (1970); *Chem. Abstr.*, **73,** 75363.
710. L. Rylski, B. Kosakiewicz, B. Dekarz, J. Lewicka, and H. Polkowska-Krajewska, *Acta Polon. Pharm.*, **27,** 349 (1970); *Chem. Abstr.*, **74,** 12779.
711. French Patent No. 2 113 720.
712. C. Roussel, R. Gallo, M. Chanon, and J. Metzger, *Bull. Soc. Chim. France*, **5,** 1902 (1971); *Chem. Abstr.*, **75,** 48966.
713. C. Roussel, Thesis, University of Marseille, France, 1971.
714. C. Roussel, M. Chanon, and J. Metzger, *Tetrahedron Lett.* **21,** 1861 (1971).
715. A. Pollet, Thesis, University of Lille, France, 1971.
716. U.S. Patent No. 3 558 644; *Chem. Abstr.*, **74,** 87951.
717. K. A. Maier and O. Hromatka, *Monatsh.*, **102,** 102 (1971); *Chem. Abstr.*, **74,** 125538.
718. Japanese Patent No. 71 03 972; *Chem. Abstr.*, **74,** 141756.
    Japanese Patent No. 71 06 049; *Chem. Abstr.*, **74,** 141763.
719. Japanese Patent No. 71 03 779; *Chem. Abstr.*, **74,** 141757.
720. German Patent No. 2 053 178; *Chem. Abstr.*, **75,** 63766.
721. J. D. Modi, S. S. Sabnis, and C. V. Deliwala, *J. Med. Chem.*, **14,** 887 (1971); *Chem. Abstr.*, **75,** 129708.
722. E. D. Sych and E. K. Mikitenko, *Khim. Geterotsikl. Soedinenii*, **7,** 857 (1971); *Chem. Abstr.*, **76,** 25154.
723. V. A. Smirnov and A. E. Lipkin, *Khim. Geterotsikl. Soedinenii*, **7,** 1369 (1971); *Chem. Abstr.*, **76,** 46129.
724. M. F. A. Abdel-Lateef, E. Kowalska, and Z. Eckstein, *Bull. Acad. Polon. Sci., Ser. Sci. Chim.*, **19,** 713 (1971); *Chem. Abstr.*, **76,** 99550.
725. K. F. Rafla, *Lybian J. Sci.*, **1,** 56 (1971); *Chem. Abstr.*, **76,** 140626.
726. V. P. Bronovitskaya, A. A. Bakhmedova, A. Y. Berlin, L. M. Abksuva, and Y. Sheinker, *Khim. Geterotsikl. Soedinenii*, **7,** 1493 (1971); *Chem. Abstr.*, **77,** 19572.
727. L. M. Werbel and J. R. Battaglia, *J. Med. Chem.*, **14,** 10 (1971); *Chem. Abstr.*, **75,** 76682.
728. E. Schmitz and H. Striegler, *J. Prakt. Chem.*, 1125 (1971).
729. H. Boehme and R. Braun, *Justus Liebigs Ann. Chem.*, **744,** 20 (1971); *Chem. Abstr.*, **74,** 125962.
    H. Boehme and R. Braun, *Justus Liebigs Ann. Chem.*, **44,** 27 (1971).
730. K. Hirai and T. Ishiba, *Chem. Commun.*, **20,** 1318 (1971).
    K. Hirai, *Tetrahedron Lett.* 1137 (1971).
731. U. H. Lindberg, *Acta Pharm. Suecica*, **8,** 39 (1971).
    U. H. Lindberg, G. Bessel, and B. Ulff, *Acta Pharm. Suecica*, **8,** 49 (1971).
732. C. V. Deliwala, J. D. Modi, and S. S. Sabnis, *J. Med. Chem.*, **14,** 450 (1971); *Chem. Abstr.*, **75,** 35860.
733. J. D. Modi, S. S. Sabnis, and C. V. Deliwala, *J. Med. Chem.*, **14,** 887 (1971); *Chem. Abstr.*, **75,** 129708.

IX. References 333

734. H. G. Garg and R. A. Sharma, *Can. J. Pharm. Sci.*, **6,** 45 (1971); *Chem. Abstr.*, **75,** 76667.
735. K. A. Thakar and N. R. Manjaramkar, *J. Indian Chem. Soc.*, **48,** 621 (1971); *Chem. Abstr.*, **75,** 129706.
736. H. Beyer and S. Schmidt, *Justus Liebigs Ann. Chem.*, **748,** 109 (1971); *Chem. Abstr.*, **75,** 88538.
737. A. K. Panigrahi and M. K. Rout, *J. Indian Chem. Soc.*, **48,** 665 (1971); *Chem. Abstr.*, **75,** 129705.
738. G. Vernin, R. Jauffred, C. Ricard, H. J. M. Dou, and J. Metzger, *J. Chem. Soc. Perkin Trans 2*, 1145 (1972); *Chem. Abstr.*, **77,** 74544S.
739. A. Babadjamian, Thesis, University of Marseille, France, 1972.
740. G. Vernin, H. J. M. Dou, and J. Metzger, *Bull. Soc. Chim. France*, **3,** 1173 (1972).
741. B. Dash and G. N. Mahapatra, *Current Sci.*, **41,** 143 (1972); *Chem. Abstr.*, **76,** 140628.
742. I. Simiti and G. Hintz, *Pharmazie*, **27,** 146 (1972); *Chem. Abstr.*, **77,** 5393.
743. I. Simiti and M. Coman, *Ann. Chim. France*, **7,** 33 (1972); *Chem. Abstr.*, **77,** 75156.
744. M. S. Solanki and J. P. Trivedi, *J. Indian Chem. Soc.*, **49,** 37 (1972).
745. U.S. Patent No. 3 673 201; *Chem. Abstr.*, **77,** 88486.
746. German Patent No. 2 152 557; *Chem. Abstr.*, **77,** 88488.
747. U.S. Patent No. 3 674 871; *Chem. Abstr.*, **77,** 101581.
748. J. M. Blaquist and F. J. Goetz, *J. Heterocyclic Chem.*, **9,** 937 (1972); *Chem. Abstr.*, **77,** 126488.
749. A. Gursoy, *Istanbul Univ. Eczacilik Fak. Mecmuasi.*, **8,** 115 (1972); *Chem. Abstr.*, **79,** 66236.
    A. Gursoy, *Istanbul Univ. Eczacilik Fak. Mecmuasi.*, **8,** 75 (1972).
750. O. H. Oldenziel and A. M. Van Leusen, *Tetrahedron Lett.* **27,** 2777 (1972); *Chem. Abstr.*, **77,** 114291.
751. P. Chauvin, J. Morel, C. Paulmier, and P. Pastour, *Compt. Rend.*, **274C,** 1347 (1972).
752. V. A. Smirnov, A. E. Lipkin, and T. B. Ryskina, *Khim. Farm. Zh.*, **24** (1972); *Chem. Abstr.*, **77,** 109860.
753. I. Iwataki, *Bull. Chem. Soc. Japan*, **45,** 3218 (1972).
754. K. Arakawa, T. Miyasaka, and H. Ohtsuka, *Chem. Pharm. Bull.*, **20,** 1041 (1972).
755. A. Takeda, S. Tsuboi, M. Wada, and H. Kato, *Bull. Chem. Soc. Japan*, **45,** 1217 (1972); *Chem. Abstr.*, **77,** 48106.
756. D. Wobig, *Justus Liebigs Ann. Chem.*, **764,** 125 (1972); *Chem. Abstr.*, **78,** 84309.
757. K. Bunge, R. Huisgen, R. Raab, and H. J. Sturm, *Chem. Ber.*, **105,** 1307 (1972).
758. German Patent No. 2 132 392; *Chem. Abstr.*, **78,** 84395.
759. German Patent No. 2 158 699; *Chem. Abstr.*, **78,** 97634.
760. German Patent No. 2 235 377; *Chem. Abstr.*, **78,** 111292.
761. B. S. Drach, I. Y. Dolgushina, and A. V. Kirsanov, *Zh. Org. Khim.*, **9,** 414 (1973); *Chem. Abstr.*, **78,** 136154.
762. A. F. Oleinik, G. A. Midnikova, and K. Y. Novitskii, *Khim. Geterotsikl. Soedinenii*, **4,** 437 (1973); *Chem. Abstr.*, **79,** 18622.
763. I. Simiti, M. Coman, and I. Schwartz, *Rev. Roumaine Chim.*, **18,** 685 (1973).
764. K. Gewald, M. Hentschel, and R. Heikel, *J. Prakt. Chem.*, **315,** 539 (1973).
765. R. J. C. Kleipool and A. C. Tas, *Riechst Aromen Koeperfluger*, **23,** 326 (1973); *Chem. Abstr.*, **80,** 27158.
766. P. Chauvin, J. Morel, and P. Pastour, *Compt. Rend.*, **276C,** 1453 (1973).
767. F. Eiden and J. Iwan, *Arch. Pharm.*, **306,** 470 (1973).
768. G. Y. Sarkis and S. Al-Azawe, *J. Chem. Eng. Data*, **18,** 99 (1973); *Chem. Abstr.*, **78,** 84310.

769. U. Stauss, H. P. Harter, and O. Schindler, *Chimia*, **27,** 99 (1973); *Chem. Abstr.*, **79,** 140252.
770. C. P. Joshua and P. N. Nambisan, *Indian J. Chem.*, **11,** 118 (1973); *Chem. Abstr.*, **79,** 31973.
771. A. S. Shawali, M. I. Ali, and A. A. Fahmi, *Bull. Chem. Soc. Japan*, **46,** 1798 (1973).
772. E. C. Taylor and R. C. Portnoy, *J. Org. Chem.*, **38,** 806 (1973).
773. M. I. Ali, M. A. Abou-State, and N. M. Hassan, *Indian J. Chem.*, **11,** 4 (1973).
774. H. Neef, K. D. Kohnert, and A. Schellenberger, *J. Prakt. Chem.*, **315,** 701 (1973); *Chem. Abstr.*, **79,** 105186.
775. K. Grohe and H. Heitzer, *Justus Liebigs Ann. Chem.*, **5–6,** 1018 (1973); *Chem. Abstr.*, **79,** 78673.
776. K. Hartke and G. Golz, *Justus Liebigs Ann. Chem.*, **10,** 1644 (1973); *Chem. Abstr.*, **80,** 47893.
777. H. Von Hartmann, H. Schaefer and K. Gewald, *J. Prakt. Chem.*, **315,** 497 (1973).
778. P. H. Porsch, *Dissertation University of Gottingen*, F.R. of Germany 1973.
779. S. Al Azawe and M. A. Bakir, *Bull. Coll. Sci. Univ. Baghdad*, **12–13,** 91 (1973).
780. S. Al Azawe and A. S. Al Tai, *Bull. Coll. Sci. Univ. Baghdad*, **12–13,** 213 (1973).
781. German Patent No. 2 331 246; *Chem. Abstr.*, **80,** 95929.
782. German Patent No. 2 236 796; *Chem. Abstr.*, **80,** 120915.
783. L. Lardicci, C. Battistini, and R. Menicagli, *J. Chem. Soc., Perkin Trans.* 1, **3,** 344 (1974); *Chem. Abstr.*, **80,** 146066.
784. P. Dubs and M. Pesaro, *Synthesis*, 294 (1974); *Chem. Abstr.*, **81,** 25596.
785. O. G. Vitzthum and P. Werkhoff, *Z. Lebensm. Unters. Forsch.*, **156,** 300 (1974).
786. D. Tornetta, F. Guerrera, and G. Ronsivalle, *Ann. Chim. Italy*, **64,** 477 (1974); *Chem. Abstr.*, **84,** 17210.
787. I. Zhelgazkov, N. Todorova, and P. Manolova, *Tr. Nauchno Izled. Khim. Farm. Inst.*, **9,** 54 (1974).
788. A. Gursoy, *Instanbul Univ. Eczacilik Fak. Mecmuasi.*, **10,** 57 (1974); *Chem. Abstr.*, **82,** 125313.
789. German Patent No. 2 423 403.
790. German Patent No. 2 316 185; *Chem. Abstr.*, **82,** 4243.
791. German Patent No. 2 450 814; *Chem. Abstr.*, **83,** 97268.
792. C. B. Nerurkar and D. T. Chaudhari, *Bull. Haffkine Inst.*, **3,** 57 (1975); *Chem. Abstr.*, **83,** 193153.
793. S. N. Sawhney, J. Singh, and O. P. Bansal, *J. Indian Chem. Soc.*, **52,** 561 (1975); *Chem. Abstr.*, **83,** 193158.
794. German Patent No. 2 365 526; *Chem. Abstr.*, **83,** 179041.
795. G. P. Andronnikova, T. E. Pavlovskaya, and Z. V. Pushkareva, *Khim. Farm. Zh.*, **9,** 22 (1975).
796. German Patent No. 2 453 106; *Chem. Abstr.*, **83,** 97271.
797. German Patent No. 2 453 082; *Chem. Abstr.*, **83,** 114378.
798. E. Schroetter, A. Raddatz, and M. Oettel, *J. Prakt. Chem.*, **31,** 520 (1975); *Chem. Abstr.*, **83,** 114274.
799. S. S. Berg, B. J. Peart, and M. P. Toft, *J. Chem. Soc., Perkin Trans.* 1, **11,** 1040 (1975); *Chem. Abstr.*, **83,** 97115.
800. B. Arena, R. Gulbe, and A. Arens, *Latv. Psr Zinat. Akad. Vestis, Kim. Ser.*, **5,** 600 (1975).
801. A. Babadjamian, J. Metzger, and M. Chanon, *J. Heterocyclic Chem.*, **12,** 643 (1975).
802. J. Boedeker, H. Pries, D. Roesch, and G. Malewski, *J. Prakt. Chem.*, **317,** 953 (1975).

803. G. D. Hartman, M. Sletzinger, and L. M. Weinstock, *J. Heterocyclic Chem.*, **12,** 1081 (1975).
804. H. Schaefer and K. Gewald, *J. Prakt. Chem.*, **317,** 771 (1975).
805. R. Lantzsch and D. Arldt, *Synthesis*, **10,** 657 (1975).
806. German Patent No. 2 515 420.
807. P. Dubs, *Synthesis*, 696 (1976).
808. F. Asinger, K. Fabian, H. Vossen, and K. Hentschel, *Justus Liebigs Ann. Chem.*, **3,** 410 (1975); *Chem. Abstr.*, **83,** 58707.
809. U.S. Patent No. 3 882 110.
    German Patent No. 2 422 849.
810. R. G. Buttery, D. G. Guadagni, and R. E. Lundin, *J. Agric. Food Chem.*, **24,** 1 (1976); *Chem. Abstr.*, **84,** 59281.
811. W. Ried and L. Kaiser, *Justus Liebigs Ann. Chem.*, **3,** 395 (1976).
812. G. D. Hartman and L. M. Weinstock, *Synthesis*, **6,** 681 (1976).
813. M. Ferrey, A. Robert, and A. Foucaud, *Synthesis*, **4,** 261 (1976).
814. W. Walek, M. Pallas, and M. Augustin, *Tetrahedron*, **32,** 623 (1976).
815. M. F. A. Abd El Lateef, D. Rusek, J. Cybulski, A. Mizerski, and Z. Eckstein, *Racz. Chem. Caire*, **50,** 323 (1976).
816. Japanese Patent No. 76 48 656.
    German Patent No. 2 544 341.
817. German Patent No. 2 541 720; *Chem. Abstr.*, **85,** 78112.
818. G. Ferrand, F. Eloy, A. Cabrol, and A. St. Blancat, *Eur. J. Med. Chem.*, **11,** 49 (1976).
819. German Patent No. 2 548 505.
820. D. Wobig, *Justus Liebigs Ann. Chem.*, **6,** 1166 (1976); *Chem. Abstr.*, **85,** 108573.
821. A. Babadjamian, R. Gallo, J. Metzger, and M. Chanon, *J. Heterocyclic Chem.*, **13,** 1205 (1976).
822. German Patent No. 2 541 632.
823. G. Vernin, S. Treppendahl, and J. Metzger, *Helv. Chim. Acta*, **60,** 284 (1977).
824. P. Hassanaly, G. Vernin, and H. J. M. Dou, Unpublished Results. (1977)
825. R. Barone, M. Chanon, and J. Metzger, *Rev. Inst. Fr. Pétrole.*, **28,** 771 (1973).
826. R. Barone, M. Chanon, and J. Metzger, *Tetrahedron Lett.* 2761 (1974).
827. R. Barone, M. Chanon, and J. Metzger, Chimia, **32,** 216 (1978).

III

Alkyl, Aryl, Aralkyl, and Related Thiazole Derivatives

JEAN PIERRE AUNE

Institut de Pétroléochimie et de Synthèse Organique Industrielle, Marseille, France

HENRI JEAN-MARIE DOU

Laboratoire de Chimie Organique A, Centre Scientifique de Saint-Jérôme, Marseille, France

JACQUELINE CROUSIER

Laboratoire de Chimie de matèriaux, Centre Scientifique de Saint-Charles, Marseille, France

I. Synthesis 339
   1. Replacement of a Ring Substituent by Hydrogen 339
   2. Dehydrogenation of Thiazolines 340
   3. Reactivity of Side-Chain Functional Substituents 340
   4. Isotopic Synthesis 342
II. Nuclear Magnetic Resonance 342
   1. Chemical Shifts and Coupling Constants 342
      A. Chemical Shifts 342
      B. Coupling Constants 344
   2. Correlations with Electron Densities 344
   3. ^{13}C Chemical Shifts and Coupling Constants 345
III. Mass Spectrometry 347
   1. Alkylthiazoles 348
   2. Arylthiazoles 349

IV. Infrared Spectroscopy 349
   1. C–H Vibrations 349
   2. Combination Vibrations of the $\gamma_{(CH)}$ Modes 351
   3. Vibrations of the Thiazole Ring 351
V. Ultraviolet Spectroscopy 352
   1. Alkylthiazoles 352
   2. Styrylthiazoles 353
   3. Arylthiazoles 353
   4. Charge-Transfer Complexes 354
VI. Basicity and pK_a 355
VII. Other Physicochemical Studies 357
   1. Viscosity and Diffusion 357
   2. Thermodynamic Properties 357
   3. Theoretical Calculations 357
VIII. Gas-Liquid and Thin-Layer Chromatography 358
   1. Gas-Liquid Chromatography 358
   2. Thin-Layer Chromatography 362
IX. Radical Reactivity 364
   1. Free-Radical Reactivity of Thiazoles 364
      A. Reactivity with Aryl Radicals 364
      B. Influence of Radical Source and Medium 366
      C. Reactivity of the Conjugate Acid of Thiazole 368
      D. Comparison of Free-Radical Reactivity with Theoretical Calculations . 370
   2. Thiazolyl Radicals 370
      A. Electron Spin Resonance 373
   3. Photochemical Rearrangement 374
      A. Alkylthiazoles 374
      B. Arylthiazoles 376
X. Nucleophilic Reactivity 378
   1. Organomagnesium Compounds 378
   2. Organolithium Compounds 378
   3. Dimerization 379
   4. Isotopic Effect 379
XI. Electrophilic Reactivity 380
   1. Alkylthiazoles 380
      A. Halogenation 380
      B. Mercuration 380
      C. Nitration 381
   2. Arylthiazoles 382
XII. Reactivity of the Nitrogen Atom 386
   1. Quaternization 386
      A. Steric Effects of Substituents at the 2- and 4-Positions 386
      B. Effect of a Group Adjacent to an Ortho-Substituent 389
      C. Electronic Effects of 5-Substituents 390
      D. Quaternization of Arylthiazoles 391
   2. Thiazole N-Oxides 392
   3. Metallic Complexes of Thiazoles 392
XIII. Condensation on the Methyl Group 392
   1. Reactions with 2-Alkyl Groups 392
   2. Reaction with the Hydrogen Atom at the 2-Position 393

XIV. Substituent Effects 393
XV. Applications 395
    1. Natural Flavors 395
    2. Polymers 395
        A. Polymerization by Total Synthesis of the Thiazole Ring 396
        B. Polymerization through a Substituent on the Thiazole Ring 397
        C. Copolymerization with Dienes 398
        D. Study of Different Monomers 398
    3. Biological and Pharmacological Uses of Thiazoles 399
XVI. Tables of Products 400
XVII. References 492

In this chapter we examine in turn the properties of alkyl and arylthiazoles that do not possess functional groups bonded directly to the thiazole ring. The general trends are underlined, and the applications of certains thiazole compounds in such areas as polymers, flavorings, and pharmacological and agricultural chemicals are discussed.

I. SYNTHESIS

The most general pathways to thiazoles bearing such groups as alkyl, aryl, aralkyl, and alkenyl, substituted or not by functional groups, are the cyclization reactions described in Chapter II. A certain number of indirect methods also exist, though only a few examples of each are given here. Others are discussed in the following chapters, with the more important references cited here.

1. Replacement of a Ring Substituent by Hydrogen

This method has mainly been used to prepare thiazoles nonsubstituted in the 2-position and involves the replacement of a functional substituent (amino, halo, mercapto, hydroxy, or carboxy) by a hydrogen. In this way the often delicate cyclization of thioformamide can be avoided.

The replacement of 2-amino group by a hydrogen can be achieved by diazotization, followed by reduction with hypophosphorous acid (1–8, 13). Another method starting from 2-aminothiazole is to prepare the 2-halothiazole by the Sandmeyer reaction (prepared also from the 2-hydroxythiazole), which is then dehalogenated chemically or catalytically (1, 9, 10).

A 2-carboxy group can be easily replaced by a hydrogen upon heating (11, 12, 14–16).

Another possible route to 2-unsubstituted thiazoles is replacement of a mercapto group by a hydrogen. Various methods have been used: hydrogen peroxide in acid medium (17–19) dilute nitric acid (17), and metallic catalysts (20–22).

2. Dehydrogenation of Thiazolines

Asinger et al. have developed a simple preparative method for variously substituted Δ_3-thiazolines by the action of sulfur and ammonia on ketones.

Dehydrogenation of the Δ_3-thiazoline in the presence of sulfur gives the thiazole (23–30, 853).

A more recent method (31) is to prepare the Δ_3-thiazoline from the mercaptoacetaldehyde dimer, ammonia, and an aldehyde using Asinger's method (32).

Dehydrogenation of 2,5-dihydro-1,3-thiazoles of this type by classical methods ($FeCl_3$, S, . . .) gives low yields, but in gentler conditions using quinones such as chloranil or 2,3-dichloro-5,6-dicyano-1,4-benzoquinone, the reaction is more efficient (31). Yields vary between 67 and 81%, depending on the nature of the alkyl group at the 2-position. This method has also the advantage of being applicable to systems with sensitive groups such as a double bond or thioethers.

3. Reactivity of Side-Chain Functional Substituents

The reactivity of alkylthiazoles possessing a functional group linked to the side-chain is discussed here neither in detail nor exhaustively since it is analogous to that of classical aliphatic and aromatic compounds. These reactions are essentially of a synthetic nature. In fact, the cyclization methods discussed in Chapter II lead to thiazoles possessing functional groups on the alkyl chain if the aliphatic compounds to be cyclized, carrying the substituent on what will become the alkyl side chain, are available. If this is not the case, another functional substituent can be introduced on the side-chain by cyclization and can then be converted to the desired substituent by a classical reaction.

Haloalkylthiazoles are generally prepared by cyclization or from the corresponding hydroxyalkythiazole. They in turn can lead to numerous functional derivatives, such as aminoalkylthiazoles (33–42, 50, 854), ethers

(42–45), and thioethers and mercaptans (37, 47, 48). They can be condensed with malonic derivatives or cyanoacetic esters (49, 33–35, 50), leading, for example, to the synthesis of thiazolylalanine (51, 52). In the Sommelet reaction, they undergo condensation to thiazoleacetaldehyde (53) and are also precursors of phosphorous insecticides (54).

Hydroxyalkylthiazoles are also obtained by cyclization or from alkoxyalkyl-thiazoles by hydrolysis (36, 44, 45, 52, 55–57) and by lithium aluminium hydride reduction of the esters of thiazolecarboxylic acids (58–60) or of the thiazoleacetic acids. The Cannizzaro reaction of 4-thiazolealdehyde gives 4-(hydroxymethyl)-thiazole (53). The main reactions of hydroxyalkyl thiazoles are the synthesis of halogenated derivatives by the action of hydrobromic acid (55, 61–63), thionyl chloride (44, 45, 63–66), phosphoryl chloride (52, 62, 67), phosphorus pentachloride (58), tribromide (38, 68), esterification (58, 68–71), and elimination that leads to the alkenylthiazoles (49, 72).

The three thiazoleacetic acids and higher homologs such as the derivatives of propionic acid have been synthesized (5). They are usually prepared by the Hantzsch's method, either as the free acid or as the ester from which the free acid is obtained by hydrolysis (49, 73).

The three isomers of thiazoleacetic acid can be decarboxylated, the order of facility being $2>5>4$, though the relative stability depends on each particular compound and the reaction conditions (72–75). This reaction may be used to obtain certain alkylthiazoles (73). Malonic derivatives can also be decarboxylated to give aliphatic thiazole acids (49, 51)

The acid function of an aliphatic chain bonded to a thiazole ring can be esterified. The corresponding acid chloride can also be prepared by the action of thionyl chloride, though the reaction is often accompanied by secondary reactions and gives poor yields (49, 74).

The amide can be prepared classically from the acid chloride or by ammonolysis of the ester, prepared directly by cyclization (15, 75, 78–80).

Dehydration of thiazolylacetamide gives the corresponding nitrile in low yields. Nitriles can be reduced to 2-thiazolylethylamine (78–81).

2-(4-Phenyl-5-thiazolyl)acetic acids variously substituted in the 2-position give the corresponding naphtho[1,2]thiazoles in the presence of acetic anhydride and sodium acetate (397, 426, 857).

Oxidation of carbon side-chains has resulted in the synthesis of dithiazolyl ketone (82) and thiazolyl phenyl ketone (83). The hydrocarbon chain can also be dehydrogenated in acetic acid in the presence of mercuric acetate to give alkenyl thiazoles (84, 85). The same reaction can also be carried out using ferric chloride and hydrochloric acid (86).

4. Isotopic Synthesis

Different isotopically substituted thiazoles have been synthesized for diverse studies:

- Biological studies: ^{14}C (87–89) or ^{35}S (90).

- NMR and microwave spectroscopy: ^{13}C (91, 92, 113) or ^{14}N (95, 638).

- Reaction mechanisms, infrared spectroscopy, and mass spectrometry: ^{2}H. Carbanion studies (96–98), infrared and mass spectrometry (97, 99), hydrogen abstraction (13, 28, 349, 572, 843).

II. NUCLEAR MAGNETIC RESONANCE

The NMR spectra of thiazoles show the same behavior as those of aromatic compounds, but the chemical shifts also depend on the two heteroatoms.

1. Chemical Shifts and Coupling Constants

A. Chemical Shifts

The chemical shifts of alkylthiazoles and arylthiazoles are indicated in Tables III-1 and III-2. For all spectra the signals for the 2- and 4-positions are relatively broad in contrast to the sharp signal of the proton in the 5-position (100). Methyl substitution in all cases shifts the protons signals to a higher field, suggesting a certain degree of aromaticity for the ring. The effect of the alkyl groups is more efficiently transmitted through the nitrogen atom than through the sulfur atom. When alkyl groups are in the β-position to the thiazole nucleus their effect on other ring protons is negligible; in the same way, protons in the γ-position to the thiazole nucleus show a chemical shift identical to those observed in the aliphatic series.

TABLE III-1. CHEMICAL SHIFTS OF ALKYLTHIAZOLES

Products	Thiazole ring			Substituents							Ref.
				α-Protons			β-Protons			Other protons	
	H_2	H_4	H_5	2	4	5	2	4	5		
Thiazole	8.77	7.86	7.27	—	—	—	—	—	—	—	101, 103, 104
2-Me	—	7.44	6.97	2.68	—	—	—	—	—	—	101, 102, 103, 104
2,4-DiMe	—	—	6.50	2.61	2.33	—	—	—	—	—	100, 101, 103
2,5-DiMe	—	7.03	—	2.54	—	2.36	—	—	—	—	101
4,5-DiMe	8.26	—	—	—	2.35	2.31	—	—	—	—	101
2-Me-4-Et	—	—	6.70	2.55	2.70	—	—	1.21	—	—	105
2-Me-4-i-Pr	—	—	6.66	2.53	3.01	—	—	1.25	—	—	105
2-Me-4-t-Bu	—	—	6.52	2.63	—	—	—	1.29	—	—	100, 101
2-Et	—	7.58	7.18	3.04	—	—	—	—	—	—	101
2-Et-4-Me	—	—	6.51	2.91	2.33	—	—	—	—	—	101
2-Et-5-Et	—	7.13	—	2.88	—	2.76	1.32	—	1.27	—	101
2-Et-4-Et	—	—	6.70	2.90	2.71	—	1.27	1.23	—	—	105
2-Et-4-i-Pr	—	—	6.67	2.90	2.9	—	1.28	1.25	—	—	105
2-Et-4-t-Bu	—	—	6.55	2.90	—	—	1.35	1.29	—	—	100, 101
2-n-Pr	—	7.51	7.04	2.91	—	—	1.80	—	—	0.97	102
2-i-Pr	—	7.58	7.08	3.31	—	—	1.40	—	—	—	101, 102
2-i-Pr-4-Me	—	—	6.52	3.18	—	2.34	1.34	—	—	—	100, 101
2-i-Pr-5-Me	—	—	7.06	—	3.14	2.37	1.32	—	—	—	101
2-i-Pr-4-Et	—	—	6.70	3.20	2.72	—	1.35	1.25	—	—	105
2-i-Pr-4-i-Pr	—	—	6.65	3.18	3.00	—	1.30	1.25	—	—	105
2-i-Pr-4-t-Bu	—	—	6.53	3.31	—	—	1.40	1.29	—	—	100, 101
2-i-Bu	—	7.54	7.10	2.86	—	—	2.13	—	—	0.98	102
2-t-Bu	—	7.58	7.10	—	—	—	1.44	—	—	—	101, 102
2-t-Bu-4-Me	—	—	6.52	—	2.35	—	1.38	—	—	—	100
2-t-Bu-4-Et	—	—	6.62	—	2.71	—	1.34	1.22	—	—	105
2-t-Bu-5-i-Pr	—	7.16	—	—	—	3.10	1.38	—	1.30	—	101
2-t-Bu-5-Et	—	7.14	—	—	—	2.71	1.37	—	1.28	—	101
2-t-Bu-4-i-Pr	—	—	6.60	—	3.00	—	1.36	1.23	—	—	105
2-t-Bu-4-t-Bu	—	—	6.53	—	—	—	1.40	1.30	—	—	100, 101
2-neopentyl	—	7.61	7.11	2.88	—	—	—	—	—	0.99	102
4-Me	8.50	—	6.74	—	2.45	—	—	—	—	—	101, 103, 104
5-Me	8.38	7.39	—	—	—	2.49	—	—	—	—	101
2,3,5-TriMe	—	—	—	2.52	2.25	2.20	—	—	—	—	105
2,5-DiMe-4-Et	—	—	—	2.46	2.50	2.20	—	1.17	—	—	105
2,5-DiMe-4-i-Pr	—	—	—	2.47	2.84	2.20	—	1.17	—	—	105
2,5-DiMe-4-t-Bu	—	—	—	2.47	—	2.35	—	1.30	—	—	105
2-Cyclohexyl	—	7.67	7.15	—	—	—	—	—	—	2.2–1.1	106
2-Cyclohexyl-4-Me	—	—	6.70	—	2.39	—	—	—	—	2.2–1.2	106
2-Cyclohexyl-4,5-diMe	—	—	—	—	2.28	2.28	—	—	—	2.1–1.2	106
2-Benzyl	—	7.47	6.93	—	—	—	4.18	—	—	7.11	106
2-Benzyl-4-Me	—	—	6.63	—	2.40	—	4.18	—	—	7.20	107
2-Benzyl-5-Me	—	7.20	—	—	—	2.19	4.10	—	—	7.20	107
2-Benzyl-4,5-diMe	—	—	—	—	2.19	2.19	—	—	—	7.17	107
2-Phenethyl	—	7.65	7.03	3.15	—	—	3.15	—	—	7.31	107
2-Phenethyl-4-Me	—	—	6.57	3.14	2.38	—	3.14	—	—	7.17	107
2-Phenethy-5-Me	—	—	7.32	3.10	—	2.33	3.10	—	—	7.32	107
2-Phenethyl-4,5-diMe	—	—	—	3.07	2.23	2.23	3.07	—	—	7.20	107

TABLE III-2. CHEMICAL SHIFTS OF ARYLTHIAZOLES

Products	Thiazole ring			Substituents							Ref.
				α-Protons			β-Protons			Other protons	
	H_2	H_4	H_5	2	4	5	2	4	5		
2-Ph-thiazole	—	7.77	7.16	—	—	—	—	—	—	—	108, 109
4-Ph	8.76	—	7.38	—	—	—	—	—	—	—	108
5-Ph	8.74	—	8.09	—	—	—	—	—	—	—	109
2-Ph-4-Me	—	—	6.71	—	2.42	—	—	—	—	—	108
2-Ph-4-t-Bu	—	—	6.67	—	—	—	—	1.34	—	—	108
2-Ph-5-Me	—	7.39	—	—	—	2.31	—	—	—	—	108
2-Ph-5-Et	—	7.34	—	—	—	2.73	—	—	1.23	—	108
2-Ph-5-i-Pr	—	7.45	—	—	—	3.08	—	—	1.26	—	108
4-Ph-2-Me	—	—	7.12	2.52	—	—	—	—	—	—	108
4-Ph-2-Et	—	—	7.15	2.48	—	—	1.84	—	—	—	108
4-Ph-2-Pr	—	—	7.07	2.56	—	—	—	—	—	0.97	108
4-Ph-2-i-Pr	—	—	7.13	3.26	—	—	1.36	—	—	—	108
4-Ph-2-t-Bu	—	—	7.11	—	—	—	1.42	—	—	—	108
4-Ph-5-Me	8.48	—	—	—	—	2.42	—	—	—	—	108
4-Ph-5-Et	8.55	—	—	2.88	—	—	—	—	1.23	—	108
2,4-DiPh	—	—	7.48	—	—	—	—	—	—	—	109
2,5-DiPh	—	7.78	—	—	—	—	—	—	—	—	109
4,5-DiPh	8.40	—	—	—	—	—	—	—	—	—	109
2-o-tolyl	—	7.53	7.14	—	—	—	—	—	—	2.54	108
2-(p-NO$_2$Ph)	—	8.06	7.58	—	—	—	—	—	—	—	108
4-(p-NO$_2$Ph)	8.94	—	7.71	—	—	—	—	—	—	—	108
4-(p-BrPh)	8.96	—	7.38	—	—	—	—	—	—	—	108
4-(p-MeOPh)	8.76	—	7.47	—	—	—	—	—	—	—	108
4-p-diPh	9	—	7.33	—	—	—	—	—	—	—	109
4-Ph-2,4-diMe	—	—	—	2.62	2.62	—	—	—	—	—	109
4,5-DiPh-2-Me	—	—	—	2.67	—	—	—	—	—	—	109
2,5-DiPh-4-Me	—	—	—	—	2.53	—	—	—	—	—	109

B. *Coupling Constants*

Coupling constants are shown in Table III-3 (101, 103).

2. Correlations with Electron Densities

The correlation is difficult in the case of thiazole and substituted thiazoles because of different effects: field effect and anisotropy of heteroatoms (110), which are very difficult to describe and calculate. When the importance of these two effects is determined it is then possible to have a good correlation between π electron densities and corrected chemical

TABLE III-3. COUPLING CONSTANTS OF DIFFERENT ALKYLTHIAZOLES (101, 103)

Thiazoles	J_{2-4}	J_{4-5}	J_{2-5}
Thiazole	0	3.10	1.80
2-Methyl	—	3.35	—
2-Ethyl	—	3.20	—
2-i-Propyl	—	3.10	—
2-i-Butyl	—	3.20	—
2-t-Butyl	—	3.20	—
2-Neopentyl	—	3.40	—
4-Methyl	—	—	1.90

shifts. The results are summarized in Table III-4. This type of correlation has also been extended to phenylthiazoles (108).

It is also possible to use NMR spectroscopy in acidic solvent for analytical purposes. The difference in chemical shift induced by protonation will allow in some cases the identification of the compound [e.g., phenyl or arylthiazoles (109)].

TABLE III-4. CORRELATION BETWEEN π ELECTRON DENSITIES AND CORRECTED CHEMICAL SHIFTS (108, 110)

Compounds	Positions	Chemical shifts	π Electron densities
Thiazole	2	7.215	0.146
	4	6.636	0.012
	5	6.391	−0.042
2-Methylthiazole	4	6.381	0.014
	5	6.203	−0.052
4-Methylthiazole	2	7.156	0.149
	5	5.989	−0.067
5-Methylthiazole	2	7.016	0.135
	4	6.305	−0.009

3. ^{13}C Chemical Shifts and Coupling Constants

Chemical shifts and coupling constants are described in Tables III-5 and III-6 (112–114).

The ^{13}C–H coupling constants can also be used to calculate interorbital and internuclear angles. For thiazole, very precise results have been

TABLE III-5. ^{13}C CHEMICAL SHIFTS OF ALKYLTHIAZOLES (STANDARD EXTERNAL CS$_2$) (112, 114)

Compounds	Ring carbons			α Carbons			β Carbons	
	2	4	5	2	4	5	2	4
Thiazole	40.3	50.0	74.0	—	—	—	—	—
2-Methyl	28.5	51.4	74.4	175.6	—	—	—	—
2-Ethyl	21.8	51.1	75.2	166.7	—	—	179.3	—
2-i-Propyl	16.7	51.1	75.8	160.2	—	—	170.2	—
2-t-Butyl	13.2	51.2	75.9	155.6	—	—	162.2	—
4-Methyl	40.6	39.5	79.4	—	176.4	—	—	—
4-Ethyl	40.6	33.4	80.75	—	168.2	—	—	179.4
2,4-Dimethyl	29.2	41.0	80.6	175.1	176.9	—	—	—
2-Methyl-4-t-butyl	29.6	27.3	83.9	174.2	158.4	—	—	163.1
2-Ethyl-4-t-butyl	22.7	27.3	84.7	166.2	158.6	—	179.3	163.2
2-i-Propyl-4-i-propyl	17.7	30.6	84.3	159.9	162.0	—	170.2	170.9
2-i-Propyl-4-t-butyl	17.4	27.3	84.9	159.9	158.3	—	170.1	163.0
2-t-Butyl-4-methyl	13.9	41.3	81.4	155.8	176.1	—	162.2	—
2-t-Butyl-4-t-butyl	14.6	27.8	85.1	155.8	158.6	—	162.4	163
4,5-Dimethyl	44.4	43.5	67.1	—	179.2	182.9	—	—
2-Methyl-4-ethyl-5-methyl	33.0	39.9	69.2	174.8	171.0	179.5	—	182.8
2-Methyl-4-i-propyl-5-methyl	33.6	36.7	70.7	174.8	165.7	183.2	—	171.2

TABLE III-6. ^{13}C–H COUPLING CONSTANTS FOR RING CARBONS (112, 114)

Compounds	C$_2$–H$_2$	C$_2$–H$_4$	C$_2$–H$_5$	C$_2$–H$_\alpha$	C$_4$–H$_4$	C$_4$–H$_2$	C$_4$–H$_5$	C$_4$–H$_\alpha$	C$_5$–H$_5$	C$_5$–H$_2$	C$_5$–H$_4$	C$_5$–H$_\alpha$
Thiazole	211.5	15.1	6.8	—	184.5	7.3	15.2	—	188.6	—	15.8	—
2-Methyl	—	—	—	—	184.0	—	—	—	188.5	—	—	—
2-Ethyl	—	—	—	—	185.0	—	6.0	—	188.2	—	16.0	—
2-i-Propyl	—	—	—	—	182.0	—	5.3	—	185.5	—	15.5	—
2-t-Butyl	—	—	—	—	184.2	—	6.2	—	186.7	—	15.6	—
4-Methyl	211.5	—	—	—	—	—	—	—	185.5	—	—	—
4-Ethyl	206.5	—	6.5	—	—	—	—	—	183.0	4.2	—	—
2,4-Dimethyl	—	—	6.9	7.0	—	—	6.5	6.5	188.0	—	—	—
2-Methyl-4-t-butyl	—	—	—	—	—	—	—	—	184.2	—	—	—
2-Ethyl-4-t-butyl	—	—	—	—	—	—	—	—	182.9	—	—	—
2-i-Propyl-4-i-propyl	—	—	—	—	—	—	—	—	183.2	—	—	—
2-i-Propyl-4-t-butyl	—	—	—	—	—	—	—	—	185.0	—	—	—
2-t-Butyl-4-methyl	—	—	—	—	—	—	—	—	185.3	—	—	—
2-t-Butyl-4-t-butyl	—	—	—	—	—	—	—	—	181.7	—	—	—
4,5-Dimethyl	212.0	—	—	—	—	—	—	—	—	—	—	—
2-Methyl-4-ethyl-5-methyl	—	—	—	7.5	—	—	—	—	—	—	—	5.5
2-Methyl-4-i-propyl-5-methyl	—	—	—	7.5	—	—	7.0	—	—	—	—	6.0

TABLE III-7. CALCULATED INTERORBITAL AND INTERNUCLEAR ANGLES IN THIAZOLE (113)

	Angles			Angles	
	Interorbital	Internuclear		Interorbital	Internuclear
SC_2H	121.6	132.4	NC_4C_5	120.3	114.8
NC_2H	118.5	115.0	C_4C_5H	116.8	105.7
SC_2N	119.9	112.6	SC_5H	123.6	143.4
NC_4H	125.6	154.4	C_4C_5S	119.6	110.8
C_5C_4H	114.3	91.8	—	—	—

obtained (see Table III-7) (114). Chemical shifts may also be related to electronic densities on carbon atoms. In this case, it is necessary to take into account $(\sigma + \pi)$ electron densities. With semiempirical methods, it is not possible to obtain good correlations (113), but with CNDO and extended Hückel methods excellent results have been recorded. This confirms that ^{13}C nuclei engaged in thiazole molecules are not submitted to ring currents. (113)

III. MASS SPECTROMETRY

The first mass spectrometric study of thiazoles was carried out by Clark et al. (115), who investigated the method of fragmentation of thiazole, and 4-methyl- and 2,4-dimethylthiazole.

The mass spectra of these products are characterized by an intense molecular ion and ion fragments resulting from the cleavage of the $S-C_2$ and $N-C_4$ bonds with charge retention by the sulfur-containing fragment, which is most probably the corresponding thiirenium ion-radical (115). This fragmentation pattern has been confirmed by mass spectra of deuterated (91, 116, 120) or ^{13}C-substituted thiazoles (91).

With 2-methyl- and 2,4-dimethylthiazole, the methyl thiirenium ion (m/e 72) is obtained, which can easily lose a hydrogen radical to give the m/e 71 ion (confirmed by the metastable peak). This latter can rearrange by ring expansion to give the thietenyl cation whose structure was confirmed in certain spectra by the presence of a metastable peak corresponding to the decomposition of the m/e 71 ion to give the thioformyl cation m/e 45, probably by elimination of acetylene.

1. Alkylthiazoles

The mass spectra of more substituted thiazoles, or those with larger alkyl groups are more complex and involve other fragmentation patterns (117, 118, 374). The molecular ion is still abundant but decreases with increasing substitution past the ethyl group.

For 4,5-dialkylthiazoles, the molecular ion decomposes by two competitive pathways, either loss of HCN followed by elimination of the radical R˙ in the position β to the double bond of the resulting substituted thiirene, or by β cleavage followed by elimination of HCN (119).

Alkylthiazoles, with side-chains of three or more carbons in the 2- or 4-positions, give an intense ion, m-(R-15), as the result of the McLafferty rearrangement (117). Such is the case with 4-propyl-, 4,5-dimethyl-2-propyl-, and 4-butyl-2,5-dimethylthiazole.

With a side-chain of three or more carbon atoms in the 5-position an intense peak is obtained at m-(R-14), resulting from β cleavage of the side-chain. Thus the highest intensity peak for 2,4-dimethyl-5-propylthiazole occurs at m/e 126 (m-29). Secondary ions have been demonstrated by Clark (115), especially at m/e 45.

An important application of these results lies in the analysis of food flavorings using a combination of gas-phase chromatography and mass spectrometry (121, 122). Similarly, metabolic products of "chlomethiazole" have been characterized (123).

2. Arylthiazoles

The mass spectra of phenylthiazoles are characterized by the presence of intense molecular ion peaks, due to the aromatic nature of the molecules, which represent 35, 41, and 44% of the total ionization for 2-, 4-, and 5-phenylthiazoles, respectively.

As in the case of thiazole and the alkylthiazoles, cleavage of the thiazole ring takes place at the 1,2 and 3,4 bonds, confirmed by a metastable peak. The other important peaks result from fragmentation of thiirenium ion, in the case of 4- and 5-phenylthiazole and of the phenyl ring. These latter are generally present in the spectra of all compounds with benzene-ring substituents, they occur at m/e 77, 76, 75, 51, 50, 39 (124). The ion m/e 45 (HCS^+) is always present.

The spectra of alkylarylthiazoles generally possess fragmentation patterns similar to those previously mentioned for alkyl- and arylthiazoles. In this case, scission of the $S-C_2$ and C_3-C_4 bonds of the thiazole ring can occur in ion fragments as well as in the molecular ion (124).

The mass spectra of arylthiazoles with functional substituents on the benzene ring have also been studied (125, 126). They possess the fragmentation pattern of the aromatic derivative corresponding to the substituent together with that of the thiazole ring described previously (126).

The mass spectra of some thiazolylpyridines have also been recorded (127).

IV. INFRARED SPECTROSCOPY

The infrared and Raman spectra of many alkyl and arylthiazoles have been recorded. Band assignment and more fundamental work has been undertaken on a small number of derivatives. Several papers have been dedicated to the interpretation of infrared spectra (128–134, 860), but they are not always in agreement with each other. However, the work of Chouteau (99, 135) is noteworthy. The infrared spectrum of thiazole consists of 18 normal vibrations as well as harmonic and combination bands.

1. C–H Vibrations

The valency vibrational frequencies are in the order: $\nu_{C_5H} > \nu_{C_4H} \geqslant \nu_{C_2H}$. The ν_{C_2H} and ν_{C_4H} bands are almost equivalent, and they are split

TABLE III-8. INFRARED VIBRATION FREQUENCIES OF ALKYL AND ARYLTHIAZOLES (99, 102, 126, 135)

Compounds	ν_{C_5H}	ν_{C_4H}	ν_{C_2H}	δ_{CH_2}	δ_{CH_4}	δ_{CH_5}	γ_{C_4H}	γ_{C_2H}	γ_{C_5H}
Thiazole	3118	3083	3083	1239	1121	1040	881	802	726
2-Methyl	3114	3079	—	—	1157	1053	882		718
4-Methyl	3110	—	3080	1233	—	1136	—	809	728
5-Methyl	—	3078	3078	1235	1105	—	850	787	—
5-i-Propyl	—	3080	3080	1234	1103	—	853	787	—
2-Phenyl	3112	3079	—	—	1138	1050	880	—	720
4-Phenyl	3110	—	3075	—^{a}	—	1040	—	814	727
5-Phenyl	—	3080	3080	1217	1115	—	855	787	—
2-Methyl-4-phenyl	3108	—	—	—	—	1030	—	—	718
2-Phenyl-5-methyl	—	3075	—	—	1135	—	910	—	—
4-Phenyl-5-methyl	—	—	3078	1258	—	—	—	800	—
2-Ethyl	3112	3080	—	—	—^{a}	—^{a}	875	—	715
2-i-Propyl	3116	3080	—	—	—^{a}	—^{a}	877	—	715
2-t-Butyl	3114	3079	—	—	—^{a}	—^{a}	872	—	715
2-n-Propyl	3112	3079	—	—	—^{a}	—^{a}	875	—	715
2-i-Butyl	3112	3078	—	—	—^{a}	—^{a}	875	—	715
2-Neopentyl	3115	3079	—	—	—^{a}	—^{a}	878	—	715

^{a} Not attributed.

only in the case of 5-isopropylthiazole in solution. The order of the out-of-plane vibrational frequencies is $\gamma_{C_4H} > \gamma_{C_2H} > \gamma_{C_5H}$, and the 2- and 4-protons, which appeared equivalent in the valency vibrations, are seen to differ in the case of the out-of-plane deformation frequencies, which are separated by 80 cm^{-1}. These assignments have been verified for methylthiazoles (99) and phenylthiazoles (126) (see Table III-8).

TABLE III-9. HARMONIC AND COMBINATION BANDS FOR THE γ_{CH} VIBRATION (99)

Compounds		$2\gamma_{C_4H}$	$\gamma_{C_4H}+\gamma_{C_2H}$	$\gamma_{C_4H}+\gamma_{C_5H}$	$2\gamma_{C_2H}$	$\gamma_{C_2H}+\gamma_{C_5H}$	$2\gamma_{C_5H}$
Thiazole	a^a	1772	1680	1616	—	—	—
	b	1760	1670	1602	1591	1516	1436
2-Methyl	a	1755	—	1616	—	—	—
	b	1755	—	1592	—	—	1436
4-Methyl	a	—	—	—	1620	1540	[1441]
	b	—	—	—	1610	1524	[1444]
5-Methyl	a	1692	—	—	1570	—	—
	b	1694	1634	—	1568	—	—

^{a} Note: (a) pure liquid; (b) in solution. The frequencies in brackets have also been assigned to another vibration.

2. Combination Vibrations of the $\gamma_{(CH)}$ Modes

The spectra of substituted thiazole derivatives, especially the methyl derivatives, can be used to determine harmonic and combination bands of the $\gamma_{(CH)}$ modes: $2\gamma_{(C_4H)}$, $2\gamma_{(C_2H)}$, $2\gamma_{(C_5H)}$, $\gamma_{(C_4H)} + \gamma_{(C_5H)}$, and $\gamma_{(C_2H)} + \gamma_{(C_5H)}$ (see Table III-9).

It should be noted that several authors have assigned the band at 1616 cm^{-1} to a vibration of the thiazole ring (128–131, 134).

3. Vibrations of the Thiazole Ring

The vibrations of the thiazole ring have been assigned in the case of thiazole, using the appellations ω_1 to ω_7 for the in-plane vibrations and Γ_1 and Γ_2 for the out-of-plane vibrations (135). For the substituted derivatives, they are classified in series numbered I to X (see Table III-10) (99).

It can be seen that of the series numbered I to VIII, five (I, II, III, IV, VII) are relatively constant in position, and because of the similarity of their frequencies they seem to correspond to the ω_1, ω_2, ω_3, ω_4, and ω_7 modes of thiazole. The frequencies classified as V and VI are variable, and it can be noted that:

- For 2- and 5-substituted derivatives the frequencies of the series VI are close to the ω_6 vibration of thiazole.
- For the 4-substituted derivatives the frequencies of the series V are close to the ω_5 vibration of thiazole.

It is possible that for the methyl derivatives of thiazole the variable frequencies of the series V and VI correspond to the modes resulting from a coupling of the $\nu_{(C-CH_3)}$ vibration with the ω_5 vibration in the case of 2- or 5-substituted derivatives and with the ω_6 vibration in the case of 4-substituted derivatives (99).

TABLE III-10. VIBRATIONAL FREQUENCIES OF THE THIAZOLE RING (99)

Compounds	ω_1	ω_2	ω_3	ω_4	ω_5	ω_6	ω_7		Γ_1	Γ_2
Thiazole	1479	1380	1318	864	811R	756	610R	—	606	463R
	I	II	III	IV	V	VI	VII	VIII	IX	X
2-Methyl	1505	1432	1310	874	1175	764	642	567	601	492
4-Methyl	1516	1409	1305	876	830	930	671	553	634	484
5-Methyl	1521	1399	1304	867	1171	742	665	554	604	483

For phenylthiazoles the assignments are complicated by the presence of the benzene ring whose vibrations fall in the same region as those of the thiazole ring. The resulting spectra are analogous to those described above [126].

V. ULTRAVIOLET SPECTROSCOPY

Thiazole shows a first absorption band assigned to a $\pi \rightarrow \pi^*$ type transition, situated at 233.5 nm in ethanol and in water (136, 137). For alkyl or aryl derivatives, the spectra are identical, though sometimes complicated by the presence of substituents.

1. Alkylthiazoles

The data are given in Table III-11.

TABLE III-11. ULTRAVIOLET ABSORPTION OF ALKYL-THIAZOLES IN ETHANOL

Compounds	λ_{max}(nm)	ε	Ref.
Thiazole	233	3650	137, 136
2-Methyl	241	5230	140
2-Ethyl	242.5	5630	140
2-i-Propyl	242.5	5910	140
2-t-Butyl	244	6080	140
4-Methyl	242	5940	140, 858
4-Ethyl	242	6000	858
4-i-Propyl	241	—	858
4-t-Butyl	241	4960	858
2,4-Dimethyl	246	3390	100, 139
4-Methyl-2-ethyl	244.2	4110	100
4-Methyl-2-i-propyl	244.4	3780	100
4-Methyl-2-t-butyl	244.5	4060	100
4-t-Butyl-2-ethyl	242.7	3840	100
4-t-Butyl-2-i-propyl	242.6	3861	100
2,4-Di-t-butyl	242	4050	100
2,5-Dimethyl	245	3710	139
5-Ethyl-2-t-butyl	246	3777	139
5-i-Propyl-2-t-butyl	246	3410	139

2. Styrylthiazoles

The ultraviolet spectra of these compounds are similar to those of *trans* stilbene or of 2- and 4-stilbazole. The effect on the ultraviolet spectrum of various substituents have been found to parallel in many respects the effects produced by the corresponding group in derivatives of aromatic hydrocarbons (142).

3. Arylthiazoles

Arylthiazoles substituted by functional groups follow the same pattern as aromatic hydrocarbons.

For nonsubstituted phenyl thiazoles or for alkylarylthiazoles, one of the problems investigated is the determination of the angle between the aryl and the thiazole rings. In the case of 4,5-diphenylthiazole the problem is complicated by the interaction of the two phenyl rings (126).

Calculations show that the deviation from planarity leads to greater conformational stability for the phenylthiazoles (143, 145). In particular, the potential energy minimum is achieved at a twist angle of about 30° for 4-phenylthiazole, 40° for 2-phenylthiazole, and 45° for 5-phenylthiazole.

Ultraviolet absorption data of different arylthiazoles are indicated in Table III-12.

TABLE III-12. ULTRAVIOLET ABSORPTION OF ARYLTHIAZOLES (126, 134)

	Solvents			
	Ethanol		Cyclohexane	
Compounds	λ_{max} (nm)	log ε	λ_{max} (nm)	log ε
2-Phenyl	287	4.14	285	4.15
2-Phenyl-4-methyl	296	4.10	296	4.12
2-Phenyl-4-t-butyl	294	4.11	296	4.13
2-Phenyl-5-methyl	295	4.16	296	4.13
2-Phenyl-5-ethyl	295	4.16	297	4.19
2-Phenyl-5-i-propyl	294	4.18	296	4.20
4-Phenyl	253	4.18	255	4.16
2-Methyl-4-phenyl	259	4.14	261	4.11
2-Ethyl-4-phenyl	259	4.14	261	4.14
2-i-Propyl-4-phenyl	259	4.14	261	4.13
2-t-Butyl-4-phenyl	260	4.12	261	4.14

TABLE III-12. (Continued)

Compounds	Ethanol λ_{max} (nm)	Ethanol log ε	Cyclohexane λ_{max} (nm)	Cyclohexane log ε
4-Phenyl-5-methyl	250	4.06	254	4.10
4-Phenyl-5-ethyl	250	4.05	254	4.08
4-(p-Methoxyphenyl)	261	4.26	263	4.24
4-(p-Nitrophenyl)	{ 314	4.21	304	—
	221	4.09	221	—
4-(p-Bromophenyl)	{ 298	4.16	299	—
	262	4.30	263	—
5-Phenyl	275	4.11	274	4.10
2-(o-Tolyl)	274	3.98	279	4.02
2,4-Diphenyl	{ 312	3.97	312	3.93
	250	4.41	251	4.41
2,5-Diphenyl	{ 322	4.39	322	4.40
	225	4.07	225	4.07
2,4,5-Triphenylthiazole	{ 323	4.24	327	4.21
	244	4.35	247	4.38
2-Phenyl-4-biphenylyl	292.5	4.41	—	—
4-Phenyl-2-biphenylyl	268	4.39	326	4.2
2,4-bis(Diphenylyl-4)	291	4.61	—	—
2,2'-(p-Phenylene)-bis-(4-methyl)	236	4.16	341	4.57
4,4'-(p-Phenylene)-bis-(2-phenyl)	298.5	4.63	308.5	4.67
2,2'-(p-Phenylene)-bis-(4-phenyl)	261.5	4.85	353.5	4.62
4,4'-(p-Phenylene)-bis-2-(4-biphenylyl)	295	—	—	—
2,2'-(p-Phenylene)-bis[-4-(4-biphenylyl)]	288	—	358	—
Oxidi-2,2'-(p-phenylene)-bis-4-phenyl	262.5	4.10	323.5	3.93

4. Charge-Transfer Complexes

Arylthiazoles derivatives are good subjects for the study of these transfers. Thus the absorption wavelengths and the enthalpies of formation of a series of charge-transfer complexes of the type arylthiazole–TCNE, have been determined (147). The results are given in Table III-13.

TABLE III-13. ULTRAVIOLET ARYLTHIAZOLES-TCNE COMPLEXES (147)

Complexes (thiazole-TCNE)	λ_{max} (nm)	K (liter mole^{-1} cm^{-1})
2-Phenyl	492	1.23
4-Phenyl	512	1.58
5-Phenyl	505	—
2,4-Diphenyl	575	0.75
2,5-Diphenyl	580	0.256
4,5-Diphenyl	525	0.156
2,4,5-Triphenyl	612	0.095
2-Phenyl-4-biphenylyl	609	2.93
4-Phenyl-2-biphenylyl	590	2.46
2,4-bis(4-Diphenylyl)	620	—

VI. BASICITY AND pK_a

The measurement of pK_a for bases as weak as thiazoles can be undertaken in two ways: by potentiometric titration and by absorption spectrophotometry. In the cases of thiazoles, the second method has been used (140, 148–150). A certain number of anomalies in the results obtained by potentiometry in aqueous medium using Henderson's classical equation directly have led to the development of an indirect method of treatment of the experimental results, while keeping the Henderson equation (144).

The pK_a values of the main alkylthiazoles are given in Table III-14.

Parallel to the determination of pK_a, the thermodynamic constants of the ionization reaction

$$BH^+_{aq} + H_2O \rightleftharpoons B_{aq} + H_3O^+_{aq}$$

were measured (150). The results are shown in Table III-15.

Substitution by a methyl group increases ΔG°_i and ΔH°_i, and this increase is attributed to polar effects. As can be seen from Table III-15, there is an increase in ΔG°_i and ΔH°_i of roughly 1 kcal/mole for each methyl group. Similar effects have been observed with picolines and lutidines (151). There is only a slight difference for the isomeric compounds, the substituent effect being weakest for the 5-derivative.

The increments of ΔH°_i in passing from thiazole to methylthiazole are 1.19 ± 0.04 for the 2-isomer, 0.97 ± 0.04 for the 4-isomer, and 0.86 ± 0.04 kcal/mole for the 5-isomer.

Table III-16 shows the estimations of ΔH°_i for di- and trimethylthiazoles based on measured values for thiazole and methylthiazoles (150).

TABLE III-14. pK_a VALUES OF ALKYLTHIAZOLES

Compounds	pK_a (25°C)	Ref.
Thiazole	2.52, 2.55, 2.53	144, 150, 153
2-Methyl	3.42, 3.40	152, 150
2-Ethyl	3.37, 3.33	144, 150
2-n-Propyl	3.35	144, 152
2-i-Propyl	3.28, 3.22	152, 140
2-i-Butyl	3.15, 3.12	144, 140
2-Neopentyl	3.37	144, 152
2,4-Dimethyl	3.8, 3.98	144, 150
2,5-Dimethyl	3.91	150
4,5-Dimethyl	3.73	150
4-Methyl	3.15, 3.16	144, 150
4-Ethyl	3.20, 3.17	144, 148
4-i-Propyl	3.00	148
4-t-Butyl	3.06, 2.87	144, 148
5-Methyl	3.11, 3.03, 3.27	144, 150, 149
5-Ethyl	3.36	149
5-i-Propyl	3.37	149
5-t-Butyl	3.37	149
2,4,5-Trimethyl	4.55	150

These results show that the measured values of H_i^v are close to those calculated from the thiazole value and the increments. That compounds substituted *ortho-ortho* to nitrogen (2,4-dimethyl and 2,4,5-trimethylthiazole) also obey this rule shows that the methyl groups do not interact sterically (150). The same conclusion had been reached in the case of the picolines and lutidines (151).

TABLE III-15. THERMODYNAMIC CONSTANTS OF IONIZATION FOR ALKYLTHIAZOLES (151)

Compounds	$\Delta G°$ (kcal/mole)	$\Delta H°$ (kcal/mole)	$\Delta S°$ (e.u.)
Thiazole	3.48	2.02	−4.9
2-Methyl	4.64	3.21	−4.8
4-Methyl	4.31	2.99	−4.4
5-Methyl	4.13	2.88	−4.2
2,4-Dimethyl	5.43	4.21	−4.1
2,5-Dimethyl	5.33	3.99	−4.5
4,5-Dimethyl	5.15	3.97	−4.0
2,4,5-Trimethyl	6.24	4.95	−4.3
2-t-Butyl	4.09	3.71	−1.3
4-t-Butyl	4.15	3.86	−1.0

TABLE III-16. ESTIMATIONS OF ΔH_i° FOR DI- AND TRIMETHYLTHIAZOLES (150)

Compounds	ΔH_i°(kcal/mole)		$\Delta(\Delta H_i^\circ)$
	Obsd	Estd	
2,4-Dimethyl	4.21±0.03	4.18±0.06	−0.03
2,5-Dimethyl	3.99±0.04	4.07±0.06	+0.08
4,5-Dimethyl	3.97±0.05	3.85±0.06	−0.012
2,4,5-Trimethyl	4.95±0.02	5.04±0.07	+0.09

VII. OTHER PHYSICOCHEMICAL STUDIES

1. Viscosity and Diffusion

The object of these studies has been the determination of the degree of association in thiazole and its alkyl derivatives. Various solvents have been used: cyclohexane (154), carbon tetrachloride (155, 156), benzene and nitrobenzene (157).

For example, a thiazole–cyclohexane solution at 25°C is less viscous than the ideal system, and the deviation from ideality can be explained assuming that in solution there is a breakage between the existing association of the thiazole molecules in pure state (157).

In other cases, association has been demonstrated by means of phase diagrams of binary mixtures (158).

2. Thermodynamic Properties

The determination of specific heats (159) has led to the conclusion that thiazole is associated intermolecularly. The measurements can be carried out by adiabatic calorimetry (159) or by using the observed fundamental vibration frequencies and molecular parameters (160, 161).

3. Theoretical Calculations

Bond orders, charges, dipole moments, and reaction orders have been calculated for thiazole and alkylthiazoles. The order of electron density is $2 < 4 < 5$. Different methods of calculation include LCAO SCF (162)

TABLE III-17. MOLECULAR DIAGRAMS FOR METHYLTHIAZOLES (164)

Compounds	π Charges					Free valences			Bond orders				
	1	2	3	4	5	2	4	5	1,2	2,3	3,4	4,5	5,1
2-Methyl	1.558	0.903	1.412	1.058	1.070	0.453	0.416	0.452	0.594	0.685	0.531	0.785	0.495
4-Methyl	1.541	0.951	1.403	1.021	1.084	0.451	0.413	0.448	0.596	0.684	0.536	0.783	0.501
5-Methyl	1.552	0.958	1.390	1.078	1.021	0.456	0.416	0.443	0.586	0.690	0.536	0.781	0.508

and ω (101, 163). In the case of phenylthiazoles, theoretical studies have been concerned mainly with the determination of electron densities and the degree of planarity of the rings (165, 166).

Some results obtained by the HMO method (one of the more useful for reactivities comparisons) are summarized in Tables III-17 and III-18.

TABLE III-18. POLARIZATION ENERGIES OF METHYLTHIAZOLES (164)

Compounds	Positions	Polarization Energies in β Units		
		Radical	Electrophilic	Nucleophilic
2-Methyl	2	2.04	2.07	2.00
	4	2.52	2.24	2.79
	5	2.25	2.01	2.49
4-Methyl	2	2.12	2.04	2.21
	4	2.39	2.23	2.56
	5	2.20	1.94	2.46
5-Methyl	2	2.12	2.02	2.22
	4	2.46	2.14	2.77
	5	2.14	2.03	2.26

VIII. GAS-LIQUID AND THIN-LAYER CHROMATOGRAPHY

1. Gas-Liquid Chromatography

The relative retention volumes

$$\alpha_r = (t'_R)_{ThX}/(t'_R)_{ThH}$$

for 2-, 4-, and 5-alkylthiazoles (168), 2,4- and 2,5-dialkylthiazoles (168), 2-arylthiazoles (176), and thiazolylpyridines (178) have been measured on weakly polar (DEGS, PMPE, carbowax 20 M, Ucon Polar 50 MB) or nonpolar (silicon SE 30, Apiezon L on support impregnated or not with potassium hydroxide) columns (167).

On nonpolar columns, the compounds of a homologous series separate as a function of their boiling points, and linear relationships have been established between the logarithms of the retention volumes and the number of carbon atoms in the 2-, 4-, and 5-positions (see Fig. III-1).

A comparison of the molar volumes of 2-, 4-, and 5-alkylthiazoles with their relative retention volumes shows that these values also vary in the same direction (see Fig. III-2).

The Kovats' indices (173) of thiazole on various columns are given in Table III-19.

The Kovats' indices of various alkyl, dialkyl, and other 2-substituted thiazoles are given in Table III-20 (174, 175).

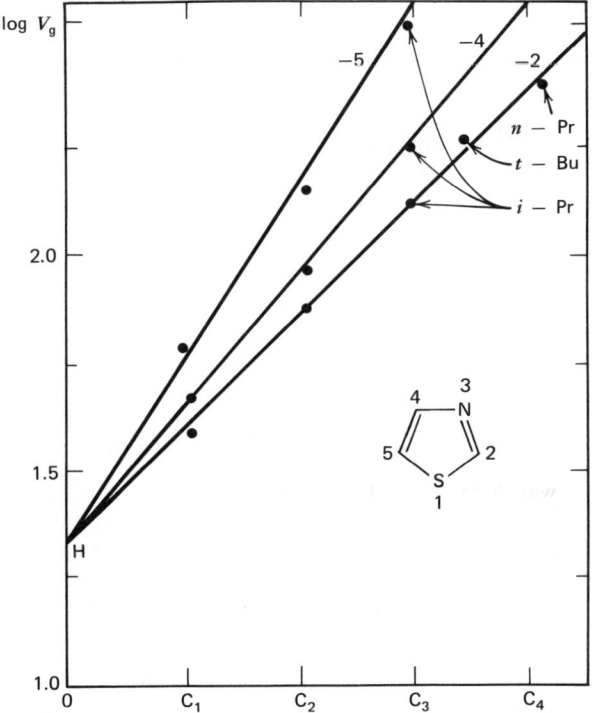

Fig. III-1. Logarithm of the specific retention volumes of 2-, 4-, and 5-alkylthiazoles as a function of carbon atoms in the sidechain (on silicon SE 30 at 70°C).

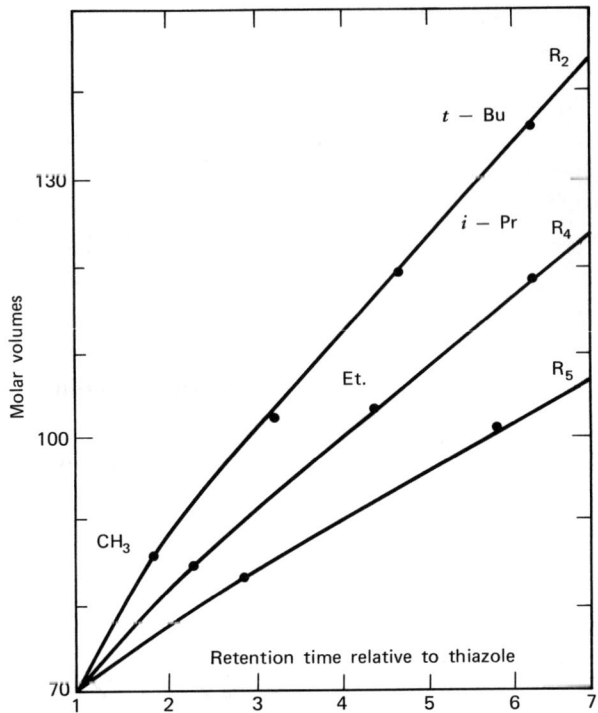

Fig. III-2. Comparison between molar volumes and retention volumes for 2-, 4-, and 5-alkylthiazoles.

TABLE III-19. COMPARISON OF KOVATS' INDICES FOR THIAZOLE (173)

Column[a]	Temperature (°C)	Kovats' Indices for Thiazole
SE 30	85	730
Apiezon L	125	750
Carbowax 20 M	125	1260
DEGS	100	1410
Benzyldiphenyl	100	945
Reoplex	100	1250

[a] 10% on chromosorb W HMDS 60/80 or 80/100; length, 2 m.

TABLE III-20. KOVATS' INDICES FOR THIAZOLES DERIVATIVES[a] (174, 175)

R	Ring positions				
	2	4	5	2,4	2,5
Methyl	795	805	840	—	—
Ethyl	870	900	940	—	—
i-Propyl	920	945	1000	—	—
n-Propyl	970	1000	—	—	—
i-Butyl	1020	—	—	—	—
n-Butyl	1070	—	—	—	—
t-Butyl	940	970	1040	—	—
Methyl, Methyl	—	—	—	870	910
Methyl, Ethyl	—	—	—	965	1000
Methyl, i-Propyl	—	—	—	—	1060
Methyl, t-Butyl	—	—	—	1030	—
Ethyl, Methyl	—	—	—	955	990
Ethyl, Ethyl	—	—	—	1035	1080
Ethyl, i-Propyl	—	—	—	—	1145
Ethyl, t-Butyl	—	—	—	1120	—
t-Butyl, Methyl	—	—	—	1025	1060
t-Butyl, Ethyl	—	—	—	1135	1150
t-Butyl, i-Propyl	—	—	—	—	1200
t-Butyl, t-Butyl	—	—	—	1175	—

[a] Column SE 30, 10% on chromosorb W HMDS 60/80; temperature, 85°C.

From these results the increments of the Kovats' indices (ΔI) of the alkyl groups (R) in the 2- and 4-positions can be calculated using the expression

$$\Delta I = (I)_{\text{ThR}} - (I)_{\text{ThH}}$$

which has been verified to within 5% for 2,4- and 2,5-dialkylthiazoles (see Table III-21).

TABLE III-21. INCREMENTS OF KOVATS' INDICES ΔI (167, 170, 172)

Group R	Alkylthiazoles			Alkylbenzenes
	2	4	5	
Methyl	65	75	110	105
Ethyl	140	170	210	200
i-Propyl	190	215	270	260
t-Butyl	210	240	315	320

These values are practically temperature independant, and they are very close to those found for the Apiezon L column. Comparison with the values of a series of alkybenzenes shows that the 5-position of thiazole possesses behavior analogous to that of a benzenic position in gas-liquid chromatography.

This additivity of incremental indices has enabled the identification of a large number of arylthiazoles variously substituted in the ortho, para, and meta positions of the aromatic ring (174, 175).

The accurate determination of relative retention volumes and Kovats' indices is of great utility to the analyst, for besides being tools of identification, they can also be related to thermodynamic properties of solutions (measurements of vapor pressure and heats of vaporization on nonpolar columns) and activity coefficients on polar columns by simple relationships (179).

2. Thin-Layer Chromatography

A large number of variously 2-, 4-, and 5-substituted thiazoles with alkyl, aryl, hydroxy, methylthio, mercapto, halo, and nitro groups have been analyzed by thin-layer chromatography on silica and alumina by the Stahl's technique (167, 170, 172). Among the many systems recommended for the elution of these compounds are the following:

Compounds	Chromatographic System
Alkyl and polyalkyl	Alumina–CCl_4
	Silica–CH_2Cl_2
Aryl and polyaryl	Alumina–heptane: CH_2Cl_2, 8:2
2-(Methylthio)	Alumina–benzene

Most of the thiazoles studied absorb in the ultraviolet above 254 nm, and the best detection for these compounds is an ultraviolet lamp (with plates containing a fluorescent indicator). Other indicator systems also exist, among which 5% phosphomolybdic acid in ethanol, diazotized sulfanilic acid or Pauly's reagent (Dragendorff's reagent for arylthiazoles), sulfuric anisaldehyde, and vanillin sulfuric acid followed by Dragendorff's reagent develop alkylthiazoles. Iodine vapor is also a useful wide-spectrum indicator.

Linear relationships have been established on one hand between the Rf and pK_a values of these azaaromatic bases (in the absence of steric hindrance of the ring nitrogen) and on the other hand, between the R_M

($R_M = \log[(1/Rf)-1)]$ and the number of CH_2 groups on a side-chain in the 2- or 4-position of the thiazole ring (168).

The steric effects of alkyl substituents (R = methyl, ethyl, i-propyl, t-butyl) on the nitrogen have been related to the steric factors of these same groups as measured in kinetic studies (152).

Application of Snyder's theory of linear chromatographic adsorption (171) gives the variation in adsorption energy of the thiazole nitrogen atom as a function of this steric hindrance for silica and alumina (see Table III-22). These results show that alumina is more sensitive toward steric effects while silica shows a higher selectivity in the case of polar effects.

Thus it would appear that the adsorption site to which the nitrogen atom is attached is smaller in alumina than in silica. Furthermore, the sensitivity of azaaromatic molecules to steric effects is proportional to the adsorption energy of the nitrogen atom in a given chromatographic system. There is also an additivity of overall substituent effects of alkyl groups in different positions of the thiazole ring (with the condition that steric hindrance around the nitrogen atom is not too important) for adsorption on silica and alumina.

TABLE III-22. VARIATION OF THE ADSORPTION ENERGY OF THE THIAZOLE NITROGEN ATOM WITH SUBSTITUENTS [$\Delta Q°$ (–N=)] (171)

Compounds	Alumina[a]	Silica[b]
Thiazole	1	1
2-Methyl	−0.7	0
2-Ethyl	−0.9	−0.1
2-i-Propyl	−1.3	−0.4
2-t-Butyl	−2.0	−0.8
2,4-Dimethyl	−0.8	0
2,4-Di-t-butyl	−4.5	−1.8
2,5-Di-t-butyl	−2.1	−0.9
2-Phenyl	−0.9	−0.4
4-Phenyl	−0.3	−0.2
2,4-Diphenyl	−1.8	−1.2

[a] On alumina GF 254 eluted with CCl_4. The chromatographic parameters determined by Snyders' method are $\alpha = 0.4$ and $p = -1.65$. The energy of adsorption of thiazole is $S° = 6$ and $Q°(-N=) = 4.5$.

[b] On silica GF 254 eluted with benzene. The chromatographic parameters are $\alpha = 0.6$ and $p = -1.2$. The energy of adsorption of thiazole is $S° = 6.8$ and $Q°(-N=) = 5.3$.

IX. RADICAL REACTIVITY

1. Free-Radical Reactivity of Thiazoles

The free-radical reactivity of thiazoles has been well studied with various radicals such as methyl, phenyl, substituted phenyl, cyclohexyl, and aromatic-heterocyclic, in nonpolar solvent or strong acids (180–182).

A. *Reactivity with Aryl Radicals*

Benzoyl peroxide has been the most common source of phenyl radicals. But in reaction with thiazoles the benzoyloxy radical abstracts a hydrogen atom from the thiazole nucleus or from a methyl group in the case of methylthiazoles, giving by-products such as dithiazolyls or 2,2'-dithiazolylethane (183). The results obtained with benzoyl peroxide are summarized in Tables III-23, III-24, and III-25.

TABLE III-23. PHENYLATION OF THIAZOLE WITH BENZOYL PEROXIDE (183)

Compound	Yield (moles)[a]
Phenylthiazoles[b]	0.12
2-Phenylthiazole (47%)	
4-Phenylthiazole (11.5%)	
5-Phenylthiazole (41.5%)	
Secondary Products[c]	
Dithiazolyls[b]	0.023
2,2'-Dithiazolyl (25%)	
2,5'-Dithiazolyl (40%)	
5,5'-Dithiazolyl (15%)	
Other dithiazolyls (20%)	
2,5-Diphenylthiazole	0.005
2,4,5-Triphenylthiazole	Traces
Thiazolyl benzoate	0.2
Benzoic acid	1.45
Diphenyl	0.005
Tars	~0.1

[a] Concentration expressed in moles per mole of benzoyl peroxide used.
[b] Percent of all phenylthiazoles or dithiazolyls.
[c] Secondary products detected in the mixture.

TABLE III-24. PRODUCTS FORMED BY THE DECOMPOSITION OF BENZOYL PEROXIDE IN THIAZOLE, 4-METHYLTHIAZOLE, AND 2,4-DIMETHYLTHIAZOLE WITH A MOLAR RATIO OF $\frac{1}{25}$ AND A REACTION TIME OF 20 hr AT 78°C (184)

Reactants[a]	Thiazole	4-Methylthiazole	2,4-Dimethylthiazole
$(C_6H_5CO_2)_2$	10	12.6	19.36
Thiazole	85	129	224
Products			
Carbon dioxide	0.5(0.3)	1.4(0.62)	1.6(0.45)
Benzoic acid (and phenylbenzoic)	7(1.45)	7.3(1.15)	13(1.3)
Total acids	8.3(1.7)	8.5(1.35)	14.6(1.5)
Basic fraction (*total* 100%)			
Phenylthiazoles	80%	45%	50%
Dithiazoles	15% 1(0.15)	42% 3.5(0.5)	20% 5.2(0.35)
Others	5%	13%	30%
Tars	1%	1.35%	1.5%

[a] The quantities of reactants and products are expressed in grams, the figures in parentheses are the yields in mole per mole of benzoyl peroxide.

TABLE III-25. ACTION OF BENZOYL PEROXIDE ON 4-METHYLTHIAZOLE AT 80°C UNDER NITROGEN ($\frac{1}{25}$ MOLE OF PEROXIDE/MOLE OF THIAZOLE) (184, 185)

Product obtained	Medium			
	Neutral (% of Isomers)	Yield[a]	Acetic acid (% of Isomer)	Yield[a]
4-Methyl-2-phenyl	60	0.22	75	0.35
4-Methyl-5-phenyl	40		25	
4-Methyl-2,5-diphenyl		Traces		
Dithiazolyls		0.18		0.018
2,2'-Dithiazolyl	43.5	—		—
2,5'-Dithiazolyl	44	—		—
5,5'-Dithiazolyl	12.5	—		—
4,4'-Dithiazolylethane	—	Traces	—	—
Diphenyl	—	0.004	—	0.004
Esters	—	0.15	—	0.035
Benzoic acid	—	1.15	—	1.4
Tars	—	0.05	—	low

[a] Yield in mole reported to one mole of benzoic peroxide.

TABLE III-26. PHENYLATION OF ALKYLTHIAZOLES (BENZOYL PEROXIDE, 110°C)[a] (182, 186)

Substrates	Isomers obtained			$_{PhH}^{ThX}K$	Partial rate constants		
	2	4	5		2	4	5
2-Methyl	—	30	70	0.60	—	1	2.4
2-Ethyl	—	30	70	0.60	—	1	2.4
2-i-Propyl	—	27	73	0.55	—	0.89	2.4
2-t-Butyl	—	15	85	0.5	—	0.45	2.55
4-Methyl	40	—	60	1.20	2.9	—	4.3
4-Ethyl	45	—	55	1.15	3.1	—	3.8
4-i-Propyl	65	—	35	0.82	3.2	—	1.7
4-t-Butyl	85	—	15	0.70	3.5	—	0.6
5-Methyl	80	20	—	0.80	3.84	0.96	—
5-Ethyl	80	20	—	0.80	3.84	0.96	—
5-i-Propyl	90	10	—	0.75	4.05	0.45	—
5-t-Butyl	100	—	—	0.70	4.2	—	—
2,4-Dimethyl	—	—	100	1	—	—	6
2-t-Butyl-4-methyl	—	—	100	0.95	—	—	5.7
4-t-Butyl-2-methyl	—	—	100	0.5	—	—	3
2,5-Dimethyl	—	100	—	0.2	—	1	—
4,5-Dimethyl	100	—	—	0.75	4.5	—	—

[a] 1/25 mole of peroxide/mole of thiazole.

The results of phenyl substitution of different alkyl and arylthiazoles are given in Tables III-26 and III-27.

In the case of substituted aryl radicals, the results may be slightly different, depending on the polarity of the radicals. With electrophilic radicals the overall reactivity of the thiazole nucleus will decrease and the percentage of 5-substituted isomer (electron-rich position) will increase, in comparison with phenyl radicals. The results are indicated in Table III-28.

B. Influence of Radical Source and Medium

In the intermediate complexe of free radical arylation, it is necessary to oxidize the reaction intermediate to avoid dimerization and disporportionation (190–193, 346) In this case isomer yield and reactivity will be highest with radical sources producing very oxidative radicals or in solvents playing the role of oxidants in the reaction. The results are summarized in Tables III-29 and III-30.

TABLE III-27. PHENYLATION OF MONO- AND DIARYLTHIAZOLES (187, 188)

Substrates	Isomers obtained	
	On the thiazole ring (in %)	On the benzene ring (in %)
2-Phenyl	45 (5) 5 (4)	30 (ortho) 7.5 (meta) 12.5 (para)
4-Phenyl	14.5 (2) 49.5 (5)	25 (ortho) 3 (meta) 8 (para)
5-Phenyl	20 (2) 20 (5)	60 (undetermined)
2,4-Diphenyl	17	—
2,5-Diphenyl	23	—
4,5-Diphenyl	25	—

^{a} Positions in parentheses.

TABLE III-28. PERCENTAGE OF ARYLISOMERS FROM THE ARYLATION OF THIAZOLE AND ALKYLTHIAZOLES WITH p-SUBSTITUTED BENZOYL PEROXIDES (182, 189)

Substrates	Radicals	Conditions of reaction	Isomers (%)				Reactivity $(^{ThH}_{PhH}K)$
			2	5	5+4	4	
Thiazole	p-BrC$_6$H$_4$	Nitrogen, 98°C	54	—	46	—	—
		Air, 115°C	45.5	34.5	—	20	1.25
	p-NO$_2$C$_6$H$_4$	Nitrogen, 98°C	64.7	—	35.3	—	—
		Air, 115°C	68	16	—	16	1.35
	p-MeOC$_6$H$_4$	Nitrogen, 98°C	57	—	43	—	2
4-Methyl	p-BrC$_6$H$_4$	Air, 115°C	42.5	57.5	—	—	1.55
	p-NO$_2$C$_6$H$_4$	Air, 115°C	45	55	—	—	—
4-Ethyl	p-BrC$_6$H$_4$	Air, 115°C	44.5	55.5	—	—	—
	p-NO$_2$C$_6$H$_4$	Air, 115°C	48	52	—	—	—
4-t-Butyl	p-BrC$_6$H$_4$	Air, 115°C	100	Traces	—	—	—
2-Methyl	p-BrC$_6$H$_4$	Air, 110°C	—	59	—	—	41
	p-NO$_2$C$_6$H$_4$	Air, 110°C	—	57	—	—	43

TABLE III-29. VARIATION IN THE PERCENTAGES OF PHENYL ISOMERS AND REACTIVITY OF THIAZOLE WITH DIFFERENT PHENYL RADICAL SOURCES (182, 184).

Radical source	Molar concentration	Temperature (°C)	Percentage of phenyl isomers			Reactivity ($_{PhH}^{ThH} K$)
			2	4	5	
$(PhCOO)_2$	$\frac{1}{50}$	110	51	39	10	0.75
$(PhCOO)_2$ + oxidizing agents	$\frac{1}{20}$	110	62.5	26.5	11	1.6
$(p\text{-}NO_2C_6H_4COO)_2$	$\frac{1}{20}$	110	65	35		—
$PhN{=}NCl$	$\frac{1}{10}$	0	80	20		—
	$\frac{1}{20}$	40	75	25		2
$PhN(NO)COCH_3$	$\frac{1}{20}$	60–80	63	37		—

TABLE III-30. VARIATION IN THE PERCENTAGE OF PHENYL ISOMERS AND REACTIVITY OF 4-METHYLTHIAZOLE WITH DIFFERENT RADICAL SOURCES (184).

Radical source	Molar concentration	Temperature (°C)	Percentage of isomers		Reactivity ($_{PhH}^{4\text{-Me}} K$)
			2	5	
$(PhCOO)_2$	$\frac{1}{20}$	110	40	60	1.2
$(PhCOO)_2$	$\frac{1}{20}$	80^a	60	40	—
$(p\text{-}NO_2C_6H_4COO)_2$	$\frac{1}{20}$	110	43	57	1.55
$(PhCOO)_2$ + nitroderivatives	$\frac{1}{20}$	110	40.5	59.5	—
$(PhCOO)_2$ + Cu	$\frac{1}{20}$	110	55	45	—
$PhN{=}NCl$	$\frac{1}{20}$	40	58	42	2.4
$PhN(NO)COCH_3$	$\frac{1}{20}$	80	40	60	1.8

^{a} Under nitrogen.

C. Reactivity of the Conjugate Acid of Thiazole

The protonation of the nitrogen atom of thiazole induces a large increase in reactivity of the 2-position (193, 194). This is in contrast to the pyridine series, where the reactivity of both positions adjacent to the nitrogen atom are enhanced (194). The phenylation of conjugate acid of 5-alkylthiazoles may then be considered as a preparative route to alkylthiazoles. The results (isomer percent and overall reactivity) are indicated in Tables III-31 (196) and III-32 (196).

TABLE III-31. INFLUENCE OF ACIDITY ON THE REPARTITION OF PHENYL ISOMERS OF THIAZOLE AND 4-METHYLTHIAZOLE AT 115°C (196).

	Thiazole		4-Methylthiazole	
Acids	2	4+5	2	5
Acetic	82.5	17.5	75	25
Trifluoroacetic	96.5	3.5	90	10
Trichloroacetic	94	6	—	—
Trichloracetic+nitromethane	72	28	—	—
Trichloroacetic+acetic	87.5	12.5	—	—
Acetic+hydrochloric+ acetic anhydride	96	4	91.5	8.5

In the case of alkyl radicals [e.g., methyl radical (197, 198) and cyclohexyl radical (198)], their nucleophilic behaviour enhances the reactivity of the 2-position. Here it is necessary to have full protonation of the nitrogen atom and to use specific solvents and radical sources.

In Table III-33 results for the methylation of thiazoles in acetic acid are given (lead tetraacetate is used as radical source), but in this case some discrepancies appear, the acidic medium being too weak, and the heterocyclic base not fully protonated. Thiazole has also been methylated by the DMSO–H_2O_2 method (201), and the results are in agreement with those described previously.

TABLE III-32. PHENYLATION OF THIAZOLE AND ALKYL-THIAZOLES IN TRIFLUOROACETIC ACID SOLUTION WITH BENZOYLPEROXIDE (COMPETITION WITH BENZENE) AT 80°C (196).

Substrate	Total reactivity	Partial rate constants[a]	
Thiazole	5.5	2–	30
		5–	3
		4–	0
2-Methylthiazole	0.6	5–	2.5
		4–	1.05
4-Methylthiazole	7.5	2–	40.5
		5–	4.5
5-Alkylthiazole	7	2–	41.5
		4–	0.5

[a] For each position of thiazole ring.

TABLE III-33. FREE-RADICAL METHYLATION IN ACIDIC MEDIUM OF AL-KYLTHIAZOLE[a] (197)

Substrates	Reaction time (hr)	Substitution (%)	Methylated products obtained (total 100%)
Thiazole	1	17	86.5 (2)[b]
			13.5 (5)
Thiazole	4	25	76.5 (2)
			23.5 (5)
2-Methyl	2	2	Not determined
4-Methyl	2	22	77.5 (2)
			16.5 (5)
5-Methyl	2	15	95 (2)
			5 (unidentified)
2,5-Dimethyl	2	0	—
4,5-Dimethyl	2	17	100 (2)
2,4-Dimethyl	2	2	100 (5)

[a] All reactions carried out at 100°C.
[b] Number in parenthesis is thiazole ring position.

D. Comparison of Free-Radical Reactivity with Theoretical Calculations

Free-radical reactivity of thiazole has been calculated by semiempirical methods, and results (free valence and localization energy) have been compared with experimental data. For mono- and dimethylthiazoles the radical localization energy of the unsubstituted position may be correlated with the logarithm of experimental reactivity (180, 200). The value of the slope shows that a Wheland-type complex is involved in the transition state.

2. Thiazolyl Radicals

The thiazolyl radicals are, in comparison to the phenyl radical, electrophilic as shown by isomer ratios obtained in reaction with different aromatic and heteroaromatic compounds. Sources of thiazolyl radicals are few: the corresponding peroxide and 2-thiazolylhydrazine (202, 209, 210) (see Table III-34) are convenient reagents, and it is the reaction of an alkyl nitrite (isoamyl) on the corresponding (2-, 4-, or 5-) amine that is most commonly used to produce thiazolyl radicals (203–206). The yields of substituted thiazole are around 40%. These results are summarized in Tables III-35 and III-36.

TABLE III-34. PRODUCTS FROM 2-THIAZOLYLHYDRAZINE AND THIAZOLE CARBONYL PEROXIDES[a] (211)

R group	Solvent ArH	Temperature (°C)	Acids RCO$_2$H	CO$_2$	Thiazole	Arylthiazole (R–Ar)	Ester (RCO$_2$Ar)	Other
2-Thiazolyl-NHNH$_2$ / (RCO$_2$)$_2$	PhH	—	—	—	30.7	12.1	—	—
	PhBr	—	—	—	18.7	10.9 ($o:m:p$ = 51:31:18)	—	—
	Cumene	—	—	—	23.4	9.8 ($o:m:p$ = 40:35:25)	—	1.3 Bicumyl
R = 2-methylthiazolyl	PhH	80	—	—	18.2	36.9	1.5	4 RCO$_2$R(?)
	PhBr	100	—	—	21.1	54.5 ($o:m+p$ = 49:51)	—	—
	Cumene	100	—	—	72.1	23.1 ($o:m:p$ = 43:32:25)	—	1.3 Bicumyl
R = 4-thiazolyl	PhH	80	90	—	1.9	21.2	13.3	1 Ph$_2$
	PhBr	80	119.5	55	1.0	28.4 ($o:m+p$ = 69:31)	—	—
	Cumene	155	64	—	—	—	—	40.7 Bicumyl
R = 5-thiazolyl	PhH	80	115	—	0.5	—	7.2	7.1 Ph$_2$
	PhBr	80	112.6	18.7	0.5	—	—	25?
	Cumene	155	37	—	—	—	—	13.8 Bicumyl

[a] Yields are in moles percent moles of starting material.

TABLE III-35. ISOMER RATIOS AND RELATIVES RATES FOR HOMOLYTIC 2-THIAZOLYLATION OF ALKYLBENZENES (207)

Substrates	Radical and Source[a]	Isomer (%) o	m	p	Reactivity $(_{PhH}^{PhX}K)$
Toluene	2-Th ($h\nu$)	69	19	12	2.1
	2-Th (G)	85	15		2.2
	Ph ($h\nu$)	65.5	20	14.4	1.65
Ethylbenzene	2-Th ($h\nu$)	66	21	13	1.8
	2-Th G)	64	23.5	12.5	1.7
	Ph ($h\nu$)	57	26.5	16.5	1.2
i-Propylbenzene	2-Th ($h\nu$)	45	35.5	19.5	1.5
	2-Th (G)	45	36.5	18.5	1.3
	Ph ($h\nu$)	36	41.7	22.3	0.85
t-Butylbenzene	2-Th ($h\nu$)	31	45.5	23.5	1.05
	2-Th (G)	19	51	30	1
	Ph ($h\nu$)	22.5	51.5	26	0.75

[a] G, Gomberg reaction.

TABLE III-36. EXPERIMENTAL ISOMER DISTRIBUTION, RELATIVE REACTIVITIES (k), AND f_m/f_p RATIOS, FOR HOMOLYTIC THIAZOL-2-YLATION AND 5-SUBSTITUTED THIAZOL-2-YLATION OF ALKYLBENZENES (176)

Substrates	5-Substituents	Isomer ratios (±1) o-	m	p	k_M	f_m/f_p
Toluene	H	65.8	19.3	14.9	2.2	0.65
	Me	66.5	19.5	14.0	2.1	0.7
	Br	64.5	19.4	16.1	2.3	0.60
	NO$_2$	62.5	19.7	17.8	2.6	0.55
Ethylbenzene	H	56.5	25.4	18.1	1.8	0.7
	Me	56.5	26.8	16.7	1.6	0.8
	Br	52.5	27.7	19.8	1.9	0.7
	NO$_2$	51.0	27.7	21.3	2.2	0.65
i-Propylbenzene	H	46.0	33.2	20.8	1.45	0.8
	Me	43.5	36.4	20.1	1.4	0.9
	Br	41.5	35.3	23.2	1.55	0.76
	NO$_2$	40.2	35.3	24.5	1.9	0.72
t-Butylbenzene	H	21.5	50.5	28	1.12	0.9
	Me	20.2	52.4	27.6	1.08	0.95
	Br	22.5	47.8	28.7	1.25	0.85
	NO$_2$	21.0	48.6	30.4	1.6	0.8

TABLE III-37. COMPARISON OF THE RELATIVE REACTIVITIES (WITH RESPECT TO BENZENE) OF VARIOUS PYRIDINE SUBSTRATES TOWARDS PHENYL AND 2-THIAZOLYL RADICALS AT 70 TO 80°C (208)

Pyridine Substrate	Peroxide (Source)	Aniline+NOOR[a] A	B	2-Aminothiazole +NOOR[b]
Pyridine	1.10	1.50	1.2	0.45
4-Methylpyridine	1.35	3.30	2.0	0.70
4-Methylpyridine-N-oxide	—	—	45	23.0
4-Propylpyridine	—	—	—	0.70
4-t-Butylpyridine	—	—	—	0.60
4-Cyanopyridine	6.40	10.0	6.0	1.60
4-Acetylpyridine	6.75	11.00	6.50	1.80
2,6-Dimethylpyridine	1.35	2.70	2.0	7.50
2,4,6-Trimethylpyridine	2.0	3.80	3.15	7.0
3,5-Dimethylpyridine	2.2–3.4	11.00	10.0	5.0

[a] This study (A) in the presence of a stoichiometric quantity of nitrite with respect to aniline, (B) in the presence of five times the stoichiometric quantity of nitrite.
[b] These values represent the average of at least two competitive experiments carried out in the presence of a slight excess of nitrite with respect to 2-aminothiazole.

With heteroaromatic substrates it is possible to prepare, for example, thiazolylpyridines. It is noteworthy that basic solvents (e.g., heterocyclic nitrogen compounds) increase the yield of substitution by a cage effect (see Tables III-37 and III-38) (208).

TABLE III-38. REACTIVITY OF ALKYLPYRIDINES TOWARDS 2-THIAZOLYL RADICALS (208)

Substrates	Isomerization[a]		Yield vs 2-Aminothiazole (%)
Pyridine	54.5 (2)	45.5 (3+4)	15
4-Methylpyridine	47 (2)	53 (4)	12
2,4,6-Trimethylpyridine	100 (3+5)		40
2,6-Dimethylpyridine	95 (3+5)	5 (4)	36

[a] Per cent, position on the pyridine ring.

A. *Electron Spin Resonance*

Several ESR spectra of thiazolyl and benzothiazolyl radicals have been recorded (667, 859), and this type of studies may be used in elucidation of the reaction mechanism.

3. Photochemical Rearrangement

Alkyl- and arylthiazoles rearrange under ultraviolet irradiation in different solvents to yield the corresponding isothiazoles or isomeric thiazoles. With alkylthiazoles the overall yields are very low, and it is not possible to use this method preparatively. For arylthiazoles it is possible; 2-arylthiazoles, for instance, can be used to prepare 3-arylisothiazoles that are otherwise very difficult to obtain.

A. *Alkylthiazoles* (213, 214)

The isomerization of alkylisothiazoles has been studied and leads to alkylthiazoles. The isomerization seems to occur by a zwitterion mechanism (Scheme 1).

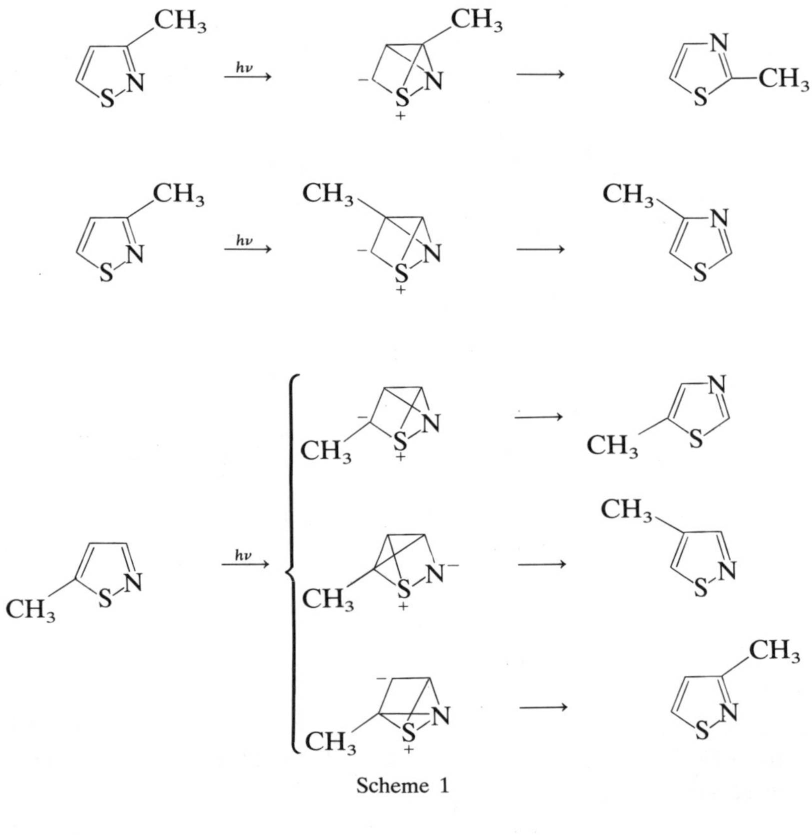

Scheme 1

TABLE III-39. RELATIVE PERCENTAGE OF ISOMERS BY IRRADIATION OF PHENYLTHIAZOLES AND ISOTHIAZOLES (217)

Starting Products	Additive	Yield in isomers (%)						Transposition (%)
		1	2	3	4	5	6	
2-Phenylthiazole (1)	—	1.65	—	14.8	12.4	1.15	—	70
	I$_2$	24.5	—	1.75	14.1	0.75	—	59
5-Phenyl (2)	—	—	14.5	—	1.05	1.05	13.5	70
	I$_2$	—	15.2	—	0.75	0.75	21.3	62
4-Phenyl (3)	—	0.4	—	74	1.1	—	—	24.5
	I$_2$	—	—	—	82	—	—	18
3-Phenylisothiazole (4)	—	—	—	16.7	31.1	—	—	52
	I$_2$	—	—	4.0	76.0	—	—	20
5-Phenyl (5)	—	0.1	—	5.85	10.9	33.15	—	50
	I$_2$	—	—	0.84	22.7	56.6	—	20
4-Phenyl (6)	—	—	—	—	—	—	45	54
	I$_2$	—	—	—	—	—	83	17

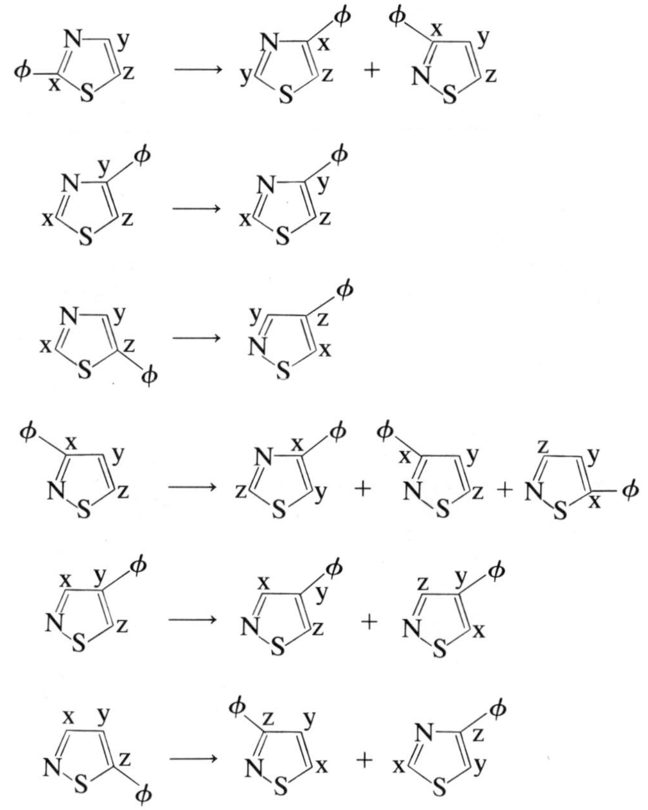

Scheme 2

B. Arylthiazoles (220, 221)

Irradiation with ultraviolet light of arylthiazoles in different solvents gave the transpositions described in Table III-39. (215, 216).

Reactions with substituted aryl- and alkylthiazoles and deuterated thiazoles (98) gave evidence of the general pathway of the rearrangement (Scheme 2).

In the isomerization of 2-phenylthiazole the selectivity of the reaction may be increased by the addition of iodine (98) (see Table III-40).

TABLE III-40. EFFECT OF IODINE ON THE ISOMERIZATION OF 2-PHENYLTHIAZOLE (98)

Additives	3-Phenylisothiazole	4-Phenylthiazole	Yields[a]
Air	34.5	65.5	12
Air+I_2	90	10	11.5
N_2	36	64	13
O_2	20	80	15

[a] Yields with respect of 2-phenylthiazole.

The same isomerization also occurs with diarylthiazoles, but when two adjacent phenyl groups are present, even in the final product, a photochemical cyclization gives rise to a polycyclic benzothiazole (Scheme 3) (213, 218, 219).

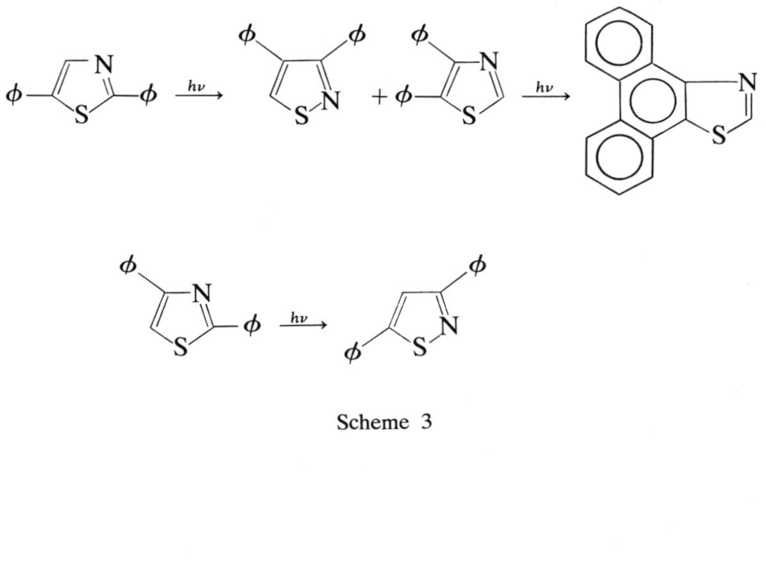

Scheme 3

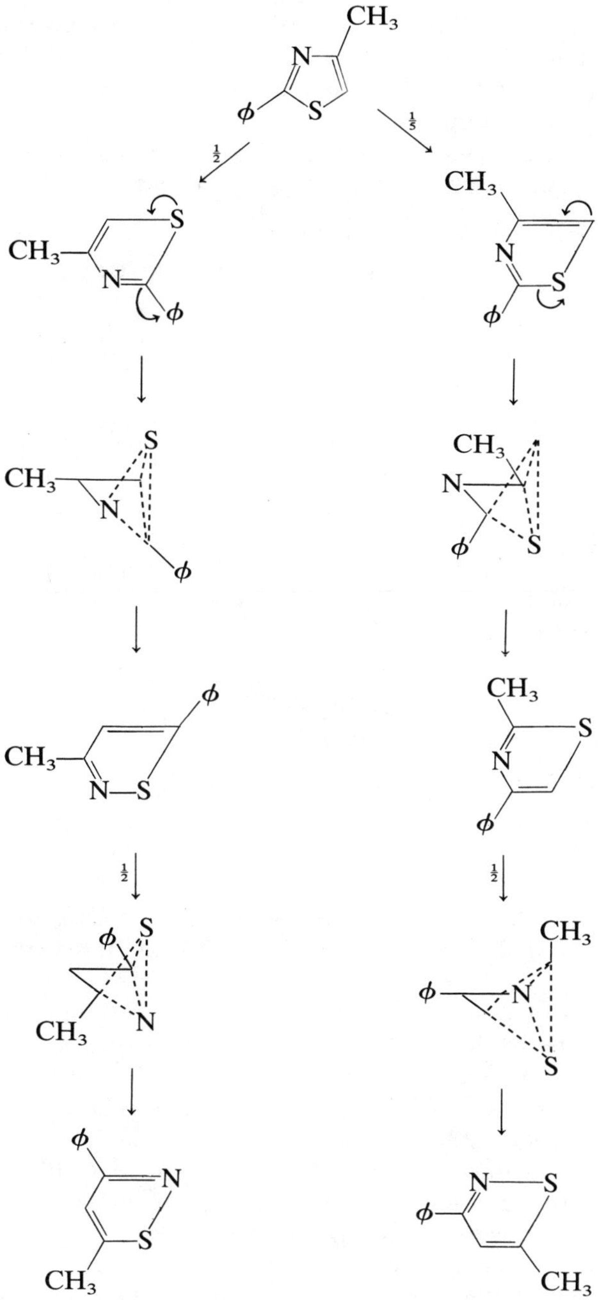
CH3
N
S
ϕ
½
⅕
S
CH3
N
ϕ
CH3
N
S
ϕ
S
CH3
N
ϕ
CH3
N
ϕ
S
ϕ
CH3
N—S
CH3
S
N
ϕ
½
½
S
ϕ
CH3
N
CH3
ϕ
N
S
ϕ
N
S
CH3
N—S
ϕ
CH3

Scheme 4

The general mechanism of the rearrangement of aryl and diarylthiazoles seems to exclude the zwitterion route. Instead it takes place through bending of thiazoles bonds (98, 213). Moreover, tricyclic sulfonium cation intermediates, after irradiation of deuterated phenylthiazoles, have been suggested by several workers (98).

Scheme 4 summarizes the photoisomers obtained from 4-methyl-2-phenylthiazole, involving bending of the bonds adjacent to sulfur.

X. NUCLEOPHILIC REACTIVITY

1. Organomagnesium Compounds

Alkylthiazoles react with ethylmagnesium bromide to give thiazolylmagnesium compounds, as demonstrated for 4- and 5-methylthiazoles, 4-ethylthiazole, and 4,5-dimethylthiazole. The resulting addition compounds do not decompose at high temperature and pressure to yield alkylthiazoles as do the addition compounds obtained with pyridine.

The high mobility of the 2-hydrogen towards Grignard reagents has been demonstrated (222).

2. Organolithium Compounds

The reaction of n-butyllithium with 2-methylthiazole leads to three different lithium derivatives following attack on three positions of the molecule (223). Attack on the 4-position gives less than 2% lithium derivative.

Deuterolysis of the organolithium compounds was used to characterize the three deuterated thiazoles corresponding to the three lithium derivatives.

The reaction of 2,4-dimethylthiazole with butyllithium shows that, in contrast to 2-methylthiazole, the benzyl position (the 2-position) is the most reactive. The effect of the substituent in the 4-position may well be steric: 4-t-butyl-2-methylthiazole in the same reaction gives no 5-substituted product (223).

Reaction of various reagents (CH_3I, C_2H_5I, PhCHO) on the organolithium products obtained by reaction of butyl-lithium with 2-methyl-4-phenylthiazole gives approximately 90% 5-substitution. The increased reactivity of the hydrogen in the 5-position can be explained by the fact that the $+I$ effect of a 4-methyl group would increase the electron density on the 5-position and hence favor the formation of the 2-lithium derivative, whereas the $-I$ effect of a 4-phenyl group favors attack in the 5-position (224, 225).

2-Benzylthiazole reacts with n-butyllithium to give 2- and 5-substituted products, but as expected from the particular properties of the 2-methylene group, the proportion of 2-lithium derivatives is much more important (223).

3. Dimerization

If the organolithium derivative of 2,4-dimethylthiazole or 2-methyl-4-phenylthiazole (prepared at $-78°C$) is allowed to warm to room temperature, the 2-lithium compound reacts with the nonmetallated thiazole (Scheme 5) (225).

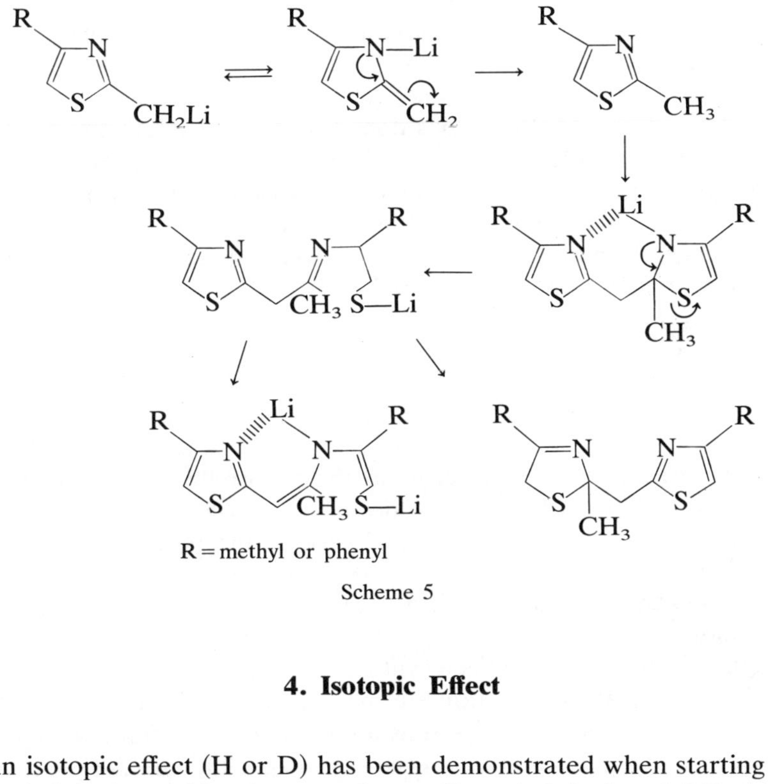

R = methyl or phenyl

Scheme 5

4. Isotopic Effect

An isotopic effect (H or D) has been demonstrated when starting from 2-methyl-4-phenylthiazole or from 2-methyl-4-phenyl-5-D-thiazole (224) in the dimerisation reaction.

XI. ELECTROPHILIC REACTIVITY

The electronic densities on the thiazole nucleus or on the thiazolium ion suggest that electrophilic substitution should occur preferentially at the 5-position. The order of π electronic densities is $2 < 4 < 5$ (163, 164).

The most widely studied electrophilic substitution reactions are halogenation and nitration. Two main types of substrates are possible: alkylthiazoles and arylthiazoles.

1. Alkylthiazoles

A. *Halogenation*

Halogenation (e.g., bromination) takes place in chloroform for the 2,4-dialkylthiazoles, and the majority of studies have been of 2,4-dimethylthiazole (227, 228). In other cases and in acetic or stronger acids, substitution occurs at the 5-position and is promoted by electron-releasing groups in the 2-position. When the releasing group is in the 4-(or 5-)-position, steric hindrance may decrease the yield of substitution at the 5- (or 4-) position. Nevertheless, the thiazole nucleus is not very reactive since 4-methylthiazole and 2,5-dimethylthiazole are inert in dilute sulfuric acid with bromine (229–231).

B. *Mercuration*

In acetic acid, mercuration occurs at the 5-position at room temperature. At 60 to 70°C in the case of arylthiazoles, where the 5-position is unsubstituted, mercuration also takes place at the para-position of the phenyl ring (861).

Mercurated thiazoles also yield 5-halothiazoles by the replacement of Hg by halogen.

5-Bromo derivatives have been synthetized in this way (229, 233, 234).

Results of thiazole mercuration are summarized in Table III-41 (861).

Mercuration has been also used as an extractive technique to separate thiazoles from complex mixtures such as crude petroleum (273, 861–863).

TABLE III-41. MERCURATION OF 2,4-DIPHENYL-THIAZOLE[a] (861)

Reaction Temperature (°C)	Position of mercuration	
	on thiazole	on benzene
Room temperature	5	no
60–70	5	para

[a] The substrate was 2,4-diphenyl; the 5-position on the thiazole nucleus is mercurated.

C. *Nitration*

This aspect of electrophilic reactivity has been studied with several alkylthiazoles, and it is noteworthy that reduction of 4-(5-) nitrothiazoles yields the amino derivatives that are good starting compounds for synthesis. Reaction takes place at the 5-position or at the 4-position if the 5-position is blocked, on the 4- and 5-positions at the same time but with poorer yields, if the 2-position is substituted (228, 231, 235–239, 244).

For 2-alkylthiazoles, substitution ratios are indicated in Table III-42.

TABLE III-42. ISOMERS OBTAINED DURING THE NITRATION OF 2-ALKYLTHIAZOLES

Substrates	5-Nitro%	4-Nitro%	Ref.
2-Methyl	77	23	240
2-*n*-Propyl	71	29	241

The overall reactivity of the 4- and 5-positions compared to benzene has been determined by competitive methods, and the results agreed with kinetic constants established by nitration of the same thiazoles in sulfuric acid at very low concentrations (242). In fact, nitration of alkylthiazoles in a mixture of nitric and sulfuric acid at 100°C for 4 hr gives nitro compounds in preparative yield, though some alkylthiazoles are oxidized. Results of competitive nitrations are summarized in Table III-43 (241, 243). For 2-alkylthiazoles, reactivities were too low to be measured accurately.

TABLE III-43. COMPETITIVE NITRATION OF ALKYLTHIAZOLES (241, 243)

Thiazoles	Reactivity
2,4-Dimethyl	1
2,5-Dimethyl	0.5
4-Alkylthiazoles	0.07
5-Alkylthiazoles	0.04

In Tables III-44 and III-45 kinetic constants of nitration in sulfuric acid are given (242).

It is also possible to use more powerful reagents to nitrate thiazoles; for example, 2-methylthiazole has been nitrated using nitronium tetrafluoroborate and the complex nitrogen dioxide–boron trifluoride (240). The overall yield is about 50 to 60%.

The NMR determination in strongly acidic medium (trifluoroacetic acid) of the chemical shifts of the protons in the 4- or 5-position can be used to establish a reactivity scale. If the proton appears at low field, this indicates that this substitution site will be poorly or not at all nitrated (111).

2. Arylthiazoles

Arylthiazoles undergo electrophilic substitution mainly in the para-position of the phenyl ring. In some cases, it is also possible to substitute the 5-position of the thiazole nucleus (244–246).

Bromination in acetic acid has been the more widely studied reaction, and substitution occurs readily at the 5-position of the thiazole nucleus (247). For instance, it is possible to study the influence of groups X and R on the reactivity of the 5-position towards Br^+ (Scheme 6) (249).

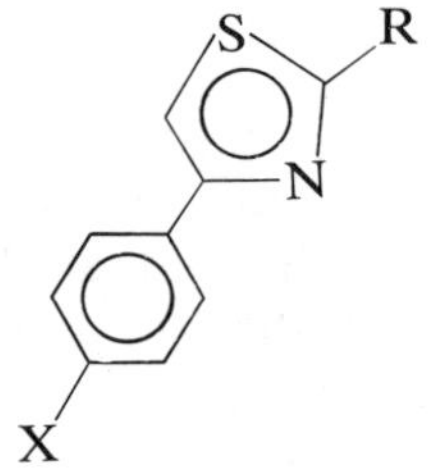

R = $-CH_2Cl$, $-CH_2OH$, $-CO_2H$
X = $-OEt$, H, Cl, Br, $-NO_2$

Scheme 6

TABLE III-44. STANDARD RATE CONSTANTS FOR THIAZOLES (242)

Compound	Range (%)	Temperature (°C)	Range (H_0)	$d\|\log k_2\|/d\|H_0\|$	$\log k_2$ (at $H_0 = 6.6$)	$\log k_2(25°C)$	$\log k_2^0$
2,4-Dimethyl-thiazole	70–88	70	5.3–7.6	−1.96	−3.53	−6.90	−6.90
2,5-Dimethyl-thiazole	83–87	80	6.9–7.4	−2.05	−3.60	−7.60	−7.60
2,3,4-Trimethyl-thiazolium	81–87	80	6.5–7.3	−2.58	−3.52	−7.52	−7.52
2,3,5-Trimethyl-thiazolium	76–88	80	5.3–7.5	−2.00	−3.72	−7.72	−7.72
5-Ethyl-2-t-butyl-thiazole	74–82	108	5.2–6.3	−1.94	−1.37	−6.96	−6.96
5-i-Propyl-2-t-butylthiazole	77–82	132	5.2–5.9	−1.93	−0.65	−7.43	−7.43

TABLE III-45. RATE PROFILE SLOPES FOR NITRATION OF THIAZOLE (242).

Compound	Position of Nitration	Low Acidity				High Acidity		
		Temperature (°C)	Slope[b]	Correlation Coefficient	Species Reacting	Temperature (°C)	Slope[b]	Species Reacting
2,4-Dimethyl-thiazole	5	70	0.80	0.992	CA[a]	70	0.29	CA
2,5-Dimethyl-thiazole	4	100	1.06	0.999	CA	80	0.28	CA
2,3,4-Trimethyl-thiazolium perchlorate	5	80	0.83	0.996	CA	70	0.13	CA
2,3,5-Trimethyl-thiazolium perchlorate	4	80	0.77	0.998	CA	80	0.15	CA
5-Ethyl-2-t-butyl-thiazole	4	108	0.89	0.995	CA	60	0.33	CA
5-i-Propyl-2-t-butyl-thiazole	4	132	0.84	1.000	CA	75	0.43	CA

[a] CA = conjugate acid.
[b] $d[\log k_2]/d[-H_0]$, $H_0 > 9.0$

In the case of 4-(chloromethyl)-2-phenylthiazole, chloration in acetic acid yields only the product substituted at the 5-position of the thiazole ring (247).

Arylthiazoles are nitrated on the phenyl ring. The yields in sulfuric acid average 60 to 85% (248). Results are indicated in Table III-46 (250).

TABLE III-46. ISOMER DISTRIBUTION OBTAINED DURING THE NITRATION OF PHENYLTHIAZOLES (250)

Thiazoles	Isomer Ratio (%)		
	o	m	p
2-Phenyl	3	8	89
4-Phenyl	7	4	89
5-Phenyl[a]	15	2	83

[a] In this case, 5% dinitrated products are also obtained.

In the literature, nevertheless, most of the authors indicate para-nitro products (126, 251, 252).

For 2-R-substituted 4-phenylthiazoles, the influence of R groups on the yield of substitution has been reported (see Table III-47). Substitution occurred at the para-position of the phenyl ring (249).

TABLE III-47. INFLUENCE OF 2-R GROUP DURING THE NITRATION OF 2-R-4-PHENYLTHIAZOLES

R	Yield (%)
$-CH_2NHCOC_6H_5$	80
$-CH_2OCOC_6H_5$	75
$-CH_2OH$	65
$-CH(OH)CH_3$	60
$-CH=CHCOOH$	86
$-CH=CHCOOEt$	84
$-COCH_3$	80

For benzylthiazoles and phenethylthiazoles, the influence of one and two $-CH_2-$ groups has been determined for the 2-position; the results are summarized in Table III-48 (250).

TABLE III-48. NITRATION OF BENZYL- AND PHENETHYLTHIAZOLES (250)

Compounds	Isomer Ratio (%)		
	o	m	p
2-Benzyl	11	25	64
4-Methyl-2-benzyl	13	27	60
5-Methyl-2-benzyl	18	35	47
4,5-Dimethyl-2-benzyl	18	37	45
2-Phenethyl	30	5	65
4-Methyl-2-phenethyl	23	9	68
5-Methyl-2-phenethyl	25	8	67
4,5-Dimethyl-2-phenethyl	27	8	65

XII. REACTIVITY OF THE NITROGEN ATOM

1. Quaternization

The quaternization of the nitrogen atom of the thiazole ring (the Menschutkin's reaction) by alkyl halide or methyl tosylate can be used to measure the reactivity of this atom and thus to evaluate steric and electronic effects of ring substituents.

The first kinetic results in the area were obtained by studying the quaternization of 4-alkyl-, 5-alkyl-, and 2-alkylthiazoles with methyl iodide (253–255). A deeper and more exhaustive study of this reaction has been carried out recently with more elaborate substrates (152).

The quaternization reaction of the thiazole nitrogen has been used to evaluate the steric effect of substituents in heterocyclic compounds since thiazole and its alkyl derivatives are good models for such study. In fact, substituents in the 2- and 4-positions of the ring only interact through their steric effects (inductive and resonance effects were constant in the studied series). The thiazole ring is planar, and the geometries of the ground and transition states are identical. Finally, the 2- and 4-positions have been shown to be different (259, 260).

A. *Steric Effects of Substituents at the 2- and 4-Positions*

A 2-Alkyl group contributes to the basicity of the thiazole ring. The only significant fall in pK_a (for 2-i-propyl and 2-t-butyl thiazole) is not attributed to a variation in the donor ability of the substituent but to a decrease in solvatation (256, 257). In Table III-49 the pK_a of 2-alkylthiazoles are indicated.

TABLE III-49. pK_a OF 2-ALKYLTHIAZOLES IN WATER AT 25°C (256, 257)

R	pK_a
H	2.52
Methyl	3.42
Ethyl	3.37
i-Propyl	3.28
t-Butyl	3.15
n-Propyl	3.35
i-Butyl	3.38
Neopentyl	3.37

If the rate constants for quaternization of 2-alkylthiazoles depended on electronic factors, they would all be greater than that of thiazole, which has the lowest pK_a, and all of the same order. The decrease in rate constants that is observed is attributed wholly to steric effects. In Table III-50 we report the main parameters for the reaction of 2-alkylthiazoles with methyl iodide.

TABLE III-50. RATE CONSTANTS AND ACTIVATION PARAMETERS FOR THE REACTION OF 2-ALKYLTHIAZOLES WITH METHYL IODIDE IN NITROBENZENE (256)

R	$k_{25°C} \times 10^6$	E^a	log A	ΔH^a	$-\Delta S^b$
Methyl	11.83	15.7±0.1	6.6±0.1	15.1±0.1	30.2±0.8
Ethyl	9.25	15.6±0.1	6.4±0.1	15.0±0.2	31.0±1.2
i-Propyl	4.32	16.2±0.4	6.5±0.3	15.6±0.3	30.7±1.9
t-Butyl	0.29	17.5±0.4	6.3±0.2	16.9±0.6	31.7±2.5
n-Propyl	9.15	15.9±0.1	6.6±0.1	15.3±0.1	29.9±0.8
i-Butyl	8.56	15.9±0.3	6.5±0.2	15.3±0.1	30.1±1.1
Neopentyl	6.02	15.7±0.5	6.3±0.3	15.1±0.4	31.6±2.2

[a] E and ΔH in kilocalories.
[b] $-\Delta S$ in calories per degree.

A good correlation between $\log k/k_0$ (k_0 is the rate constant for methylthiazole), and the Taft's parameter E_s (261) has been found for this series. Table III-51 indicates this correlation.

TABLE III-51. PLOT OF $\log(k/k_0) = \delta E_s$ FOR 2-ALKYLTHIAZOLES (256)

R	E_s (Taft)	$\log k/k_0$ at 25°C
Methyl	0	0
Ethyl	−0.07	−0.10
i-Propyl	−0.47	−0.43
t-Butyl	−1.54	−1.61

The value of δ obtained by linear regression is 0.96 with a correlation coefficient of 0.9985. For 2-alkylpyridines δ is 2.030 (256), which leads to the conclusion that 2-alkylpyridines are twice as sensitive to steric effects as their thiazole analogs.

2- and 4-alkyl substituents contribute similarly to the basicity of the thiazole ring. However, in quaternization, the dissymetric structure of the ring leads to 2- or 4-substituents possessing different steric effects in the neighborhood of the nitrogen atom that undergoes electrophilic attack.

A good correlation is obtained by

$$\log\left(\frac{k}{k_0}\right)_{4\text{-alkyl}} = \delta \log\left(\frac{k}{k_0}\right)_{2\text{-alkyl}}$$

with δ = 1.189.

Table III-52 indicates the kinetic data for quaternization of 2- and 4-alkylthiazoles. The δ value shows that the 4-position is more sensitive than the 2-position to steric effects, the bond angle $\widehat{H_2C_2N}$ (123°6) being greater than that of $\widehat{H_4C_4N}$ (119°4). This result has been confirmed for all solvents and leaving groups (256).

TABLE III-52. KINETIC DATA FOR THE QUATERNIZATION OF 2- AND 4-ALKYLTHIAZOLES WITH METHYL IODIDE IN NITROBENZENE AT 25°C (256)

R	4-R Thiazole		2-R Thiazole	
	$k \times 10^6$	$\log k/k_0$	$k \times 10^6$	$\log k/k_0$
Methyl	9.78	0	11.83	0
Ethyl	7.85	−0.096	9.25	−0.108
i-Propyl	2.58	−0.582	4.32	−0.436
t-Butyl	0.12	−1.91	0.29	−1.611

XII. Reactivity of the Nitrogen Atom

The quaternization of 16 2,4-dialkylthiazoles with methyl tosylate in nitrobenzene at 25°C has been studied (256) in order to examine the simultaneous influence of two substituents grouped around a reaction center (257).

Table III-53 indicates the rate constants for the quaternization of 2-alkyl- and 2,4-dialkylthiazoles.

TABLE III-53. RATE CONSTANTS ($k \times 10^6$) FOR THE QUATERNIZATION OF 2-ALKYL AND 2,4-DIALKYLTHIAZOLES WITH METHYL IODIDE IN NITROBENZENE AT 25°C (256)

R	2-R Thiazole	2-R-4-Methyl-thiazole	2-R-4-Ethyl-thiazole	2-R-4-i-Propyl-thiazole	2-R-4-t-Butyl-thiazole
Methyl	4.54	4.23	2.75	0.59	0.0121
Ethyl	3.47	2.82	1.86	0.54	0.0068
i-Propyl	1.53	1.01	0.517	0.131	0.0023
t-Butyl	0.098	0.0154	0.0143	0.003	—

Here again it is possible to find a linear relationship between the log (k/k_0) (k_0 = methyl) values of 2-alkyl- and 2,4-dialkylthiazoles and between the latter value and Tafts' E_S parameter (256). The value of δ for 2,4-dialkylthiazoles is 1.472 with a correlation coefficient of 0.9994. Thus the sensitivity to substituent effects is more marked than in the case of a single substituent in the 2-position. Furthermore, the 4-position is again more sensitive than the 2-position.

B. *Effect of a Group Adjacent to an Ortho-Substituent*

The results of kinetic studies of the quaternization of 4-alkyl-2,5-dimethylthiazoles do not give a linear correlation between log (k/k_0) and Tafts' E_S parameter, such as is found for 2-alkyl-, 4-alkyl-, and 2,4-dialkylthiazoles (258).

The log (k/k_0) value for 4-isopropyl-2,5-dimethylthiazole is twice that expected if the curve were linear, which implies a rate constant 6.5 times smaller than expected. This result can be explained by the existence of a privileged conformation, induced by the presence of the methyl group in the 5-position and that has a lower reactivity (258). This result also leads to a limitation in the use of Tafts' E_S parameter to cases where the environment of a substituent does not induce particular conformation for this latter (258).

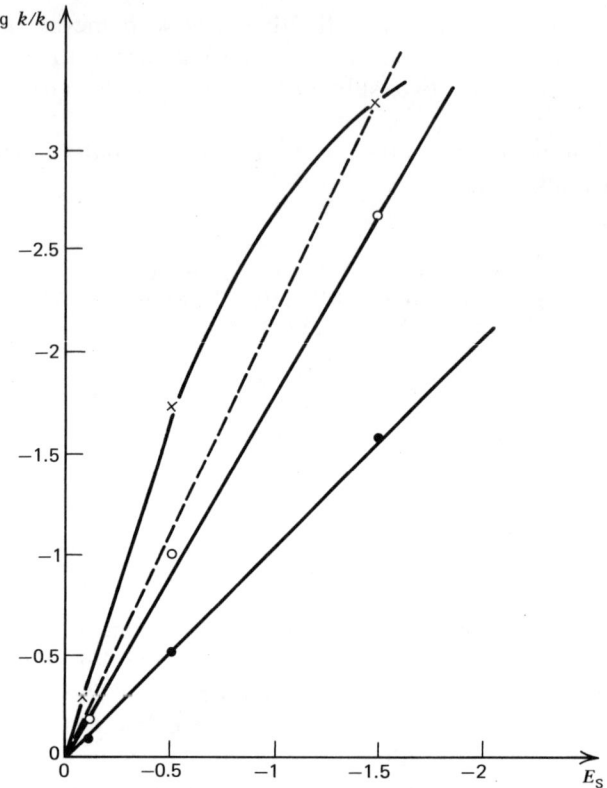

Fig. III-3. Log k/k_0 for quaternization of alkylthiazoles versus Taft's E_S parameters: (●) 2-alkylthiazoles, (○) 2-methyl-4-alkylthiazoles, (X) 2,5-dimethyl-4-alkylthiazoles.

C. *Electronic Effects of 5-Substituents*

The quaternization of 5-alkylthiazoles with methyl iodide in nitrobenzene has been studied (254); results are summarized in Table III-54.

The effect of alkyl groups in the 5-position on the reactivity of the thiazole nitrogen is analogous to that found for 3-alkylpyridines, in other words, a simple inductive effect. In passing from the unsubstituted heterocycle to the methyl derivative, the rate constant doubles; a further increase in substitution produces a much less pronounced variation.

The application of the Taft equation, derived from the Hammett equation (262)

$$\log\left(\frac{k}{k_0}\right) = \rho^* \sigma^* \qquad k_0 = k \text{ methyl}$$

can be verified. The value of ρ^* is -0.5.

TABLE III-54. RATE AND ACTIVATION CONSTANTS FOR THE REACTION OF 5-ALKYLTHIAZOLES WITH METHYL IODIDE IN NITROBENZENE. (254).

R	Temperature (°C)	$k \times 10^6$	E	log A	$\Delta H^{\neq a}$	$-\Delta S^{\neq b}$
H	30	21.95	15.5	6.50	14.9	30.9
	50	106.69				
Methyl	30	42.48	15.6	6.88	15.0	29.1
	50	211.14				
Ethyl	30	46.41	15.6	6.89	15.10	29.1
	50	229.79				
i-Propyl	30	52.23	15.2	6.69	14.6	30.0
	50	249.32				
t-Butyl	30	55.17	15.4	6.83	14.8	29.4
	50	267.91				

[a] $\Delta H^{\neq}$ in kilocalories.
[b] $-\Delta S^{\neq}$ in calories per degree.

Table III-55 shows the verification of the Taft equation. The result is analogous to that obtained for 3-alkylpyridines.

TABLE III-55. VERIFICATION OF THE TAFT EQUATION FOR THE KINETICS OF QUATERNIZATION OF 5-ALKYLTHIAZOLES (254)

R	σ^*	$\log (k/k_0)$
H	0.49	−0.286
Methyl	0	0
Ethyl	−0.10	0.039
i-Propyl	−0.19	0.095
t-Butyl	−0.30	0.117

D. *Quaternization of Arylthiazoles*

Some studies on the quaternization of arylthiazoles have been published, among them the quaternization of 2-methyl-4-phenyl thiazole in various solvents (263). The order of reactivity is the following: 2-methyl-4-phenyl > 2-methyl-4-(3-nitrophenyl) > 2-methyl-4-(2-chlorophenyl) > 2-methyl-4-(4-nitrophenyl). Introduction of a phenyl group in the

5-position slows down the reaction by comparison with 2-methyl-4-phenylthiazole (263).

2. Thiazole N-Oxides

Little work has been carried out on thiazole N-oxides. These products are unstable and breakdown by autoxidation to give thiazolium-N-oxide sulfates and other decomposition products (264). They are prepared by direct oxidation with hydrogen peroxide, or by tungstic acid (264, 265) or peracetic acid (265–267).

Only one reaction of thiazole N-oxides has been studied in detail. The rearrangement in acetic anhydride of 2,4-dimethylthiazole-3-oxide gave 2-acetoxy-4-methylthiazole and 4-acetoxymethyl-2-methylthiazole in a ratio of about 4.5 to 1(264).

4-Methylthiazole-3-oxide in the same conditions gave both 4-acetoxymethylthiazole and 2-acetoxy-4-methylthiazole but in low yields. The reaction of 2,4-dimethylthiazole-3-oxide with p-tolyl chloride has also been studied (268), substitution occurring on the 2- and 4-methyl groups.

3. Metallic Complexes of Thiazoles

Aza-aromatic compounds can give rise to metallic complexes, and various complexes of thiazole have been studied:

- Rhodium III for their bacteriostatic activity (269).
- Copper, nickel, cobalt, iron, and zinc (270) for their physical properties using ultraviolet and infrared spectrometry (271).
- With different anions such as SCN^-, $(ClO_4)_2^{2-}$ (272, 273).

XIII. CONDENSATION ON THE METHYL GROUP

It is well known that in nitrogen-containing heterocyclic compounds the reactivity of alkyl groups is enhanced. In the thiazole series, alkyl groups in the 2-position are reactive towards carbonyl compounds and condensations may be realized.

1. Reactions with 2-Alkyl Groups

2-Alkylthiazoles heated with an aromatic aldehyde in the presence of zinc chloride undergo condensation. For example, 2-methylthiazole gives

2-styrylthiazole with benzaldehyde (274, 275). The substituents in the 4- and 5-positions have little influence on the reactivity of 2-alkyl groups (275). With nonaromatic aldehydes (e.g., cinnamaldehyde or trichloracetaldehyde) the reaction occurs, but with lower yields (275, 276). In some cases, the carbinol (trichloracetaldehyde) may be obtained in 20 to 30% yield (49, 279). Acetic anhydride may also be used as catalyst, for example, with heptaldehyde. In this case, reaction with 2-methylthiazole at 250°C yields 2-(1-octenyl)-thiazole (280, 281).

This reaction may also be extended to 2-ethylthiazole (276).

Various substituted benzaldehydes react with 2,4-dimethylthiazole at the 2-methyl group (6, 282). Results have been obtained with

$$X-\phi-C\overset{O}{\underset{H}{}}$$

X = o-OH, p-OH, o-OMe, p-OMe, 3,4-(OH)$_2$, p-NO$_2$, p-CH=CH-C$_6$H$_5$, p-C$_6$H$_5$O, p-Br, m-NO$_2$, p-N(Me)$_2$, p-(2-furyl).

This reaction also occurs with ketones and may be carried out in a EtOH–EtONa medium. For instance, 5-diethylamino-2-pentanone reacts with 2-benzylthiazole in this way (64).

2. Reaction with the Hydrogen Atom at the 2-Position

The 2-position of thiazole reacts in the same conditions as with aldehydes previously discussed. In this case, the carbinol is obtained in poor yields with zinc chloride and with improved yields in acetic anhydride. But, in this reaction, the 2-position seems to be more sensitive to the influence of substituents at the 4- or 5-positions, or both as exemplified in the following case.

4,5-dimethylthiazole reacts with benzaldehyde to give the corresponding carbinol in 30% yield (14, 83), whereas with 4-methylthiazole only a 8% yield is obtained (83). Similarly, 4,5-dimethylthiazole reacts with aqueous formaldehyde at 120 to 170°C to give about 30% 2-hydroxymethyl-4,5-dimethylthiazole (83).

XIV. SUBSTITUENT EFFECTS

The transmission of the effects of substituents in the 2- and 4-positions across the thiazole ring has been determined from the rates of solvolysis of the halo group in 5-(1-chloroethyl)thiazole or 4-(1-chloroethyl)thiazole, the substituents under consideration (X) being at the 2-position, X = H, Cl, Me, –SMe, –OMe, –C$_6$H$_5$ (283, 284, 285). Tables III-56 and III-57 indicate the rate of solvolysis for the considered reaction.

TABLE III-56. RATE OF SOLVOLYSIS OF 2-SUBSTITUTED 4-(1-CHLOROETHYL)-THIAZOLES (286)

Substituents	$k(10^{-5} \text{ sec}^{-1})^a$
H	80.9 ± 0.7
Cl	23.4 ± 0.5
Methyl	6070
–S–Methyl	79700
–O–Methyl	5.5 × 10^5

^{a} At 45°C.

TABLE III-57. RATE OF SOLVOLYSIS OF 2-SUBSTITUTED 4-(1-CHLOROETHYL)-THIAZOLES (283)

Substituents	$k(10^{-5} \text{ sec}^{-1})^a$
H	11.1 ± 0.3
Br	1.28 ± 0.003
Methyl	72.3 ± 1
Phenyl	28 ± 0.5
–S–Methyl	67.2 ± 0.1

^{a} At 45°C.

A similar study has been carried out with the 1-chloroethyl group in the 2-position and the X group in the 4-position. The results are given in Table III-58 (283, 286).

These results are used to calculate the following Hammett σ^+ parameters of the 2-, 4-, and 5-thiazolyl groups (taking a value of -5.12 for ρ for the reaction of solvolysis): 5-thiazolyl = -0.18, 4-thiazolyl = -0.01, 2-thiazolyl = $+0.26$.

TABLE III-58. RATE OF SOLVOLYSIS OF 4-SUBSTITUTED 2-(1-CHLOROETHYL)-THIAZOLES (283, 286)

Substituents	$k(10^{-5} \text{ sec}^{-1})^a$
H	0.464
Methyl	5.15 ± 0.2
Phenyl	1.54 ± 0.05
4,5-Dimethyl	593

^{a} At 45°C.

XV. APPLICATIONS

1. Natural Flavors

Analytical chemistry has in recent years been equipped with a number of powerful means of investigation. Their application, especially that of gas-phase chromatography coupled with a mass spectrometer, has demonstrated the presence of a certain number of thiazoles in natural products such as fruits or cereals (287, 288, 297). The many results are shown in Table III-59.

TABLE III-59. THIAZOLES IN NATURAL FLAVORS

Products	Thiazoles	Ref.
Cereals, meat	Various	292
Beef, Chicken	Various	293, 298
Tomato	2-i-Butyl, 10–50 ppb	294, 295 303, 304, 584
Roasted peanuts	Various	296, 864
Roasted coffee	More than 20 alkylthiazoles	297, 302
Tea	More than 4 alkylthiazoles	300, 301
Hazel nuts	4-Methyl-5-vinyl	291, 296, 297

Moreover, certain thiazoles are used in canned food to enhance fruit aroma (e.g., 2-i-butylthiazole for tomatoes).

2. Polymers

The use of thiazoles in polymer chains has received a certain amount of attention, especially where thermostable polymers are demanded. Several directions of study are possible:

- Polymerization of a noncyclic monomer leading to a total synthesis of the ring or the polymerization of a substituent on the thiazole ring, for example, a vinyl group.
- A study of structural units, where a monomer is examined by physicochemical methods to determine its thermostability, its chemical and physical properties, and its sites of degradation.

Furthermore, if the structural unit to be polymerized is a vinyl group, mixed reactions can be carried out with a second compound such as a vinyl derivative or maleic anhydride.

A. *Polymerization by Total Synthesis of the Thiazole Ring*

The condensation of a dithioamide with a dibromodiketone (305–311) leads to a polymer according to Scheme 7. Polymers formed by thiazole rings bonded by arylene groups have also been obtained (Scheme 8). The ultraviolet spectra of the polymers and their models are given in Table III-60.

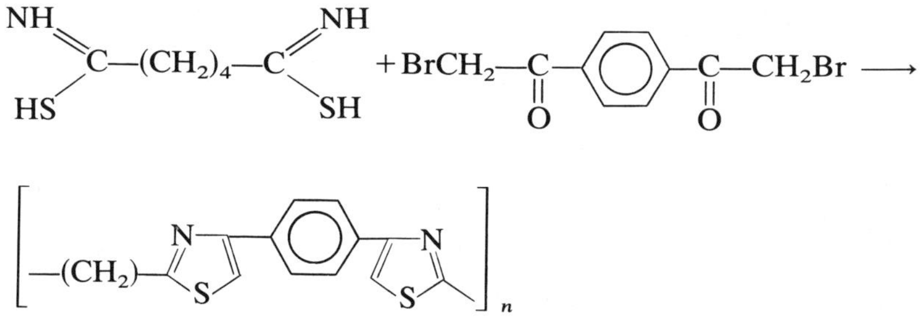

Scheme 7

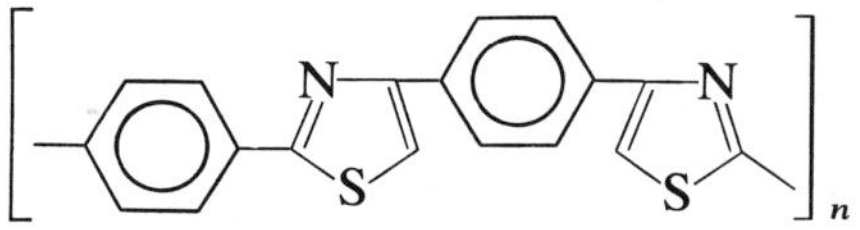

Scheme 8

TABLE III-60. ULTRAVIOLET SPECTRA OF THIAZOLE POLYMERS (305–311)

Compound	λ_{max}	ε	λ_{max}	ε
Polymer 1	282	—	—	—
	279	25200	—	—
Polymer 2	282	—	375	—
	261	23000	377	26700

The best solvent for this type of polymerization is dimethylformamide. The molecular weight is between 5300 and 5800, and the products have melting points up to 350°C.

B. *Polymerization through a Substituent on the Thiazole Ring*

The monomer bearing the vinyl group is synthesized and then polymerized. The most common monomers are of the type shown in Scheme 9 (312, 313).

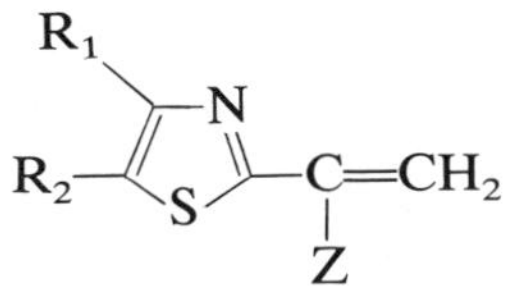

$$R_2 \underset{S}{\overset{R_1}{\underset{N}{\bigg[}}} C(Z)=CH_2$$

Z = Methyl, H,
R_1, R_2 = methyl, ethyl, propyl, *i*-propyl, butyl, phenyl

Scheme 9

Polymerization takes place in the following manner: in the presence of suitable peroxide catalyst these compounds polymerize with themselves (homopolymerization) in aqueous emulsion. When the reaction is complete, the emulsified polymer may be used directly or the emulsion coagulated to yield the solid polymer (312). A typical polymerization mixture is total monomer (2-vinylthiazole), 100; sodium stearate, 5; potassium persulfate, 0.3; laurylmercaptan, 0.4 to 0.7; and water, 200 parts.

In the case of anionic polymerization (with 2-isopropenylthiazole) there is a chain-monomer equilibrium. Furthermore, lowering the temperature of polymerization increases the conversion of monomer to polymer (314).

C. Copolymerization with Dienes

Copolymerization can be carried out with styrene, acetonitrile, vinyl chloride, methyl acrylate, vinylpyridines, 2-vinylfurans, and so forth. The addition of 2-substituted thiazoles to different dienes or mixtures of dienes with other vinyl compounds often increases the rate of polymerization and improves the tensile strength and the rate of cure of the final polymers. This allows vulcanization at lower temperature, or with reduced amounts of accelerators and vulcanizing agents.

A typical example is total monomers, 100; sodium stearate, 5; potassium persulfate, 0.3; lauryl mercaptan, 0.4 to 0.7; and water, 200 parts. In this formula, 75 parts of 1,3-butadiene and 25 parts of 4-methyl-2-vinylthiazole give 86% conversion to a tacky rubber-like copolymer in 15 hr at 45°C. The polymer contains 62% benzene-insoluble gel. Sulfur analysis indicates that the polymer contains 21 parts of combined 4-methyl-2-vinylthiazole (312). Butadiene alone in the above reaction normally requires 25 hr to achieve the same conversion, thus illustrating the acceleration due to the presence of 4-methyl-2-vinylthiazole.

Similarly, the addition of low quantities of vinyl or polyvinylthiazoles in the synthesis of aromatic polyesters increases the rate of polymerization (315).

A number of reviews discuss the polymerization of thiazole monomers in a general context (316–319).

D. Study of Different Monomers

Various monomers have been studied for their physicochemical properties and electronic structures (320, 321). For example, a series of monomers can be synthesized following Mulvaney et al. (310) and then theoretical diagrams and degradation sites are studied (Table III-61) (134).

The pyrolysis thresholds of this series of compounds show that the thiazole structure is thermostable, decomposition temperatures being generally between 450 and 510°C. Other observations that can be made are:

- Thermal stability increases with increasing molecular weight.
- Methyl derivatives are much less resistant.
- The presence of –O– bonds in the case of diphenylether derivatives slightly increases the thermal stability.

Thus the study of polymers containing thiazole rings, linked by alkyl or

TABLE III-61. PYROLYSIS THRESHOLDS OF VARIOUS MONOMERS (134)

Monomer	Pyrolysis thresholds (°c)
Thiazole	530
2,4-Diphenylthiazole	431
2-Phenyl-4-biphenylthiazole	442
4-Phenyl-2-biphenylthiazole	452
2-(Diphenylyl-4)bis-thiazole	510
2,2'-(p-Phenylene)bis-4-methylthiazole	403
4,4'-(p-Phenylene)bis-2-phenylthiazole	484
2,2'-(p-Phenylene)bis-4-phenylthiazole	495
4,4'-(p-Phenylene)bis-2-(4-biphenyl)thiazole	505
2,2'-(p-Phenylene)bis-4-(4-biphenyl)thiazole	513
2,2'-(p-Phenylene)oxidi-bis-4-phenylthiazole	476
2,2'-(p-Phenylene)oxidi-bis-4-(4-biphenyl)thiazole	495

phenyl or heterocyclic rings, shows that these compounds are resistant to thermal degradation (322–325, 327) and even to γ-irradiation (319, 326). It would appear then that if the cost price of the monomer were competitive, such polymers would find their place in the future market.

3. Biological and Pharmacological Uses of Thiazoles

The thiazole ring can be found in numerous molecules that possess biological activity: the thiamine (vitamin B_1), penicillins, antiinflamatory and bactericidals compounds, and so forth.

On the other hand, the thiazole is also found in the structures of numerous pesticides, fungicides, herbicides, and nematocides. Association of thiazoles with other heterocyclic compounds has been widely used in this field.

These preceding properties imply that the thiazole has to be introduced in various molecules, by direct cyclization or with precursors already bearing the thiazole ring. Among these last products the clomethiazole, nitrothiazole, and aryl or alkylthiazoles with the functional group on the aryl or alkyl substituent have been widely used.

The prime importance of these biological applications, far beyond the scope of this book, has in recent years focused interest on biological applications of thiazoles instead of on typical chemical research (at least for those described in Chapter 3). In the tables of products, thiazoles that are of biological interest are indicated.

XVI. TABLES OF PRODUCTS

Alkylthiazoles
Table III-62	2-Alkylthiazoles	402
Table III-63	4-Alkylthiazoles	403
Table III-64	5-Alkylthiazoles	403

Dialkyl- and Trialkylthiazoles
Table III-65	2,4-Dialkylthiazoles	404
Table III-66	2,5-Dialkylthiazoles	406
Table III-67	4,5-Dialkylthiazoles	407
Table III-68	2,4,5-Trialkylthiazoles	408

Arylthiazoles
Table III-69	2-Arylthiazoles	409
Table III-70	4-Arylthiazoles	410
Table III-71	5-Arylthiazoles	410

Diraryl- and Triarylthiazoles
Table III-72	2,4-Diarylthiazoles	410
Table III-73	2,5-Diarylthiazoles	411
Table III-74	4,5-Diarylthiazoles	411
Table III-75	2,4,5-Triarylthiazoles	411

Arylalkylthiazoles
Table III-76	2-Aryl-4-alkylthiazoles	412
Table III-77	2-Alkyl-4-arylthiazoles	412
Table III-78	2-Aryl-5-alkylthiazoles	413
Table III-79	2-Alkyl-5-arylthiazoles	413
Table III-80	4-Aryl-5-alkylthiazoles	413
Table III-81	4-Alkyl-5-arylthiazoles	413
Table III-82	Dialkyl-arylthiazoles	414
Table III-83	Alkyl-diarylthiazoles	414

Alkenylthiazoles
Table III-84	2-Position	415
Table III-85	4-Position	416
Table III-86	5-Position	416
Table III-87	2-Position with a heterocyclic group	417
Table III-88	4-Position with a heterocyclic group	419

Aralkylthiazoles
Table III-89	2-Aralkylthiazoles	419
Table III-90	4-Aralkylthiazoles	420
Table III-91	5-Aralkylthiazoles	420

Aralkylthiazoles with One Substituent on the Aryl Group
Table III-92	2-Aralkylthiazoles with one substituent on the aryl group	421
Table III-93	4-Aralkylthiazoles with one substituent on the aryl group	422
Table III-94	5-Aralkylthiazoles with one substituent on the aryl group	422

XVI. Tables of Products

Arylthiazoles with Substituents on the Aryl group
- Table III-95   2-Arylthiazoles with one substituent on the aryl group 422
- Table III-96   2-Arylthiazoles with two substituents on the aryl group . . . 424
- Table III-97   4-Arylthiazoles with one substituent on the aryl group 424
- Table III-98   4-Arylthiazoles with substituents on the aryl group 426
- Table III-99   5-Arylthiazoles with one substituent on the aryl group 426

Diarylthiazoles with Substituents on the Aryl Groups
- Table III-100   2,4-Diarylthiazoles with substituents on the aryl groups 427
- Table III-101   2,5-Diarylthiazoles with substituents on the aryl groups . . . 428
- Table III-102   4,5-Diarylthiazoles with substituents on the aryl groups . . . 428

Alkylthiazoles with One Function on the Alkyl Group
- Table III-103   2-Alkylthiazoles with one function on the alkyl group 429
- Table III-104   4-Alkylthiazoles with one function on the alkyl group 439
- Table III-105   5-Alkylthiazoles with one function on the alkyl group 452

Thiazoles Substituted by Polyfunctional Groups
- Table III-106   Thiazoles substituted by one polyfunctional group 460
- Table III-107   Thiazoles substituted by several polyfunctional groups 464

Thiazolylthiazoles
- Table III-108   2-Thiazolylthiazoles 468
- Table III-109   4-Thiazolylthiazoles 468
- Table III-110   5-Thiazolylthiazoles 469
- Table III-111   Benzothiazolylthiazoles 469

Thiazoles Substituted by One Heterocyclic Group
- Table III-112   Furylthiazoles 469
- Table III-113   Thienylthiazoles 470
- Table III-114   Pyrolyl and benzopyrrolyl thiazoles 471
- Table III-115   Oxazolylthiazoles 471
- Table III-116   Thiazolidinyl and thiazolynylthiazoles 472
- Table III-117   Imidazolyl and benzimidazolylthiazoles 473
- Table III-118   1,3,4-Oxadiazolylthiazoles 476
- Table III-119   Triazolinylthiazoles 476

Pyridylthiazoles
- Table III-120   2-Pyridylthiazoles 477
- Table III-121   4-Pyridylthiazoles 478
- Table III-122   5-Pyridylthiazoles 478
- Table III-123   2,4-Dipyridylthiazoles 478

Thiazoles Substituted by Several Heterocyclic Groups
- Table III-124   Thiazoles substituted by several heterocyclic groups 479

Thiazoles Substituted by Heterocyclic Groups Nonadjacent to the Thiazole Ring
- Table III-125   Thiazoles substituted by heterocyclic groups nonadjacent to the thiazole ring 481

Nonadjacent Dithiazoles
- Table III-126   Nonadjacent dithiazoles 482

Miscellaneous Substituted Thiazoles
- Table III-127   Miscellaneous substituted thiazoles 487

TABLE III-62. 2-ALKYLTHIAZOLES

	b.p. (°C)	Ref.[a]
Me	129/760	23, 75, 96, 97, 99 iq, 103 r, 111 r, 121, 128 i, 129 i,
	126–127/760	130 i, 131 i, 135 i, 152, 150 t,
	126.5/729	157 u, 168 g, 197 g, 222,
	65–67/80	239, 240, 258, 275, 278, 378, 434, 454, 493, 494, 501, 508 i, 524 g, 525, 537, 562 t, 570, 600, 601, 641 g, 642 u, 684 r, 717, 758 h
Et	158/760	96, 97 k, 107 r, 119 m, 128 i,
	66.5–67/46	152 t, 168 g, 257, 258, 278, 501, 526, 537, 600, 785 m
n-Pr	172/760	97 k, 119 m, 152 t, 168 g, 240, 600, 601, 640, 785 m
n-Bu	—	640
n-Amyl	198/760	168, 600, 640, 686
	142/4	
n-Hexyl	—	685
i-Pr	160/760	31, 97 k, 119 m, 128 i, 152 t,
	75/46	168 g, 257, 258, 501, 600, 640, 643, 684 r, 685
sec-Bu	—	640
sec-Amyl	—	640
i-Bu	180/760	31, 152 t, 168 g, 257, 304, 371 k, 600, 640, 643, 685, 687
tert-Bu	175/760	97 k, 119, 150 t, 152 t, 257,
	180/760	271, 501, 579, 600, 601, 640,
	57/46	684 r
neo-Pentyl	46/0.06	31, 152

[a] For the meaning of letters following references, cf. p. 2.

XVI. Tables of Products

TABLE III-63. 4-ALKYLTHIAZOLES

	b.p. (°C)	Ref.
Me	133/760 130–133/760 70/90	17, 55, 99 iq, 103 r, 111 r, 115 m, 121, 128 i, 129 i, 130 i, 131 i, 135 i, 150 t, 197, 198, 222, 239, 272, 350, 359, 379, 392, 403 416 h, 417, 418, 435, 454, 455, 472, 479, 493, 508 i, 524 g, 526, 537, 549 t, 552 562 t, 570, 602, 603, 642, 644, 645, 646, 684 r, 688
Et	145/760 151–152/743	128 i, 168 g, 436, 602, 716 r
n-Pr		117
i-Pr	171/760	128 i, 168 g
i-Bu	—	716r
tert-Bu	—	119 m, 128 i, 150 t, 602

TABLE III-64. 5-ALKYLTHIAZOLES

	b.p. (°C)	Ref.
Me	141–142/760 82.8/100 57/32	14, 99 iq, 119 m, 128 i, 130 i, 131 i, 135 i, 150 t, 168 g, 199, 222, 239, 285, 379, 399, 404, 454, 479, 493, 508, 516, 562 t, 570, 602, 603, 642, 647, 684 r.
Et	162/760 99.7/100	119 m, 128 i, 168 g, 222, 360, 399, 516, 602, 641 g, 717
n-Pr	—	119 m, 587
n-Bu	79/11	16, 119 m, 587
i-Pr	109.5/100	128 i, 516, 587, 602
tert-Bu	117/100	128 i, 516, 684 r.

TABLE III-65. 2,4-DIALKYLTHIAZOLES

Position		b.p. (°C)	Ref.
2	4		
Me	Me	150–151/760	23, 39 u, 55,
		145/760	96, 100 ur,
		145–147/760	103 r, 111 r,
		70–73/50	118 m, 121, 130 i,
		67–58/50	131 i, 150 t, 168,
			197 g, 228, 229,
			268, 280, 354,
			372, 379, 394,
			405, 416 h, 437,
			454, 455, 473 i,
			475, 482, 493,
			508 i, 517, 526,
			537, 602, 603,
			604, 645, 647,
			648, 717 s, 718,
			763, 791
Me	Et	165–169/719	117, 121, 304, 438, 785 m.
Me	n-Pr	—	117, 785 m.
Me	n-Bu	—	117
Me	cyclohexyl		24
Me	i-Pr	79–81/0.5	105, 117
Me	tert-Bu	113/50	100 ur, 105, 168

	(CH₂)ₙ			
	CH—(CH₂)ₙ—C—			
	(CH₂)ₙ			

R₁	R₂	R₃	bp	Refs
Me	—	—	—	605
Et	Me		160/728.5	96, 119 m, 411, 532, 582 ur
Et	Et		116/173	105, 117
Et	n-Pr		97.5–98/15	117, 527, 689
Et	i-Pr		—	105
Et	i-Bu		89–90/11.5	528
Et	tert-Bu		108/50	24, 105
n-Pr	Me		76–78/14	582 ur.
n-Pr	Et		—	119 m
n-Pr	n-Pr		—	117
n-Bu	Me		—	117
i-Pr	Me		92/50	517
i-Pr	Et		—	117, 168 g, 582
i-Pr	i-Pr		113/50	105
i-Pr	tert-Bu		—	105
i-Bu	Me		96/50	105, 582 ur
tert-Bu	Me		—	18
tert-Bu	Et		—	119 m, 168, 582 ur
tert-Bu	i-Pr		114/50	105
tert-Bu	tert-Bu		—	105, 168, 582 ur, 602

TABLE III-66. 2,5-DIALKYLTHIAZOLES

Position		b.p. (°C)	Ref.
2	5		
Me	Me	86/80 90/100 149/760 151–153/760	96, 107 r, 111 r, 119 m, 128 i, 129 i, 130 i, 131 i, 150 t, 168 g, 197 g, 278, 378, 379, 439, 454, 456, 472, 475, 493, 508 i, 516, 602, 717 s, 719, 763
Me	Et	107.6/100	128 i, 516, 602, 641 g, 717, 785 m
Me	n-Pr	—	119 m
Me	i-Pr	118.9/100	128 i, 516
Me	tert-Bu	126.8/100	128 i, 516
Et	Me	106.2/100	119 m, 128 i, 222, 516, 602
Et	Et	123/100	117, 128 i, 516
Et	n-Pr	—	119 m
Et	n-Bu	—	119 m
Et	i-Pr	134/100	128 i
n-Pr	Me	185/760	466
n-Pr	n-Pr	—	117
n-Bu	Me	—	119 m
n-Bu	n-Pr	—	119 m
n-Hex	n-Hex	—	808
i-Pr	Me	114.9/100	128 i, 516
i-Pr	Et	131/100	128 i, 516, 602
i-Pr	i-Pr	141.5/100	128 i, 516
tert-Bu	Me	120/100	128 i, 516, 602
tert-Bu	Et	135.8/100	128 i, 516
tert-Bu	i-Pr	155.5/100	128 i, 516

TABLE III-67. 4,5-DIALKYLTHIAZOLES

Position 4	Position 5	b.p. (°C)	Ref.
Me	Me	55–57/13 52–53/14 75–77/47 158/760	14, 17, 26, 118 m, 119 m, 128 i, 131 i, 150 t, 168 g, 197 g, 198, 222, 364, 393, 428, 454, 475, 479, 493, 505, 603, 688, 716 r, 776 m
Me	Et	39.5–44/2 78–79/25 173–174/755	9, 117, 119 m, 121, 392, 466, 716 r
Me	n-Pr		606 l, 716 r
Me	n-Bu		374
Me	i-Pr	—	716 r
Me	i-Bu	—	649, 716 r
Me	tert-Bu	50–52/3	8
Et	Me	36–38/0.3 35–36/1 44–46/1 169/70	23, 26, 29, 119 m, 502
Et	Et	186/751 178–179/760	440, 509, 650
n-Pr	Et	82–83/15	118 m, 440
	–(CH$_2$)$_3$–	80/11	406
	–(CH$_2$)$_4$–	68–70/1 94/11 126–127/22	26, 30, 380, 406
i-Pr	Me	187–188/760	466

TABLE III-68. 2,4,5-TRIALKYLTHIAZOLES

	Position		m.p. (°C)	b.p. (°C)	Ref.
2	4	5			
Me	Me	Me		45–47/6	23, 55, 96,
				57/12	117, 118 m,
				48–50/14	119 m, 121 m,
				65–67/20	130 i, 131 i,
				64–66/20	150 t, 258,
				164–166/760	412, 454,
					462, 475,
					493, 502,
					508 i, 510,
					532, 553 g,
					641, 645,
					690 rgl,
					720, 785 m
Me	Me	Et		185/760	117, 119 m,
					315, 466,
					641 g, 717,
					720 m, 785 m
Me	Me	n-Pr		—	117
Me	Et	Me		32/11	26, 29,
					105, 117,
					119 m, 258,
					720, 785 m
Me	Et	n-Pr		—	117
Me	n-Pr	Me		90–100/12	544
Me	n-Pr	Et		93–95/12	117, 544
Me	n-Bu	Me		—	117
Me	n-Undecyl	Me		125–129/16	544
Me	n-Tetradecyl	Me		122–126/0.1	544
Me	n-Hexadecyl	Me	34 m.p.	125–130/0–0.2	544
Me	i-Pr	Me		196–197/760	105, 258, 466
Me	i-Bu	Me	36–38 m.p.	106/8 90–100/12	258, 510, 544
Me	–(CH$_2$)$_3$–			100/13	406
Me	–(CH$_2$)$_4$–			102/11 230/715	30, 355, 380
Me	–(CH$_2$)$_5$–			115–120/16 115–118/18	486, 487
Et	Me	Me		—	117, 532
Et	Me	Et		81.5/12 82–90/13	117, 503, 528, 689
Et	Me	i-Pr		89.5–90/12	528, 689
Et	Et	Me		52–53/1	119 m, 498,

TABLE III-68. (Continued)

	Position		m.p. (°C)	b.p. (°C)	Ref.
2	4	5			
				58–60/15	502
Et	n-Pr	Et		70–73/16	544
Et	i-Bu	Et		95–105/12	544
Et	–(CH$_2$)$_4$–			120/10	30
				121/16	
Et	–(CH$_2$)$_5$–			80–82/0.5	30
n-Pr	Me	Me		—	117
n-Pr	Me	Et		—	117
n-Pr	Et	Me		62–63/1	26, 29
				102–103/18	
n-Bu	Et	Me		—	119 m
n-Pentyl	i-Bu	n-Pentyl		115–117/0.15	544
n-Heptyl	i-Bu	n-Heptyl		111–115/0.15	544
n-Undecyl	n-Pr	n-Undecyl		165–170/0.15	544
n-Heptadecyl	i-Bu	n-Heptadecyl	42–43		544
i-Pr	Me	Me		—	121 m, 809 mg
i-Pr	Me	Et		—	121 m
i-Pr	Et	Me		53–54/0.8	29
i-Bu	Me	Me		—	121 m, 720 m

TABLE III-69. 2-ARYLTHIAZOLES

	b.p. (°C)	Ref.
Ph	135–140/5	108 r, 109, 124 m, 130 i, 143,
	135/18	145, 147 u, 184 r, 216, 217,
	266/743	245, 250, 251, 283, 438, 457,
		469, 505, 508 i, 565, 575,
		607, 642 u, 691 hd, 734, 733 r
p-MeC$_6$H$_4$	—	176 m, 607
m-MeC$_6$H$_4$	—	176 m, 607
o-MeC$_6$H$_4$	—	108 r, 176 m, 607
p-EtC$_6$H$_4$	—	176 m
m-EtC$_6$H$_4$	—	176 m
o-EtC$_6$H$_4$	—	176 m
p-BuC$_6$H$_4$	—	176 m
m-BuC$_6$H$_4$	—	176 m
o-BuC$_6$H$_4$	—	176 m
p-i-PrC$_6$H$_4$	—	176 m
m-i-PrC$_6$H$_4$	—	176 m
o-i-PrC$_6$H$_4$	—	176 m
o-PhC$_6$H$_4$	—	187 g

TABLE III-70. 4-ARYLTHIAZOLES

	m.p. (°C)	b.p. (°C)	Ref.
Ph		273/760	21, 98, 108 r, 109, 124 m,
	51–55		145, 147, 216, 217, 219 u,
	52		246, 251, 351, 351, 428, 470,
	55		495, u, 505, 554, 575, 607,
	51–53		642 u, 691 hd, 692, 733 r
	49–52		
p-MeC$_6$H$_4$	—		607
m-MeC$_6$H$_4$	—		607
o-MeC$_6$H$_4$	—		607
p-PhC$_6$H$_4$	—		109, 336 g
o-PhC$_6$H$_4$	—		336 g

TABLE III-71. 5-ARYLTHIAZOLES

	m.p. (°C)	Ref.
Ph	45–46	108 r, 109, 124 m, 145,
	45	147 u, 216, 217, 251, 252,
	40–41	457, 505, 575, 642, 691 hd,
		733 r

TABLE III-72. 2,4-DIARYLTHIAZOLES

Position				
2	4	m.p. (°C)	b.p. (°C)	Ref.
Ph	Ph	93–94		108 hd, 109,
		92–93		118 m, 134 tiu,
		93–93.5		187 g, 219 u,
				329, 351, 439,
				470, 518, 566,
				575, 651
Ph	p-MeC$_6$H$_4$	116		538
Ph	p-PhC$_6$H$_4$		159/60	134 tiu, 329,
				575
p-MeC$_6$H$_4$	Ph	132		538
m-MeC$_6$H$_4$	Ph	65.3		538
p-PhC$_6$H$_4$	Ph	162		134 tiu, 147 u
				329, 575
p-PhC$_6$H$_4$	p-PhC$_6$H$_4$	207–208		134 tiu, 329
α-Naphthyl	Ph	88		470

TABLE III-73. 2,5-DIARYLTHIAZOLES

Position			
2	5	m.p. (°C)	Ref.
Ph	Ph	103–104	7, 108 hd, 109, 130 i, 147 u, 187 g, 218, 456, 508 i, 547, 575
Ph	p-MeC$_6$H$_4$	120–121	547
Ph	p-PhC$_6$H$_4$	191–192	453
Ph	β-Naphthyl	145–146.5	489 u
p-MeC$_6$H$_4$	p-MeC$_6$H$_4$	—	810 g
p-PhC$_6$H$_4$	Ph	190–191	489 u
α-Naphthyl	Ph	93.5–96.5	489 u

TABLE III-74. 4,5-DIARYLTHIAZOLES

Position			
4	5	m.p. (°C)	Ref.
Ph	Ph	53–54 57 63–64	109, 147 u, 187 g, 218, 252, 336, 479, 575, 691 dh
p-PhC$_6$H$_4$	Ph	156	538
p-PhC$_6$H$_4$	p-PhC$_6$H$_4$	145.5–147	467
β-Naphthyl	β-Naphthyl	117	467

TABLE III-75. 2,4,5-TRIARYLTHIAZOLES

Position				
2	4	5	m.p. (°C)	Ref.
Ph	Ph	Ph	86–87	109, 147 u, 439, 691 hd, 811 u, 812, 813

TABLE III-76. 2-ARYL-4-ALKYLTHIAZOLES

Position				
2	4	m.p. (°C)	b.p. (°C)	Ref.
Ph	Me	145–16 97–98	111/6 140–141/12 284/760	24, 98, 108 r, 109, 118 m, 124 m, 185 g, 245, 352, 395, 418, 439, 455, 607, 686, 756
Ph	Et	117–118		189 g, 245, 274, 418, 851
Ph	tert-Bu		158–159/15	24, 185 g, 189 g
p-MeC$_6$H$_4$	Me	34		25
m-MeC$_6$H$_4$	Me		165/18	25
o-MeC$_6$H$_4$	Me		162/23	25
p-PhC$_6$H$_4$	Me	151–152		686
α-Naphthyl	Me		190–192/1.5	25

TABLE III-77. 2-ALKYL-4-ARYLTHIAZOLES

Position				
2	4	m.p. (°C)	b.p. (°C)	Ref.
Me	Ph	68 68.5	280–284/760 284/760	73, 98, 108 r, 109, 124 m, 131 i, 185 g, 189 g, 352, 418, 439, 455, 482, 495 u, 607, 647, 652, 756, 814, 815, 816, 817
Me	p-PhC$_6$H$_4$		363/747	275
Et	Ph		295–296/729	27, 108 r, 109, 124 m, 439
n-Pr	Ph		—	108 r, 124 m
i-Pr	Ph		—	109
tert-Bu	Ph		—	108 r, 124 m

XVI. Tables of Products

TABLE III-78. 2-ARYL-5-ALKYLTHIAZOLES

Position			
2	5	b.p. (°C)	Ref.
Ph	D	—	756
Ph	Me	254/734 283.5/750	108 r, 124 m, 456, 756, 818 l.
Ph	Et	—	108 r, 124 m, 819
Ph	i-Pr	—	108 r, 124 m

TABLE III-79. 2-ALKYL-5-ARYLTHIAZOLES

Position			
2	5	m.p. (°C)	Ref.
Me	Ph	81 84	109, 185 g, 189 g, 456, 482, 490
Me	p-MeC$_6$H$_4$	187	467
Me	β-Naphthyl	127–128	467

TABLE III-80. 4-ARYL-5-ALKYLTHIAZOLES

Position			
4	5	b.p. (°C)	Ref.
Ph	D	—	756
Ph	Me	111/2 146/11 142/12	108 r, 109 m, 124 m, 246, 495 u, 756
Ph	Et	152.5/11	108 r, 109 m, 495 u

TABLE III-81. 4-ALKYL-5-ARYLTHIAZOLES

Position			
4	5	b.p. (°C)	Ref.
Me	Ph	134–135/25	184 r, 185 g, 189 g, 756, 820
Et	Ph	—	189 g
tert-Bu	Ph	—	185 g, 189 g

TABLE III-82. DIALKYL-ARYLTHIAZOLES

Position			b.p. (°C)	Ref.
2	4	5		
Me	Me	Ph	148/11	109, 278
			135/11	457, 488 u,
			155/14	495 u
			270–271/755	
Me	Ph	Me	153–162/19–21	109, 275,
			296/725	511, 607
Me	Ph	Et	157/11	495 u, 608
Ph	Me	Me	126–128/6	109, 118 m,
			148/14	245, 509,
			268/760	607, 650
Ph	Et	Me	124/1.6	23, 29, 502
			138/2	
Ph	–(CH$_2$)$_4$–		149/2	30
Ph	–(CH$_2$)$_5$–		128–129/0.05	30

TABLE III-83. ALKYL-DIARYLTHIAZOLES

Position			m.p. (°C)	b.p. (°C)	Ref.
2	4	5			
Me	Ph	Ph	50		109, 246,
			51–2		275, 439,
					652, 693 l,
					764
Me	p-MeC$_6$H$_4$	Ph	—		589 u
Me	p-EtC$_6$H$_4$	Ph	—		589 u
Me	2,5-Me$_2$C$_6$H$_4$	Ph	—		589 u
Et	Ph	Ph	62.5		477
Ph	Ph	Me	—		109
Ph	p-MeC$_6$H$_4$	Me	82		747
Ph	m-MeC$_6$H$_4$	Me	198		747
Ph	o-MeC$_6$H$_4$	Me	45		747
Ph	Me	Ph		115/10	109

TABLE III-84. 2-POSITION

$$\begin{array}{c} R_2 \\ \diagdown \\ C=CH- \\ \diagup \\ R_2' \end{array} \begin{array}{c} N R_4 \\ \| \\ SR_5 \end{array}$$

Substituent					
R_2	R_2'	R_4	R_5	m.p. (°C) b.p. (°C)	Ref.
Ph	H	p-$NO_2C_6H_4$	H	158	519 u
Ph	H	H	p-$NO_2C_6H_4$	142	519 u
Ph	H	p-$MeOC_6H_4$	H	163	519 u
Ph	H	H	p-$MeOC_6H_4$	170–171	519 u
p-(MeCONH)C_6H_4	H	Me	H	153	555 l
p-(MeCONH)C_6H_4	H	Ph	H	213	555 l
p-ClC_6H_4	H	Me	H	98	555 l
				112–113	609, 694
				122–123	
p-ClC_6H_4	H	Ph	H	173	555 l
m-ClC_6H_4	H	Me	H	58	555 l
m-ClC_6H_4	H	Ph	H	133	555 l
o-ClC_6H_4	H	Me	H	56	555 l
o-ClC_6H_4	H	Ph	H	100	555 l
p-BrC_6H_4	H	Me	H	100	277
p-$NMe_2C_6H_4$	H	H	H	124	282
p-$NMe_2C_6H_4$	H	Me	H	100	282
p-$NEt_2C_6H_4$	H	Me	H	—	282
p-$NO_2C_6H_4$	H	Me	H	129	282
p-$NO_2C_6H_4$	H	Ph	H	168	519 u
p-$NO_2C_6H_4$	H	p-$MeOC_6H_4$	H	152	519 u
p-$NO_2C_6H_4$	H	H	Ph	192	519 u
p-$NO_2C_6H_4$	H	H	p-$MeOC_6H_4$	182	519 u
m-$NO_2C_6H_4$	H	Me	H	104	277
p-OHC_6H_4	H	Me	H	201	277
o-OHC_6H_4	H	Me	H	200	277
p-MeC_6H_4	H	Me	H	50	277,
				55	555 l
p-$MeOC_6H_4$	H	Ph	H	143	519 u,
					555 l
p-$MeOC_6H_4$	H	p-$NO_2C_6H_4$	H	161	519 u
p-$MeOC_6H_4$	H	H	Ph	142	519 u
p-$MeOC_6H_4$	H	H	p-$NO_2C_6H_4$	—	519 u
o-$MeOC_6H_4$	H	Me	H	—	277
p-$PhOC_6H_4$	H	Me	H	—	277
![3,4-methylenedioxyphenyl]	H	Me	H	100	277
3,4-$(OH)_2C_6H_3$	H	Me	H	—	277
3-Me-4-OHC_6H_3	H	Me	H	109–111	277
3,4-$diMeOC_6H_3$	H	Me	H	102	277
CO (o-ClC_6H_4)	H	H	H	—	610
CO_2H	H	H	H	187–190	49
CO_2H	H	p-$NO_2C_6H_4$	H	218	248, 533

TABLE III-84. (Continued)

	Substituent			m.p. (°C)	b.p. (°C)	Ref.
R_2	R'_2	R_4	R_5			
CO_2Et	H	$p\text{-}NO_2C_6H_4$	H		132/3	248
NH_2	Me	H	H	52–53		26
					98/0.5	
CO_2Me	Me	Ph	H	—		821 r
$p\text{-}ClC_6H_4$	Ph	Me	H	119–120		609
$(CH_2)_3NEt_2$	Me	Me	H		145/0.1	64

TABLE III-85. 4-POSITION

$$\underset{R_2}{\underset{\|}{\text{N}}}\!\!\!\!\!\overset{\text{CH}=C\overset{R_4}{\underset{R'_4}{}}}{\underset{S}{\bigcirc}}\!\!R_5$$

	Substituent			m.p. (°C)	Ref.
R_2	R_4	R'_4	R_5		
Cyclohexyl	CO_2H	H	H	148–149	822 1
Ph	CO_2H	H	H	191–192	529 i, 534,
				192–194	569
$p\text{-}MeC_6H_4$	CO_2H	H	H	203–204	569
$m\text{-}MeC_6H_4$	CO_2H	H	H	179–180	569
$p\text{-}ClC_6H_4$	CO_2H	H	H	—	695 1, 765
$p\text{-}NO_2C_6H_4$	CO_2H	H	H	250	534
Ph	CO_2Et	H	H	126	534
Cyclohexyl	CO_2NH_2	H	H	179–180	822 1
H	NO_2	Me	H	99–100	441
$p\text{-}NO_2C_6H_4$	CO_2Et	COMe	H	176	533
$p\text{-}NO_2C_6H_4$	CO_2H	CO_2H	H	150	533
$p\text{-}NO_2C_6H_4$	CO_2Et	CO_2Et	H	—	530 i, 533

TABLE III-86. 5-POSITION

$$\underset{R_2}{\underset{\|}{\text{N}}}\!\!\!\!\!\overset{R_4}{\underset{S}{\bigcirc}}\!\!\text{CH}=C\overset{R_5}{\underset{R'_5}{}}$$

	Substituent			m.p. (°C)	Ref.
R_2	R_4	R_5	R'_5		
H	Me	H	H	142–143	121, 721
Me	Me	H	H	—	121
H	H	NO_2	H	114	523
H	Me	NO_2	H	128–130	523
H	H	NO_2	Me	106–107	441
H	Me	NO_2	Me	104–107	523
H	H	CO_2H	NHCOPh	194–195	442

TABLE III-87. 2-POSITION WITH A HETEROCYCLIC GROUP

	Substituent				
R_2	R'_2	R_4	R_5	m.p. (°C)	Ref.
2-Oxazolyl	H	Me	H	71	2771, 555, 7371
2-Oxazolyl	H	Me	Me	—	7371
2-Oxazolyl	H	Me	Et	—	7371
2-Oxazolyl	H	Ph	H	110	555
2-(5-NO$_2$-oxazolyl)	H	H	H	—	7351
2-(5-NO$_2$-oxazolyl)	H	Me	Me	—	266
2-(5-NO$_2$-oxazolyl)	H	Me	Et	—	266, 7361
2-(5-NO$_2$-oxazolyl)	H	Me	Bu	—	266, 7361
2-(5-NO$_2$-oxazolyl)	H	Me	—(CH$_2$)$_n$—	—	266
2-(5-NO$_2$-oxazolyl)	H	Me	CH$_2$CONHNH$_2$	—	6551
2-(5-NO$_2$-oxazolyl)	H	Me	CH$_2$CO$_2$Et	—	6551, 754
2-(5-NO$_2$-oxazolyl)	H	Et	Me	—	6551, 7361
2-Thiophenyl	H	Me	H	75	5551
2-Thiophenyl	H	Ph	H	138	5551
2-(5-NO$_2$-thiophenyl)	H	CH$_2$Cl	H	—	611, 654
2-(5-NO$_2$-thiophenyl)	H	—CH$_2$—N(morpholino)	H	—	611
2-(4-Me-thiazolyl)	H	Me	H	—	84, 861
2-(4,5-DiPh-thiazolyl)	H	Ph	Ph	—	84, 85, 861
2-pyridyl	H	Me	H	—	5551
2-pyridyl	H	Ph	H	99	5551
3-pyridyl	H	Ph	H	99	5551

TABLE III-87. (*Continued*)

	Substituent				
R_2	R_2'	R_4	R_5	m.p. (°C)	Ref.
4-pyridyl	H	Ph	H	125	555 l
2-Benzimidazolyl	2-OHC$_6$H$_4$	H	H	230–231	565
2-Benzimidazolyl	2-Oxazolyl	H	H	206–207	565
![structure: CEtO-C(=O)-C(=S)-N-CH$_2$CO$_2$Et thiazolidine]	H		H	—	611

TABLE III.-88. 4-POSITION WITH A HETEROCYCLIC GROUP

$$\underset{R_2}{\overset{N}{\diagdown}}\underset{S}{\diagdown}\overset{CH=C}{\underset{R_5}{\diagdown}}\overset{R_4}{\diagdown}R_4'$$

Substituent				m.p./b.p.	
R_2	R_4	R_4'	R_5	(°C)	Ref.
Me	2-Oxazolyl	H	H	—	612 l
Me	2-Oxazolyl	Me	H	—	612 l
Me	2-Oxazolyl	Cl	H	—	612 l

TABLE III-89. 2-ARALKYLTHIAZOLES

	Position		m.p. (°C)	b.p. (°C)	Ref.
2	4	5			
–CH$_2$Ph	H	H		104/2	78, 107 r
–CH$_2$Ph	Me	H		150/14	14, 64, 107 r,
				94–96/0.1	417, 455,
				149–150/12	567 g
				150/14–15	
–CH$_2$Ph	H	Me		149–150/12	83, 107 r, 567 g
–CH$_2$Ph	Me	Me		150–151/13	14, 107 r
–CH$_2$Ph	Ph	Ph	99.5		477
–CH(Me)Ph	H	H	—		107 r
9-(9,10-dihydroanthracenyl)	Ph	H	121–122		580
–(CH$_2$)$_2$Ph	H	H	—		107 r, 567 g
–(CH$_2$)$_2$Ph	Me	H	—		107 r, 595, 648
–(CH$_2$)$_2$Ph	H	Me	—		107 r
–(CH$_2$)$_2$Ph	Me	Me	—		107 r

TABLE III-90. 4-ARALKYLTHIAZOLES

Position 2	4	5	b.p. (°C)	Ref.
H	-CH$_2$Ph	H	—	603
H	-CH$_2$Ph	Me	147.5/11	488 u, 495 u
H	-CH$_2$Ph	Et	155–160/12	495 u
Me	-CH$_2$Ph	H	—	567 g
Me	-CH$_2$Ph	Me	152/11 159/19	488 u, 511
H	-CH—CHPh \\CH$_2$/	H	143–145/2	419
H	-C(Ph)CH$_2$ \\CH$_2$/	H	80/0.5	420

TABLE III-91. 5-ARALKYLTHIAZOLES

Position 2	4	5	m.p. (°C)	b.p. (°C)	Ref.
H	Me	-CH$_2$Ph		148.5/11	495 u 567 gr
Me	Me	-CH$_2$Ph		157/11	495 u
Ph	Me	-CH$_2$Ph	56–57		25
p-MeC$_6$H$_4$	Me	-CH$_2$Ph	73–75		25
Ph	Me	-CH$_2$(p-MeC$_6$H$_4$)	43–45		25
p-MeC$_6$H$_4$	Me	-CH$_2$(p-MeC$_6$H$_4$)	85–86		25
α-Napthyl	Me	-CH$_2$(α-Naphthyl)	142		25
Me	H	—CH—CH$_2$ (naphthyl ring)	75		467

TABLE III-92. 2-ARALKYLTHIAZOLES WITH ONE SUBSTITUENT ON THE ARYL GROUP

2	Position 4	5	m.p. (°C)	Ref.
$-CH(Ph)(p-MeOC_6H_4)$	Me	H	66	5631
$-CH(p-MeCOC_6H_4)_2$	H	H	119–120	6561
$-CH(p-MeCOC_6H_4)_2$	Me	H	123–124	6561
$-CH(p-MeCOC_6H_4)_2$	Et	Me	105	6561
$-CH(p-MeCOC_6H_4)_2$	tert-Bu	H	152	6561
$-CH(p-MeCOC_6H_4)_2$	Ph	Me	88	6561
$-CH(p-MeCOC_6H_4)_2$	Ph	Ph	138	6561
$-CH(p-OHC_6H_4)_2$	H	H	230	5631, 6561
$-CH(p-OHC_6H_4)_2$	Me	H	230	6561
$-CH(p-OHC_6H_4)_2$	Et	H	216	6561
$-CH(p-OHC_6H_4)_2$	Et	Me	251–252	6561
$-CH(p-OHC_6H_4)_2$	n-Pr	H	219	6561
$-CH(p-OHC_6H_4)_2$	$-(CH_2)_4-$		—	6561
$-CH(p-OHC_6H_4)_2$	tert-Bu	H	232	6561
$-CH(p-OHC_6H_4)_2$	Ph	H	208	6561
$-CH(p-OHC_6H_4)_2$	Ph	Me	228	6561
$-CH(p-OHC_6H_4)_2$	Ph	Ph	288–290	6561
$-CH(p-MeOC_6H_4)_2$	Me	H	84	5631, 6561
$-CH(p-MeOC_6H_4)_2$	Et	H	57	5631, 6561
$-CH(p-MeOC_6H_4)_2$	Et	Me	—	563, 6561
$-CH(p-MeOC_6H_4)_2$	n-Pr	H	—	5631
$-CH(p-MeOC_6H_4)_2$	$-(CH_2)_4-$	H	—	6561
$-CH(p-MeOC_6H_4)_2$	tert-Bu	H	84	5631, 6561
$-CH(p-MeOC_6H_4)_2$	Ph	H	115 113	5631 6561
$-CH(p-MeOC_6H_4)_2$	Ph	Me	148	5631, 6561
$-CH(p-MeOC_6H_4)_2$	Ph	Ph	96	5631, 6561
$-CH(o-MeOC_6H_4)(p-MeOC_6H_4)$	Me	H	—	5631
$-CH(p-MeCO_2C_6H_4)_2$	H	H	119–120	5631
$-CH(o-MeCO_2C_6H_4)(p-MeCO_2C_6H_4)$	Me	H	97	5631
$-CH(OH)(p-MeC_6H_4)$	Ph	H	—	738
$-CH(o-CH_2Ph)(p-MeC_6H_4)$	Ph	H	—	738
$-C(OH)(p-MeOC_6H_4)_2$	H	H	147	6561
$-C(OH)(p-MeOC_6H_4)_2$	Me	H	—	6561

TABLE III-93. 4-ARALKYLTHIAZOLES WITH ONE SUBSTITUENT ON THE ARYL GROUP

	Position				
2	4	5	m.p. (°C)	b.p. (°C)	Ref.
H	$-CH_2(2\text{-}OH\text{-}3\text{-}ClC_6H_3)$	H		130/4	56
Me	$-CH_2(p\text{-}CO_2MeC_6H_4)$	Me	64–65		544
				120–130/16	
Me	$-CH_2(p\text{-}OHC_6H_4)$	H	285–286		5631
n-Pr	$-CH_2(p\text{-}OHC_6H_4)$	H	—		5631
Me	$-CH_2(p\text{-}MeOC_6H_4)$	H	81		5631
H	$-CH(p\text{-}OHC_6H_4)_2$	H	263		5631
H	$-CH(p\text{-}MeOC_6H_4)_2$	H	107		5631
H	$-CH(p\text{-}MeCO_2C_6H_4)_2$	H	117		5631

TABLE III-94. 5-ARALKYLTHIAZOLES WITH ONE SUBSTITUENT ON THE ARYL GROUP

	Position			
2	4	5	m.p. (°C)	Ref.
$p\text{-}Me_2NC_6H_4$	Me	$CH_2(p\text{-}Me_2NC_6H_4)$	131–133	25

TABLE III-95. 2-ARYLTHIAZOLES WITH ONE SUBSTITUENT ON THE ARYL GROUP

	Position				
2	4	5	m.p. (°C)	b.p. (°C)	Ref.
$p\text{-}MeCOC_6H_4$	Ph	H	124		538
$m\text{-}MeCOC_6H_4$	Ph	H	79		538
$p\text{-}MeCONHC_6H_4$	Me	H	—		395
$p\text{-}MeCONHC_6H_4$	Ph	H	—		356
$m\text{-}MeCONHC_6H_4$	Me	H	130–131		395
$o\text{-}MeCONHC_6H_4$	Me	H	116–117		395
$p\text{-}MeCO_2C_6H_4$	Me	H	72–74		25
$p\text{-}FC_6H_4$	H	Ph	137–138.5		489 u
$p\text{-}ClC_6H_4$	H	H	—		339 l, 607
$p\text{-}ClC_6H_4$	Me	H	119–120		125 m, 565 l, 607 l, 686
$p\text{-}ClC_6H_4$	Ph	H	110.5		538
$p\text{-}ClC_6H_4$	H	Ph	144.5–145		489 u
$p\text{-}ClC_6H_4$	$-CH_2Cl$	H	—		125
$m\text{-}ClC_6H_4$	Ph	H	75		538
$o\text{-}ClC_6H_4$	H	Ph	77–78		489 u

TABLE III-95. (*Continued*)

Position 2	4	5	m.p.	b.p. (°C)	Ref.
p-BrC$_6$H$_4$	Et	H	—		189 g
p-BrC$_6$H$_4$	*tert*-Bu	H	—		189 g
p-BrC$_6$H$_4$	–CH$_2$Cl	H	—		779 iu, 780 h
m-BrC$_6$H$_4$	Ph	H	81		538
m-IC$_6$H$_4$	H	Ph	117–117.5		489 u
p-NH$_2$C$_6$H$_4$	H	H	123–124		245, 251
p-NH$_2$C$_6$H$_4$	Me	H	112.5–113.5		245, 395, 823
p-NH$_2$C$_6$H$_4$	Et	H	106.5–107		245
p-NH$_2$C$_6$H$_4$	Me	Me	130.5–131.5		245
p-NH$_2$C$_6$H$_4$	Ph	H	146–146.5		356, 789
p-NH$_2$C$_6$H$_4$	Ph	Me	—		789
m-NH$_2$C$_6$H$_4$	Me	H	46–48		395.
p-NMe$_2$C$_6$H$_4$	H	Ph	160–162		489 u
p-NMe$_2$C$_6$H$_4$	Me	H	110–112		25
p-NO$_2$C$_6$H$_4$	H	H	147–148		245, 25 l, 607
p-NO$_2$C$_6$H$_4$	H	Et	—		807
p-NO$_2$C$_6$H	Me	H	106		245, 395
p-NO$_2$C$_6$H$_4$	Me	Me	169–170		245, 505
p-NO$_2$C$_6$H$_4$	Et	H	79.5–80		189, 245
p-NO$_2$C$_6$H$_4$	*tert*-Bu	H	—		189 g
p-NO$_2$C$_6$H$_4$	–CH$_2$Cl	H	—		780 h
o-NO$_2$C$_6$H$_4$	H	H	—		565 l
p-HOC$_6$H$_4$	H	H	163–165		357, 588 l
p-HOC$_6$H$_4$	Me	H	220–221		357, 588 l
p-MeOC$_6$H$_4$	H	H	—		357, 607
p-MeOC$_6$H$_4$	Me	H	56–57		357
p-MeOC$_6$H$_4$	Me	Me	65–66		505
p-MeOC$_6$H$_4$	Ph	H	101		538, 805
p-EtOC$_6$H$_4$	H	H		139–141/6	245
p-EtOC$_6$H$_4$	Me	H		160/6	357
p-EtOC$_6$H$_4$	Me	Me	84–86		245
p-EtOC$_6$H$_4$	Et	Me		150–152/6	245
⟨C$_6$H$_4$⟩NHCSNH⟨C$_6$H$_4$⟩OMe	H	H	—		805
p-AsO(OH)$_2$C$_6$H$_4$	Ph	H	—		356
p-PO(EtO)$_2$C$_6$H$_4$	Me	H	—		588 l
p-PS(MeO)$_2$C$_6$H$_4$	Me	H	—		588 l
p-PS(MeO)$_2$C$_6$H$_4$	H	H	—		588 l
p-PS(EtO)$_2$C$_6$H$_4$	H	H	—		588 l
p-PS(EtO)$_2$C$_6$H$_4$	Me	H	—		588 l
p-(2-Benzothiophenyl)-C$_6$H$_4$	Me	H	—		657
m-[2(4-Ph)thiazolyl]-C$_6$H$_4$	Ph	H	270–272		325 p
p-[2(4,5-diPh)thiazolyl]-C$_6$H$_4$	Ph	Ph	—		755

TABLE III-96. 2-ARYLTHIAZOLES WITH TWO SUBSTITUENTS ON THE ARYL GROUP

Position			m.p. (°C)	Ref.
2	4	5		
2-MeCO$_2$-4-NH$_2$-C$_6$H$_3$	H	H	169–70	443
2-MeCO$_2$-4-NO$_2$-C$_6$H$_3$	H	H	174–6	443
2-OH-4-NH$_2$-C$_6$H$_3$	H	H	138–40	443
2-OH-4-NO$_2$-C$_6$H$_3$	H	H	182–4	443
3-NH$_2$-2-OH-C$_6$H$_3$	H	H	96.5–7.5	245
3-NH$_2$-4-EtO-C$_6$H$_3$	Me	H	126–7	245
3-NH$_2$-4-EtO-C$_6$H$_3$	Me	Me	163.5–4.5	245
3-NH$_2$-4-EtO-C$_6$H$_3$	Et	Me	109	245
3-NO$_2$-4-EtO-C$_6$H$_3$	H	H	—	245
3-NO$_2$-4-EtO-C$_6$H$_3$	Me	H	130–2	245
3-NO$_2$-4-EtO-C$_6$H$_3$	Me	Me	140.2–1.2	245
3-NO$_2$-4-EtO-C$_6$H$_3$	Et	H	71–71.5	245
3-NO$_2$-4-OH-C$_6$H$_3$	H	H	—	245
3-OH-4-OH-C$_6$H$_3$	Me	H	250	357
Piperonyl	Me	H	93–5	357

TABLE III-97. 4-ARYLTHIAZOLES WITH ONE SUBSTITUENT ON THE ARYL GROUP

Position			m.p. (°C)	Ref.
2	4	5		
H	o-MeOC$_6$H$_4$	H	65–66	443
Me	o-HCO$_2$C$_6$H$_4$	H	143–147	413
H	p-CH$_3$CONHC$_6$H$_4$	H	165	246
H	o-HOC$_6$H$_4$	H	64–66	443
H	p-HOC$_6$H$_4$	H	146	373
H	p-MeOC$_6$H$_4$	H	—	108, 607, 789
Ph	p-MeOC$_6$H$_4$	H	135.5	538
Me	p-PhOC$_6$H$_4$	H	71	458
Me	p-FC$_6$H$_4$	H	81	361
Me	p-FC$_6$H$_4$	Ph	—	589 u
Et	p-FC$_6$H$_4$	H	160	361
H	p-ClC$_6$H$_4$	H	—	607
Me	p-ClC$_6$H$_4$	H	122–123	358, 652
Et	p-ClC$_6$H$_4$	H	164	358
H	p-BrC$_6$H$_4$	H	—	108
Me	p-BrC$_6$H$_4$	H	—	189
Me	p-BrC$_6$H$_4$	Ph	—	589 u
Et	p-BrC$_6$H$_4$	H	82–84	613
Ph	p-BrC$_6$H$_4$	H	135	538
Ph	m-BrC$_6$H$_4$	H	70	538
Ph	p-IC$_6$H$_4$	H	149	538
H	p-NH$_2$C$_6$H$_4$	H	99	246, 251, 554
H	p-NH$_2$C$_6$H$_4$	Me	80	982

TABLE III-97. (Continued)

	Position		m.p. (°C)	Ref.
2	4	5		
H	p-$NH_2C_6H_4$	Et	—	803
n-Pr	p-$NH_2C_6H_4$	H	—	614
Ph	p-$NH_2C_6H_4$	H	—	651, 685
n-Pr	m-$NH_2C_6H_4$	H	—	614
$CH_2C_6H_5$	p-CH_2=NC_6H_4	H	—	696
$CH_2C_6H_5$	p-CH_2=NC_6H_4	Me	—	696
Me	p-$(CH_2CH_2Cl)_2NC_6H_4$	H	—	806
Me	p-$(CH_2CH_2OH)_2NC_6H_4$	H	—	806
H	p-$NO_2C_6H_4$	H	178 180	108, 124, 248, 468r 533, 554, 607, 740, 781
H	p-$NO_2C_6H_4$	Me	98	246
Me	p-$NO_2C_6H_4$	H	145	189, 482, 614, 652
Ph	p-$NO_2C_6H_4$	H	122	538, 641, 747
Ph	p-$NO_2C_6H_4$	Me	152	747
Me	m-$NO_2C_6H_4$	H	—	652
Ph	m-$NO_2C_6H_4$	H	136	538
Ph	m-$NO_2C_6H_4$	Me	145	747
p-MeC_6H_4	m-$NO_2C_6H_4$	Me	163–164	538
m-MeC_6H_4	m-$NO_2C_6H_4$	Me	122–123	538
Ph	o-$NO_2C_6H_4$	Me	244	747
Me	p-$SCNC_6H_4$	Me	—	614, 739
Et	p-$SCNC_6H_4$	Me	64–66	739
n-Pr	p-$SCNC_6H_4$	Me	—	614
i-Pr	p-$SCNC_6H_4$	Me	—	739
H	m-$SCNC_6H_4$	Me	60–62	739
Me	m-$SCNC_6H_4$	Me	51–54	739, 741, 742, 745, 746
Me	m-$SCNC_6H_4$	n-Bu	—	739
n-Pr	m-$SCNC_6H_4$	Me	—	739
i-Pr	m-$SCNC_6H_4$	Me	—	739
H	p-$[(p$-$MeOC_6H_4)NHCSNH]C_6H_4$	H	—	789
H	p-$[(CH_2CH_2Cl)_2N$-C$_6$H$_4$-CH=N$]C_6H_4$	H	—	803
H	(cyclohexyl with OMe)–N=CH–(C$_6$H$_4$)–N(ClCH$_2$CH)$_2$	H	—	804
H	p-$[(OH)_2OAs]C_6H_4$	H	—	373
Me	p-$(2$-Me-4-thiazolyl$)C_6H_4$	H	223–224	329 p
Ph	p-$(2$-Ph-4-thiazolyl$)C_6H_4$	H	241–242	329 p
Me	m-$(2$-Me-4-thiazolyl$)C_6H_4$	H	66–67	329 p
Ph	m-$(2$-Me-4-thiazolyl$)C_6H_4$	H	178–195	329 p

TABLE III-98. 4-ARYLTHIAZOLES WITH SUBSTITUENTS ON THE ARYL GROUP

____	Position	____	m.p. (°C)	Ref.
H	2-Cl-4-$NO_2C_6H_3$	H	—	554
H	2,4-$(MeO)_2C_6H_3$	H	129	5811
Me	3-SCN-4-ClC_6H_3	H	120–121	739
Me	3-SCN-4-BrC_6H_3	H	123–125	739
Ph	3,4-$(OH)_2C_6H_3$	H	164–165	353
Ph	3-NH_2-4-$MeOC_6H_3$	H	—	804
Me	2,4,6-$(NO_2)_3C_6H_2$	H	172	379
Me	2,4,6-$(MeO)_3C_6H_2$	H	—	5811

TABLE III-99. 5-ARYLTHIAZOLES WITH ONE SUBSTITUENT ON THE ARYL GROUP

	Position		m.p. (°C)	Ref.
2	4	5		
H	H	p-(p-ClC_6H_4)CH=CHC$_6$H$_4$	—	658
Me	H	p-(p-$NO_2C_6H_4$)C$_6$H$_4$	203	482
H	H	p-ClC_6H_4	40	252
Me	Ph	o-ClC_6H_4	—	589u
Me	H	p-BrC_6H_4	—	189g
H	Me	p-BrC_6H_4	—	189g
H	Et	p-BrC_6H_4	—	189g
H	tert-Bu	p-BrC_6H_4	—	189g
Me	Ph	2,4-$Cl_2C_6H_3$	—	589u
H	H	p-$NH_2C_6H_4$	149	251, 252,
Me	H	p-$NH_2C_6H_4$	160	490
Me	H	p-$MeCONHC_6H_4$	186	490
H	H	p-$NO_2C_6H_4$	145–146	251, 252, 740r
Me	H	p-$NO_2C_6H_4$	—	189g
H	Et	p-$NO_2C_6H_4$	—	189g
H	tert-Bu	p-$NO_2C_6H_4$	—	189g
Me	H	p-$NO_2C_6H_4$	146	189g, 482
H	H	m-$NO_2C_6H_4$	59	252
Me	H	p-HOC_6H_4	225	490
Me	Ph	p-$MeOC_6H_4$	—	589u

TABLE III-100. 2,4-DIARYLTHIAZOLES WITH SUBSTITUENTS ON THE ARYL GROUPS

Position				
2	4	5	m.p. (°C)	Ref.
p-MeC$_6$H$_4$	m-NO$_2$C$_6$H$_4$	H	163–164	538
m-MeC$_6$H$_4$	m-NO$_2$C$_6$H$_4$	H	122–123	538
p-MeCOC$_6$H$_4$	p-BrC$_6$H$_4$	H	187	538
p-MeCOC$_6$H$_4$	m-BrC$_6$H$_4$	H	120	538
p-MeCOC$_6$H$_4$	p-IC$_6$H$_4$	H	—	538
p-MeCOC$_6$H$_4$	p-NO$_2$C$_6$H$_4$	H	169	538
p-MeCOC$_6$H$_4$	m-NO$_2$C$_6$H$_4$	H	174	538
m-MeCOC$_6$H$_4$	m-NO$_2$C$_6$H$_4$	H	160	538
p-ClC$_6$H$_4$	p-NH$_2$C$_6$H$_4$	H	—	790
p-ClC$_6$H$_4$	o-RCH=NC$_6$H$_4$	H	—	696
p-ClC$_6$H$_4$	o-RCH=NC$_6$H$_4$	Me	—	696
m-BrC$_6$H$_4$	m-NO$_2$C$_6$H$_4$	H	147	538
p-NH$_2$C$_6$H$_4$	p-MeC$_6$H$_4$	H	—	789
p-NH$_2$C$_6$H$_4$	p-MeC$_6$H$_4$	Me	—	789
p-NH$_2$C$_6$H$_4$	p-MeCONHC$_6$H$_4$	H	—	789
p-NH$_2$C$_6$H$_4$	p-MeCONHC$_6$H$_4$	Me	—	789
p-NH$_2$C$_6$H$_4$	p-ClC$_6$H$_4$	H	—	789
p-NH$_2$C$_6$H$_4$	p-ClC$_6$H$_4$	Me	—	789
p-NH$_2$C$_6$H$_4$	p-BrC$_6$H$_4$	H	—	789
p-NH$_2$C$_6$H$_4$	p-BrC$_6$H$_4$	Me	—	789
p-NH$_2$C$_6$H$_4$	p-NH$_2$C$_6$H$_4$	Me	—	789
p-NH$_2$C$_6$H$_4$	p-MeOC$_6$H$_4$	Me	—	789
p-NH$_2$C$_6$H$_4$	p-MeOC$_6$H$_4$	Ph	—	789
p-NH$_2$C$_6$H$_4$	p-EtOC$_6$H$_4$	H	—	789
p-NH$_2$C$_6$H$_4$	p-EtOC$_6$H$_4$	Me	—	789
p-NO$_2$C$_6$H$_4$	p-NO$_2$C$_6$H$_4$	H	213	651, 747
p-NO$_2$C$_6$H$_4$	p-NO$_2$C$_6$H$_4$	Me	233	747
p-NO$_2$C$_6$H$_4$	m-NO$_2$C$_6$H$_4$	Me	209	747
p-NO$_2$C$_6$H$_4$	o-NO$_2$C$_6$H$_4$	Me	151	747
p-NO$_2$C$_6$H$_4$	3-NO$_2$-4-MeC$_6$H$_3$	Me	218	747
p-HOC$_6$H$_4$	3,4-(OH)$_2$C$_6$H$_3$	H	212–213	357
p-MeOC$_6$H$_4$	3,4-(OH)$_2$C$_6$H$_3$	H	154–155	357
p-(EtNHCO)C$_6$H$_4$	p-NO$_2$C$_6$H$_4$	H	204	248
p-(4p-NH$_2$C$_6$H$_4$-2-thiazolyl)C$_6$H$_4$	p-NH$_2$C$_6$H$_4$	H	280–292	828p
m-(4p-NH$_2$C$_6$H$_4$-2-thiazolyl)C$_6$H$_4$	p-NH$_2$C$_6$H$_4$	H	230–232	325p
p-[4(3,5-(NH$_2$)$_2$C$_6$H$_4$)-2-thiazolyl]C$_6$H$_4$	3,5-(NH$_2$)$_2$C$_6$H$_3$	H	—	325p
m-(4p-NO$_2$C$_6$H$_4$-2-thiazolyl)C$_6$H$_4$	p-NO$_2$C$_6$H$_4$	H	—	324p
3,5-(OH)$_2$C$_6$H$_3$	3,4-(OH)$_2$C$_6$H$_3$	H	135–140	357

TABLE III-100. (Continued)

	Position			
2	4	5	m.p. (°C)	Ref.
(benzodioxole)	3,4-(OH)$_2$C$_6$H$_3$	H	185	357
R_1^a	p-RCH=NC$_6$H$_4^a$	R_2^a		696

[a] R_1 = Ph, PhCH$_2$, 4-MeOC$_6$H$_4$, 4-MeC$_6$H$_4$, Me, 3,4,5-(MeO)$_3$C$_6$H$_2$, 4-ClC$_6$H$_4$, 3,4-(MeO)$_2$C$_6$H$_3$, or pyridyl. R = 2,5-(OH)XC$_6$H$_3$, where X = Br, Cl, or NO$_2$. R_2 = H or Me.

TABLE III-101. 2,5-DIARYLTHIAZOLES WITH SUBSTITUENTS ON THE ARYL GROUPS

	Position			
2	4	5	m.p./b.p. (°C)	Ref.
p-CNC$_6$H$_4$	H	p-CNC$_6$H$_4$	—	547
p-ClC$_6$H$_4$	H	p-ClC$_6$H$_4$	—	547
p-BrC$_6$H$_4$	H	p-BrC$_6$H$_4$	—	547
p-NH$_2$C$_6$H$_4$	H	p-NH$_2$C$_6$H$_4$	—	547
p-NMe$_2$C$_6$H$_4$	H	p-NMe$_2$C$_6$H$_4$	—	547
p-NO$_2$C$_6$H$_4$	H	p-NO$_2$C$_6$H$_4$	—	547
p-HOC$_6$H$_4$	H	p-HOC$_6$H$_4$	—	547
p-SHC$_6$H$_4$	H	p-SHC$_6$H$_4$	—	547
p-SO$_2$NH$_2$C$_6$H$_4$	H	p-SO$_2$NH$_2$C$_6$H$_4$	—	547

TABLE III-102. 4,5-DIARYLTHIAZOLES WITH SUBSTITUENTS ON THE ARYL GROUPS

	Position			
2	4	5	m.p. (°C)	Ref.
Me	p-NO$_2$C$_6$H$_4$	p-NO$_2$C$_6$H$_4$	183	246
H	p-MeOC$_6$H$_4$	p-MeOC$_6$H$_4$	—	615

TABLE III-103. 2-ALKYLTHIAZOLES WITH ONE FUNCTION ON THE ALKYL GROUP

Substituent					
R_2	R_4	R_5	m.p. (°C)	b.p. (°C)	Ref.
$CH_2CH=CH_2$	H	H	—		785m
CH_2COMe	H	H		95/14	75
CH_2COMe	Me	Me		85–86/1	26
CH_2COMe	Et	Me		94/1	26
CH_2COMe	—$(CH_2)_4$—			112–114/1	26
CH_2COMe	Ph	H		—	821, 822r
CH_2COEt	Me	Me		102–103/1	26
CH_2COEt	Et	Me		96/1	26
CH_2COEt	—$(CH_2)_4$—			122–125/1	26
CH_2COPh	Me	H	96		512
CH_2COPh	Ph	H	106		512
CH_2COCl	Me	H	—		15
CH_2CONH_2	H	H	115–116		75
CH_2CONH_2	Me	H	131		15, 444
			130–132		
$CH_2CONHCH_2CH_2OH$	Me	H	80–82	176–178/12	421
CH_2CONEt_2	Me	H	99–100		15
$CH_2CONHNH_2$	Me	H	186–187		15
CH_2CO_2H	H	H	109–110		75, 444,
CH_2CO_2H	Me	H	90–91		15
CH_2CO_2H	Ph	H	93		73
CH_2CO_2H	p-ClC$_6$H$_4$	H	—		766
CH_2CO_2H	p-BrC$_6$H$_4$	H	—		77

TABLE III-103. (*Continued*)

	Substituent				
R_2	R_4	R_5	m.p. (°C)	b.p. (°C)	Ref.
CH(Me)CO$_2$H	p-ClC$_6$H$_4$	H	—		697
CH(Me)CO$_2$H	p-BrC$_6$H$_4$	H	—		77, 697
CH(Et)CO$_2$H	p-BrC$_6$H$_4$	H	—		77
CH(i-Pr)CO$_2$H	p-BrC$_6$H$_4$	H	—		77
C(Me)$_2$CO$_2$H	p-BrC$_6$H$_4$	H	—		77
CH$_2$CO$_2$Et	H	H		120–122/15	75, 722
CH$_2$CO$_2$Et	Me	H	—		722
CH$_2$CO$_2$Et	p-ClC$_6$H$_4$	H	—		697
CH$_2$CN	H	H	—		560
CH$_2$CN	Ph	H	—		599, 698
CH$_2$CN	Ph	Me	—		599, 698
CH$_2$CN	Ph	Ph	—		599, 698
CH$_2$F	H	H	—		445
CH$_2$Cl	H	H		62/5	66, 524g
				65–70/2	659, 8241
				48/3	54, 64,
				92/20	824
CH$_2$Cl	Me	H		—	757
CH$_2$Cl	Me	Me			44, 52,
CH$_2$Cl	Ph	H		184/16	62, 249
CH$_2$Cl	p-MeC$_6$H$_4$	H	—		249
CH$_2$Cl	p-ClC$_6$H$_4$	H	—		249
CH$_2$Cl	p-BrC$_6$H$_4$	H	—		249
CH$_2$Cl	p-NO$_2$C$_6$H$_4$	H	140–143		44, 52,
			157		249
CH$_2$Cl	p-EtOC$_6$H$_4$	H	—		249

CH$_2$Cl	Ph	H	—	459
CH$_2$Cl	p-NO$_2$C$_6$H$_4$	H	103	459
CH(Me)Cl	H	H	—	284, 285
CH(Me)Cl	H	Me	—	286
CH(Me)Cl	H	Ph	—	286
CH(Me)Cl	H	Me	—	285
CH(Me)Cl	Me	Me	—	286
CH$_2$Br	Ph	H	140–145/0.05	38, 62
			195/15	
CH$_2$Br	Me	Me	45–46	55
CH$_2$NH$_2$	H	H	93–95/14	49, 66, 722
CH$_2$NH$_2$	Me	H	83/5	414, 429, 722
			82–84/5	
CH$_2$NH$_2$	Ph	H	—	460
CH$_2$NH$_2$	3,4-(OH)$_2$C$_6$H$_3$	H	—	353
CH$_2$NHMe	3,4-(OH)$_2$C$_6$H$_3$	H	110–111	353
CH$_2$NH(p-MeC$_6$H$_4$)	Ph	H	—	460
CH$_2$NMe$_2$	Ph	H	—	38
CH$_2$NEt$_2$	Ph	H	—	38
CH$_2$N⟨piperidine⟩	Me	H	110–111	507
CH$_2$N⟨piperidine⟩	p-BrC$_6$H$_4$	H	110–111	507
CH$_2$N⟨piperidine⟩	p-MeOC$_6$H$_4$	H	88–89	507
CH(Me)NH$_2$	Me	H	98–100/12	415
CH(Me)NH$_2$	Ph	H	—	508, 513
CH(i-Pr)NH$_2$	Ph	H	201–202	508

TABLE III-103. (Continued)

	Substituent				
R_2	R_4	R_5	m.p. (°C)	b.p. (°C)	Ref.
$C(Me)_2NH_2$	Me	H	—		508
$C(Me)_2NH_2$	$3,4-(OH)_2C_6H_3$	H	210–215		353
$CH_2NHCOMe$	$3,4-(OH)_2C_6H_3$	H	—		353
$CH_2NHCOCH_2Ph$	$3,4-(OH)_2C_6H_3$	H	98		430
$CH_2NHCOPh$	Me	H	114–115		414, 429
$CH_2NHCOPh$	Ph	H	148		460
$CH_2NHSO_2(p-NH_2C_6H_4)$	Me	H	124–126		414
$CH_2N(Me)COMe$	$3,4-(OH)_2C_6H_3$	H	—		353
![phthalimidomethyl structure] CH_2N(phthalimide)	H	H	117–118		66
$CH(Me)NHCOPh$	Me	H	112		429
$CH(Me)NHCOPh$	Ph	H	162–164		508
$CH(Me)NHSO_2(p-NHCOMeC_6H_4)$	Me	H	180–181		429
$CH(i-Pr)NHCOPh$	Ph	H	126		508
$C(Me)_2NHCOMe$	$3,4-(OH)_2C_6H_3$	H	198–200		353
$C(p-MeOC_6H_4)NHCSNH(p-EtOC_6H_4)$	H	H	156–157		594
CH_2OH	H	H		70–80/0.2 123/5.5	66, 69, 8241
CH_2OH	Me	H		86/0.1	64
CH_2OH	Ph	H	89		38, 62, 249
CH_2OH	$p-MeC_6H_4$	H	—		249
CH_2OH	$p-ClC_6H_4$	H	—		249

Compound	R	mp	bp/p	Refs
CH$_2$OH	p-BrC$_6$H$_4$	H	—	249
CH$_2$OH	p-NO$_2$C$_6$H$_4$	H	200	44, 52, 248, 249, 533
			192–194	
			196–197	
CH$_2$OH	p-EtOC$_6$H$_4$	H	—	249
CH$_2$OH	H	Ph	205–208/12	459
CH$_2$OH	Me	Me	44–45	14, 55
CH$_2$OH	Ph	Ph	113–117	688, 825
CH(Me)OH	H	H		278, 284, 517
			110–120/12	
			112–115/13–15	
CH(Me)OH	Ph	H	76	36
CH(Me)OH	p-NO$_2$C$_6$H$_4$	H		548
CH(Me)OH	3,4-(MeO)$_2$C$_6$H$_3$	H	101–102	356
CH(Me)OH	H	Me		
CH(Me)OH	Me	Me		286
CH(Et)OH	H	H	124/12–13	517
CH(n-Pr)OH	H	H	136/12	517
CH(n-Pr)OH	Me	H	60	517
CH(i-Pr)OH	H	H	127/13	517
CH(n-Hexyl)OH	H	H	62	417, 723
CH(Ph)OH	Me	H	96	14, 83
CH(Ph)OH	H	Me	89	14
CH(Ph)OH	Me	Me	124–125	793
CH(p-ClC$_6$H$_4$)OH	Ph	H	—	793
CH(p-ClC$_6$H$_4$)OH	p-MeC$_6$H$_4$	H	—	793
CH(p-ClC$_6$H$_4$)OH	p-MeOC$_6$H$_4$	H	—	793
CH(2,4-Cl$_2$C$_6$H$_3$)OH	Ph	H		793
CH(p-NH$_2$C$_6$H$_4$)OH	H	H	117	827
CH(p-NO$_2$C$_6$H$_4$)OH	H	H	131	827
C(Me)$_2$OH	H	H	—	96
C(Me)$_2$OH	p-BrC$_6$H$_4$	H	73	616, 826
C(Ph)$_2$OH	H	H	112	517
C(p-MeOC$_6$H$_4$)$_2$OH	H	H	147	563

TABLE III-103. (Continued)

	Substituent				
R_2	R_4	R_5	m.p. (°C)	b.p. (°C)	Ref.
CH_2OCOMe	H	H		70/0.2	69
CH_2OCOMe	Ph	H	40	193/4	62
CH_2OCOPh	Me	H	—	125–127/0.1	64, 653
CH_2OCOPh	$CH_2C_6H_5$	H	73–74		653
CH_2OCOPh	Ph	H	161		62
CH_2OCOPh	$p\text{-}NO_2C_6H_4$	H	156–157 155–156		44, 52, 248
CH_2OCOPh	Me	Me	—		653
CH_2OCOPh	H	CH_2Ph	—		653
CH_2OCOPh	Ph	Ph	157/60		688, 825
$CH_2OCO\triangle\!\!\!\!\!\!\!\!\!{}^{(Me)_2}\!\!CH_2CH=CH_2$			—		653
$CH_2OCO\triangle\!\!\!\!\!\!\!\!\!{}^{(Me)_2}_{(Me)_2}$	$-(CH_2)_4-$		—		653
$CH_2OCO\ [3,4\text{-}(NO_2)_2C_6H_3]$	Me	Me	117–118		14
$CH(Me)OCOPh$	H	H		135/139/1.5 137–142/2	446, 447
$CH(Me)OCOPh$	Me	H		155/4	446
$CH(Me)OCOPh$	Ph	H		252–254/14	36
$CH(Me)OCOPh$	$3,4\text{-}(MeO)_2C_6H_3$	H	86		356
$CH(Me)OCOPh$	Me	Me		148/1	446

CH(Me)OCOPh	Ph	Me		194/1	446
CH(Ph)OCOMe	Me	H		198–202/22	417
CH(Ph)OCOMe	Me	Me		135–137/0.2	14
CH(Ph)OCO(p-NO$_2$C$_6$H$_4$)	Me	Me	103–104		14
C(Me$_2$)OCOPh	Me	H		110–130/1	446
CH$_2$OMe	Me	Me		94/13	14
CH$_2$OEt	H	H		187/15	62
CH$_2$OEt	p-MeCONHC$_6$H$_4$	H	—		44
CH$_2$OEt	p-NH$_2$C$_6$H$_4$	H	149		44
CH$_2$OEt	p-NO$_2$C$_6$H$_4$	H	96–97		44
CH$_2$OEt	H	Ph		153–155/5	459
CH$_2$OEt	H	p-NH$_2$C$_6$H$_4$	—		459
CH$_2$OEt	H	p-NO$_2$C$_6$H$_4$	65		459
CH$_2$OBu	p-AcNHC$_6$H$_4$	H	147		44
CH$_2$OBu	p-NO$_2$C$_6$H$_4$	H	60–61		44
CH$_2$OBu	H	Ph		190–192/8	459
CH$_2$OBu	p-NH$_2$C$_6$H$_4$	H	—		459
CH$_2$OBu	p-NO$_2$C$_6$H$_4$	H	—		459
CH$_2$OPh	p-NH$_2$C$_6$H$_4$	H	174–175		44
CH$_2$OPh	p-AcNHC$_6$H$_4$	H	134		44
CH$_2$OPh	p-NO$_2$C$_6$H$_4$	H	65		44
CH$_2$OPh	H	Ph			459
CH$_2$OPh	H	p-NH$_2$C$_6$H$_4$	114–115		459
CH$_2$OPh	H	p-NO$_2$C$_6$H$_4$	153–154		459
CH$_2$O(p-ClC$_6$H$_4$)	p-NH$_2$C$_6$H$_4$	H	162–163		44
CH$_2$O(p-ClC$_6$H$_4$)	p-AcNHC$_6$H$_4$	H	195		44
CH$_2$O(p-ClC$_6$H$_4$)	p-NO$_2$C$_6$H$_4$	H	160		44
CH$_2$O(p-ClC$_6$H$_4$)	H	p-NH$_2$C$_6$H$_4$	113–114		459
CH$_2$O(p-ClC$_6$H$_4$)	H	p-NO$_2$C$_6$H$_4$	158		459
CH$_2$O(p-BrC$_6$H$_4$)	p-NH$_2$C$_6$H$_4$	H	—		790
CH$_2$O(p-MeOC$_6$H$_4$)	p-NH$_2$C$_6$H$_4$	H	139		44
CH$_2$O(p-MeOC$_6$H$_4$)	p-AcNHC$_6$H$_4$	H	202		44
CH$_2$O(p-MeOC$_6$H$_4$)	p-NO$_2$C$_6$H$_4$	H	136–137		44

TABLE III-103. (Continued)

Substituent with structure:

![structure: R2CH2-C(=N)-C(R4)=C(R5)-S ring (thiazole) with R4 and R5 substituents]

R_2	R_4	R_5	m.p. (°C)	b.p. (°C)	Ref.
CH$_2$O(o-MeOC$_6$H$_4$)	p-NH$_2$C$_6$H$_4$	H	—		44
CH$_2$O(o-MeOC$_6$H$_4$)	p-AcNHC$_6$H$_4$	H	149		44
CH$_2$O(o-MeOC$_6$H$_4$)	p-NO$_2$C$_6$H$_4$	H	113		44
CH$_2$O(o-MeOC$_6$H$_4$)	H	Ph	90		459
CH$_2$O(2-Pr-5-MeC$_6$H$_3$)	p-NH$_2$C$_6$H$_4$	H	—		790
CH$_2$O(3-Me-4-ClC$_6$H$_3$)	p-NH$_2$C$_6$H$_4$	H	—		790
CH$_2$O(2,4-Cl$_2$C$_6$H$_3$)	p-NH$_2$C$_6$H$_4$	H	—		790
CH$_2$O(2,4-Br$_2$C$_6$H$_3$)	p-NH$_2$C$_6$H$_4$	H	—		791
CH(Ph)OEt	Me	H	—		807
CH$_2$SMe	H	H	—		8241
CH$_2$SP(OMe)$_2$=O	Me	H	—		541
CH$_2$SP(OEt)$_2$=O	Me	H	112		541
CH$_2$SP(Me)$_2$=S	Me	H		123/0.01	541
CH$_2$S(OEt)(Et)=S	Me	H		118/0.01	541
CH$_2$S(OMe)$_2$=S	Me	H		—	541
CH$_2$S(OEt)$_2$=S	Me	H		125/0.01	541
CH$_2$CH=CMe$_2$	H	H		52/0.03	31, 685
CH$_2$CONH$_2$	H	H	106–109		49
CH$_2$CONH$_2$	Me	H	—		590
CH$_2$CO$_2$H	H	H	126–127 129–132		49, 278
CH$_2$CO$_2$H	Ph	Ph	52–58		6881, 825

CH(Ph)CN	Me		133/0.1	64
CH$_2$NH$_2$	H			49
CH$_2$NH$_2$	Me	—	102–104/10	49, 415
CH$_2$NH$_2$	Ph	—		49
CH$_2$NH$_2$	3,4-(HO)$_2$C$_6$H$_3$	—		57
CH$_2$NMe$_2$	Me		90–92/3	468, 484
CH$_2$NEt$_2$	Me		96–97/2	468
CH$_2$NHCOC$_6$H$_5$	Me	76		415, 429
CH$_2$NHCOC$_6$H$_5$	Me	204		248
CH$_2$N(CO)$_2$C$_6$H$_4$ (phthalimido)	3,4-(OH)$_2$C$_6$H$_3$	H	203–205	57
CH(Me)NHCOMe	Me			422
CH$_2$NHSO$_2$(p-NH$_2$C$_6$H$_4$)	H	237–238		423
CH$_2$NHSO$_2$(p-NH$_2$C$_6$H$_4$)	Me	241–242		423
CH$_2$NHSO$_2$(p-MeCONHC$_6$H$_4$)	Me	—		423
CH$_2$OH	H		145/3	96, 517
CH$_2$OH	Me	—		96, 418
CH$_2$OH	p-NO$_2$C$_6$H$_4$	—		5351
CH(Me)OH	Me		133/22	517
CH(Me)OH	Et		148/32	517
CH(Pr^i)OH	Et		190/80	517
CH(Ph)OH	Me			609, 694
CH(p-ClC$_6$H$_4$)OH	Me	122–123		609
C(n-Pr)$_2$OH	Me		133–134/5	609
C(CH$_2$)$_4$OH	Me		114–116/3	609, 694
C(CH$_2$)$_5$OH	Me		142–144/8	694
C(Me)(Ph)OH	Me	71.5–73		609, 694
C(Ph)$_2$OH	H	151–152		609
C(Ph)(p-ClC$_6$H$_4$)OH	Me	119–120		609, 694
CH(Ph)CO$_2$Me	Me		127/0.05	64
C(—NHPh)=NH	H		—	560
CH$_2$SCH$_3$	H		82–4/0.09	31

TABLE III-103. (*Continued*)

	Substituent				
R_2	R_4	R_5	m.p. (°C)	b.p. (°C)	Ref.
$(CH_2)_3NH_2$	Me	H		120–122/12	415
$(CH_2)_3NH_2$	3,4-$(OH)_2C_6H_3$	H	226–228		57
$CH(Ph)CH_2CH_2NMe_2$	H	H	—		448
$CH(Ph)CH_2CH_2NMe_2$	Me	H		128–130/0.1	64, 448
$CH(Ph)CH_2CH_2NMe_2$	Ph	H	—		448
$CH(Ph)CH_2CH_2NEt_2$	H	H	—		448
$CH(Ph)CH_2CH_2NEt_2$	Me	H		125/0.1	448
$CH(Ph)CH_2CH_2$piperidinyl	H	H	—		448
$CH(Ph)CH_2CH_2$piperidinyl	Me	H		140–143/0.1	64
$CH(Ph)CH_2CH_2N\underset{CO}{\overset{CO}{\diagup}}$	Me	H		145–148/0.05	64
$(CH_2)_3N\underset{CO}{\overset{CO}{\diagup}}$	Me	H	80		415, 429
$(CH_2)_3N\underset{CO}{\overset{CO}{\diagup}}$	3,4-$(OH)_2C_6H_3$	H	114–115		57
$(CH_2)_3NHSO_2(p$-$NH_2C_6H_4)$	Me	H	130–132		415
$(CH_2)_3NHSO_2(p$-$MeCONHC_6H_4)$	Me	H	91		415
$(CH_2)_{10}CO_2H$	H	H	40–41		280
$(CH_2)_{10}NH_2$	Me	H	58–60	166–170/0.5	414, 429
$(CH_2)_{10}NHCOC_6H_4$	Me	H	108		414
$(CH_2)_{10}NHSO_2(p$-$NH_2C_6H_4)$	Me	H	102–104		414
$(CH_2)_{10}NHSO_2(p$-$MeCONHC_6H_4)$	Me	H	58–59		414
$(CH_2)_9CH(CH_2Me)NHCOC_6H_4$	Me	H			429

TABLE III-104. 4-ALKYTHIAZOLES WITH ONE FUNCTION ON THE ALKYL GROUP

Substituent: 4-alkylthiazole with R_2, R_4, R_5 substituents; $R_4 = -CH_2COCH_2-$

R_2	R_4	R_5	m.p. (°C)	b.p. (°C)	Ref.
p-ClC$_6$H$_4$	CH$_2$CHO	H	—		339 1
H	CH$_2$COCl	H	87–88		520
p-ClC$_6$H$_4$	CH$_2$COCl	H	—		829
Ph	CH(Me)COCl	H	—		699
H	CH$_2$CONH$_2$	H	122–123		78, 81, 403
Me	CH$_2$CONH$_2$	H	139–140		78, 403
CH$_2$C$_6$H$_5$	CH$_2$CONH$_2$	H	132–133		78
Cyclohexyl	CH$_2$CONH$_2$	H	136–137		8221
p-MeOC$_6$H$_4$	CH$_2$CONH$_2$	H	—		699
H	CH$_2$CONHMe	H	—		560
Et	CH$_2$CONHEt	H	—		560
p-ClC$_6$H$_4$	CH$_2$CONHNH$_2$	H	—		125 m
Cyclohexyl	CH$_2$CONH(CH$_2$)$_2$NMe$_2$	H	—	146–150/0.15	8221
Me	CH$_2$CONH(CH$_2$)$_2$OH	H	130		421
			93–94		
Me	CH$_2$CONH-(decalin)	H	—		617

439

TABLE III-104. (Continued)

	Substituent		m.p. (°C)	b.p. (°C)	Ref.
R_2	R_4	R_5			
Me	CH$_2$C(Et)(CO—NH)(CO—NH)CO (cyclic)	H	264–265		50
H	CH$_2$N(CO)(CO)C$_6$H$_4$	H	156–157, 157–158		49, 66
Me	CH$_2$N(CO)(CO)C$_6$H$_4$	H	146–146.5		66
Ph	CH$_2$N(CO)(CO)C$_6$H$_4$	H	151–152		33
o-MeC$_6$H$_4$	CH$_2$N(CO)(CO)C$_6$H$_4$	H	186–187		33
Ph	CH(Me)CONH$_2$	H	—		699
Ph	CH(Me)CONH(CH$_2$)$_2$NMe$_2$	H	—		699
Ph	CH(Me)CO-piperidinyl	H	—		699
Me	CH$_2$CSNH$_2$	H	132		774
H	CH$_2$CO$_2$H	H	134, 138, 139		393, 403, 424, 5961, 618, 829
Me	CH$_2$CO$_2$H	H	121, 125–126		403, 618, 829

CH₂C₆H₅	CH₂CO₂H	H	126.5–128		78
Cyclohexyl	CH₂CO₂H	H	132–133		8221
Cyclohexyl	CH₂CO₂H	Ph	162/3		8221
Ph	CH₂CO₂H	H	90		33, 6881
					6921, 699
Ph	CH₂CO₂H	Ph	171	202/5	688, 825
Ph	CH₂CO₂H	p-ClC₆H₄			688, 6921
o-MeC₆H₄	CH₂CO₂H	H	110–111		6921
α-Naphthyl	CH₂CO₂H	H	153–154		6921
β-Naphthyl	CH₂CO₂H	H	140–141		688, 6921
p-ClC₆H₄	CH₂CO₂H	H	—		76, 77,
					125 m, 3391
					619, 620,
					660, 699,
					724, 7251
					7261
p-ClC₆H₄	CH₂CO₂H	Ph	—	200/1	6881, 8251
p-BrC₆H₄	CH₂CO₂H	H	174–175		76
p-HOC₆H₄	CH₂CO₂H	H	202		125 m, 7251
3-Cl-4-HOC₆H₃	CH₂CO₂H	H	208–212		7251
4-Cl-2-HOC₆H₃	CH₂CO₂H	H	—		7251
4-Cl-3-HOC₆H₃	CH₂CO₂H	H	—		125 m, 7251
p-ClC₆H₄	CH₂CO₂H	H	—		6611
Cyclohexyl	CH(Me)CO₂H	H	—		8221
Ph	CH(Me)CO₂H	H	—	176/6	699
p-ClC₆H₄	CH(Me)CO₂H	H	—		77, 697
p-ClC₆H₅	CH₂CO₂Me	H	—		78
p-ClC₆H₄	CH₂CO₂Me	H	—		125 m
H	CH₂CO₂Et	H	—	122–123/11	78, 393,
					403
Me	CH₂CO₂Et	H	—	127/13	78, 403
CH₂C₆H₅	CH₂CO₂Et	H	—		78

TABLE III-104. (Continued)

	Substituent				
R_2	R_4	R_5	m.p. (°C)	b.p. (°C)	Ref.
Cyclohexyl	CH_2CO_2Et	H	—	120–122/0.4	699, 8221
Ph	CH_2CO_2Et	Ph	—		688, 699, 8251
Ph	CH_2CO_2Et	p-ClC$_6$H$_4$		108/9	688, 8251
p-ClC$_6$H$_4$	CH_2CO_2Et	H	—		699
p-MeOC$_6$H$_4$	CH_2CO_2Et	H	—		699
Cyclohexyl	CH_2CO_2Bu	H		134–136/0.5	8221
Me	$CH_2CO_2(CH_2)_2Cl$	H		110/4	403
Me	$CH_2CO_2(CH_2)_2NEt_2$	H		131/4	403
Me	CH_2CO_2–△(Me$_2$)–$CH_2C(Me)\!=\!CH_2$	H	—		653
Et	CH_2CO_2–△(Me$_2$)–$CH_2C(Me)\!=\!CH_2$	H	—		653
n-Pr	CH_2CO_2–△(Me$_2$)–$CH_2C(Me)\!=\!CH_2$	H	—		653
i-Pr	CH_2CO_2–△(Me$_2$)–$CH_2C(Me)\!=\!CH_2$	H	—		653
$CH_2C_6H_5$	CH_2CO_2–△(Me$_2$)–$CH_2C(Me)\!=\!CH_2$	H	—		653

Ph	CH₂CO₂△Me₂ CH₂C(Me)=CH₂	H	—	653
Me	CH₂CO₂△Me₂	H	—	653
Et	CH₂CO₂△Me₂	H	—	653
n-Pr	CH₂CO₂△Me₂	H	—	653
i-Pr	CH₂CO₂△Me₂	H	—	653
CH₂C₆H₅	CH₂CO₂△Me₂	H	—	653
Ph	CH₂CO₂△Me₂	H	—	653
Ph	CH(Me)CO₂Et	H	—	699
Ph	CH(Me)CO₂(CH₂)₂NMe₂	H	—	699
Ph	CH(Me)CO₂(o-HCO₂C₆H₄)	H	—	622
H	CH₂CN	H	30, 110/5	78, 81, 560, 749
Me	CH₂CN	H	146/26	78
n-Pr	CH₂CN	H	—	560

TABLE III-104. (Continued)

	Substituent				
R_2	R_4	R_5	m.p. (°C)	b.p. (°C)	Ref.
$CH_2C_6H_5$	CH_2CN	H	43–44		78
Cyclohexyl	CH_2CN	H	—		8221
Ph	CH_2CN	H	43–44		381, 688, 6921
Ph	CH_2CN	H		180–185/4	33
p-MeOC$_6$H$_4$	CH_2CN	H	73		33, 692
H	CH_2Cl	H		64–65/1	55, 396, 524 g, 552, 5561, 564, 8241
Me	CH_2Cl	H		65–67/3	38, 50, 58, 653
Et	CH_2Cl	H		85/10	653
n-Pr	CH_2Cl	H	—		653
$CH_2C_6H_5$	CH_2Cl	H	—		653, 662
Cyclohexyl	CH_2Cl	H	—		700
i-Bu	CH_2Cl	H	—		8221
Ph	CH_2Cl	H	31		653
			50–51	155–156/4	33, 34, 38, 50, 53, 247, 381, 545, 623, 653, 700, 779 iu, 780 h, 824, 830

p-MeC$_6$H$_4$	CH$_2$Cl	H	95–96		53, 779 iu 780
m-MeC$_6$H$_4$	CH$_2$Cl	H	47		53, 623, 663
o-MeC$_6$H$_4$	CH$_2$Cl	H	—		53
p-BrC$_6$H$_4$	CH$_2$Cl	H	119		247, 623
p-NO$_2$C$_6$H$_4$	CH$_2$Cl	H	120		43, 245, 247, 534, 623
p-MeOC$_6$H$_4$	CH$_2$Cl	H	55–56		33, 35, 692
3-Me-4-NO$_2$C$_6$H$_3$	CH$_2$Cl	H	—		663
3-NO$_2$-4-ClC$_6$H$_3$	CH$_2$Cl	H	—		125 m
4,5-(MeO)$_2$C$_6$H$_3$	CH$_2$Cl	H	88–89		33
α-Naphthyl	CH$_2$Cl	H	—		6921
Piperonyl	CH$_2$Cl	H	106–107		33
H	CH$_2$Br	H	—		55
Me	CH$_2$Br	Me	—	88–90/15	55
CH$_2$C$_6$H$_5$	CH$_2$I	H	104–105		662
Ph	CH$_2$I	H	—		53
H	CH(Me)Cl	H	—		284, 286
Me	CH(Me)Cl	H	—		286
Ph	CH(Me)Cl	H	—		286
H	CH$_2$NH$_2$	H	—	90–93/12	49, 66, 759
Me	CH$_2$NH$_2$	H	—	109–107/22	66
Ph	CH$_2$NH$_2$	H	205–206		33
p-OHC$_6$H$_4$	CH$_2$NH$_2$	H	224.5		33
p-MeOC$_6$H$_4$	CH$_2$NH$_2$	H	—		33
Ph	CH$_2$N(Me)$_2$	H		15/0.01	38
Me	CH$_2$N(Me)$_2$	H		170–180/0.05	38
Ph	CH$_2$N(Et)$_2$	H			38

TABLE III-104. (Continued)

	Substituent				
R_2	R_4	R_5	m.p. (°C)	b.p. (°C)	Ref.
p-$NO_2C_6H_4$	$CH_2N(Et)_2$	H	202–204		245
Me	$CH_2N(CH_2CH_2Cl)_2$	H	—		39
Me	$CH_2N(CH_2CH_2OH)_2$	H	—		39
H	CH_2OH	H		85–87/2 115–117/11 123/15	55, 374, 524, 552, 5681, 8241
Me	CH_2OH	H		104/5	58
Ph	CH_2OH	H		67–68/4	53, 381, 700, 824, 830
p-MeC_6H_4	CH_2OH	H	116–117		53
m-MeC_6H_4	CH_2OH	H	—		53
p-ClC_6H_4	CH_2OH	H	—		339
p-$NO_2C_6H_4$	CH_2OH	H	169–170		43
Ph	CH_2OMe	H		116/1	381
p-$MeCONHC_6H_4$	CH_2OMe	H	121		43
p-$NH_2C_6H_4$	CH_2OMe	H	77		43
p-$NO_2C_6H_4$	CH_2OMe	H	143		43
Me	CH_2OEt	H		93.5/13	514
Ph	CH_2OEt	H		172–174/12	43
p-$MeCONHC_6H_4$	CH_2OEt	H	150		43
p-$NO_2C_6H_4$	CH_2OEt	H	90–90.5		43
Ph	CH_2OPr	H		183–185/12	43
p-$MeCONHC_6H_4$	CH_2OPr	H	146–147		43
p-$NH_2C_6H_4$	CH_2OPr	H	75		43
p-$NO_2C_6H_4$	CH_2OPr	H	84–84.5		43

Ph	CH$_2$OBu		173–174/5	43
p-MeCONHC$_6$H$_4$	CH$_2$OBu	H		43
p-NO$_2$C$_6$H$_4$	CH$_2$OBu	H	206–208/5	43
Ph	CH$_2$O-cyclohexyl	H		43
p-NO$_2$C$_6$H$_4$	CH$_2$OCH$_2$Ph	H		43
Ph	CH$_2$O-i-Pr	H	147	43
Ph	CH$_2$O-i-Bu	H	165–167/5	43
p-MeCONHC$_6$H$_4$	CH$_2$O-i-Bu	H	183–184/10	43
p-NH$_2$C$_6$H$_4$	CH$_2$O-i-Bu	H		43
p-NO$_2$C$_6$H$_4$	CH$_2$O-i-Bu	H		43
Ph	CH$_2$O-i-Amyl	H	179–181/5	43
Ph	CH$_2$OPh	H	67–68	50
p-MeCONHC$_6$H$_4$	CH$_2$OPh	H	141–142	43
p-NH$_2$C$_6$H$_4$	CH$_2$OPh	H	98	43
p-NO$_2$C$_6$H$_4$	CH$_2$OPh	H	112	43
p-AcNHC$_6$H$_4$	CH$_2$O(p-MeCONHC$_6$H$_4$)	H	259	43
p-AcNHC$_6$H$_4$	CH$_2$O(o-MeCONHC$_6$H$_4$)	H	217	43
p-NO$_2$C$_6$H$_4$	CH$_2$O(p-NO$_2$C$_6$H$_4$)	H	198	43
p-NO$_2$C$_6$H$_4$	CH$_2$O(o-NO$_2$C$_6$H$_4$)	H	216–217	43
p-MeCONHC$_6$H$_4$	CH$_2$O(p-ClC$_6$H$_4$)	H	171	43
p-NH$_2$C$_6$H$_4$	CH$_2$O(p-ClC$_6$H$_4$)	H	119	43
p-NO$_2$C$_6$H$_4$	CH$_2$O(p-ClC$_6$H$_4$)	H	120	43
p-MeCONHC$_6$H$_4$	CH$_2$O(p-MeOC$_6$H$_4$)	H	170–171	43
p-NH$_2$C$_6$H$_4$	CH$_2$O(p-MeOC$_6$H$_4$)	H	144–145	43
p-NO$_2$C$_6$H$_4$	CH$_2$O(p-MeOC$_6$H$_4$)	H	152	43
p-MeCONHC$_6$H$_4$	CH$_2$O(o-MeOC$_6$H$_4$)	H	161–161.5	43
p-NH$_2$C$_6$H$_4$	CH$_2$O(o-MeOC$_6$H$_4$)	H	130–131	43
p-NO$_2$C$_6$H$_4$	CH$_2$O(o-MeOC$_6$H$_4$)	H	109.5–110	43
H	CH(Me)OH	H	—	284, 286
Me	CH(Me)OH	H	—	286
Ph	CH(Me)OH	H	—	286
p-NO$_2$C$_6$H$_4$	CH(OH)CH(NH$_2$)CH$_2$OH	H	175–180	5311

TABLE III-104. (Continued)

![Thiazole structure: N-C(R2)=S-C(R5)=C(CH2-R4), a thiazole with R2 at 2-position, CH2-R4 at 4-position, R5 at 5-position]

	Substituent				
R_2	R_4	R_5	m.p. (°C)	b.p. (°C)	Ref.
Ph	CH(OH)CH(NHCONHCCl$_2$)CH$_2$OH	H	—	131/3	5311
H	CH$_2$OCOMe	H	—	76–78/8	55, 56,
				115/14	524
Ph	CH$_2$OCOMe	H	42–43	117–119/14	381
Ph	CH$_2$OCOC$_6$H$_5$	H	77		53
p-MeCONHC$_6$H$_4$	CH$_2$O(o-MeOC$_6$H$_4$)	H	161–161.5		43
Me	CH$_2$SH	H	—	89–91/3	47
Ph	CH$_2$SH	H	42		50
p-ClC$_6$H$_4$	CH$_2$SH	H	—		125 m
Me	CH$_2$SCH$_2$C$_6$H$_5$	H	—	131–135/0.75	47
H	CH$_2$NCS	H	—		7621
p-ClC$_6$H$_4$	CH$_2$COCl	H	—		665
Cyclohexyl	CH$_2$CONH$_2$	H	108–109		822
Ph	CH$_2$CONHNH$_2$	H	142–143		34
p-MeOC$_6$H$_4$	CH$_2$CONHNH$_2$	H	158–159		35
Ph	CH$_2$CON$_3$	H	72		34
p-MeOC$_6$H$_4$	CH$_2$CON$_3$	H	78–79		35
Cyclohexyl	CH$_2$CO$_2$H	H	—	155/7	822 I
Ph	CH$_2$CO$_2$H	H	83–84		34

p-MeOC$_6$H$_4$	CH$_2$CO$_2$H	H	126–127		35
Ph	CH$_2$CO$_2$Et	H	42–43		34
p-MeOC$_6$H$_4$	CH$_2$CO$_2$Et	H	53–54		35
H	CH$_2$CN	H	—		560
Me	CH$_2$CN	H	—		560
H	CH$_2$Cl	Me	—	116/16	78
H	CH$_2$Cl	H		97/0.1	63
Ph	CH$_2$Cl	H			41
p-MeOC$_6$H$_4$	CH$_2$Cl	Me	79–80		41
H	CH$_2$Br	H		122–123/12	63
H	CH$_2$NH$_2$	H		117–119/11	49,81, 496, 6661
Me	CH$_2$NH$_2$	H		82–83/2–3	49, 78
CH$_2$C$_6$H$_5$	CH$_2$NH$_2$	H		162–165/2	78
Ph	CH$_2$NH$_2$	H		146–147/2–3	34, 41
p-HOC$_6$H$_4$	CH$_2$NH$_2$	H	—		35
p-MeOC$_6$H$_4$	CH$_2$NH$_2$	H		139/0.1	35, 41
				292–293/3–4	
Ph	CH$_2$NHMe	H		150/0.1	41
p-MeOC$_6$H$_4$	CH$_2$NHEt	H		139/0.1	41
H	CH$_2$NHCH$_2$Ph	H		145/0.3	41
Me	CH$_2$NHCH$_2$Ph	H		154–155/1	78
H	CH$_2$NMe$_2$	H		138/60	41i
CH$_2$C$_6$H$_5$	CH$_2$NMe$_2$	H	—		38
Ph	CH$_2$NMe$_2$	H	—		38, 41i
p-MeOC$_6$H$_4$	CH$_2$NMe$_2$	H		141/0.3	41
Me	CH$_2$NEt$_2$	H	—		38
CH$_2$C$_6$H$_5$	CH$_2$NEt$_2$	H	—		38
Ph	CH$_2$NEt$_2$	H	—		38, 41
p-MeOC$_6$H$_4$	CH$_2$NEt$_2$	H	—		41
Me	2-Piperidinoethyl	H	—		38
CH$_2$C$_6$H$_5$	2-Piperidinoethyl	H	—		38

TABLE III-104. (Continued)

	Substituent				
R_2	R_4	R_5	m.p. (°C)	b.p. (°C)	Ref.
Ph	2-Piperidinoethyl	H	—		38
Ph	$-CH_2N(Me)CH_2-$		87		41i
p-MeOC$_6$H$_4$	$-CH_2N(Me)CH_2-$		117–118		41i
H	CH$_2$N(Me)CH$_2$C$_6$H	H	—		41i
H	CH$_2$N(Me)NH$_2$	H		110–112/12	441
H	CH$_2$N(phthalimido)	H	101–102		49
Me	CH$_2$N(phthalimido)	H	102–103		49
Ph	CH$_2$N(phthalimido)	H	113–114		34
p-MeOC$_6$H$_4$	CH$_2$N(phthalimido)	H	120–121		35
H	CH(Me)N(phthalimido)	H	107–108		441
H	C(NHα-naphtyl)=NH	H	—		560
Me	CH$_2$N=CHPh	H	—	152–154/1	78
H	CH$_2$OH	H	—		61
H	CH$_2$OH	Me	—		63, 667
p-ClC$_6$H$_4$	CH$_2$OH	H	—		125 m, 3391
					6191

H	CH(Me)OH	H	392
p-ClC₆H₄	C(Me)₂OH	H	125 m
H	CH₂OCOMe	H	61
Me		H	664
Me		H	664

[a] R = Me or CO₂Et. R' = H or OH. R'' = H or Me.

[b] R = H, Me, CO₂Et, or CO₂H. R' = H or OH.

TABLE III-105. 5-ALKYLTHIAZOLES WITH ONE FUNCTION ON THE ALKYL GROUP

Substituent					
R_2	R_4	R_5	m.p. (°C)	b.p. (°C)	Ref.
H	H	$-CH_2COCH_2-$	87–88		520
H	H	CH_2CONH_2	114		431
H	Me	CH_2CONH_2	136		80, 362
Me	Me	CH_2CONH_2	173		78, 412
Ph	Me	CH_2CONH_2	168–169		78
Ph	Ph	CH_2CONH_2	209–210		7271
Ph	p-ClC$_6$H$_4$	CH_2CONH_2	223–224		7271
H	Me	$CH_2CONHNH_2$	94–95		362, 375, 721
			111		
Ph	p-ClC$_6$H$_4$	$CH_2CONHOH$	176–177		7271
H	H	CH_2CO_2H	153–154		55, 431
H	Me	CH_2CO_2H	192–193		362, 365, 382, 620, 624, 625, 721, 786, 787
			189		
Me	Ph	CH_2CO_2H	202–203		397, 425
Me	p-MeC$_6$H$_4$	CH_2CO_2H	200–202		397, 425
Me	p-EtC$_6$H$_4$	CH_2CO_2H	155		425
H	Ph	CH_2CO_2H	154–156		425
Me	Me	CH_2CO_2H	183–184		78, 412, 620
Me	p-EtC$_6$H$_4$	CH_2CO_2H	155		397

Me	2,4-Me$_2$C$_6$H$_3$	CH$_2$CO$_2$H	189–190 199–200	397, 425
Me	p-i-PrC$_6$H$_4$	CH$_2$CO$_2$H	173–174	397, 425
Me	α-Naphthyl	CH$_2$CO$_2$H	212–213	397, 426
Me	β-Naphthyl	CH$_2$CO$_2$H	226–229	397, 425
Me	p-ClC$_6$H$_4$	CH$_2$CO$_2$H	202–204	397, 425
Me	p-MeOC$_6$H$_4$	CH$_2$CO$_2$H	189–190 177–179	397, 425
Me	p-EtOC$_6$H$_4$	CH$_2$CO$_2$H	188–190 169–190	397, 425
Ph	Me	CH$_2$CO$_2$H	—	78, 620
Ph	Ph	CH$_2$CO$_2$H	152–153	7271
Ph	p-MeC$_6$H$_4$	CH$_2$CO$_2$H	168–169	7271
Ph	p-CF$_3$C$_6$H$_4$	CH$_2$CO$_2$H	168–169	626, 627
Ph	p-FC$_6$H$_4$	CH$_2$CO$_2$H	173–174	7271, 750
Ph	p-ClC$_6$H$_4$	CH$_2$CO$_2$H	—	627, 668, 7021, 7031 7271, 750
Ph	o-ClC$_6$H$_4$	CH$_2$CO$_2$H	178–180	727
Ph	p-BrC$_6$H$_4$	CH$_2$CO$_2$H	178–180	627, 7271
Ph	p-HOC$_6$H$_4$	CH$_2$CO$_2$H	214–215	7271
Ph	p-MeOC$_6$H$_4$	CH$_2$CO$_2$H	180–181	7271
Ph	3,4-(MeO)$_2$C$_6$H$_3$	CH$_2$CO$_2$H	147–149	7271
Ph	α-Naphthyl	CH$_2$CO$_2$H	166–167	7271
Ph	β-Naphthyl	CH$_2$CO$_2$H	168–169	7271
p-MeC$_6$H$_4$	Ph	CH$_2$CO$_2$H	173–175	7271
p-MeC$_6$H$_4$	p-ClC$_6$H$_4$	CH$_2$CO$_2$H	188–189	7271
m-MeC$_6$H$_4$	p-ClC$_6$H$_4$	CH$_2$CO$_2$H	123–125	7271
m-MeC$_6$H$_4$	p-ClC$_6$H$_4$	CH$_2$CO$_2$H	152–153	7031, 7271
o-MeC$_6$H$_4$	Ph	CH$_2$CO$_2$H	172–173	7271
o-MeC$_6$H$_4$	p-ClC$_6$H$_4$	CH$_2$CO$_2$H	174–175	7271
o-MeC$_6$H$_4$	β-naphthyl	CH$_2$CO$_2$H	171–172	7271

TABLE III-105. (*Continued*)

	Substituent				
R_2	R_4	R_5	m.p. (°C)	b.p. (°C)	Ref.
p-iPrC$_6$H$_4$	Ph	CH$_2$CO$_2$H	146–147		7271
p-CycloPrC$_6$H$_4$	p-ClC$_6$H$_4$	CH$_2$CO$_2$H	—		6691
m-CF$_3$C$_6$H$_4$	Ph	CH$_2$CO$_2$H	143–145		7271
p-CO$_2$HC$_6$H$_4$	p-ClC$_6$H$_4$	CH$_2$CO$_2$H	267–269		7271
p-ClC$_6$H$_4$	Me	CH$_2$CO$_2$H	—		620
p-ClC$_6$H$_4$	Ph	CH$_2$CO$_2$H	157–158 153–155		626, 7271
p-ClC$_6$H$_4$	p-FC$_6$H$_4$	CH$_2$CO$_2$H	194–196		7271
p-ClC$_6$H$_4$	p-ClC$_6$H$_4$	CH$_2$CO$_2$H	199–201		7271
p-ClC$_6$H$_4$	p-BrC$_6$H$_4$	CH$_2$CO$_2$H	206–207		7271
p-ClC$_6$H$_4$	p-MeOC$_6$H$_4$	CH$_2$CO$_2$H	199–201		626
m-ClC$_6$H$_4$	p-ClC$_6$H$_4$	CH$_2$CO$_2$H	165–166		7271
o-ClC$_6$H$_4$	Ph	CH$_2$CO$_2$H	169–171		7271
p-BrC$_6$H$_4$	Ph	CH$_2$CO$_2$H	174–175		7271
p-BrC$_6$H$_4$	p-FC$_6$H$_4$	CH$_2$CO$_2$H	192–193		7271
p-Me$_2$NC$_6$H$_4$	Ph	CH$_2$CO$_2$H	154–156		7271
p-Me$_2$NC$_6$H$_4$	p-ClC$_6$H$_4$	CH$_2$CO$_2$H	205–207		7271
p-MeOC$_6$H$_4$	Ph	CH$_2$CO$_2$H	151–152		7271
p-MeOC$_6$H$_4$	p-ClC$_6$H$_4$	CH$_2$CO$_2$H	173–174		7271
p-MeOC$_6$H$_4$	p-MeOC$_6$H$_4$	CH$_2$CO$_2$H	176–178		626, 7271
o-MeOC$_6$H$_4$	Ph	CH$_2$CO$_2$H	179–181		7271
o-MeOC$_6$H$_4$	p-MeOC$_6$H$_4$	CH$_2$CO$_2$H	140–141		626, 7271
p-NO$_2$Ph-hydrazone	Me	CH$_2$CO$_2$H	173		550
2,3-Me$_2$C$_6$H$_3$	Ph	CH$_2$CO$_2$H	143–145		7271
2,6-Me$_2$C$_6$H$_3$	Ph	CH$_2$CO$_2$H	203–205		7271
2-Me-4-ClC$_6$H$_3$	Ph	CH$_2$CO$_2$H	175–177		7271

2-Me-6-ClC$_6$H$_3$	Ph	CH$_2$CO$_2$H	217–219		7271
2-Me-4-MeOC$_6$H$_3$	Ph	CH$_2$CO$_2$H	136–138		7271
2,4-Cl$_2$C$_6$H$_3$	Ph	CH$_2$CO$_2$H	158–160		7271
2,6-Cl$_2$C$_6$H$_3$	Ph	CH$_2$CO$_2$H	210–212		7271
2-MeO-4-ClC$_6$H$_3$	Ph	CH$_2$CO$_2$H	204–205		7271
2,4-(MeO)$_2$C$_6$H$_3$	Ph	CH$_2$CO$_2$H	157–159		7271
α-Naphthyl	Ph	CH$_2$CO$_2$H	145–148		7271
β-Naphthyl	Ph	CH$_2$CO$_2$H	171–172		7271
Me	Ph	CH(Me)CO$_2$H	172–173		397, 425
p-ClC$_6$H$_4$	Ph	CH(Me)CO$_2$H	—		6691
Ph	Ph	CH(Me)CO$_2$H	142–144		7271
p-ClC$_6$H$_4$	Ph	CH(Me)CO$_2$H	135–136		7271
H	Me	CH$_2$CO$_2$Me	208		61, 362, 382
Me	Ph	CH$_2$CO$_2$Me	132–133		425
Ph	Ph	CH$_2$CO$_2$Me	122–123		7271
p-ClC$_6$H$_4$	Ph	CH$_2$CO$_2$Me	—	125.7/11	668
H	H	CH$_2$CO$_2$Et	—	80–81/0.2	75, 431
Me	Me	CH$_2$CO$_2$Et	114		59, 61, 362, 365
Me	Me	CH$_2$CO$_2$Et	—	150/18	412
Ph	Me	CH$_2$CO$_2$Et	—		78
Ph	Ph	CH$_2$CO$_2$Et	95–96		7271
p-ClC$_6$H$_4$	Ph	CH$_2$CO$_2$Et	69–70		7271
p-MeOC$_6$H$_4$	Ph	CH$_2$CO$_2$Et	68–69		7271
p-MeOC$_6$H$_4$	Ph	CH$_2$CO$_2$Et	65–67		7271
p-ClC$_6$H$_4$	Ph	CH$_2$CO$_2$Bu	68–70		7271
H	H	CH$_2$CN	—		560
H	Me	CH$_2$CN	87	92–93/2	80
Me	Me	CH$_2$CN	104.5–106		78, 412
Ph	Me	CH$_2$CN	—		78
H	H	CH$_2$Cl	—		66, 3471

455

TABLE III-105. (*Continued*)

	Substituent				
R_2	R_4	R_5	m.p. (°C)	b.p. (°C)	Ref.
H	Me	CH_2Cl	32–33	65/3	3471, 368,
Me	H	CH_2Cl		62–64/15	58
H	Me	CH_2Br	—	93.5/13	61
Me	Me	CH_2Br	—		55
Ph	H	CH_2Br	72–73		38
H	H	$CH(Me)Cl$	—		283, 284, 285
Me	H	$CH(Me)Cl$	—		285
Ph	H	$CH(Me)Cl$	—		283
H	Me	CH_2NH_2		94–95/5	61
Ph	H	CH_2NMe_2	—		38
Ph	H	CH_2NEt_2	—		38
H	H	CH_2OH	91–92	138–141/10	66, 3471
H	Me	CH_2OH		105/1	59, 61,
				113/3	3471, 368,
				48–49/2	628
H	Et	CH_2OH	65.8–66	120/3	849
Me	H	CH_2OH		117.5/3	58, 96
Me	Me	CH_2OH	43–45	123–125/4	55, 96, 462
Ph	H	CH_2OH	76–77		38
H	H	$CH(Me)OH$	—		284
H	Me	$CH(Me)OH$	128–129	146/18	364, 384, 388, 392, 721

R1	R2	R3	mp	bp	Refs
Me	H	CH(Me)OH	—	—	285
Me	Me	CH(Me)OH	—	156/70	462, 517
Ph	H	CH(Me)OH	—	—	667
Me	Me	CH(Et)OH	—	—	96
Me	Me	CH(n-Pr)OH	—	165/40–42	517
Me	H	C(Me)$_2$OH	—	—	96
Me	H	CH$_2$OEt	—	73/7	58
H	Me	CH$_2$OMe	125–126	—	721
H	Me	CH$_2$OEt	127–128	—	721
H	Me	CH$_2$OCOMe	—	—	61
Ph	H	CH(Me)OCO-p-NO$_2$C$_6$H$_4$	—	—	667
H	Me	(CH$_2$)$_2$CONH$_2$	97–98	—	392
H	Me	(CH$_2$)$_2$CO$_2$H	179	—	392, 551
Ph	Ph	(CH$_2$)$_2$CO$_2$H	150–151	—	727l
p-ClC$_6$H$_4$	Ph	(CH$_2$)$_2$CO$_2$H	143–144	—	727l
o-MeC$_6$H$_4$	Ph	(CH$_2$)$_2$CO$_2$H	107–109	—	727l
p-ClC$_6$H$_4$	Ph	(CH$_2$)$_2$CO$_2$H	177–178	—	727l
p-MeOC$_6$H$_4$	Ph	(CH$_2$)$_2$CO$_2$H	174–175	—	727l
H	Me	(CH$_2$)$_2$CO$_2$Me	—	130–132/7	392
H	Me	(CH$_2$)$_2$CO$_2$Et	124	—	365, 551l
H	Me	(CH$_2$)$_2$CN	—	99–101/10	551l
H	H	(CH$_2$)$_2$Cl	—	74–75/3	347
H	Me	(CH$_2$)$_2$Cl	—	100–105/4, 92/7	65, 123 m, 363, 385, 497l, 539, 551l, 597l, 598 mg, 625, 630, 670, 671, 672, 673, 674, 704l, 721, 728l, 760l, 767, 769, 788, 794l, 831, 832

TABLE III-105. (*Continued*)

Substituent			m.p. (°C)	b.p. (°C)	Ref.
R_2	R_4	R_5			
H	Me	$CH_2CH(Me)Cl$		100–104/2	3471
H	Me	$(CH_2)_2Br$		95/3	9, 68, 392, 398
H	Me	$(CH_2)_2NH_2$		100–105/5	61, 80, 3471
				79–80/1	369, 389
Me	Me	$(CH_2)_2NH_2$	218–220	84/2	78, 412
Ph	Me	$(CH_2)_2NH_2$			78
H	Me	$(CH_2)_2N=CHC_6H_5$	—		850
H	H	$(CH_2)_2OH$	—		374
H	Me	$(CH_2)_2OH$	—		4, 9, 71, 123 m, 130 i, 347, 449, 455, 461 l, 471, 478, 480 l, 491 l, 499, 504, 506, 508 i, 521, 539, 540, 541, 551, 571, 576, 585, 625, 631, 632, 633, 644, 645, 667 s, 675, 676 urimg, 677, 721, 729, 748 l, 768, 771, 772, 776 m, 777, 834 gm, 835
H	Et	$(CH_2)_2OH$		127–128/3	845, 846, 847, 848, 852
H	n-Pr	$(CH_2)_2OH$	—		374
H	i-Pr	$(CH_2)_2OH$	—		374
Me	H	$(CH_2)_2OH$			374, 500 g

458

			mp	Ref
Me	Me	(CH$_2$)$_2$OH	130–135/6	364, 365, 368, 399, 455, 462, 541
Et	Me	(CH$_2$)$_2$OH	130–132/6	368
n-Pr	Me	(CH$_2$)$_2$OH	130–131/7	368
Ph	Me	(CH$_2$)$_2$OH	117–121/2	368
H	Me	CH$_2$CH(Me)OH	116–121/2	392, 721
Me	Me	(CH$_2$)$_2$OPh	170–180/5	836
Me	Me	CH$_2$CH–CH$_2$ _O_/	140/6	386
H	Me	(CH$_2$)$_2$OCOMe	105–108/4	461, 550, 721
Me	Me	(CH$_2$)$_2$OCOMe	—	837
H	Me	(CH$_2$)$_2$OCOC$_7$H$_{15}$	181/4	838
H	Me	(CH$_2$)$_2$OCOC$_{11}$H$_{23}$	198/4	838
H	Me	(CH$_2$)$_2$OCOC$_{13}$H$_{27}$	219/4	838
H	Me	(CH$_2$)$_2$OCOC$_{15}$H$_{31}$	230–233/4	838
H	Me	(CH$_2$)$_2$OCO(o-CO$_2$HC$_6$H$_4$)	145	68, 721, 777
H	Me	(CH$_2$)$_2$OCO(o-HOC$_6$H$_4$)	63	845, 846
H	Et	(CH$_2$)$_2$OCOPh	127–128/3	631
H	Me	(CH$_2$)$_2$OP(O)(OH)$_2$	—	399
H	Me	(CH$_2$)$_2$SH	75/0.5	

H	H	(CH$_2$)$_4$CO$_2$H	108–109	5
H	Me	(CH$_2$)$_3$Br	—	844
H	Me	(CH$_2$)$_3$OH	141/7	392
H	Me	(CH$_2$)$_4$OH	233	374
H	Me	(CH$_2$)$_6$OH	225	374

TABLE III-106. THIAZOLES SUBSTITUTED BY ONE POLYFUNCTIONNAL GROUP

	Substituent				
R_2	R_4	R_5	m.p. (°C)	b.p. (°C)	Ref.
CH(CO)₂ (phthaloyl)	Ph	H	257		352
CH(CO₂Me)₂	H	H	173–174		586
CH(CO₂Et)₂	H	H	161.5		586
CH(CO₂Et)₂	p-ClC₆H₄	H	—		697, 766
CH(CO₂Et)₂	p-BrC₆H₄	H	—		697
C(Me)(CO₂Me)₂	H	p-ClC₆H₄	129–130		586
C(Me)(CO₂Et)₂	p-ClC₆H₄	H	—		697
C(Me)(CO₂Et)₂	p-BrC₆H₄	H	45–46		586, 697
CH(CN)COMe	Ph	H	—		599
CH(CN)COPh	Ph	H	—		599
CH(CN)CO₂CH₂Ph	Ph	H	—		599
CH₂NHCO₂Et	Ph	H	59–61		460
CH₂NHCONH₂	H	H	190		460
CH₂ONHCONH₂	H	H	—		839
CH₂ONHCSNHR	H	H	—		839
CH₂ONHCONHR	H	H	—		839
CH(NH₂)CO₂Et	Me	H	—		840
CH(NH₂)CH₂CO₂H	H	H	—		730
CH(Ph)OCONHPh	Me	H	154–155		417
CH(CO₂Et)NHCOCH₂Ph	H	H	—		430 u
C(CO₂Et)(Ph)CH₂CH₂NEt₂	Me	H		160–163/0.1	64
C(CO₂Et)(Ph)CH₂CH₂N(morpholino)	Me	H		181–183/0.05	64
C(CO₂Et)(Ph)CH(Me)CH₂NEt₂	Me	H		165/0.2	64
C(CO₂Et)(Ph)CH(Me)CH₂NEt₂	Ph	H		197–200/0.05	64

R_2	R_4	mp (°C)	bp (°C/mm)	Ref.
C(CO$_2$Et)(Ph)CH(Me)CH$_2$N(O) [morpholine]	Me	—	187–190/0.1	64
CH$_2$C(NHC$_6$H$_5$)=NH	H	—		560
CH(OH)CH(CH$_2$OH)NHCOMe	p-NO$_2$C$_6$H$_4$	120		5351
CH$_2$CH(OH)CCl$_3$	H	126–128 124–126		49, 278
CH$_2$CH(OH)CCl$_3$	Me	124.5		277
CH$_2$C(Ph)(COMe)CH$_2$CH$_2$NEt$_2$	Me		152.5/0.01	64
CH$_2$C(Ph)(COMe)CH$_2$CH$_2$N(O) [morpholine]	Me		180/0.1	64
CH$_2$CH(NH$_2$)CO$_2$H	H	197–198		49, 266, 634, 635, 722
CH$_2$CH(NH$_2$)CO$_2$H	Me			722
CH$_2$CH(NH$_2$)CO$_2$H	Ph	222–223		52
CH$_2$CH(NH$_2$)CO$_2$H	p-NO$_2$C$_6$H$_4$	240–245		52
CH(NH$_2$)CO$_2$Et	H	—		722
CH(NH$_2$)CO$_2$Et	Me	—		722
CH(CO$_2$Et)NHCOCH$_2$Ph	H	—		722
CH(CO$_2$Et)NHCOCH$_2$Ph	Me	—		722
CH$_2$C(=NH)NHPh	H	—		560
CH$_2$C(=NH)NHPh	n-Pr	—		560
CH$_2$C(CN)(NHCOMe)CO$_2$H	Ph	73–75		52
CH$_2$C(CN)(NHCOMe)CO$_2$H	p-NO$_2$C$_6$H$_4$	190–196		52
CH(CH$_2$NMe$_2$)CH$_2$NMe$_2$	Me	—	98–100/3	463
CH$_2$SP(S)(OEt)$_2$	Me	—		757
CH$_2$SCO—N(piperidine)	H	—		743
H	CH(p-MeOC$_6$H$_4$)$_2$	133		5631

TABLE III-106. (Continued)

Substituent					
R_2	R_4	R_5	m.p. (°C)	(b.p. (°C))	Ref.
Ph	$CH_2CH(CO_2H)_2$	H	141–142		366, 367
p-MeOC$_6$H$_4$	$CH_2CH(CO_2H)_2$	H	97		35
Ph	$CH_2C(Et)(CO_2H)_2$	H	145		50
p-ClC$_6$H$_4$	$CH(CO_2Me)_2$	H	46		586
2,4-Cl$_2$C$_6$H$_3$	$C(Me)(CO_2Et)_2$	H	—		697
Ph	$CH_2CH(CO_2Et)_2$	H	30–31		366, 367
p-MeOC$_6$H$_4$	$CH_2CH(CO_2Et)_2$	H	55–56		35
Me	$CH_2C(Et)(CO_2Et)_2$	H		235–293/3–4	50
Ph	$CH_2C(Et)(CO_2Et)_2$	H		168–174/4–5	50
H	$CHCl_2$	H		208/4–5	552, 5681
Me	$CHBr_2$	H		63–65/1	577
p-NH$_2$C$_6$H$_4$	$CH_2NH(CH_2)_2NEt_2$	H	94.5		578
p-NO$_2$C$_6$H$_4$	$CH_2NH(CH_2)_2NEt_2$	H	—		578
H	$(CH_2)_2C(NH)NHPh$	H	—		560
Me	$(CH_2)_2C(NH)NH(p$-MeC$_6$H$_5$)	H	—		560
H	$CH_2C(NHC_6H_5)=NH$	H	—		560
n-Pr	$CH_2C(NHC_6H_5)=NH$	H	—		560
H	$CH_2C(=NH)NH$-α-Naphthyl	H	—		560
H	$CH_2C(=NH)NH$-β-Naphthyl	H	—		560
H	$(CH_2)_2C(=NH)NHC_6H_5$	H	—		560
Me	$(CH_2)_2C(=NH)NHC_6H_5$	H	—		560
H	$(CH_2)_2C(=NH)NH(p$-MeC$_6$H$_4$)	H	—		560
H	$(CH(NH_2)CO_2H$	H	155–156		6781
Me	$CH(NH_2)CO_2H$	H	157–158		6781
Me	$CH_2C(Ph)(COMe)CH_2CH_2CO_2H$	H	137–138		485
Me	$CH(NH_2)CO_2H$	H	—		678
H	$CH_2CH(NH_2)CO_2H$	H	237–238		49, 51
p-NO$_2$C$_6$H$_4$	$CH_2CH(NH_2)CO_2H$	H	—		529i
H	$CH_2C(CO_2Et)_2NHCOMe$	H	103–104		49, 51
			104–105		

Ph	CH₂C(Et)(CONH-CO-CONH) (ring)	H	210–211	50
p-ClC₆H₄	C(Br)(CO₂Me)₂	H	110–111	586
Me	CH(OH)p-MeOC₆H₄	H	104	563i
p-NO₂C₆H₄	CH(OH)CH(CH₂OH)NHCOCHCl₂	H	—	531i
Me	CH₂SCH₂CO₂H	H	104.5 105.5	450
Ph	CH₂SCH₂CO₂H	H	78–79	450
p-ClC₆H₄	CH₂SCH₂CO₂H	H	—	125 m

Thiazole structure with R₂, R₄, R₅ (N in ring, S in ring)

R₂	R₃	R₄	R₅	mp	ref
H		CH(NH₂)CO₂H		155–157	6781
Me		CH(NH₂)CO₂H		202–203	6781
Me		CH(NH₂)CO₂H		197–199	320, 6781
H		CH₂CH(NH₂)CO₂H		240	369, 370, 376
H		CH₂CH(NH₂)CO₂H		227–228	442
Me		CH₂CH(NH₂)CO₂Me		187	376
Me		CH₂CH(NHCOMe)CO₂H		191	376
Me		CH(COMe)CO₂CN		—	78
p-ClC₆H₄		CH(CO₂Me)₂		140–141	586
p-ClC₆H₄		C(CO₂Me)₂N(piperidinyl)		90–91	586
Me		CF₂CF₂CF₃		151–153	835
H		CH₂C(=NH)NHC₆H₅		—	560
H		CH₂C(=NH)NH(p-SO₃HC₆H₄)		—	560
Me		(CH₂)₂NH(CH₂)₃NEt₂		198–199	3471
Ph		CH₂CHOHCH₂Cl		95	386
H		(CH₂)₂OCO(CH₂)₂CO₂H		97–100	71, 465
H		(CH₂)₂SC(=NH)NH₂		—	65

TABLE III-107. THIAZOLES SUBSTITUTED BY SEVERAL POLYFUNCTIONAL GROUPS

	Substituent				
R_2	R_4	R_5	m.p. (°C)	b.p. (°C)	Ref.
CH_2CONH_2	CH_2CONH_2	H	218–219		75
CH_2CO_2H	CH_2CO_2H	H	124		75
CH_2CO_2Et	CH_2CO_2Et	H	—	143–147/20	75
CH_2NH_2	$CH_2N(Et)_2$	H	—		377
$(CH_2)_2NH_2$	$(CH_2)_2NH_2$	H	—		49
$(CH_2)_3NH_2$	CH_2NEt_2	H	—	138–139/4	40
$(CH_2)_4NH_2$	CH_2NEt_2	H	—	149–150/6	481
$CH_2NHCOPh$	CH_2CO_2H	H	184–186		414, 429
$CH_2NHCOPh$	CH_2CO_2Et	H	86–88		414, 429
CH_2–phthalimidyl	CH_2Cl	H	133–134		377
$CH_2CH_2NHCOPh$	CH_2CH_2–phthalimidyl	H	107		49
CH_2CH_2–phthalimidyl	CH_2Cl	H	115–116		40

Structure	R	R'	mp/bp	Ref
CH₂CH₂CH₂N(phthalimide)	CH₂NEt₂	H	204–208/4	40
(CH₂)₄N(phthalimide)	CH₂Cl	H	82–83	481
(CH₂)₄N(phthalimide)	CH₂NEt₂	H	—	481
	CH(Me)OH	Me	—	636
	CH₂OPh	H	182–184	432
	CH₂OPh	H	172–173	432
	CH₂OPh	H	173–174	432
	CH₂CH₂OH	H	205–207	432
	CH₂CH₂NEt₂	H	166–168	432
	CH₂CH₂NPr₂	H	193–195	432
	CH₂CH₂NBu₂	H	184–186	432
	CH₂CH₂N(morpholine)	H	164–166	432
CH₂OPh				
CH₂O(p-MeC₆H₄)	CH₂CH₂NPr₂			
CH₂O(p-MeC₆H₄)	CH₂CH₂N(morpholine)			
CH₂O(o-iPrC₆H₄)	CH₂CH₂NEt₂			
CH₂O(o-iPrC₆H₄)	CH₂CH₂NPr₂			
CH₂O(o-iPrC₆H₄)	CH₂CH₂N(morpholine)	H	207–209	432

TABLE III-107. (Continued)

	Substituent			
R_2	R_4	R_5	m.p. (°C)	b.p. (°C)	Ref.
$CH_2O(p\text{-}PhC_6H_4)$	$CH_2CH_2NEt_2$	H	183–185		432
$CH_2O(p\text{-}PhC_6H_4)$	$CH_2CH_2NPr_2$	H	—		432
$CH_2O(p\text{-}ClC_6H_4)$	$CH_2CH_2NEt_2$	H	196–197		432
$CH_2O(p\text{-}ClC_6H_4)$	$CH_2CH_2NPr_2$	H	182–184		432
$CH_2O(p\text{-}ClC_6H_4)$	$CH_2CH_2NBu_2$	H	105–107		432
$CH_2O(p\text{-}ClC_6H_4)$	$CH_2CH_2N\smallsetminus O$	H	222–224		432
$CH_2O(m\text{-}ClC_6H_4)$	$CH_2CH_2NEt_2$	H	188–190		432
$CH_2O(m\text{-}ClC_6H_4)$	$CH_2CH_2NPr_2$	H	125–127		432
$CH_2O(m\text{-}ClC_6H_4)$	$CH_2CH_2N\smallsetminus O$	H	193–195		432
$CH_2O(2,5\text{-}Me_2C_6H_3)$	$CH_2CH_2NEt_2$	H	172–174		432
$CH_2O(2,5\text{-}Me_2C_6H_3)$	$CH_2CH_2NPr_2$	H	152–154		432
$CH_2O(2,5\text{-}Me_2C_6H_3)$	$CH_2CH_2N\smallsetminus O$	H	208–210		432
$CH_2O(5\text{-}Me\text{-}2\text{-}i\text{-}PrC_6H_3)$	$CH_2CH_2NEt_2$	H	202–204		432
$CH_2O(5\text{-}Me\text{-}2\text{-}i\text{-}PrC_6H_3)$	$CH_2CH_2NPr_2$	H	159–161		432
$CH_2O(5\text{-}Me\text{-}2\text{-}iPrC_6H_3)$	$CH_2CH_2N\smallsetminus O$	H	199–200		432

R_2	R_4	R_5	mp	bp	Ref
CH$_2$CO$_2$Ph	Me	CH$_2$CH=CH$_2$	—	—	653
CH$_2$OH	Me	CH$_2$CH$_2$OH	81–82	—	5501
CH(Me)OH	Me	CH$_2$CH$_2$OH	83–84	—	5501, 5731
CH(Me)OH	Me	CH$_2$OCOMe	—	150/0.04	5501
CH(Me)OH	Me	CH$_2$CH$_2$OCOMe	—	—	5501, 5731
CH(Et)OH	Me	CH$_2$CH$_2$OH	88–89	—	5501
CH(i-Pr)OH	Me	CH$_2$CH$_2$OH	—	—	5501
CH(i-Pr)OH	Me	CH$_2$OCOMe	—	148–50/0.01	5501
CH(i-Pr)OH	Me	CH$_2$CH$_2$OCOMe	—	—	5501
CH(CH$_2$C≡CH)OH	Me	CH$_2$OCOMe	—	145/0.01	5501
CH(CH$_2$C≡CH)OH	Me	(CH$_2$)$_2$OCOMe	—	—	5501
CH(CH$_2$CH$_2$Ph)OH	Me	(CH$_2$)$_2$OCOMe	—	162–164/0.002	5501
CH(CH$_2$CH$_2$CH$_2$Ph)OH	Me	CH$_2$CH$_2$OH	114–115	—	5501
CH(Ph)OH	Me	CH$_2$CH$_2$OH	101–102	—	5501, 777, 778
CH(Ph)OH	Me	(CH$_2$)$_2$OCOPh	—	—	777
CH$_2$CH$_2$OH	Me	CH$_2$CH$_2$OH	—	—	5731
CH$_2$CH(Me)OH	Me	CH$_2$OCOMe	—	155/0.04	5501
CH$_2$CH(Me)OH	Me	CH$_2$CH$_2$OCOMe	—	—	5501
CH$_2$CH(Ph)OH	Me	CH$_2$CH$_2$OCOMe	—	155–158/0.002	5501
CH(Me)CH$_2$OH	Me	CH$_2$CH$_2$OH	104	—	5501
CH(Me)CH$_2$OH	Me	CH$_2$CH$_2$OCOMe	—	—	5501

[Structure: thiazole ring with R_2 at 2-position, R_4 at 4-position, R_5 at 5-position]

R_2	R_4	R_5	mp	bp	Ref
Me	CH$_2$CO$_2$Et	CH$_2$CO$_2$Et	106–107	116–117/0.1	520
Me	CH$_2$OH	CH$_2$OH	—	—	60

Table III-108. 2-THIAZOLYLTHIAZOLES

	Position			
2	4	5	m.p. (°C)	Ref.
4-Me-2-thiazolyl	Me	H	—	184 re
5-Ph-2-thiazolyl	H	Ph	240–241	489 u
4-Ph-2-thiazolyl	Ph	H	222	479, 841uv
4-(p-NH$_2$C$_6$H$_4$)-2-thiazolyl	p-NH$_2$C$_6$H$_4$	H	280–283	828
4-(p-NO$_2$C$_6$H$_4$)-2-thiazolyl	p-NO$_2$C$_6$H$_4$	H	—	324 p
4-Me-5-[(CH$_2$)$_2$Cl]-2-thiazolyl	Me	Cl(CH$_2$)$_2$	151–152	721
4-Me-5-[(CH$_2$)$_2$OH]-2-thiazolyl	Me	(CH$_2$)$_2$OH	—	777
4,5-Ph$_2$-2-thiazolyl	Ph	Ph	238	479
4-(CH$_2$)$_2$OCOMe-5-Me-2-thiazolyl	Me	(CH$_2$)$_2$OAc	116	427 f
4-(5-NO$_2$-2-furyl)-2-thiazolyl	H	H	290	824 l
4-Thiazolyl	CH$_2$Br	H	—	508 i
4-Thiazolyl	CH$_2$OH	H	155–158	130 i, 508, 513
2-NHCH$_2$CH=CH$_2$-4-thiazolyl	p-NO$_2$C$_6$H$_4$	H	—	338 l
2-NHCH$_2$CH=CH$_2$-4-thiazolyl-4-thiazolyl	p-NO$_2$C$_6$H$_4$	H	—	338 l
5-Thiazolyl	p-NO$_2$C$_6$H$_4$	H	228–229	535 l
4-(p-PhC$_6$H$_4$)-2-thiazolyl	p-PhC$_6$H$_4$	H	—	841 u
4-(2-Fluorenyl)-2-thiazolyl	2-Fluorenyl	H	—	841 u
4-(5-Indenyl)-2-thiazolyl	5-Indenyl	H	—	841 u

TABLE III-109. 4-THIAZOLYLTHIAZOLES

	Position			
2	4	5	m.p. (°C)	Ref.
CH(p-MeOC$_6$H$_4$)$_2$	2-CH(p-MeOC$_6$H$_4$)$_2$-4-thiazolyl	H	180	563
H	2-C$_6$H$_4$R-4-Me-5-thiazolyl[a]	H	—	705
CH$_2$CH=CH$_2$	2-C$_6$H$_4$R-4-Me-5-thiazolyl[a]	H	—	707
Ph	2-Ph-4-OH-5-thiazolyl	H	—	536

[a] R = H, 2-Me, 3-Me, 4-Me, 4-Cl, 4-MeO, or 4-NO$_2$.

TABLE III-110. 5-THIAZOLYLTHIAZOLES

Position				
2	4	5	m.p. (°C)	Ref.
Me	Me	4-Thiazolyl	84	515
H	Me	2-Me-4-thiazolyl	42	515
Me	Me	2-Me-4-thiazolyl	62.5	515
H	H	5-Thiazolyl	94–95	6
H	Me	4-Me-5-thiazolyl	—	184 r

TABLE III-111. BENZOTHIAZOLYTHIAZOLES

Position				
2	4	5	m.p. (°C)	Ref.
Me	2-Benzothiazolyl	H	176	492
Me	H	2-Benzothiazolyl	122	492

TABLE III-112. FURYLTHIAZOLES

Position					
2	4	5	m.p. (°C)	b.p. (°C)	Ref.
2-Furyl	H	H	—		641g, 752
2-Furyl	Me	H		88/1	5571, 558
2-Furyl	Et	H		96–99/3	558
2-Furyl	tert-Bu	H		133–135/15	24
2-Furyl	CH$_2$Cl	H	74	114–118/3	5571, 558
2-Furyl	H	p-ClC$_6$H$_4$	152		453
5-NO$_2$-2-furyl	Me	H	142		5571
5-NO$_2$-2-furyl	Et	H	100		558
5-NO$_2$-2-furyl	CH$_2$Cl	H		131/2	5571, 558, 654
5-NO$_2$-2-furyl	Ph	H	168–169		701, 842 l
H	2-Furyl	H	80.6–81.5		751u

TABLE III-112. (*Continued*)

	Position				
2	4	5	m.p. (°C)	b.p. (°C)	Ref.
Me	2-Furyl	H	—		591
Ph	2-Furyl	H	65–67		751u
Me	5-p-ClC$_6$H$_4$-2-furyl	H	—		680
Me	5-p-BrC$_6$H$_4$-2-furyl	H	—		680
Me	5-p-NO$_2$C$_6$H$_4$-2-furyl	H	—		680
H	3-Benzofuryl	H	135–136		751u
Ph	3-Benzofuryl	H	128–129		751u

TABLE III-113. THIENYLTHIAZOLES

	Position				
2	4	5	m.p. (°C)	b.p. (°C)	Ref.
2-Thienyl	H	Ph	93.5–95.5		489 u
5-NO$_2$-2-thienyl	CH$_2$NH-cyclohexyl	H	—		611
5-NO$_2$-2-thienyl	CH$_2$NEt$_2$	H	—		611
5-NO$_2$-2-thienyl	CH$_2$N(Me)CH$_2$CH$_2$OH	H	—		611
5-NO$_2$-2-thienyl	CH$_2$Cl	H	—		611, 637, 654
5-NO$_2$-2-thienyl	p-ClC$_6$H$_4$	H	—		611
2-NHC$_6$H$_5$-4-NH$_2$-5-COC$_6$H$_5$-3-thienyl	Ph	H	—		599
2-SCH$_2$COC$_6$H$_5$-4NH$_2$-5-COC$_6$H$_5$-3-thienyl	Ph	H	—		599
2-NH$_2$-3-benzothienyl	Ph	H	—		599, 698
H	2-Thienyl	H	104–106		751u
Me	2-Thienyl	H	60–61 63–64	140–150/15	591u 751u
Me	2-Thienyl	CH$_2$CO$_2$H	158–159		397, 425
Ph	2-Thienyl	CH$_2$CO$_2$H	154–155		727l
o-MeC$_6$H$_4$	2-Thienyl	CH$_2$CO$_2$H	136–138		626, 727l
p-ClC$_6$H$_4$	2-Thienyl	CH$_2$CO$_2$H	137–138		626, 727l
p-MeOC$_6$H$_4$	2-Thienyl	CH$_2$CO$_2$H	149–151		727l
H	CH$_2$(5-Cl-2-thienyl)	H	—		679
H	3-Me-2-benzothienyl	H	68–70		751u
Me	3-Me-2-benzothienyl	H	97–99		751u
Ph	3-Me-2-benzothienyl	H	114–115		751u

TABLE III-114. PYRROLYL AND BENZOPYRROLYL THIAZOLES

Position			m.p. (°C)	b.p. (°C)	Ref.
2	4	5			
Me	3-Pyrrolyl	H	94–95		433
H	H	2-Tetrahydropyrrolyl		171/18 135–137/14	800
H	3-Benzopyrrolyl	H	134–135		583 iu

TABLE III-115. OXAZOLYLTHIAZOLES

Position			m.p. (°C)	Ref.
2	4	5		
4-Oxazolyl	H	H	85–93	732
4-Oxazolyl	Me	H	45–47	732
4-Oxazolyl	Ph	H	112–114	732
2-Me-4-oxazolyl	H	H	83–85	732
2-Me-4-oxazolyl	Me	H	123–124	732
2-Me-4-oxazolyl	Ph	H	90–91	732
2-Me-4-oxazolyl	p-MeC$_6$H$_4$	H	126–128	732
2-Me-4-oxazolyl	p-ClC$_6$H$_4$	H	118–120	732
2-Me-4-oxazolyl	p-BrC$_6$H$_4$	H	140	732
2-Me-4-oxazolyl	p-MeOC$_6$H$_4$	H	158–160	732
2-Ph-4-oxazolyl	H	H	105–106	732
2-Ph-4-oxazolyl	Me	H	130–132	732
2-Ph-4-oxazolyl	Ph	H	184–186	732
2-Ph-4-oxazolyl	p-ClC$_6$H$_4$	H	180–181	732
2-Ph-4-oxazolyl	p-BrC$_6$H$_4$	H	203–204	732
2-Ph-4-oxazolyl	p-MeOC$_6$H$_4$	H	184–186	732
2-CH$_2$Ph-5-EtO-4-oxazolyl	H	H	138	430u
4-Me-5-oxazolyl	H	H	77–79	732
4-Me-5-oxazolyl	Me	H	110–112	732
4-Me-5-oxazolyl	Ph	H	86–87	732

TABLE III-116. THIAZOLIDINYL AND THIAZOLINYLTHIAZOLES

	Position				
2	4	5	m.p. (°C)	b.p. (°C)	Ref.
Ph	4-CO$_2$H-2-thiazolidinyl	H		108/9	569
p-MeC$_6$H$_4$	4-CO$_2$H-2-thiazolidinyl	H	130–131		569
m-MeC$_6$H$_4$	4-CO$_2$H-2-thiazolidinyl	H	128		569
Ph	4-CO$_2$Et-2-thiazolidinyl,HCl	H	146–147		569
p-MeC$_6$H$_4$	4-CO$_2$Et-2-thiazolidinyl,HCl	H	142–143		569
m-MeC$_6$H$_4$	4-CO$_2$Et-2-thiazolidinyl,HCl	H	114–115		569
Ph	CH=NHN—N (thiazolinone)	H	239–240		569
p-MeC$_6$H$_4$	CH=NHN—N (thiazolinone)	H	221		569
m-MeC$_6$H$_4$	CH=NHN—N (thiazolinone)	H	234–235		569

TABLE III-117. IMIDAZOLYL AND BENZIMIDAZOLYLTHIAZOLES

Position				
2	4	5	m.p. (°C)	Ref.
Me	Me	$C(Ph_2)$(1-imidazolyl)	—	629
H	2-Benzimidazolyl	H	203–204	543, 546, 552 1, 565, 568, 670, 706 1, 707 1, 708 1, 709 1, 710 1, 711 1, 712 1, 713 1, 773, 798
H	2-Benzimidazolyl	Me	—	559
Me	2-Benzimidazolyl	H	144.5	559, 565
n-Bu	2-Benzimidazolyl	H	—	559
Ph	2-Benzimidazolyl	H	—	559
H	6-$(CH_2)_2CO_2$H-2-benzimidazolyl	H	—	638 1
H	6-$(CH_2)_2CO_2$Et-2-benzimidazolyl	H	—	638 1
H	6-CH$(CO_2$H$)(CH_2)_9$Me-2-benzimidazolyl	H	—	638 1
H	6-CO$(CH_2)_2CO_2$Et-2-benzimidazolyl	H	—	638 1
H	6-CO_2H-2-benzimidazolyl	H	—	638 1
H	6-NHCO$_2$(CH)Me$_2$-2-benzimidazolyl	H	—	714 1, 802
H	6,7-Cl$_2$-2-benzimidazolyl	H	—	559
H	6-NO$_2$-2-benzimidazolyl	H	—	639
H	6-HO-2-benzimidazolyl	H	—	574 1
H	6-NaO-2-benzimidazolyl	H	—	574 1
H	6-SO_3NH_2-2-benzimidazolyl	H	—	574 1
H	6-NHCOC$_6$H$_5$-2-benzimidazolyl	H	—	802
H	6-NHCO$_2$Me-2-benzimidazolyl	H	—	802
H	6-NHCO$_2$Et-2-benzimidazolyl	H	—	802
H	1-Bu-2-benzimidazolyl	H	—	773
H	1-Bu-6-NO$_2$-2-benzimidazolyl	H	—	773

TABLE III-117. (*Continued*)

2	4	5	m.p. (°C)	Ref.
H	1-CH$_2$Ph-2-benzimidazolyl	H	—	559
H	1-CH$_2$(*p*-ClC$_6$H$_4$)-7-NH$_2$-2-benzimidazolyl	H	—	5741
H	1-CH$_2$(*p*-ClC$_6$H$_4$)-7-HNCOMe-2-benzimidazolyl	H	—	5741
H	1-COMe-7-NHCOMe-2-benzimidazolyl	H	—	5741
H	1-CO$_2$Et-2-benzimidazolyl	H	—	6381
H	1-NHCO$_2$CHMe$_2$,7-OH-2-benzimidazolyl	H	—	639
H	1-NH$_2$SO$_2$-2-benzimidazolyl	H	—	6381
H	1-HO-7-NHCOC$_6$H$_5$-2-benzimidazolyl	H	—	759
H	1-HO-7-NHCO$_2$Me-2-benzimidazolyl	H	—	739
H	1-HO-7-NHCO$_2$Et-2-benzimidazolyl	H	—	739
H	1-HO-7-NHCO$_2$CHMe$_2$-2-benzimidazolyl	H	—	739, 802
H	CH$_2$(2-benzimidazolyl)	H	144–145	5651
H	CH$_2$(1-COMe-2-benzimidazolyl)	H	88–89	5651
H	CH$_2$(1-CH(OH)CCl$_3$-2-benzimidazolyl)	H	116–117	5651
H	(CH$_2$)$_9$ (2-benzimidazolyl)	H	—	6381
H	(CH$_2$)$_{11}$ (2-benzimidazolyl)	H	—	6381

R_1	R_2	R_3		mp	Ref
H	H	H	H	210	565
Cl	H	H	H	—	565
NO_2	H	H	H	—	565
NO_2	H	NO_2	H	230–231	565
H	MeO	OH	H	212–213	565
=CH(2-furyl)			H	206–207	565
=CH(p-$NO_2C_6H_4$), OCOMe			H	180–181	565
=CH(m-$NO_2C_6H_4$), OCOMe			H	—	565
H	CH_2(2-benzimidazolyl)		H	—	744

TABLE III-118. 1,3,4-OXADIAZOLYTHIAZOLES

2	4	5	m.p. (°C)	Ref.
5-NH$_2$-2-oxadiazolyl	Me	H	251–254	753
5-NH$_2$-2-oxadiazolyl	Me	H	262–265	753
Ph	5-NH$_2$-2-oxadiazolyl	H	—	795
p-MeC$_6$H$_4$	5-NH$_2$-2-oxadiazolyl	H	—	795
p-BrC$_6$H$_4$	5-NH$_2$-2-oxadiazolyl	H	—	795
Me	Me	5-NH$_2$-2-oxadiazolyl	238–240	753
Ph	Me	5-NH$_2$-2-oxadiazolyl	230–232	753

TABLE III-119. TRIAZOLINYLTHIAZOLES

2	4	5	m.p. (°C)	Ref.
1-Ph-triazolin-3-yl (N–NH, N-Ph, =S)	Me	H	217–220	753
1-Ph-triazolin-3-yl (N–NH, N-Ph, =S)	Ph	H	272–273	753
Me	Me	1-Ph-triazolin-3-yl (N–NH, N-Ph, =S)	263–266	753
Ph	Me	1-Ph-triazolin-3-yl (N–NH, N-Ph, =S)	274–278	753

TABLE III-120. 2-PYRIDYLTHIAZOLES

	Position			
2	4	5	m.p. (°C)	Ref.
2-Pyridyl	H	H	—	270, 542g
2-Pyridyl	p-ClC$_6$H$_4$	CH$_2$CO$_2$H	202–203	727 l
6-HO-2-pyridyl	Me	H	124	731
6-HO-2-pyridyl	Ph	H	218	731
6-HO-2-pyridyl	p-MeC$_6$H$_4$	H	184	731
6-HO-2-pyridyl	p-ClC$_6$H$_4$	H	221	731
6-HO-2-pyridyl	p-BrC$_6$H$_4$	H	225–226	731
6-HO-2-pyridyl	p-MeOC$_6$H$_4$	H	158	731
3-Pyridyl	H	Ph	104–104.5	489 u
3-Pyridyl	p-ClC$_6$H$_4$	CH$_2$CO$_2$H	—	727 l
2-HO-6-Me-3-pyridyl	Ph	H	—	599
4-Pyridyl	H	CH$_2$CO$_2$H	—	621
4-Pyridyl	Me	CH$_2$CO$_2$H	—	620, 621
4-Pyridyl	p-ClC$_6$H$_4$	CH$_2$CO$_2$H	215–216	727 l
4-Pyridyl	H	CH$_2$CO$_2$Et	—	621
4-Pyridyl	Me	CH$_2$CO$_2$Et	—	621
4-Pyridyl	H	CH$_2$CONH$_2$	—	621
4-Pyridyl	Me	CH$_2$CONH$_2$	—	621
4-Pyridyl	CH$_2$CO$_2$Et	H	—	621
2-Quinolyl	Me	H	121.5–122.5	390
3-Quinolyl	Me	H	118–118.5	390
4-Quinolyl	Me	H	82–85	390
5-Quinolyl	Me	H	97–98	390
6-Quinolyl	Me	H	90.5–91.5	390
CH$_2$(2-pyridyl)	Me	H	—	774
CH(2-pyridyl)NH$_2$	H	H	—	594
CH(3-pyridyl)NH$_2$	H	H	—	594
CH(4-pyridyl)NH$_2$	H	H	—	594
CH(3-pyridyl)OH	H	H	75–76.5	827
C(Me)(3-pyridyl)OH	H	H	116–117	827
C(cyclohexyl)(2-pyridyl)OH	H	H	90–91.5	827
C(cyclohexyl)(3-pyridyl)OH	H	H	127–128	827
C(cyclohexyl)(4-pyridyl)OH	H	H	188–189	827
C(Ph)(3-pyridyl)OH	H	H	131–132.5	827
CH(3-pyridyl)OCOMe	H	H	68.5–69.5	827

TABLE III-121. 4-PYRIDYLTHIAZOLES

	Position			
2	4	5	m.p. (°C)	Ref.
H	2-Pyridyl	H	—	270g
CH(p-HOC$_6$H$_4$)$_2$	2-Pyridyl	H	220	6561
CH(p-MeOC$_6$H$_4$)$_2$	2-Pyridyl	H	108	5631 6561
5-NO$_2$-2-furyl	2-Pyridyl	H	213–214	8421
CH(p-HOC$_6$H$_4$)$_2$	3-Pyridyl	H	165–170	6561
CH(p-MeOC$_6$H$_4$)$_2$	3-Pyridyl	H	156	5631 6561
5-NO$_2$-2-furyl	3-Pyridyl	H	232–234	8421
CH(p-HOC$_6$H$_4$)$_2$	4-Pyridyl	H	232	6561
CH(p-MeOC$_6$H$_4$)$_2$	4-Pyridyl	H	118	5631 6561
5-NO$_2$-2-furyl	4-Pyridyl	H	184	8421
Me	4-Me-2-quinolyl	H	105–106	584
Ph	4-Me-2-quinolyl	H	149–150	584
Me	4-Ph-2-quinolyl	H	158–159	584
Ph	4-Ph-2-quinolyl	H	202–203	584
Ph	CH=NHNCO-4-pyridyl	H	208–209	569
p-MeC$_6$H$_4$	CH=NHNCO-4-pyridyl	H	205	569
m-MeC$_6$H$_4$	CH=NHNCO-4-pyridyl	H	209–210	569

TABLE III-122. 5-PYRIDYLTHIAZOLES

Position				
2	4	5	m.p. (°C)	Ref.
H	H	2-Pyridyl	—	587
Me	Me	4-Me-2-quinolyl	69–70	584
Ph	Me	4-Me-2-quinolyl	100–101	584
Me	Me	4-Ph-2-quinolyl	100–101	584
H	H	CH$_2$CH$_2$N⟨⟩NHCOPh	—	770

TABLE III-123. 2,4-DIPYRIDYLTHIAZOLES

Position				
2	4	5	m.p. (°C)	Ref.
2-Pyridyl	2-Pyridyl	H	142–144	464, 782 783, 784
3-Me-2-pyridyl	2-Pyridyl	H	—	782

TABLE III-124. THIAZOLES SUBSTITUTED BY SEVERAL HETEROCYLCIC GROUPS

	Position			
2	4	5	m.p.(°C)	Ref.
5-NO_2-2-furyl	CH_2(1-morpholinyl)	H	143–144	5571
5-NO_2-2-furyl	5-NO_2-2-furyl	H	252–253	8421
5-NH_2-2-oxadiazolyl	2-Furyl	H	250–252	753
![N—NH, S, N, Ph structure]	2-Furyl		258–261	753
5-NH_2-2-oxadiazolyl	2-Benzofuryl	H	202–203	753
![N—NH, S, N, Ph structure]	2-Benzofuryl	H	274–277	753
5-NO_2-2-thienyl	CH_2(1-pyrrolidinyl)	H	—	611
5-NO_2-2-thienyl	CH_2(1-piperidinyl)	H	—	611
5-NO_2-2-thienyl	CH_2(4-Me-1-piperazinyl)	H	—	611
5-NO_2-2-thienyl	CH_2(4-morpholinyl)	H	—	611
5-NO_2-2-thienyl	CH_2(2,6-Me_2-4-morpholinyl)	H	—	611
CH=CH(5-NO_2-2-thienyl)-	CH=N(4-morpholinyl)	H	—	637
5-NH_2-2-oxadiazolyl	CH_2(4-morpholinyl)	H	—	611
	2-Thienyl	H	259–262	753
![N—NH, O, N, Ph structure]	2-Thienyl	H	285–290	753

TABLE III-124. (*Continued*)

	Position			
2	4	5	m.p.(°C)	Ref.
2-Me-4-oxazolyl	2-(2-Me-5-oxazolyl)-4-thiazolyl	H	294–296	732
2-Ph-4-oxazolyl	2-(2-Me-5-oxazolyl)-4-thiazolyl	H	294–296	732
5,7-Diterpentyl-2-benzoxazolyl	5,7-Diterpentyl-2-benzoxazolyl	H	129–130	592
5,7-Diterpentyl-2-benzoxazolyl	5,7-Diterpentyl-2-benzoxazolyl	Me	125–127	592
5,7-Diterpentyl-2-benzoxazolyl	5,7-Diterpentyl-2-benzoxazolyl	Ph	131–133	592
5,7-Diterpentyl-2-benzoxazolyl	Ph	5,7-Diterpentyl-2-benzoxazolyl	207–210	592
Me	5,7-Diterpentyl-2-benzoxazolyl	5,7-Diterpentyl-2-benzoxazolyl	103–105	592
Ph	5,7-Diterpentyl-2-benzoxazolyl	5,7-Diterpentyl-2-benzoxazolyl	72–74	592
5,7-Diterpentyl-2-benzoxazolyl	5,7-Diterpentyl-2-benzoxazolyl	5,7-Diterpentyl-2-benzoxazolyl	139–140	592
5-NH$_2$-2-oxadiazolyl	2-Me-5-thiazolyl	H	270–272	753

TABLE III-125. THIAZOLES SUBSTITUTED BY HETEROCYCLIC GROUPS NONADJACENT TO THE THIAZOLE RING

$$R_5(CH_2)_2 \underset{S}{\overset{R_4}{\underset{\big|}{\bigsqcup}}} \overset{N}{\underset{R_2}{\big|}}$$

Substituent				
R_2	R_4	R_5	b.p. (°C)	Ref.
H	Me	1-Piperidinyl	—	681
Me	Me	1-Piperidinyl	—	681
Et	Me	1-Piperidinyl	—	681
H	Me	1-Pyrrolidinyl	—	681
Me	H	1-Pyrrolidinyl	—	681
Et	Me	1-Pyrrolidinyl	—	681
H	Me	1-Piperazinyl	—	681
H	Me	4-Me-1-piperazinyl	135–138/0.1	347
Me	Me	4-Me-1-piperazinyl	—	347
Et	Me	4-Me-1-Piperazinyl	—	347
H	Me	Morpholinyl	—	347
Me	Me	Morpholinyl	—	347
Et	Me	Morpholinyl	—	347
H	Me	Tetrahydroisoquinolyl	—	681
Me	Me	Tetrahydroisoquinolyl	—	681
Et	Me	Tetrahydroisoquinolyl	—	681
H	Me	1-Pyrazolyl	—	681
Me	Me	1-Pyrazolyl	—	681
Et	Me	1-Pyrazolyl	—	681
H	Me	1-Imidazolyl	—	681
Me	Me	1-Imiddzolyl	—	681
Et	Me	1-Imidazolyl	—	681
H	Me	$-O-\underset{O}{\overset{O}{\underset{\|}{P}}}\overset{O-}{\underset{O-}{\big<}}$	—	682

TABLE III-126. NONADJACENT DITHIAZOLES

$$R_4\underset{R_5}{\overset{N}{\diagdown}}\underset{S}{\diagup}(CH_2)_n\underset{S}{\overset{N}{\diagdown}}\underset{R_5'}{\overset{R_4'}{\diagup}}$$

		Substituent				
n	R_4	R_5	R_4'	R_5'	m.p.(°C)	Ref.
1	Me	H	Me	H	221	400
1	Ph	H	Ph	H	119–20	278,400
2	Me	H	Me	H	68–69	407
2	Ph	H	Ph	H	120	407
3	Me	H	Me	H	—	407
3	Ph	H	Ph	H	91–92	407
4	Me	H	Me	H	251	278, 391
4	Ph	H	Ph	H	288	305
4	Ph	Ph	Ph	Ph	116–117	278
5	Me	H	Me	H	—	387
5	Ph	H	Ph	H	60	387
7	Me	H	Me	H	—	401
7	Ph	H	Ph	H	78	451
8	Ph	H	Ph	H	70	401
9	Me	H	Me	H	—	401

$$R_2\underset{S}{\overset{N}{\diagdown}}\underset{R_5}{\diagup}(CH_2)_n\underset{S}{\overset{N}{\diagdown}}\underset{R_2'}{\diagup}$$

			Substituent			
n	R_2	R_5	R_2'	R_5'	m.p. (°C)	Ref.
1	H	H	H	H	35–36	82, 408
1	Me	H	Me	H	55–56	408
1	Ph	H	Ph	H	83–84	408
7	Ph	H	Ph	H	84	309
8	$MeSO_2$-C_6H_4	H	C_6H_4-SO_2Me	H	152–155	383

TABLE III-126. (*Continued*)

$$R_2 \underset{S}{\overset{N}{\diagdown}} -CH_2-\underset{A}{\overset{B}{\underset{|}{C}}}-CH_2-\underset{S}{\overset{N}{\diagdown}} R_2'$$

	Substituent				
R_2	A	B	R_2'	m.p. (°C)	Ref.
Ph	H	CONHNH$_2$	Ph	—	34
Ph	H	CO$_2$H	Ph	127–128	34
Ph	H	N(CO)$_2$C$_6$H$_4$	Ph	158–159	34
Ph	H	NH$_2$	Ph	158–159	34
Ph	CO$_2$H	CO$_2$H	Ph	156–157	34
Ph	CO$_2$Et	CO$_2$Et	Ph	116	34

$$R_4 \underset{S}{\overset{N}{\diagdown}} -CH_2-A-CH_2-\underset{S}{\overset{N}{\diagdown}} R_4'$$

	Substituent			
R_4	A	R_4'	m.p. (°C)	Ref.
p-MeC$_6$H$_4$	NH(CH$_2$)$_2$NH	p-MeC$_6$H$_4$	—	507
p-ClC$_6$H$_4$	NH(CH$_2$)$_2$NH	p-ClC$_6$H$_4$	—	507
p-BrC$_6$H$_4$	NH(CH$_2$)$_2$NH	p-BrC$_6$H$_4$	145–147	507
p-MeOC$_6$H$_4$	NH(CH$_2$)$_2$NH	p-MeOC$_6$H$_4$	116–117	507
p-MeC$_6$H$_4$	NH(COMe)(CH$_2$)$_2$(COMe)NH	p-MeC$_6$H$_4$	170–172	507
p-ClC$_6$H$_4$	NH(COMe)(CH$_2$)$_2$(COMe)NH	p-ClC$_6$H$_4$	206–208	507
p-BrC$_6$H$_4$	NH(COMe)(CH$_2$)$_2$(COMe)NH	p-BrC$_6$H$_4$	214–216	507
p-MeOC$_6$H$_4$	NH(COMe)(CH$_2$)$_2$(COMe)NH	p-MeOC$_6$H$_4$	180–183	507
p-BrC$_6$H$_4$	—N(CH$_2$CH$_2$)$_2$N—	p-BrC$_6$H$_4$	252–258	507
p-MeOC$_6$H$_4$	—N(CH$_2$CH$_2$)$_2$N—	p-MeOC$_6$H$_4$	214–217	507
Ph	CH$_2$SCH$_2$	Ph	68–69	452
Ph	CH(Me)S(Me)CH	Ph	66–67	452
p-BrC$_6$H$_4$	CH(Me)S(Me)CH	p-BrC$_6$H$_4$	132–133	452
Ph	CH(Me)S(Me)CH ↓ O	Ph	—	452

TABLE III-126. (*Continued*)

$$\underset{R_2}{\overset{N}{\underset{S}{\|}}}-(CH_2)_2-A-(CH_2)_2-\underset{S}{\overset{N}{\underset{S}{\|}}}R_2'$$

	Substituent				
R_2	A	R_2	m.p. (°C)	b.p. (°C)	Ref.
Ph	NHCONH	Ph	176–177		34
Me	NH	Me		150–210/2	78
3,4-(MeO)$_2$C$_6$H$_3$	NH	3,4-(MeO)$_2$C$_6$H$_3$	165–166		843

$$\underset{S}{\overset{N}{\|}}\overset{R_4}{\underset{CH_2-A-CH_2}{-}}\overset{R_4'}{\underset{S}{\|}}\overset{N}{}$$

	Substituent			
R_4	A	R_4'	m.p. (°C)	Ref.
Me	O	Me	180	61
Me	CH$_2$SSCH$_2$	Me	34–37	65

$$\left[\underset{S}{\overset{N}{\underset{Ph-}{\|}}}-CH_2\right]_2 CHNHCONHCH \left[\overset{CH_2}{\underset{S}{\|}}-Ph\right]_2 \quad 182–183 \quad 34$$

XVI. Tables of Products

TABLE III-126. (*Continued*)

$$R_4 \underset{S}{\boxed{}} N-A-N \underset{S}{\boxed{}} R_4'$$

	Substituent			
R_4	A	R_4'	m.p. (°C)	Ref.
Me	1,4-C_6H_4	Me	166 169–170	134 tiu, 307
Ph	1,4-C_6H_4	Ph	225 229–230 230.5–231	134 tiu, 307, 310
p-PhC$_6$H$_4$	1,4-C_6H_4	p-PhC$_6$H$_4$	350	134 tiu
Ph	1,3-C_6H_4	Ph	170	310
Me	—⟨C$_6$H$_4$⟩—⟨C$_6$H$_4$⟩—	Me	208	402
Ph	—⟨C$_6$H$_4$⟩—O—⟨C$_6$H$_4$⟩—	Ph	230–231.5	134 tiu
p-PhC$_6$H$_4$	—⟨C$_6$H$_4$⟩—O—⟨C$_6$H$_4$⟩—	p-PhC$_6$H$_4$	343–344.5	134 tiu

$$R_2 \underset{S}{\boxed{}}^{N} -A- \underset{S}{\boxed{}}^{N} R_2'$$

	Substituent			
R_2	A	R_2'	m.p. (°C)	Ref.
Et	1,4-C_6H_4	Et	114–116	310
Ph	1,4-C_6H_4	Ph	232–233	134 tiu
p-PhC$_6$H$_4$	1,4-C_6H_4	p-PhC$_6$H$_4$	350	134 tiu
Ph	1,3-C_6H_4	Ph	178–195	329
Me	$C_6H_4OC_6H_4$	Me	146–147	458
Me	$C_6H_4OC_6H_4$	Me	154–156	458

TABLE III-126. (*Continued*)

Structure	m.p. (°C)	Ref.
![structure] bis-thiazole bridged by (CH$_2$)$_4$ groups	182–183	409, 410
CH$_2$CH$_2$OH–[4-Me-thiazol-2-yl]–CH(Ph)OCH$_2$CH$_2$–[4-Me-thiazol-2-yl]–CHOHPh	—	777
1,3,5-tris(2-phenylthiazol-4-yl... / thiazol-5-yl)benzene	195	306

TABLE III-127. MISCELLANEOUS SUBSTITUTED THIAZOLES

Substituent			m.p./	
R_2	R_4	R_5	b.p. (°C)	Ref.
R = H	Me	H	—	723
R = NH$_2$ (C(Ph)=CH-pyrimidine-Me)	Me	H	—	723
R = NH$_2$	Me	Me	—	723
R = NHMe	Me	H	—	723
R = NMe$_2$	Me	H	—	723
R = NMe$_2$	Me	CH$_2$CH$_2$OH	—	777, 778
C(OH)(Ph)CH$_2$-(aminopyrimidine-Me)	Me	Me	—	777
C(OH)(Ph)CH$_2$-(aminopyrimidine-Me)	Me	CH$_2$CH$_2$OH	—	777
H	(2-methyl-4-oxoquinazolin-3-yl)	H	201–202	5481

TABLE III-127. (Continued)

	Substituent		m.p./b.p. (°C)	Ref.
R_2	R_4	R_5		
H	(3-amino-2-methylquinazolin-4(3H)-one)	H	191	548
H	(pyrido-fused pyrimidinone)	H	281–282	548
Ph	(barbiturate-type structure)	H	—	569
p-MeC$_6$H$_4$	(barbiturate-type structure)	H	—	569
m-MeC$_6$H$_4$	(barbiturate-type structure)	H	261	569

p-NO$_2$C$_6$H$_4$	CH$_2$CH(C=O)O-N-Et (structure)	H	235/6	534 l
p-NO$_2$C$_6$H$_4$	CH$_2$CH(C=O)O-N-Ph (structure)	H	240	5341
p-ClC$_6$H$_4$	Me	C(CO$_2$Me)$_2$-2-morpholinyl	124–126	586
p-ClC$_6$H$_4$	Me	C(CO$_2$Me)$_2$-1-morpholinyl	125–126	586
p-NO$_2$C$_6$H$_4$	CH$_2$-C(=O)-N(Ph)-N=Me (structure)	H	209–210	7271
H	CH$_2$-(N=N-NH triazole)	p-ClC$_6$H$_4$	209–210	7271
H	(chromone R, Me) ^{a}	H	—	715, 749
Me	(chromone R, Me) ^{b}	H	—	749
Ph	(chromone R, Me) ^{b}	H	—	749

TABLE III-127. (*Continued*)

Substituent			m.p./	
R_2	R_4	R_5	b.p. (°C)	Ref.
H	CH(NH$_2$)CONH—[β-lactam-S-C(Me$_2$)-CO$_2$H]	H	—	320, 6781
Me	CH(NH$_2$)CONH—[β-lactam-S-C(Me$_2$)-CO$_2$H]	H	—	320, 6781
p-ClC$_6$H$_4$	CH(NH$_2$)CONH—[β-lactam-S-C(Me$_2$)-CO$_2$H]	H	—	665
H	CH$_2$CONH—[β-lactam-S-C(Me$_2$)-CO$_2$K]	H	—	8291
Me	CH$_2$CONH—[β-lactam-S-C(Me$_2$)-CO$_2$K]	H	—	8291
H	H	CH(NH$_2$)CONH—[β-lactam-S-C(Me$_2$)-CO$_2$H]	—	320, 6781

H	Me	CH(NH₂)CONH—[β-lactam-S-CMe₂-CO₂H]	—	320, 6781
Me	Me	CH(NH₂)CONH—[β-lactam-S-CMe₂-CO₂H]	—	320, 6781
H	H	CH(NH₂)CONH—[β-lactam-S-C⁺Me₂-CONEt₂]	—	6781
H	Me	CH(NH₂)CONH—[β-lactam-S-C⁺Me₂-CONEt₂]	—	6781
Me	Me	CH(NH₂)CONH—[β-lactam-S-C⁺Me₂-CONEt₂]	—	6781
H	H	CHRCONH—[cepham-CO₂H] ^{c}	—	683

^{a} R = H, Me, COMe, CH₂CO₂Et, or OMe.
^{b} R = H, Me, COMe, or CH₂CO₂Et.
^{c} R = H, NH₂, or OH.

XVII. REFERENCES

1. K. Ganapathi and A. Venkataraman, *Proc. Indian Acad. Sci.*, **22A,** 362 (1945); *Chem. Abstr.*, **40,** 4059.
2. P. Schatzmann, *Annalen*, **261,** 1 (1891).
3. G. Popp, *Annalen*, **250,** 273 (1889).
4. Japanese Patent No. 175 910; *Chem. Abstr.*, **44,** 8960.
5. G. Swain, *J. Chem. Soc.*, 2898 (1949); *Chem. Abstr.*, **44,** 3976.
6. M. Erne, L. Herzfeld, B. Prijs, and H. Erlenmeyer, *Helv. Chim. Acta*, **36,** 354 (1953); *Chem. Abstr.*, **49,** 1012.
7. E. C. Taylor, Jr., J. A. Anderson, and G. A. Berchtold, *J. Amer. Chem. Soc.*, **77,** 5444 (1955); *Chem. Abstr.*, **50,** 6434.
8. V. A. Krasovskii and S. I. Burmistrov, *Khim. Geterotsikl. Soedinenii*, **1,** 56 (1969); *Chem. Abstr.* **70,** 115051.
9. K. Takeda, K. Tamamura, and H. Tone, *J. Pharm. Soc. Japan*, **74,** 290 (1954); *Chem. Abstr.*, **49,** 3162.
10. G. Vernin and J. Metzger, *Bull, Soc. Chim. France*, 2498 (1963); *Chem. Abstr.*, **60,** 4122.
11. H. Schenkel, E. Marbet, and H. Erlenmeyer, *Helv. Chim. Acta*, **27,** 1437 (1944); *Chem. Abstr.*, **39,** 3281.
12. H. Erlenmeyer, R. Marbet, and H. Schenkel, *Helv. Chim. Acta*, **28,** 924 (1945); *Chem. Abstr.*, **40,** 869.
13. H. A. J. Schoutissen, *J. Amer. Chem. Soc.*, **55,** 45 (1933).
14. M. Erne and H. Erlenmeyer, *Helv. Chim. Acta*, **31,** 652 (1948); *Chem. Abstr.*, **42,** 4575.
15. H. Erlenmeyer, J. Junod, W. Guex, and M. Erne, *Helv. Chim. Acta*, **31,** 1342 (1948); *Chem. Abstr.*, **43,** 5021.
16. L. Herzfeld, B. Prijs, and H. Erlenmeyer, *Helv. Chim. Acta*, **36,** 1842 (1953); *Chem. Abstr.*, **49,** 1012.
17. E. R. Buchman, A. O. Reims, and H. Sargent, *J. Org. Chem.*, **6,** 764 (1941); *Chem. Abstr.*, **36,** 1606.
18. U.S. Patent No. 2 179 984; *Chem. Abstr.*, **34,** 1687.
19. British Patent No. 492 637; *Chem. Abstr.*, **33,** 1760.
20. A. H. Cook, I. H. Heilbron, and A. L. Levy, *J. Chem. Soc.*, 1594 (1947); *Chem. Abstr.*, **42,** 8795.
21. C. D. Hurd and B. Rudner, *J. Amer. Chem. Soc.*, **73,** 5157 (1951); *Chem. Abstr.*, **47,** 587.
22. U.S. Patent No. 2 402 642; *Chem. Abstr.*, **40,** 5758.
23. U.S. Patent No. 2 870 158; *Chem. Abstr.*, **53,** 11409.
24. F. Asinger, M. Thiel, G. Peschel, and K. H. Meinicke, *Annalen*, **619,** 145 (1958); *Chem. Abstr.*, **53,** 17105.
25. F. Asinger, M. Thiel, and K. Gewald, *Annalen*, **639,** 133 (1961); *Chem. Abstr.*, **55,** 23495.
26. F. Asinger, L. Schroeder, and S. Hoffmann, *Annalen*, **648,** 83 (1961); *Chem. Abstr.*, **56,** 7297.
27. F. Asinger, M. Thiel, P. Puechel, F. Haaf, and W. Schaefer, *Annalen*, **660,** 85 (1962); *Chem. Abstr.*, **58,** 9058.
28. R. A. Olofson, J. M. Landesberg, K. N. Houk, and J. S. Michelman, *J. Amer. Chem. Soc.*, **88,** 4265 (1966).

XVII. References

29. F. Asinger, M. Thiel, and H. G. Hauthal, *Annalen*, **634,** 131 (1959); *Chem. Abstr.*, **55,** 2616.
30. F. Asinger, M. Thiel, H. Usbeck, K. H. Grobe, H. Grundmann, and S. Trankner, *Annalen*, **634,** 144 (1959); *Chem. Abstr.*, **55,** 2617.
31. P. Dubs and M. Pesaro, *Synthesis*, 294 (1974).
32. M. Thiel, F. Asinger, and K. Schmiedel, *Annalen*, **619,** 133 (1958); *Chem. Abstr.*, **53,** 17102.
33. C. M. Suter, and T. B. Johnson, *Rec. Trav. Chim.*, **49,** 1066 (1930); *Chem. Abstr.*, **25,** 952.
34. W. S. Hinegardner and T. B. Johnson, *J. Amer. Chem. Soc.*, **52,** 3724 (1930); *Chem. Abstr.*, **24,** 5038.
35. W. S. Hinegardner and T. B. Johnson, *J. Amer. Chem. Soc.*, **52,** 4139 (1930); *Chem. Abstr.*, **24,** 5751.
36. J. F. Olin and T. B. Johnson, *J. Amer. Chem. Soc.*, **53,** 1473 (1931); *Chem. Abstr.*, **25,** 2722.
37. J. M. Sprague, A. H. Land, and C. Ziegler, *J. Amer. Chem. Soc.*, **68,** 2155 (1946); *Chem. Abstr.*, **41,** 960.
38. R. Dahlbom, *Acta Chem. Scand.*, **7,** 885 (1953); *Chem. Abstr.*, **49,** 4626.
39. H. T. Liang, H. Y. Ch'in, and L. C. Pi, *Yao Hsueh Hsueh Pao*, **7,** 218 (1959); *Chem. Abstr.*, **54,** 10996.
40. Chi Yuon Fong and Tshin Chi Yuan, *Sci. Sinica Peking*, **5,** 449 (1956); *Chem. Abstr.*, **52,** 3776.
41. G. Palazzo and M. Tavella, *Gazz. Chim. Ital.*, **92,** 1084 (1962); *Chem. Abstr.*, **58,** 12543.
42. K. M. Murav'eva and M. N. Shchukina, *Zh. Obshchei Khim.*, **33,** 3723 (1963); *Chem. Abstr.*, **60,** 8012.
43. S. G. Fridman, *Zh. Obshchei Khim.*, **24,** 909 (1954); *Chem. Abstr.*, **49,** 8250.
44. S. G. Fridman, *Zh. Obshchei Khim.*, **24,** 1059 (1954); *Chem. Abstr.*, **49,** 8923.
45. S. G. Fridman, *J. Gen. Chem. U.S.S.R.*, **25,** 933 (1955); *Chem. Abstr.*, **51,** 11327.
46. H. J. M. Dou, W. J. Spillane, and J. Metzger, *Communication Orsay France*, 1975.
47. F. Kipnis, I. Levy, and J. Ornfelt, *J. Amer. Chem. Soc.*, **71,** 3570 (1949); *Chem. Abstr.*, **44,** 610.
48. J. Harley-Mason, *J. Chem. Soc.*, 323 (1947); *Chem. Abstr.*, **41,** 5510.
49. R. G. Jones, E. C. Kornfeld, and K. C. MacLaughlin, *J. Amer. Chem. Soc.*, **72,** 4526 (1950); *Chem. Abstr.*, **45,** 3383.
50. F. E. Hooper and T. B. Johnson, *J. Amer. Chem. Soc.*, **56,** 484 (1934); *Chem. Abstr.*, **28,** 1700.
51. W. T. Caldwell and S. M. Fox, *J. Amer. Chem. Soc.*, **73,** 2935 (1951); *Chem. Abstr.*, **46,** 2538.
52. S. Tatsuoka and H. Hitomi, *J. Pharm. Soc. Japan*, **73,** 334 (1953); *Chem. Abstr.*, **48,** 3343.
53. A. Silberg, I. Simiti, and H. Mantsch, *Chem. Ber.*, **94,** 2887 (1961); *Chem. Abstr.*, **56,** 7295.
54. German Patent No. 1 161 275; *Chem. Abstr.*, **60,** 10717.
55. B. M. Mikhailov and V. P. Bronovitskaya, *Zh. Obshchei Khim.*, **27,** 726 (1957); *Chem. Abstr.*, **51,** 16436.
56. Japanese Patent No. 8569 ('56); *Chem. Abstr.*, **52,** 11951.
57. J. F. Olin and T. B. Johnson, *J. Amer. Chem. Soc.*, **53,** 1475 (1931); *Chem. Abstr.*, **25,** 2722.
58. V. M. Zubarovskii and R. N. Moskaleva, *Zh. Obshchei Khim.*, **32,** 570 (1962); *Chem. Abstr.*, **58,** 2525.

59. A. J. Eusebi, E. V. Brown, and L. R. Cerecedo, *J. Amer. Chem. Soc.*, **71,** 2931 (1949); *Chem. Abstr.*, **44,** 610.
60. L. H. Conover and D. S. Tarbell, *J. Amer. Chem. Soc.*, **72,** 5221 (1950); *Chem. Abstr.*, **45,** 5148.
61. E. R. Buchman and H. Sargent, *J. Amer. Chem. Soc.*, **67,** 400 (1945); *Chem. Abstr.*, **39,** 1872.
62. J. F. Olin and T. B. Johnson, *J. Amer. Chem. Soc.*, **53,** 1470 (1931); *Chem. Abstr.*, **25,** 2722.
63. British Patent No. 792 158; *Chem. Abstr.*, **52,** 20199.
64. D. J. Brown, A. H. Cook, and I. Heilbron, *J. Chem. Soc.*, **106S** (1949); *Chem. Abstr.*, **44,** 145.
65. Y. Sawa and T. Ishida, *J. Pharm. Soc. Japan*, **76,** 337 (1956); *Chem. Abstr.*, **50,** 13875.
66. A. Y. Berlin and V. P. Bronovitskaya, *Zh. Obshchei Khim.*, **31,** 1356 (1961); *Chem. Abstr.*, **55,** 24719.
67. H. Yonemoto, *Yakugaku Zasshi*, **79,** 143 (1959); *Chem. Abstr.*, **53,** 13168.
68. German Patent No. 927 031; *Chem. Abstr.*, **51,** 14833.
69. S. Fallab, *Helv. Chim. Acta*, **35,** 215 (1952); *Chem. Abstr.*, **46,** 8647.
70. G. Leichssenring and J. Schmidt, *Chem. Ber.*, **95,** 767 (1962); *Chem. Abstr.*, **57,** 798.
71. Japanese Patent No. 13 225 ('61); *Chem. Abstr.*, **56,** 7326.
72. G. B. Bachman and L. V. Heisey, *J. Amer. Chem. Soc.*, **71,** 1985 (1949); *Chem. Abstr.*, **43,** 7018.
73. C. Scherschener, *Pr Montevideo*, **9–10,** 45 (1959); *Chem. Abstr.*, **58,** 5655.
74. W. R. Boon, *J. Chem. Soc.*, 601 (1945); *Chem. Abstr.*, **40,** 573.
75. R. Mory and H. Schenkel, *Helv. Chim. Acta*, **33,** 405 (1950); *Chem. Abstr.*, **44,** 5873.
76. R. G. Button and P. J. Taylor, *J. Chem. Soc. Perkin Trans 2*, 557 (1973); *Chem. Abstr.*, **78,** 123592.
77. P. J. Taylor, *J. Chem. Soc. Perkin Trans 2*, 1077 (1972); *Chem. Abstr.*, **77,** 74520F.
78. A. Burger and G. E. Ullyot, *J. Org. Chem.*, **12,** 342 (1947); *Chem. Abstr.*, **41,** 4489.
79. H. Gregory and L. F. Wiggins, *J. Chem. Soc.*, 590 (1947); *Chem. Abstr.*, **41,** 5865.
80. D. Price and F. D. Pickel, *J. Amer. Chem. Soc.*, **63,** 1067 (1941); *Chem. Abstr.*, **35,** 3635.
81. H. Erlenmeyer and M. Muller, *Helv. Chim. Acta*, **28,** 922 (1945); *Chem. Abstr.*, **40,** 869.
82. A. Wartburg, *Helv. Chim. Acta*, **32,** 1097 (1949); *Chem. Abstr.*, **43,** 7931.
83. M. Erne, *Helv. Chim. Acta*, **32,** 2205 (1949); *Chem. Abstr.*, **44,** 2981.
84. British Patent No. 623 428; *Chem. Abstr.*, **44,** 4510.
85. Swiss Patent No. 260 230; *Chem. Abstr.*, **44,** 4511.
86. U.S. Patent No. 2 483 392; *Chem. Abstr.*, **44,** 2036.
87. P. F. Torrence and H. Tieckelmann, *Biochim. Biophys. Acta*, **158,** 183 (1968); *Chem. Abstr.*, **68,** 101794.
88. U. H. Lindberg, *Acta Chem. Scand.*, **20,** 917 (1966); *Chem. Abstr.*, **65,** 5451G.
89. D. J. Tocco, R. P. Buhs, H. D. Brown, A. R. Matzuk, H. E. Mertel, R. E. Harman, and N. R. Trenner, *J. Med. Chem.*, **7,** 399 (1964); *Chem. Abstr.*, **61,** 16627.
90. I. Kh. Fel'dman, G. S. Semicheva, and R. D. Podorova, *Mechenye Biol. Aktivn. Veshchestva, St. Statei*, 89 (1971); *Chem. Abstr.*, **76,** 25234M.
91. I. N. Bojesen, J. H. Hoeg, J. T. Nielsen, I. B. Petersen, and K. Schaumburg, *Acta Chem. Scand.*, **25,** 2739 (1971); *Chem. Abstr.*, **76,** 24291.
92. E. J. Vincent, R. Phan Tan Luu, and J. Metzger, *Bull. Soc. Chim. France*, 3530 (1966).
93. U.S. Patent No. 2 221 147; *Chem. Abstr.*, **35,** 2033.
94. M. Witanowski, L. Stefaniak, H. Januszewski, and Z. Grabowski, *Tetrahedron*, **28,** 637 (1972); *Chem. Abstr.*, **76,** 79007H.

XVII. References

95. J. P. Kintzinger and J. M. Lehn, *Mol. Phys.*, **14**, 133 (1968); *Chem. Abstr.*, **69**, 6961.
96. J. Crousier and J. Metzger, *Bull. Soc. Chim. France*, **11**, 4134 (1967); *Chem. Abstr.*, **69**, 10391S.
97. R. Cottet, R. Gallo, J. Metzger, and J. M. Surzur, *Bull. Soc. Chim. France*, **12**, 4502 (1967); *Chem. Abstr.*, **69**, 2397H.
98. C. Riou, G. Vernin, H. J. M. Dou, and J. Metzger, *Bull. Soc. Chim. France*, **7**, 2673 (1972); *Chem. Abstr.*, **78**, 15316.
99. G. Davidovics, P. Roepstorff, J. Chouteau, and C. Garrigou-Lagrange, *Spectrochim. Acta*, **23A**, 2669 (1967); *Chem. Abstr.*, **68**, 7766.
100. A. Babadjamian and J. Metzger, *Bull. Soc. Chim. France*, **12**, 4878 (1968); *Chem. Abstr.*, **70**, 68234.
101. E. J. Vincent, R. Phan Tan Luu, J. Metzger, and J. M. Surzur, *Bull. Soc. Chim. France*, **11**, 3524 (1966); *Chem. Abstr.*, **66**, 64912.
102. R. Cottet, R. Gallo, and J. Metzger, *Bull. Soc. Chim. France*, **12**, 4499 (1967); *Chem. Abstr.*, **69**, 2894.
103. A. Taurins and W. G. Schneider, *Can. J. Chem.*, **38**, 1237 (1960); *Chem. Abstr.*, **55**, 3196.
104. G. Borgen and S. Gronowitz, *Acta Chem. Scand.*, **20**, 2593 (1966); *Chem. Abstr.*, **66**, 60541.
105. C. Roussel, A. Babadjamian, M. Chanon, and J. Metzger, *Bull. Soc. Chim. France*, **3**, 1087 (1971); *Chem. Abstr.*, **75**, 20263.
106. R. Vivaldi, H. J. M. Dou, G. Vernin, and J. Metzger, *Bull. Soc. Chim. France*, **11**, 4014 (1969); *Chem. Abstr.*, **72**, 55313.
107. R. Vivaldi, H. J. M. Dou, and J. Metzger, *C.R. Acad. Sci. Ser. C*, **264**, 1652 (1967); *Chem. Abstr.*, **67**, 64280.
108. J. P. Aune, R. Phan Tan Luu, E. J. Vincent, and J. Metzger, *Bull. Soc. Chim. France*, 2679 (1972); *Chem. Abstr.*, **78**, 3389.
109. G. Vernin, J. P. Aune, H. J. M. Dou, and J. Metzger, *Bull. Soc. Chim. France*, 4523 (1967); *Chem. Abstr.*, **69**, 19062T.
110. E. J. Vincent, R. Phan Tan Luu, and J. Metzger, *Bull. Soc. Chim. France*, **11**, 3537 (1966); *Chem. Abstr.*, **66**, 104623.
111. H. J. M. Dou and J. Metzger, *Bull. Soc. Chim. France*, 2395 (1966); *Chem. Abstr.*, **65**, 18467G.
112. E. J. Vincent, R. Phan Tan Luu, and J. Metzger, *C.R. Acad Sci. Ser. C*, **270**, 666 (1970).
113. R. Garnier, R. Faure, A. Babadjamian, and E. J. Vincent, *Bull. Soc. Chim. France*, **3**, 1040 (1972); *Chem. Abstr.*, **77**, 11964H.
114. E. J. Vincent and J. Metzger, *C.R. Acad. Sci. Ser. C*, **261**, 1964 (1965); *Chem. Abstr.*, **63**, 17826.
115. G. M. Clarke, R. Grigg, and D. H. Williams, *J. Chem. Soc. B*, **4**, 339 (1966); *Chem. Abstr.*, **64**, 17389.
116. R. Graham, I. Cooks, S. Howe, W. Tam, and D. H. Williams, *J. Amer. Chem. Soc.*, **90**, 4064 (1968).
117. R. G. Buttery, L. C. Ling, and R. E. Lundin, *J. Agric. Food Chem.*, **21**, 488 (1973); *Chem. Abstr.*, **79**, 31976.
118. R. A. Khmelnitskii, E. A. Kunina, S. L. Gusinskaya, and V. Y. Telly, *Khim. Geterotsikl. Soedinenii*, **7**, 1372 (1971); *Chem. Abstr.*, **76**, 98611S.
119. R. Tabacchi, *Helv. Chim. Acta*, **57**, 324 (1974).
120. G. Buttery, R. M. Seifert, and L. C. Ling, *Agri. Food Sci.*, **23**, 516 (1975).
121. R. G. Buttery and L. C. Ling, *J. Agric. Food Chem.*, **22**, 912 (1974).
122. O. G. Witzthum and P. Werkhoff, *J. Food Sci.*, **39**, 1210 (1974).

123. R. Bonnichsen, R. Hjalm, T. Marde, M. Moller, and R. Ryhage, *Z. Rechtsmed.*, **73,** 225 (1973).
124. J. P. Aune and J. Metzger, *Bull. Soc. Chim. France*, **9,** 3536 (1972); *Chem. Abstr.*, **78,** 3421.
125. M. J. Rix and B. R. Webster, *Org. Mass Spectrom.*, **5,** 311 (1971); *Chem. Abstr.*, **75,** 34719.
126. J. P. Aune, *Thesis* 1969. University of Marseille, France.
127. G. Vernin and J. Metzger, *J. Chim. Phys.*, **6,** 865 (1974).
128. D. E. Ryan, *Can. J. Chem.*, **39,** 2389 (1961); *Chem. Abstr.*, **56,** 12306.
129. A. Taurins, J. G. E. Fenyes, and R. N. Jones, *Can. J. Chem.*, **35,** 423 (1957); *Chem. Abstr.*, **52,** 883.
130. M. P. V. Mijovic and J. Walker, *J. Chem. Soc.*, 3381 (1961); *Chem. Abstr.*, **56,** 14252.
131. P. Bassignana, C. Cogrossi, and M. Gandino, *Spectrochim. Acta*, **19,** 1885 (1963); *Chem. Abstr.*, **59,** 13482.
132. French Patent No. 1 356 908; *Chem. Abstr.*, **61,** 3115.
133. G. Sbrana, E. Castellucci, and M. Ginanneschi, *Spectrochim. Acta*, **23A,** 751 (1967); *Chem. Abstr.*, **66,** 109841.
134. R. Arnaud, M. Gelus, J. C. Malet, and J. M. Bonnier, *Bull. Soc. Chim. France*, **9,** 2857 (1966); *Chem. Abstr.*, **66,** 37815.
135. G. Davidovics, C. Garrigou Lagrange, J. Chouteau, and J. Metzger, *Spectrochim. Acta*, **23A,** 1477 (1967); *Chem. Abstr.*, **67,** 37861.
136. B. Ellis and P. J. F. Griffiths, *Spectrochim. Acta*, **21,** 1881 (1965); *Chem. Abstr.*, **64,** 171.
137. D. Bouin and A. Friedman, *C.R. Acad. Sci. Ser. C*, **269,** 1343 (1969); *Chem. Abstr.*, **72,** 78116.
138. A. Friedmann, A. Cormons, and J. Metzger, *C.R. Acad. Sci. Ser. C*, **271,** 17 (1970).
139. A. R. Katritzky, C. O. Ogetir, H. O. Tarhan, H. J. M. Dou, and J. Metzger, *J. Chem. Soc. Perkin Trans* 2, 1614 (1975).
140. M. Azzaro, *Thesis Marseille*, 1962.
141. U.S. Patent No. 2 226 153; *Chem. Abstr.*, **35,** 2084.
142. P. L. Southwick and D. Sapper, *J. Org. Chem.*, **19,** 1926 (1954); *Chem. Abstr.*, **50,** 963.
143. N. Bodor, M. Farkas, and N. Trinajstic, *Croat. Chem. Acta*, **43,** 107 (1971); *Chem. Abstr.*, **75,** 98074.
144. R. Phan Tan Luu, J. M. Surzur, J. Metzger, J. P. Aune, and C. Dupuy, *Bull. Soc. Chim. France*, **9,** 3274 (1967); *Chem. Abstr.*, **68,** 38910.
145. V. Galasso and N. Trinajstic, *Tetrahedron*, **28,** 2799 (1972); *Chem. Abstr.*, **77,** 47764G.
146. J. P. Aune, *Thesis* 1969. University of Marseille, France.
147. J. M. Bonnier and R. Arnaud, *Bull. Soc. Chim. France*, 243 (1968); *Chem. Abstr.*, **68,** 99212.
148. M. Carrega, D. L. Nealy, and J. Metzger, *J. Chim. Phys.*, **56,** 820 (1959).
149. M. Poite, *Thesis* 1961. University of Marseille, France.
150. P. Goursot and I. Wadso, *Acta Chem. Scand.*, **20,** 1314 (1966); *Chem. Abstr.*, **65,** 17781B.
151. L. Sacconi, P. Paoletti, and M. Ciampolini, *J. Amer. Chem. Soc.*, **82,** 3831 (1960).
152. R. Cottet, R. Gallo, J. Metzger, and J. M. Surzur, *Bull. Soc. Chim. France*, 4502 (1967); *Chem. Abstr.*, **69,** 2397.
153. A. Albert, R. Goldagre, and J. Phillips, *J. Chem. Soc.*, 2240 (1948).
154. R. Meyer and J. Metzger, *Bull. Soc. Chim. France*, **12,** 4465 (1967); *Chem. Abstr.*, **68,** 108142.

XVII. References

155. M. Meyer, R. Meyer, B. Brun, J. Salvinien, and J. Metzger, *Bull. Soc. Chim. France*, 3132 (1968); *Chem. Abstr.*, **69**, 100011.
156. R. Meyer and M. Meyer, *C.R. Acad. Sci. Ser. C*, **266**, 1664 (1968); *Chem. Abstr.*, **69**, 46253N.
157. R. Meyer, M. Meyer, and J. Metzger, *Bull. Soc. Chim. France*, **7**, 2692 (1968); *Chem. Abstr.*, **69**, 89832T.
158. R. Meyer, G. Bourrelly, and J. Metzger, *C.R. Acad. Sci., Ser. C*, **267**, 114 (1968); *Chem. Abstr.*, **69**, 70354S.
159. P. Goursot and E. F. Westrum, Jr., *C.R. Acad. Sci., Ser. C*, **266**, 667 (1968); *Chem. Abstr.*, **69**, 13527B.
160. T. R. Manley and D. A. Williams, *Spectrochim. Acta*, **24A**, 361 (1968); *Chem. Abstr.*, **68**, 99276.
161. B. Soptrajanov, *Croat. Chem. Acta*, **39**, 229 (1967); *Chem. Abstr.*, **68**, 108663.
162. M. Gelus, P. M. Vay, and G. Berthier, *Theor. Chim. Acta*, **9**, 182 (1967); *Chem. Abstr.*, **68**, 33422.
163. R. Phan Tan Luu, L. Bouscasse, E. J. Vincent, and J. Metzger, *Bull. Soc. Chim. France*, **9**, 3283 (1967); *Chem. Abstr.*, **68**, 44384.
164. J. Vitry Raymond, *Thesis* 1965. University of Marseille, France.
165. J. M. Bonnier, R. Arnaud, and M. Maurey-Mey, *C.R. Acad. Sci. Ser. C*, **267**, 10 (1968); *Chem. Abstr.*, **69**, 64067J.
166. M. Gelus and J. M. Bonnier, *J. Chim. Phys.*, **64**, 1602 (1967); *Chem. Abstr.*, **68**, 100244.
167. U.S. Patent No. 2 242 237; *Chem. Abstr.*, **35**, 5518.
168. G. Vernin and J. Metzger, *Bull. Soc. Chim. France*, 846 (1967); *Chem. Abstr.*, **67**, 7782.
169. G. Vernin, R. Cottet, R. Gallo, J. M. Surzur, and J. Metzger, *Bull. Soc. Chim. France*, 4492 (1967).
170. G. Vernin, "La Chromatographie en Couche Mince," Dunod, Paris, 1970, p. 111.
171. L. R. Snyder, "Principles of Adsorption Chromatography," J. C. Giddings, Ed., Marcel Dekker, New York, 1968.
172. G. Vernin and G. Vernin, *J. Chromatogr.*, **46**, 48 (1970); *Chem. Abstr.*, **72**, 59532.
173. K. Kovats, *Helv. Chim. Acta*, **41**, 1915 (1958).
174. G. Vernin, R. Jauffred, H. J. M. Dou, and J. Metzger, *J. Chem. Soc. B*, 1678 (1970).
175. G. Vernin, R. Jauffred, C. Ricard, H. J. M. Dou, and J. Metzger, *J. Chem. Soc. Perkin Trans* 2, 1145 (1972); *Chem. Abstr.*, **77**, 74544S.
176. G. Vernin, H. J. M. Dou, and J. Metzger, *J. Chem. Soc. Perkin Trans* 2, 1093 (1973); *Chem. Abstr.*, 79, 31217.
177. G. Vernin, M. A. Lebreton, H. J. M. Dou, and J. Metzger, *Bull. Soc. Chim. France*, 1085 (1974).
178. G. Vernin, M. A. Lebreton, H. J. M. Dou, and J. Metzger, *Tetrahedron*, **30**, 4171 (1974).
179. G. Vernin, Personal Communication,
180. H. J. M. Dou, G. Vernin, and J. Metzger, *Bull. Soc. Chim. France*, 4189 (1971).
181. H. J. M. Dou, G. Vernin, and J. Metzger, *Bull. Soc. Chim. France*, 4593 (1971).
182. G. Vernin, H. J. M. Dou, and J. Metzger, *Bull. Soc. Chim. France*, 1173 (1972).
183. H. J. M. Dou, G. Vernin, and J. Metzger, *Tetrahedron Lett.*, **23**, 2223 (1967); *Chem. Abstr.*, **67**, 54069.
184. G. Vernin, H. J. M. Dou, and J. Metzger, *Bull. Soc. Chim. France*, **8**, 3280 (1968); *Chem. Abstr.*, **70**, 3913.
185. H. J. M. Dou and J. Metzger, *C.R. Acad. Sci., Ser. C*, **262**, 687 (1966); *Chem. Abstr.*, **64**, 19591.

186. H. J. M. Dou, G. Vernin, and J. Metzger, *C.R. Acad. Sci., Ser. C*, **263,** 1243 (1966); *Chem. Abstr.*, **66,** 54840.
187. G. Vernin, H. J. M. Dou, and J. Metzger, *C.R. Acad. Sci., Ser. C*, **263,** 1310 (1966); *Chem. Abstr.*, **66,** 75942.
188. G. Vernin, H. J. M. Dou, and J. Metzger, *Bull. Soc. Chim. France*, 4514 (1967); *Chem. Abstr.*, **69,** 10049M.
189. G. Vernin, H. J. M. Dou, and J. Metzger, *C.R. Acad. Sci., Ser. C*, **264,** 336 (1967); *Chem. Abstr.*, **67,** 2530.
190. D. R. Augood, D. H. Hey, and G. H. Williams, *J. Chem. Soc.*, 2094 (1952).
191. E. L. Eliel, S. Meyerson, and Z. Welwart, *Tetrahedron Lett.*, 479 (1962).
192. H. J. M. Dou, G. Vernin, and J. Metzger, *Tetrahedron Lett.*, 953 (1968).
193. H. J. M. Dou and B. M. Lynch, *Tetrahedron Lett.*, 897 (1965).
194. H. J. M. Dou and B. M. Lynch, *Bull. Soc. Chim. France*, **12,** 3815 (1966); *Chem. Abstr.*, **66,** 104545.
195. J. M. Bonnier and J. Court, *C.R.Acad. Sci., Ser, C*, **265,** 133 (1967).
196. G. Vernin and H. J. M. Dou, *C.R. Acad. Sci., Ser. C*, **266,** 822 (1968); *Chem. Abstr.*, **69,** 43233V.
197. H. J. M. Dou, *Bull. Soc. Chim. France*, 1678 (1966); *Chem. Abstr.*, **65,** 8711F.
198. M. Baule, G. Vernin, H. J. M. Dou, and J. Metzger, *Bull. Soc. Chim. France*, **6,** 2083 (1971); *Chem. Abstr.*, **75,** 76668.
199. H. J. M. Dou and B. M. Lynch, *Bull. Soc. Chim. France*, 3820 (1966).
200. H. J. M. Dou, G. Vernin, and J. Metzger, *Bull. Soc. Chim. France*, 4521 (1967); *Chem. Abstr.*, **69,** 2411H.
201. U. Rudquist and K. Torssell, *Acta Chem. Scand.*, **25,** 2183 (1971); *Chem. Abstr.*, **76,** 3146.
202. D. MacKay, *Can. J. Chem.*, **44,** 2881 (1966).
203. J. I. G. Cadogan, *J. Chem. Soc.*, 4257 (1962).
204. Shu Huang, *Acta Chim. Sinica*, **25,** 171 (1959).
205. P. Hassanaly, G. Vernin, H. J. M. Dou, and J. Metzger, *Bull. Soc. Chim. France*, 560 (1974).
206. G. Fillipi, G. Vernin, H. J. M. Dou, J. Metzger, and M. J. Perkins, *Bull. Soc. Chim. France*, 1075 (1974).
207. H. J. M. Dou, G. Vernin, and J. Metzger, *J. Chem. Soc. A*, 1678 (1970).
208. G. Vernin, M. A. Lebreton, H. J. M. Dou, and J. Metzger, *Tetrahedron*, **30,** 4171 (1974).
209. M. C. Ford and D. MacKay, *J. Chem. Soc.*, 1294 (1958).
210. M. C. Ford and D. MacKay, *J. Chem. Soc.*, 4620 (1957).
211. A. L. Lee, D. MacKay, and E. L. Manery, *Can. J. Chem.*, **48,** 3554 (1970); *Chem. Abstr.*, **74,** 31707.
212. K. Torssell, *Tetrahedron*, **26,** 2769 (1970).
213. A. Lablache-Combier and M. Remy, *Bull. Soc. Chim. France*, 679 (1971).
214. A. Couture and A. Lablache-Combier, *Chem. Commun.* 891 (1971).
215. G. Vernin, H. J. M. Dou, and J. Metzger, *C.R. Acad. Sci., Ser. C*, **271,** 1616 (1970).
216. G. Vernin, J. C. Poite, J. Metzger, J. P. Aune, and H. J. M. Dou, *Bull. Soc. Chim. France*, **3,** 1103 (1971); *Chem. Abstr.*, **75,** 20260.
217. G. Vernin, C. Riou, H. J. M. Dou, L. Bouscasse, J. Metzger, and G. Loridan, *Bull. Soc. Chim. France*, **5,** 1743 (1973); *Chem. Abstr.*, **79,** 91365.
218. M. Kojima and M. Maeda, *J. Chem. Soc. D*, 386 (1970); *Chem. Abstr.*, **72,** 120787.
219. M. Ohashi, A. Iio, and T. Yonezawa, *J. Chem. Soc. D*, 1148 (1970); *Chem. Abstr.*, **73,** 120547.

220. M. Maeda and and M. Kojima, *Tetrahedron Lett.*, 3523 (1973).
221. M. Maeda, A. Kawahara, M. Kai, and M. Kojima, *Heterocycles*, **3,** 389 (1975).
222. J. Metzger and B. Koether, *Bull. Soc. Chim. France*, **20,** 702 (1953); *Chem. Abstr.*, **48,** 10738.
223. J. Crousier and J. Metzger, *Bull. Soc. Chim. France*, 4134 (1967); *Chem. Abstr.*, **69,** 10391.
224. G. Knaus and A. I. Meyers, *J. Org. Chem.*, **39,** 1192 (1974).
225. G. Knaus and A. I. Meyers, *J. Org. Chem.*, **39,** 1189 (1974).
226. J. Vitry Raymond, *Thesis* 1965. University of Marseille, France.
227. G. M. Dyson, *J. Indian Chem. Soc.*, **8,** 147 (1931).
228. K. Ganapathi and A. Venkataraman, *Proc. Indian Acad. Sci.*, **22A,** 343 (1945); *Chem. Abstr.*, **40,** 4056.
229. G. Schwarz, *Org. Synth.*, **25,** 35 (1945); *Chem. Abstr.*, **40,** 330.
230. H. Nagasawa, *J. Pharm. Soc. Japan*, **59,** 471 (1939); *Chem. Abstr.*, **34,** 102.
231. H. Nagasawa, *J. Pharm. Soc. Japan*, **60,** 219 (1940).
232. S. Gusinskaya, V. Telly, and N. L. Ovchinnikova, *Uzbek. Khim. Zh.*, **15,** 47 (1971).
233. G. Travagli, *Gazz. Chim. Ital.*, **78,** 592 (1948); *Chem. Abstr.*, **43,** 2615.
234. C. D. Hurd and H. L. Wehrmeister, *J. Amer. Chem. Soc.*, **71,** 4007 (1949).
235. Y. Nagasawa, *Chem. Zentralbl.*, **2,** 199 (1941).
236. B. Prijs, J. Ostertag, and H. Erlenmeyer, *Helv. Chim. Acta*, **30,** 2110 (1947); *Chem. Abstr.*, **42,** 1934.
237. H. Babo and B. Prijs, *Helv. Chim. Acta*, **33,** 306 (1950); *Chem. Abstr.*, **44,** 5872.
238. K. Ganapathi and K. D. Kulkarni, *Current Sci.*, **21,** 314 (1952); *Chem. Abstr.*, **48,** 2046.
239. K. Ganapathi and K. D. Kulkarni, *Proc. Indian Acad. Sci.*, **37A,** 758 (1953); *Chem. Abstr.*, **48,** 10006.
240. G. Asato, *J. Org. Chem.*, **33,** 2544 (1968); *Chem. Abstr.*, **69,** 27317V.
241. H. J. M. Dou, G. Vernin, and J. Metzger, *J. Heterocyclic Chem.*, **6,** 575 (1969); *Chem. Abstr.*, **71,** 91361.
242. A. R. Katritzky, C. O. Ogretir, H. O. Tarhan, H. J. M. Dou, and J. Metzger, *J. Chem. Soc. Perkin Trans.* 2, 1614 (1975).
243. H. J. M. Dou, A. Friedmann, G. Vernin, and J. Metzger, *C.R. Acad. Sci., Ser. C*, **266,** 714 (1968); *Chem. Abstr.*, **69,** 43182.
244. E. Ochiai and K. Kokeguti, *J. Pharm. Soc. Japan*, **60,** 271 (1940).
245. B. S. Friedman, M. Spark, and R. Adams, *J. Amer. Chem. Soc.*, **59,** 2262 (1937); *Chem. Abstr.*, **32,** 556.
246. E. Ochiai, T. Kakuda, I. Nakayama, and G. Masuda, *J. Pharm. Soc. Japan*, **59,** 462 (1939); *Chem. Abstr.*, **34,** 101.
247. I. Simiti and M. Farkas, *Chem. Ber.*, **98,** 3446 (1965); *Chem. Abstr.*, **64,** 5061.
248. A. Silberg, A. Benko, and A. Panczel, *Rev. Roumaine Chim.*, **10,** 617 (1965); *Chem. Abstr.*, **64,** 723.
249. I. Simiti, M. Farkas, and I. Schwartz, *Stud. Univ. Babes Bolyai, Ser. Chem.*, **16,** 141 (1971); *Chem. Abstr.*, **76,** 126183.
250. M. Baule, R. Vivaldi, J. C. Poite, H. J. M. Dou, G. Vernin, and J. Metzger, *Bull. Soc. Chim. France*, 4310 (1971); *Chem. Abstr.*, **76,** 85184E.
251. H. Erlenmeyer, C. Becker, E. Sorkin, H. Bloch, and E. Suter, *Helv. Chim. Acta*, **30,** 2058 (1947); *Chem. Abstr.*, **42,** 2253.
252. M. Ohta, *J. Pharm. Soc. Japan*, **71,** 869 (1951); *Chem. Abstr.*, **46,** 4002.
253. M. Carrega, *Thesis* 1959. University of Marseille, France.
254. M. Poite, *Thesis* 1961. University of Marseille, France.

255. M. Azzaro and J. Metzger, *Bull. Soc. Chim. France*, **7,** 1575 (1964); *Chem. Abstr.*, **61,** 10554.
256. R. Gallo, *Thesis* 1972. University of Marseille, France.
257. R. Gallo, M. Chanon, H. Lund, and J. Metzger, *Tetrahedron Lett.*, **36,** 3857 (1972); *Chem. Abstr.*, **77,** 151308.
258. A. Babadjamian, M. Chanon, R. Gallo, and J. Metzger, *J. Amer. Chem. Soc.*, **95,** 3807 (1973); *Chem. Abstr.*, **79,** 31338.
259. L. W. Deady, *Austr. J. Chem.*, **26,** 1949 (1973).
260. L. W. Deady and J. A. Zoltewicz, *J. Amer. Chem. Soc.*, **93,** 5475 (1971).
261. R. W. Taft, *J. Amer. Chem. Soc.*, **74,** 3120 (1952).
262. R. W. Taft, "Steric Effects in Organic Chemistry," Wiley, New York, 1956.
263. G. B. Behera, J. N. Kar, R. C. Acharya, and M. K. Rout, *J. Org. Chem.*, **38,** 2164 (1973); *Chem. Abstr.*, **79,** 31173.
264. H. J. Anderson, D. J. Barnes, and Z. M. Khan, *Can. J. Chem.*, **42,** 2375 (1964).
265. E. Ochiai and E. Hayashi, *J. Pharm. Soc. Japan*, **67,** 34 (1947); *Chem. Abstr.*, **45,** 9533.
266. Japanese Patent No. 72 42 662; *Chem. Abstr.*, **78,** 72123.
267. A. Fujita, J. Aramoto, S. Minami, and H. Takamatsu, *Yakugaku Zasshi*, **86,** 427 (1966); *Chem. Abstr.*, **65,** 3870C.
268. E. Matsumura, T. Hirooka, and K. Imagawa, *Nippon Kagaku Zasshi*, **82,** 616 (1961); *Chem. Abstr.*, **57,** 12466.
269. A. W. Addison, K. Dawson, R. D. Gillard, B. T. Heaton, and H. Shaw, *J. Chem. Soc. C*, 589 (1972); *Chem. Abstr.*, **76,** 93965A.
270. W. J. Eilbeck, F. Holmes, T. W. Thomas, and G. Williams, *J. Chem. Soc. A*, 2348 (1968); *Chem. Abstr.*, **69,** 113015.
271. M. N. Hughes and K. J. Rutt, *Spectrochim. Acta*, **A27,** 924 (1971); *Chem. Abstr.*, **75,** 55841.
272. M. N. Hughes and K. J. Rutt, *Inorg. Chem.*, **10,** 414 (1971); *Chem. Abstr.*, **74,** 60322.
273. J. A. Weaver, Thesis, 1970; *Chem. Abstr.*, **76,** 30254X.
274. W. H. Mills and J. L. B. Smith, *J. Chem. Soc.*, **121,** 2724 (1922); *Chem. Abstr.*, **17,** 1023.
275. J. L. B. Smith, *J. Chem. Soc.*, **123,** 2288 (1923); *Chem. Abstr.*, **18,** 264.
276. H. Kondo and F. Nagasawa, *J. Pharm. Soc. Japan*, **57,** 249 (1937).
277. T. Matsuo, *J. Pharm. Soc. Japan*, **71,** 684 (1951); *Chem. Abstr.*, **46,** 8093.
278. H. Erlenmeyer, O. Weber, P. Schmidt, G. Kung, C. Zinsstag, and B. Prijs, *Helv. Chim. Acta*, **31,** 1142 (1948); *Chem. Abstr.*, **42,** 7291.
279. F. M. Hamer, *J. Chem. Soc.*, 3197 (1952).
280. F. Brody and M. T. Bogert, *J. Amer. Chem. Soc.*, **65,** 1080 (1943); *Chem. Abstr.*, **37,** 4733.
281. A. Schuftan, *Berichte*, **27,** 1009 (1894).
282. D. M. Brown and G. A. R. Kon, *J. Chem. Soc.*, 2147 (1948); *Chem. Abstr.*, **43,** 3426.
283. D. S. Noyce and S. A. Fike, *J. Org. Chem.*, **38,** 2433 (1973); *Chem. Abstr.*, **79,** 52486.
284. D. S. Noyce and S. A. Fike, *J. Org. Chem.*, **38,** 3316 (1973); *Chem. Abstr.*, **79,** 114769.
285. D. S. Noyce and S. A. Fike, *J. Org. Chem.*, **38,** 3318 (1973); *Chem. Abstr.*, **79,** 114768.
286. D. S. Noyce and S. A. Fike, *J. Org. Chem.*, **38,** 3321 (1973); *Chem. Abstr.*, **79,** 114764.
287. British Patent No. 1 247 829; *Chem. Abstr.*, **76,** 2719J.
288. French Patent No. 2 065 062; *Chem. Abstr.*, **77,** 163257.

289. German Patent No. 2 152 557; *Chem. Abstr.*, **77,** 88488P.
290. German Patent No. 2 026 735; *Chem. Abstr.*, **77,** 87105Z.
291. S. Kinoshita and T. Tamanishi, *Nippon Nogei Kagaku Zasshi,* **11,** 737 (1973).
292. British Patent No. 1 247 829.
293. French Patent No. 2 065 062.
294. R. E. Naipawer, R. Potter, P. Vallon, and R. E. Erickson, *Flavour Ind.*, **2,** 465 (1971); *Chem. Abstr.*, **75,** 128560.
295. U.S. Patent No. 3 660 112.
296. J. P. Walradt, A. O. Pittet, T. E. Kinlin, R. Muralidhara, and A. Sanderson, *J. Agric. Food. Chem.*, **19,** 972 (1971); *Chem. Abstr.*, **75,** 108666.
297. O. G. Vitzthum and P. Werkhoff, *J. Food Sci.*, **39,** 1210 (1974).
298. R. A. Wilson, C. J. Mussinan, J. Katz, and A. Sanderson, *J. Agric. Food Chem.*, **21,** 873 (1973).
299. T. E. Kinlin, R. Muralidhara, A. O. Pittet, and A. Sanderson, *J. Agric. Food Chem.*, **20,** 1021 (1972).
300. W. Renold, R. Naef Mueller, R. Keller, B. William, and G. Ohloff, *Helv. Chim. Acta,* **57,** 1301 (1974).
301. G. Sanderson and H. N. Graham, *J. Agric. Food Chem.*, **21,** 576 (1973).
302. M. Stoll, M. Winter, F. Gautschi, I. Flament, and B. Willham, *Helv. Chim. Acta,* **50,** 628 (1967).
303. S. J. Kazeniac and R. M. Hall, *J. Food Sci.*, **35,** 519 (1970); *Chem. Abstr.*, **74,** 63278.
304. R. G. Buttery, R. M. Seifert, D. G. Guadagni, and L. C. Ling, *J. Agric. Food Chem.*, **19,** 524 (1971); *Chem. Abstr.*, **75,** 18920.
305. H. Lehr and H. Erlenmeyer, *Helv. Chim. Acta,* **27,** 489 (1944); *Chem. Abstr.*, **38,** 6275.
306. G. Bischoff, O. Weber, and H. Erlenmeyer, *Helv. Chim. Acta,* **27,** 947 (1944); *Chem. Abstr.*, **39,** 1641.
307. H. Erlenmeyer, W. Buchler, and H. Lehr, *Helv. Chim. Acta,* **27,** 969 (1944); *Chem. Abstr.*, **39,** 1641.
308. H. Erlenmeyer and M. Erne, *Helv. Chim. Acta,* **29,** 275 (1946); *Chem. Abstr.*, **40,** 2829.
309. H. Erlenmeyer and W. Buchler, *Helv. Chim. Acta,* **29,** 1924 (1946); *Chem. Abstr.*, **41,** 2041.
310. J. E. Mulvaney and C. S. Marvel, *J. Org. Chem.*, **26,** 95 (1961); *Chem. Abstr.*, **55,** 19902.
311. C. S. Marvel and G. E. Hartzell, *J. Amer. Chem. Soc.*, **81,** 448 (1959).
312. U.S. Patent No. 2 515 328.
313. Y. Iwakura, F. Toda, H. Suzuki, N. Kusakawa, and K. Yagi, *J. Polymer Sci., Part A* 1, *Polymer Chem.*, **10,** 1133 (1972); *Chem. Abstr.*, **77,** 62340J.
314. K. Yagi, T. Miyazaki, H. Okitsu, F. Toda, and Y. Iwakura, *J. Polymer Sci., Part A* 1, **10,** 1149 (1972); *Chem. Abstr.*, **77,** 20119P.
315. D. R. Stevenson, *Thesis,* 1972; *Chem. Abstr.*, **77,** 35025D.
316. H. Ho Wei, *Thesis,* 1965. *Chem. Abstr.*, **64,** 6757.
317. V. V. Korshak and A. D. Maksimov, *Itogi Nauki Khim. Teknol. Vysokomol. Soedin,* **3,** 60 (1971); *Chem. Abstr.*, **77,** 48823N.
318. K. Takemoto, N. Ueda, and T. Kawabata, *Technol. Reports Osaka Univ.*, **21,** 757 (1971); *Chem. Abstr.*, **77,** 35011W.
319. J. Huml. and J. Kepecki, *Spec. Plast. Hmoty Nar. Hospod,* 121 (1971); *Chem. Abstr.*, **76,** 127448.
320. J. M. Bonnier, M. Gelus, and B. Papoz, *Bull. Soc. Chim. France,* 2485 (1965).

321. I. B. John, E. McHill, and J. O. Smith, *Technical Report*, 582 (1959).
322. J. Preston and W. B. Black, *Monsanto Tech. Rev.*, **12,** 32 (1967); *Chem. Abstr.*, **67,** 22784.
323. J. Preston and W. B. Black, *J. Polymer Sci., Part A* 1, **5,** 2429 (1967); *Chem. Abstr.*, **67,** 109051.
324. J. Preston, *J. Heterocyclic Chem.*, **2,** 441 (1965); *Chem. Abstr.*, **64,** 8175.
325. U.S. Patent No. 3 335 149; *Chem. Abstr.*, **67,** 73603S.
326. W. C. Sheehan, T. B. Cole, and L. G. Picklesimer, *J. Appl. Polymer Sci.*, **9,** 1455 (1965); *Chem. Abstr.*, **62,** 13319.
327. M. H. Ting, *Hua Hsueh Pao*, 385 (1965); *Chem. Abstr.*, **64,** 6769.
328. H. Ho-Wei Un, *Diss. Abstr.*, 2486 (1965); *Chem. Abstr.*, **64,** 6757.
329. A. De Rooker and P. De Radzitky, *Bull. Soc. Chim. Belges*, **75,** 641 (1966); *Chem. Abstr.*, **66,** 37814.
330. C. L. Schilling and J. E. Mulvaney, *Amer. Chem. Soc., Div. Polym Chem.*, **8,** 363 (1967); *Chem. Abstr.*, **66,** 105221.
331. C. L. Schilling and J. E. Mulvaney, *Macromolecules*, **1,** 452 (1968); *Chem. Abstr.*, **69,** 107142.
332. C. L. Schilling and J. E. Mulvaney, *Macromolecules*, **1,** 445 (1968); *Chem. Abstr.*, **69,** 107146.
333. P-J. Wang, *Ko Fenizu T Ung Hsun Chine*, **8,** 79 (1966); *Chem. Abstr.*, **65,** 20217B.
334. G. C. Barrett and J. R. Chapman, *Chem. Commun.*, **6,** 335 (1968); *Chem. Abstr.*, **69,** 10704.
335. H. P. Gurtner, *Bibl. Nutr. et Dieta*, **8,** 97 (1966); *Chem. Abstr.*, **66,** 36277.
336. U.S. Patent No. 3 355 426; *Chem. Abstr.*, **68,** 22805.
337. T. Severin and F. Ledl, *Chem. Mikrobiol. Technol. Lebensm*, **1,** 135 (1972); *Chem. Abstr.*, **77,** 152524.
338. A. Benko and I. Borus, *Chem. Ber.*, **100,** 2188 (1967); *Chem. Abstr.*, **67,** 82149.
339. R. Howe, R. H. Moore, B. S. Rao, and A. H. Wood, *J. Med. Chem.*, **15,** 1040 (1972); *Chem. Abstr.*, **78,** 12020.
340. W. Hepworth, B. B. Newbould, D. S. Platt, and G. J. Stagey, *Nature*, **221,** 582 (1969).
341. S. J. Alcok, *Excerpta Medica Internat. Congress*, **220,** 184 (1970).
342. U. H. Lindberg, *Acta Pharm. Suecica*, **3,** 161 (1966).
343. U. H. Lindberg, G. Bexell, J. Pedersen, and S. Ross, *Acta Pharm. Suecica*, **7,** 423 (1970).
344. S. N. Sawhney and J. Singh, *Indian J. Chem.*, **8,** 882 (1970); *Chem. Abstr.*, **74,** 13045.
345. U. H. Lindberg, *Acta Pharm. Suecica*, **8,** 647 (1971); *Chem. Abstr.*, **76,** 148733.
346. U. H. Lindberg, *Acta Pharm. Suecica*, **8,** 39 (1971).
347. U. H. Lindberg, J. Pedersen, and B. Ulff, *Acta Pharm. Suecica*, **4,** 269 (1967); *Chem. Abstr.*, **68,** 105069.
348. K. Brown, D. P. Cater, J. Cavalla, D. Green, R. Newberry, and A. B. Wilson, *J. Med. Chem.*, **17,** 1177 (1974).
349. W. J. Spillane, H. J. M. Dou, and J. Metzger, *Tetrahedron Lett.*, **26,** 2269 (1976).
350. A. Hantzsch, *Berichte*, **60B,** 2537 (1927); *Chem. Abstr.*, **22,** 1158.
351. K. Auwers and E. Waltraut, *Z. Phys. Chem. Frankfurt*, **122,** 217 (1926); *Chem. Abstr.*, **20,** 3385.
352. W. H. Mills and J. L. Smith, *J. Chem. Soc.*, **121,** 2724 (1922); *Chem. Abstr.*, **17,** 1024.
353. T. B. Johnson and E. Gatewood, *J. Amer. Chem. Soc.*, **51,** 1815 (1929); *Chem. Abstr.*, **23,** 3470.
354. G. M. Dyson, R. F. Hunter, J. W. T. Jones, and E. R. Styles, *J. Indian Chem. Soc.*, **8,** 147 (1931); *Chem. Abstr.*, **25,** 4880.

XVII. References

355. J. L. B. Smith and R. H. Sapiro, *Trans. Roy. Soc. S. Africa*, **18,** 229 (1929); *Chem. Abstr.*, **24,** 2130.
356. D. W. MacCorquodale and T. B. Johnson, *Rec. Trav. Chim.*, **51,** 483 (1932); *Chem. Abstr.*, **26,** 3794.
357. C. M. Suter and T. B. Johnson, *J. Amer. Chem. Soc.*, **52,** 1585 (1930); *Chem. Abstr.*, **24,** 2460.
358. J. P. Wetherill and R. M. Hann, *J. Amer. Chem. Soc.*, **56,** 970 (1934); *Chem. Abstr.*, **28,** 3072.
359. M. L. Tomlinson, *J. Chem. Soc.*, 1607 (1936); *Chem. Abstr.*, **31,** 685.
360. W. H. Schopfer,*C.R. Acad., Sci.*, **205,** 445 (1937); *Chem. Abstr.*, **31,** 8819.
361. J. P. Wetherill and R. M. Hann, *J. Amer. Chem. Soc.*, **57,** 1752 (1935); *Chem. Abstr.*, **29,** 7328.
362. L. R. Cerecedo and J. G. Tolpin, *J. Amer. Chem. Soc.*, **59,** 1660 (1937); *Chem. Abstr.*, **31,** 7873.
363. E. R. Buchman, *J. Amer. Chem. Soc.*, **58,** 1803 (1936); *Chem. Abstr.*, **30,** 7572.
364. B. C. Knight and H. MacIlwain, *Biochem. J.*, **32,** 1241 (1938); *Chem. Abstr.*, **32,** 9170.
365. W. J. Robbins and F. Kavanagh, *Proc. Nat. Acad. Sci., U.S.A.*, **24,** 145 (1938); *Chem. Abstr.*, **32,** 5022.
366. German Patent No. 663 305; *Chem. Abstr.*, **32,** 9099.
367. A. Abd el Salaam and P. C. Leong, *Biochem. J.*, **32,** 958 (1938); *Chem. Abstr.*, **32,** 9160.
368. A. G. Pesina, *J. Gen. Chem. U.S.S.R.*, **9,** 804 (1939); *Chem. Abstr.*, **34,** 425.
369. C. R. Harington and R. C. G. Moggridge, *J. Chem. Soc.*, 443 (1939); *Chem. Abstr.*, **33,** 4242.
370. E. R. Buchman and E. M. Richardson, *J. Amer. Chem. Soc.*, **61,** 891 (1939); *Chem. Abstr.*, **33,** 4242.
371. G. O. Doak, H. G. Steinmam, and H. Eagle, *J. Amer. Chem. Soc.*, **62,** 3012 (1940); *Chem. Abstr.*, **35,** 86.
372. A. Schoberl, and M. Stock, *Chem. Ber.*, **73B,** 1240 (1940); *Chem. Abstr.*, **35,** 3636.
373. E. Ochiai and O. Suzuki, *J. Pharm. Soc. Japan*, **60,** 353 (1940); *Chem. Abstr.*, **34,** 7289.
374. F. Schultz, *Z. Physiol. Chem.*, **265,** 113 (1940); *Chem. Abstr.*, **35,** 1839.
375. D. Price, E. L. May, and F. D. Pickel, *J. Amer. Chem. Soc.*, **62,** 2818 (1940); *Chem. Abstr.*, **34,** 7903.
376. C. R. Harington and R. C. G. Moggridge, *Biochem. J.*, **34,** 685 (1940); *Chem. Abstr.*, **34,** 5447.
377. Y. F. Chi and S. Y. Tshin, *J. Amer. Chem. Soc.*, **64,** 90 (1942); *Chem. Abstr.*, **36,** 1322.
378. R. Manzoni Ansidei and G. Travagli, *Gazz. Chim. Ital.*, **71,** 680 (1941); *Chem. Abstr.*, **36,** 3021.
379. J. McLean and G. D. Muir, *J. Chem. Soc.*, 383 (1942); *Chem. Abstr.*, **36,** 5815.
380. H. Erlenmeyer and M. Simon, *Helv. Chim. Acta*, **25,** 362 (1942); *Chem. Abstr.*, **36,** 5816.
381. E. H. Huntress and K. Pfister, *J. Amer. Chem. Soc.*, **65,** 1667 (1943); *Chem. Abstr.*, **37,** 6263.
382. M. Kofler and L. Sternbach, *Helv. Chim. Acta*, **24,** 1014 (1941); *Chem. Abstr.*, **36,** 2556.
383. U.S. Patent No. 2 269 472; *Chem. Abstr.*, **36,** 3068.
384. P. Baumgarten, A. Dornow, K. Gutschmidt, and H. Krehl, *Berichte*, **75B,** 442 (1942); *Chem. Abstr.*, **37,** 3091.
385. J. Weijlard, *J. Amer. Chem. Soc.*, **64,** 2279 (1942); *Chem. Abstr.*, **37,** 379.

386. H. Beyer, *Chem. Ber.*, **74B**, 1100 (1941); *Chem. Abstr.*, **36**, 4819.
387. R. Ammon, *Deutsch Zahnarztl. Wochschr.*, **45**, 85 (1942); *Chem. Abstr.*, **37**, 1742.
388. Y. M. Slobodin and E. E. Hel'ms, *Compte Rend. Acad. Sci. U.R.S.S.* (*in English*), **39**, 145 (1943); *Chem. Abstr.*, **38**, 1239.
389. W. Huber, *J. Amer. Chem. Soc.*, **66**, 876 (1944); *Chem. Abstr.*, **38**, 3983.
390. H. Coates, A. H. Cook, I. H. Heilbron, and F. B. Lewis, *J. Chem. Soc.*, 419 (1943); *Chem. Abstr.*, **38**, 106.
391. H. Erlenmeyer and G. Bischoff, *Helv. Chim. Acta*, **27**, 412 (1944); *Chem. Abstr.*, **38**, 6275.
392. E. R. Buchman and E. M. Richardson, *J. Amer. Chem. Soc.*, **67**, 395 (1945); *Chem. Abstr.*, **39**, 1871.
393. H. Erlenmeyer and C. J. Morel, *Helv. Chim. Acta*, **28**, 362 (1945); *Chem. Abstr.*, **40**, 1501.
394. K. Ganapathi and A. Venkataraman, *Proc. Indian Acad. Sci.*, **21A**, 34 (1945); *Chem. Abstr.*, **39**, 3524.
395. A. H. Cook, I. H. Heilbron, and K. J. Reed, *J. Chem. Soc.*, 182 (1945); *Chem. Abstr.*, **39**, 2978.
396. B. Beilenson, F. M. Hamer, and R. J. Rathbone, *J. Chem. Soc.*, 222 (1945); *Chem. Abstr.*, **39**, 3290.
397. E. B. Knott, *J. Chem. Soc.*, 455 (1945); *Chem. Abstr.*, **39**, 4869.
398. U.S. Patent No. 2 366 189; *Chem. Abstr.*, **39**, 1737.
399. E. M. Gibbs and F. A. Robinson, *J. Chem. Soc.*, 925 (1945); *Chem. Abstr.*, **40**, 1830.
400. H. Lehr, W. Guex, and H. Erlenmeyer, *Helv. Chim. Acta*, **27**, 970 (1944); *Chem. Abstr.*, **39**, 1641.
401. H. Lehr, H. J. Micheels, and H. Erlenmeyer, *Helv. Chim. Acta*, **28**, 165 (1945); *Chem. Abstr.*, **40**, 1500.
402. P. Karrer and J. Schukri, *Helv. Chim. Acta*, **28**, 820 (1945); *Chem. Abstr.*, **40**, 1501.
403. E. R. H. Jones, F. A. Robinson, and M. N. Strachan, *J. Chem. Soc.*, 87 (1946); *Chem. Abstr.*, **40**, 3441.
404. H. Erlenmeyer and P. Schmidt, *Helv. Chim. Acta*, **29**, 1957 (1946); *Chem. Abstr.*, **41**, 1667.
405. M. Q. Doja and J. C. Banerjee, *J. Indian Chem. Soc.*, **23**, 217 (1946); *Chem. Abstr.*, **41**, 2901.
406. H. Erlenmeyer and G. Bischoff, *Helv. Chim. Acta*, **29**, 280 (1946); *Chem. Abstr.*, **40**, 2447.
407. H. Lehr, W. Guex, and H. Erlenmeyer, *Helv. Chim. Acta*, **28**, 1281 (1945); *Chem. Abstr.*, **40**, 2146.
408. P. Ruggli, A. Von-Wartburg, and H. Erlenmeyer, *Helv. Chim. Acta*, **30**, 348 (1947); *Chem. Abstr.*, **41**, 3096.
409. H. Erlenmeyer and K. Degen, *Helv. Chim. Acta*, **29**, 1080 (1946); *Chem. Abstr.*, **41**, 448.
410. H. Erlenmeyer and K. Degen, *Helv. Chim. Acta*, **30**, 592 (1947); *Chem. Abstr.*, **41**, 4487.
411. C. T. Bahner, D. Pickens, and B. D. Bales, *J. Amer. Chem. Soc.*, **70**, 1652 (1948); *Chem. Abstr.*, **42**, 5450.
412. H. Gregory and L. F. Wiggins, *J. Chem. Soc.*, 1400 (1947); *Chem. Abstr.*, **42**, 1262.
413. E. B. Knott, *J. Chem. Soc.*, 1656 (1947); *Chem. Abstr.*, **42**, 2969.
414. A. A. Goldberg and W. Kelley, *J. Chem. Soc.*, 1372 (1947); *Chem. Abstr.*, **42**, 1263.
415. A. Goldberg and W. Kelly, *J. Chem. Soc.*, 1372 (1947); *Chem. Abstr.*, **42**, 1263.
416. J. Metzger and A. Pullman, *Bull. Soc. Chim. France*, 1166 (1949); *Chem. Abstr.*, **43**, 5777.

XVII. References

417. H. Erlènmeyer, H. Baumann, and E. Sorkin, *Helv. Chim. Acta*, **31**, 1978 (1948); *Chem. Abstr.*, **43**, 3820.
418. A. Pullman and J. Metzger, *Bull. Soc. Chim. France*, 1021 (1948); *Chem. Abstr.*, **43**, 2511.
419. D. G. Markees and A. Burger, *J. Amer. Chem. Soc.*, **70**, 3329 (1948); *Chem. Abstr.*, **43**, 635.
420. D. G. Markees and A. Burger, *J. Amer. Chem. Soc.*, **71**, 2031 (1949); *Chem. Abstr.*, **43**, 9042.
421. R. G. Jones, Q. F. Soper, O. K. Behrens, and J. W. Corse, *J. Amer. Chem. Soc.*, **70**, 2843 (1948); *Chem. Abstr.*, **43**, 3362.
422. T. Ekstrand, *Acta Chem. Scand.*, **2**, 727 (1948); *Chem. Abstr.*, **43**, 4263.
423. K. Ganapathi and A. Venkataraman, *Proc. Indian Acad. Sci.*, **28A**, 556 (1948); *Chem. Abstr.*, **43**, 6180.
424. H. Erlenmeyer, J. Junod, W. Guex, and M. Erne, *Helv. Phys. Acta*, **31**, 1342 (1948); *Chem. Abstr.*, **43**, 5021.
425. British Patent No. 593 024; *Chem. Abstr.*, **43**, 8401.
426. British Patent No. 593 024; *Chem. Abstr.*, **43**, 5048.
427. Swiss Patent No. 238 517; *Chem. Abstr.*, **43**, 5807.
428. U.S. Patent No. 2 509 453; *Chem. Abstr.*, **44**, 7885.
429. British Patent No. 630 671; *Chem. Abstr.*, **44**, 4042.
430. A. H. Cook and J. A. Elvidge, *J. Chem. Soc.*, 2362 (1949); *Chem. Abstr.*, **44**, 1488.
431. M. Aeberli and H. Erlenmeyer, *Helv. Chim. Acta*, **33**, 503 (1950); *Chem. Abstr.*, **44**, 8351.
432. C. Djerassi, R. Mizzoni, and C. Scholz, *J. Org. Chem.*, **15**, 700 (1950); *Chem. Abstr.*, **44**, 9412.
433. U.S. Patent No. 2 481 674; *Chem. Abstr.*, **44**, 7173.
434. Japanese Patent No. 177 625; *Chem. Abstr.*, **45**, 7153.
435. M. Ohta and K. Sato, *J. Pharm. Soc. Japan*, **67**, 227 (1947); *Chem. Abstr.*, **45**, 9533.
436. J. Metzger and B. Koether, *Ann. Univ. Saraviensis*, **1**, 23 (1952); *Chem. Abstr.*, **47**, 10524.
437. M. Ohta, *J. Pharm. Soc. Japan*, **67**, 199 (1947); *Chem. Abstr.*, **45**, 9533.
438. P. Aubert and E. Knott, *Nature*, **166**, 1039 (1950); *Chem. Abstr.*, **45**, 8006.
439. A. A. Rosen, *J. Amer. Chem. Soc.*, **74**, 2994 (1952); *Chem. Abstr.*, **47**, 9963.
440. J. Metzger and B. Koether, *Ann. Univ. Saraviensis*, **1**, 151 (1952); *Chem. Abstr.*, **47**, 10524.
441. M. Erne, F. Ramirez, and A. Burger, *Helv. Chim. Acta*, **34**, 143 (1951); *Chem. Abstr.*, **45**, 7565.
442. F. C. Brown, H. Erlenmeyer, and E. Sorkin, *Helv. Chim. Acta*, **34**, 1654 (1951); *Chem. Abstr.*, **46**, 8099.
443. W. Voegtli, E. Sorkin, and H. Erlenmeyer, *Helv. Chim. Acta*, **33**, 1297 (1950); *Chem. Abstr.*, **45**, 2474.
444. M. Ohta and K. Kato, *J. Pharm. Soc. Japan*, **67**, 136 (1947); *Chem. Abstr.*, **45**, 9532.
445. R. D. Beaty and W. K. R. Musgrave, *J. Chem. Soc. A*, 875 (1952); *Chem. Abstr.*, **46**, 9560.
446. D. Schoene, *J. Amer. Chem. Soc.*, **73**, 1970 (1951); *Chem. Abstr.*, **46**, 972.
447. U.S. Patent No. 2 591 506; *Chem. Abstr.*, **46**, 11248.
448. R. Dahlbom, *Acta Chem. Scand.*, **4**, 744 (1950); *Chem. Abstr.*, **45**, 2936.
449. Japanese Patent No. 178 630; *Chem. Abstr.*, **46**, 1593.
450. U.S. Patent No. 2 580 476; *Chem. Abstr.*, **46**, 7126.
451. Y. Takata, *J. Pharm. Soc. Japan*, **72**, 588 (1952); *Chem. Abstr.*, **47**, 2751.
452. R. Dahlbom, *Acta Chem. Scand.*, **5**, 690 (1951); *Chem. Abstr.*, **46**, 430.

453. F. N. Hayes, L. C. King, and D. E. Peterson, *J. Amer. Chem. Soc.*, **74,** 1106 (1952); *Chem. Abstr.*, **47,** 12355.
454. R. P. Kurkjy and E. V. Brown, *J. Amer. Chem. Soc.*, **74,** 5778 (1952); *Chem. Abstr.*, **49,** 1013.
455. K. Miyatake and T. Yoshikawa, *J. Pharm. Soc. Japan,* **75,** 1054 (1955); *Chem. Abstr.*, **50,** 5633.
456. J. D'Amico, *J. Amer. Chem. Soc.*, **75,** 681 (1953); *Chem. Abstr.*, **48,** 2045.
457. M. Erne, *Helv. Chim. Acta,* **36,** 138 (1953); *Chem. Abstr.*, **48,** 2688.
458. M. Tomita, H. Kumaoka, and M. Takase, *J. Pharm. Soc. Japan,* **74,** 850 (1954); *Chem. Abstr.*, **49,** 9623.
459. S. G. Fridman, *Zh. Obshchei Khim.*, **25,** 970 (1955); *Chem. Abstr.*, **50,** 3410.
460. I. T. Strukov, *Dokl. Akad. Nauk SSSR,* **88,** 483 (1953); *Chem. Abstr.*, **48,** 2686.
461. K. Miyatake, G. Ohta, G. Ouchi, and S. Ichimura, *J. Pharm. Soc. Japan,* **75,** 1060 (1955); *Chem. Abstr.*, **50,** 5634.
462. B. M. Mikhailov and V. P. Bronovitskaia, *Zh. Obshchei Khim,* **26,** 66 (1956); *Chem. Abstr.*, **50,** 13874.
463. B. M. Mikhailov and I. K. Platova, *Zh. Obshchei Khim.*, **26,** 491 (1956); *Chem. Abstr.*, **50,** 13875.
464. R. Menasse, G. Klein, and H. Erlenmeyer, *Helv. Chim. Acta,* **38,** 1289 (1955); *Chem. Abstr.*, **50,** 12984.
465. S. Yoshida, M. Nagawa, and M. Kataoka, *Takamine Kenkyujo Nempo,* **12,** 48 (1960); *Chem. Abstr.*, **55,** 6487.
466. J. Metzger and J. Beraud, *C.R. Acad Sci.*, **242,** 2362 (1956); *Chem. Abstr.*, **51,** 2741.
467. E. D. Sych, *Ukr. Khim. Zh.*, **22,** 217 (1956); *Chem. Abstr.*, **51,** 372.
468. J. Gotze and O. Riester, *Mitt. Forschungslab. Agfa Leverkusen Munchen,* **1,** 56 (1955); *Chem. Abstr.*, **51,** 7354.
469. A. Lawson and C. E. Searle, *J. Chem. Soc.*, 1556 (1957); *Chem. Abstr.*, **51,** 12078.
470. G. M. Badger and N. Kowanko, *J. Chem. Soc.*, 1652 (1957); *Chem. Abstr.*, **51,** 13849.
471. R. Charonnat, P. Lechat, and J. Chareton, *Therapie,* **11,** 261 (1956); *Chem. Abstr.*, **51,** 9936.
472. H. Makayama, *Vitamins Kyoto,* **11,** 20 (1956); *Chem. Abstr.*, **51,** 18091.
473. J. M. Waisvisz, M. G. Vanderhoeven, and B. Nijenhuis, *J. Amer. Chem. Soc.*, **79,** 4524 (1957); *Chem. Abstr.*, **52,** 357.
474. German Patent No. 927 527; *Chem. Abstr.*, **52,** 939.
475. Y. K. Yur'ev and I. G. Zhukova, *Zh. Obshchei Khim.*, **28,** 7 (1958); *Chem. Abstr.*, **52,** 11821.
476. H. Beyer, U. Hess, and W. Liebenow, *Chem. Ber.*, **90,** 2372 (1957); *Chem. Abstr.*, **52,** 10057.
477. F. S. Babichev and P. N. Taran, *Nauk, Zapiski. Kiiv. Univ.*, **14,** 125 (1955); *Chem. Abstr.*, **52,** 20131.
478. Japanese Patent No. 5718 ('56); *Chem. Abstr.*, **52,** 11953.
479. H. Beyer, G. Berg, and D. Behrens, *Chem. Ber.*, **90,** 2080 (1957); *Chem. Abstr.*, **53,** 1310.
480. R. Charonnat, P. Lechat, J. Chareton, and P. Volat, *Therapie,* **12,** 954 (1957); *Chem. Abstr.*, **53,** 10554.
481. Io-Fon Chi and Shi-Yun Zin, *Zh. Obshchei Khim.*, **28,** 1492 (1958); *Chem. Abstr.*, **53,** 1312.
482. E. D. Sych, *Ukr. Khim. Zh.*, **25,** 344 (1959); *Chem. Abstr.*, **54,** 5619.
483. U.S. Patent No. 2 904 540; *Chem. Abstr.*, **54,** 2384.
484. U.S. Patent No. 2 900 299; *Chem. Abstr.*, **54,** 2361.

XVII. References

485. E. J. Cragoe, A. M. Pietruszkiewicz, and C. M. Robb, *J. Org. Chem.*, **23,** 971 (1958); *Chem. Abstr.*, **54,** 4372.
486. U.S. Patent No. 2 916 487; *Chem. Abstr.*, **54,** 8386.
487. P. Bassignana and M. Gandino, *Compt. Rend. Congr. Intern. Chim. Ind.*, 31 ieme Liege, **2,** 518 (1958); *Chem. Abstr.*, **54,** 14232.
488. H. Bredereck, R. Gompper, and F. Reich, *Chem. Ber.*, **93,** 723 (1960); *Chem. Abstr.*, **54,** 15360.
489. V. N. Kerr, F. N. Hayes, D. G. Ott, and E. Hansbury, *J. Org. Chem.*, **24,** 1861 (1959); *Chem. Abstr.*, **54,** 9891.
490. E. D. Sych, *Ukr. Khim. Zh.*, **25,** 767 (1958); *Chem. Abstr.*, **54,** 13144.
491. H. Laborit, R. Coirault, R. Damasio, R. Gaujard, G. Laborit, and F. Fabrizy, *Anesthes. Analges.*, **14,** 384 (1957); *Chem. Abstr.*, **54,** 11271.
492. V. M. Zubarovskii, *Zh. Obshchei Khim.*, **29,** 2018 (1959); *Chem. Abstr.*, **54,** 8792.
493. G. Travagli and G. Mazzoli, *Studi Urbinati, Fac. Farm.*, **30,** 101 (1956); *Chem. Abstr.*, **54,** 24661.
494. T. P. Sycheva, I. V. Lebedeva, and M. N. Shchukina, *Zh. Vzesoyuz. Khim. Obshch. Im. D. T. Mendeleeva*, **5,** 234 (1960); *Chem. Abstr.*, **54,** 21049.
495. H. Bredereck, R. Gompper, and F. Reich, *Chem. Ber.*, **93,** 1389 (1960); *Chem. Abstr.*, **54,** 21105.
496. F. Holmes and F. Jones, *J. Chem. Soc.*, 2398 (1960); *Chem. Abstr.*, **54,** 21080.
497. P Lechat and J. Chareton, *Proc. Intern. Congr. Neuro Pharm.*, 1st Rome, 344 (1958); *Chem. Abstr.*, **54,** 17726.
498. F. Asinger, M. Thiel, and H. G. Hauthal, *Annalen*, **634,** 131 (1959); *Chem. Abstr.*, **55,** 2616.
499. Y. Imai, J. Suzuoki, and A. Kobata, *J. Biochem. Japan*, **48,** 341 (1960); *Chem. Abstr.*, **55,** 2884.
500. S. Ogawa and H. Nishimura, *Bitamin*, **14,** 423 (1958); *Chem. Abstr.*, **55,** 3010.
501. R. Didier and J. Metzger, *C.R. Acad. Sci.*, **252,** 1619 (1961); *Chem. Abstr.*, **55,** 16516.
502. M. Thiel, F. Asinger, W. Schafer, and H. G. Hauthal, *Annalen*, **638,** 174 (1960); *Chem. Abstr.*, **55,** 15464.
503. A. Schellenberger and K. Winter, *Z. Physiol. Chem.*, **322,** 164 (1960); *Chem. Abstr.*, **55,** 15465.
504. I. A. Rubtsov, M. V. Balyakina, and E. S. Zhdanovich, *Tr. Vsesoyuz. Nauch. Issledovatel. Vitamin. Inst.*, **6,** 27 (1959); *Chem. Abstr.*, **55,** 12388.
505. J. Vitry, *C.R. Acad. Sci.*, **250,** 139 (1960); *Chem. Abstr.*, **55,** 22288.
506. I. M. Lisnyanskii and E. S. Zhdanovich, *Tr. Vsesoyuz. Nauch. Issledovatel. Vitamin. Inst.*, **6,** 28 (1959); *Chem. Abstr.*, **55,** 19903.
507. H. A. Braun, H. Kuhne, and B. Prijs, *Helv. Chim. Acta*, **43,** 659 (1960); *Chem. Abstr.*, **55,** 25917.
508. B. M. Dean, M. P. V. Mijovic, and J. Walker, *J. Chem. Soc.*, 3394 (1961); *Chem. Abstr.*, **56,** 14253.
509. S. L. Gusinskaya, T. P. Adylova, and D. M. Elenskaya, *Dokl. Akad. Nauk Uzbeck S.S.R.*, **29** (1961); *Chem. Abstr.*, **56,** 11877.
510. T. Takahashi and M. Hayami, *Yakugaku Zasshi*, **81,** 1419 (1961); *Chem. Abstr.*, **56,** 11713.
511. T. Takahashi and M. Hayami, *Yakugaku Zasshi*, **81,** 1426 (1961); *Chem. Abstr.*, **56,** 11713.
512. B. Tornetta, *Ann. Chim. Italy*, **51,** 930 (1961); *Chem. Abstr.*, **56,** 7298.
513. P. Brookes, R. J. Clark, B. Majhofer, M. P. V. Mijovic, and J. Walker, *J. Chem. Soc.*, 925 (1960); *Chem. Abstr.*, **56,** 14390.

514. Z. El-Hewehi and F. Runge, *J. Prakt. Chem.*, **16,** 297 (1962); *Chem. Abstr.*, **58,** 5671.
515. E. D. Sych and E. D. Smaznaya-Il'ina, *Zh. Obshchei Khim.*, **32,** 984 (1962); *Chem. Abstr.*, **58,** 6953.
516. M. Poite and J. Metzger, *Bull. Soc. Chim. France*, 2078 (1962); *Chem. Abstr.*, **58,** 11340.
517. J. Beraud and J. Metzger, *Bull. Soc. Chim. France*, 2072 (1962); *Chem. Abstr.*, **58,** 13930.
518. W. Laessig, *Ber. Bunsenges. Phys. Chem.*, **95,** 2792 (1962); *Chem. Abstr.*, **58,** 9256.
519. E. D. Sych and L. P. Umanskaya, *Zh. Obshchei Khim.*, **33,** 80 (1963); *Chem. Abstr.*, **59,** 591.
520. M. Muehlstaedt and E. Bordes, *J. Prakt. Chem.*, **20,** 285 (1963); *Chem. Abstr.*, **59,** 10019.
521. French Patent No. 1 310 062; *Chem. Abstr.*, **58,** 13963.
522. Japanese Patent No. 13 476 ('62); *Chem. Abstr.*, **59,** 10061.
523. Japanese Patent No. 13 477 ('62); *Chem. Abstr.*, **59,** 10061.
524. Belgian Patent No. 623 148; *Chem. Abstr.*, **60,** 10687.
525. M. Koral, D. Bonis, A. J. Fusco, P. Dougherty, A. Leifer, and J. E. LuValle, *J. Chem. Eng. Data*, **9,** 406 (1964); *Chem. Abstr.*, **61,** 9487.
526. D. H. Reid, F. S. Skelton, and W. Bonthrone, *Tetrahedron Lett.*, 27 (1964); *Chem. Abstr.*, **61,** 9486.
527. W. Firshein and W. Braun, *J. Bacteriol.*, **87,** 1245 (1964); *Chem. Abstr.*, **61,** 2211.
528. F. Asinger, W. Schaefer, M. Baumann, and H. Roemgens, *Annalen*, **672,** 103 (1964); *Chem. Abstr.*, **61,** 1854.
529. E. M. Peresleni and Y. N. Sheinker, *Zh. Fiz. Khim.*, **38,** 2152 (1964); *Chem. Abstr.*, **61,** 14499.
530. A. Silberg, E. Hamburg, Z. Frenkel, and L. Cormos, *Rev. Roumaine Chim.*, **9,** 215 (1964); *Chem. Abstr.*, **61,** 14045.
531. A. Silberg and A. Benko, *Berichte*, **97,** 1915 (1964); *Chem. Abstr.*, **61,** 8290.
532. T. Pyl, H. Gille, and D. Nusch, *Annalen*, **679,** 139 (1964); *Chem. Abstr.*, **62,** 5262.
533. A. Silberg, Z. Frenkel, and L. Cormos, *Stud. Univ. Babes Bolyai, Ser. Chem.*, **8,** 135 (1963); *Chem. Abstr.*, **61,** 16060.
534. A. Silberg, Z. Frenkel, and L. Cormos, *Stud. Univ. Babes Bolyai, Ser. Chem.*, **7,** 23 (1962); *Chem. Abstr.*, **62,** 1643.
535. A. Silberg and A. Benko, *Berichte*, **97,** 3045 (1964); *Chem. Abstr.*, **62,** 1642.
536. H. Behringer and D. Weber, *Annalen*, **682,** 201 (1965); *Chem. Abstr.*, **62,** 14646.
537. R. M. Archeson, M. W. Foxton, and G. R. Miller, *J. Chem. Soc.*, 3200 (1965); *Chem. Abstr.*, **63,** 4300.
538. J. Okamiya, *Nippon Kagaku Zasshi*, **86,** 315 (1965); *Chem. Abstr.*, **63,** 4123.
539. P. Lechat, G. Streichenberger, A. Boime, and M. Lemeignan, *Ann. Pharm. Franc.*, **23,** 179 (1965); *Chem. Abstr.*, **63,** 7507.
540. East German Patent No. 35 205; *Chem. Abstr.*, **63,** 1791.
541. Japanese Patent No. 10 413 ('65); *Chem. Abstr.*, **63,** 5656.
542. H. Bartels and H. Erlenmeyer, *Helv. Chim. Acta*, **48,** 285 (1965); *Chem. Abstr.*, **63,** 2421.
543. British Patent No. 988 956; *Chem. Abstr.*, **63,** 11568.
544. N. Gerencevic, A. Castek, and M. Prostenik, *Bull. Sci., Conseil Acad. R.S.F., Yougoslavie*, **10,** 34 (1965); *Chem. Abstr.*, **63,** 16326.
545. A. Silberg and Z. Frenkel, *Studia Univ. Babes Bolyai, Ser. Chem.*, **10,** 27 (1965); *Chem. Abstr.*, **63,** 16387.
546. French Patent No. 1 413 905; *Chem. Abstr.*, **64,** 6658H.

547. British Patent No. 1 008 631; *Chem. Abstr.*, **64,** 7578F.
548. Netherlands Appl. No. 6 507 580; *Chem. Abstr.*, **64,** 12698E.
549. M. Mansson and S. Sunner, *Acta Chem. Scand.*, **20,** 845 (1966); *Chem. Abstr.*, **65,** 6385C.
550. Belgian Patent No. 657 236; *Chem. Abstr.*, **64,** 17610A.
551. Netherlands Appl. No. 6 510 389; *Chem. Abstr.*, **65,** 2268B.
552. Netherlands Appl. No. 6 413 477; *Chem. Abstr.*, **65,** 8919C.
553. R. G. Glushkov, *Zh. Obshchei Khim.*, **36,** 948 (1966); *Chem. Abstr.*, **65,** 8891C.
554. Netherlands Appl. No. 6 515 006; *Chem. Abstr.*, **65,** 13721G.
555. G. Pappalardo, B. Tornetta, and G. Scapini, *Farmaco Pavia Ed. Sci.*, **21,** 740 (1966); *Chem. Abstr.*, **66,** 46363.
556. B. Vollmert, *Kunst. Plastics*, **56,** 680 (1966); *Chem. Abstr.*, **66,** 11181.
557. Japanese Patent No. 16 787 ('66); *Chem. Abstr.*, **66,** 10927B.
558. Japanese Patent No. 19 340 ('66); *Chem. Abstr.*, **66,** 3791P.
559. French Patent No. 1 438 566; *Chem. Abstr.*, **65,** 20136F.
560. French Patent No. 1 451 284; *Chem. Abstr.*, **66,** 85785V.
561. U.S. Patent No. 3 299 082; *Chem. Abstr.*, **66,** 85784U.
562. R. Meyer and J. Metzger, *Bull. Soc. Chim. France*, **5,** 1711 (1967); *Chem. Abstr.*, **67,** 57626.
563. S. Geiger, M. Joannic, M. Pesson, H. Techer, E. Legrange, and M. Aurousseau, *Chim. Ther.*, 425 (1966); *Chem. Abstr.*, **67,** 11413.
564. H. Watanabe, S. Kuwata, T. Sakata, and K. Matsumura, *Bull. Chem. Soc. Japan*, **39,** 2473 (1966); *Chem. Abstr.*, **67,** 22130.
565. Netherlands Appl. No. 6 603 614; *Chem. Abstr.*, **67,** 82214T.
566. R. Huisgen, E. Funke, F. C. Schaefer, H. Gotthardt, and E. Brunn, *Tetrahedron Lett.*, **19,** 1809 (1967); *Chem. Abstr.*, **67,** 82142.
567. R. Vivaldi, H. J. M. Dou, and J. Metzger, *C.R. Acad. Sci., Ser. C*, **264,** 1862 (1967); *Chem. Abstr.*, **67,** 90710.
568. I. T. Kashkaval, *Farmatsevt Zh. Kiev*, **22,** 11 (1967); *Chem. Abstr.*, **67,** 82151.
569. I. Simiti and H. Demian, *Farmacia Bucharest*, **14,** 735 (1966); *Chem. Abstr.*, **67,** 90711.
570. British Patent No. 1 077 529; *Chem. Abstr.*, **68,** 21927.
571. M. Stoll, P. Dietrich, E. Sundt, and M. Winter, *Ind. Eng. Chem.*, **50,** 2065 (1967); *Chem. Abstr.*, **68,** 12890.
572. W. J. Spillane, P. Hassanaly, and H. J. M. Dou, *C.R. Acad. Sci.*, **283,** 289 (1976).
573. J. Kiss, R. D'Souza, and H. Spiegelberg, *Helv. Chim. Acta*, **51,** 325 (1968); *Chem. Abstr.*, **68,** 78184.
574. French Patent No. 1 476 529; *Chem. Abstr.*, **68,** 87289.
575. J. M. Bonnier, R. Arnaud, and M. Maurey-Mey, *C.R. Acad. Sci. Ser. C*, **267,** 10 (1968); *Chem. Abstr.*, **69,** 64067.
576. U.S.S.R. Patent No. 213 028; *Chem. Abstr.*, **69,** 96699.
577. H. Baganz and J. Rueger, *Chem. Ber.*, **101,** 3872 (1968); *Chem. Abstr.*, **70,** 19965.
578. W. Hampel and I. Muller, *J. Prakt. Chem.*, **38,** 320 (1968); *Chem. Abstr.*, **70,** 37690.
579. Japanese Patent No. 68 22865; *Chem. Abstr.*, **70,** 57890.
580. U.S. Patent No. 3 418 332; *Chem. Abstr.*, **70,** 57827.
581. A. Boucherle, G. Carraz, M. I. Hicter, H. Beriel, and M. Trivin, *Chim. Ther.*, **3,** 360 (1968); *Chem. Abstr.*, **70,** 106424.
582. Japanese Patent No. 69 05228; *Chem. Abstr.*, **71,** 3403.
583. Y. I. Smushkevich, T. M. Babueva, and N. N. Suvorov, *Khim. Geterotsikl. Soedinenii*, 91 (1969); *Chem. Abstr.*, **70,** 115049.

584. G. Kempter, H. Schaefer, and G. Sarodnick, *Z. Chem.*, **9,** 186 (1969); *Chem. Abstr.*, **71,** 30388.
585. B. I. Shapira, A. A. Malina, G. Z. Yakovleva, Z. V. Repina, and I. A. Rubstov, *Khim. Farm. Zh.*, **3,** 46 (1969); *Chem. Abstr.*, **71,** 91369.
586. German Patent No. 1 913 472; *Chem. Abstr.*, **72,** 43659.
587. German Patent No. 1 953 954; *Chem. Abstr.*, **73,** 35358.
588. U.S. Patent No. 3 518 279; *Chem. Abstr.*, **73,** 130990.
589. K. Mukherjee, S. Misra, C. S. Panda, G. B. Behera, and M. K. Rout, *J. Indian Chem. Soc.*, **47,** 323 (1970); *Chem. Abstr.*, **73,** 36528.
590. Japanese Patent No. 70 09 378; *Chem. Abstr.*, **72,** 132714.
591. S. P. Misra and P. K. Jesthi, *Current Sci.*, **39,** 417 (1970); *Chem. Abstr.*, **73,** 120551.
592. British Patent No. 1 189 008; *Chem. Abstr.*, **73,** 26624.
593. German Patent No. 2 003 190; *Chem. Abstr.*, **73,** 98930.
594. V. Ermolaeva and M. Shchukina, *Khim. Geterotsikl. Soedinenii*, **1,** 84 (1967); *Chem. Abstr.*, **67,** 54064.
595. L. J. Altman and S. L. Richkeimer, *Tetrahedron Lett.*, 4709 (1971).
596. J. L. Garraway, *Pesticide Sci.*, **1,** 240 (1970); *Chem. Abstr.*, **74,** 75415.
597. G. Herbertz, T. Metz, H. Reinauer, and W. Staib, *Biochem. Pharmacol.*, **22,** 1541 (1973); *Chem. Abstr.*, **79,** 87311.
598. R. Bonnichsen, R. Hjalm, Y. Marde, M. Moller, and R. Ryhage, *Z. Rechtsmed.*, **73,** 225 (1973); *Chem. Abstr.*, **81,** 20733.
599. H. Schaefer and K. Gewald, *J. Prakt. Chem.*, **316,** 684 (1974); *Chem. Abstr.*, **81,** 169467.
600. R. Cottet, R. Gallo, and J. Metzger, *Bull. Soc. Chim. France*, 4499 (1967); *Chem. Abstr.*, **69,** 2894T.
601. J. II. Bowie, P. F. Donaghue, II. J. Rodda, and B. K. Simons, *Tetrahedron*, **24,** 3965 (1968); *Chem. Abstr.*, **69,** 2357V.
602. A. Friedmann, D. Bouin, and J. Metzger, *Bull. Soc. Chim. France*, **8–9,** 3155 (1970); *Chem. Abstr.*, **74,** 31389.
603. G. Vernin, H. J. M. Dou, and J. Metzger, *C.R. Acad. Sci. Ser. C.* **272,** 272 (1971); *Chem. Abstr.*, **74,** 125546.
604. Japanese Patent No. 71 24 260; *Chem. Abstr.*, **75,** 121432.
605. Japanese Patent No. 71 04 370; *Chem. Abstr.*, **75,** 5888.
606. U. H. Lindberg, G. Bexell, J. Pedersen, and S. Ross, *Acta Pharm. Suecica*, **7,** 423 (1970); *Chem. Abstr.*, **73,** 118624.
607. G. Vernin, H. J. M. Dou, and J. Metzger, *C.R. Acad. Sci. Ser. C.* **271,** 271 (1970); *Chem. Abstr.*, **74,** 75865.
608. Japanese Patent No. 71 24 259; *Chem. Abstr.*, **75,** 121429.
609. C. Ivanov, V. Dryanska, and I. Arnaudova, *Dokl. Bolg. Akad. Nauk*, **22,** 891 (1969); *Chem. Abstr.*, **72,** 31674.
610. R. Laliberte, J. Manson, H. Warwick, and G. Medawar, *Can. J. Chem.*, **46,** 1952 (1968); *Chem. Abstr.*, **69,** 35632J.
611. German Patent No. 1 943 155; *Chem. Abstr.*, **74,** 13140.
612. N. Saldabols, S. Hillers, L. N. Alekseeva, and L. Kruzmetra, *Khim. Farm. Zh.*, **5,** 15 (1971); *Chem. Abstr.*, **76,** 14410Q.
613. German Patent No. 1 929 963; *Chem. Abstr.*, **72,** 90500.
614. German Patent No. 2 053 178; *Chem. Abstr.*, **75,** 63766.
615. U.S. Patent No. 3 558 644; *Chem. Abstr.*, **74,** 87951.
616. British Patent No. 1 180 268; *Chem. Abstr.*, **72,** 132713.
617. French Patent No. 1 491 242; *Chem. Abstr.*, **69,** 35586X.
618. J. L. Garraway, *Pesticide Sci.*, **1,** 240 (1970); *Chem. Abstr.*, **74,** 75415.
619. British Patent No. 1 183 850; *Chem. Abstr.*, **73,** 33905.

620. P. Lechat, M. Freyss-Beguin, A. Boime, N. Mathieu-Levy, and E. Van Brussel, *Therapie*, **26**, 497 (1971); *Chem. Abstr.*, **75**, 108274.
621. German Patent No. 2 023 425; *Chem. Abstr.*, **74**, 87964.
622. French Patent No. 1 564 081; *Chem. Abstr.*, **73**, 14837.
623. I. Simiti, I. Schwartz, L. Proinov, M. Farkas, and S. Silberg, *J. Electroanal. Chem. Interfacial Electrochem.*, **30**, 517 (1971); *Chem. Abstr.*, **75**, 29304.
624. H. Teubner, W. Weuffen, and H. Hoeppe, *Arch. Exp. Veterinaermed*, **24**, 849 (1970); *Chem. Abstr.*, **74**, 108556.
625. G. Herbertz and H. Reinauer, *Naunyn Schmiedebergs Arch. Pharmakol.*, **270**, 192 (1971); *Chem. Abstr.*, **75**, 107943.
626. German Patent No. 1 917 432; *Chem. Abstr.*, **72**, 43655.
627. U.S. Patent No. 3 539 585; *Chem. Abstr.*, **74**, 53775.
628. Japanese Patent No. 70 36 908; *Chem. Abstr.*, **74**, 88044.
629. German Patent No. 1 949 013; *Chem. Abstr.*, **74**, 139880.
630. J. P. Ehrhardt, J. C. Wissocq, and P. Niaussat, *Compt. Rend. Soc. Biol.*, **164**, 1984 (1970); *Chem. Abstr.*, **75**, 95925.
631. H. Mitsuda, T. Tanaka, Y. Takii, and F. Kawai, *J. Vitaminol.*, **17**, 89 (1971); *Chem. Abstr.*, **75**, 106126.
632. R. A. Neal, *Methods Enzymol.*, **18**, 133 (1970); *Chem. Abstr.*, **75**, 2398.
633. W. H. Amos and R. A. Neal, *J. Biol. Chem.*, **245**, 5463 (1970); *Chem. Abstr.*, **74**, 1786.
634. N. L. Couse, D. J. Cummings, V. A. Chapman, and S. S. Delong, *Virology*, **42**, 590 (1970); *Chem. Abstr.*, **74**, 39709.
635. A. Carere and S. Russi, *G. Bot. Ital.*, **104**, 349 (1970); *Chem. Abstr.*, **75**, 31655.
636. K. Tengo and K. Murata, *Bitamin*, **43**, 172 (1971); *Chem. Abstr.*, **75**, 30454.
637. German Patent No. 2 034 961; *Chem. Abstr.*, **74**, 141767.
638. French Patent No. 1 491 244; *Chem. Abstr.*, **69**, 43912R.
639. German Patent No. 2 110 396; *Chem. Abstr.*, **76**, 14541H.
640. U.S. Patent No. 3 660 112; *Chem. Abstr.*, **77**, 33120G.
641. E. J. Mulders, *Z. Lebensm. Untersuch.*, **152**, 193 (1973); *Chem. Abstr.*, **79**, 90591.
642. M. Y. Kornilov, E. D. Sych, and L. I. Smeshko, *Ukr. Khim. Zhur.*, **39**, 353 (1973); *Chem. Abstr.*, **79**, 17689.
643. German Patent No. 2 262 471; *Chem. Abstr.*, **79**, 78785.
644. German Patent No. 2 253 774; *Chem. Abstr.*, **79**, 32395.
645. I. T. Depeshko, *Strukt. Mekh. Deistviya Fiziol. Akativ Veshchestv.*, **67** (1972); *Chem. Abstr.*, **78**, 123477.
646. U.S. Patent No. 3 740 400; *Chem. Abstr.*, **79**, 42505.
647. A. I. Meyers and G. N. Knaus, *J. Amer. Chem. Soc.*, **95**, 3408 (1973); *Chem. Abstr.*, **79**, 18789.
648. L. J. Altman and S. L. Richheimer, *Tetrahedron Lett.*, 4709 (1971); *Chem. Abstr.*, **76**, 33896U.
649. V. A. Krasovskii and S. I. Burmistrov, *Khim. Geterotsikl. Soedinenii*, **168** (1971); *Chem. Abstr.*, **78**, 71979.
650. L. Lewicki, *Pr. Nauk. Inst. Chem. Teknol. Nafty. Wegla. Politech. Wroclaw*, **10**, 157 (1972); *Chem. Abstr.*, **78**, 74522.
651. A. Silberg, V. Farcasan, A. Donea, and A. Tomoaia, *Stud. Univ. Babes Bolyai, Ser. Chem.*, **18**, 59 (1973); *Chem. Abstr.*, **79**, 126375.
652. G. B. Behera, J. N. Kar, R. C. Acharya, and M. K. Rout, *J. Org. Chem.*, **38**, 2164 (1973); *Chem. Abstr.*, **79**, 31173.
653. K. Sota, M. Aida, K. Noda, and A. Hayashi, *Agr. Biol. Chem. Tokyo*, **36**, 2287 (1972); *Chem. Abstr.*, **78**, 111193.
654. British Patent No. 1 317 257; *Chem. Abstr.*, **79**, 66346.

655. German Patent No. 2 162 468; *Chem. Abstr.*, **77,** 126612.
656. French Patent No. 1 601 733; *Chem. Abstr.*, **78,** 58400.
657. German Patent No. 2 162 439; *Chem. Abstr.*, **79,** 127404.
658. German Patent No. 2 038 408; *Chem. Abstr.*, **77,** 7320C.
659. Swedish Patent No. 319 577; *Chem. Abstr.*, **77,** 66190K.
660. S. J. Alcock, *Proc. Eur. Soc. Study. Drug Toxicity*, **12,** 184 (1971); *Chem. Abstr.*, **77,** 14344S.
661. G. Ciaceri, G. Pennisi, and L. Gialdi, *Minerva Med.*, **63,** 2409 (1972); *Chem. Abstr.*, **77,** 121995.
662. I. Simiti and G. Hintz, *Pharmazie*, **27,** 146 (1972); *Chem. Abstr.*, **77,** 5393E.
663. I. Simiti and M. Farkas, *Acta Chim. Acad. Sci. Hung.*, **76,** 107 (1973); *Chem. Abstr.*, **79,** 18628.
664. V. P. Khilya, L. G. Grishko, and V. Szabo, *Khim. Geterotsikl. Soedinenii*, 1321 (1972); *Chem. Abstr.*, **78,** 43338.
665. British Patent No. 1 292 603; *Chem. Abstr.*, **78,** 58408.
666. J. L. Ambrus and C. M. Ambrus, *Res. Commun. Chem. Pathol. Pharmacol.*, **3,** 265 (1972); *Chem. Abstr.*, **77,** 14224C.
667. W. Damerau, D. Schwarz, G. Lassmann, G. Huebner, and A. Schellenberger, *Tetrahedron Lett.*, **17,** 1545 (1973); *Chem. Abstr.*, **79,** 65446.
668. German Patent No. 2 234 675; *Chem. Abstr.*, **78,** 111289.
669. French Patent No. 2 096 994; *Chem. Abstr.*, **77,** 140036.
670. M. O. Kolosova, E. V. Pereverzeva, N. N. Ozeretskovskaya, O. M. Gein, and M. E. Pudel, *Med. Parazitol. Parazit. Bolez.*, **40,** 540 (1971); *Chem. Abstr.*, **76,** 148730.
671. G. Herbertz, T. Metz, H. Reinauer, and W. Staib, *Biochem. Pharmacol.*, **22,** 1541 (1973); *Chem. Abstr.*, **79,** 87311.
672. K. Naber, K. Mueller, and W. Rehbehn, *Deut. Med. Wochschr.*, **98,** 226 (1973); *Chem. Abstr.*, **78,** 106220.
673. H. Lauressergues, F. Boismare, G. Streichenberger, and N. Hugon, *Therapie*, **26,** 741 (1971); *Chem. Abstr.*, **76,** 68149C.
674. K. Naber, D. Maroske, K. H. Bichler, and T. H. Thaut, *Arzneim. Forsch.*, **21,** 1612 (1971); *Chem. Abstr.*, **76,** 68117T.
675. B. K. Dwivedi and R. G. Arnold, *J. Food Sci.*, **37,** 886 (1972); *Chem. Abstr.*, **78,** 56713.
676. B. K. Dwivedi, R. G. Arnold, and L. M. Libbey, *J. Food Sci.*, **37,** 689 (1972); *Chem. Abstr.*, **78,** 28194.
677. B. V. Passet, N. G. Volikova, and K. I. Novitskii, *Khim. Farm. Zh.*, **7,** 29 (1973); *Chem. Abstr.*, **79,** 92080.
678. M. Hatanaka and T. Ishimaru, *J. Med. Chem.*, **16,** 978 (1973); *Chem. Abstr.*, **79,** 133253.
679. U.S. Patent No. 3 668 212; *Chem. Abstr.*, **77,** 88468G.
680. A. F. Oleinik, G. A. Midnikova, and K. Y. Novitskii, *Khim. Geterotsikl. Soedinenii*, **4,** 437 (1973); *Chem. Abstr.*, **79,** 18622.
681. French Patent No. 2 073 427; *Chem. Abstr.*, **77,** 61985M.
682. T. Nguyen Tanh and P. Chabrier, *C.R. Acad. Sci. Ser. C.* **275,** 1125 (1972); *Chem. Abstr.*, **78,** 58329.
683. German Patent No. 2 301 509; *Chem. Abstr.*, **79,** 105267.
684. M. Yu. Kornilov and A. V. Turov, *Ukr. Khim. Zhur.*, **40,** 215 (1974); *Chem. Abstr.*, **80,** 127661.
685. P. Dubs and M. Pesaro, *Synthesis*, 294 (1974); *Chem. Abstr.*, **81,** 25596.
686. M. O. Lozinskii, S. N. Kukota, P. S. Pel'kis, and T. N. Kudrya, *Khim. Geterotsikl. Soedinenii*, 757 (1969); *Chem. Abstr.*, **72,** 43547.
687. R. Viani, J. Bricout, J. P. Marion, F. Mueggler Chavan, D. Reymond, and R. H. Egli,

Helv. Chim. Acta, **52,** 887 (1969); *Chem. Abstr.,* **71,** 11875.
688. German Patent No. 065 078; *Chem. Abstr.,* **71,** 49929.
689. F. Asinger, W. Schaefer, and F. Haaf, *Annalen,* **672,** 134 (1964); *Chem. Abstr.,* **61,** 1857.
690. P. Dubs and M. Pesaro, *Synthesis,* 294 (1974); *Chem. Abstr.,* **81,** 25596.
691. J. M. Bonnier and R. Arnaud, *C.R. Acad. Sci. Ser. C.* **270,** 885 (1970); *Chem. Abstr.,* **72,** 126203.
692. British Patent No. 1 137 529; *Chem. Abstr.,* **71,** 30469.
693. U.S. Patent No. 3 769 413; *Chem. Abstr.,* **80,** 87507.
694. I. Chavdar, V. Dryanska, and I. Arnaudova, *Dokl. Bolg. Akad. Nauk,* **22,** 891 (1969); *Chem. Abstr.,* **72,** 31674.
695. U.S. Patent No. 3 769 412; *Chem. Abstr.,* **80,** 87509.
696. T. S. Malvankar, S. S. Sabnis, M. R. Patel, and C. V. Deliwala, *Bull. Haffkine Inst.,* **2,** 28 (1974); *Chem. Abstr.,* **82,** 57594.
697. British Patent No. 1 337 661; *Chem. Abstr.,* **80,** 82942.
698. H. Schaefer and K. Gewald, *J. Prakt. Chem.,* **316,** 684 (1974); *Chem. Abstr.,* **81,** 169467.
699. French Patent No. 1 561 433; *Chem. Abstr.,* **72,** 43654.
700. I. Simiti and G. Hintz, *Pharmazie,* **29,** 443 (1974); *Chem. Abstr.,* **81,** 105384.
701. British Patent No. 1 328 549; *Chem. Abstr.,* **79,** 137130.
702. F. Rossi, R. Di Nola, M. Cazzola, and R. Spadaro, *Gazz. Med. Ital.,* **133,** 87 (1974); *Chem. Abstr.,* **81,** 130971.
703. K. Brown, D. Cater, J. F. Cavalla, D. Green, R. A. Newberry, and A. B. Wilson, *J. Med. Chem.,* **17,** 1177 (1974); *Chem. Abstr.,* **82,** 38473.
704. S. H. M. Nystrom, E. J. R. Riihimaki, M. Riihimaki, and J. Vainio, *Ircs Libre Compend.,* **2,** 1236 (1974); *Chem. Abstr.,* **82,** 11211.
705. G. Csavassy and Z. A. Gyorfi, *Justus Liebigs Ann. Chem.,* **8,** 1195 (1974); *Chem. Abstr.,* **81,** 169464.
706. M. V. Carter and T. V. Price, *Aust. J. Agr. Res.,* **25,** 105 (1974); *Chem. Abstr.,* **80,** 104741.
707. G. J. Wilson, *N.Z.J. Exp. Agric.,* **2,** 265 (1974); *Chem. Abstr.,* **82,** 52499.
708. T. M. Subramanian, R. Sadasivam, and N. V. Raman, *Indian J. Agric. Sci.,* **43,** 284 (1973); *Chem. Abstr.,* **80,** 104724.
709. J. H. Holmden, M. J. McMullan, G. C. Cairns, and W. H. D. Leaning, *Proc. N.Z. Soc. Anim. Prod.,* **33,** 115 (1973); *Chem. Abstr.,* **80,** 104041.
710. M. A. Hasslinger, *Dstch. Tieraerstl,* **81,** 379 (1974); *Chem. Abstr.,* **82,** 51602.
711. W. Karg, U. Burth, and A. Ramson, *Nachrichtenbl. Pflanzenschutzdienst,* **27,** 169 (1973); *Chem. Abstr.,* **80,** 56425.
712. Japanese Patent No. 73 16 163; *Chem. Abstr.,* **80,** 91409.
713. I. I. Bogolepova and L. D. Semenkov, *Veterinaria,* **11,** 66 (1973); *Chem. Abstr.,* **80,** 103835.
714. K. C. Kates, M. L. Colglazier, F. D. Enzie, I. L. Lindahl, and G. Samuelson, *Proc. Helminthol. Soc. Wash.,* **40,** 87 (1973); *Chem. Abstr.,* **80,** 91262.
715. V. P. Khilya, V. Szabo, L. G. Grishko, D. V. Vikhman, and F. S. Babichev, *Zh. Org. Khim.,* **9,** 2561 (1973); *Chem. Abstr.,* **80,** 70747.
716. U. H. Lindberg, G. Bexell, J. Pedersen, and S. Ross, *Acta Pharm. Suecica,* **7,** 423 (1970); *Chem. Abstr.,* **73,** 118624.
717. E. J. Mulders, *Versl. Landbrowk. Onderz.* N°. 798 (1973); *Chem. Abstr.,* **80,** 13787.
718. E. Occhiai and K. Kokeguti, *J. Pharm. Soc. Japan,* **60,** 271 (1940).
719. S. Kinoshita and T. Yamanishi, *Nippon Nogei Kagaku Zasshi,* **47,** 737 (1973); *Chem. Abstr.,* **80,** 106964.
720. R. G. Buttery, R. M. Seifert, and L. C. Ling, *Agr. Food Sci.,* **23,** 516 (1975).

721. U. H. Lindberg, *Acta Pharm. Suecica*, **8,** 39 (1971).
722. M. Hatanaka and T. Ishimaru, *Bull. Chem. Soc. Japan*, **46,** 3600 (1973).
723. A. Takamizawa and H. Harada, *Chem. Pharm. Bull.*, **22,** 2818 (1974).
724. S. J. Alcock, *Exc. Med. Intern. Long. Sci.*, **220,** 184 (1970).
725. D. M. Foulkes, *J. Pharm. Exp. Ther.*, **172,** 115 (1970); *Chem. Abstr.*, **72,** 109596.
726. W. Hepworth, B. B. Newbould, D. S. Platt, and G. J. Stacey, *Nature*, **221,** 582 (1969).
727. K. Brown, D. P. Cater, J. F. Cavalla, D. Green, R. A. Newberry, and A. B. Wilson, *J. Med. Chem.*, **17,** 1177 (1974); *Chem. Abstr.*, **72,** 109596.
728. U. H. Lindberg, *Acta Pharm. Suecica*, **3,** 161 (1966).
729. P. E. Linnett and J. Walker, *Biochim. Biophys. Acta*, **184,** 381 (1969).
730. Y. Seto, K. Torii, K. Bori, K. Inabata, S. Kuwata, and H. Watanabe, *Bull. Chem. Soc. Japan*, **47,** 151 (1974).
731. I. Steffan and B. Prijs, *Helv. Chim. Acta*, **44,** 1429 (1961).
732. T. Rinderspacher and B. Prijs, *Helv. Chim. Acta*, **43,** 1522 (1960).
733. M. Maeda and M. Kojima, *Tetrahedron Lett.*, 3523 (1973); *Chem. Abstr.*, **80,** 2847.
734. H. J. M. Dou, G. Vernin, and J. Metzger, *Chim. Acta Turc.*, **2,** 82 (1974); *Chem. Abstr.*, **81,** 152078.
735. D. W. Henry, V. H. Brown, M. Cory, J. G. Johanson, and E. Bueding, *J. Med. Chem.*, **16,** 1287 (1973); *Chem. Abstr.*, **80,** 10273.
736. Japanese Patent No. 73 03 382; *Chem. Abstr.*, **80,** 26079.
737. Japanese Patent No. 72 43 233; *Chem. Abstr.*, **80,** 36031.
738. M. R. Chaphekar and L. P. Ghalsasi, *J. Indian Chem. Soc.*, **51,** 564 (1974); *Chem. Abstr.*, **81,** 152074.
739. Swiss Patent No. 540 282; *Chem. Abstr.*, **80,** 3501.
740. F. Humm, R. Romametti. P. Tordo, L. Bouscasse, and R. Phan Tan Luu, *Org. Magn. Reson.*, **5,** 365 (1973); *Chem. Abstr.*, **80,** 2810.
741. Swiss Patent No. 540 285; *Chem. Abstr.*, **80,** 3502.
742. Swiss Patent No. 540 283; *Chem. Abstr.*, **80,** 3503.
743. French Patent No. 2 188 606; *Chem. Abstr.* **81,** 13514.
744. French Patent No. 1 604 908; *Chem. Abstr.*, **81,** 13516.
745. Swiss Patent No. 540 284; *Chem. Abstr.*, **80,** 3504.
746. Swiss Patent No. 540 281; *Chem. Abstr.*, **80,** 3505.
747. A. Silberg, Z. Frenkel, L. Szotyori, and E. Stanciu, *Rev. Roumaine Chim.*, **12,** 905 (1967).
748. H. Yamasaki, H. Sanemori, K. Yamada, and T. Kawasaki, *J. Bacteriol.*, **116,** 1280 (1973); *Chem. Abstr.*, **80,** 34993.
749. V. P. Khilya, V. Szabo, L. G. Grishko, D. V. Vikhman, and F. S. Babichev, *Zh. Org. Khim.*, **9,** 2561 (1973); *Chem. Abstr.*, **80,** 70747.
750. E. Marmo, A. P. Caputi, C. Vacca, and M. Cazzola, *Riv. Farmacol. Ter.*, **5,** 279 (1974); *Chem. Abstr.*, **81,** 130820.
751. G. Kempter, G. Sarodnick, H. Schafer, H. J. Fiebig, and J. Spindler, *Z. Chem.*, **8,** 339 (1968).
752. R. J. C. Kleipool and A. C. Tas, *Riechst Aromen Koeperfluger*, **23,** 326 (1973); *Chem. Abstr.*, **80,** 27158.
753. G. Kempter, H. Schafer, and G. Sarodnick, *Z. Chem.*, **10,** 460 (1970).
754. Japanese Patent No. 73 25 499; *Chem. Abstr.*, **81,** 21795.
755. German Patent No. 2 232 260; *Chem. Abstr.*, **81,** 14722.
756. C. Riou, J. C. Poite, G. Vernin, and J. Metzger, *Tetrahedron*, **30,** 879 (1974); *Chem. Abstr.*, **81,** 90786.
757. Japanese Patent No. 74 00 442; *Chem. Abstr.*, **81,** 22289.

XVII. References

758. T. Avignon, E. J. Vincent, J. Raymond, and M. Chaillet, *J. Mol. Structure*, **21**, 319 (1974); *Chem. Abstr.*, **81**, 36912.
759. Canadian Patent No. 939 356; *Chem. Abstr.*, **81**, 49681.
760. T. Dobrazanski, *Endokrynol. Pol.*, **24**, 171 (1973); *Chem. Abstr.*, **81**, 21875.
761. Japanese Patent No. 73 91 090; *Chem. Abstr.*, **81**, 25657.
762. H. Teubner, W. Weuffen, and H. Hoeppe, *Arch. Exp. Veterinaermed*, **28**, 249 (1974); *Chem. Abstr.*, **81**, 21586.
763. S. Kinoshita and T. Yamanishi, *Nippon Nogei Kagaku Kaishi*, **47**, 737 (1973); *Chem. Abstr.*, **80**, 106964.
764. U.S. Patent No. 3 769 413; *Chem. Abstr.*, **80**, 87507.
765. U.S. Patent No. 3 769 412; *Chem. Abstr.*, **80**, 87509.
766. British Patent No. 1 337 661; *Chem. Abstr.*, **80**, 82942.
767. M. Fischler, E. P. Frisch, and B. Ortengren, *Acta Pharm. Suecica*, **10**, 483 (1973); *Chem. Abstr.*, **80**, 115941.
768. C. P. Heinrich, D. Schmidt, and K. Noack, *Eur. J. Biochem.*, **41**, 555 (1974); *Chem. Abstr.*, **80**, 117593.
769. M. Lemeignan, A. Rodallec, N. Hugon, and P. Lechat, *J. Pharmacol.*, **4**, 535 (1973); *Chem. Abstr.*, **80**, 128116.
770. J. L. Archibald and G. A. Benke, *J. Med. Chem.*, **17**, 736 (1974); *Chem. Abstr.*, **81**, 114405.
771. G. A. Kochetov and A. E. Izotova, *Biokhimiya*, **38**, 954 (1973); *Chem. Abstr.*, **80**, 92579.
772. H. Mitsuda, Y. Takii, and T Tanaka, *Vitamin*, **48**, 19 (1974); *Chem. Abstr.*, **80**, 106030.
773. East German Patent No. 99 787; *Chem. Abstr.*, **80**, 95953.
774. F. Andrasi, J. Borsi, and L. Farkas, *Acta Pharm. Hung.*, **43**, 116 (1973); *Chem. Abstr.*, **80**, 44021.
775. B. K. Dwivedi and R. G. Arnold, *J. Food Sci.*, **21**, 54 (1973); *Chem. Abstr.*, **78**, 70257.
776. B. K. Dwivedi, R. G. Arnold, and L. M. Libbey, *J. Food Sci.*, **38**, 450 (1973); *Chem. Abstr.*, **78**, 134700.
777. A. Takamizawa, and H. Harada, *Chem. Pharm. Bull.*, **21**, 770 (1973); *Chem. Abstr.*, **79**, 66287.
778. A. Takamizawa, *Vitamins Kyoto*, **47**, 1 (1973); *Chem. Abstr.*, **78**, 97514.
779. I. Simiti, M. Coman, and I. Schwartz, *Rev. Roumaine Chim.*, **18**, 685 (1973).
780. I. Schwartz, *Stud. Univ. Babes Bolyai, Ser. Chem.*, **16**, 51 (1971); *Chem. Abstr.*, **76**, 24487.
781. A. Benko and A. Botar, *Stud. Univ. Babes Bolyai, Ser. Chem.*, **18**, 29 (1973); *Chem. Abstr.*, **81**, 105378.
782. H. A. Goodwin and D. W. Mather, *Austr. J. Chem.*, **25**, 715 (1972); *Chem. Abstr.*, **76**, 119009.
783. H. A. Goodwin and R. N. Sylva, *Austr. J. Chem.*, **21**, 2881 (1968); *Chem. Abstr.*, **70**, 33894.
784. E. Konig and G. Ritter, *Chem. Phys.* **1**, 17 (1973).
785. F. Ledl and T. Severin, *Chem. Mikrobiol. Technol. Lebensm*, **2**, 155 (1973); *Chem. Abstr.*, **80**, 119312.
786. J. P. Famaey and M. W. Whitehouse, *Biochem. Pharmacol.*, **22**, 2707 (1973).
787. J. P. Famaey, *Biochem. Pharmacol.*, **22**, 2693 (1973); *Chem. Abstr.*, **80**, 33719.
788. M. Kuhnert-Brandstatter, G. Kramer, and P. D. Lark, *Microchem. J.*, **17**, 739 (1972).
789. C. B. Nerurkar and D. T. Chaudhari, *Bull. Haffkine Inst.*, **3**, 57 (1975); *Chem. Abstr.*, **83**, 193153.
790. D. R. Shekawat, S. S. Sabnis, and C. V. Deliwala, *Bull. Haffkine Inst.*, **1**, 35 (1973).

791. A. Babadjamian, J. Metzger, and M. Chanon, *J. Heterocyclic Chem.*, **12,** 643 (1975).
792. H. Kishi, S. Nishiyama, and E. Hiraoka, *Bitamin,* **42,** 81 (1970); *Chem. Abstr.,* **74,** 29241.
793. M. R. Chaphekar and L. P. Ghalsasi, *J. Indian Chem. Soc.,* **51,** 564 (1974).
794. H. Wallgren, P. Nikander, P. Von Bogulawsky and J. Linkola, *Acta Physiol. Scand.,* **91,** 83 (1974); *Chem. Abstr.,* **81,** 73083.
795. I. Simiti and H. Demian, *Farmacia Bucharest,* **21,** 713 (1973); *Chem. Abstr.,* **81,** 63506.
796. German Patent No. 2 126 811; *Chem. Abstr.,* **78,** 85929.
797. German Patent No. 2 223 648; *Chem. Abstr.,* **78,** 43533.
798. South African Patent No. 6 700 194; *Chem. Abstr.,* **70,** 77975.
799. P. Chauvin, J. Morel, C. Paulmier, and P. Pastour, *C.R. Acad. Sci., Ser. C,* **274,** 1347 (1972); *Chem. Abstr.,* **77,** 48125.
800. H. Erlenmeyer and R. Marbet, *Helv. Chim. Acta,* **29,** 1946 (1946); *Chem. Abstr.,* **41,** 1668.
801. I. G. Leder, *Methods Enzymol.,* **18,** 166 (1970); *Chem. Abstr.,* **75,** 5767.
802. Canadian Patent No. 953 726; *Chem. Abstr.,* **82,** 43421.
803. S. S. Sabnis, *Indian J. Chem.,* **5,** 619 (1967).
804. C. V. Deliwala, J. D. Modi, and S. S. Sabnis, *J. Med. Chem.,* **14,** 450 (1971); *Chem. Abstr.,* **75,** 35860.
805. B. S. Kulkarni, B. S. Fernandez, M. R. Patel, R. A. Bellare, and C. V. Deliwala, *J. Pharm. Sci.,* **58,** 852 (1969).
806. J. D. Modi, S. S. Sabnis, and C. V. Deliwala, *J. Med. Chem.,* **14,** 887 (1971).
807. French Patent No. 824 837.
808. British Patent No. 490 571; *Chem. Abstr.,* **33,** 813.
809. R. G. Buttery and L. C. Ling, *J. Agr. Food Chem.,* **22,** 912 (1974); *Chem. Abstr.,* **81,** 168003.
810. A. Takamizawa, K. Hirai, T. Ishiba, and S. Hayakawa, *J. Vitaminol. Kyoto,* **11,** 204 (1965); *Chem. Abstr.,* **64,** 6718.
811. T. Matsuura and I. Saito, *Bull. Chem. Soc. Japan,* **42,** 2973 (1969); *Chem. Abstr.,* **72,** 39061.
812. K. Masaharu and M. Minoro, *J. Chem. Soc. D,* **6,** 386 (1970); *Chem. Abstr.,* **72,** 120787.
813. M. M. Bursey and R. L. Nunnally, *J. Org. Chem.,* **37,** 3032 (1972); *Chem. Abstr.,* **77,** 139031.
814. J. Roberts, *Chem. Ind.,* 658 (1947); *Chem. Abstr.,* **43,** 6206.
815. I. T. Strukov, *Zh. Obshchei Khim.,* **22,** 1025 (1952); *Chem. Abstr.,* **47,** 8063.
816. R. McKee and J. Thayer, *J. Org. Chem.,* **17,** 1494 (1952); *Chem. Abstr.,* **47,** 9319.
817. E. D. Sych and A. I. Kiprianov, *Zh. Obshchei Khim.,* **31,** 3926 (1961); *Chem. Abstr.* **57,** 9834.
818. K. E. Schulte, J. Reisch, and M. Sommer, *Arch. Pharm.,* **299,** 107 (1966); *Chem. Abstr.,* **64,** 14190.
819. H. Kato, T. Shiba, and K. Yamane, *J. Chem. Soc. D,* **23,** 1592 (1970); *Chem. Abstr.,* **74,** 53617.
820. H. Erlenmeyer and M. Simon, *Helv. Chim. Acta,* **25,** 528 (1942); *Chem. Abstr.,* **36,** 6539.
821. T. R. Govindachari, S. Rajappa, and K. Nagarajan, *Helv. Chim. Acta,* **51,** 2102 (1968); *Chem Abstr.,* **70,** 28861.
822. German Patent No. 1 804 306; *Chem. Abstr.,* **71,** 81351.
823. D. T. Chaudhari, S. S. Sabnis, and C. V. Deliwala, *Bull. Haffkine Inst.,* 81 (1975).

XVII. References

824. U.S. Patent No. 3 438 992; *Chem. Abstr.*, **71,** 61386.
825. British Patent No. 1 147 626; *Chem. Abstr.*, **71,** 49930.
826. British Patent No. 1 180 268; *Chem. Abstr.*, **72,** 132713.
827. V. G. Ermolaeva and M. N. Shchukina, *Zh. Obshchei Khim.*, **33,** 2716 (1963); *Chem. Abstr.*, **60,** 514.
828. U.S.S.R. Patent No. 173 409; *Chem. Abstr.*, **64,** 892.
829. U.S. Patent No. 3 296 250; *Chem. Abstr.*, **66,** 55485.
830. K. Brown and R. A. Newberry, *Tetrahedron Lett.*, 2797 (1969); *Chem. Abstr.*, **71,** 91367.
831. British Patent No. 1 147 731; *Chem. Abstr.*, **70,** 118116.
832. H. I. Vapaatalo and H. Karppanen, *Ag. Actions*, **1,** 206 (1970); *Chem. Abstr.*, **74,** 110113.
833. P. E. Linnett and J. Walker, *Biochem. J.*, **109,** 161 (1968).
834. R. M. Seifert and C. F. Hiller, *J. Ass. Offic. Anal. Chem.*, **56,** 1273 (1973); *Chem. Abstr.*, **80,** 25969.
835. L. M. Yagupol'skii and A. G. Galushko, *Zh. Obshchei Khim.*, **39,** 2087 (1969); *Chem. Abstr.*, **72,** 31676.
836. A. R. Todd, F. Bergel, and A. Jacob, *J. Chem. Soc.*, 1555 (1936); *Chem. Abstr.*, **31,** 401.
837. A. R. Todd and F. Bergel, *J. Chem. Soc.*, 1504 (1937); *Chem. Abstr.*, **32,** 159.
838. S. Yoshida, M. Nagawa, and M. Kataoka, *Takamine Kenkyujo Nempo*, **12,** 48 (1960); *Chem. Abstr.*, **55,** 6487.
839. G. H. Hamor and H. Andon, *J. Pharm. Sci.*, **59,** 276 (1970); *Chem. Abstr.*, **72,** 100586.
840. M. Hatanaka and T. Ishimaru, *Bull. Chem. Soc. Japan*, **46,** 3600 (1973); *Chem. Abstr.*, **80,** 71072.
841. J. Shukri and S. Alazawe, *J. Indian Chem. Soc.*, **45,** 1056 (1968); *Chem. Abstr.*, **70,** 57718.
842. British Patent No. 969 031; *Chem. Abstr.*, **61,** 14679.
843. H. J. M. Dou, *Chim. Actual. Paris*, 41 (1976).
844. British Patent No. 456 751; *Chem. Abstr.*, **31,** 2232.
845. Swiss Patent No. 204 688.
846. British Patent No. 471 416.
847. F. Schultz, *Z. Physiol. Chem.*, **265,** 113 (1940); *Chem. Abstr.*, **35,** 1839.
848. U.S. Patent No. 2 386 766; *Chem. Abstr.*, **40,** 1284.
849. British Patent No. 456 735; *Chem. Abstr.*, **31,** 2232.
850. German Patent No. 703 775; *Chem. Abstr.*, **36,** 784.
851. K. Hubacher, *Justus Liebigs Ann. Chem.*, **259,** 228 (1890).
852. French Patent No. 824 837.
853. F. Asinger, M. Thiel, and H. G. Hauthal, *Annalen*, **634,** 131 (1959); *Chem. Abstr.*, **55,** 2616.
854. F. E. Hooper and T. B. Johnson, *J. Amer. Chem. Soc.*, **56,** 484 (1934); *Chem. Abstr.*, **28,** 1700.
855. E. B. Knott, *J. Chem. Soc.*, 455 (1945); *Chem. Abstr.*, **39,** 4869.
856. British Patent No. 593 024; *Chem. Abstr.*, **43,** 5048.
857. U.S. Patent No. 2 423 709; *Chem. Abstr.*, **41,** 6582.
858. M. Carrega, *Thesis* 1959. University of Marseille, France.
859. W. Damerau, D. Schwarz, G. Lassmann, G. Huebner, and A. Schellenberger, *Z. Chem.*, **13,** 179 (1973); *Chem. Abstr.*, **79,** 136222.
860. M. Conte, *Thesis* 1975. University of Lyon, France.

861. S. L. Gusinskaya, V. Yu. Telly, and N. L. Ovchinnikova, *Uzbek, Khim. Zhur.*, **15,** 47 (1971); *Chem. Abstr.*, **76,** 72617.
862. V. Yu. Telly and S. L. Gusinskaya, *Neftekhimiya*, **11,** 902 (1971); *Chem. Abstr.*, **76,** 88123V.
863. G. Travagli, *Stud. Urbinati, Fac. Farm.*, **36,** 24 (1962); *Chem. Abstr.*, **64,** 6410.
864. East German Patent No. 65 078; *Chem. Abstr.*, **71,** 49929.

IV

Thiazolecarboxylic Acids, Thiazolecarboxaldehydes, and Thiazolyl Ketones

ROGER MEYER

Laboratoire de Chimie Physique Organique, Faculté des Sciences et Techniques de Saint-Jérôme, Marseille, France

I. Thiazolecarboxylic Acids	520
1. Syntheses	520
A. Ester Hydrolysis	520
B. Oxidation	521
C. Carbonating Organometallic Compounds	522
D. Other Methods	522
2. Properties and Reactions	522
A. Acidity	522
B. Decarboxylation	523
C. Reduction	524
3. Functional Derivatives	525
A. Esters	525
B. Thiazolecarboxylic Acid Chlorides	528
C. Thiazolecarboxamides	529
D. Thiazolecarboxylic Acid Hydrazides	530
E. Cyanothiazoles	531
II. Thiazolecarboxaldehydes	532
1. Syntheses	532
2. Reactions	534
III. Thiazolyl Ketones	535
1. Syntheses	535
2. Reactions	537
A. Reduction	537
B. Oxidation	537
C. Reactions of Condensation and Others	537
IV. Tables of Compounds	538
V. References	557

I. THIAZOLECARBOXYLIC ACIDS

As noted by Robinson and Strachan (1), after considerable activity in the period 1885 to 1895 thiazolecarboxylic acids received little attention until 1935. Isolation of 4-methyl-5-thiazolecarboxylic acid after degradation of vitamin B_1 gave new interest to the chemistry of these compounds.

1. Syntheses

They are easily accessible, and all possible substituted carboxylic acids are known.

A. *Ester Hydrolysis*

The most widely used method for the preparation of carboxylic acids is ester hydrolysis. The esters are generally prepared by heterocyclization (cf. Chapter II), the most useful and versatile of which is the Hantzsch's synthesis, that is the condensation of an halogenated α- or β keto ester with a thioamide (1–20). For example ethyl 4-thiazole carboxylate (**3**) was prepared by Jones et al. from ethyl α-bromoacetoacetate (**1**) and thioformamide (**2**) (1). Hydrolysis of the ester with potassium hydroxide gave the corresponding acid (**4**) after acidification (Scheme 1).

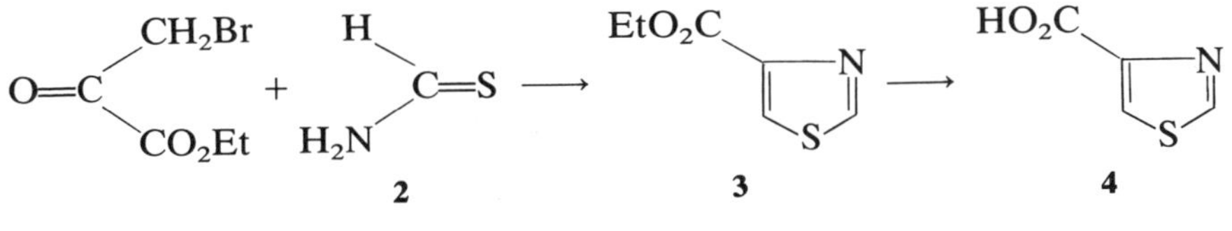

Scheme 1

The 2,4- or 2,5-diacids and 2,4,5-triacids have been prepared in good yields (40 to 50%) by hydrolysis of the esters obtained by heterocyclization (7, 12).

B. Oxidation

On account of the high degree of stability of the thiazole ring a large variety of substituted derivatives yield thiazolecarboxylic acids upon oxidation. The oxidation of a methyl group or a substituted methyl group to a carboxyl group has been accomplished in a few instances.

The alkyl derivatives of thiazoles can be catalytically oxidized in the vapor phase at 250 to 400°C to afford the corresponding formyl derivatives (21). Molybdenum oxide, V_2O_5, and tin vanadate are used as catalysts either alone or with a support. The resulting carbonyl compounds can be selectively oxidized to the acids.

Thiazolecarboxylic acids can also be prepared by catalytic dehalogenation of 2-halo derivatives (22, 23, 24).

4-Thiazolecarboxylic acid (24) has been synthesized in this way by the action of concentrated H_2SO_4 and HNO_3 upon 4-(dichloromethyl)thiazole.

Derivatives in which the substituents are already in a higher oxidation state than alkyl groups can be good precursors of acids. Acids can be prepared by the oxidation of alcohols.

Thus 2-phenyl-4(hydroxymethyl) thiazole upon oxidation with aqueous chromic acid yields 2-phenyl-4-thiazolecarboxylic acid (25, 26).

An other example is the oxidation of 4-(hydroxymethyl)thiazole with a mixture of nitric and sulfuric acids at −5°C (24, 27).

A reaction carried out by Haruki et al. is the energic oxidation of 5-(1-hydroxyethyl)2-phenylthiazole (5) in dioxane by aqueous KI-iodine to give 2-phenyl-5-thiazolecarboxylic acid (6) (Scheme 2) (28).

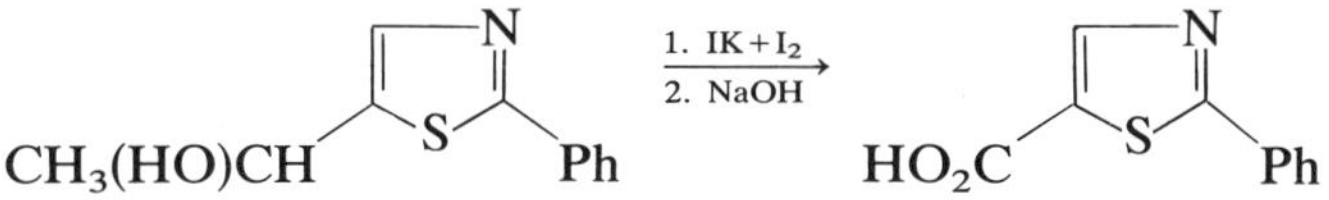

Scheme 2

Aldehydes are among the most readily oxidized classes of thiazolyl compounds. They are converted to thiazolecarboxylic acids by numerous oxidizing agents, including not only the usual reagents such as permanganate and dichromate but also relatively weak oxidizing agents such as Ag_2O.

For example, 2-phenyl-5-formylthiazole treated with Ag_2O in water and dioxane under basic conditions gives 2-phenyl-5-thiazolecarboxylic acid (29).

Ozonolysis of 2-styryl-4-methylthiazole followed by oxidation of the intermediate carbonyl compound with peracetic acid yields 4-methyl-2-thiazolecarboxylic acid (30).

C. Carbonating Organometallic Compounds

2- and 5-thiazolecarboxylic acids are obtained with good yields by carbonating the corresponding lithium derivatives (31–34).

2- and 5-bromothiazole treated at −70°C with BuLi stirred and poured onto dry ice give after acidification 2- or 5-thiazolecarboxylic acid in 40% yield (Scheme 3). The formation of ketones was not observed (31–32).

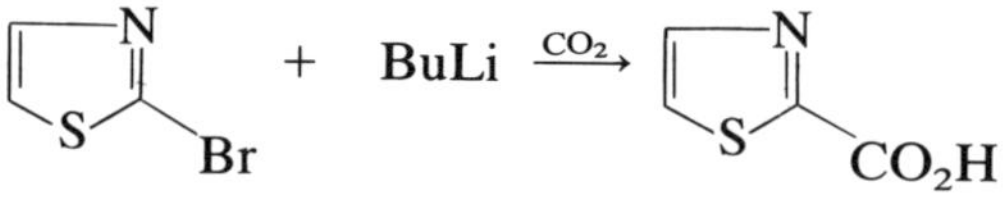

Scheme 3

D. Other Methods

The haloform reaction has occasional utility in the preparation of acids from the corresponding methyl ketone. Note that the reaction mixture must be treated with sodium thiosulfate before acidification (35). These acids are identical to those obtained from corresponding aldehydes by the Cannizzaro reaction (35, 36). Thiazolecarboxylic acids can also be prepared by hydrolysis of their derivatives (5, 11, 12, 28, 37, 39–41). Thus cyanothiazoles can be hydrolyzed to give acids in good yield (37). Some thiazolecarboxylic acids are also found in low yield in the antibiotics, micrococcin P (16, 42) thiostrepton (43), siomycin (44), and saramycetin (44). They are derived biogenetically from N-acylated cysteines by ring closure followed by dehydrogenation (16).

These interesting reactions, however, have limited interest for the synthesis of acids.

2. Properties and Reactions

A. Acidity

The thiazolecarboxylic acids are stronger acids than their homologs in the pyridine series (45).

I. Thiazolecarboxylic Acids

As already noted by Sprague and Land (46), no accurate data are available for 2-thiazolecarboxylic acid, but it appears to be a considerably stronger acid than either 4- or 5-thiazolecarboxylic acid (7). It is difficult to study experimentally the acidity of these compounds because of their amphoteric character.

To evaluate the dissociation constants it would be necessary to measure the equilibrium constant for the reaction in Scheme 4. The dissociation of thiazolecarboxylic acids has been studied principally by Erlenmeyer et al. (47, 48). It seems that no systematic and reliable determination of the acidity dissociation constants have been realized until now.

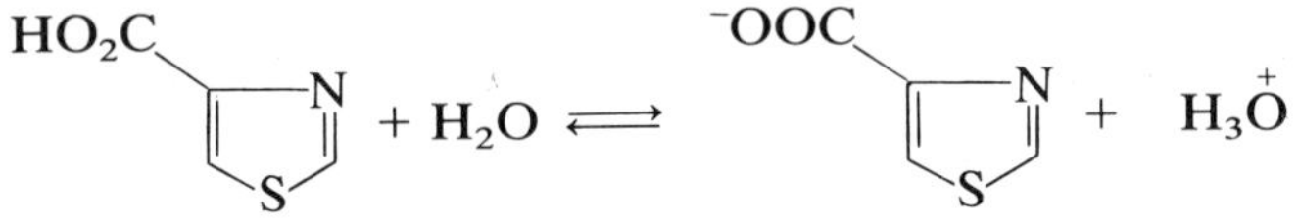

Scheme 4

B. *Decarboxylation*

Decarboxylation of 2-thiazolecarboxylic acids occurs readily (49), 5-carboxylic acids can also be decarboxylated without difficulty (50), but 4-thiazolecarboxylic acids are relatively stable (7, 38). For example,

2,5-diacid $\xrightarrow{85°C}$ 5-Acid $\xrightarrow{250°C}$ thiazole

2,4-diacid $\xrightarrow{140°C}$ 4-Acid

As with pyridine acids, the presence of a positive center or an electron accepting group facilitates decarboxylation that must proceed by way of the anion.

On the other hand, with 4,5-thiazoledicarboxylic acid the 5-carboxyl group is the more labile (3, 5, 51) contrary to results found by other workers (52). For example, Huntress et al. (5) have shown that pyrolysis of 2-phenyl-4,5-thiazoledicarboxylic acid (**7**) gives evolution of carbon dioxide and the formation of 2-phenyl-4-thiazolecarboxylic acid (**8**) (Scheme 5).

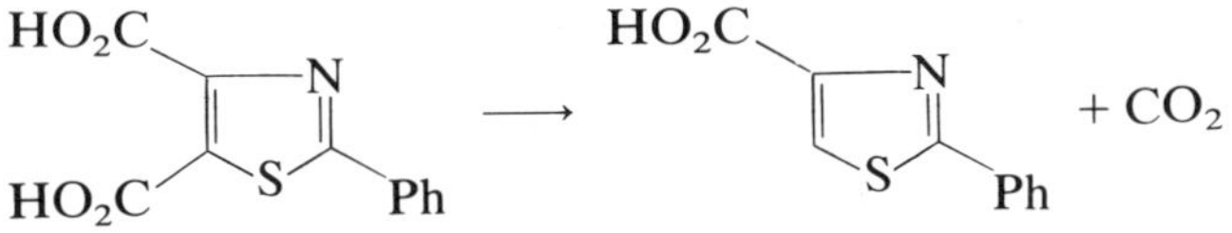

Scheme 5

Hydrolysis of ethyl 4-methyl-2,5-thiazole dicarboxylate (**9**) or dicarboxylic acid dichloride gives an excellent yield of 4-methyl-5 thiazole carboxylic acid (**10**) instead of the dicarboxylic acid (Scheme 6). This lability is a general Property of 2-thiazolecarboxylic acids.

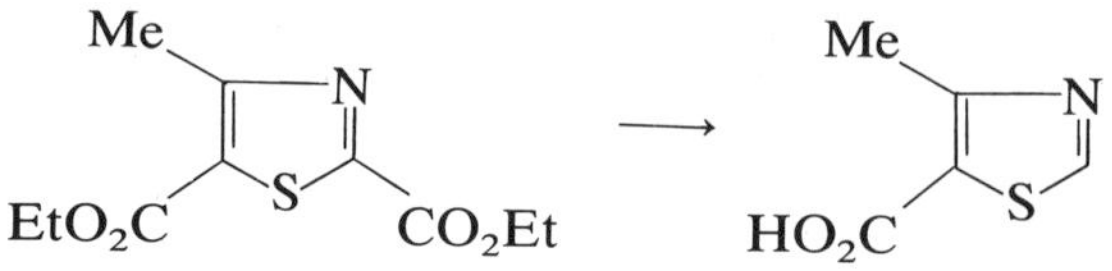

Scheme 6

The 4-carboxylic acid in all instances is the more stable. Heated to 220°C 2-(α-thienyl)-4,5-thiazoledicarboxylic acid lost 1 mole of carbon dioxide to give the 4-monoacid.

The observed order of decreasing ease of decarboxylation for the three thiazolecarboxylic acids is $2 > 5 > 4$.

This is verified by the stepwise decarboxylation of 2,4,5-thiazoletricarboxylic acid (12, 48).

C. Reduction

Thiazole acids may undergo many different types of reduction. Chemical reduction of thiazolecarboxylic acids and of their derivatives to yield the corresponding alcohols can be accomplished with lithium aluminium hydride in ether solution (53).

Reduction with sodium in alcohol was unsuccessful (54). The introduction of lithium aluminium hydride has provided an elegant method for the reduction of thiazole esters to hydroxythiazoles; for example, ethyl 2-methyl-4-thiazolecarboxylate (**11**) with lithium aluminium hydride in diethyl ether gives 2-methyl-4-(hydroxymethyl)thiazole (**12**) in 66 to 69% yield (Scheme 7) (53).

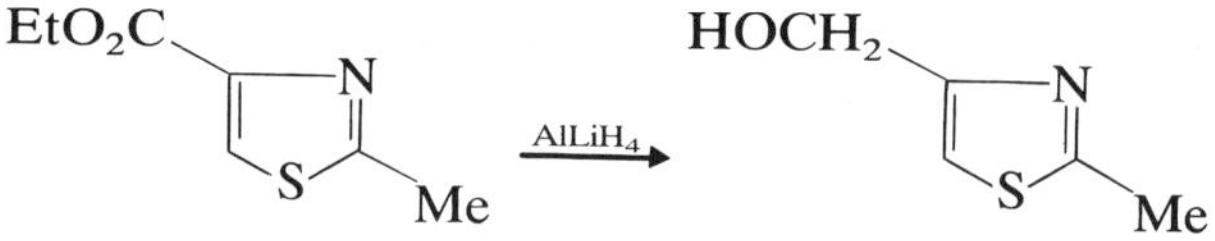

Scheme 7

I. Thiazolecarboxylic Acids 525

Reductions carried out with lithium aluminium hydride are not always so successful. As noted by Sprague (46) the esters of 2-aminothiazole carboxylic acids behave somewhat differently with $AlLiH_4$ (55).

It is possible to reduce thiazole carbonyl derivatives with aluminium isopropoxide (56, 57). For example, 4-methyl-5-acetylthiazole gives 4-methyl-5-(α-hydroxyethyl)thiazole. With aluminium amalgam one obtains a duplication like to the pinacol reaction, and the yield is generally poor (57). The preparation of thiazole alcohols by reduction is also discussed in Chapter VIII.

Iversen and Lund have realized the polarographic reduction of 2-thiazolecarboxylic acid and some of its derivatives in acid solution. They obtained 2-thiazolecarboxaldehyde (58).

Aldehydes are more generally prepared by electrolytic reduction of amides, the reduction of carboxylic acids being possible only when they are activated by a strongly electron-withdrawing group (58).

The reduction of the carboxylic group is easier in the 2-position than in the 4- or 5-positions. These differences in reductibility run parallel to chemical reactivity (58).

The activated position is the most vulnerable to nucleophilic attack (58).

The thiazolecarboxylic acids and especially 2-alkyl-5-thiazolecarboxylic acids have hypolimic vasodilatory and antiinflammatory activity (19, 59, 60).

3. Functional Derivatives

A. *Esters*

a. SYNTHESES

As previously mentioned the esters of thiazolecarboxylic acids are generally prepared by the Hantzsch's synthesis (17, 20).

Usually no difficulties are encountered in the esterification of thiazole acids. Direct esterification with alcohol and acid in the presence of an acid catalyst (7, 61, 62), or prior conversion to the acid chloride (6, 63, 64) followed by reaction with an alcohol in basic conditions give good yields.

These methods were not applicable to all acids, and various alternative routes were investigated. The conversion of an acid to its methyl ester by diazomethane is a method of choice when other methods are unsatisfactory (6, 30, 61, 65). With appropriate alcohols thiazoleanhydrides give the esters or diesters in good yield (64), dimethyl 2-phenyl-4,5-thiazoledicarboxylate (**13**) has been prepared in this way (Scheme 8) (17).

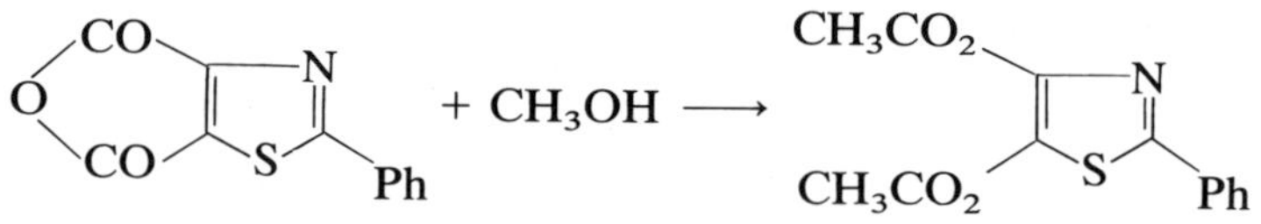

Scheme 8

Erlenmeyer et al. showed that it is possible to realize a moderated saponification of diethyl 4,5 thiazoledicarboxylate in mild conditions with an ethanolic potassium solution. The attack begins at the 5-position, and finally monoesters in the 4-position are obtained (Scheme 9) (17).

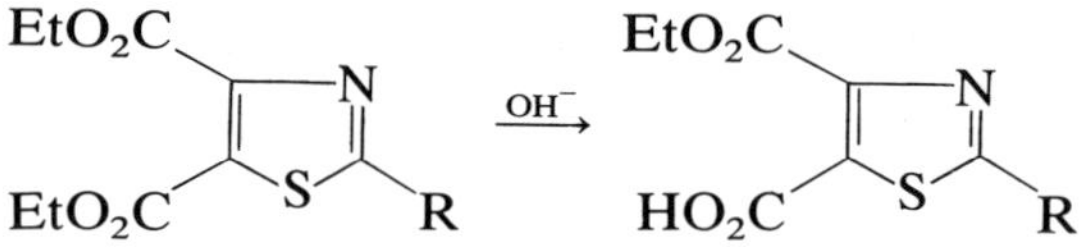

R = Ph, α-thineyl

Scheme 9

Some esters of substituted alcohols have been synthesized by transesterification. Treatment of 4-methyl-5-thiazolecarboxylic acid (**14**) with β-chloroethyldiethylamine in acetone in the presence of anhydrous potassium carbonate gives the desired ester (**15**) in good yield (60%) (Scheme 10) (163).

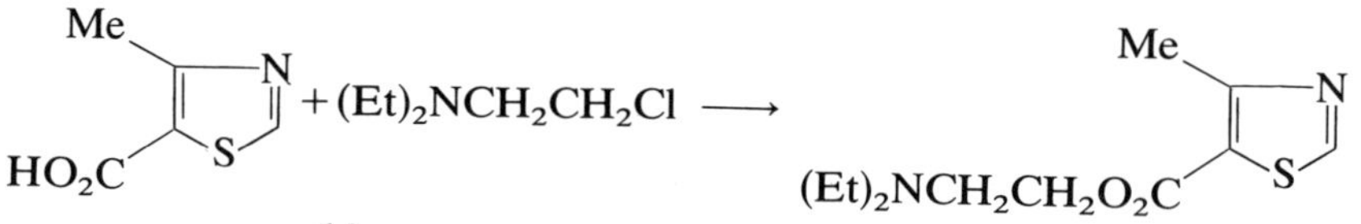

Scheme 10

b. REACTIONS

Thiazolyl esters react with various nucleophilic compounds such as alcohols, bases, sodamide, and its derivatives (6, 16).

Thus esters were converted to hydrazides with an excess of hydrazine hydrate in alcohol with a yield of 84% (66).

Heating diethyl 2-phenyl-4,5-thiazoledicarboxylate (**16**) in alcohol with an excess of hydrazine hydrate gave a mixture of cyclohydrazide (**17**) and the open-chain 2-phenyl-4,5-thiazoledicarboxylic dihydrazide (**18**) (Scheme 11) (5). The relative amounts of these two products depended on

I. Thiazolecarboxylic Acids 527

the duration of the reaction, the proportion of cyclohydrazide increasing with time. Hydrolysis of **16** with methanolic potassium hydroxide gave 2-phenyl-4,5-thiazoledicarboxylic acid (**7**). Pyrolysis of the diacid (**7**) caused the formation of the monoacid (**8**). Reduction of diethyl 2-amino-4,5-thiazoledicarboxylate (**19**) with lithium aluminium hydride (53) gave unexpected results. With a small excess of LiAlH$_4$ in alcohol only the 4-ester group (**20**) is reduced, whereas a large excess of reducing agent gives 5-methyl-4-hydroxymethyl-2-aminothiazole (**21**) (Scheme 12) (53). All the aminoalkylesters tested pharmacologically exhibit spasmolytic and analeptic properties (1).

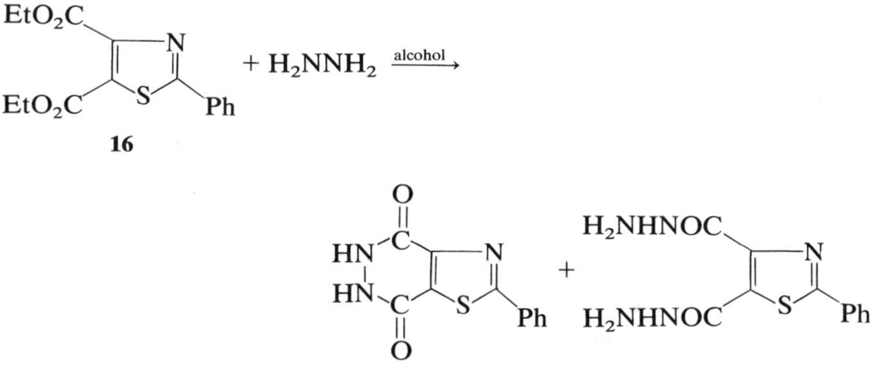

Scheme 11

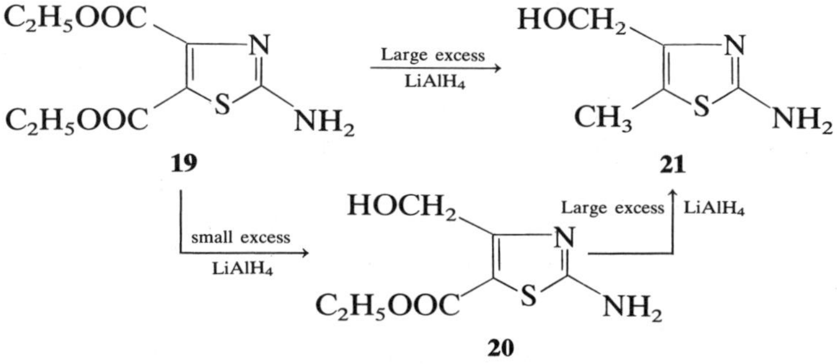

Scheme 12

Recently, new esters possessing physiological properties have been synthesized (67).

They have adrenalin and vasodilating properties that can be utilized in the treatment of hypertension.

B. *Thiazolecarboxylic Acid Chlorides*

a. SYNTHESES

Thiazolecarboxylic acid chlorides, generally prepared from the corresponding acid and thionyl chloride (Scheme 13), are relatively unstable (2, 7, 23, 25, 63, 64, 68–78).

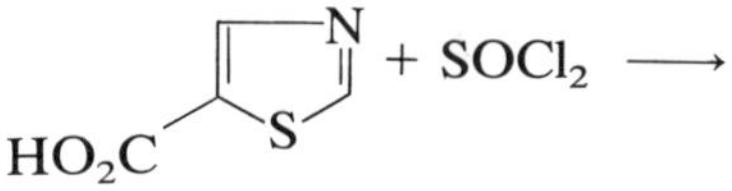

Scheme 13

The 2-thiazolecarboxylic acids are too readily decarboxylated to be converted to acid chlorides (6, 12, 79).

Boon has observed that when 2,4-dimethyl-5-thiazolecarboxylic acid (**22**) is heated with an excess of thionyl chloride, the main product (76%) is 4-methyl-2,5-thiazoledicarboxylic acid dichloride (**23**) (Scheme 14) (7).

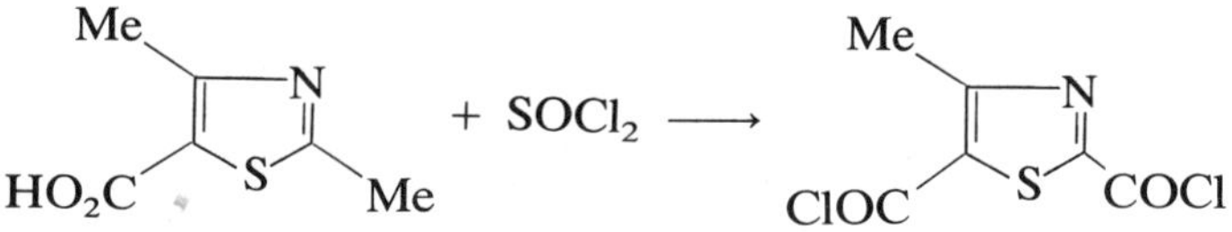

Scheme 14

The normal acid chloride is obtained by reaction between equimolar quantities of acid and thionyl chloride in the presence of pyridine at 0°C.

b. REACTIONS

These compounds are used in the preparation of esters (63), amides (7, 71, 77, 80, 81), and hydrazides.

Treatment of acid chlorides with ammonia or amines in benzene gives thiazolecarboxamides (71, 77, 78).

The potassium salt of 4,5-thiazoledicarboxylic acid with thionyl chloride gives the corresponding anhydride (Scheme 15) (5, 17).

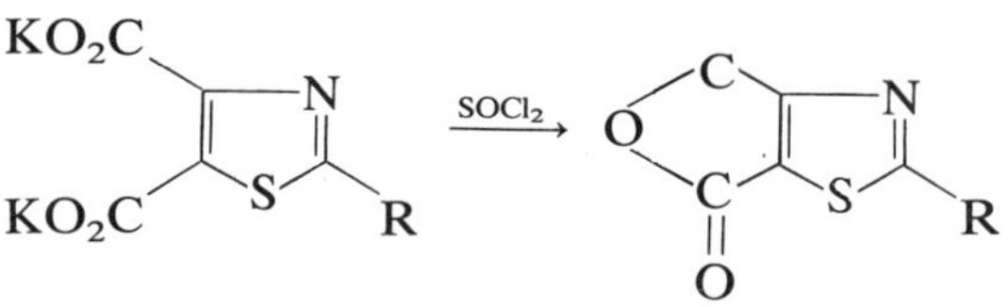

R = Et, α-furyl, α-thienyl, or β-thienyl

Scheme 15

I. Thiazolecarboxylic Acids   529

Thiazole acid chlorides react with diazomethane to give the diazoketone. The later reacts with alcoholic hydrogen chloride to give chloroacetylthiazole (Scheme 16). However, the Wolff rearrangement of the diazoketone is not consistently satisfactory (82). Heated with alcohol in the presence of copper oxide the 5-diazomethylketone (**24**) gives ethyl 5-thiazoleacetate (**25**) instead of the expected ethoxymethyl 5-thiazolyl ketone (Scheme 17) (83).

Scheme 16

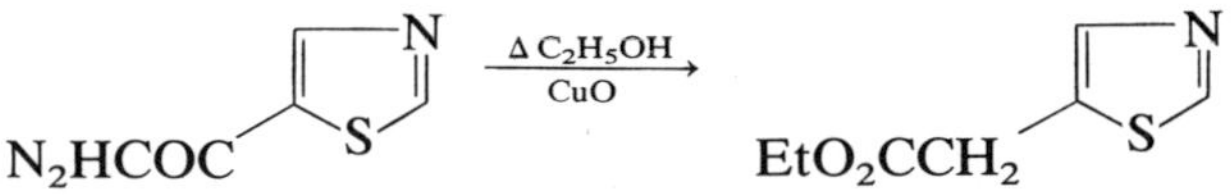

Scheme 17

C. *Thiazolecarboxamides*

Thiazolecarboxamides are generally prepared by the reaction of a thiazolecarboxylic acid, ester, or acid chloride with ammonia or an amine (1, 5, 7, 23, 37, 62, 68, 71, 74, 75, 77, 80, 81, 84, 88–90). For example, the treatment of 4-thiazolecarboxylic acid chloride with dry NH_3 in the presence of benzene gives 4-thiazolecarboxamide (71, 75, 77, 81).

Diamides and triamides have been obtained from the action of an aqueous saturated solution of ammonia on the corresponding ester (Scheme 18) (88). Amides can also be obtained by the Curtius (16) or Hofmann reactions (80). Thus the Curtius reaction with 2-substituted 4-thiazolecarboxylic acids gives the 4-acetamido compounds (16).

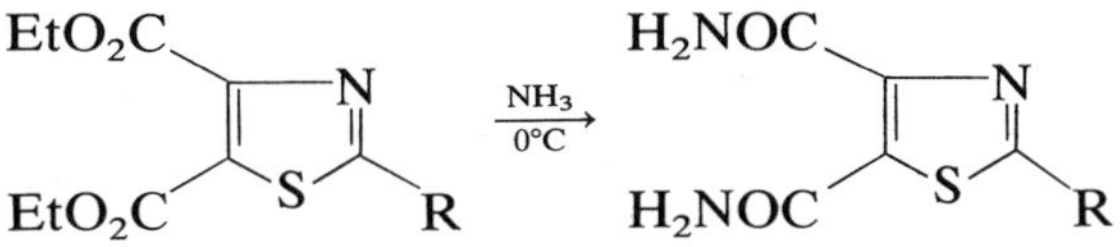

R = Cl, Et, α-furyl, α-thienyl, or β-thienyl

Scheme 18

Thiazole carboxamides are readily dehydrated to nitriles in good yields by heating with phosphorus oxychloride (91), phosphorus pentoxide (87, 71), or phosphoryl chloride (16) (Scheme 19).

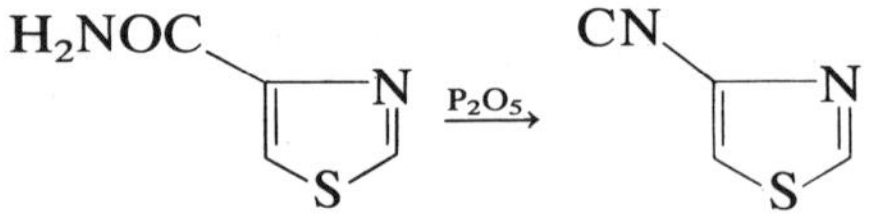

Scheme 19

Treatment of thiazole-4,5-dicarboxamides (**26**) with phosphorus oxychloride in the presence of pyridine afforded 4,5-dicyanothiazoles (**27**) (Scheme 20) (91).

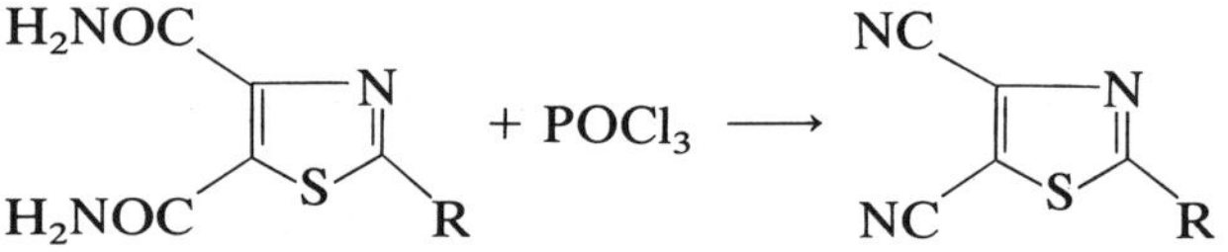

R = Cl, Me, Et, Ph, α-furyl, α-thienyl, or β-thienyl

Scheme 20

D. *Thiazolecarboxylic Acid Hydrazides*

Thiazolecarboxylic acid hydrazides are prepared by the same general methods used to prepare amides, that is, by treating acids, esters, amides, anhydrides, or acid halides with hydrazine or substitued hydrazines. For example, see Scheme 21 (92). The dihydrazides are obtained in the same way (88). With diethyl 2-chloro-4,5-thiazoledicarboxylate this reaction gives the monohydrazide monoester of 2-hydrazine-4,5-thiazoledicarboxylic acid as a result of the great mobility of the chlorine in the 2-position and not, as expected, the dihydrazides.

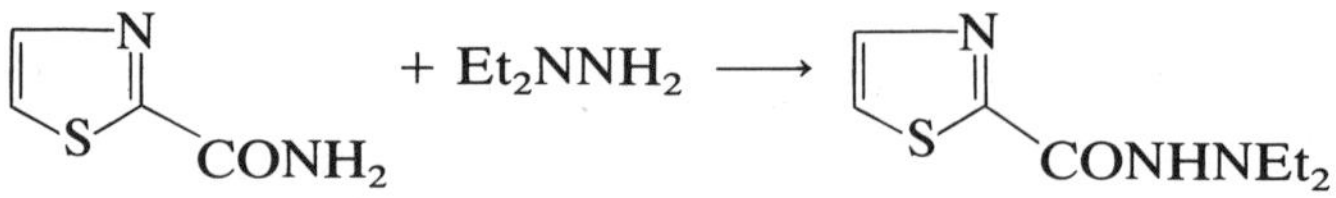

Yield = 83%

Scheme 21

I. Thiazolecarboxylic Acids

E. *Cyanothiazoles*

a. SYNTHESES

Cyanothiazoles can be prepared by different methods. Dehydration of amides to give nitriles has already been mentioned. The yields are good (37, 38, 71, 87, 93, 94).

Alkylthiazoles can be oxidized to nitriles in the presence of ammonia and a catalyst. For example, 4-cyanothiazole was prepared from 4-methylthiazole by a one-step vapor-phase process (94) involving reaction with a mixture of air, oxygen, and ammonia at 380 to 460°C. The catalyst was MoO_3 and V_2O_5 or MoO_3, V_2O_5, and CoO on an alumina support.

In the same way, 4-cyanothiazole is obtained from the action of a mixture of N_2/NO on the methyl derivative at 500°C with Re catalysts (Scheme 22) (95).

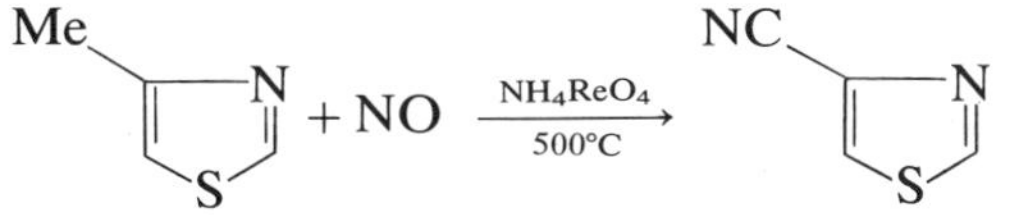

Scheme 22

The cyanothiazoles are not obtained by replacement of a nuclear halogen with cyanide.

b. REACTIONS

The Raney nickel reduction of cyanothiazoles leads to the corresponding amino compounds (96).

The Stephen's method allows the reduction of nitriles by stannous chloride in acid medium. If the amine chlorhydrate initially formed is hydrolyzed, the corresponding aldehyde is obtained (37, 91). Harington and Moggridge (37) have reduced 4-methyl-5-cyanothiazole by this method (Scheme 23). However, Robba and Le Guen (91) did not obtain the expected products with 4,5-dicyanothiazole and 2-methyl-4,5-dicyanothiazole. These compounds have been reduced with diisobutylaluminium hydride with very low yields (3 to 6%) (Scheme 24). In other conditions the reaction gives a thiazole nitrile aldehyde with the same yield as that of the dialdehyde.

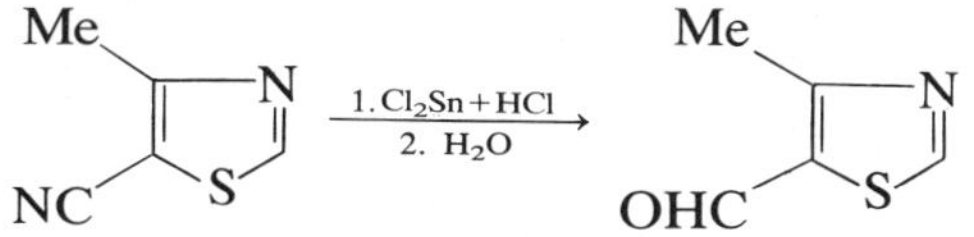

Scheme 23

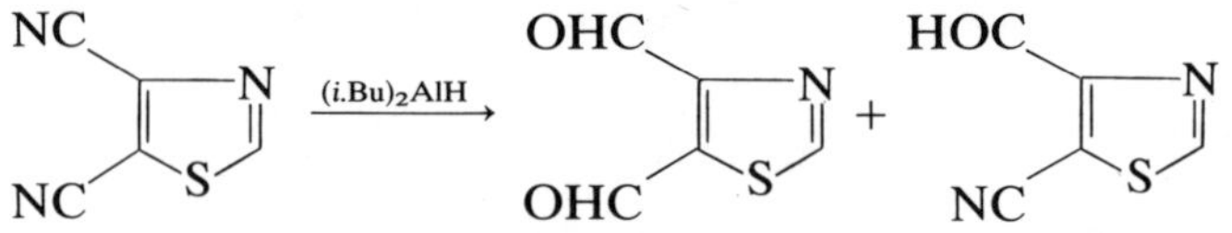

Scheme 24

It has been shown that the cyano group in the 4-position is more easily reduced than at other positions (91).

Organometallic reagents react with cyanothiazoles to give thiazolyl ketones (87).

The conversion of 4,5-dicyanothiazoles to diketones has been attempted (91). A difference in reactivity between the two cyano groups has been observed; the least labile is the group in the 5-position. These Grignard reactions are limited and lead to 4-acetyl-5-cyanothiazole (Scheme 25).

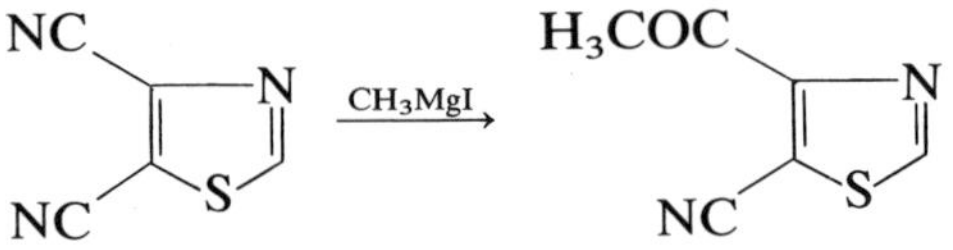

Scheme 25

Nitriles react with ammonia, or primary or secondary amines in the presence of an acid catalyst to give amidines (Scheme 26) (75, 77, 81). The catalysts used are hydrochloric acid and aluminium chloride. The amidines are anthelmintics for animals such as sheep, goats, cattle, horses, and swine.

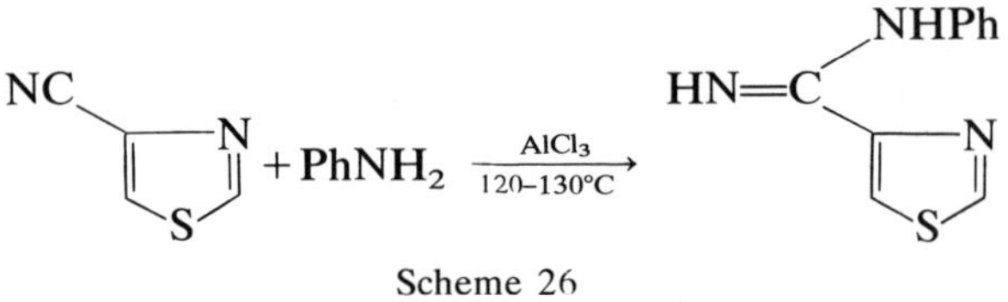

Scheme 26

II. THIAZOLECARBOXALDEHYDES

1. Syntheses

Different preparative methods of thiazolecarboxaldehyde have been reported by Iversen and Lund (97):

- Chromic oxidation of the corresponding carbinols (29, 98).
- The McFadyen-Stevens synthesis (66, 99) and its Newman-Caflish modification.

II. Thiazolecarboxaldehydes

The conversion of esters to hydrazides and of hydrazides to the sulfonyl derivatives occurs in good yield in the McFadyen-Stevens synthesis, but the decomposition of sulfonyl derivatives gives low yields of the desired products, for example, thiazole hydrazide (**28**) with 10% excess of $PhSO_2Cl$ in pyridine gave a 75% yield of 1-phenylsulfonyl-2-(4-methyl-5-thiazolecarbonyl)hydrazine (**29**) (66). The Newman-Caflish modification of the McFadyen-Stevens synthesis gave 37% 4-methyl-5-thiazolecarboxaldehyde (**30**) (Scheme 27).

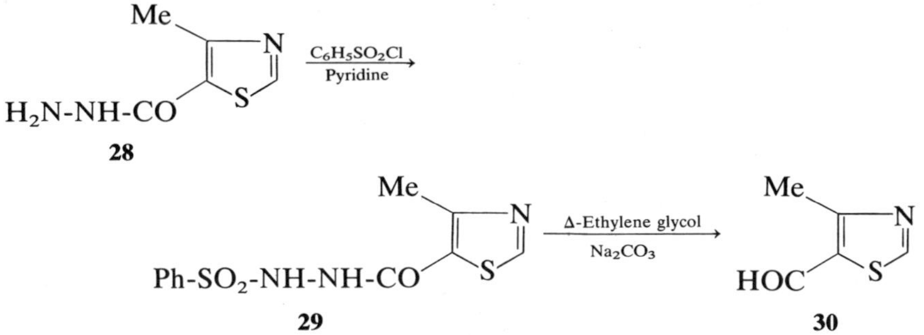

Scheme 27

As mentioned previously, aldehydes can be prepared by Stephen's method of reduction of nitriles by stannous chloride (37, 91). Polarographic reduction of thiazolecarboxylic acids and their derivatives gives lower yields of aldehydes (58). Ozonolysis of styrylthiazoles, for example, 2-styryl-4-methylthiazole, followed by catalytic reduction gives aldehyde with 47% yield of crude product (30).

Catalytic oxidation of alkylthiazoles in the vapor phase (21) or oxidation of halomethylthiazoles (24, 27, 36) gives the carbonyl compound in low yield.

Reaction of 2-thiazolyl diazonium salts with formaldoxime (100) is of little use.

A novel synthesis of 2-thiazolecarboxaldehyde uses the reaction of 2-thiazolyllithium with *N*-methylformanilide with a yield of 46% (99). Similarly, 5-thiazolecarboxaldehyde was obtained in 5% yield, the poor yield probably being a result of the low stability of 5-thiazolyllithium. These unsatisfactory yields led Iversen and Lund to investigate four new methods (97).

1. Addition of *N,N*-dimethylformamide to lithium derivatives (Scheme 28).

$$RLi \xrightarrow[\text{2. } H_2O]{\text{1. DMF}} RCHO$$

Scheme 28

2. Selenium dioxide oxidation of carbinols (Scheme 29).

$$RCH_2OH \xrightarrow[\text{100°C, dioxane}]{SeO_2} RCHO$$

Scheme 29

3. Selenium dioxide oxidation of methyl derivatives (Scheme 30).

$$RCH_3 \xrightarrow[\text{100°C, dioxane}]{SeO_2} RCHO$$

Scheme 30

4. Nitric acid oxidation of carbinols (Scheme 31).

$$RCH_2OH \xrightarrow{HNO_3} RCHO + RCOOH$$

Scheme 31

2. Reactions

Thiazolecarboxaldehydes exhibit many reactions typical of aldehydes. However, they give no aldolization reaction (no α-hydrogen), but they do react with different compounds such as acetic anhydride, hippuric acid, acetylglycine, and so for (37, 101, 102). Thus 2-phenyl-4-formylthiazole (**31**) mixed with hippuric acid and treated with AcO_2 and anhydrous NaOAc gives the azalactone (**32**) (Scheme 32).

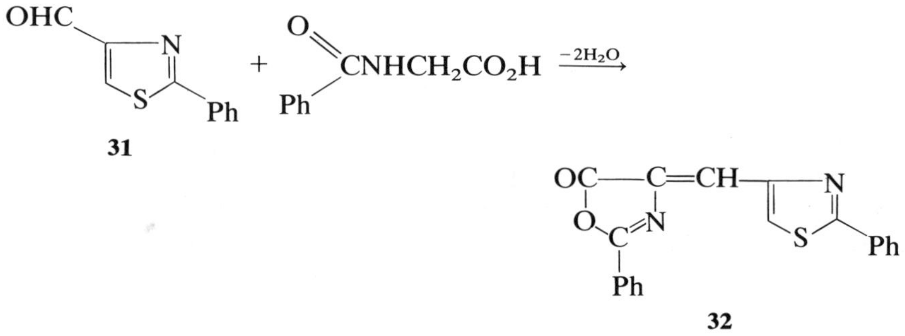

Scheme 32

With malonic acid in a mixture of pyridine and piperidine 2-phenyl-4-formyl-5-chlorothiazole yields 2-phenyl-5-chloro-4-thiazoleacrylic acid (103).

Thiazolecarboxaldehydes in the presence of a strong base (103) give equal amounts of the corresponding alcohol and carboxylic acid (Cannizaro reaction). In the presence of potassium cyanide thiazolecarboxaldehyde undergoes the benzoin condensation (104, 105).

Oximes, hydrazones and semicarbazones are readily prepared from the aldehydes (37, 106).

The reduction of thiazole carboxaldehydes with isopropanol and aluminium isopropoxide gives the corresponding alcohols (37, 107, 108).

Catalytic reduction over a platinum catalyst fails because of poisoning of the catalyst (101).

Thiazolecarboxaldehydes are very easily oxidized to carboxylic acids by most oxidizing agents, the most common being $KMnO_4$ in cold pyridine or boiling acetone. Thiazolecarboxylic acids are obtained in 50% yield (29). Other oxidizing agents such as Ag_2O in dioxane and water (29, 103), chromic acid, and so forth are also used.

III. THIAZOLYL KETONES

1. Syntheses

Alcohols are easily oxidized to ketones by acid dichromate practically at room temperature (28, 98, 105, 109–113). For example, oxidation of 2-(α-hydroxyethyl)-4-methylthiazole (**33**) gives 2-acetyl-4-methylthiazole (**34**) (Scheme 33) (111). Although this is the most common reagent, many other strong oxidizing agents, for example, $KMnO_4$, MnO_2, or SeO_2 have also been employed. Thus oxidation of 2-benzyl-5-methylthiazole with selenium dioxide in dioxane gives a 75% yield of phenyl 5-methyl-2-thiazolyl ketone (11). Thiazole ketones can be obtained from the Friedel-Crafts acylation reaction using more than one equivalent of $AlCl_3$ catalyst (114). 4,5-Dimethylthiazole (**35**) and benzaldehyde with $ZnCl_2$ at 160°C give the corresponding carbonyl compound (**36**) (Scheme 34).

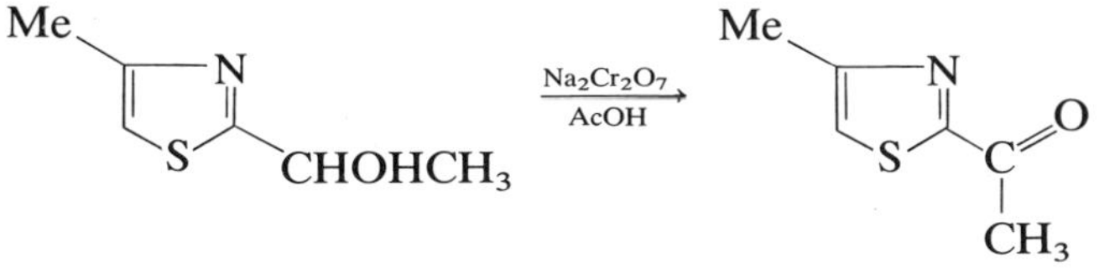

Scheme 33

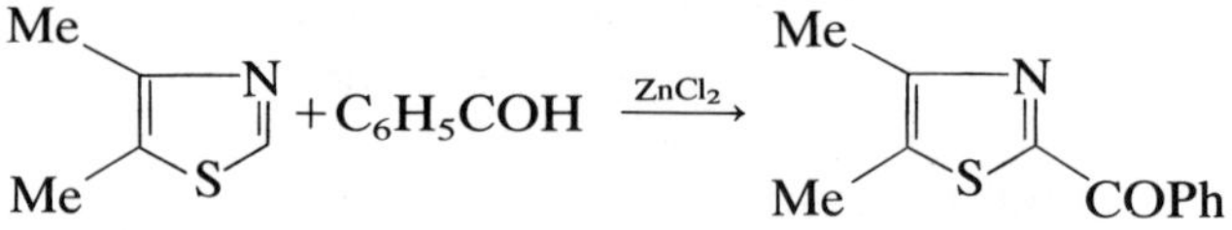

536 Thiazolecarboxylic Acids, Thiazolecarboxaldehydes, and Thiazolyl Ketones

$$\text{35} + C_6H_5COH \xrightarrow{ZnCl_2} \text{36}$$

Scheme 34

The ketones can be obtained by addition of organometallic reagents to acids and their derivatives.

2-Thiazolyl magnesium halide reacts at low temperature with acetic anhydride to give acetylthiazole in low yield (10 to 20%) (31).

It should be noted that Grignard reagents obtained from 2-methylthiazole do not yield acetyl derivatives (31). Ketones can also be obtained from cyanothiazoles and Grignard reagents (87). Thus phenyl magnesium bromide and 5-cyanothiazole give phenyl 5-thiazolyl ketone (Scheme 35).

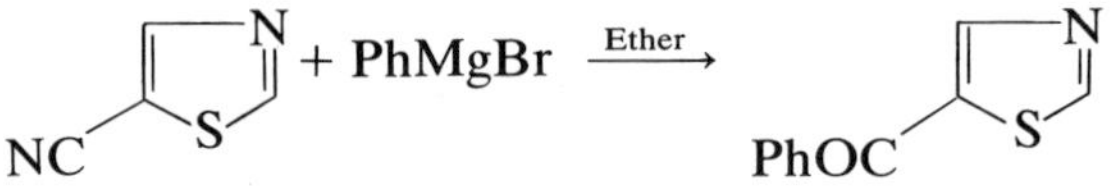

$$\text{NC-thiazole} + PhMgBr \xrightarrow{Ether} \text{PhOC-thiazole}$$

Scheme 35

Robba and Le Guen (91) have shown that 4,5-dicyanothiazole and various Grignard reagents react partially to give 4-acetyl-5-cyanothiazoles. As previously mentioned, the cyano group in the 5-position is the least reactive.

The Claisen condensation of an aliphatic ester and a thiazolic ester gives after acidic hydrolysis a thiazolylketone (56). For example, the Claisen condensation of ethyl 4-methyl-5-thiazolecarboxylate with ethyl acetate followed by acid hydrolysis gives methyl 4-methyl-5-thiazolyl ketone in 16% yield.

Aldehydes may be converted to their homologs with diazomethane (Scheme 36) (35).

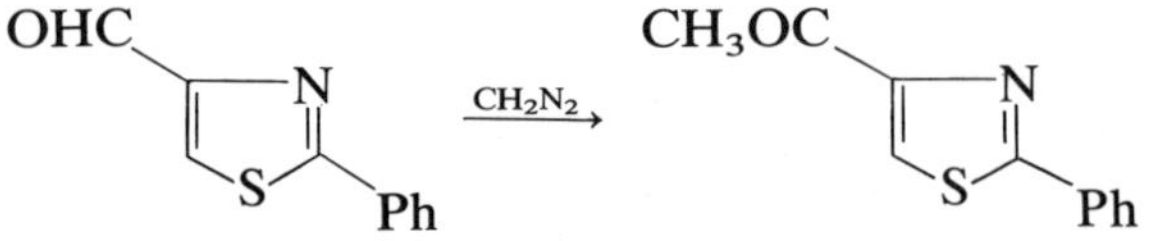

$$\text{OHC-thiazole-Ph} \xrightarrow{CH_2N_2} \text{CH}_3\text{OC-thiazole-Ph}$$

Scheme 36

2. Reactions

A. *Reduction*

The easiest large-scale reduction method for conversion of ketones to alcohols is the catalytic hydrogenation, the usual catalysts being nickel and palladium. For example, hydrogenation of 3-pyridyl-2-thiazolyl ketone in AcOH over Raney Ni at room temperature gave 3-pyridyl-2-thiazolylcarbinol (115). Reduction of thiazolyl ketones may also be effected with Grignard reagents (115).

Reduction with aluminium isopropoxide by the Meerwein-Pondorf procedure yields the alcohol (56, 57, 112).

The Clemmensen reduction of 2-acetyl-5-methyl thiazole gives 2-ethyl-5-methyl thiazole (31) (Scheme 37).

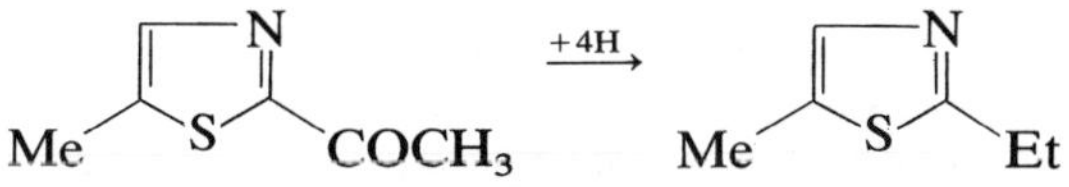

Scheme 37

B. *Oxidation*

The carbonyl group of thiazolylketones is oxidized with more difficulty than that of thiazolecarboxaldehyde.

Ketones are oxidized by potassium permanganate or by sodium hypochlorite (91) in aqueous solution to the corresponding acids. For example, oxidation of 5-acetylthiazole with aqueous $KMnO_4$ at 70°C gives 5-thiazolecarboxylic acid.

C. *Reactions of Condensation and Others*

Ammonia and a variety of its derivatives condense with thiazolylketones (28, 80, 115–117).

The oximes of ketones undergo the Beckmann rearrangement on treatment with phosphorus pentachloride (118).

The Mannich reaction can be realized with formaldehyde and secondary amines.

The haloform reaction is realized in good yield (28, 82, 112, 119, 120). Bromination of 2,4-dimethyl-5-acetylthiazole in CCl_4 or dry $CHCl_3$ under reflux gives 75% 2,4-dimethyl-5-bromomethylthiazole (Scheme 38) (120).

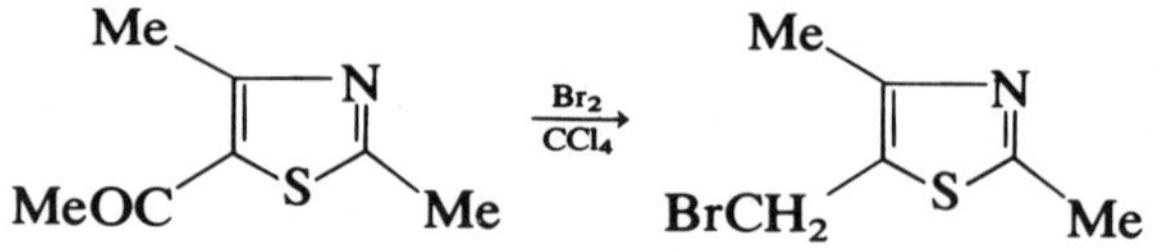

Scheme 38

IV. TABLES OF COMPOUNDS

Table IV-1	Thiazolemonocarboxylic acids	538
Table IV-2	Thiazolecarboxylates	540
Table IV-3	Thiazolecarboxylic acid chlorides	545
Table IV-4	Thiazolecarboxamides and thiazolecarboxylic acid hydrazides	546
Table IV-5	Thiazolepolycarboxylic acids and derivatives	550
Table IV-6	Cyanothiazoles	552
Table IV-7	Thiazolecarboxaldehydes	553
Table IV-8	Thiazolylketones	554
Table IV-9	Thiazole amidine and other compounds	556

TABLE IV-1. THIAZOLEMONOCARBOXYLIC ACIDS

	Substituent			
R_1	R_2	m.p. (°C)	b.p. (°C)	Ref.[a]
	2-Thiazolecarboxylic acids			
H	H	97–102	—	8, 31, 32, 34
H	n-Bu	80—81	—	13
Me	H	—	122/5mm	6, 30, 34, 661[a]
Me	Me	98	—	11
	4-Thiazolecarboxylic acids			
H	H	192–194	—	4, 12, 15, 27, 39, 42, 69, 71, 73, 76, 121 v
H	Me	—	—	3, 51
H	CH_2Ph	—	—	36

H	$CH_2CH_2CH_2CHCO_2H$
	                     |
	                     NH_2
		—	—	122

H	$CH_2CHMeCH_2NH_2$	—	—	161

IV. Tables of Compounds

TABLE IV-1. (Continued)

Substituent				
R_1	R_2	m.p. (°C)	b.p. (°C)	Ref.
H	CH_2NH_2	180	—	123, 124 l
H	$CHNH_2CH_3$	—	—	126 v
H	$CHNH_2$i-Pr	—	—	121 v
H	$CHOH–CH_3$	148–150	—	121
H	$(NH_2)CH(OH)$-$MeC(OH)CHMe$	—	—	43n, v
H	Ph	—	—	5, 16 l, 25, 52, 127, 128
H	p-$NO_2C_6H_4$	228	—	106
H	α-Thienyl	168	—	17 i
H	CHO	—	—	43 m, n, r, v
H	COMe	169–171	—	42 i, u
H	COEt	169–171	—	42 u, 121, 129 e
H	Br	—	—	130
Ph	H	—	—	131 l
Ph	Ph	140–141	—	41
COC_6H_5	Me	—	—	132
5-Thiazolecarboxylic acids				
H	H	218	107–108	2, 32, 39, 49, 66, 133, 134
H	Me	—	—	1, 39, 47, 61, 66, 69 125, 135–138
H	CH_2CO_2H	189–190	—	12
H	Ph	203–204	—	139
Me	H	209	—	3, 138, 140–142
Me	Me	230–231	—	1, 33, 51, 141
Me	Ph	216	—	10, 132
Et	H	—	—	19, 59 l, 60 l, 143
n-Pr	H	—	—	19 l, 59 l, 60 l, 143 l
n-Bu	H	—	—	19 l, 59 l, 60 l, 143 l
Pentyl	H	—	—	19 l, 59 l, 60 l, 143 l
n-Hexyl	H	—	—	19 l, 59 l, 60 l, 143 l
Undecyl	H	—	—	19 l, 59 l, 60 l, 143
Ph	H	187.5–188.5 186	— —	5, 29, 52
Ph	Me	204	—	64, 144, 145
Ph	CO_2Et	126	—	52
m-HOC_6H_4	H	—	—	146 l
p-$C_6H_4SO_2Me$	Me	—	—	18
CH_2CO_2H	Me	235	—	147
CH_2NH_2	Me	—	—	123
$CH_2NHCOPh$	Me	—	—	123
Cl	H	—	—	148
Br	H	175–176	—	148

[a] For the meaning of letters, please cf. p. 2.

TABLE IV-2. THIAZOLECARBOXYLATES

	Substituent				
R_1	R_2	R_3	m.p. (°C)	b.p. (°C)	Ref.
2-Thiazolecarboxylate					
Me	Me	H	65	115/2 mm	6, 30
Et	H	H	48	117–120/15 mm	8, 551
Et	H	Me	—	—	149
Et	H	n-Bu	—	156–158/11 mm	13
Et	Me	H	—	—	6, 66
Et	Me	Me	—	—	11
4-Thiazolecarboxylate					
Me	H	H	—	—	121v
Me	H	Ac	78–80	—	121v
Me	H	CHOHMe	89.92	—	121v
Et	H	H	51.5–53	—	23, 701, 150
			57		151, 152
Et	H	Me	58	—	1, 153, 154
Et	H	Ph	—	—	52
Et	H	$CH_2NHCOPh$	82–83	—	52
Et	H	$CHCH_3NHCOPh$	124	—	126
Et	H	$-CH(NHCOPh)CH(Me)CH_3$	119–120	—	1261
Et	Ph	Ph	110–111	—	41
$CH_2CH_2N(Et)_2$	H	Me	—	—	1
$p-NO_2C_6H_4$	H	H	—	—	76
5-Thiazolecarboxylate					
Me	H	H	68–69	—	155, 156
Me	H	Me	73–74	—	66, 157

Me	H	CH$_2$CO$_2$H	147–148	—	12
Me	H	o-CO$_2$HC$_6$H$_4$	—	—	158
Me	Me	H	88	—	7
Et	H	H	—	—	23, 39, 66, 135
Et	H	Me	—	98–99	1, 7, 39, 61, 66, 69 117, 125, 135, 137
Et	H	CH$_2$CO$_2$Et	—	—	12
Et	H	Ph	89–91	—	139
Et	H	o-CO$_2$HC$_6$H$_4$	—	—	158
Et	Me	H	—	117–120/9 mm	3, 149
Et	Me	Me	50–51	—	14, 59, 159
Et	Me	CO$_2$Et	—	—	53
Et	Et	Me	208–209	—	14
Et	Pr	Me	192–193	—	14
Et	i-Pr	Me	189–190	—	14
Et	Bu	Me	162–163	—	14
Et	i-Bu	Me	184–185	—	14
Et	Am	Me	146–147	—	14
Et	i-Am	Me	165–167	—	14
Et	CH$_2$NHCOPh	Me	—	—	123
Et	CH$_2$CONH$_2$	Me	91–92	—	160
Et	Ph	H	58–60	—	5
Et	Ph	Me	43	—	144
Et	CH$_2$Ph	Me	—	141–144/0.02 mm	161
Et	p-C$_6$H$_4$SO$_2$CH$_3$	Me	151	—	181
Et	NAcNH$_2$	Me	—	—	162
Et	Cl	H	—	—	148
Et	Br	H	175–176	—	148
Ph	H	H	—	—	86
CH$_2$CH$_2$N(Me)$_2$	H	H	—	—	11
CH$_2$CH$_2$N(Et)$_2$	H	H	—	101–105/10 mm	63
CH$_2$CH$_2$N(Et)$_2$	H	Me	—	—	1, 163, 164
CH$_2$CH$_2$N(Et)$_2$	Me	Me	—	—	1

TABLE IV-2. (Continued)

	Substituent			m.p. (°C)	b.p. (°C)	Ref.
R₁	R₂	R₃				
CH₂CH₂N(CH₂—CH₂)₂N—CH₂	H	Me		—	—	11
CH₂CH₂N(CH₂—CH₂)₂O	H	Me		—	—	1
CH₂CHMeN(Et)₂	H	Me		—	—	11
CH₂CMe₃	H	H		—	130–132/15 mm	63
CH₂CH₂Cl	H	Me		59	150–160/20 mm	1
CH₂CH₂I	H	Me		—	—	1
(3-pyridyl)CH₂	H	H		64	—	165
CH₂CH₂CH₂N(Et)₂	H	Me		—	—	11
(2-ClC₆H₄)O—CMe₂—CH₂	H	H		77	—	63
CH₂CH₂—N(CH₂CH₂)₂N—Ph	Pr	H		—	—	671, u, i

![structure] 4-Me-C6H4-piperazine-CH2CH2-	Pr	H	—	—	67 i, u, l
2-Me-C6H4-piperazine-CH2CH2-	Pr	H	—	—	67
PhCH2N-piperazine-CH2CH2-	Pr	H	—	—	67 i, u, l
2-MeO-C6H4-piperazine-CH2CH2-	Me	H	—	—	67 i, u, l
4-MeO-C6H4-piperazine-CH2CH2-	Pr	H	—	—	67 i, u, l
2-MeO-C6H4-piperazine-CH2CH2-	Pr	H	—	—	67 l, u, v
2,6-(MeO)2-C6H3-piperazine-CH2CH2-	Pr	H	—	—	67 l, u, i
2-EtO-C6H4-piperazine-CH2CH2-	Pr	H	—	—	67

TABLE IV-2. (*Continued*)

R_1	Substituent			m.p. (°C)	b.p. (°C)	Ref.
	R_2	R_3				
CH₂CH₂—N(piperazine)—C₆H₄(2-OMe)	Bu	H		—	—	67 u, i, l
CH₂CH₂—N(piperazine)—C₆H₄(2-OMe)	Pr	H		—	—	67 u, i, l
CH₂CH₂—N(piperazine)—C₆H₄(2-Cl)	Pr	H		—	—	67 u, i, l
CH₂CH₂—N(piperazine)—C₆H₄(3-CF₃)	Pr	H		—	—	67 u, i, l

TABLE IV-3. THIAZOLECARBOXYLIC ACID CHLORIDES

Substituent				
R_1	R_2	m.p. (°C)	b.p. (°C)	Ref.
4-Thiazolecarboxylic acid chlorides				
H	H	85	—	23, 71, 74, 75, 77
H	Ph	97–98	—	25
H	Cl	114–116	—	70, 76, 781
5-Thiazolecarboxylic acid chlorides				
H	H	—	—	2, 63, 68
H	Me	64	—	1, 65
Me	Me	—	—	7
Ph	H	125.3–126.5	—	5
Ph	Me	—	—	166
Cl	Me	—	—	72

TABLE IV-4. THIAZOLECARBOXAMIDES AND THIAZOLECARBOXYLIC ACID HYDRAZIDES

Substituent					
R_1	R_2	R_3	m.p. (°C)	b.p. (°C)	Ref. |// structure: R_2–thiazole(N,S)–R_3, C(=O)–R_1
NH_2	H	H	118	—	8, 167
NH_2	H	n-Bu	135–137	—	13
NH_2	Me	H	152	—	6
$NHNH_2$	H	H	175–176	—	92, 1681
NHNHEt	H	H	—	—	92
NHNHPh	H	H	—	—	92
$NHNHC_6H_3(NO_2)2,4$	H	H	—	—	92
$NHN(Et)_2$	H	H	—	—	92
$NHN=C(Me)_2$	Me	Ph	—	—	92
$NHN=CMeCO_2H$	H	H	—	—	92

structure: R_2–thiazole(N,S)–R_3 (2-position), C(=O)–R_1 (at 4-position)

R_1	R_2	R_3	m.p. (°C)	b.p. (°C)	Ref.
NH_2	H	H	152–153	—	23, 71, 75, 77, 81
NH_2	H	Me	152	—	1
NH_2	H	Ph	143	—	25
$N(Et)_2$	H	H	155	—	23

546

R	R₃		mp		Refs
o-NHC₆H₄NH₂	H	H	106.5–109	—	701, 78, 781
o-NHC₆H₄NH₂	H	Cl	148–150	—	78, 781
o-NHC₆H₄NO₂	H	H	145–146.5	—	70
o-NHC₆H₄NO₂	H	Cl	160–161	—	78, 781
m-NHC₆H₄NO₂	H	H	145–146	—	169
p-NHC₆H₄N(Me)₂	H	H	181–183 (HCl)	—	74
NH–C₆H₄–C₆H₄Fp (p)	H	H	212–213	—	73, 731
NH–C₆H₃(NO₂)–NMe₂	H	H	210.5–211	—	74
NH–C₆H₃(NO₂)–Ph	H	H	215–217	—	731
NH–C₆H₃(NO₂)–C₆H₄Fp	H	H	—	—	731
N-(thiazolyl)indole	H	H	—	—	170
HNNH₂	H	H	144	—	171

Structure: 2-R₂, 4-R₃, 5-C(=O)R₁ thiazole

TABLE IV-4. (Continued)

	Substituent					
R_1	R_2	R_3	m.p. (°C)	b.p. (°C)	Ref.	
NH_2	H	H	196	—	68, 87	
NH_2	H	Me	148	—	37, 80, 84, 172	
NH_2	Me	Me	—	—	1, 7, 80	
NH_2	Ph	H	213.7–214.5	—	5	
NHMe	Me	Me	—	—	7	
NHEt	Me	Me	69	—	7	
NHPr	Me	Me	39.5	195/24 mm	7	
NHBu	Me	Me	17.6	190/14 mm	7	
NH-amyl	Me	Me	—	116/0.83	7	
$NHCH_2Ph$	Me	Me	102	—	7	
NHPh	Me	Me	—	—	170	
NHPh	Ph	Me	—	—	64	
NH-pyridyl	H	H	143–144	—	173	
$NHCH_2CH_2$—⌬—SO_2NHR	H	Me	—	—	174	
$N(Me)_2$	H	Me	—	164/23 mm	1	
$N(Me)_2$	Me	Me	—	169/25 mm	9	
$N(Me)_2$	Ph	Pyridyl	—	—	175	
$N(Et)_2$	H	H	28	152/11 mm	2, 86, 89, 176	
$N(Et)_2$	H	Me	—	169–170/23 mm	1	
$N(Et)_2$	Me	Me	—	174/15 mm	7, 90	
$N(Pr)_2$	H	Me	—	—	177	
$N(Bu)_2$	H	Me	—	—	1, 50	
$N(Allyl)_2$	H	Me	—	—	1, 177	
N⟨CH_2-CH_2⟩$_2$	H	Me	—	197	1	

Structure	R1	R2	mp	bp	Refs
N(CH₂-CH₂)₂ (piperidine)	Me	Me	—	—	7, 90
N(CH₂-CH₂)₂O (morpholine)	H	Me	—	—	1
N(CH₂-CH₂)₂O (morpholine)	Me	Me	66	192/14 mm	7
[N(CH₂-CH₂)₂N]₂ (bis)	H	Me	—	—	1
NHNH₂	H	H	157–159	—	155, 178
NHNH₂	H	Me	166	—	66, 80, 101
HNH₂	Me	Me	—	—	7, 80, 179
NHNHSO₂C₆H₅	H	Me	170	—	1, 66
HNNHSO₂-C₆H₄-OMe	H	Me	164.3–164.9	—	66
HNNHSO₂-C₆H₄-NO₂	H	Me	214.5–215.5	—	66
HNNHSO₂-C₆H₄-Br	H	Me	194–193.3	—	66
N₃	H	H	103	—	155, 178
N₃	H	Me	83	—	80
N₃	Me	Me	—	—	80, 179

TABLE IV-5. THIAZOLEPOLYCARBOXYLIC ACIDS AND DERIVATIVES

	Substituent				
R_1	R_2	R_3	m.p. (°C)	b.p. (°C)	Ref.

[Structure: 2,4-diacyl thiazole with R_3 at 5-position, R_2CO at 4, R_1CO at 2]

R_1	R_2	R_3	m.p. (°C)	b.p. (°C)	Ref.
OH	OH	H	—	—	12, 38, 42, 121 v
OEt	OEt	H	43–44	—	38

[Structure: 2,5-diacyl thiazole with R_2 at 4-position, R_3CO at 5, R_1CO at 2]

R_1	R_2	R_3	m.p. (°C)	b.p. (°C)	Ref.
OH	H	OH	—	—	49
OH	Me	OH	—	—	7
OEt	Me	OEt	59	—	7
NH_2	Me	NH_2	—	—	7
NHMe	Me	NHMe	217	152–154/10 mm	7
NHEt	Me	NHEt	158	—	7
$N(Me)_2$	Me	$N(Me)_2$	—	—	7
$N(Et)_2$	Me	$N(Et)_2$	—	173/0.4 mm	7
morpholino	Me	morpholino	118	—	7
$NHNH_2$	Me	OEt	166	—	7
Cl	Me	Cl	—	—	7

[Structure: 4,5-diacyl thiazole with R_3 at 2-position, R_1CO at 4, R_2CO at 5]

R_1	R_2	R_3	m.p. (°C)	b.p. (°C)	Ref.
OH	OH	H	—	—	2, 4, 12, 39, 48
OH	OH	Me	—	—	3, 51
OH	OH	Et	212	—	17
OH	OH	Ph	—	—	5, 52
OH	OH	α-Furyl	225	—	17, 17 i
OH	OH	α-Thienyl	255	—	17 i
OH	OH	β-Thienyl	218	—	17 i
OEt	OH	α-Furyl	151	—	88
OEt	OMe	Ph	121	—	17 i

TABLE IV-5. (Continued)

	Substituent		m.p. (°C)	b.p. (°C)	Ref.
R_1	R_2	R_3			
OEt	OEt	H	—	175/12 mm	2
OEt	OEt	Me	—	—	3, 51
OEt	OEt	Et	—	152–157/ 2mm	
OEt	OEt	Ph	104	—	5, 17 i
OEt	OEt	CH=CHPh	125	—	17 i
OEt	OEt	α-Furyl	85	—	17 i
OEt	OEt	α-Thienyl	70	—	17 i
OEt	OEt	β-Thienyl	70	—	17
OEt	OMe	α-Thienyl	87	—	17
OEt	OH	α-Furyl	151	—	17
NH_2	NH_2	Me	200	—	7
NH_2	NH_2	Et	212	—	88 i
NH_2	NH_2	α-Furyl	303	—	88
NH_2	NH_2	α-Thienyl	323	—	88 i
NH_2	NH_2	β-Thienyl	291	—	88
$N(Et)_2$	$N(Et)_2$	H	44	—	2
NH_2	OH	Ph	290	—	88 i
NH_2	OH	α-Thienyl	268	—	88 i
NH_2	OH	α-Furyl	285	—	88 i
$NHCH_3$	OH	α-Thienyl	248	—	88 i
$NHNH_2$	$NHNH_2$	Me	243	—	7
$NHNH_2$	$NHNH_2$	Et	167	—	88 i
$NHNH_2$	$NHNH_2$	Ph	—	—	5
$NHNH_2$	$NHNH_2$	α-Thienyl	253	—	88 i
$NHNH_2$	$NHNH_2$	β-Thienyl	245	—	88 i
$NHNH_2$	OH	Ph	260	—	88
$NHNH_2$	OH	α-Thienyl	250	—	88 i
NHNHMe	OH	α-Thienyl	200	—	88 i, r
NHNHPh	OH	α-Thienyl	238	—	88 i, r
$NHN(Me)_2$	OH	α-Thienyl	208	—	88
$NHNH=C(Me)_2$	$NHN=C(Me)_2$	Ph	253	—	5
Cl	Cl	H	—	—	2
Cl	Cl	Me	—	—	7

OH	OH	OH	—	—	38
OEt	OEt	OEt	—	—	38
NH_2	NH_2	NH_2	—	—	38

TABLE IV-6. CYANOTHIAZOLES

Substituent				
R_1	R_2	m.p. (°C)	b.p. (°C)	Ref.
2-Cyanothiazole				
H	H	31	—	167
4-Cyanothiazole				
H	H	57–61	100–120/20 mm	71, 73, 75, 77, 81, 91, 94, 95, 121 v, 180, 181
H	Me	—		182
5-Cyanothiazole				
H	H	53	90–93/15 mm	87, 91
H	Me	—	—	37
H	Ac	125	—	91 i
H	COEt	113	—	91
Me	Ac	84	—	91 i
Me	COEt	80	—	91 i
Cl	Cl	—	—	183
4-5-Dicyanothiazole				
	H	—	—	91
	Me	98	—	91 i
	Et	76	—	91 i
	Ph	147	—	91 i
	α-Furyl	77	—	91 i
	α-Thienyl	183	—	91 i
	β-Thienyl	150	—	91 i
	Cl	51	—	91 i
	CN	127	—	38

TABLE IV-7. THIAZOLECARBOXALDEHYDES

Substituent				
R_1	R_2	m.p. (°C)	b.p. (°C)	Ref.
2-Thiazolecarboxaldehyde				
H	H	—	63/10 mm	97, 99 p, 100, 139
H	Me	—	49.5/3 mm	100
			80–80.5/10 mm	97
Me	H	—	73–76/11 mm	30, 66, 97
Me	Me	—	62/3 mm	100
Me	CH_2CO_2H	—	—	96
Ph	H	66.5–67.5	—	98, 184
Ph	Me	—	—	163
Ph	Ph	—	—	170
p-$NO_2C_6H_4$	H	—	—	29
p-$NO_2C_6H_4$	CO_2H	—	—	29
4-Thiazolecarboxaldehyde				
H	H	63–65	99–104/14 mm	21, 24 l, 66, 185
H	Ph	—	—	35, 186
H	$PhCH_2$	—	—	36
H	p-$CH_3C_6H_4$	—	—	35
H	m-$CH_3C_6H_4$	—	—	35
H	p-BrC_6H_4	—	—	35
H	p-$NO_2C_6H_4$	—	—	35, 106, 186
Cl	Ph	91–29	—	103 i
Cl	p-$NO_2C_6H_4$	162–163	—	103 i
Br	C_6H_5	—	—	35
5-Thiazolecarboxaldehyde				
H	H	90–94	—	187
H	Me	74.5–75.3	—	66 l, 101, 188
		72.5	112–118/21 mm	37, 108
Me	H	99–101	110–116/18 mm	187
Ph	H	93–94	—	29, 187, 189
$PhCH_2$	H	—	136–138	187, 189
o-ClC_6H_4	H	100–101	—	187, 189
4-(2-ethylpyridyl)	H	61–62	—	189
4,5-Thiazoledicarboxaldehyde				
	H	65–66	—	91 i
	Me	71–72	—	91 i

TABLE IV-8. THIAZOLYLKETONES

2-Thiazolylketones

R₁	R₂	R₃	m.p.	b.p.	Ref.
Me	H	H	—	95–105/15 mm	119
Me	H	Me	—	—	190
Me	H	Et	—	—	190
Me	Me	H	34	90–93/12 mm	111 u, 190
Me	Me	Me	—	65–75/6 mm	190, 191
Me	Et	Et	—	—	190
Me	CH₂CH₂OH	Me	—	118–134/6 mm	191
Me	Ph	H	78–79	—	105, 110, 192
Me	p-NO₂C₆H₄	H	185	—	112
Ph	Me	Me	88–89	—	113, 191
Ph	CH₂OH	Me	84	—	113
p-NO₂C₆H₄	H	H	118–119	—	115
Pyridyl-3	H	H	74–75	—	115
CH₂Br	H	H	—	—	119
CH₂Br	p-NO₂C₆H₄	H	—	195–196	112
CH₂Cl	Ph	H	—	—	105
CH₂Cl	(3,4-dihydroxyphenyl)	H	—	—	193
CH₂CH₂NH₂	Ph	H	—	—	110
CH₂CH₂N(Me)₂	Ph	H	—	—	110
CH₂CH₂N(Et)₂	Ph	H	—	—	110
CH₂CH₂N(Pr)₂	Ph	H	—	—	110
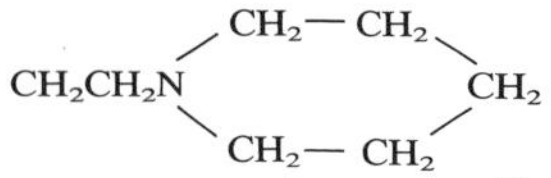 CH₂CH₂N(morpholinyl)	Ph	H	—	—	110
N₃	p-NO₂C₆H₄	H	—	—	194
N₃	p-NC₂C₆H₄	H	—	—	194
N₃	p-BrC₆H₄	H	—	—	194
N₃	p-MeC₆H₄	H	—	—	194

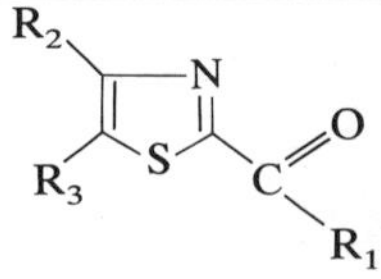

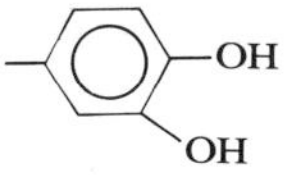

4-Thiazolylketone

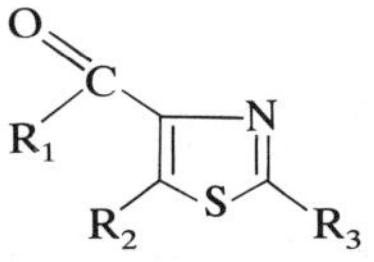

R₁	R₂	R₃	m.p.	b.p.	Ref.
Me	H	H	56	—	116, 195 a
Me	H	Ph	90–92	—	28, 35
Me	H	m-MeC₆H₄	—	—	35
Me	H	p-MeC₆H₄	—	—	35

IV. Tables of Compounds

TABLE IV-8. (Continued)

R₁	R₂	R₃	m.p.	b.p.	Ref.
Me	H	p-NO$_2$C$_6$H$_4$	—	—	35
Me	H	p-BrC$_6$H$_4$	—	—	35
Me	Me	Me	—	—	195
Me	OMe	OMe	—	—	195
Me	OEt	OEt	—	—	195
Me	Br	Ph	—	—	35

5-Thiazolylketone

$$\underset{R_1}{\overset{R_3}{\underset{}{\text{O=C}}}}\underset{S}{\overset{N}{\diagdown}}R_2$$

R₁	R₂	R₃	m.p.	b.p.	Ref.
Me	H	H	—	—	134
Me	H	Me	28	88/6 mm	120
			27.5	—	56, 57, 117
Me	Me	Me	—	—	80, 118, 120
Me	Ph	H	151–152	—	28
Ph	Ph	H	—	—	196
p-MeOC$_6$H$_4$	(R–S ring, 4-methyl)	H	—	—	197
p-BrC$_6$H$_4$	p-MeC$_6$H$_4$	H	—	—	196
F[19]	H	H	—	—	198
CH$_2$CO$_2$H	H	Me	—	—	56
(CH$_2$)$_3$OMe	H	H	—	154–155/14 mm	87
CH$_2$Cl	H	H	—	—	82
CH$_2$Cl	H	Me	76	—	69
CH$_2$Br	H	Me	60	—	120
CH$_2$	Me	Me	—	—	120
CHN$_2$	H	H	—	—	82
CHN$_2$	H	Me	—	—	69

TABLE IV-9. THIAZOLE AMIDINE AND OTHER COMPOUNDS

	Substituent				
R_1	R_2	R_3	m.p.	b.p.	Ref.

2-Thiazole amidine

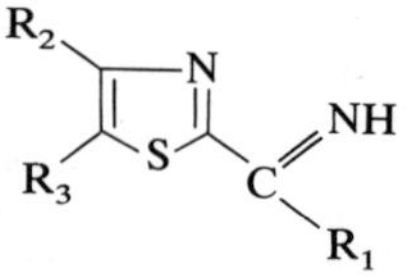

R_1	R_2	R_3	m.p.	b.p.	Ref.
NH_2	H	H	—	—	182
NH_2	Me	H	—	—	182

4-Thiazole amidine

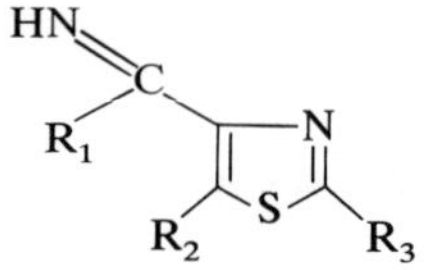

R_1	R_2	R_3	m.p.	b.p.	Ref.
NH_2	H	H	—	—	182
NH_2	H	Me	—	—	182
NHPh	H	H	255–227 (HCl)	—	71, 75 l, 77, 81
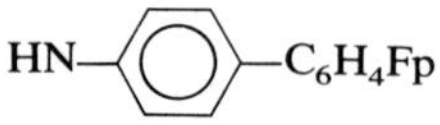 HN—⟨○⟩—C_6H_4Fp	H	H	157	—	73
NClPh	H	H	—	—	81
NBrPh	H	H	—	—	81

Arylhydrazones of ethyl ester and amide of 4-phenylthiazolylglyoxalic acid

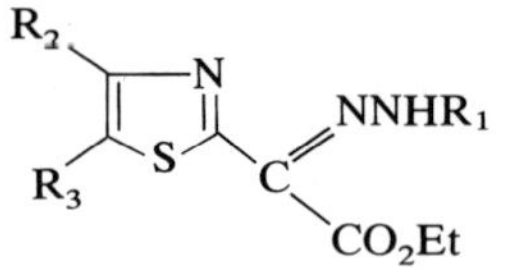

R_1	R_2	R_3	m.p.	b.p.	Ref.
Ph	Ph	H	161–162	—	119 u
p-MeC$_6$H$_4$	Ph	H	132–133	—	199 u
o-EtOC$_6$H$_4$	H	Ph	164–165	—	119 u
o-PrOC$_6$H$_4$	Ph	H	105–106	—	199 u
p-BrC$_6$H$_4$	Ph	H	174–175	—	199 u
m-ClC$_6$H$_4$	Ph	H	149–150	—	199 u
OEt	H	Me	—	—	182

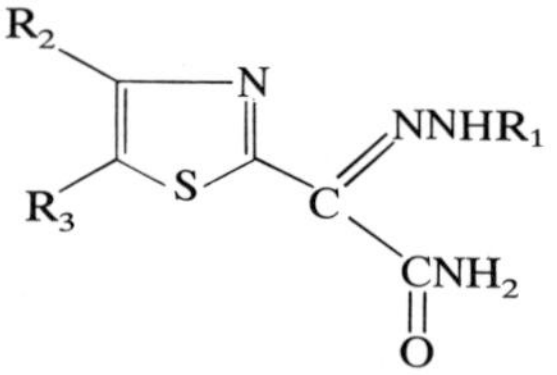

R_1	R_2	R_3	m.p.	b.p.	Ref.
o-MeOC$_6$H$_4$	Ph	H	203–204	—	199 u
o-EtOC$_6$H$_4$	Ph	H	226–227	—	199 u
o-PrOC$_6$H$_4$	Ph	H	213–214	—	199 u
o-ClC$_6$H$_4$	Ph	H	218–219	—	199 u
o-BrC$_6$H$_4$	Ph	H	209–210	—	199 u
5,2-Cl(MeO)C$_6$H$_3$	Ph	H	—	—	199 u

V. REFERENCES

1. E. R. H. Jones, F. A. Robinson, and M. N. Strachan, *J. Chem. Soc.*, 87 (1946); *Chem. Abstr.*, **40,** 3441.
2. H. Erlenmeyer and H. Meyenburg, *Helv. Chim. Acta*, **20,** 204 (1937); *Chem. Abstr.*, **31,** 3916.
3. A. Schoberl and M. Stock, *Chem. Ber.*, **73B,** 1240 (1940); *Chem. Abstr.*, **35,** 3636.
4. H. Erlenmeyer and C. J. Morel, *Helv. Chim. Acta*, **25,** 1073 (1942); *Chem. Abstr.*, **37,** 1702.
5. E. H. Huntress and K. Pfister, *J. Amer. Chem. Soc.*, **65,** 2167 (1943); *Chem. Abstr.*, **38,** 98.
6. H. Schenkel, E. Marbet, and H. Erlenmeyer, *Helv. Chim. Acta*, **27,** 1437 (1944); *Chem. Abstr.*, **39,** 3281.
7. W. R. Boon, *J. Chem. Soc.*, 601 (1945); *Chem. Abstr.*, **40,** 573.
8. H. Erlenmeyer, R. Marbet, and H. Schenkel, *Helv. Chim. Acta*, **28,** 924 (1945); *Chem. Abstr.*, **40,** 869.
9. U.S. Patent No. 2 396 893; *Chem. Abstr.*, **40,** 3779.
10. E. B. Knott, *J. Chem. Soc.*, 1656 (1947); *Chem. Abstr.*, **42,** 2969.
11. M. Erne and H. Erlenmeyer, *Helv. Chim. Acta*, **31,** 652 (1948); *Chem. Abstr.*, **42,** 4575.
12. H. Erlenmeyer, J. Junod, W. Guex, and M. Erne, *Helv. Chim. Acta*, **31,** 1342 (1948); *Chem. Abstr.*, **43,** 5021.
13. L. Herzfeld, B. Prijs, and H. Erlenmeyer, *Helv. Chim. Acta*, **36,** 1842 (1953); *Chem. Abstr.*, **49,** 1012.
14. A. L. Mndzhoyan, A. A. Aroyan, M. A. Kaldrikyan, T. R. Ovsepyan, and R. H. Arshakyan, *Izvest. Akad. Nauk Armyan. S.S.R., Khim. Nauk*, **17,** 204 (1964); *Chem. Abstr.*, **61,** 8298.
15. East German Patent No. 38 188; *Chem. Abstr.*, **63,** 13269.
16. G. E. Hall and J. Walker, *J. Chem. Soc. C*, **6,** 1357 (1966); *Chem. Abstr.*, **65,** 12189G.
17. M. Robba and Y. Le Guen, *Bull. Soc. Chim. France*, **5,** 1762 (1969); *Chem. Abstr.*, **71,** 81255.
18. H. S. Forrest, A. T. Fuller, and J. Walker, *J. Chem. Soc.*, 1501 (1948).
19. South African Patent No. 69 05 123; *Chem. Abstr.*, **75,** 49065.
20. K. A. Maier and O. Hromatka, *Monatsh.*, **102,** 102 (1971); *Chem. Abstr.*, **74,** 125538.
21. French Patent No. 1 471 783; *Chem. Abstr.*, **68,** 69004.
22. R. V. Jones and H. R. Henze, *J. Amer. Chem. Soc.*, **64,** 1669 (1942); *Chem. Abstr.*, **36,** 5822.
23. H. Erlenmeyer and C. J. Morel, *Helv. Chim. Acta*, **28,** 362 (1945); *Chem. Abstr.*, **40,** 1501.
24. Netherlands Appl. No. 6 413 477; *Chem. Abstr.*, **65,** 8919C.
25. E. H. Huntress and K. Pfister, *J. Amer. Chem. Soc.*, **65,** 1667 (1943); *Chem. Abstr.*, **37,** 6263.
26. E. N. Odintsova, *Compt. Rend. Acad. Sci. U.R.S.S. (in English)*, **41,** 250 (1943); *Chem. Abstr.*, **38,** 4644.
27. U.S. Patent No. 3 374 207; *Chem. Abstr.*, **66,** 287580.
28. E. Haruki, S. Izumita, and E. Imoto, *Nippon Kagaku Zasshi*, **86,** 942 (1965); *Chem. Abstr.*, **65,** 13688A.
29. A. Silberg, A. Benko, and G. Csavassy, *Berichte*, **97,** 1684 (1964); *Chem. Abstr.*, **61,** 16061.

30. H. Kondo and F. Nagasawa, *J. Pharm. Soc. Japan*, **57**, 249 (1937).
31. J. Metzger and B. Koether, *Bull. Soc. Chim. France*, **20**, 708 (1953); *Chem. Abstr.*, **48**, 10738.
32. H. Beyerman, P. Berben, and J. Bontekoe, *Rec. Trav. Chim.*, **73**, 325 (1954); *Chem. Abstr.*, **49**, 6230.
33. B. Mikhailov and V. Bronovitskaya, *Zh. Obshchei Khim.*, **26**, 66 (1956); *Chem. Abstr.*, **50**, 13874.
34. P. E. Iversen, *Acta Chem. Scand.*, **22**, 694 (1968); *Chem. Abstr.*, **69**, 77145.
35. I. Simiti and M. Farkas, *Bull. Soc. Chim. France*, **9**, 3862 (1968); *Chem. Abstr.*, **70**, 3914.
36. I. Simiti and G. Hintz, *Pharmazie*, **27**, 146 (1972); *Chem. Abstr.*, **77**, 5393.
37. C. R. Harington and G. Moggridge, *J. Chem. Soc.*, 443 (1939); *Chem. Abstr.*, **33**, 4242.
38. H. Erlenmeyer, J. Junod, W. Guex, and M. Erne, *Helv. Phys. Acta*, **31**, 1342 (1948); *Chem. Abstr.*, **43**, 5021.
39. German Patent No. 658 353.
40. H. Carrington, *J. Chem. Soc.*, 1619 (1948); *Chem. Abstr.*, **43**, 1766.
41. A. H. Cook, J. D. Downer, and I. H. Heilbron, *J. Chem. Soc.*, 2028 (1948); *Chem. Abstr.*, **43**, 3408.
42. P. Brookes, A. T. Fuller, and J. Walker, *J. Chem. Soc.*, 689 (1957); *Chem. Abstr.*, **51**, 9590.
43. M. Bodanszky, J. Alicino, C. A. Birkhimer, and N. J. Williams, *J. Amer. Chem. Soc.*, **84**, 2003 (1962); *Chem. Abstr.*, **57**, 11181.
44. A. Aszalos, A. I. Cohen, J. Alicino, and B. T. Keeler, *Antimicrob. Agents Chemother.*, 456 (1967); *Chem. Abstr.*, **70**, 58259.
45. British Patent No. 456 735; *Chem. Abstr.*, **31**, 2232.
46. J. M. Sprague and A. H. Land, Heterocyclic Compounds, R. Elderfield Ed., 1957, Vol. V, p. 484.
47. H. Erlenmeyer, A. Epprecht, and H. Meyenburg, *Helv. Chim. Acta*, **20**, 310 (1937); *Chem. Abstr.*, **31**, 3765.
48. H. Schenkel and M. Schenkel, *Helv. Chim. Acta*, **31**, 924 (1948); *Chem. Abstr.*, **42**, 5906.
49. M. Erne, *Helv. Chim. Acta*, **36**, 138 (1953); *Chem. Abstr.*, **48**, 2688.
50. J. M. Sprague, R. M. Lincoln, and C. Ziegler, *J. Amer. Chem. Soc.*, **68**, 266 (1946); *Chem. Abstr.*, **40**, 2126.
51. P. Aubert and E. Knott, *Nature*, **166**, 1039 (1950); *Chem. Abstr.*, **45**, 8006.
52. H. Erlenmeyer, P. Buchmann, and H. Schenkel, *Helv. Chim. Acta*, **27**, 1432 (1944); *Chem. Abstr.*, **39**, 3281.
53. L. Conover and D. Tarbell, *J. Amer. Chem. Soc.*, **72**, 5221 (1950); *Chem. Abstr.*, **45**, 5148.
54. L. R. Cerecedo and J. G. Tolpin, *J. Amer. Chem. Soc.*, **59**, 1660 (1937); *Chem. Abstr.*, **31**, 7873.
55. S. Fallab, *Helv. Chim. Acta*, **35**, 215 (1952); *Chem. Abstr.*, **46**, 8647.
56. P. Baumgarten, A. Dornow, K. Gutschmidt, and H. Krehl, *Berichte*, **75B**, 442 (1942); *Chem. Abstr.*, **37**, 3091.
57. E. R. Buchman and E. M. Richardson, *J. Amer. Chem. Soc.*, **67**, 395 (1945); *Chem. Abstr.*, **39**, 1871.
58. P. E. Iversen and H. Lund, *Acta Chem. Scand.*, **21**, 389 (1967); *Chem. Abstr.*, **67**, 17255.
59. French Patent No. 2 047 876; *Chem. Abstr.*, **76**, 14325.

V. References

60. British Patent No. 1 264 197; *Chem. Abstr.*, **76,** 140780.
61. M. L. Tomlinson, *J. Chem. Soc.*, 1030 (1935); *Chem. Abstr.*, **29,** 6596.
62. British Patent No. 536 951; *Chem. Abstr.*, **36,** 1806.
63. German Patent No. 1 936 751; *Chem. Abstr.*, **72,** 79023.
64. C. C. Kartha, *Current Sci.*, **41,** 635 (1972); *Chem. Abstr.*, **77,** 164581.
65. E. H. Wells, *J. Ass. Offic. Agr. Chem.*, **25,** 747 (1942); *Chem. Abstr.*, **36,** 6750.
66. E. Campaigne, R. L. Thompson, and J. E. Van Werth, *J. Med. Pharm. Chem.*, **1,** 577 (1959); *Chem. Abstr.*, **54,** 22573.
67. French Patent No. 2 247 243; *Chem. Abstr.*, **83,** 206330.
68. H. Erlenmeyer, E. H. Schmid, and A. Kleiber, *Helv. Chim. Acta*, **25,** 375 (1942); *Chem. Abstr.*, **36,** 5816.
69. P. Karrer and W. Graf, *Helv. Chim. Acta*, **28,** 824 (1945); *Chem. Abstr.*, **40,** 1501.
70. D. J. Tocco, R. P. Buhs, H. D. Brown, A. R. Matzuk, H. E. Mertel, R. E. Harman, and N. R. Trenner, *J. Med. Chem.*, **7,** 399 (1964), *Chem. Abstr.*, **61,** 16627.
71. British Patent No. 988 784; *Chem. Abstr.*, **63,** 16357.
72. Belgian Patent No. 624 259; *Chem. Abstr.*, **59,** 12956.
73. Netherlands Appl. No. 6 412 728; *Chem. Abstr.*, **64,** 3560E.
74. Netherlands Appl. No. 6 413 475; *Chem. Abstr.*, **64,** 8191E.
75. U.S. Patent No. 3 236 855; *Chem. Abstr.*, **64,** 15891F.
76. U.S. Patent No. 3 294 811; *Chem. Abstr.*, **66,** 55481A.
77. U.S. Patent No. 3 299 082; *Chem. Abstr.*, **66,** 85784U.
78. Hungarian Patent No. 155 098; *Chem. Abstr.*, **70,** 47432.
79. Y. Tamura, H. Hayashi, E. Saeki, J.-H. Kim, and M. Ikeda, *J. Heterocyclic Chem.*, **11,** 459 (1974).
80. K. Ganapathi and A. Venkataraman, *Proc. Indian Acad. Sci.*, **22A,** 343 (1945); *Chem. Abstr.*, **40,** 4056.
81. British Patent No. 1 009 807; *Chem. Abstr.*, **64,** 8191C.
82. H. Erlenmeyer and J. Ostertag, *Helv. Chim. Acta*, **31,** 26 (1948); *Chem. Abstr.*, **42,** 4166.
83. M. Aeberli and H. Erlenmeyer, *Helv. Chim. Acta*, **33,** 503 (1950); *Chem. Abstr.*, **44,** 8351.
84. F. Lipman and G. Perlmann, *J. Amer. Chem. Soc.*, **60,** 2574 (1938); *Chem. Abstr.*, **33,** 597.
85. H. Erlenmeyer and W. Wurgler, *Helv. Chim. Acta*, **25,** 249 (1942); *Chem. Abstr.*, **37,** 377.
86. H. Erlenmeyer and K. Degen, *Helv. Chim. Acta*, **29,** 1080 (1946); *Chem. Abstr.*, **41,** 448.
87. H. Erlenmeyer and R. Marbet, *Helv. Chim. Acta*, **29,** 1946 (1946); *Chem. Abstr.*, **41,** 1668.
88. M. Robba and Y. Le Guen, *Bull. Soc. Chim. France*, **6,** 2152 (1969); *Chem. Abstr.* **71,** 81254.
89. German Patent No. 668 874.
90. British Patent No. 1 211 889; *Chem. Abstr.*, **74,** 139879.
91. M. Robba and Y. Le Guen, *Bull. Soc. Chim. France*, **11,** 4026 (1969); *Chem. Abstr.*, **72,** 66856.
92. British Patent No. 768 482; *Chem. Abstr.*, **51,** 15590.
93. A. E. Lipkin and V. A. Smirnov, *Khim. Geterotsikl. Soedinenii*, **3,** 571 (1968); *Chem. Abstr.*, **69,** 96555.
94. Netherlands Appl. No. 6 509 377; *Chem. Abstr.*, **65,** 13715G.
95. E. Fisher, *Chimia*, **22,** 437 (1968); *Chem. Abstr.*, **70,** 37624.

96. H. Gregory and L. F. Wiggins, *J. Chem. Soc.*, 1400 (1947); *Chem. Abstr.*, **42,** 1262.
97. P. E. Iversen and H. Lund, *Acta Chem. Scand.*, **20,** 2649 (1966); *Chem. Abstr.*, **66,** 104953.
98. J. F. Olin and T. B. Johnson, *J. Amer. Chem. Soc.*, **53,** 1470 (1931); *Chem. Abstr.*, **25,** 2722.
99. J. Tirouflet, E. Laviron, J. Metzger, and J. Boichard, *Coll. Czech. Chem. Commun.*, **25,** 3277 (1960); *Chem. Abstr.*, **56,** 13956.
100. H. Beyer, U. Hess, and W. Liebenow, *Chem. Ber.*, **90,** 2372 (1957); *Chem. Abstr.*, **52,** 10057.
101. E. R. Buchman and E. M. Richardson, *J. Amer. Chem. Soc.*, **61,** 891 (1939); *Chem. Abstr.*, **33,** 4242.
102. A. Silberg, I. Simiti, and H. Mantsch, *Acad. Rep. Populare Romine, Stud. Cercetari Chim.*, **12,** 315 (1961); *Chem. Abstr.*, **59,** 591.
103. A. Silberg and Z. Frenkel, *Stud. Univ. Babes Bolyai, Ser. Chem.*, **10,** 27 (1965); *Chem. Abstr.*, **63,** 16327.
104. J. F. Olin and T. B. Johnson, *J. Amer. Chem. Soc.*, **53,** 1475 (1931); *Chem. Abstr.*, **25,** 2722.
105. J. F. Olin and T. B. Johnson, *J. Amer. Chem. Soc.*, **53,** 1473 (1931); *Chem. Abstr.*, **25,** 2722.
106. A. Silberg, Z. Frenkel, and L. Cormos, *Stud. Univ. Babes Bolyai, Ser. Chem.*, **8,** 273 (1963); *Chem. Abstr.*, **61,** 16060.
107. B. C. Jansen, *Z. Vitaminforsch.*, **7,** 239 (1938); *Chem. Abstr.*, **33,** 3834.
108. C. R. Harington and R. C. G. Moggridge, *Biochem. J.*, **34,** 685 (1940); *Chem. Abstr.*, **34,** 5447.
109. F. F. Heyroth, *Bull. Basic. Sci. Res.* **4,** 1 (1932); *Chem. Abstr.*, **27,** 2183.
110. G. A. Levy and H. B. Nisbet, *J. Chem. Soc.*, 1053 (1938); *Chem. Abstr.*, **32,** 7453.
111. K. Daigo and L. J. Reed, *J. Amer. Chem. Soc.*, **84,** 659 (1962); *Chem. Abstr.*, **57,** 11182.
112. A. Silberg and A. Benko, *Berichte*, **97,** 3045 (1964); *Chem. Abstr.*, **62,** 1642.
113. E. Ochiai, *J. Pharm. Soc. Japan*, **60,** 55 (1940).
114. E. Ochiai and H. Nagasawa, *Berichte*, **72B,** 1470 (1939); *Chem. Abstr.*, **33,** 7783.
115. V. G. Ermolaeva and M. N. Shchukina, *Zh. Obshchei Khim.*, **33,** 2716 (1963); *Chem. Abstr.*, **60,** 514.
116. H. Erlenmeyer and H. Ueberwasser, *Helv. Chim. Acta*, **23,** 197 (1940).
117. K. Ganapathi and A. Venkataraman, *Proc. Indian Acad. Sci.*, **22A,** 362 (1945).
118. J. L. Smith and R. H. Sapiro, *Trans. Roy. Soc. S. Africa*, **18,** 229 (1929); *Chem. Abstr.*, **24,** 2130.
119. H. Erlenmeyer, O. Weber, P. Schmidt, G. Kung. C. Zinsstag, and B. Prijs, *Helv. Chim. Acta*, **31,** 1142 (1948); *Chem. Abstr.*, **42,** 7291.
120. E. D. Sych and E. D. Smaznaya-Ll'Ina, *Zh. Obshchei Khim.*, **32,** 984 (1962); *Chem. Abstr.*, **58,** 6953.
121. P. Brookes, R. J. Clark, A. T. Fuller, M. P. V. Mijovic, and J. Walker, *J. Chem. Soc.*, 916 (1960); *Chem. Abstr.*, **56,** 14389.
122. J. D'A. Jeffery, P. F. Abraham, and G. G. F. Newton, *Biochem. J.*, **75,** 216 (1960); *Chem. Abstr.*, **54,** 17373.
123. A. Goldberg and W. Kelly, *J. Chem. Soc.*, 1372 (1947); *Chem. Abstr.*, **42,** 1263.
124. D. F. W. Cross, G. W. Kenner, R. C. Sheppard, and C. E. Stehr, *J. Chem. Soc.*, 2143 (1963); *Chem. Abstr.*, **59,** 741.
125. H. T. Clarke and S. Gurin, *J. Amer. Chem. Soc.*, **57,** 1876 (1935).
126. D. M. Dean, M. P. V. Mijovic, and J. Walker, *J. Chem. Soc.*, 3394 (1961); *Chem. Abstr.*, **56,** 14253.

V. References

127. J. Laudon and B. Sjogren, *Svensk Kem. Tidskr.*, **52,** 64 (1940); *Chem. Abstr.*, **34,** 4734.
128. A. Berlin and V. Maimind, *Dokl. Akad. Nauk S.S.S.R.*, **60,** 1181 (1948); *Chem. Abstr.*, **42,** 7292.
129. G. W. Kenner, R. C. Sheppard, and C. E. Stehr, *Tetrahedron Lett.*, **1,** 23 (1960); *Chem. Abstr.*, **54,** 13104.
130. G. O. Doak, H. G. Steinmam, and H. Eagle, *J. Amer. Chem. Soc.*, **62,** 3012 (1940); *Chem. Abstr.*, **35,** 86.
131. A. Cook, G. Harris, J. Pollock, and J. M. Swan, *J. Chem. Soc.*, 1947 (1950); *Chem. Abstr.*, **45,** 1999.
132. K. A. Maier and O. Hromatka, *Monatsh.*, **102,** 102 (1971); *Chem. Abstr.*, **74,** 125538.
133. Swiss Patent No. 192 849.
134. N. K. Kochetkov, E. E. Nifant'Ev, and N. V. Molodtsov, *Zh. Obshchei Khim.*, **29,** 2330 (1959); *Chem. Abstr.*, **54,** 14230.
135. J. J. D'Amico and T. W. Bartram, *J. Org. Chem.*, **25,** 1336 (1960); *Chem. Abstr.*, **54,** 24662.
136. E. R. Buchman, R. R. Williams, and J. C. Keresztesy, *J. Amer. Chem. Soc.*, **57,** 1849 (1935).
137. A. Wohl and K. Jaschinowski, *Berichte*, **54,** 476 (1921).
138. A. Pullman and J. Metzger, *Bull. Soc. Chim. France*, 1021 (1948); *Chem. Abstr.*, **43,** 2511.
139. P. E. Iversen, *Acta Chem. Scand.*, **24,** 2459 (1970); *Chem. Abstr.*, **74,** 76364.
140. H. Erlenmeyer and H. M. Weber, *Helv. Chim. Acta*, **21,** 863 (1938); *Chem. Abstr.*, **32,** 7915.
141. H. Erlenmeyer, H. M. Weber, and P. Wiessmer, *Helv. Chim. Acta*, **21,** 1017 (1938); *Chem. Abstr.*, **33,** 603.
142. German Patent No. 705 432; *Chem. Abstr.*, **36,** 1950.
143. German Patent No. 1 936 695; *Chem. Abstr.*, **74,** 53774.
144. K. Hubacher, *Annalen*, **259,** 228 (1890).
145. German Patent No. 1 959 307; *Chem. Abstr.*, **75,** 99256.
146. German Patent No. 2 045 818; *Chem. Abstr.*, **75,** 36009.
147. M. Ohta, R. Sudo, and K. Kato, *J. Pharm. Soc. Japan*, **68,** 40 (1948); *Chem. Abstr.*, **44,** 3974.
148. G. Klein and B. Prijs, *Helv. Chim. Acta*, **37,** 2057 (1954); *Chem. Abstr.*, **50,** 307.
149. H. Erlenmeyer and P. Schmidt, *Helv. Chim. Acta*, **29,** 1957 (1946); *Chem. Abstr.*, **41,** 1667.
150. H. Erlenmeyer and D. Markees, *Helv. Chim. Acta*, **29,** 1229 (1946).
151. C. Zee, Y. C. Kwang, C. Chia, *J. Heterocyclic Chem.*, **7,** 1439 (1970); *Chem. Abstr.*, **74,** 53623.
152. P. P. Didenko, *Tr. Vses. Inst. Gel Mintol.* **15,** 103 (1969); *Chem. Abstr.*, **75,** 47793.
153. V. M. Zubarovskii and R. N. Moskaleva, *Zh. Obshchei Khim.*, **32,** 570 (1962); *Chem. Abstr.*, **58,** 2525.
154. K. Ganapathi and A. Venkataraman, *Proc. Indian Acad. Sci.*, **22A,** 359 (1945); *Chem. Abstr.*, **40,** 4059.
155. H. Erlenmeyer, W. Mengisen, and B. Prijs, *Helv. Chim. Acta*, **30,** 1865 (1947); *Chem. Abstr.*, **43,** 221.
156. H. Erlenmeyer, W. Buchler, and H. Lehr, *Helv. Chim. Acta*, **27,** 969 (1944); *Chem. Abstr.*, **39,** 1641.
157. Y. M. Slobodin and M. S. Ziegel, *J. Gen. Chem. U.S.S.R.*, **11,** 1019 (1941); *Chem. Abstr.*, **36,** 6542.

158. V. Barkane and E. Gudriniece, *Latv. Psr Zinat. Akad. Vestis, Kim. Ser.*, **6,** 695 (1970); *Chem. Abstr.*, **74,** 53619.
159. R. Metze, *Chem. Berichte*, **89,** 2056 (1956); *Chem. Abstr.*, **51,** 7383.
160. M. Ohta and K. Kato, *J. Pharm. Soc. Japan*, **67,** 136 (1947); *Chem. Abstr.*, **45,** 9532.
161. D. Brown, A. Cook, and I. Heilbron, *J. Chem. Soc.*, 5106 (1949); *Chem. Abstr.*, **44,** 145.
162. I. Simiti and M. Coman, *Ann. Chim. France*, **8,** 373 (1973).
163. J. L. Smith, *J. Chem. Soc.*, **123,** 2288 (1923); *Chem. Abstr.*, **18,** 264.
164. J. M. Sprague and L. W. Kissinger, *J. Amer. Chem. Soc.*, **63,** 578 (1941); *Chem. Abstr.*, **35,** 2144.
165. U.S. Patent No. 3 491 107; *Chem. Abstr.*, **72,** 79011.
166. Japan Patent No. 65 13504; *Chem. Abstr.*, **64,** 9858D.
167. D. D. Libman and R. Slack, *J. Chem. Soc.*, 2253 (1956); *Chem. Abstr.*, **51,** 1957.
168. British Patent No. 768 481; *Chem. Abstr.*, **51,** 15590.
169. French Patent No. 1 413 905; *Chem. Abstr.*, **64,** 6658H.
170. German Patent No. 2 415 351; *Chem. Abstr.*, **82,** 4241.
171. M. C. Dodson and W. R. Todd, *J. Lab. Clin. Med.*, **30,** 891 (1945); *Chem. Abstr.*, **40,** 916.
172. P. Karrer, W. Graf, and J. Schukri, *Helv. Chim. Acta*, **28,** 1523 (1945); *Chem. Abstr.*, **40,** 1500.
173. H. Erlenmeyer, H. Bloch, and H. Kiefer, *Helv. Chim. Acta*, **25,** 1066 (1942); *Chem. Abstr.*, **37,** 1460.
174. German Patent No. 2 410 003; *Chem. Abstr.*, **82,** 4242.
175. German Patent No. 2 350 222; *Chem. Abstr.*, **81,** 29516.
176. Swiss Patent No. 192 930.
177. M. Soodak and L. R. Cerecedo, *J. Amer. Chem. Soc.*, **66,** 1988 (1944); *Chem. Abstr.*, **39,** 519.
178. M. Aeberli and H. Erlenmeyer, *Helv. Chim. Acta*, **91,** 470 (1948); *Chem. Abstr.*, **42,** 4574.
179. K. A. Jensen and O. Rosenlund Hansen, *Dansk Tidsskr. Farm.*, **17,** 189 (1943); *Chem. Abstr.*, **39,** 2058.
180. British Patent No. 988 956; *Chem. Abstr.*, **63,** 11568.
181. M. Mansson and S. Sunner, *Acta Chem. Scand.*, **20,** 845 (1966); *Chem. Abstr.*, **65,** 6385C.
182. French Patent No. 1 487 304; *Chem. Abstr.*, **69,** 19142.
183. German Patent No. 2 213 865; *Chem. Abstr.*, **80,** 27243.
184. T. Uno and S. Akihama, *Yakugaku Zasshi*, **81,** 579 (1961); *Chem. Abstr.*, **55,** 19904.
185. U.S. Patent No. 3 262 939; *Chem. Abstr.*, **65,** 20131G.
186. A. Silberg, Z. Frenkel, and L. Cormos, *Stud. Univ. Babes Bolyai, Ser. Chem.*, **7,** 23 (1962); *Chem. Abstr.*, **62,** 1643.
187. German Patent No. 1 182 234; *Chem. Abstr.*, **62,** 7764.
188. E. R. Buchman and H. Sargent, *J. Amer. Chem. Soc.*, **67,** 400 (1945); *Chem. Abstr.*, **39,** 1872.
189. Belgian Patent No. 612 839; *Chem. Abstr.*, **59,** 2831.
190. J. Metzger and B. Koether, *Bull. Soc. Chim. France*, **20,** 702 (1953); *Chem. Abstr.*, **48,** 10738.
191. Japanese Patent No. 71 03 779; *Chem. Abstr.*, **74,** 141757.
192. K. Makino and T. Imai, *Z. Physiol. Chem.*, **239,** 1 (1936); *Chem. Abstr.*, **30,** 2970.
193. U.S. Patent No. 1 743 083; *Chem. Zbt* (1930), 1973.

194. A. Benko and A. Botar, *Stud. Univ. Babes Bolyai, Ser. Chem.*, **18,** 29 (1973).
195. German Patent No. 2 152 557; *Chem. Abstr.*, **77,** 88488.
196. J. C. Meslin and H. Quiniou *Synthesis*, **4,** 298 (1974).
197. J. P. Pradere, *C.R. Acad. Sci., Ser. C,* **281,** 119 (1975).
198. K. Schaumburg, *Can. J. Chem.*, **49,** 1146 (1971); *Chem. Abstr.*, **74,** 132807.
199. R. G. Dubenko, *Zh. Org. Khim.*, **2,** 485 (1966), *Chem. Abstr.*, **65,** 8890D.

V

Halo- and Nitrothiazoles

L. FORLANI and P. E. TODESCO

Istituto di Chimica Organica, Università, Bologna, Italy

I. Monohalogenothiazoles 565
   1. Preparative Methods 565
   2. Properties 567
      A. Nucleophilic Substitution Reactions on Halogenothiazoles 567
      B. Substituent Effect on the Reactivity of Halogenothiazoles 571
      C. Other Reactions of Halogenothiazoles 573
II. Polyhalogenothiazoles 575
   1. Preparative Methods 575
   2. Properties 576
III. Mononitrothiazoles 576
   1. Preparative Methods 576
   2. Properties 577
IV. Polynitrothiazoles 578
V. Halogenonitrothiazoles 578
   1. Preparative Methods 578
   2. Properties and Reactivity 578
VI. References 582

I. MONOHALOGENOTHIAZOLES

1. Preparative Methods

All four 2-halogenothiazoles are known. 2-Chloro (1–11), 2-bromo (1, 2, 5, 8–10, 12–24), and 2-iodothiazole (8, 10, 25, 26) can be prepared from 2-aminothiazole by Sandmeyer reactions, with yields between 30 and 70%. 2-Aminothiazole is a commercially available product, being prepared by easy cyclization reaction between thiourea and α-haloacetaldehyde, or a precursor. The 2-bromothiazole can also be prepared by direct bromination of thiazole in the gaseous phase (21); bromination in a solvent such as chloroform fails.

The bromination with N-bromosuccinimide in CCl_4 can be accomplished yielding to 2-bromothiazole (12.5% yield) (27).

2-Iodothiazole also can be prepared by an exchange reaction from 2-bromothiazole and I_2 in presence of Li, with 90% yield of conversion (25). 2-Fluorothiazole has recently been prepared by two different syntheses: thermal decomposition of 2-thiazolydiazoniumfluoroborate (28) (no yields are reported) or by exchange reaction of 2-nitrothiazole and KF in anhydrous dimethylpyrrolidone (20% yields) (29) (Scheme 1). Only the 4-chloro- and 4-bromothiazole are known, and they are prepared by selective dehalogenation of the 2,4-dihalogenothiazoles or 4,5-dihalogenothiazoles with Zn (3, 30, 31). The halogen in the 4 position- is more resistent to reductive dehalogenation (Scheme 2). 5-Chlorothiazole can be obtained in relative high yields (65%) from selective dehalogenation of 2,5-dichlorothiazole with Zn (Scheme 3) (3). 5-Bromothiazole is obtained by catalytic reduction of 2,5-dibromothiazole with Ni/Raney (40% yield) (13). In an alternative method, 2-amino-5-bromothiazole eliminates the amino group in position 2 by diazotization/reduction (44% yield), leading to 5-bromothiazole (13, 18).

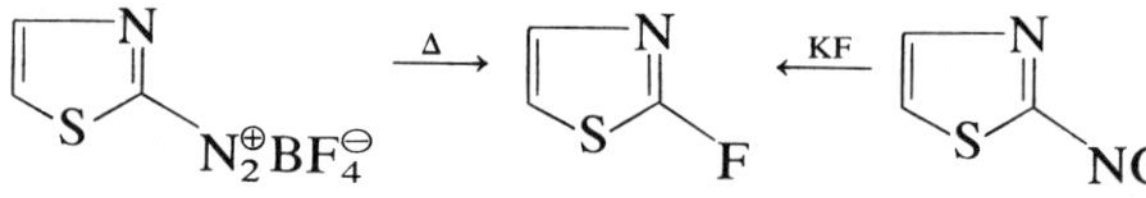

Scheme 1

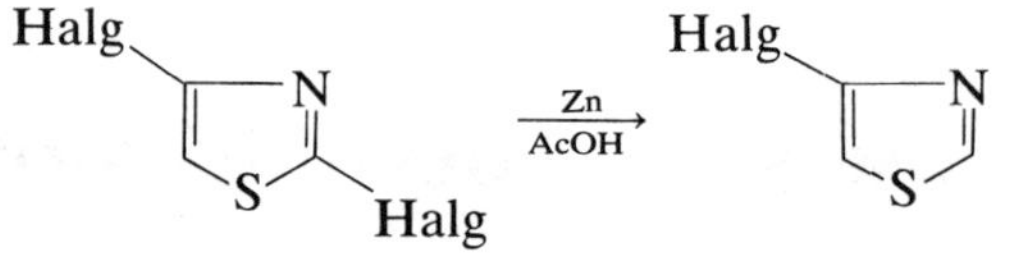

Scheme 2

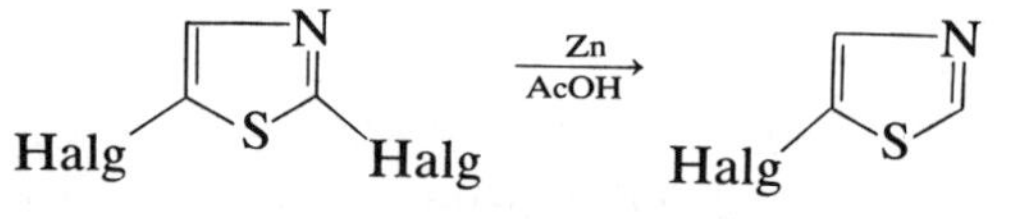

Scheme 3

2. Properties

The halothiazoles are distillable oils under reduced pressure. The 2-fluorothiazole is extremely volatile at room temperature and must be handled with care.

A. *Nucleophilic Substitution Reactions on Halogenothiazoles*

The halogen bonded to the thiazole ring exhibits different degrees of reactivity depending on the position of bonding and on reagents employed. The 2-halogenothiazoles easily react with nucleophilic reagents leading to the expected products of nucleophilic substitution as shown in Scheme 4.

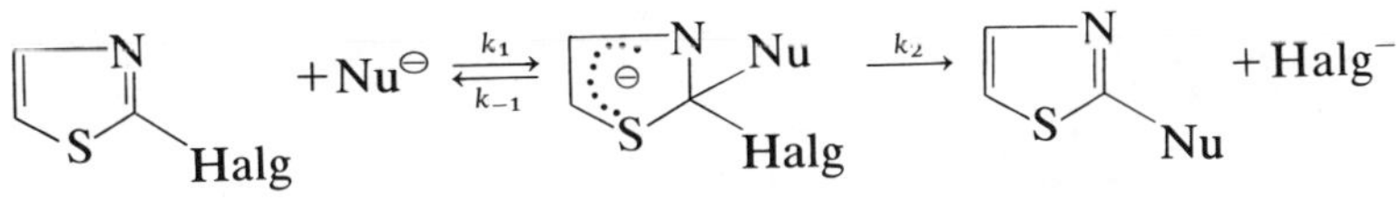

Scheme 4

The nucleophiles used are OH$^-$ (32) [the 2-hydroxythiazole can also be obtained by acidic hydrolysis with strong mineral acids (33)], OR$^-$ (5, 8, 30, 34), SR$^-$ (8, 9, 12), ArSH (35), and amines (4, 7, 14, 33). Benzamide also reacts with 2-bromothiazole, yielding 2-benzamidothiazole (36). Sulfonamide also reacts with 2-halogenothiazoles in presence of a base and copper powder, yielding 2-sulfonamidothiazoles (37, 38).

2-Halogenothiazoles react with salts of sulfinic acids leading to the corresponding sulfones (6, 37, 39). A comparison of reactivity of some 2-halogenothiazoles with some nucleophiles, with respect to other aza-activated heteroaromatic substrates, is made in Table V-1. The higher or comparable reactivity of 2-fluorothiazole with regard to other 2-halogenothiazoles indicates that the bond breaking is not affected in the slow step of reaction, confirming the proposed SnAr mechanism. The data in Table V-1 indicate also that the 2-halogenothiazoles, with all the nucleophiles used, are more activated than the corresponding 2-halogenopyridines.

TABLE V-1. RELATIVE REACTIVITIES OF SOME HALO-AZA-ACTIVATED AROMATIC SUBSTRATES WITH NUCLEOPHILES[a]

Compound	Nucleophile			Piperidine in Piperidine
	MeO$^-$ in MeOH	PhS$^-$ in MeOH	PhSH in MeOH	
2-Fluorothiazole	2.22×10^{3b}	2.67×10^b	—	—
2-Chlorothiazole	$1.00^{b,c}$	1.00^d	1.00^e	1.00^f
2-Bromothiazole	1.30^b	3.91^b	4.89^e	—
2-Iodothiazole	4.91×10^{-1b}	2.51^b	—	—
2-Chloropyridine	4.09×10^{-3g}	—	—	3.73×10^{-2h}
2-Bromopyridine	—	—	—	2.26×10^{-1k}
2-Chloroquinoline	3.21^i	3.73×10 (at 86°C)^j	1.59×10^2 (at 30°C)^j	1.86^k
2-Bromoquinoline	—	—	—	3.79×10^k
2-Chlorobenzothiazole	4.16×10^{2l}	4.22×10^{2l}	—	to fast^k

[a] For a fixed nucleophile the rate constant of each compound at 50°C (different temperatures indicated) was divided by the following values of 2-chlorothiazole (k in sec$^{-1} M^{-1}$ unless otherwise indicated): $k_{MeO^-} = 8.1 \times 10^{-6}$, $k_{PhS^-} = 4.5 \times 10^{-6}$, $k_{PhSH} = 4.7 \times 10^{-6}$, $k_{piperidine} = 1.5 \times 10^{-5}$ (sec^{-1}).
[b] Data from Ref. 40.
[c] Data from Ref. 8.
[d] Data from Ref. 9.
[e] Data from Ref. 35.
[f] L. Forlani and P. E. Todesco, unpublished work.
[g] Extrapolated value from Ref. 41.
[h] Data from Ref. 42.
[k] Data from Ref. 4.
[i] Extrapolated value from Ref. 41.
[j] Data from Ref. 44.
[l] Data from Ref. 45.

The higher reactivity of 2-halogenothiazoles with respect to halogenopyridines can be related to the different aromaticity of the two systems, less for thiazole than for pyridine, for example, the relatively stronger fixation of the π bond in the thiazole than in the case of pyridine. As the data reported in Table V-1 (footnote a) indicates, the free thiophenol is more reactive than the thiolate anion toward the 2-halogenothiazoles. This fact should be considered when one prepares the thiazolyl sulfides.

A more unusual fact observed in thiazole chemistry is that also the other positions (4 and 5) are activated toward the nucleophilic substitution, as found independently by Metzger and coworkers (46) and by Todesco and coworkers (30, 47). Some kinetic data are reported in Table V-2. As the data in Table V-2 indicate, no simple relationship between nucleophilic reactivity and charge density, or other parameters available from more or less sophisticated calculation methods, can be applied. As a

TABLE V-2. REACTIVITY RATIOS OF HALOGENOTHIAZOLES WITH DIFFERENT NUCLEOPHILES[a]

Halogen in Thiazole	MeO⁻ in MeOH	EtO⁻ in EtOH	i-PrO⁻ in i-PrOH	t-BuO⁻ in t-BuOH	PhS⁻ in MeOH	PhSH in MeOH	Piperidine in MeOH	Piperidine in Piperidine	MeS⁻ in MeOH
2-Cl	1[b]	1[c]	1[c]	1[c]	1[d]	1[e]	1[f]	1[d]	—
4-Cl	0.074[b]	0.59[c]	3.9[c]	42[c]	—	—	0.080[d]	0.17[d]	—
5-Cl	2.3[b]	—	—	—	—	—	—	—	1[f]
2-Br	1.3[b]	—	—	—	3.9[d]	4.9[e]	—	—	—
4-Br	0.16[b]	—	—	—	0.36[f]	No reaction	—	—	—
5-Br	2.8[b]	—	—	—	0.20[f]	0.4[e]	—	—	0.16[f]

[a] For a fixed nucleophile the rate constant of each compound at 50°C was divided by following values of 2-chlorothiazole: $k_{MeO^-} = 0.81$; $k_{EtO^-} = 0.44$; $k_{i\text{-}PrO^-} = 0.74$; $k_{t\text{-}BuO^-} = 0.18$, $k_{PhS^-} = 0.45$, $k_{PhSH} = 0.47$; $k_{Piper} = 0.10$ (in MeOH), $k_{Piper} = 1.5$ (sec⁻¹, in piperidine); 2-bromothiazole, $k_{MeS^-} = 11$; k values in sec⁻¹ M^{-1}, at 50° unless otherwise indicated.
[b] Values from Ref. 8 and 30.
[c] Values from Ref. 34.
[d] Values from Ref. 9.
[e] Values from Ref. 35.
[f] Values from Ref. 48.

matter of fact the thiazole residue displays a net electron-withdrawing action in all positions as indicated by the fact that the thiazole carboxylic acids are all more acidic than the benzoic acid (49). Analogously, the thiazolylphenylsulfoxides are oxidized to the corresponding sulfones more slowly than the *p*-nitrodiphenylsulfoxide, indicating that, considering only the polar effect, the 2-thiazolyl residue is more electron-withdrawing than a *p*-nitrophenyl residue (48, 50); the experimental order of electron-withdrawing efficiency of the different positions is 2>5 (48). Moreover, in the nucleophilic substitution at different positions of halogenothiazoles other factors must be taken into account. In fact, with methoxide ion in methanol the reactivity sequence is 5-chlorothiazole > 2-chlorothiazole > 4-chlorothiazole, while with thiophenoxide ion or piperidine the sequence 2>4>5 is observed (48).

The order of reactivity is completely reversed using more crowded and basic alcohols and nucleophiles, as *t*-butoxide in *t*-butanol, the 4-chlorothiazole being more reactive than 2-chloro isomer (34). While the reactivity of 2-halogenothiazoles can easily be interpreted by the usual SnAr mechanism, the reactivity of 4-halogeno- and 5-halogenothiazoles requires that in the intermediate complex the negative charge imported by the nucleophile can be supported by all the atoms forming the heterocycle and not only by the aza substituent. Another possibility arises from intervention of ion pairs as in Scheme 5.

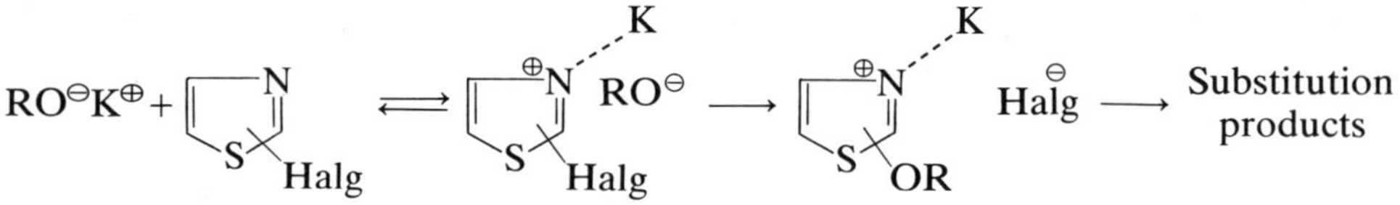

$$RO^{\ominus}K^{\oplus} + \underset{S}{\underset{\|}{\overset{N}{\bigcap}}}\text{Halg} \rightleftarrows \underset{S}{\underset{\|}{\overset{\overset{K}{\cdots}}{\overset{\oplus}{N}}}}\text{Halg}\ RO^{\ominus} \rightarrow \underset{S}{\underset{\|}{\overset{\overset{K}{\cdots}}{\overset{\oplus}{N}}}}\text{OR}\ \text{Halg}^{\ominus} \rightarrow \text{Substitution products}$$

Scheme 5

The intervention of ion pairs, more important in *t*-butanol than in methanol, can increase the substitution reaction in such cases as the 4- and 5-halogenothiazoles, which are poorly activated by the aza substituent.

The role of the quaternization of the azasubstituent in the nucleophilic substitution at 2-halogenothiazoles is in fact emphasized by the reactivity of 2-halogenothiazoles with undissociated thiophenol (35), which proceeds faster than the corresponding reaction of 2-halogenothiazoles with thiophenolate anion, through the pathways shown in Scheme 6. Moreover, the 4-halogenothiazoles do not react with undissociated thiophenols, while the 5-halogenothiazoles react well (48).

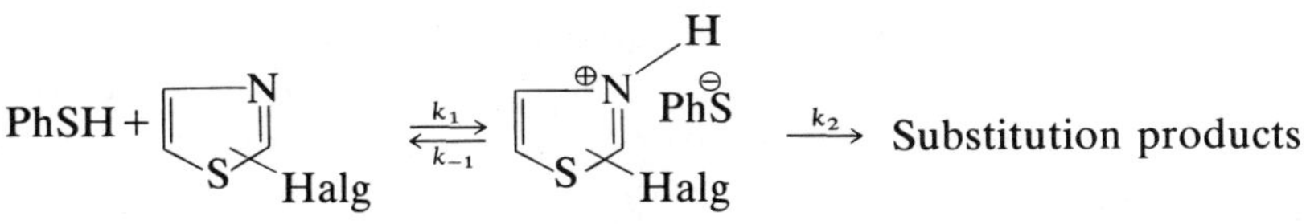

PhSH + [thiazole-Halg] $\underset{k_{-1}}{\overset{k_1}{\rightleftarrows}}$ [protonated thiazole-Halg with PhS⁻] $\xrightarrow{k_2}$ Substitution products

Scheme 6

B. *Substituent Effect on the Reactivity of Halogenothiazoles*

The nucleophilic reactivity of 2-halogenothiazoles is strongly affected by the substituent effect, depending on the kind of substitution reaction. Positions 4 and 5 can be considered as "meta and para", respectively, with regard to carbon 2 and to groups linked to it; consequently, it is possible to correlate the reactivity data with Hammett's relationships.

Some Hammett values for reactions in thiazole and in nucleophile are reported in Table V-3. The observed ρ values for "normal" substitution processes (methoxy and thiophenoxysubstitution) are high and positive, indicating that the substituent plays an important role in modifying the stability of the intermediate anion.

TABLE V-3. ρ VALUES FOR SOME SUBSTITUTION REACTIONS OF 2-HALOGENO-X-THIAZOLES WITH SUBSTITUTED NUCLEOPHILES

Reaction	In Thiazole	In Nucleophile
Methoxydechlorination	5.9[a]	—
Thiophenoxydechlorination	5.3[b]	—
Thiopheno1dechlorination	−3[c]	1.4[c]

[a] Data from Ref. 51.
[b] Data from Ref. 9.
[c] Data from Ref. 35.

Supporting this idea, when X is able to carry fully the negative charge and is bonded to a conjugable position, as in the case of 2-chloro-5-nitrothiazole, we observe a strong increase of reactivity, as expressed by the experimental σ_{para} that reaches, in thiophenoxysubstitution, the very high value of 1.6. In fact, the 5-nitro-2-chlorothiazole is more reactive than the 2-chlorothiazole by a factor of 10^8. The reaction between undissociated thiophenol and 2-halogenothiazoles has shown an opposite trend of structural effects (35). The donor groups in the thiazole ring are activators, while electron-withdrawing groups slow down the substitution rates (with the exception of nitro group). On the other hand, in the nucleophile the structural changes are opposite to that usually observed in nucleophilic substitution; the donor groups in thiophenol slow down the rate. All these phenomena are clearly explained by Scheme 6, observing that the first equilibrium plays a dominant role in determining the overall observed rate.

Regarding the substituent effect on reactivity of groups in positions 4 and 5 there is little information in the literature. The reactivity of halogen in position 5 seems to be increased when an amino group is present in position 2. Substitution products are easily obtained using neutral nucleophiles such as thiourea, thiophenols, and mercaptans (52–59).

Scheme 7, involving ionic couples, seems reasonable, when aza-amino tautomerism is considered. Moreover, the 2-acetylamino-5-halogenothiazole, in which the amino nitrogen is hardly affected by protonation, reacts well with mercaptans. Other substituents can be tested to clarify the role of electronic and tautomeric effects. On the other hand, in simple-cases, the different weight of electronic effects on the thiazole system, depending on the relative position of the reaction center and substituent, has been detected, and there is evidence that the substituent effect is more important for groups in position 2 than in positions 4 or 5. This was the first indication for the acidity of 2-carboxy-5-X-thiazoles, 4-carboxy-2-X-thiazoles, or 5-carboxy-2-X-thiazoles with ρ values 2.35, 1.34, and 0.83, respectively, reported by Imoto and coworkers (49). Another supporting observation is the oxidation reaction of phenylsulfinyl-X-thiazoles to the corresponding sulfones. In fact, the ρ value is -1.19 for 2-phenyl-sulfinyl-5-X-thiazoles, while it only attains the value of -0.5 for 5-phenylsulfinyl-2-X-thiazoles.

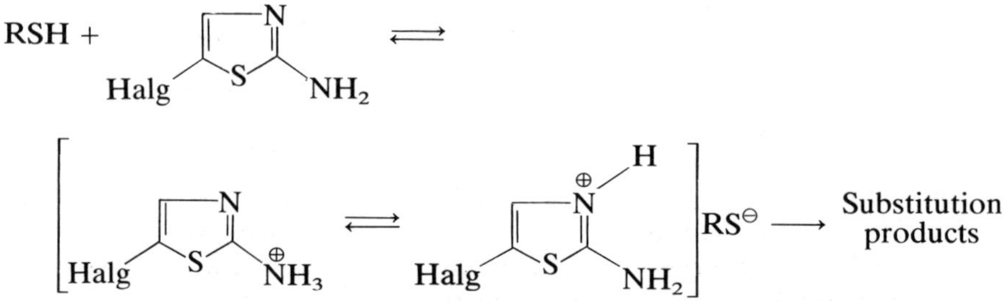

Scheme 7

Although Noyce and Fike have recently found for the solvolysis of 2-thiazolyl-ethyl chlorides analogous modality of substituent electronic effect transmission from position 2 toward position 5 and from position 5 toward position 2(60), a more general conclusion indicates that the groups bonded to position 2 are more sensitive to a substituent effect in conjugable position 5 than are the groups bonded to position 5 with regard to a substituent in position 2.

This is also confirmed by the observation that in the reaction of 2,4- or 2,5-dihalogenothiazoles with anionic nucleophiles, the halogen in position 2 reacts first (8, 9, 35). This halogen is strongly activated by the aza substituent and by the other halogen substituent. Only when the halogen in position 2 is substituted more than 90% can the substitution proceed in position 4 or 5.

The metal reduction of 2,4-dichlorothiazole or 2,5-dichlorothiazole for which the mechanism has not been fully investigated also proceeds almost exclusively in position 2 (3, 13).

C. *Other Reactions of Halogenothiazoles*

Halogenothiazoles react with *n*-butyllithium leading to the corresponding thiazolyllithium, which can react with carbonyl compounds yielding the expected addition derivative (61). The thiazole carboxylic acids can be obtained from the lithium derivatives with CO_2 (Scheme 8) (13, 62). The lithium derivatives can decompose in CH_3COOD leading to corresponding deuterated thiazoles (18). The Grignard reagents, in all thiazole positions, can be prepared using magnesium in the presence of an excess of magnesium bromide, while the normal procedure fails (26). The halogen in position 2 can be easily replaced by the hydrogen, using zinc and acetic acid (6, 53, 63–65) or tin and hydrochloric acid; hydroiodic acid and phosphorus can also be used. Catalytic reduction can be accomplished by potassium hydroxide or triethylamine with Raney nickel catalyst (66, 67).

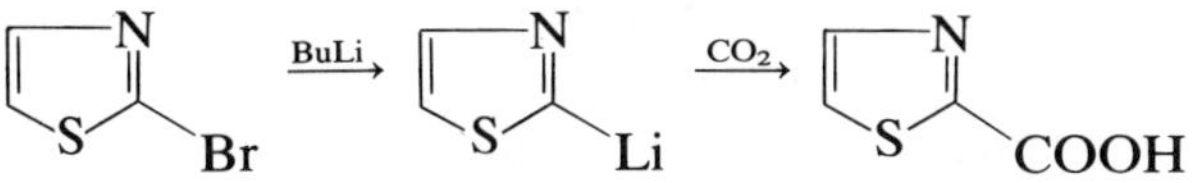

Scheme 8

The 2-bromothiazole can also be reduced to thiazole by zinc (18) in CH_3COOH, or by electrolytic reduction (68). The deuterated thiazoles can also be obtained starting from the corresponding halogenothiazoles, by Zn/Cu in CH_3COOD (18).

Treating 2-bromothiazole with copper at 170°C in cumene as solvent affords the 2,2'-bisthiazole (Scheme 9) (69). The 2-halogenothiazoles can generate complexes with mercuric chloride (21). The halogenothiazoles can also be alkylated by alkylhalides, halogen exchange being observed simultaneously as in pyridine or quinoline derivatives (Scheme 10) (70, 71). The presence of halogen, particularly in position 4, decreases the basicity of the aza substituent. As a consequence, picrate or picrolonate are difficult to obtain.

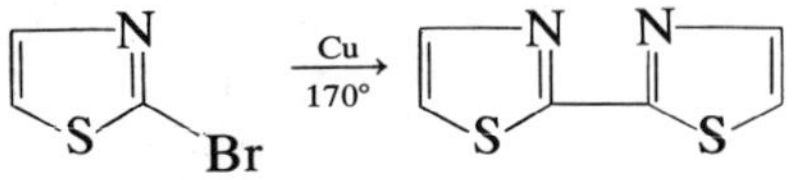

Scheme 9

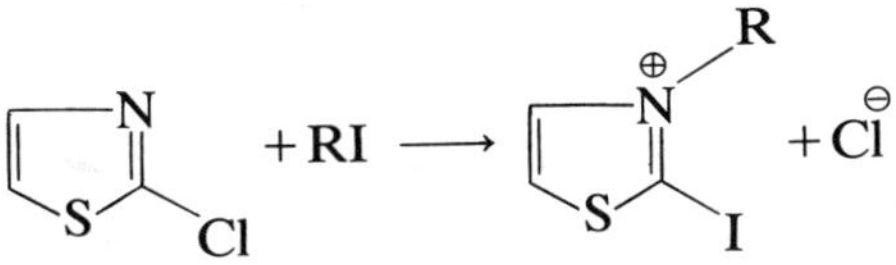

Scheme 10

Bromination of 2-bromothiazole leads to 2,5-dibromothiazole (5). 2-Bromothiazole can be used as a substrate in a malonic synthesis (72); starting from phenylacetonitrile the α phenyl-(2-thiazolyl)-acetonitrile is obtained in high yields (84%) (Scheme 11).

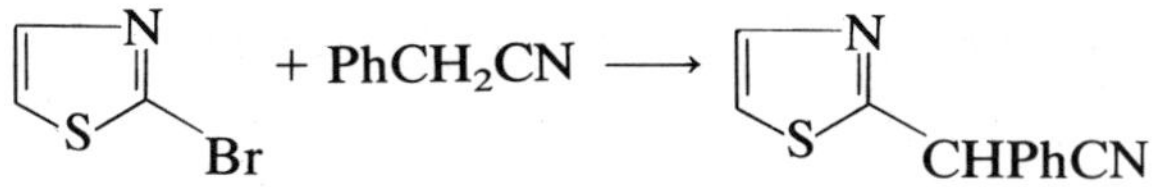

Scheme 11

All the halogenothiazoles, depending on the electron-withdrawing power of the halosubstituent, together with the electron-withdrawing power of the azasubstituent, are only slightly susceptible to electrophilic substitution reactions such as nitration, sulfonation, and so on, while the polyhalogenation reaction can take place.

The polarographic properties of the halogenothiazoles in comparison with other thiazole compounds have also been investigated (73, 74). Infrared, Raman, ultraviolet, and NMR spectra of monohalogenothiazoles have been measured (2, 3, 6, 10, 15, 17, 24, 29) (Table V-4).

TABLE V-4. MONOHALOGENOTHIAZOLES

Halogen	Physical Properties and Derivatives	Ref.
2-F	b.p. 78–80°C	28, 29
2-Cl	b.p. 145°C; n_D^{20} 1.5503	1–10, 24
2-Br	b.p. 70–71°C/20 mm; n_D^{20} 1.5913	1, 2, 5, 8, 10, 12, 13, 15–21, 23, 24, 27, 31, 68
2-J	b.p. 118°C/40 mm; n_D^{25} 1.6670	8, 10, 24–26
4-Cl	b.p. 165–66°C; methiodide, m.p. 178°C	3, 10, 24, 30
4-Br	b.p. 189–90°C; methiodide, m.p. 199–200°C	30, 31, 56
5-Cl	b.p. 139–40°C	3, 10, 24, 30
5-Br	b.p. 81°C/18 mm; n_D^{25} 1.5976, mercuric chloride, m.p. 148°C	2, 10, 13, 18, 24, 30

II. POLYHALOGENOTHIAZOLES

1. Preparative Methods

All possible dichloro- or dibromothiazoles are known. The 2,5-dihalogeno derivatives can be prepared from the 5-halogeno-2-aminothiazoles by diazotization/decomposition with CuCl or CuBr (3, 12, 13, 18, 75). The 5-halogeno-2-aminothiazoles can be easily prepared by halogenation of 2-aminothiazole (65, 76–79); 2,5-dibromothiazole can also be prepared by direct bromination of 2-bromothiazole (5).

The 2,4-isomers can be obtained by halogenation of 2,4-dihydroxythiazole by POCl₃ or POBr₃ in pyridine (3).

The 4,5-dihalogenothiazoles are obtained by cyclization-halogenation reactions as show in scheme 12 (3). 2-Acetamido-4,5-diiodothiazole has been obtained by Hurd and Wehrmeister (80). The triiodothiazole can be prepared by iodination by molecular iodine of the mercuric complex of 2-iodothiazole following the Travagli method (81).

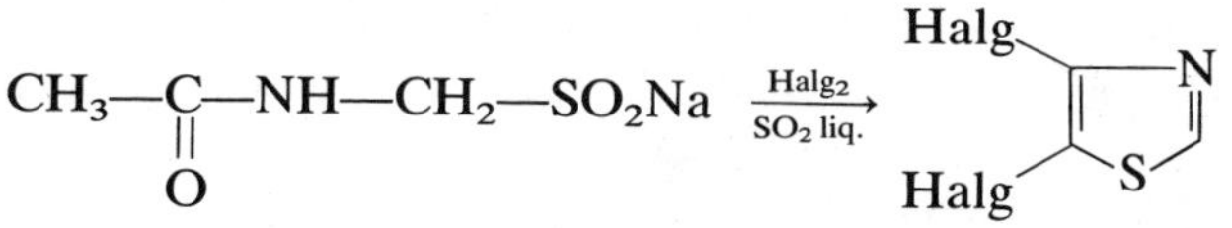

$$CH_3-\underset{\underset{O}{\|}}{C}-NH-CH_2-SO_2Na \xrightarrow[SO_2 \text{ liq.}]{Halg_2} \begin{array}{c} Halg \\ \diagdown \\ Halg \end{array} \begin{array}{c} N \\ \diagup \\ S \end{array}$$

Scheme 12

The mixed 2-chloro-5-bromothiazole has been obtained from 2-amino-5-bromothiazole, as previously indicated (75).

Trichlorothiazole, which behaves as an insecticide, can be prepared by the reaction at 150 to 200°C of sulfur with either 1,2,2,2-tetrachlorethylisocyanide dichloride (103), N-(1,2,2,2-tetrachlorethyl)formimidine (104), or by the reaction of sulfur chlorides at elevated temperature in the presence of Friedel-Crafts catalysts with 1,2-dichlorethylcarbylamine dichloride (105). Another mode of preparation results from the chlorination at 60°C of 2,4-dichlorothiazole by chlorine in the presence of $SbCl_3$ (106). Other perhalogenated thiazoles have been described, and their use as soil nematocides has been patented (106).

2. Properties

Both halogens of the dihalogenothiazoles can be replaced by nucleophiles. At any rate, the halogen in position 2 is always more reactive than those in positions 4 or 5, as previously discussed. Analogously, the halogen can be selectively removed only from position 2 by reduction with metals (Table V-5).

TABLE V-5. POLYHALOGENOTHIAZOLES

Halogens	Physical Properties and Derivatives	Ref.
2,4-Dichloro	m.p. 42–43°C, b.p. 184°C	3, 10
2,5-Dichloro	b.p. 159–161°C	3
2,4-Dibromo	m.p. 82°C	3, 10
2,5-Dibromo	m.p. 48°C; mercuric chloride, m.p. 188–9°C	3, 5, 10, 12, 13, 18, 75
2-Chloro-5-bromo	m.p. 52–53°C	10, 75
4,5-Dichloro	m.p. 38–39°C; b.p. 183°C; $n_D^{43} = 1.565$	3
4,5-Dibromo	m.p. 75°C	3
Tribromo	m.p. 32°C	31
Triiodo	m.p. 118 (decomp.)	81
Trichloro	m.p. −13°C, b.p. 195°C	106
2-Fluoro-4,5-dichloro	m.p. −34°C, b.p. 158°C	106
2-Bromo-4,5-dichloro	m.p. 18°C, b.p. 226°C	106
4-Chloro-2,5-dibromo	m.p. 6°C, b.p. 235°C	106
2-Fluoro-4-chloro-5-bromo	m.p. −42°C, b.p. 205°C	106
2-Fluoro-4,5-dibromo	m.p. −17°C, b.p. 198°C	106

III. MONONITROTHIAZOLES

1. Preparative Methods

2-Nitrothiazole and 5-nitrothiazole are known products, while 4-nitrothiazole does not appear to have been prepared. 2-Nitrothiazole (2,

82–84) can be prepared from 2-aminothiazole by diazotization/decomposition with NaNO$_2$ (20% yields). It is possible to obtain the 2-nitrothiazole starting from 2-nitro-5-bromothiazole and reducing the 5-bromo with alcoholic potassium cyanide or cuprous cyanide in pyridine; no trace of 5-cyano-2-nitro derivative is detected (85).

5-Nitrothiazole is obtained by reductive dehalogenation of 2-bromo-5-nitrothiazole (2, 14, 83).

2. Properties

The 2-nitrothiazole reacts with nucleophiles, such as CH$_3$O$^-$, leading to the normal substitution product, 2-methoxythiazole, at high rate (86). The 5-nitrothiazole with methoxide in DMSO decomposes rapidly, probably following a ring-opening reaction (87) similar to that found in the benzothiazole derivative (88) (Scheme 13). This ring-opening reaction can explain the characteristic colored solution developed by the action of alcoholic or aqueous alkali on the nitrothiazoles or R-substituted nitrothiazoles, and attributed previously to acinitroquinone-like structures (82, 85, 89) or Meisenheimer-like compounds.

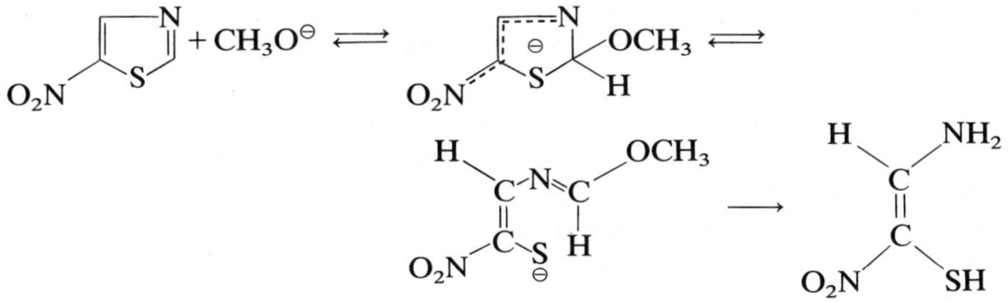

Scheme 13

The 2-nitrothiazole can be reduced to the corresponding aminothiazole by catalytic or chemical reduction (82, 85, 89). The 5-nitrothiazole can also be reduced with low yield to impure 5-aminothiazole (1, 85). All electrophilic substitution reactions are largely inhibited by the presence of the nitro substituent. Nevertheless, the nitration of 2-nitrothiazole to 2,4-dinitrothiazole can be accomplished (see Section IV).

IV. POLYNITROTHIAZOLES

Only the 2,4-dinitrothiazole is known, being prepared by nitration of 2-nitrothiazole by N_2O_4–NO_2/BF_3. The yield is 80% (90). The reduction with Raney Ni/Ac_2O of 2,4-dinitrothiazole proceeded smoothly, yielding the corresponding 2,4-diacetamidothiazole.

V. HALOGENONITROTHIAZOLES

1. Preparative Methods

Some 2-halogeno-5-nitrothiazoles and 2-nitro-5-halogenothiazoles are known. 2-Halogeno-5-nitrothiazoles can be prepared by a Sandmeyer reaction from 2-amino-5-nitrothiazole (1, 85), while 2-nitro-5-halogenothiazoles can be analogously prepared by decomposition of diazonium salts arising from 2-amino-5-halogenothiazoles in presence of nitrite anion (82, 84).

2. Properties and Reactivity

Many 2-substituted 5-nitrothiazoles are prepared (by nucleophilic substitution reactions on 2-halogeno-5-nitrothiazoles) for use as biocides or for their biological activity (31, 91–95).

Taking into account the previous discussion of the reactivities of simple halogenothiazoles or nitrothiazoles, the mechanism of nucleophilic reactivity of 2-halogeno-5-nitrothiazoles with an excess of basic nucleophile such as CH_3O^- can be proposed (Scheme 14).

Evidence of **3**, **4**, **8**, and **9** was reported by Todesco and coworkers (51, 87, 88), while **2** and **8** were found by Illuminati and coworkers (96). The ring-opening reaction is a slow process in methanol and can be made faster using a solvent such as DMSO.

If the reactions are carried out using a concentration of CH_3O^- slightly less than that of halogenothiazole, the reaction almost quantitatively yields the 2-methoxy-5-nitrothiazole. The reaction of methoxide ion in excess with regard to 2-halogeno-5-nitrothiazole, previously reported as leading to a mixture of unidentified products together with the expected 2-methoxy-5-nitrothiazole in low yields (97), can be explained on the basis of Scheme 14.

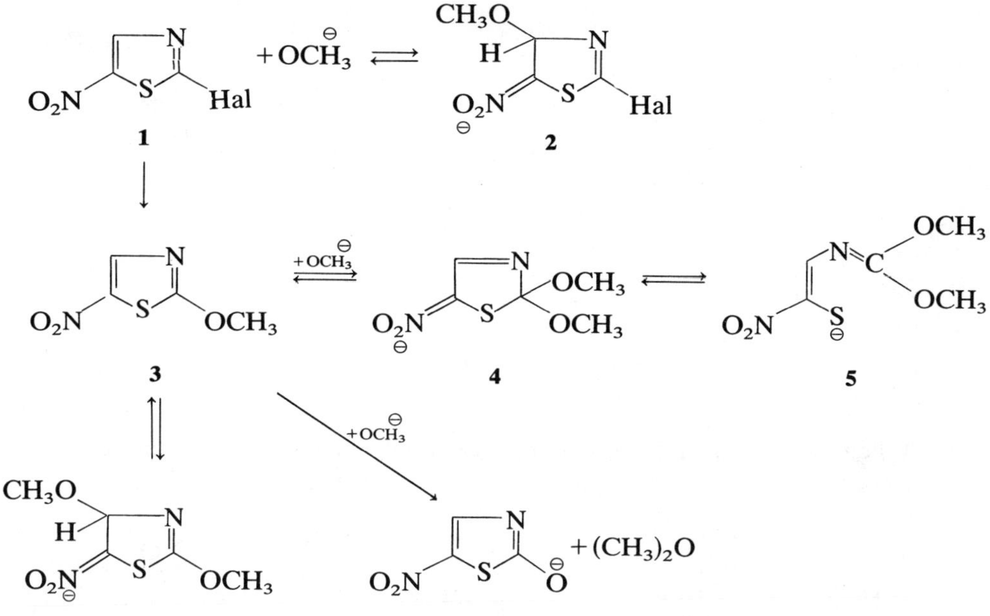

V. Halogenonitrothiazoles

Scheme 14

Operating with sodium benzenethiolate the only reaction observed is the thiophenoxydehalogenation, which is strongly favored by the activation performed by 5-nitro substituent, as previously reported (Scheme 15) (9).

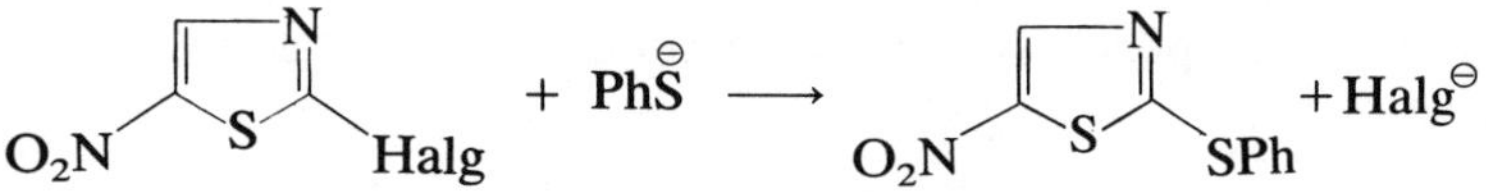

Scheme 15

The very high rate of thiophenoxy substitution, compared with low stability of Meisenheimer-like sulfurated compounds, can explain the simple behavior of the thiophenoxy-substitution reaction.

Ilvespaa (98) has demonstrated that, using some amines, the 2-chloro-5-nitrothiazole undergoes an opening reaction in a competitive reaction parallel to the normal substitution process. This confirms the sensitivity of position 4 to nucleophilic attack when a nitro group is present in position 5 (Scheme 16).

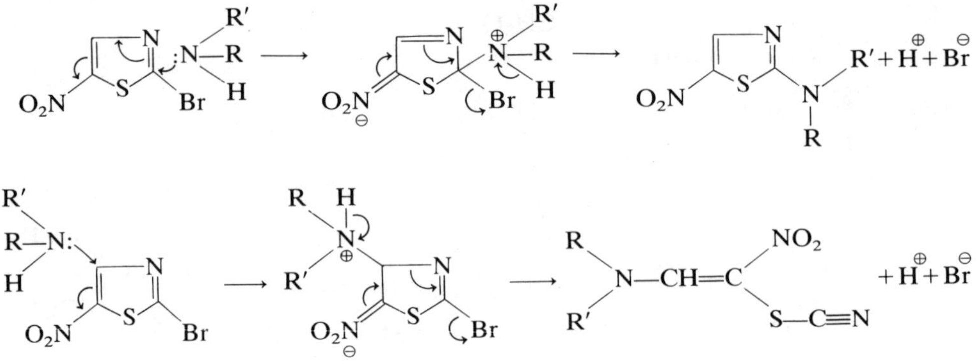

Scheme 16

When 5-halogeno-2-nitrothiazole reacts with methoxide ion, the leaving group is the nitro group in position 2, following Scheme 17 (86).

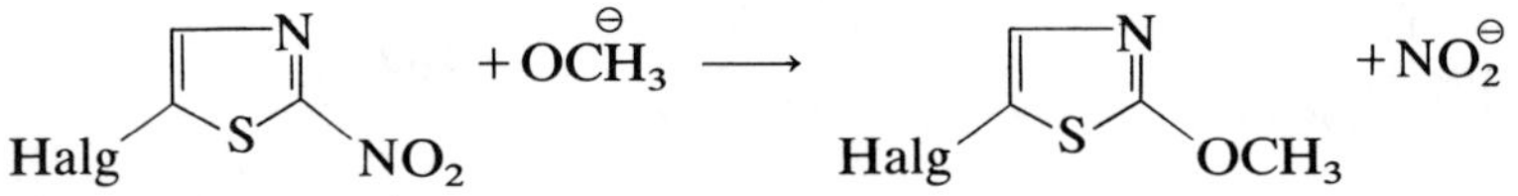

Scheme 17

This confirms the low activation of position 5 conferred by the substituent in position 2 as opposed to that observed in opposite sense, as previously reported also for different kinds of reactions. On the other hand, the nitro group in many cases acts as a very good leaving group, and this may explain the observed reactivity. For a complete understanding of nucleophilic reactivity of substituted thiazoles further investigation seems to be necessary, requiring syntheses of new thiazole derivatives, and testing the respective reactivities with different nucleophiles, solvents, and substituents.

TABLE V-6. NITROTHIAZOLES AND NITROHALOGENOTHIAZOLES

Compound	m.p. (°C)		Ref.
2-Nitrothiazole	77–78	Pale yellow plates	2, 29, 82, 83, 85, 101
5-Nitrothiazole	63–65	Pale yellow plates	2, 14, 83
2,4-Dinitrothiazole	144.5–146.5	—	90
2-Chloro-5-nitrothiazole	61	Hexagonal plates	1, 9, 82, 97, 98, 99
2-Bromo-5-nitrothiazole	91	Thick colorless plates	1, 82, 85, 91, 92, 97, 99
2-Iodo-5-nitrothiazole	75	—	80, 97
2-Nitro-5-bromothiazole	82	—	85
2,4-Dibromo-5-nitrothiazole	72	—	31

TABLE V-7. REACTIONS OF 2-HALOGENOTHIAZOLES

Halogen	Reagents and Conditions	Products (Yields, %)	Ref.
F, Cl Br, J	MeO$^-$Na$^+$ in MeOH	2-Methoxythiazole (80–90), b.p. 149–150°C; picrate, m.p. 93–94°C	2, 5, 29, 30
Cl	Piperidine in piperidine in MeOH, in DMSO	2-Piperidylthiazole (80), b.p. 130–131°C/13 mm, n_D^{25} 1.5690; picrate, m.p. 142–143°C	4, 7, 32, 102
Cl	Pyrrolidine in DMSO	2-Pyrrolidinylthiazole (63), b.p. 123°C/11 mm m.p. 41–42.5°C	7
Cl, Br	PhS$^-$Na$^+$ or PhSH in MeOH	2-Phenylthiothiazole (75), b.p. 154°/3 mm	9, 12, 33
Cl	Morpholine in DMSO	2-Morpholinylthiazole (60), b.p. 132°C/10 mm, n_D^{25} 1.5760; picrate, m.p. 135°C	7, 102
Br	4-Cl-C$_6$H$_4$S$^-$Na$^+$ in Carbitol at 180	2-(p-Chlorophenylthio)thiazole (51), m.p. 49.5–50°C	12
Br	2,4,5-Cl$_3$-C$_6$H$_2$S$^-$Na$^+$ in Carbitol at 180°	2-(2,4,5-Trichlorophenylthio)thiazole (70), m.p. 63–63.5°C	12
Br	Me$_2$NH	2-Dimethylaminothiazole, b.p. 89°C/20 mm, n_D^{25} 1.5530; picrate, m.p. 165°C	102
Br	Et$_2$N-CH$_2$CH$_2$O$^-$Na$^+$	2-(Et$_2$N-CH$_2$CH$_2$O)thiazole (66), b.p. 130–132°C/11 mm; picrate, m.p. 114–115°C	14
Br	Glyceral-1,2acetonide	(±)-3-(2-Thiazolyloxy)-1,2-propanediolacetonide, b.p. 95–100°C/0.3 mm, n_D^{25} 1.4966, λ_{max} 235 nm (log ε 3.73)	94, 95
Br	Br$_2$ in H$_2$O/HBr	2,5-Dibromothiazole, m.p. 48°C	5
Br	Butyllithium(−70°)/N-methylformanilide	2-Thiazolealdehyde (46), b.p. 64–70°C/10 mm	2
Br	Butyllithium/CO$_2$ in Et$_2$O at −70°	2-Carboxythiazole (62), m.p. 97–98°C	13, 62
Br	Butyllithium/AcOD in Et$_2$O at −60°	2-D-Thiazole (40)	18
Br	Zn/H$^+$ in AcOH	Thiazole (90)	18
Br	Cu, Zn/D$^+$ in AcOD	2-D-Thiazole (50)	67
Br	Electrolytic reduction	Thiazole hydrobromide (96), m.p. 186–187°C	68

VI. REFERENCES

1. K. Ganapathi and A. Venkataraman, *Proc. Indian Acad. Sci.*, **22A,** 343 and 362 (1945); *Chem. Abstr.*, **40,** 4056 and 4059.
2. G. Borgen and S. Gronowitz, *Acta Chem. Scand.*, **20,** 2593 (1966); *Chem. Abstr.*, **66,** 60541.
3. P. Reynaud, M. Robba, and R. C. Moreau, *Bull. Soc. Chim. France*, 1735 (1962); *Chem. Abstr.*, **58,** 6816.
4. T. E. Young and E. D. Amstutz, *J. Amer. Chem. Soc.*, **73,** 4773 (1951).
5. G. Klein and B. Prijs, *Helv. Chim. Acta*, **37,** 2057 (1954), *Chem. Abstr.*, **50,** 307.
6. J. McLean and G. D. Muir, *J. Chem. Soc.*, 383 (1942); *Chem. Abstr.*, **36,** 5815.
7. H. Grube and H. Suhr, *Chem. Ber.*, **102,** 1570 (1969).
8. M. Foa, A. Ricci, and P. E. Todesco, *Boll. Sci. Fac. Chim. Ind. Bologna*, **23,** 229 (1965); *Chem. Abstr.*, **63,** 12998.
9. P. E. Todesco, M. Bosco, L. Forlani, V. Liturri, and P. Riccio, *J. Chem. Soc. B*, **7,** 1373 (1971); *Chem. Abstr.*, **75,** 48198.
10. E. J. Vincent, R. Phan Tan Luu, J. Metzger, and J. M. Surzur, *Bull. Soc. Chim. France*, **11,** 3524 (1966); *Chem. Abstr.*, **66,** 64912.
11. K. Okada, *Ann. Rep. Takamine Lab.*, **4,** 40 (1952); *Chem. Abstr.*, **49,** 11665.
12. P. A. Van Zwieten and H. O. Huisman, *Rec. Trav. Chim.*, **81,** 554 (1962); *Chem. Abstr.*, **57,** 12465.
13. H. Beyerman, P. Berben, and J. Bontekoe, *Rec. Trav. Chim.*, **73,** 325 (1954); *Chem. Abstr.*, **49,** 6230.
14. G. Klein, B. Prijs, and H. Erlenmeyer, *Helv. Chim. Acta*, **38,** 1412 (1955); *Chem. Abstr.*, **50,** 5643.
15. I. N. Bojesen, J. H. Hoeg, J. T. Nielsen, I. B. Petersen, and K. Schaumburg, *Acta Chem. Scand.*, **25,** 2739 (1971); *Chem. Abstr.*, **76,** 24291.
16. J. A. Braun and J. Metzger, *Bull. Soc. Chim. France*, 503 (1967); *Chem. Abstr.*, **67,** 11446.
17. G. M. Clarke, R. Grigg, and D. H. Williams, *J. Chem. Soc. B*, **4,** 339 (1966) and 4597 (1965); *Chem. Abstr.*, **64,** 17389.
18. P. Roussel and J. Metzger, *Bull. Soc. Chim. France*, 2075 (1962); *Chem. Abstr.*, **58,** 13932.
19. P. E. Iversen and H. Lund, *Acta Chem. Scand.*, **20,** 2649 (1966); *Chem. Abstr.*, **66,** 104953.
20. R. Handley, E. F. G. Herington, M. Azzaro, and J. Metzger, *Bull. Soc. Chim. France*, 1904 (1963).
21. P. Wibaut and H. E. Jansen, *Rec. Trav. Chim.*, **53,** 77 (1934) and **56,** 699 (1937); *Chem. Abstr.*, **28,** 4417 and **31,** 6232.
22. S. G. Friedmann, *Zh. Obshchei Khim.*, **23,** 778 (1953); *Chem. Abstr.*, **48,** 33401.
23. P. E. Iversen, H. Lund, A. Berg, and S. R. Johansen, *Sin. Geterotsikl. Soedin.*, **9,** 27 (1972); *Chem. Abstr.*, **79,** 126379.
24. G. Davidovics, P. Roepstorff, J. Chouteau, and C. Garrigou-Lagrange, *Spectrochim. Acta*, **23A,** 2669 (1967); *Chem. Abstr.*, **68,** 7766.
25. P. E. Iversen, *Acta Chem. Scand.*, **22,** 1690 (1968); *Chem. Abstr.*, **70,** 3912.
26. G. Travagli, *Gazz. Chim. Ital.*, **78,** 592 (1948); *Chem. Abstr.*, **43,** 2615.
27. B. M. Mikhailov and V. P. Bronovitskaya, *Zh. Obschei Khim.*, **27,** 726 (1957); *Chem. Abstr.*, **51,** 16436.
28. C. Gruenert, H. Schellong, and K. Wiechert, *Z. Chem.*, **10,** 116 (1970); *Chem. Abstr.*, **72,** 132594.

29. G. Bartoli, A. Latrofa, F. Naso, and P. E. Todesco, *J. Chem. Soc., Perkin Trans.* 1, 2671 (1972).
30. M. Bosco, L. Forlani, P. E. Todesco, and L. Troisi, *Chem. Commun.*, 1093 (1971).
31. M. Robba and R. C. Moreau, *Ann. Pharm. France.*, **22**, 201 (1964); *Chem. Abstr.*, **61**, 3086.
32. M. Wohmann, *Annalen*, **259**, 277 (1890).
33. H. Erlenmeyer, N. P. Buchman, and H. Schenkel, *Helv. Chim. Acta*, **27**, 1432 (1944); *Chem. Abstr.*, **39**, 3281.
34. M. Bosco, L. Forlani, P. E. Todesco, and L. Troisi, *Chim. Ind.*, **55**, 931 (1973).
35. M. Bosco, V. Liturri, L. Troisi, L. Forlani, and P. E. Todesco, *J. Chem. Soc. Perkin Trans* 2, **5**, 508 (1974); *Chem. Abstr.*, **81**, 62773.
36. K. A. Jensen and T. Thorsteinsson, *Dansk Tidsskr. Farm.*, **15**, 41 (1941); *Chem. Abstr.*, **35**, 5109.
37. K. Ganapathi and A. Venkataraman, *Proc. Indian Acad. Sci.*, **28A**, 556 (1948); *Chem. Abstr.*, **43**, 6180.
38. Brit Patent No. 517 272 and U.S. Patent No. 2 433 388; *Chem. Abstr.*, **35**, 6741.
39. H. J. Backer and J. A. Keverling Buisman, *Rec. Trav. Chim.*, **64**, 102 (1945); *Chem. Abstr.*, **40**, 3414.
40. G. Bartoli, L. DiNunno, L. Forlani, and P. E. Todesco, *Int. J. Sulfur Chem.*, Part C, **6**, 77 (1971).
41. M. Liveris and J. Miller, *J. Chem. Soc.*, 3486 (1963).
42. K. R. Brower, W. P. Samuel, J. W. Way, and E. D. Amstutz, *J. Org. Chem.*, **19**, 1830 (1954).
43. M. L. Belli, G. Illuminati, and G. Marino, *Tetrahedron*, 345 (1967).
44. G. Illuminati, P. Linda, and G. Marino, *J. Amer. Chem. Soc.*, **80**, 3521 (1967).
45. A. Ricci, P. E. Todesco, and P. Vivarelli, *Gazz. Chim. Ital.*, **95**, 101 (1965).
46. J. Metzger, 2nd Congress Int. Chimie Heterocyclique, Montpellier, 1969.
47. M. Bosco, L. Forlani, and P. E. Todesco, *Chim. Ind.* **50**, 813 (1968).
48. M. Bosco, L. Forlani, and P. E. Todesco, Unpublished Results (1973).
49. Y. Otsuji, T. Kimura, Y. Sugimoto, E. Imoto, Y. Omori, and T. Okawara, *Nippon Kagaku Zasshi*, **80**, 1024 (1959); *Chem. Abstr.*, **55**, 5467.
50. M. Bosco, L. Forlani, D. Sapone, and P. E. Todesco, *Boll. Sci. Fac. Chim. Ind. Bologna*, **23**, 83 (1969); *Chem. Abstr.*, **71**, 60287.
51. G. Bartoli, O. Sciacovelli, M. Bosco, L. Forlani, and P. E. Todesco, *J. Org. Chem.*, **40**, 1275 (1975); *Chem. Abstr.*, **83**, 95894.
52. Y. Garreau, *Bull. Soc. Chim. France*, 816 (1948); *Chem. Abstr.*, **43**, 632.
53. E. M. Gibbs and F. A. Robinson, *J. Chem. Soc.*, 925 (1945); *Chem. Abstr.*, **40**, 1830.
54. R. Dahlbom and T. Ekstrand, *Svensk Kem. Tidsskr.*, **57**, 229 (1945); *Chem. Abstr.*, **40**, 4061.
55. Swedish Patent No. 115 581 and 118 143; *Chem. Abstr.*, **41**, 160 and **41**, 447.
56. K. A. Jensen and O. R. Hansen, *Dansk Tidsskr. Farm.*, **20**, 226 (1946); *Chem. Abstr.*, **41**, 447.
57. E. Hoggarth, *J. Chem. Soc.*, 110 (1947); *Chem. Abstr.*, **41**, 3458.
58. C. S. Mahajanshetti and L. D. Basanagoudar, *Can. J. Chem.*, **45**, 1807 (1967); *Chem. Abstr.*, **67**, 63461.
59. C. S. Mahajanshetti, M. G. Dhaplapur, and S. Siddappa, *J. Karnatak Univ.*, **8**, 1 (1963); *Chem. Abstr.*, **61**, 8291.
60. D. S. Noyce and S. A. Fike, *J. Org. Chem.*, **38**, 2433 (1973); *Chem. Abstr.*, **79**, 52486.
61. R. P. Kurkjy and E. V. Brown, *J. Amer. Chem. Soc.*, **74**, 6260 (1952); *Chem. Abstr.*, **48**, 1340.

62. P. E. Iversen, H. Lund, A. Berg, and S. R. Johansen, *Sin. Geterotsikl. Soedin.*, **9,** 27 (1972); *Chem. Abstr.*, **79,** 126379.
63. H. Erlenmeyer and J. Ostertag, *Helv. Chim. Acta*, **31,** 26 (1948); *Chem. Abstr.*, **42,** 4166.
64. H. Andersag and K. Westphal, *Berichte*, **70,** 2035 (1937); *Chem. Abstr.*, **32,** 2127.
65. G. M. Dyson, R. F. Hunter, J. W. Jones, and E. R. Styles, *J. Indian Chem. Soc.*, **8,** 147 (1931); *Chem. Abstr.*, **25,** 4881.
66. J. Gregory and R. Mathes, *J. Amer. Chem. Soc.*, **74,** 1719 (1952); *Chem. Abstr.*, **48,** 164.
67. H. Erlenmeyer and C. J. Morel, *Helv. Chim. Acta*, **25,** 1073 (1942) and **28,** 362 (1945); *Chem. Abstr.*, **37,** 1702 and **40,** 1501.
68. P. Iversen, *Synthesis*, **9,** 484 (1972).
69. H. Erlenmeyer and E. H. Schmid, *Helv. Chim. Acta*, **22,** 698 (1939).
70. J. Kendall and H. Suggate, *J. Chem. Soc.*, 1503 (1949); *Chem. Abstr.*, **44,** 466.
71. H. L. Bradlow and C. A. Van Der Werf, *J. Org. Chem.*, **16,** 1143 (1951), *Chem. Abstr.*, **46,** 3535B.
72. Y. Mizuno, K. Adachi, and K. Ikeda, *Pharm. Bull. Japan*, **2,** 225 (1954).
73. J. Tirouflet and E. Laviron, *Compt. Rend.*, **244,** 2306 (1957).
74. E. Laviron *Abandl. Deut. Akad. Wiss. Berlin, Kl. Chem.*, *Geol. Biol.*, **63,** (1964); *Chem. Abstr.*, **62,** 8674.
75. H. Erlenmeyer and H. Kiefer, *Helv. Chim. Acta*, **28,** 985 (1945); *Chem. Abstr.*, **40,** 1500.
76. J. P. English, J. M. Clark, J. W. Clapp, D. Seeger, and R. H. Ebel, *J. Amer. Chem. Soc.*, **68,** 453 (1946); *Chem. Abstr.*, **40,** 2808.
77. Y. Garreau, *Compt. Rend.*, **222,** 963 (1946); *Chem. Abstr.*, **40,** 4374.
78. Y. Garreau, *Compt. Rend.*, **218,** 597 (1944); *Chem. Abstr.*, **40,** 2445.
79. E. Pedley, *J. Chem. Soc.*, 431 (1947); *Chem. Abstr.*, **41,** 5509.
80. C. D. Hurd and H. L. Wehrmeister, *J. Amer. Chem. Soc.*, **71,** 4007 (1949); *Chem. Abstr.*, **45,** 155.
81. G. Travagli, *Gazz. Chim. Ital.*, **85,** 926 (1955); *Chem. Abstr.*, **51,** 13851.
82. B. Prijs, J. Ostertag, and H. Erlenmeyer, *Helv. Chim. Acta*, **30,** 1200 (1947); *Chem. Abstr.*, **41,** 7397.
83. D. Bouin and A. Friedmann, *C.R. Acad. Sci., Ser. C*, **269,** 1343 (1969); *Chem. Abstr.*, **72,** 78116.
84. W. Kirk, J. Johnson, and A. Blomquist, *J. Org. Chem.*, **8,** 557 (1943).
85. H. Babo and B. Prijs, *Helv. Chim. Acta*, **33,** 306 (1950); *Chem. Abstr.*, **44,** 5872.
86. M. Bosco, L. Forlani, V. Liturri, and P. E. Todesco, *Chim. Ind.*, **54,** 266 (1972).
87. G. Bartoli, L. Forlani, and P. E. Todesco, Unpublished Results, (1972).
88. G. Bartoli, F. Ciminale, M. Fiorentino, and P. E. Todesco, *Chem. Commun.*, 732 (1974).
89. B. Prijs, J. Ostertag, and H. Erlenmeyer, *Helv. Chim. Acta*, **30,** 2110 (1947); *Chem. Abstr.*, **42,** 1934.
90. R. A. Parent, *J. Org. Chem.*, **27,** 2282 (1962); *Chem. Abstr.*, **57,** 5901.
91. Japanese Patent No. 22,885 '63; *Chem. Abstr.*, **60,** 4154.
92. L. M. Werbel, E. F. Elslager, A. A. Phillips, D. F. Worth, P. J. Islip, and M. C. Neville, *J. Med. Chem.*, **12,** 521 (1969); *Chem. Abstr.*, **71,** 37377.
93. A. Silberg, A. Ursu-Donea and L. Fazakas, *Stud. Univ. Babes Bolyai, Ser. Chem.*, **11,** 29 (1966); *Chem. Abstr.*, **67,** 3018.
94. A. P. Roszkowski, A. M. Strosberg, L. M. Miller, J. A. Edwards, B. Berkoz, G. S. Lewis, O. Halpern, and J. H. Fried, *Experientia*, **28,** 1336 (1972); *Chem. Abstr.*, **78,** 52718.

95. J. A. Edwards, B. Berkoz, G. S. Lewis, O. Halpern, J. H. Fried, A. M. Strosberg, L. M. Miller, S. Urich, F. Liu, and A. P. Roszkowski, *J. Med. Chem.*, **117,** 200 (1974); *Chem. Abstr.*, **80,** 140991.
96. G. Illuminati and F. Stegel, Unpublished Results, (1970).
97. A. Friedmann and J. Metzger, *C.R. Acad. Sci., Ser. C,* **270,** 502 (1970); *Chem. Abstr.,* **72,** 100585.
98. A. O. Ilvespaa, *Helv. Chim. Acta,* **51,** 1723 (1968); *Chem. Abstr.,* **70,** 3911.
99. French Patent No. 2 015 434; *Chem. Abstr.,* **74,** 31748.
100. E. J. Vincent and J. Metzger, *Compt. Rend.,* **261,** 1964 (1965); *Chem. Abstr.,* **63,** 17826.
101. G.F. Pedulli, P. Zanirato, H. Alberti, and M. Tiecco, *J. Chem. Soc. Perkin Trans.* 2, 293 (1975).
102. A. Friedmann and J. Metzger, *C.R. Acad. Sci., Ser. C,* **269,** 1000 (1969); *Chem. Abstr.,* **72,** 43551.
103. German Patent No. 2 294 174.
104. German Patent No. 2 294 175.
105. German Patent No. 2 289 503.
106. U.S. Patent No. 3 907 819.

Subject Index

Ab initio method, 28, 52
5-Acetamino-2-methylthiazoles, from α-acylaminonitriles, 285
5-Acetaminothiazoles, unsubstituted derivatives in 4-position synthesis of, 285
S-Acetonyl-N-methylisothiourea, from α-thiocyanatoacetone and methylamine, 278
N-Acetonylthiazolium salts, ring expansion, 141
Acetophenone and derivatives, bromination of, 180
2-Acetoxymethylthiazole, from 4-methylthiazole-3-oxide, 392
2-Acetoxymethyl-2-methylthiazole, from 2,4-dimethylthiazole-3-oxide, 392
2-Acetoxy-4-methylthiazole, from 2,4-dimethylthiazole-3-oxide, 392
4-Acetoxymethylthiazole, from 4-methylthiazole-3-oxide, 392
N-Acetylsarcosinethioamide, condensation, with α-bromoketones, 197
5-Acetyl-4-thiazolecarboxylic acids, synthesis of, 204
Acyclic intermediate, in condensation of, β-acylvinylphosphonium salts with thiourea, 299
  ethylcyanoacetate with ethylthioglycolate, 294
  α-mercaptoacids with nitriles, 294
  thiobenzamide with bromosuccinic acid, 295
  thiourea with α-halonitrile, 297
  in Hantzsch's synthesis, 269
  see also Hantzsch's synthesis
Acyclic intermediate enamines, in condensation, of β-amino crotonic ester with aminothiocyanogen, 298
Acylaminoketones, synthesis of, from acid anhydrides and α-aminoacids, 282
  from acids and chlorhydrate of α-aminoketones, 282
Acylhalides, reaction, with diazomethane, 180
4-Adamanthylthiazole, synthesis of, 179
Aldehydes, bromination of, 174
  halogenation of, 174
  trimerization of, 174
Alkenylthiazoles, by dehydrogenation, 342
  from hydroxy alkylthiazoles, 341
  synthesis of, 339
  table of products, 415
2-Alkenylthiazoles, table of products, 415, 417
4-Alkenylthiazoles, table of products, 416, 419
5-Alkenylthiazoles, table of products, 416
Alkoxyalkenylthiazoles, hydrolysis of, 341
2-(α-Alkoxycarbonyl)-4-methyl-5-(β-hydroxyethyl)thiazoles, from thioamides and 4-hydroxy-3-bromo-2-pentanone, 188
5-Alkoxy-2-phenylthiazoles, synthesis of, 283
2-Alkoxythiazoles, 5-methyl or 5-phenyl derivatives, 282
  synthesis of, from esters of thiocarbamic acid (or thiourethanes) and α-halocarbonyl compounds, 259
  from 2-halogenothiazoles and metallic alcoholates, 260
5-Alkoxythiazoles, synthesis of, from α-acylaminoesters and phosphorus pentasulfide, 282, 283
2-Alkylamino-5-nitrothiazoles, ring cleavage, 134
2-Alkylaminothiazoles, synthesis, 23
Alkylarylthiazoles, mass spectrometry of, 349

2-Alkyl-4-arylthiazoles, synthesis of, 186, 191
   table of products, 412
4-Alkyl-2-arylthiazoles, from arylthioamides and α-halomethylketones, 186
2,5-Alkylarylthiazoles, synthesis of, 172
2-Alkyl-5-arylthiazoles, tables of products, 413
4,5-Alkylarylthiazoles, synthesis of, 179
4-Alkyl-5-arylthiazoles, table of products, 413
Alkylbenzene, thiazolylation of, 372
N-Alkylchloromethylketones, preparation of, 264
Alkyl-diarylthiazoles, table of products, 414
4-Alkyl-2,5-dimethylthiazoles, quaternization of, kinetic studies, 389
3-Alkyl-4-hydroxythiazolidine-2-thiones, dealkylation of, 270
   in equilibrium, with acyclic isomers, 270
3-Alkyl-2-iminothiazolines, synthesis, 23
Alkylphenacyl chlorides, synthesis of, 265
2-Alkyl-N-phenacylthiazolium salts, ring expansion, 141
2-Alkylpyridines, 388
3-Alkylpyridines, 390, 391
Alkyl radicals, reactivity of, 369
1-Alkylsulfonylketones, preparation, from α-chloroketones and sodium alkylsulfonates, 264
2-Alkyl-4-thiazolecarboxylic acid, synthesis of, 204
Alkylthiazoles, application of, 395
   arylation with *para*substituted benzoyl peroxides, 367
   chemical shift, carbon-13, 76
      proton, 68
   competitive nitration, 382
   coupling constants H-H, 74
   derivatives, 337
   dehydrogenation of carbon side-chain, 341
   electrophilic reactivity, 380
   as flavouring agents, 167
   free radical arylation, 109
   gas-liquid chromatography, Kovat's indices, 359, 361
   gas-liquid and thin-layer chromatography, 358
   halogenation of, 380
   with heterocyclic group on alkyl chain,
      table of products, 481
   infrared spectroscopy, 53, 349
   infrared spectra, C-H vibrations, 349
      harmonic and combination bands, 351
      interpretation of, 349
      vibration frequencies, 350
      vibrations of thiazole ring, 351
   ionization potential, 52, 83
   isomerization of, 374
   mass spectrometry of, 81, 347
   mercurated derivatives, halogenation of, 380
   mercuration of, 380
   methylation of, 369
   nitration of, 104, 381
      kinetics constants, 383, 384
   N.M.R. of, 342
   nucleophilic reactivity, 378
   organolithium compounds, 378
   organomagnesium compounds of, 378
   oxidation of carbon side-chain, 341
   phase diagrams of binary mixture, 357
   phenylation of, 366
   photochemical rearrangement, 374
   pKa, 356
      and basicity, 355
   with polyfunctional group, table of products, 460, 464
   quaternization of, 386
   radical reactivity, 364
   Raman spectroscopy, 349
   reaction with ethyl magnesium bromide, 378
   reactivity with aryl radical, 364
   reactivity of nitrogen atom, 386
   refraction index, 90
   relation between RF and pKa, 362
   specific heats, determination of, 357
   steric effect, on adsorption energy, 363
   synthesis of, 339
   theoretical calculations, 357
   thermodynamic constants of ionization reaction, 355, 356
   from thiazole acetic acids, 341
   with thiazolyl group on alkyl chain, table of products, 482
   thin-layer chromatography, 362
   ultraviolet spectroscopy, 352
   variation of adsorption-energy with substituents, 363

Subject Index

viscosity in solution, for determination of degree of association, 357
2-Alkylthiazoles, nitration of, 381
   with one function on alkyl group, table of products, 428
   pKa in water, 387
   quaternization of, 386
      kinetic data, 388
      reaction with methyl iodide, correlation of rate constants with Taft's parameters, 388
      rate constants and activation parameters, 387
   reactivity of alkyl group, with aldehydes, 392
      with ketones, 393
   synthesis of, 170, 172
   tables of products, 402
4-Alkylthiazoles, with one function on alkyl group, table of products, 439
   preparation of, 179
   quaternization of, 386
      kinetic data, 388
   tables of products, 403
5-Alkylthiazoles, with one function on alkyl group, table of products, 452
   phenylation of conjugate acid, 368
   quaternization of, 386
      correlation of rate constants with Taft's parameter, 391
      kinetic data, 391
      with methyl iodide, 390
   tables of products, 403
   from thioformamide, and α-haloaldehydes, 172
3-Alkyl-Δ-4-thiazoline-2-thiones, preparation of, 271
   steric effects in, 270
Alkylthioamides, condensation with, ω-bromoacetophenones, 188
   α,β-dibromoether, 172
   sym, dichloroacetone, 185
   α-haloalkylketones, 180
2-Alkylthio-4-benzylidenethiazol-5-ones, from *N*-thioacylaminoacids and benzaldehyde, 303
2-Allylaminothiazoles, from allylthiourea and α-halocarbonyl compounds, 233, 237
Althiomycine, thiazolic structure, 1

2-Amino-4-acetylthiazole, from α-diketone and thiourea, 215
α-Aminoacids, *N*-benzoyl derivatives of, 283
2-Amino-4-(4-alkyl-selenophenyl) thiazoles, synthesis of, 216
Aminoalkylthiazoles, from haloalkylthiazoles, synthesis of, 340
2-Amino-4-alkylthiazoles, alkylation in 5-position, 103
2-Amino-5-β-aminoethylthiazoles, from isomerization of 2-amino-5-β-aminoethylidene Δ_2-thiazolines, 300
2-Amino-4-arylthiazoles, synthesis of, 216, 217
   from α-thiocyanatoacetophenones and ammonium chloride, 278
2-Amino-4-aryl-5-(*p*-aminophenyl)thiazoles, from phenylthiosemicarbazides and ω-bromoacetophenones (or acetophenones and iodine), 230
5-Amino-2-benzylthiazole, from potassium thiophenylacetate, 285
2-Amino-5-bromothiazole, deamination of, 566
4-Amino-5-carbethoxy-2-methylthiazole, from methyl *N*-cyanoiminodithiocarbamic ester and ethyl-α-mercaptoacetate, 301
5-Amino-4-carboxyethyl-2-methylthiazole, from ethyl-2-acetamido-2-thiocarbamoylacetate and $POCl_3$, 284
4-Amino-5-cyanothiazoles, 2-substituted derivatives, synthesis of, 302
2-Amino-4,5-disubstituted thiazoles, synthesis of, from thiourea and α-halogenoketones, 224, 225
2-Amino-4-(2'-furyl)thiazole, bromination, 100
2-Amino-5-halogenothiazoles, nucleophilic substitution reaction, 572
   with mercaptans, 572
   with thiophenol, 572
   with thiourea, 572
2-Amino-4-hetarylthiazoles, synthesis of, 216, 217
2-Amino-4-hydroxythiazole, from thiourea and ethyl α-chloroacetate, 215. See *also* Thiohydantoin
5-Amino-2-hydroxythiazoles, 2-alkoxy-5-benzylideneamino derivatives of, 288

Subject Index

from α-aminonitriles and carbon oxysulfide, 288
5-benzylideneamino derivatives of, 288
α-Aminoketones, synthesis of, from α-halogenoketones, 282
  by reduction of α-nitrosoketones, 282
5-Amino-2-mercaptothiazoles, benzylideneamino derivatives of, 287
  from carbon disulfide and α-aminonitriles, 286
2-Amino-4-methoxycarbonyl-5-(1-methoxyalkyl) thiazoles, from 2,4-di-chloro-2,4, epoxyalkanoates and thiourea, 231
m-Aminomethylthiazole, see 2-Amino-4-methylthiazole, 9
2-Amino-4-methylthiazole, from α-thiocyanotoacetone and ammonia, 278
  structure of, 9, 22
2-Aminomethylthiazoles, acetyl and benzoyl derivatives of, 184
  from α-haloketones and diacetylaminothioacetamide, 184
2-Aminomethyl-4-thiazole carboxylic acid, synthesis of, 204
2-Amino-4-(4-methyl-2-thiazolyl)-5-(p-aminophenyl) thiazoles, synthesis of, 230
2-Amino-4-naphthyl-5(p-aminophenyl) thiazoles, synthesis of, 230
α-Aminonitriles, condensation, with salts or esters of dithioformic and dithiophenacetic acids, 284
2-Amino-5-nitrothiazoles, Schistosomicidal activity of, 167
2-Amino-4-phenylthiazoles, from α-diazoacetophenone and thiourea, 231
2-Amino-5-phenylthiazole, diuretic properties of Schiff's bases of, 167
  synthesis of, by condensation of ammonia with N-(aryl-1,3-oxathiol-2-ylidene) tertiary iminium salts, 300
2-(p-Aminophenyl) thiazoles, syntheses of, 4,5-disubstituted derivatives of, 191
2-Amino-5-substituted thiazoles, from α-haloaldehydes and thiourea, 224, 225
N,N^1-bis (2-Amino-4 substituted thiazolyl)-p-biphenylene, from p-biphenylene dithiourea and α-halocarbonyl compounds, 243

2-Amino-4,5,6,7-tetrahydrothiazolo 5,4c pyridine, from 3-bromo-4-piperidone hydrochloride and thiourea, 230
2-Aminothiazole, 566
  deamination of, 272
  diazo coupling, 103
  labelled from ^{14}C-thiourea, 214
  mechanism of formation, 232
  pKa, 92
  preparation of, 566
  preparation and structure, 22
  reduction of diazo compound, 339
  Sandmeyer reaction, 566
  synthesis of, from ethyl α β-dihalogeno ethers and thiourea, 214
5-Aminothiazole, derivatives, from α-aminonitriles and salts or esters of dithioacids, 284
  2,4-disubstituted derivatives by cyclization of thioamides, with polyphosphoric acid, 285
  synthesis of, from ethylformate, 285
  from 2-formamidothioacetamide, 284
2-Amino-4-thiazolecarboxylic acid esters, from 3-arylamino-2-chloropro-2-enoic esters and thiourea, 205
  decarboxylation of, 215
  hydrolyse of, 215
  from thiourea and γ-bromo acetoacetic esters, 215
Aminothiazoles, chemical shift, proton, 68
  coupling constants H-H, 74
  dipole moment, 38
  electronic structure, 44
  ionization potential, 52, 83
  substitution effects, 45
2-Aminothiazoles, antimicrobic and fungicidal properties of, 213
  conversion into unsubstituted derivatives: in the 2-position, 232
  derivatives, from enamines, sulfur and cyanamide, 297
  mitodepressive and mitostatic properties of, 167
  ring cleavage, by phenylhydrazinolysis, 229
    by transition metals salts, 136
  2-substituted derivatives, from thioureas and 1,2 dichloroethylisothiocyanate, 301

Subject Index

4-substituted derivatives, from α-halomethyl ketones and thiourea, 215, 216
N-substituted derivatives, from N-substituted thioureas and α-halocarbonyl compounds, 234
synthesis, 233
from α-thiocyanatoketones and ammonium chloride or amine hydrochloride, 278
sulfonation, 100
synthesis of, 23
from β-acylvinylphosphonium salts and thiourea, 299
from aminothiocyanogen and ketones, 298
from cyanamide and α-mercaptoketones, 293
from 3-halogenoalkynes and thiourea, 230
from α-haloketones or aldehydes and thiourea, 213
from ketones, cyanamid and sulfur, 214
from thiourea, iodine and ketones, 213
from thiourea and iodomercuriketones, 214
4-Aminothiazoles, 2,5-disubstituted derivatives, by cyclization of halogeno compounds with cyanamide derivatives, 302
synthesis of, 302
2-substituted derivatives, from α-chloroalcanoic acids and sulfide, 302
synthesis of, from α-cyanoalkylthiocyanates and hydrogen iodide, 275
from α-halonitriles and thioamides, 301
from 1,3-oxathiolium salts and cyanamide, 303
5-Aminothiazoles, 2-benzylthio derivatives of, 286
unsubstituted in 2-position, synthesis by oxidation of 5-amino-2-mercaptothiazoles, 286
2-Amino-4-thiazolones, as tautomers of, 2-amino-4-hydroxythiazoles, 215
2-Amino-4(p-2-thiazolylphenyl) thiazole, from thiourea and p-2-thiazolylphenacylbromide, 216
2-Aminothieno[2,3d] thiazoles, from 3-thiocyanatoacetophenones, 278
Ammonium dithiocarbamate, condensation,
with 3-bromobutan-2-one, 264
with chloracetone, 264
decomposition of, 260
with α-halocarbonyl compounds, 258
preparation of, 258
Analeptic, aminoalkylesters as, 527
2-Anilino-4-propyl-5-isopropylthiazole, from 2-methyl-3-isothiocyanato-4-thiocyanato-hept-3-ene and aniline, 299
Anthelmintics, amidines as, 532
Antiinflammatory, 2-alkyl-5-thiazolecarboxylic as, 525
Aralkylthiazoles, and related thiazole derivatives, 337
synthesis of, 339
2-Aralkylthiazoles, table of products, 419, 421
4-Aralkylthiazoles, table of products, 420, 422
5-Aralkylthiazoles, table of products, 420, 422
Aromaticity, 33, 67
1-Aroyl-2-(2-thiazolyl)hydrazines, synthesis of, from arylthiosemicarbazides, 253
2,5-Arylalkylthiazoles, preparation of, 175
2-Aryl-4-alkylthiazoles, table of products, 412
2-Aryl-5-alkylthiazoles, table of products, 413
4-Aryl-5-alkylthiazoles, table of products, 413
2-Arylamino-5-nitrothiazoles, ring cleavage, 134
2-Arylaminothiazoles, synthesis of arsenic derivatives, 243
synthesis of 4-or 5- substituted derivatives from α,β-dihalogenoethers and N-arylthioureas, 238
2-Aryl-4-hydroxythiazoles, synthesis of, from α-mercaptoacids and nitriles, 294
2-Arylidenehydrazino-4-chloromethylthiazoles, from sym-dichloroacetone and thiosemicarbazones, 256
2-Arylimino-3,4-diarylthiazolines, preparation of, 231
2-Aryloxazoles, by photochemical conversion of 2-arylthiazoles, 309
2-Aryloxymethylthiazoles, 5-methyl or phenyl derivatives, synthesis of, 282

Aryl radicals, reactivity with thiazoles and alkylthiazoles, 364
2-Aryl-4-thiazolecarboxylic acids, synthesis of, 204
5-Aryl-4-thiazolecarboxylic acids, synthesis of, 204
Arylthiazoles, applications of, 395
  bromination of, 382
  charge transfer complexes with TCNE, 354
    ultraviolet spectra, 355
  dipole moment, 38
  electrophilic reactivity, 380
  electrophilic substitution, 382
  with functional substituents in benzene ring, mass spectra of, 349
  gas-liquid and thin-layer chromatography, 358
  infrared spectra, vibration frequencies, 350
  infrared spectroscopy, 349
  mass spectrometry of, 347, 349
  mechanism of rearrangement, 378
  N.M.R. of, 342
  nucleophilic reactivity, 378
  phenylation of, 367
  photochemical rearrangement, 374, 376
  pyrolysis of, 399
  quaternization of, 391
    order of reactivity, 391
  Raman spectroscopy, 349
  and related thiazole derivatives, 337
  synthesis of, 339
  thin-layer chromatography, 362
  ultraviolet spectroscopy, 352, 353
2-Arylthiazoles, photochemical isomerization into 3-arylisothiazoles, 310
  photorearrangement of, 112
  table of products, 409, 422, 424
4-Arylthiazoles, from arylacyl bromide, and thioformamide, 179
  table of products, 410, 424, 426
5-Arylthiazoles, synthesis of, from ω-formamidoacetophenones and P_2S_5, 279
  from ω-thioformamidoketones and H_2SO_4, 279
  from thioformamide and α-haloaldehydes, 172
  table of products, 410, 426
Arylthioamides, cyclization of, with chloroacetaldehyde, 171
2-Arylvinylthiazoles, preparation of, 184

Association, 88
  of thiazole and alkyl derivatives, 357
4-Azidothiazoles, from 4-carboxythiazoles, 204

Basicity, of alkylthiazoles, 355
  of thiazoles, 91
2-Benzamidothiazole, 43
2-Benzhydrylthiazoles, synthesis of, 185
Benzimidazolylthiazoles, table of products, 473
Benzocycloheptathiazoles, from thioamides and 6-bromo-1-benzosuberone, 200
Benzonitril-4-nitrobenzylide, cycloadditions of, with dithiobenzoate and dimethyl trithiocarbonate, 307
Benzopyrrolylthiazoles, table of products, 471
Benzothiazolyl radicals, electron spin resonance of, 373
Benzothiazolylthiazoles, table of products, 469
Benzoyl peroxides, source of phenyl radicals, 364
5-Benzoyl-4-thiazolecarboxylic acids, synthesis of, 204
5-Benzylideneamino-2-mercaptothiazoles, from α-aminonitrile, carbon disulfide and benzaldehyde, 286
2-Benzylthiazole, reaction with n-butyllithium, 379
Benzylthiazoles, nitration of, 385
N-Benzylthiazolium chloride, thermolysis of, 170
Binary mixtures, 85
p-Bis (2-amino-4-thiazolyl) phenylene ethers and thioethers derivatives, from thiourea and bis-α-chloroketone, 224
Bis-(4-aryl-2 thiazolyl) sulfides, as by-products of the 4-aryl-2 mercaptothiazoles synthesis from α-thiocyanatoacetophenones and thiourea, 276
Bis-(4-phenyl-2-thiazolyl)sulfide, from 2-mercapto-4-phenyl-2-isothiazolyl isothiouronium chloride, 276
Bis-(2-thiazolyl)amine, from dithiobiuret and aromatic α-haloketones, 244
m- and p- Bis (2-thiazolyl)benzenes, from thioamides and bis-(haloacetyl) benzenes, 193

Subject Index

m- and *p*- Bis (4-thiazolyl)benzenes, 193
2,2'-Bithiazoles, from rubeanic acid and bromoketones, 197
Boiling point, 85
Bond lengths, thiazole, 39, 45
Bond orders, methyl thiazoles, 41
  thiazole, 30, 32
Bromination, of arylthiazoles, 382
Bromine, as reactant in 2-aminothiazoles syntheses, 214
α-Bromoacetals, preparation of, 175
2-Bromo-4-acetylaminothiazoles, preparation of, 276
2-Bromoacetylthiazoles, condensation, with thiourea, 224
α-Bromoaldehydes, use of, 174
2-Bromo-4-aminothiazoles, from α-cyanoalkylthiocyanates and hydrogen bromide, 275
Bromomethylketones, synthesis, 180
2-Bromothiazole, bromination of, 575
  preparation of, 565
  reaction with copper, 573
  reduction of, 573
  sulfonation, 100
4-Bromothiazole, preparation of, 566
5-Bromothiazole, 380
  preparation of, 566
2-Bromothiazoles, synthesis of, from α-thiocyanatoketones and hydrogen bromide, 274, 275
1,3-Butadiene, copolymerization with 4-methyl-2-vinyl thiazole, 398
4-Butyl-2,5-dimethylthiazole, mass spectrometry of, 348
Butyllithium, reaction with alkylthiazoles, 378
2-*t*-Butyl-5-methylthiazole, rate of nitration, 104
2-*i*-Butylthiazole, in tomatoes aromas, 395
  pKa, 92
2-*t*-Butylthiazole, free enthalpy of ionization, 92
  pKa, 92
4-*t*-Butylthiazole, free enthalpy of ionization, 92
  pKa, 92
5-*t*-Butylthiazole, synthesis of, 174

2-Carbethoxy-4-hydroxythiazole, from chloroacetonitrile and ethyl thiooxamate, 295
Carbon disulfide, condensation, on diamines, 300
  with ethylisocyanoacetate, 307
4-Carboranylmethylthiazole, synthesis of, 179
2-Carboxythiazole, replacement of carboxy group by hydrogen, 339
4-Carboxythiazoles, from α-bromopyruvic acid, 204
4-Carboxythiazoles derivatives, synthesis of, 207
2-Carboxy-5-X-thiazoles, 572
4-Carboxy-2-X-thiazoles, 572
5-Carboxy-2-X-thiazoles, 572
Charge transfer complexes, of arylthiazoles, 354
Chemical shift, ^{13}C, 77
  of alkyl thiazoles, 345, 346
  empirical calculations, 70
  nitrogen, 77
  proton, 70
    of alkylthiazoles, 71, 342, 343
    of arylthiazole, 342, 344
    correlation with π electron densities, 344
    difference induced by protonation, 345
    methylthiazoles, 71
    thiazole derivatives, 71
    related to electronic densities on carbon atoms, 347
    theoretical calculations, 70, 77
    of thiazole, 76
    of thiazole derivatives, 76
Chlorine, as reactant in 2-aminothiazole synthesis, 214
Chloroacetaldehyde, condensation with thioamides, 171
  preparation of, 171
  reaction with thioamides, 170
2-Chloro-4-aryl-5-bromothiazoles, preparation of, 274
2-Chloro-4-arylthiazoles, action of thiourea on, 276
  preparation of, 274
2-Chloro-5-bromothiazole, 575
  preparation of, 575
2-Chloro-4,5-dialkylthiazoles, preparation of, 274

2-Chloro-4,5-dimethylthiazoles, preparation of, 274
5-(2-Chloroethyl)-4-methylthiazoles, preparation of, 179
1-Chloroethylthiazole, kinetics of solvolyse, 393
2-(1-Chloroethyl) thiazoles, substituted at 4-position, rate of solvolysis of, 395
4-(1-Chloroethyl) thiazoles, substituted at 2-position, rate of solvolysis of, 394
5-(1-Chloroethyl) thiazoles, substituted at 2-position, rate of solvolysis of, 394
2-Chloro-4-methyl-5-(β-acetoxyethyl) thiazole, preparation of, 274
2-Chloro-4-methyl-5-(β-carbethoxyethyl) thiazole, preparation of, 274
4-Chloromethyl-2-phenylthiazole, chloration of, 385
2-Chloro-4-methylthiazole, preparation of, 274
4-Chloromethylthiazoles, from alkylthioamides and sym dichloroacetone, 185
2-substituted derivatives, synthesis of, 185
Chlorosulfonic acid, as reactant in 2-aminothiazole syntheses, 214
2-Chlorothiazole, preparation of, 565
4-Chlorothiazole, preparation of, 566
5-Chlorothiazole, preparation of, 566
2-Chlorothiazoles, synthesis of, from 2-hydroxythiazoles and phosphorylchloride, 274
from α-thiocyanatoketones and by hydrochloric acid, 273, 274
α-Chlorothioacetic acid ethyl ester, condensation, with potassium salt of thioacetylcyanamide, 302
Clomethiazole, 399
CNDO method, 28, 50, 52
Cobalt, complexes with thiazoles, U.V. and I.R., 392
Cook-Heilbron's synthesis, 5-aminothiazoles derivatives, 284
Conformation, effect on kinetics of quaternization, 389
Copper, complexes with thiazole, U.V. and I.R., 392
Coupling constants, C-C, 80
  C-H, 79
  ^{13}C-H, 345
    alkylthiazoles, 346
  used to calculate inter orbital and internuclear angles of thiazole, 345
  C-N, 80
  H-H, 76
    alkyl thiazoles, 342
    arylthiazoles, 342
    thiazole, 73
    thiazole derivatives, 74
  N-H, 80
Cyanamid, condensation, with ketones and sulfur, 214
α-Cyano-α-acetylthioacetamide, condensation, with α-chloroacetonitrile, 301
α-Cyanoalkylthiocyanates, from α-chloroalkyl and α-(toluenesulfonyloxy) alkylcyanides, 275
Cyanoamido-thiocarbamates, from cyanamide and isothiocyanates, 297
Cyanohydrin, benzene sulfonates, synthesis of, from sodium cyanide and halobenzaldehyde, 297
5-Cyano-4-hydroxythiazoles, synthesis of, 306
S-Cyanomethyl isothiouronium salt, as acyclic intermediate in reaction of thiourea with α-halonitrile, 297
Cyanothiazoles, from dehydration of amides, 530, 531
  from oxidation of alkylthiazoles, 531
  physical properties of, table of, 552
  reaction of with, ammonia or amines, 532
    organometallic reagents, 532, 536
  reduction by, diisobutyl-aluminium hydride, 531, 532
    Raney nickel, 531
    Stephen's method, 531
4-Cyanothiazole, synthesis of, 306
4-Cyanothiazoles, from α-metalled isocyanides and thionoesters, 306

Del Re method, 27
Deuterated thiazoles, mass spectra of, 347
Deuterium exchange, in thiazoles, 113
Deuterothiazoles, infrared absorption, 57
N,N'-Diacetyl N,N'-bis (4-phenyl-2-thiazolylmethyl-ethylene diamine), synthesis of, 195
N,N'-Diacetylethylene diamine N,N'-dithioacetamide, reaction, with bromoacetophenone, 195

Subject Index

N,N-Dialkyl-2-aminothiazoles, preparation of, 244, 246
Dialkylarylthiazoles, table of products, 414
2,2'-Dialkyl-4,4'-bithiazoles, from dibromobiacetyl and thioamides, 195
2,4-Dialkylthiazoles, by Dub's method, 305
  preparation of, 181
  quaternization of, 389
    kinetic data, 389
    with methyl tosylate, 389
    reaction with methyliodide, correlation of rate constants with Taft's parameters, 389
  table of products, 404
2,5-Dialkylthiazoles, preparation of, 175
  table of products, 406
4,5-Dialkylthiazoles, syntheses of, 179
  table of products, 407
N,N-Dialkylthioureas, condensation, with chloroacetaldehyde or dibromoether, 244
2,4-Diamino-5-arylthiazoles, from substituted phenylbromoacetonitrile and thiourea, 297
2,4-Diamino-5-(4-halogenophenyl)thiazole benzene sulfonate, synthesis of, 297
2,4-Diaminothiazole hydrochloride, from thiourea and chloroacetonitrile, 215
2,4-Diaminothiazoles, synthesis of, from α-halonitriles and thiourea, 296
  substituted derivatives, 297
2,5-Diaminothiazoles, acyclic intermediate in synthesis of, 289
  alkaline isomerization, 134
  5-benzylideneamino derivatives of, 290
  2,5-disubstituted derivatives from acylisothiocyanates, 291
  synthesis of, from isothiocyanates and α-aminonitriles, 290
2,4-Diaryl-5-thiazole acetic acid, from dehydration of 4-hydroxy-2-aryl-2-thiazoline-5-acetic acid, 208
Diarylthiazoles, photochemical cyclization to polycyclic benzothiazoles, 376
2,4-Diarylthiazoles, synthesis of, from thiobenzamide and ω-bromoacetophenones, 192
  table of products, 410, 427
2,5-Diarylthiazoles, preparation of, 175
  table of products, 411, 428

4,5-Diarylthiazoles, table of products, 411, 428
α,α-Diarylthioacetamides, condensation, with α-haloketones, 185
α-Diazoketones, in place of α-halogenoketones in Hantzsch's synthesis, 231
Diazomethane, preparation of, 180
2,5-Dibromothiazole, catalytic reduction of, 576
  preparation of, 575
  sulfonation, 100
2,5-Dichlorothiazole, 566
  reductive dehalogenation of, 566
Dienic character, 33
4,5-Diethylthiazole, metallation, 119
α,β-Dihalogenoethylacetate, preparation, from vinylacetate and halogen, 172
α,β-Dihalogenoethylethers, condensation, with thioamides, 175
  syntheses of, 175
2,4-Dihalogenothiazoles, preparation of, 575
2,5-Dihalogenothiazoles, preparation of, 575
  reduction with metals, 576
4,5-Dihalogenothiazoles, synthesis of, from sodium acetylaminomethane sulfonate and thionyl halide, 304
9,10-Dihydro-8H-benzo-4,5-cyclohepta (d)-thiazol-2-yl hydrazine, from thiosemicarbazide and 7-bromobenzo (a) cyclohept-2-ene-1-one, 249
2,5-Dihydrothiazoles, dehydrogenation of, 340
2,4-Dihydroxythiazole, 575
  halogenation of, 575
  structure of, 10
4-Dimethylaminoethylthiazoles, from thioamides and 1-bromo-4-dimethylamino-2-butanone, 184
2,4-Dimethyl-5-benzylthiazole, from thioacetamide and 1-phenyl-2-chloro-2,3 epoxybutane, 192
2,4-Dimethyl-N-benzylthiazolium chloride, thermolysis of, 111
2,4-Dimethyl-5-carboxythiazole, hydrogen-deuterium exchange, 144
2,4-Dimethyl-5-(β-hydroxyethyl) thiazole, from 3-acetyl-3-mercaptopropanol and acetaldehyde, 293
2,4-Dimethyl-3-phenylthiazolium chloride, synthesis of, 211

4,5-Dimethyl-2-propylthiazole, mass spectrometry of, 348
2,4-Dimethylthiazole, action of benzoyl peroxide on, 365
   azepine adduct with dimethylacetylene dicarboxylate, 98
   free enthalpy of ionization, 92
   halogenation of, 380
   mass spectrometry of, 347
   organolithium products, reaction with the non metalled thiazole, 379
   perfluoroalkylation, 103
   pKa, 92
   rate of nitration, 104, 105
   reactivity of the 2-methyl group, with substituted benzaldehyde, 393
   synthesis of, 180, 181, 184
      from α-mercaptoacetone and acetamide, 293
      by vapor phase reaction of sulfur with dialkylamines, 305
2,5-Dimethylthiazole, cycloaddition with dimethylacetylene dicarboxylate, 97
   free enthalpy of ionization, 92
   pKa, 92
   rate of nitration, 104, 105
4,5-Dimethylthiazole, free enthalpy of ionization, 92
   metallation, 119
   pKa, 92
   reaction, with benzaldehyde, 393
      with ethylmagnesium bromide, 378
      with formaldehyde, 393
2,4-Dimethylthiazole-3-oxide, 392
2,4-Dimethylthiazoles, 355
   reaction with butyllithium, 378
3,4-Dimethylthiazol-2-one, rate of nitration, 105
2,4-Dinitrothiazole, preparation of, 578
   reduction of, 578
Dioxane, as solvent in preparation, of 2,4-dialkylthiazoles, 184, 191
Dioxane dibromide, as bromination agent, 180
   as halogenating agent, 174
Dioxythiazole, see 2,4-Dihydroxythiazole, structure of
2,5-Diphenyl-4-aminothiazole, from benzene sulfonic ester of mandelonitrile and thiobenzamide, 301

3,5-Diphenylisothiazole, photoisomerization, 137
2,4-Diphenylthiazole, mercuration of, 381
   from photochemical isomerization of 3,5-diphenyl-isothiazole, 547
   photoisomerization, 136
   synthesis of, 192
2,5-Diphenylthiazole, photoisomerization, 136
   synthesis of, from sulfur, ammonia and acetophenone, 309
4,5-Diphenylthiazole, from desylchloride and thioformamide, 179
   photoisomerization, 136
Dipole moment, of aminothiazoles, 38
   of arylthiazoles, 38
   of halogenothiazoles, 38
   of methylthiazoles, 38
   of thiazole, 37, 39
   of thiazoles, 89
2,4-Dipyridylthiazoles, table of products, 478
2,2'-Dithiazolyl, metallic complexes, with Cu(II) salts, 129
   with Ni(II) salts, 127
Dithioazolylamines, synthesis of, from dithiobiuret and α-halocarbonyl compounds, 245
Dithiazolylketone, by oxidation, 341
Dithiobiuret, condensation, with aromatic α-haloketones, 244

π-Electron densities, correlation with proton chemical shift, of alkylthiazoles, 344
   of phenylthiazoles, 345
Electronic charges, of thiazole, 31
   of aminothiazoles, 44
   of chlorothiazoles, 44
   of methylthiazoles, 40
Electronic effects, of 5-substituents on quaternization, 390
Electronic paramagnetic resonance, 84
   coupling constants for nitrothiazoles, 84
   coupling constants for 2-thiazolyl radical, 84
Electronic structure, 26
   aminothiazoles, 44
   halogenothiazoles, 44
   methylthiazoles, 40
   thiazole, 27

Subject Index

Electrophilic reactivity, of alkyl and aryl thiazoles, 380
Electrophilic substitution, 99
  in thiazole series, 101
Enol acetate, preparation of, 175
Enthalpy of fusion, 85
Esters, from diazomethane, 525
  from direct esterification, 525
  from heterocyclization, 520
  reaction with nucleophilic compounds, 526, 527
  from thiazole anhydrides, 525, 526
  from trans esterification, 526
Esters of thiazolecarboxylic acids, from Hantzsch's synthesis, 525
5-Ethoxy-2-phenyl-Δ^2-thiazoline, as intermediate in 2-phenylthiazole synthesis, 283
5-Ethyl-2-t-butylthiazole, rate of nitration, 105
Ethylmagnesium bromide, reactivity with alkylthiazoles, 378
4-Ethyl-5-methyl-2-mercaptothiazole, from 2-mercapto-3-pentanone and ammonium thiocyanate, 293
2-Ethyl-4-methylthiazole, synthesis of, 180
2-Ethyl-5-methylthiazole, rate of nitration, 104
4-Ethyl-5-methylthiazole, metalation, 119
2-Ethylthiazole, pKa, 92
  reactivity with aldehydes, 393
4-Ethylthiazole, pKa, 92
  reaction with ethylmagnesium bromide, 378
Ethyl 4-thiazolecarboxylate, from ethyl-α-bromoacetoacetate and thioformamide, 520
Ethylvinylether, bromination of, 175

Ferric chloride, as dehydrogenating agent of Δ^3-thiazolines, 308
2-Fluorothiazole, preparation of, 566
Food flavoring, analysis by mass-spectrometry, 348
Formamidine disulfide, as intermediate in 2-aminothiazole synthesis, 214
5-Formylthiazole carboxylic acid, from chloromalonic dialdehyde, 204
Free radical substitution, 108
Furylthiazoles, table of products, 469

Gabriel's synthesis, 278, 282
2-(D-Galactopentaacetoxypentyl) thiazoles, synthesis of, deacetylation of, 188
Gas-liquid chromatography, of alkylthiazoles, relative retention volumes, 358
  of arylthiazoles, relative retention volumes, 358
  Kovat's indices, 359
  of thiazolyl pyridines, relative retention volumes, 358
Geometrical structure, 45
  approximation from J_{C-H}, 79

Haloalkylthiazoles, condensation with malonic derivatives and cyanoacetic esters, 341
  precursors of phosphorous insecticides, 341
  reactivity of, 340
  in Sommelet reaction, 341
Halogenation, of alkylthiazoles, 380
α-Halogenoaldehydes, instability of, 175
2-Halogeno-5-nitrothiazoles, nucleophilic substitution reactions, 578
  opening reaction with amines, 579
  preparation of, 578
  reaction with MeO$^-$, 578
  reaction with PhS$^-$, 579
Halogenothiazoles, chemical shift, proton, 68
  coupling constants H-H, 68
  dipole moment, 38
  electronic structure, 44
  infrared absorption, 53
  ionization potential, 52, 83
  substitution effects, 45
  ultraviolet absorption, 48
2-Halogenothiazoles, from 2-aminothiazole, 339
  dehalogenation of, 339
  reduction of, 172
Hammett's σ^+ parameter, of thiazolyl groups, calculation from rate of solvolysis, 394
Hantzsch's synthesis, activation energies of reaction, of phenacyl bromide with thiosemicarbazide, 256
  activation energy in, 210, 232
  acylic intermediate in, 209
  in 2-hydroxythiazoles synthesis, 273

mechanism of, from α-halocarbonyl compounds and thiourea, 232
from thioamides and α-halogeno carbonyl compounds, 209
rate of reaction, correlation with Hammett σ constants, 210, 232, 256
of α-halocarbonyl compounds and thiourea, 232
of phenacylbromide with thiosemicarbazide, 256
in reaction of thiobenzoic acid with α-thiocyanatoacetophenone, 276
Heat capacity, 86
2-Hetaryl-4-thiazole, carboxylic acids, synthesis of, 204
2-Heteroarylthiazoles, syntheses of, 196
Hexamethylene tetraminium salts, hydrolysis of, 282
HMO method, 27, 89
Hydrazinothiazole carboxylic acids, synthesis of, from α-bromoacetoacetic ester arylhydrazones and thioamides, 206
2-Hydrazinothiazoles, synthesis of, from thiosemicarbazides and α-halocarbonyl compounds, 251, 256
synthesis of N,N-disubstituted derivatives, from disubstituted thiosemicarbazides, 255
synthesis of N-substituted derivatives, from substituted thiosemicarbazides, 250
synthesis of N-sulfonamido derivatives, 250
from thiosemicarbazides and α-halocarbonyl compounds, 249
4-Hydrazinothiazoles, from 4-carboxy thiazoles, 204
Hydrazinylthiazoles, from thiosemicarbazides of nicotinic and isonicotinic acids and α-halocarbonyl compounds, 255
2,2'-Hydrazo-bis-thiazoles, from 1-(2-thiazolyl) semicarbazides (or from bisthiourea), and α-halocarbonyl compounds, 255
Hydroxyalkylthiazoles, to alkenyl thiazoles, 341
from alkoxyalkylthiazoles, 341
esterification of, 341

reactivity of, to halogenated derivatives, 341
by reduction of the esters of thiazole acetic acids, 341
2-(α-Hydroxyalkyl)thiazoles, from α-haloketone and α-benzoyloxythiopropionamide and α-benzoyloxy-α-benzoylthioacetamide, 188
2-Hydroxy-5-aminothiazoles, synthesis of, from α-aminonitrile and carbon oxysulfide, 289
4-Hydroxy-5-arylthiazoles, synthesis of, from oxiranes and thioamides, 295
2-Hydroxy-4-arylthiazoles, synthesis of, from α-thiocyanatoacetophenones, 273
4-Hydroxy-2-(α-cyano-α-acetylmethylthiazole, from α-cyano-α-acetyl-thioacetamide and ethyl-α-bromoacetate, 294
2-Hydroxy-4,5-dimethyl-thiazole, synthesis of, 271
2-Hydroxy-4-hetarylthiazoles, from α-thiocyanato-heteroarylketones, 273
2-Hydroxymethyl-4,5-dimethylthiazole, from 4,5-dimethylthiazole, 393
4-(Hydroxymethylthiazole), from 4-thiazole aldehyde, 341
2-Hydroxy-4-methylthiazole, from barium thiocyanate and chloroacetone, 271
formylation reaction, 103
from thiocyanoacetone, 11
structure of, 10
2-Hydroxy-4-methyl-5 thiazolecarboxylic acid ethyl esters, from α-chloroacetoacetate and ammonium thiocarbamate, 258
2-Hydroxy-4-phenylthiazole, from thiocyanoacetophenone, 11
4-Hydroxy-2-phenylthiazoles, from α-chlorothioacids and thiobenzamide, 295
2-Hydroxythiazole, from chloroacetaldehyde and ammonium thiocarbamate, 258
4-Hydroxy-2-thiazolecarboxylic acid ethyl ester, self condensation of, 294
synthesis of, from ethylcyanoacetate and ethylthioglycolate, 294
from thioamides and α-haloacids or esters, 294
2-Hydroxythiazoles, halogenation of, with phosphorus oxychloride, 259

mechanism of formation of, 273
synthesis of, from α-chloroketones and ammonium thiocarbamate, 258
   from salts and esters of thiocarbamic acid and α-halocarbonyl compounds, 258, 259
   from α-thiocyanato ketones, 271, 272
4-Hydroxythiazoles, synthesis of, from α-mercaptoacids and nitriles, 293
5-Hydroxythiazoles, 2-substituted derivatives, cyclization of N-thioacylamino acids with phosphorus tribromide in acetic anhydride, 303
4-Hydroxythiazolidine-2-thiones, characteristic absorption of, 270
   as cyclic intermediates in condensation of ammonium dithiocarbamate and α-halocarbonyl compounds, 270
   dehydratation of, 270
   as intermediates in reaction of 1-alkyl-sulfonyl-3-bromo-2-propanones and ammonium dithiocarbamate, 264
4-Hydroxy-2-thiazolidinones, as more probable intermediates in 2-hydroxythiazoles syntheses, 273
Hydroxythiazolines, treatment by acid, 210
4-Hydroxythiazolines, in equilibrium, with α-thioketones, 210
   as intermediate in Hantzsch's synthesis, 209
4-Hydroxythiazolium salts, from N-methyl-(p-dimethylamino)thiobenzamide and α-haloketones, 211
2-Hydroxy-4-thiocyanatomethylthiazole, from ammonium thiocyanate and sym dichloroacetone, 273
Hypolimic vasodilatory, 2-alkyl-5-thiazolecarboxylic acids as, 525

Imidazolythiazoles, table of products, 473
2-Imino-3-allyl-4-thiazolone, from allyl-thiourea and ethyl chloroacetate, 233
2-Imino-4-methylthioxole, wrong structure of hydroxymethylthiazole, 12
2-Imino-4-oxathiazolidines, as tautomers of 2-amino-4-hydroxythiazoles, 215
2-Imino-3-oxa-4-thiolenes, as by-products in 2-hydroxythiazoles synthesis, 273
2-Iminothiazoles, as tautomers of 2-aminothiazoles, 248

2-Imino-Δ-4-thiazoline, as tautomers of 2-aminothiazoles, 213
3-Iminothiobutyramide, condensation, with phenacylbromide, 191
Infrared absorption, 53
   deuterothiazoles, 57
   thiazole derivatives, 63
Infrared spectra, of alkylthiazoles, 349
   of arylthiazoles, 349
   of metallic complexes of thiazole, 392
   of thiazole and alkylthiazole, interpretation of, 349
2-Iodothiazole, photolysis of, 112
   preparation of, 565
Ionization, of alkyl thiazoles, thermodynamics of, 92, 355
   of thiazoles, thermodynamics of, 92
Ionization potential, 51, 82
Iron, complexes with thiazoles, U.V. and I.R., 392
Isomerization, of phenylisothiazoles, by irradiation, 375
4-Isopropenylthiazole, synthesis of, 179
5-Isopropylidene amino-2-mercapto-thiazoles, from α-aminonitrile, carbon disulfide and acetone, 287
Isoxazole, infrared absorption, 60
Isothiazole, infrared absorption, 60
   photoisomerization to thiazole, 24, 136
   rate of hydrogen exchange, 116
   rate of quaternization, 126
Isothiocyanates, reaction, with 1,4-diamino-2-butynes, 300
1-Isothiocyanato-4-amino-2-butynes, condensation, with ammoniac or primary or secondary amines, 300
Isothiocyanoacetic acid, thiazolic structure of, 16. See also 2,4-Dihydroxythiazole
Isothiourea, as acyclic intermediate in reaction between α-halocarbonyl compounds and thiourea, 232
Isotopic effect, in lithiation, 379
   mass spectra of, 347
   of thiazoles, 342

2-Ketonylthiothiazoles, as by-products in 2-mercaptothiazoles synthesis, 264
   from 2-mercaptothiazoles and bromo-ketones, 266

Kinetics, of nitration, 381
  of quaternization, of alkylthiazole, 388
  of 2-alkylthiazoles by methyl iodide, 387
  of 5-alkyl thiazoles, 391
  of 1-chloroethyl thiazoles, 393
  of dialkyl thiazoles, 389

Linear free-energy relationships, 147, 149
2-Lithiomethylthiazole, reactivity, 121

Magnetic susceptibility, of thiazoles, 89
2-(N-Maleylsulphanilamido)thiazoles, preparation of, 233
Mass spectra, of alkylthiazoles, 82, 347
  application in analysis of food flavorings, 348
  of aryl thiazoles, 347
  of thiazolyl pyridines, 349
Mass spectrometry, 81
  ionization potentials of thiazoles, 82
  mode of cleavage, of 2-alkylthiazoles, 82
  of 5-alkylthiazoles, 83
  McLafferty rearrangement, 81
  of thiazole, 81
Mass spectrum, of thiazole, 81
Melting point, 85
2-Mercapto-4-(N-alkyl-2-benzimidazolyl)-thiazoles, from ammonium dithiocarbamate and N-alkyl-2-benzimidazolyl chloromethylketones, 265
2-Mercapto-5-aminothiazoles, from α-aminonitriles and carbon disulfide, 286
2-Mercapto-4-arylthiazoles, from ammonium dithiocarbamate and aromatic chloro (or bromo) methylketones, 265
  synthesis of, 264
2-Mercapto-4-hetarylthiazoles, synthesis of, 265
2-Mercapto-4-hydroxythiazole, from ammonium thiocyanate, 15. See also Rhodanine
Mercaptoketones, condensation, with nitriles (or α-mercapto acids), 168
2-Mercapto-4-methyl-5-carbethoxythiazole, from ammonium dithiocarbamate and α-chloroacetoacetate, 264
2-Mercapto-4-methylthiazole, from chloracetone yield of, 264

2-Mercapto-4-methyl-5-thiazolylmethyl-ketone, from ammonium dithiocarbamate and 3-chloro-2,4 pentanedione, 264
2-Mercapto-4-phenylthiazole, synthesis of, 22
  from phenacylchloride and ammonium dithiocarbamate, 266
  from α-thiocyanato-acetophenone and thioacetic acid, 276
2-Mercaptothiazole, preparation and structure, 21
  replacement of mercapto group by hydrogen, 340
  synthesis of, from ammonium dithiocarbamate and chloracetaldehyde or α,β-dihalogenoethers, 260
2-Mercapto-4-thiazole-carboxylic acid ethylester, from ammonium dithiocarbamate and ethylbromopyruvate, 264
2-Mercaptothiazoles, as accelerators in vulcanization of rubber, 260
  alkylation of, 266
  mechanism of formation, 269
  from α-bromoalkynes, 265
  oxidation of, 172
  synthesis of, from ammonium dithiocarbamate and α-bromoalkynes, 264
  from ammonium dithiocarbamate and α-halocarbonyl compounds, 260, 261
  from ammonium dithiocarbamate and α-thiocyanatoketones, 276, 277
  from hydrogen sulfide and α-thiocyanatoketones, 276
  from α-mercaptoketones and ammonium thiocyanate, 293
  from thiourea and α-thiocyanatoketones, 276
  in rubber vulcanization, 167
  tautomeric forms of, 260
2-Mercaptothiazoline, 4- and 5-methyl derivatives, preparation, 22
2-Mercaptothiazolines, from bromoamines, 21
Mercuration, of alkylthiazoles, 380
  as extractive technique, 380
Metallic complex, of thiazoles, 392
ω^* Method, 27
ω'' Method, 27

2-Methoxy-4-methylthiazole, rate of nitration, 105
4-Methyl-5-acetylthiazole, reduction of, 525
2-Methyl-4-(2-benzothiazolyl)thiazole, from 2-benzothiazolylbromomethylketone and thioacetamide, 199
2-Methyl-4-*t*-butylthiazole, reaction with butyllithium, 378
4-Methyl-5-carbethoxythiazole, pKa, 92
4-Methyl-2-carboxythiazole, pKa, 92
4-Methyl-5-carboxythiazole, pKa, 92
2-Methyl-4-(2-chlorophenyl)thiazole, quaternization of, 391
2-Methyl-5-ethylthiazole, rate of nitration, 104
5-Methyl-4-ethylthiazole, metalation, 119
Methyl group, reactivity with carbonyl compounds, 392
2-Methyl-4-heteroarylthiazoles, syntheses of, 195
4-Methyl-5-(β-hydroxyethyl)thiazole, preparation of, from α-chloro-α-acetyl-α-butyrolactone or 4-acetyl-4-chloro-1-pentanone, 179
2-Methyl-3-isothiocyanato-hept-3-ene, from thiocyanogen and oxoalkylene phosphorane, 299
2-Methyl-4-(3-nitrophenylthiazole), quaternization of, 391
2-Methyl-4-(4-nitrophenylthiazole), quaternization of, 391
4-Methyl-2-methylaminothiazole, from *S*-acetonyl-*N*-methylisothiourea, 278
2-Methyl-4-phenylthiazole, organolithium products, reaction with various reagents, 240
quaternization, 391
4-Methyl-2-phenylthiazole, synthesis of, from mercaptoacetone, benzaldehyde and ammonia, 293
α-Methylrhodim, *see* 2-Imino-4-methylthioxole
2-Methylthiazole, bromination, 100
free enthalpy of ionization, 92
lithiation, 120
mass spectrometry of, 347
nitration of, by nitrogen dioxide-boron trifluoride, 382
by nitronium tetrafluoroborate, 382
orientation of nitration, 105

pKa, 92
preparation of, by vapor phase reaction of sulfur with dialkyl-amines, 305
reactivity of methyl group, with benzaldehyde, 392
4-Methylthiazole, action of benzoyl peroxide on, 365
amination, 113, 124
from chloracetone and thioformamide, 179
free enthalpy of ionization, 92
mass spectrometry of, 347
metalation, 119
phenylation of, with different phenyl radicals sources, 368
pKa, 92
preparation of, 171
radical reactivity, influence of acidity, 369
rate of nitration, 104
reaction, with benzaldehyde, 393
with ethylmagnesium bromide, 378
by reduction of 2-hydroxy-4-methylthiazole, 11
5-Methylthiazole, free enthalpy of ionization, 92
pKa, 92
rate of nitration, 104
reaction with ethylmagnesium bromide, 378
yield of, 174
4-Methylthiazole-3-oxide, 392
Methylthiazoles, basicity, 93
chemical shift, carbon-[13], 76
proton, 68
coupling constants, H-H, 74
cycloaddition with dimethylacetylene dicarboxylate, 97
dipole moment, 38
electronic structure, 26, 40
free radical methylation of, 370
free radical reactivity, comparison with theoretical calculation, 370
infrared spectra, C-H vibrations assignments, 350
harmonic and combination bandes, 350, 351
vibrational frequencies of thiazole ring, 351
vibration frequencies, 350
nitration of, kinetics constants, 382

phenylation of, in trifluoroacetic acid solution, 369
reaction with benzaldehyde, 393
reactivity of methyl group, 392
substitution effects, 42
theoretical calculations, molecular diagrams, 358
polarization energies, 358
thermodynamic constants of the ionization reaction, 356
ultraviolet absorption, 48
2-Methylthiazoles, organolithium compounds, deuterolysis of, 378
reaction with n-butyllithium, 378
N-Methyl-Δ^4-thiazolines, from N-methylaminothiocyanogen, 298
3-Methyl-Δ^4-thiazoline-2-thione, from methylaminoacetonitrile and carbon disulfide, 287
4-Methylthiazol-2-one, rate of nitration, 105
2-Methylthieno[3,2d]thiazole, synthesis of, 200
4-Methyl-2-vinylthiazole, copolymerization with 1,3-butadiene, 398
4-Methyl-5-vinylthiazole, in natural flavors, 395
Micrococcin, thiazolic structure, 1
Miscellaneous thiazoles, table of products, 487
Molar refraction, 90
Monohalogenothiazoles, 567
nucleophilic substitution reactions, 322
with amines, 567
with ArSH, 567
with benzamide, 567
with OH⁻, 567
with OR⁻, 567
relative reactivities, 568
with SR⁻, 567
with sulfinic acid salts, 567
with sulfonamide, 567
reactions, with alkyl halides, 574
with n-butyllithium, 573
with Grignard reagents, 573
reeduction of, with nickel, 573
with zinc, 573

Natural flavors, presence of thiazoles, 395
Nickel, complexes with thiazoles, U.V. and I.R., 392

Nitration, of alkylthiazoles, kinetics of, 381, 384
by nitronium fluoroborate and complexe nitrogen-boron trifluoride, 382
(5-Nitrofuryl)vinylthiazole-N-oxides, preparation, 131
2-Nitro-5-halogenothiazoles, 578
preparation of, 578
reaction with MeO⁻, 580
2-(4-Nitrophenylimino)-4-thiazolidone, from 4-nitrophenylthiourea and chloroacetic acid, 296
2-Nitrothiazole, 576
nitration of, 577
preparation of, 576
reaction with KF, 566
reaction with MeO⁻, 577
reduction of, 577
5-Nitrothiazole, 576
preparation of, 577
reduction of, 577
ring opening reaction, 577
Nitrothiazoles, chemical shift, proton, 68
coupling constants in EPR, 84
ionization potential, 52
reduction of, 381
ultraviolet absorption, 48
Nuclear magnetic resonance, 66
of alkyl thiazoles, 342
of aryl thiazoles, 342
correlation with theoretical calculations, 70
coupling constants, of alkyl thiazoles, 342
of arylthiazoles, 342
of disubstituted thiazoles, 71
$J_{13C-13C}$ coupling constants in thiazole, 81
J_{CH} coupling constants in thiazoles, 78
J_{HH} coupling constants, in methylthiazoles, 75
in thiazoles, 74
J_{15N-H} coupling constants in thiazole, 81
J_{NH} coupling constants of thiazole, 80
line width of thiazole lines, 80
of monosubstituted thiazoles, 68
proton chemical shifts, of alkylthiazoles, 342
of arylthiazoles, 342
ring current in thiazole, 67
of substituents in substituted thiazoles, 72

^{13}C, 77
of thiazole, effect of solvents, 67, 73
^{13}C, 76
^{14}N, 77
Nucleophilic reactivity, of alkylthiazoles and arylthiazoles, 378
Nucleophilic substitution, 113

1,3,4-Oxadiazolylthiazoles, table of products, 476
1,3-Oxathiolium salts, in condensation, with cyanamide, 303
Oxazole, infrared absorption, 60
Oxazoles, heating of, with phosphorus pentasulfide, 309
Oxazolylthiazoles, table of products, 471
Oxidation, of thiazoles, by hydrogen peroxide, 392
by tungstic acid, 392
Oxiranes, condensation, with thioamides, 295
2,2'-Oxydi-p-phenylene-bis (4-phenyl and biphenyl-4-yl)thiazoles, syntheses of, 195
m-Oxymethylthiazole, see 2-Hydroxy-4-methylthiazole, structure of

Pentaacetyl-D-gluconic acid thioamide, cyclization, with chloroacetone, 188
with phenacyl bromides, 188
2-neo-Pentylthiazole, pKa, 92
Phenanthro[3,4d] thiazoles, preparation, from 3-phenanthrenylamine, 199
Phenethylthiazoles, nitration of, 385
Phenylacetaldehyde, polymerization of, 174
2-Phenyl-4-arylthiazoles, Schiff's bases derived from, 193
Phenylation, of alkylthiazole, 366
of arylthiazole, 367
N,N'-ortho (and para) Phenylene-bis-(2-amino-4-R thiazole), from ortho (and para) phenylene dithioureas, 243
meta (or para) Phenylene bis-(4-thiazolyl-2-amino), from thiourea and bis (meta (or para) haloacetyl) benzenes, 216
3-Phenylisothiazole, photoisomerization, 137
5-Phenylisothiazole, photoisomerization, 137
Phenylisothiazoles, isomerization of, by irradiation, 375
2-Phenyl-4-methylthiazole, photoisomerization of, 378
2-Phenyl-5-methylthiazole, from N-propynylbenzamide and phosphorus pentasulfide, 282
4-Phenyl-5-methylthiazole, from α-bromopropiophenone and thioformamide, 179
2-Phenyl-4-α-(or β)naphthylthiazoles, synthesis of, 193
2-Phenylthiazole, isomerization of, effect of iodine, 376
mass spectrometry of, 349
photoisomerization, 137
synthesis of, from α-bromoacetal and thiobenzamide, 172
from thiobenzamidoacetal, 283
twist angle, 353
4-Phenylthiazole, mass spectrometry of, 349
from photochemical isomerization of 3-phenylisothiazole, 310
2-R-substituted, nitration of, 385
twist angle, 353
5-Phenylthiazole, mass spectrometry of, 349
from phenylacetaldehyde and thioformamide, 174
twist angle, 353
2-Phenyl-4-thiazole carboxylic acid, synthesis of, 204
Phenylthiazoles, angle between the aryl and the thiazole rings, 353
infrared spectra, C-H vibration assignments, 350
isomerization of, by irradiation, 375
mass spectra of, 349
nitration of, 385
theoretical studies, 358
2-Phenyl-Δ^2-thiazoline-4-one, from chloroacetic acid and thiobenzamide, 294
2-(4-Phenyl-5-thiazolyl) acetic acids, to naphto[1,2] thiazoles, 341
(4-Phenylthiazol-2-yl) acetone, synthesis of, 191
Phenylthiosemicarbazide, condensation, with dichloroether or chloroacetone, 250
Phosphorence spectra, of thiazole, 51
Phosphorus insecticides, from haloalkylthiazoles, 341

Photochemical rearrangement, of alkylthiazoles, 374
of arylthiazoles, 374
Photoelectron spectroscopy, 51
Photorearrangements, of isothiazoles, 139
of thiazoles, 139
2-Phthalimidomethylthiazoles, preparation of, 233
Physical properties, 45
Piperazine N,N'-dithio-acetamide, condensation, with ω-bromoacetophenones, 197
Pka, of alkylthiazoles, 355
correlation with electronic structure, 93
of thiazoles, 91
Polarization energies, of methylthiazoles, 358
Polymerization, of vinyl thiazoles, 397
Polymers, use of thiazole in polymer chains, 395
Polymers containing thiazole ring, resistance to thermal degradation, 398
Potassium ferrocyanide, as deshydrogenating agent of Δ^3-thiazolines, 308
PPP Method, 27, 50, 89
4-i-Propyl-2,5-dimethylthiazoles, quaternization of, 389
5-i-Propyl-2-t-butylthiazole, rate of nitration, 105
2-n-Propylthiazole, orientation of nitration, 105
pKa, 92
2-i-Propylthiazole, anionic polymerization of, 397
pKa, 92
4-Propylthiazole, mass spectrometry of, 348
5-i-Propylthiazole, 350
Protonation, 48
Pseudothiohydantoine, as acyclic intermediate, in reaction of α-halogenoacids or esters with thiourea, 232. See also 2-Amino-4-hydroxythiazole
Pyridazines (4,7-dioxo-4,5,6,7-tetrahydrothiazolo[4,5d]), preparation of, 206
Pyridine, electronic structure, 36, 39, 46
ultraviolet absorption, 47
thiazolylation of, 373
2-(4-Pyridyl)-4-carboxyethylthiazole, from thioisonicotinamide and ethyl bromopyruvate, 198

2-(4-Pyridyl)-4-(4-ethoxy carbonyl-2-thiazolyl hydrazinocarbonyl) thiazole, synthesis of, 198
2-Pyridylthiazoles, table of products, 477
4-Pyridylthiazoles, table of products, 478
5-Pyridylthiazoles, table of products, 478
Pyrolysis, of aryl thiazole, 399
Pyrrolothiazine, from thiazole ylide, 95
Pyrrolyl thiazoles, table of products, 471

Quaternary salts, ultraviolet absorption, 50
Quaternization, of alkylthiazoles, by alkyl halide or methyl tosylate, 386
of 5-alkylthiazoles, kinetic data, 391
of dialkyl thiazoles, kinetic data, 389
kinetic data, 388
Quinones, as dehydrogenating agent of 2,5-dihydrothiazoles, 309

Radical reactivity, of alkylthiazoles, 364
of conjugate acid of thiazole, 368
influence of radical source and medium, 366
of thiazolyl radicals, with alkyl benzene, 372
with pyridines, 373
Raman diffusion, 53
thiazole, 57
thiazole derivatives, 63
Reactivity indices, 33
Reactivity of nitrogen atom, of thiazole and alkyl thiazole, 386
Relaxation time, 80
Rhodanines, as intermediates, in synthesis of amino acids peptides and purines, 167
Rhodaninic acid, see 2-Mercapto-4-hydroxythiazole; Rhodanine
Rhodanine, modes of preparation, 19
N-phenyl, preparation, 19
N-substituted, preparation, 20
structure of, 19
Rhodanines, condensation with aldehydes, 20
5-substituted, preparation, 20
Rhodaninic acid, see 2-Mercapto-4-hydroxythiazole; Rhodanine
Rhodanpropimin, see α-Thiocyanoacetonimine
Rhodium, complex with thiazole, U.V. and I.R., 392

Subject Index

Side chain functional substituents, reactivity of, 340
Solvolysis, of 1-chloroethylthiazole, determination of rates, 393
Spasmolytic, aminoalkylesters as, 527
Specific heats, of alkylthiazoles, 357
Steric effects of substituents, at 2-and 4-positions, evaluated by quaternization reaction, 386
  on reactivity of nitrogen atom, 386
Steroids, with thiazole ring, preparation of, 242
trans-Stilbene, 353
Styryl thiazoles, ultraviolet spectra, effect of substituents, 353
2-Styrylthiazoles, from 2-methylthiazole, 393
Substituent effect on reactivity of halogenothiazoles, 571
  Hammett's relationships, 571
Substituents effects, electronic, 390
  steric, 386
  transmission of effects, across thiazole ring, 393
Substitution effects, aminothiazoles, 43
  halogenothiazoles, 43
  methylthiazoles, 42
Sulfathiazole, 214
2-Sulfonamidothiazole, 567
2-Sulfonamidothiazoles, preparation of, 216, 233
Sulfur, as dehydrogenating agent, of Δ^3-thiazolines, 308
Sulfur chloride, as reactant in 2-aminothiazoles synthesis, 214
Sulfur trioxide, as reactant in 2-aminothiazole synthesis, 214

Theoretical calculations, 357
Theoretical model, 26
Thermal stability, 87
Thermodynamic data, 85-88
  cryoscopic data for thiazoles, 86
  heat capacity of, 2-methylthiazole, 87
    thiazole, 86
  liquid-vapor equilibrium of thiazole-solvent mixtures, 88
  molar excess functions for thiazole-solvent mixtures, 88
  pyrolysis temperatures of thiazoles, 87

standard entropy of, 2-methylthiazole, 87
  thiazole, 86
thiazole-water azeotrope, 85
vapor-liquid equilibrium of, 2-methylthiazole, 85
  thiazole, 85
Thiadiazole, electronic structure, 46
Thiamine, 212, 399
  hydrogen exchange, 114
  preparation of, from 4-methyl-5-(β-hydroxy-ethyl)thiazole, 179
1,4-Thiazines, from thiazolium salts, 139
Thiazole, addition reactions, 94
  from 2-aminothiazole, 24
  aromaticity, 67
  arylation with *para*substituted benzoyl peroxides, 367
  auto-association, 88
  basicity, 91
  binary mixture, 85
  boiling point, 85
  bond lengths, 39, 45
  bromination of, 566
  -2-^{13}C, preparation, 25
  -4-^{13}C, preparation, 25
  -5-^{13}C, preparation, 25
  -2-^{14}C, preparation, 72
  chemical shift, carbon-13, 76
    proton, 68
  complexes of cobalt (II), 129
  coupling constants C-C, 81
  coupling constants H-H, 73
  coupling constants N-H, 81
  by cyclization, 24
  by decarboxylation of thiazolecarboxylic acids, 24
  2-deutero, preparation, 24
  5-deutero, preparation, 24
  dicyanomethyl ylide of, preparation, 94
    reaction with dimethylacetylene dicarboxylate, 94
  2,5-dideutero, preparation, 24
  dipole moment, 26, 38, 39, 89
  discovery of, 8
  as electron donor, 125
  electronic charges, 29
  electronic structure, 26, 31
    ab initio method, 26
    CNDO method, 26
    HMO method, 26

parameterization, 27
PPP method, 26
ω* method, 26
ω" method, 26
enthalpy of fusion, 86
free enthalpy of ionization, 92
free radical, arylation, 108
  bromination, 108
  cyclohexylation, 111
  methylation, 110, 370
free radical reactivity, comparison with theoretical calculations, 370
gas-liquid chromatography, Kovat's indice, 359, 360
geometrical structure, 45
hydrogen exchange by phase-transfer catalysis, 119
hydrogen-metal exchange reactions, 119
infrared absorption, 53
  influence of physical state, 61
interorbital and internuclear angles, calculated from ^{13}C-H coupling constants, 347
ionization potential, 52, 83
from isothiazole by photoisomerization, 136
linear free-energy relationships, 147
mass spectrometry, 81
mechanism of, deuteration, 106
  hydrogen exchange, 117
melting point, 85
mercuration, 100
metallic complexes, 126
  mass spectra of, spectroscopic properties, 129
  structure of, 129
methylation of, 369
molecular structure of, 46
^{15}N, preparation, 25
N.M.R., carbon-13, 77
  proton, 66
numbering of, 7
one-electron reduction, 135
oxidation, 130
partial rate factors for arylation, 108
phenylation of, with benzoyl peroxide, 364
  with different phenyl radicals source, 368
  in trifluoroacetic acid solution, 369
phosphorescence, 50

photoelectron spectroscopy, 51
by photorearrangement of isothiazole, 24
physical properties, 45
pKa, 92
proton acidity, 118
protonation, 48
quaternization, steric effect, 126
radical reactivity, influence of acidity, 369
Raman diffusion, 56
rate of, deuteration, 106
  quaternization, 125
reactivity with aryl radical, 364
reactivity indices, 31
reduced system of, 167
refraction index, 90
ring cleavage, 134
  synthetic applications, 134
ring expansion, 139
substitution effects, 43
synthesis of, from chloroacetaldehyde and thioformamide, 170
  from ethylisocyanoacetate and carbon disulfite, 307
  from formylaminoacetal, 279
thermal stability, 87
from Δ^3-thiazoline, 24
transmission of substituent effects, 107
trideutero, preparation, 25
ultraviolet absorption, 46, 48
velocity of sound, 87
viscosity, 88
Thiazoleacetic acids, decarboxylation of, to alkyl thiazoles, 341
  syntheses of, 207
  by Hantzsch method, 341
2-Thiazoleacetic acids, from ethyl thiomalonamate or monothiomalonamide, 208
4-Thiazole aldehyde, Cannizzaro reaction of, 341
Thiazolealiphatic acid chlorides, from acids, 341
Thiazolealiphatic acids, esterification of, 341
  from malonic derivatives, 341
Thiazole alkanoic acids, synthesis of, 201
4-Thiazole alkanoic acids and salts, from thioamides and γ-chloro (or γ-bromo) aceta acetic or their α-alkyl derivatives, 207
5-Thiazole alkanoic acids and salts, from

thioamides and β-halo γ-keto-acids, 208
Thiazole amidine, from cyanothiazoles, 532
  physical properties of table of, 556
  from polarographic reduction, 525
  source of thiazolyl radicals, 371
Thiazolecarboxaldehyde, reaction with, acetic anhydride, 534
  acetylglycine, 534
  Cannizzaro, 535
  hippurique acid, 534
  malonic acid, 534
Thiazolecarboxaldehydes, conversion to, hydrazones, 535
  oximes, 535
  semicarbazones, 535
  oxidation of, 522
  physical properties of, table of, 553
  preparation of, by addition of N,N-dimethylformamide to lithium derivatives, 533
    by McFadyen-Stevens synthesis, 532
    by oxidation of alkylthiazoles, 533
    by oxidation of alcohols, 532
    via oxidation of carbinols or methyl derivatives by SeO_2, 534
    by ozonolysis of styrylthiazoles, 522, 533
    by polarographic reduction, 533
    by reaction of thiazolyllithium with N-methylformanilide, 533
    by Stephen's method, 533
  reaction of, benzoination, 535
    oxidation, 535
    reduction, 535
  reaction with, diazomethane, 536
Thiazolecarboxamide, from acid chloride and amines, 528, 529
  preparation of, from Curtius reaction, 529
  from Hofmann reaction, 529
Thiazolecarboxamides, dehydration of, 530, 531
  physical properties of, table of, 546
  preparation of, 341
    from acid chloride, 529
    from esters, 529
    from thiazolecarboxylic acid, 529
Thiazolecarboxylates, physical properties of, table of, 540
5-Thiazolecarboxylates, rate of decarboxylation, 118
5-Thiazolecarboxylic acid, from decarboxylation of 2,5-thiazoledicarboxylic acids, 205
Thiazolecarboxylic acid chlorides, from corresponding acid and thionyl chloride, 528
  physical properties of, table of, 545
  in preparation of: amides, esters, hydrazides, 528
  reaction with, ammonia or amines, 528
    diazomethane, 529
Thiazolecarboxylic acid esters, correlation between rate of alkaline hydrolysis and Hammett's σ values, 253
4-Thiazolecarboxylic acid ethyl ester, from ethylbromopyruvate and thioamides, 204
Thiazolecarboxylic acid hydrazide, preparation of, from acids, 530
  from amides, 530
  from esters, 530
Thiazolecarboxylic acid hydrazides, physical properties of, table of, 547
  preparation of, from acid halides, 530
  from anhydrides, 530
5-Thiazolecarboxylic acid methyl (or ethyl) ester, from ethyl α-formyl-chloroacetate and thioamides, 204
Thiazolecarboxylic acids, acidity of, 522, 523
  from carbonating organometallic compounds, 522
  from catalytic dehalogenation of 2-halo derivatives, 521
  correlation between pKa and Hammett σ values, 147
  decarboxylation of, 523, 527, 528
  from ester hydrolysis, 520, 521, 524, 527
  from haloform reaction, 522
  from hydrolysis of their derivatives, 522
  from oxidation, of alcohols, with aqueous chromic acid, aqueous KI-iodine mixture of nitric and sulfuric acids, 521
    of aldehydes, 521, 522, 535
    of methyl or substituted methyl group, 521
    of thiazolyl ketones, 537
  polarographic reduction of, 525

reduction of, with lithium aluminium
    hybride, 524
  with sodium in alcohol, 524
  synthesis of, 201
Thiazole-2,4-diacetic acid diethyl ester, from
    ethyl-γ-bromoacetoacetate, 209
Thiazole-4,5-dicarboxylic acid ethyl esters,
    from diethyl α-chloro β-keto-succinate
    and thioamides, 206
Thiazole-2,4-dicarboxylic acids or esters,
    synthesis of, 204
Thiazole-4,5-dicarboxylic acids, synthesis
    of, 204
Thiazole esters, reduction of, 524, 527
Thiazolemonocarboxylic acids, physical
    properties of, table of, 538
Thiazole-N-oxides, 392
  by oxidation, with hydrogen peroxide, 392
    with tungstic acid, 392
  preparation and stability, 131
  reaction of, 392
Thiazolepolycarboxylic acids and derivatives, physical properties of, table of, 550
Thiazole polymers, ultra-violet spectra, 397
Thiazoles, addition with dimethylacetylene
    dicarboxylate, 95
  alkylation, 101
  biological and pharmacological uses, 399
  bromination, 101
  chemical shift, carbon-13, 76
    proton, 71
  complexes with metals, 392
  coupling constants H-H, 74
  ^{13}C substituted, mass spectra, 347
  deuterium exchange, 113
  dipole moment, 89
  2,4-disubstituted derivatives, from acyl-
    amidocarbonyl compounds and
    P_2S_5, 278
    from α-mercaptoketones and nitriles,
    291
  4-and 4,5-disubstituted derivatives, synthesis, 176
  electrophilic substitution orientation, 101
  in food flavoring, analysis by mass-spectrometry, 348
  ionization potential, 52
  isotopics derivatives, spectroscopy of, 342
    synthesis of, 342

kinetics of nitration, 104
mass spectrometry, 82
mercuration, 101
meso-ionic, 167
  photorearrangement of, 139
in natural aromas, 1, 395
nitration, 101
  correlation with proton NMR, 99
  in petroleums, 1
  photorearrangement, mechanism of, 137
Prij's card index of, 167
pyrolysis of, 399
reaction with aldehyde, 393
substituted by several heterocyclic groups,
    table of products, 479
2-substituted derivatives, from 2,5-dihydro-
    thiazoles, 309
sulfonation, 101
synthesis, from α-acylaminocarbonyl
    compounds and P_2S_5, 168
  from α-amino-nitriles (or α-amino-
    amides) and carbon disulfide (or
    thioacids), 168
  from α-mercaptoketones (or acids) and
    nitriles or aldehyde oximes, 292
  4-monosubstituted derivatives, 179
  from oxazoles and phosphorus penta-
    sulfide, 309
  from photochemical isomerization of
    isothiazoles, 310
  from Δ^3-thiazolines, 308, 340
  from thioacetic acid and aminoacids (or
    N-acetyl-and N-thioacyl derivatives),
    306
  from α-thiocyanatoketones, 168, 271
survey, 167
thermostable polymers, 395
2,4,5-trisubstituted derivatives from α-
    acylaminoketones and P_2S_5, 280
  synthesis of, 189, 192, 307
  synthesis by Dub's method, 305
ultraviolet absorption, 48
unsubstituted in 2-position, synthesis, 339
2-Thiazolethiones, *see* 2-Mercaptothiazoles
Thiazolidine, 2-phenylamino-3-phenyl,
    preparation, 22
  from β-mercaptoalkylamines, 168
Thiazolidinylthiazoles, table of products,
    472
Thiazolines, rate of nitration, 104

2-Thiazolines, from β-mercaptoalkylamines, 168
Δ³-Thiazolines, chemistry of, 167
  dehydrogenation of, 340
  synthesis of, 309
    from ammonia and α-mercaptoketones or aldehydes, 308
Thiazoline-2-thione, derivatives, preparation, 21
Thiazoline-2-thiones, by dehydration of 4-hydroxythiazolidine-2-thiones, 264
Δ⁴-Thiazoline-2-thiones, from dealkylation of 3-alkyl-4-hydroxythiazolidine-2-thiones, 270
  derivatives of, 289
  mechanism of formation, from ammonium dithiocarbamate and α-halocarbonyl compounds, 270
  N-alkyl (or aryl) derivatives, from salts of N-alkyl (or aryl) dithiocarbamic acid, 269
Thiazolinylthiazoles, table of products, 472
Thiazolium chloride, syntheses of 2,3,4-trimethyl derivatives, from N-methyl-thio-acetamide and chloracetone, 211
Thiazolium ion, reaction with acetone ketyl radical, 135
Thiazolium-N-oxide sulfates, 392
Thiazolium salts, deuterium exchange, 114
  lability of H-2, 115
  preparation from 4-hydroxy thiazolium salts, 211
    from N-monosubstituted thioamides and α-halocarbonyl compounds, 211
    from N-substituted thioamides and α-halocarbonyl compounds, 212
  protodetritiation, 118
  pseudo-bases, 114
  reduction by NaBH₄, 132
  synthesis of, from N,N,N'-trimethyl-thiourea, 248
  2-thiomethyl-3-ethyl derivatives, synthesis of, 212
Thiazolium ylide, intermediate for H exchange, 114
  stabilization by d-σ overlap, 115
Thiazol-2-ones, by cyclization of enamines in acidic medium, 298
  synthesis of 5-aryl (or alkyl) thio derivatives, from p-bromophenacylaryl (or alkyl) sulfides and ethyl thiocarbamate, 259
4-Thiazolones, as tautomers of 4-hydroxythiazoles in condensation of α-mercaptoacids with nitriles, 293
Thiazol-5-ones, (2,4-substituted derivatives) synthesis of, 304
5-H-Thiazolo[3,2-a] pyridine, preparation, 94
Thiazolopyrimidines, from 6-amino-1,3-dimethyl uraciles, 200
Thiazolo[5,4d] pyrimidines, from 5-amino 4-mercapto pyrimidines and formic acid, 200
  synthesis of, 200
Thiazolylacetamide, to corresponding nitrile, 341
Thiazolylalanine, synthesis of, 341
Thiazolylalkyl thiazoles, table of products, 482
2-(4-Thiazolyl) benzimidazole (thiabendazole), from thioformamide and *ortho* nitro bromoacetylaniline, 198
5-Thiazolylbenzimidazoles, synthesis of, 200
Thiazolyl ethers, from haloalkylethers, synthesis of, 340
2-Thiazolylethylamine, by reduction of nitrile, 341
Thiazolylethyl chlorides, rate of solvolysis, correlation with electronic structure, 106
2-Thiazolyl-ethylchlorides, solvolysis of, 572
1-Thiazolylethyl chlorides, correlation of reactivity with π net charges, 146
  first order rate of solvolysis, 146
1-(5-Thiazolyl)ethyl chlorides, 2-substituted, rate of solvolysis, 106
2-Thiazolyl group, Hammett's σ^+ parameter, 394
4-Thiazolyl groups, Hammett's σ^+ parameters, 394
5-Thiazolyl groups, Hammett's σ^+ parameters, 394
2-Thiazolyl hydrazine, 370
1-(2-Thiazolyl)-2-imidazolones, synthesis of derivatives from 1-thiocarbamoyl-2-imidazolones and α-bromoketones, 197
Thiazolyl ketones, Beckmann rearrangement of oximes, 537

oxidation of, 537
physical properties of, table of, 554
preparation of, by addition of organometallic reagents to acids and their derivatives, 536
  from aldehyde and diazomethane, 536
  by Claisen condensation, 536
  from cyanothiazoles, 531, 532, 536
  by Friedel-Crafts acylation, 535
  by oxidation of alcohols, 535
reaction of, condensation with ammonia, 537
  haloform, 538
  Mannich, 537
reduction of, from catalytic hydrogenation, 537
  Clemmensen reduction, 537
  with Grignard reagents, 537
  by Meerwein-Pondorf procedure, 537
Thiazol-2-yllithium, preparation, 119
Thiazol-2-ylmagnesium bromide, preparation, 119
Thiazolyl mercaptans, from haloalkylthiazoles, synthesis of, 341
2-Thiazolylphenacyl-bromide, condensation, with thiobenzamide, 194
Thiazolyl phenylketone, by oxidation, 341
Thiazolylphenylsulfoxides, Hammett's relationships, 572
  oxidation of, 570
2-(Thiazolylphosphoramido)thiazoles, preparation of, 233
Thiazolyl pyridine, gas-liquid chromatography of, relative retention volume, 358
  mass spectra of, 349
  preparation of, 373
Thiazolyl radical, 84
  reactivity of, 370
Thiazolyl radicals, electron spin resonance of, 373
  reactivity with alkyl benzene, relative reactivity of, 372
  reactivity with pyridines, 373
  from thiazolecarbonyl peroxides, 112
Thiazol-2-yl radicals, 5-substituted, reactivity, 113
  by thermal decomposition of diazonium derivatives, 113
  from thiazol-2-ylhydrazine, 112

1-(2-Thiazolyl)semicarbazides, from N-amidothiosemicarbazide and α-halocarbonyl compounds, 255
2-Thiazolyl sulfides, antifungal activity of, 266
  synthesis of, from esters of dithiocarbamic acid and α-halocarbonyl compounds, 266
  from esters or salts of dithiocarbamic acid and α-halocarbonyl compounds, 266
  from 2-mercaptothiazoles and α-bromo compounds, 266
2-Thiazolylthiazoles, table of products, 468
4-Thiazolylthiazoles, table of products, 468
5-Thiazolylthiazoles, table of products, 469
Thiazolylthioethers, from haloalkylthiazoles, synthesis of, 341
2-Thiazolylthiophosphates, from 2-hydroxythiazoles and ethyl chlorothiophosphate, 258
2-Thiazolylthiourea, from dithiobiuret and aromatic α-haloketones, 244
(α-Thienyl)-thiazole-4,5-dicarboxylic acid, pyrolysis of, 206
Thienylthiazoles, table of products, 470
Thin-layer chromatography, of alkyl thiazoles, 362
  of aryl thiazoles, 362
Thioacetamide, condensation, with bis (α-chloroketone), 194
  with bromomethylheteroarylketones, 195
  as generator of hydrogen sulfide, 171
  preparation of, 171
N-Thioacylaminoacetic acids, cyclization, with benzaldehyde in presence of acetic anhydride, 304
2-Thioalkylthiazoles, from alkylisothiocyanates and α-mercaptoketones, 293
Thioamides, as acyclic intermediates in Cook-Heilbron's synthesis, 284
  condensation, with 3-aroyl-3-bromopropionic acid, 208
  with γ-bromoacetoacetic ester arylhydrazines, 207
  with bromomethyl-1-adamantyl ketones, 195
  with 1-chloro-1,2-epoxydes, 192
  with diethyl α-chloro β-ketosuccinate, 206

with ethylbromopyruvate, 204
with ethyl α-formyl chloroacetate, 204
with γ-halo acetoacetic acid or their
  α alkyl derivatives, 208
with β-halo γ-keto acids, 208
with 4-hydroxy-3-bromo-2-pentanone,
  188
with nitriles and hydrogen sulfide, 285
with oxiranes, 295
cyclohexyl, benzyl and phenethyl derivatives of, preparation, 171
N-substituted derivatives, from P_2S_5
  and N-substituted amides, 171
reaction, with α-halocarbonyl compounds,
  180
α-halogenoaldehydes, 174
Thioarylamides, condensation, with 3-aroyl-
  3-bromo propionic acid, 208
Thiobenzamide, condensation, with benzene
  sulfonic ester of mandelonitrile, 301
with α-bromoacetal, 172
with α-chlorothioacids, 295
with α-haloaldehydes, 175
with α or β-naphthylbromomethylketone, 193
Thiobenzamides, rates of reaction, with
  phenacylbromides, 210
α-Thiocyanatoacetophenone, condensation,
  with thioacids, 276
α-Thiocyanatoketones, condensation, with
  ammonium chloride or amine hydrochloride, 278
cyclization of, 271
from α-haloketones and metal thiocyanates, 168
preparation of, from α-carbonyl compounds and metal thiocyanates, 271
treatment of, with dry hydrogen chloride,
  273
Thiocyanoacetic acid, structure of, 15
Thiocyanoacetophenone, precursor of
  thiazole ring, 11
α-Thiocyanoacetone, isomerization into
  2-hydroxy-4-methylthiazole, 11
α-Thiocyanoacetoneimine, thiazolic structure of, 8
Thioformamide, cyclization, with diethyl
  (bromooxoalkyl) phosphonates, 179
with α-haloaldehydes, 172
with α-haloketones, 175

instability of, 171
preparation of, 171, 174
Thioformamidoacetophenone, from aminoacetophenone and potassium dithioformate, 279
Thioheteroarylamides, reaction, with haloketones or aldehydes, 197
Thiohydantoin, acidic properties of, 16
basic properties of, 16
correct structure of, 17
derivatives, synthesis of, 18
monophenyl, preparation, 16
structure of derivatives, 18
N,N'-diphenyl, preparation, 17
preparation, 17
reactivity of, 16
structure of, 15, 16
α-Thioketone, as acyclic intermediate in the
  Hantzsch's synthesis, 211
in reaction of chloroacetone and thiosemicarbazide, 249
Thiophene, electronic structure, 34, 35, 46
Thiosemicarbazide, rate of reaction, with
  phenacyl bromide in Hantzsch's
  synthesis, 256
Thiosemicarbazone, as intermediate in reaction of chloroacetone with thiosemicarbazide under acidic conditions, 249
Thiosemicarbazones, condensation, with α-
  halocarbonyl compounds, 256
Thiourea, condensation, with β-acylvinylphosphonium salts, 299
with 3-arylamino-2-chloroprop-2-enoic
  esters, 205
with benzene sulfonates of cyanohydrin,
  297
with α-haloketones or aldehydes, 213
with α-halonitrile, 296
N-substituted derivatives, condensation
  with α-haloacids and esters, 242
rate of reaction with phenacylbromide,
  232
Thioureas, cyclization of N-substituted
  derivatives with α-carbonyl compounds, 232
4-Tosylthiazoles, from tosylisocyanide and
  carboxymethyldithioates, 305
2,4,5-Tri(acetoxymercury)thiazole, by mercuration of thiazole, 100
2,4,5-Trialkylthiazoles, table of products, 408

2,4,5-Triarylthiazoles, table of products, 411
Triazolinylthiazoles, table of products, 476
2,4,5-Trichlorothiazoles, synthesis of, 304, 575
Triiodothiazole, preparation of, 575
Trimethylphenylammonium bromide, as bromination agent, 174
Trimethylpyrrolo[2,1-*b*] thiazole-5,6,7-tricarboxylate, from thiazole, 97
2,4,5-Trimethylthiazole, 355
2,4,5-Trimethylthiazole, pKa, 92
2,3,4-Trimethylthiazolium, rate of nitration, 105
2,3,5-Trimethylthiazolium, rate of nitration, 105
N,N,N'-Trimethylthiourea, reaction, with chloroacetone, 248
Triphenylthiazole, photo-oxidation, 130

Ultraviolet absorption spectra, of alkylthiazoles, 352
of aryl thiazoles, 352
of halogeno thiazoles, 48
of metallic complexes of thiazole, 392
of methyl thiazoles, 48
of nitro thiazoles, 48
of thiazole derivatives, 48
of thiazolium salts, 51
Ultraviolet absorption spectrum, of thiazole, 46, 48

Vasodilating, esters as, 527
Velocity of sound, in liquid thiazole, 87
Vibrations, normal modes, 53
  ring, 60
  valence, 56
Vinylchloride, as starting material in preparation of chloroacetaldehyde, 171
4-Vinylthiazole, synthesis of, 179
Vinylthiazoles, copolymerization of, 398
  polymerization of, 397
2-Vinylthiazole, polymerization of, 397
Viscosity, 88
  of thiazole and alkylthiazoles in solution, 357
Vitamine, B_1, *see* Thiamine

Zinc, complexes with thiazoles, U.V. and I.R., 392